Animal Physiology

FOURTH EDITION

Animal

Malcolm S. Gordon
University of California, Los Angeles

IN COLLABORATION WITH

George A. Bartholomew
University of California, Los Angeles

Alan D. Grinnell
University of California, Los Angeles

C. Barker Jørgensen
University of Copenhagen

Fred N. White
University of California, San Diego

Physiology:

PRINCIPLES AND ADAPTATIONS

MACMILLAN PUBLISHING CO., INC.
NEW YORK

COLLIER MACMILLAN PUBLISHERS
LONDON

Earlier edition entitled *Animal Function: Principles and Adaptations,* copyright © 1968 by Malcolm S. Gordon. Earlier edition entitled *Animal Physiology: Principles and Adaptations,* copyright © 1972 and 1977 by Malcolm S. Gordon.

Macmillan Publishing Co., Inc.
866 Third Avenue, New York, New York 10022

Collier Macmillan Canada, Inc.

Library of Congress Cataloging in Publication Data

Gordon, Malcolm S.
  Animal physiology.

  Includes bibliographies and index.
  1. Physiology, Comparative.   2. Adaptation (Physiology).   3. Animal ecology.   I. Title.
QP33.G65   1982   591.1   81–8227
ISBN 0–02–345320–6   AACR2

Printing: 1 2 3 4 5 6 7 8   Year: 2 3 4 5 6 7 8 9

A cautious man should above all be on his guard against resemblances; they are a very slippery sort of thing.

—Plato, ''The Sophist,'' translated by F. M. Cornford (*Plato's Theory of Knowledge,* London, 1935, p. 180).

Preface

This book is an effort to bring to undergraduate students an awareness of major features of the current state of animal biology at the level of the functional features of whole organisms and their component organs and organ systems. Emphasis is on function as it is related to the survival of organisms in their natural environments. The book might well be called a textbook of comparative ecological physiology. Phenomena at the submolecular to tissue levels are discussed only as they are needed to provide perspective at the organ to whole-animal levels. It is assumed that students will have additional exposure in their curriculum to courses in cellular or general physiology, as well as to anatomy (from ultrastructure to gross) and embryology.

Emphasis throughout is on the vertebrates, the fundamental approaches being zoological, comparative, and evolutionary. Material on ontogeny of functions is considered wherever possible. Material on invertebrates is discussed where it contributes to an appreciation of the diversity of functional adaptations shown by organisms, or where such examples will clarify the situation. The level of discussion presumes a basic familiarity, such as would be obtained from a good college level introductory biology or zoology course.

Comments we have received from some users of the previous editions make it seem worthwhile to include here a short statement giving the reasons for our strong belief that beginning students of physiology, whatever may be their personal career goals, should start with a broad zoologically based, comparative approach. We feel that it does a serious disservice to students to start them out with concentration on only mammalian, especially human, physiology.

Our reasons involve both theoretical considerations and also the realities of present-day university and professional school curricula. (1) A large part of new medical and veterinary practice derives originally from studies on other kinds of animals. A lack of understanding of the important ways in which other animals resemble or differ from the forms of direct concern to humans will make it difficult to evaluate the significance or relevance of such studies. (2) Present-day concerns with important issues such as the ecological bases for the food webs supporting animals of direct importance to humans, the indirect effects of environmental pollution, the control of destructive pest species, and the need for maximizing sustained yields of valuable animal species, all require for their solution detailed knowledge of the physiological properties of many other kinds of animals. (3) Students intending to go on to professional schools, if they succeed in obtaining entrance, will be required to take physiology courses strongly focused upon the aspects of physiology of greatest relevance to their curricula. Many of these courses are likely to be relatively intensive and short. We know from experience that a prior solid basis of understanding of principles will make these later courses both more meaningful and easier.

We have tried to make this book as complete as possible in terms of physiological principles and orientations, but it is purposely designed not to be exhaustive in terms of facts. Its field is so large and diverse that a choice had to be made between trying to cover all major topics and selecting for discussion only the most significant (to the authors) of the major topics. The latter has been the choice. The effort is made, in discussing each subject area, to give an understanding not only of the important facts but also of the experimental basis for these facts and, wherever possible, some of the reasoning that led to the performance of the experiments. We hope in this way to provide some feeling for the nature of biological enquiry along with a fundamental theoretical orientation in the given subfield that will permit independent study with comprehension of related topics not specifically considered here. To ease the process of independent study, each chapter concludes with a list of relevant recent books and review articles. These relate not only to the topics considered in the chapter but also to associated areas not adequately covered.

We hope that courses in which this text will be used will make use of additional reading materials, especially on the invertebrates. Laboratory exercises using invertebrate materials are suggested as other means for expanding the range of factual coverage.

The organization of this book emphasizes physiological processes. The introduction provides necessary theoretical, philosophical, evolutionary, and taxonomic background. The energy animals use, where they get it, and how they make it available for use in their own tissues are discussed in the first two chapters—on nutrition and metabolism. The use of the greatest part of this energy to produce movement is considered next. This is followed by descriptions of the processes producing the internal environments within organisms: respiration,

circulation, water and solute metabolism, and temperature adaptation and regulation. The final five chapters consider how animals obtain necessary information about the world both outside and inside themselves, how they correlate and control their activities in the light of this information (both neurophysiologically and endocrinologically), and how they reproduce themselves.

The first three editions of this book have been sufficiently widely used that both the authors and the publisher felt it important to try to keep the contents reasonably up-to-date. This edition involves both an updating and some expansion of coverage.

All the authors are only too aware of many limitations, shortcomings, and problems relating to their contributions. We hope, however, that despite these we have produced a text that will be useful, informative, and stimulating for students and other interested readers. We earnestly solicit any comments and suggestions for future improvements.

All of the authors have profited greatly from comments and suggestions relating to the previous editions. These comments have been generated by colleagues, friends, and students in many places. We wish to particularly thank the following: Drs. G. C. Grigg, L. O. Larsen, J. E. Minnich, L. C. Oglesby, C. R. Tracy, and P. Rosenkilde. Once again, we earnestly solicit any comments and suggestions for further improvements.

Wherever previously published photographs have been used for illustrations, the authors cited in the figure captions have kindly provided original prints. The cooperation of the many authors and publishers who have given permission to reproduce copyright material is gratefully acknowledged. Finally, the revision of the book would not have been possible without the assistance of a number of people, especially Christine Kulia, Thomas Haglund, Dana Gordon, and Alexandra Zaugg-Haglund.

M.S.G.

Contents

Detailed Contents

Malcolm S. Gordon

1

INTRODUCTION

1.1 GENERAL COMMENTS

Biology today is in the midst of one of the most active and exciting periods in its long history of activity and excitement. There are multiple causes for the electricity in the biological atmosphere, including the mundane but necessary one of the availability of funds to support research. Also there is an increasingly rapid breakdown of traditional barriers between scientific fields. The decline of scientific compartmentalism has led, in biology, to a major infusion of new talent, ideas, theoretical approaches, and techniques. Biology is rapidly changing from a largely descriptive, cataloging science to an experimental, analytic, and synthetic field based increasingly upon detailed understanding of underlying physical, chemical, and mathematical principles. In many areas biologists are beginning to develop enough perspective on the nature of the theoretical trees they have been studying to begin, at least, to perceive the nature of the forest in which they are all working.

These fundamental changes in biology are reflected in the nature of the field at every level of consideration of the biological world—from the submolecular to the ecosystem. This is as it should be, as there is certainly no logical basis for any claim that understanding the nature of life at one level of organization is more fundamental to overall understanding than comprehension at any other level.

Another major change in perspective derives from the addition of a strong time dimension in biological thinking. Modern methods for studying fossils and the rocks from which they come now permit making meaningful statements about important aspects of the finer structure, chemical composition, and physicochemical and biological environments in which ancient organisms lived. This information, combined with wider use of comparative studies on the diversity of living forms (each one of which is the current result of a continuing natural experiment that began far in the past), is leading to multidimensional pictures of living things evolving in time as well as adapting to their current environments. Evolutionary biology is making its own strong contribution to all the fields of biology one can create from a hierarchy of levels of complexity.

The living world is an object of study of a complexity and diversity not even remotely approached by the subject matter of any other field of science. It is an unending frontier full of great and small challenges, containing within itself something for literally all people with some curiosity about the world around them. To appreciate most fully the nature of the living world, one must approach it with an open mind and a broad perspective. We hope this book will aid students to make such an approach—and, perhaps, eventually to contribute to what we might call the new natural history.

1.2 A TAXONOMY OF PHYSIOLOGY

The facts resulting from scientific research are, in themselves, essentially neutral. They only take on meaning when viewed in the light of some theoretical construct or point of view. One of the most challenging and interesting features of studies of animal function and structure (all of which we shall include in the general field of physiology from here on) is that they are almost always open to interpretation in more than one way. These differing interpretations are sometimes diametrically opposed to one another, but more often they complement one another and add breadth to our understanding. It seems worthwhile to take a little time at this point to indicate the major value orientations used in physiology.

One of the oldest, most popular, and certainly most practical orientations is that of the relation of the work to human health and welfare. Medically oriented physiology, involving work primarily with man himself or other mammals, has been, is, and will continue to be the largest, best supported, and most widely known part of the field. Workers in this area continue to add to a vast literature, the greater part of which is directed toward understanding all aspects of human function in health and disease, the mode of action of drugs on people, and so on. A trend that has developed rapidly in recent years has been the realization that studies on animals other than mammals can also frequently be of considerable medical significance.

Physiological work on nonmammalian forms (or on mammals less "ordinary" than dogs, cats, monkeys, and so on) can also be of direct value to human welfare when it involves such topics as the control of reproduction in economically important species and understanding environmental effects on the survival and distribution of pest species. Increasingly, as the growing human population requires more food, more water, more recreation, and less pollution, physiological studies on other organisms become of greater value in the formulation of rational policies of environmental use. The current prospect of large-scale environmental modifications resulting from artificial rainmaking and other types of weather control make understanding the biological consequences of such changes crucial. The nature of many of these consequences will be determined in large degree by the physiologies of the organisms involved.

A third way of considering physiology is from the viewpoint of just plain curiosity. Animals do a great many things—how do they do them? How do they survive and reproduce in the broad range of natural environments in which they live? Biological nature seems to abhor an absence of life, so we find organisms living virtually every place that has conditions not absolutely beyond the range of adaptation of life as we know it. If the gross environment is beyond that range, we usually find there are microenvironments within it that organisms can exploit. What are the physiological mechanisms that permit this adaptability?

To make the issue more concrete, consider the range of some of the environmental variables tolerated by animals: air temperatures ranging from $-70°C$ (in the Arctic and Antarctic in winter) to above $45°C$ (shade temperatures in the Sahara Desert in summer); water temperatures from the freezing point of seawater (near $-2°C$ in the Antarctic) to above $42°C$ (hot springs in the American Southwest); light conditions from total darkness in deep caves to brilliant glaring sun on almost white coral sand on many tropical beaches; atmospheric pressures from the usual 1 atmosphere (atm) at sea level (a bit more in places such as Death Valley, California, below sea level) to less than half that at altitudes near 5000 meters (m) in mountains like the Himalayas and Andes; hydrostatic pressures from near 1 atm at the sea surface to near 1110 atm at the bottoms of the deepest submarine trenches (such as the Marianas Trench in the Pacific, 11,100 m deep); salt concentrations in water from near zero (in meltwater from inland snows) to nearly three times seawater concentration (in enclosed bays along the Texas coast in summer) or to saturated magnesium sulfate (some mineral springs in the deserts of the American Southwest). The list can be made much longer, but the adaptability of animals has been illustrated.

The desire for physiological understanding for its own sake can be pursued from several other points of view. However, because understanding how organisms function as organisms under their natural conditions of life is the ultimate goal of all approaches, the others may be classified as subdivisions of the environmental perspective.

There are several ways to define these subdivisions. One is to study the physiology of particular systematic categories of animals—a taxonomic approach. Another division is based upon the fact that, with present knowledge, biological understanding fits into what philosophers call an emergent type of classification. There are different levels of organizational complexity at which one can approach the physiological study of organisms (ranging from the submolecular to at least the population). Each level has properties that are unique to it and not presently predictable on the basis of a summation of the properties of all the simpler levels. Thus we have cell physiologists, organsystem physiologists, and so on.

A third division, perhaps philosophically the most significant, might be stated as the difference between the "lumpers" and the "splitters." A physiologist might be called a lumper (more ordinarily a general physiologist) if he feels that the diversity of life is underlain by broad uniformities in function and that most of the diversity is essentially a series of variations on a limited number of basic themes. Another physiologist could be called a splitter (more ordinarily a comparative physiologist) if he feels that the visible diversity of life is a reflection of equivalent underlying functional diversity. He will recognize that there are broad areas of uniformity, perhaps even a few across-the-board generalities, but he will emphasize that functional adaptations of animals to the natural world are not only complex but in most cases have probably arisen a number of times in different kinds of animals, frequently as a result of radically different evolutionary histories. This physiologist will often view supposed generalities as artifacts of our own relative ignorance, resulting from the study of only a few, frequently quite similar, species. He will also

be likely to feel that, if it is theoretically possible for living things to survive natural conditions by means of more than one mechanism, it is only a matter of time until study of a variety of species will demonstrate that all the logical possibilities have been used. The metaphor might be that, if there is more than one way to skin a cat, nature will have used all of them.

The comparative approach also lends itself easily to the evolutionary point of view. One of the more important categories of comparative studies emphasizes the unraveling of the ways in which homologous organs in different groups have adapted to varying functions as a consequence of the challenges presented by changing environments during the course of evolution.

Those users of this book who come to feel that understanding how organisms function is what they would like to devote themselves to will have to decide for themselves which of these basic points of view is most compatible with their own *Weltanschauung*. The distinctions drawn here are perhaps artificially clearly stated. A single physiologist can easily take all these points of view in different research projects at different times in his career. There is no necessary conflict between them, as long as the experiments are done carefully, properly, and honestly. If these conditions are met, everyone makes a real contribution to the body of human scientific understanding.

1.3 BACKGROUND

There are two major items that can help place the later discussions in this book in proper biological perspective: (1) the geological time scale and the relations to it of the major events in vertebrate evolutionary history, and (2) the classification of the major groups of living vertebrates. Tables 1-1 and 1-2 summarize present understanding in these two areas.

The ultimate sources for the information in this book are the research papers written by the people who did the experiments and published their results (usually after some amount of intellectual and editorial stress) in scientific journals. The number of such journals is very large and is increasing rapidly (first-class libraries devoted to the life sciences,

TABLE 1-1. Geological Time Scale and Some Important Events in Animal Evolution

Approximate Time Since Beginning of Period, Millions of Years	Era	Period or Epoch*	Some Important Events in the Evolution of Animals
0.01		Recent	
2.5	Cenozoic	Pleistocene	First true men
7	(age of mammals)	Pliocene ⎫	Culmination of mammals;
26		Miocene ⎭	radiation of apes
38		Oligocene ⎫	Modernization of
54		Eocene ⎭	mammalian faunas
65		Paleocene	Expansion of mammals
136	Mesozoic	Cretaceous	Last dinosaurs
190	(age of reptiles)	Jurassic	First birds
225		Triassic	First dinosaurs and mammals
280		Permian	Expansion of primitive reptiles
345		Carboniferous	First reptiles
395	Paleozoic	Devonian	First amphibians and insects
430		Silurian	
500		Ordovician	Earliest known fishes
570		Cambrian	Marine invertebrates abundant
3000+	Precambrian	Period names not well established	First known fossils

* An epoch is technically a subdivision of a period. The names in this column for the Cenozoic are technically epochs and those for the Mesozoic and Paleozoic are periods.

Source: W. T. Keeton. *Biological Science,* 3rd ed., W. W. Norton & Co., New York, 1980, Table 20.1, p. 906.

including medicine, currently subscribe to over 9000 different periodicals). Additional research papers appear in increasing numbers of symposium volumes. A rough estimate of the number of research papers published worldwide each year in the life sciences alone would be well over 200,000.

Finding specific items in this vast mass of literature is increasingly becoming a job for electronic computers. The computers, however, must be instructed most carefully in what to look for or essential papers will be overlooked. Considering the difficulties of such detailed programing, it seems possible that for the foreseeable future students and researchers will have to go personally to their libraries and look at the journals of particular interest to them. The short and incomplete list of abstract journals, review journals, and research jour-

nals that follows is intended to assist readers in gaining entry to the basic literature in organ and whole-animal physiology. The country of publication for each journal is listed in each case.

Abstract Journals

Berichte über die Wissenschaftliche Biologie (Germany)
Biological Abstracts (U.S.A.)
Bulletin Signalétique (France)
Referativny Zhurnal, Biologiya (U.S.S.R.)
Zoological Record (U.K.)

Review Journals

American Zoologist (U.S.A.)
Année Biologique (France)
Annual Reviews of Physiology (U.S.A.)

Biological Reviews (U.K.)
Physiological Reviews (U.S.A.)
Quarterly Reviews of Biology (U.S.A.)

Research Journals

Acta Physiologica Scandinavica (Scandinavia)
American Journal of Physiology (U.S.A.)
Archives Internationale de Physiologie et Biochimie (Belgium)
Australian Journal of Zoology (Australia)
Biological Bulletin (U.S.A.)
Brain Research (U.S.A.)
Canadian Journal of Biochemistry and Physiology (Canada)
Canadian Journal of Zoology (Canada)
Comparative Biochemistry and Physiology (U.K.)
Comptes Rendus de l'Académie des Sciences (France)
Comptes Rendus de la Société de Biologie (France)
Condor (U.S.A.)
Copeia (U.S.A.)
Die Naturwissenschaften (Germany)
Doklady Akademia Nauk U.S.S.R. (U.S.S.R.)
Endocrinology (U.S.A.)
Evolutionnoi Biokhimii i Fisiologii (U.S.S.R.)
Experientia (Switzerland)
Experimental Eye Research (U.S.A.)
Experimental Neurology (U.S.A.)
Forma et Functio (U.S.A.)
General and Comparative Endocrinology (U.S.A.)
Japanese Journal of Physiology (Japan)
Journal de Physiologie (France)
Journal of Applied Physiology (U.S.A.)
Journal of Auditory Research (U.S.A.)
Journal of Chemical Ecology (U.S.A.)
Journal of Comparative Pathology (U.S.A.)
Journal of Comparative Physiology (Germany)
Journal of Endocrinology (U.K.)
Journal of Experimental Biology (U.K.)
Journal of Experimental Marine Biology and Ecology (Holland)
Journal of Experimental Zoology (U.S.A.)
Journal of General Physiology (U.S.A.)
Journal of Insect Physiology (U.S.A.)
Journal of Mammalogy (U.S.A.)
Journal of Marine Biological Association of the United Kingdom (U.K.)
Journal of Muscle Research and Cell Motility (U.K.)
Journal of Neurophysiology (U.S.A.)
Journal of Physiology (U.K.)

Journal of Reproduction (U.K.)
Laboratory Animal Care (U.S.A.)
Marine Biology (Germany)
Molecular Physiology (Holland)
Nature (U.K.)
Physiological Zoology (U.S.A.)
Research on Physiology (U.S.A.)
Respiration Physiology (U.S.A.)
Science (U.S.A.)
The Auk (U.S.A.)
Vision Research (U.S.A.)

Although it is not a periodical, mention must be made of the ongoing large-scale publishing effort of the American Physiological Society, the *Handbook of Physiology*. As volumes appear, they become major reference sources and benchmarks against which research progress can be measured. The sections that have been published serve as invaluable keys to the knowledge making up many of the fields discussed in this book.

Recent years have also seen the development of several major, continuing series of books on important aspects of physiology published by commercial publishers. Some examples are

International Review of Physiology (University Park Press, Baltimore, Md.)
International Review of Biochemistry (University Park Press, Baltimore, Md.)
Comprehensive Biochemistry (American Elsevier Publishing Co., New York)
Horizons in Biochemistry and Biophysics (Addison-Wesley Publishing Co., Reading, Mass.)

1.4 PRACTICAL MATTERS

Physiology is preeminently an experimental subject involving the development and application of many kinds of techniques and equipment to a wide variety of organisms. Whatever aspect of physiology they decide to concentrate in, students immediately find that they not only have to become familiar with and proficient in several fields of theory but they also have to become various kinds of practical people, good with their hands and adept at solving new and unpredictable experimental situations. A certain amount of book knowledge is essential to

TABLE 1-2. Major Groups of Living Vertebrates (Phylum Chordata)*

Subphylum	Superclass	Class	Order	Vertebrates
Hemichordata		Enteropneusta		Acorn worms
		Pterobranchia		Pterobranchs
Cephalochordata				Lancelets
Tunicata		Ascidiacea		Sea squirts
		Thaliacea		Salps
		Larvacea		Larvacea
Vertebrata	Agnatha	Cyclostomata	Petromyzontiformes	Lampreys
			Myxiniformes	Hagfishes
	Gnathostomata	Elasmobranchii	Squaliformes	Sharks
			Rajiformes	Skates, rays
			Chimaeriformes	Chimaeras
		Actinopterygii	Coelacanthiformes	Coelacanths
			Dipteriformes	Lungfishes
			Polypteriformes	Bichirs
			Acipenseriformes	Sturgeons, paddlefishes
			Amiiformes	Bowfins
			Lepisosteiformes	Gars
			Anguilliformes	Eels and relations
			Clupeiformes	Herrings and relations
			Salmoniformes	Salmon, trout, and relations
			Cypriniformes	Minnows, carps, and relations
			Siluriformes	Catfishes
			Lophiiformes	Anglerfishes
			Gadiformes	Codfishes and relations
			Atheriniformes	Flying fishes, toothcarps, and relations
			Gasterosteiformes	Sticklebacks and relations
			Scorpaeniformes	Sculpins and relations
			Perciformes	All perchlike fishes
			Pleuronectiformes	Flatfishes
			Tetraodontiformes	Puffers and relations
		Amphibia	Urodela	Salamanders
			Anura	Frogs, toads, and relations
			Apoda	Caecilians
		Reptilia	Chelonia	Turtles
			Rhynchocephalia	Tuataras
			Squamata	Lizards, snakes
			Crocodilia	Crocodiles, alligators, and relations
		Aves	Paleognathiformes	Ostriches and relations
			Impenniformes	Penguins
			Gaviiformes	Loons
			Procellariiformes	Petrels, albatrosses
			Pelecaniformes	Cormorants, pelicans, and relations
			Ciconiiformes	Storks, herons, flamingoes
			Anseriformes	Ducks, geese, swans
			Falconiformes	Hawks and relations
			Galliformes	Chickens and relations
			Charadriiformes	Waders and gulls

TABLE 1-2. Major Groups of Living Vertebrates (Phylum Chordata)* (continued)

Subphylum	Superclass	Class	Order	Vertebrates
Vertebrata (cont'd)		Mammalia	Columbiformes	Pigeons
			Strigiformes	Owls
			Piciformes	Woodpeckers
			Passeriformes	Perching birds
			Monotremata	Platypus and relations
			Marsupialia	Kangaroos and relations
			Insectivora	Insectivores
			Chiroptera	Bats
			Primates	Primates
			Rodentia	Rodents
			Lagomorpha	Rabbits and relations
			Cetacea	Whales and relations
			Carnivora	Carnivores
			Proboscidea	Elephants and relations
			Perissodactyla	Horses and relations
			Artiodactyla	Even-toed ungulates

* Minor orders of fishes, birds, and mammals omitted.

Source: This classification derived from classifications given by P. H. Greenwood, et al., *Bull. Amer. Mus. Nat. Hist.*, **131,** 339–456 (1966); K. F. Lagler, et al., *Ichthyology: The Study of Fishes*, J. Wiley & Sons, Inc., New York, 1962; J. Z. Young, *The Life of Vertebrates*, 2nd ed., Oxford University Press, Fair Lawn, NJ, 1962.

the process of becoming an experimenter, but the point is quickly reached beyond which reading and talking are of no further use; it is necessary to get into the laboratory or field and try it one's self.

This book cannot even superficially discuss the book learning aspects of experimental design, methodology, equipment, and animal care and use involved in the various fields of physiology. The Additional Reading list at the end of this chapter provides some basic references in most of the important areas.

1.5 UNDERSTANDING PHYSIOLOGY: PATTERNS, MECHANISMS, CONTROLS

What constitutes an answer to a physiological question? Section 1.2 pointed out that an important approach to physiology is one that emphasizes the differences in physiological properties one encounters when organisms are studied at different levels of organizational complexity. These emergent properties of the different levels of organizational complexity also provide help in understanding the nature of physiological explanation.

Ecological physiological studies usually begin at the whole-animal level. A question is asked along the following lines: How do intact individuals of species A respond physiologically to changes in environmental factor B? Experiments are then performed to determine the pattern of responses shown by A to ecologically relevant changes in B. Once this pattern is established, second-order questions immediately arise concerning the mechanisms used by species A to produce this pattern. Which organ systems are involved in the response, and what changes occur in their functioning? Third-order questions also must be asked concerning how species A controls and coordinates the changes occurring in the involved organ systems.

A complete answer to the initial question thus involves three parts: the pattern of response at the original level of organizational complexity (the whole organism in the example); the mechanisms operating one level of complexity down, which produce the pattern (the organ-system level in the example); and the controls that integrate the mechanisms to produce a coordinated pattern.

The process can be continued: the changes in organ-system function that were found also constitute patterns of response for each organ system. Each of these patterns can be explained on the basis of mechanisms involving changes in the functioning of the individual organs in each organ system. There are also controls on the organs. Changes in organ function can then be explained on the basis of changes in the tissues making up the organs—and so on.

The ultimate, truly complete answer to the original question posed at the organismic level thus may be seen to involve patterns, mechanisms, and controls at all levels of organizational complexity down to the submolecular. In practice most physiologists feel that they have satisfactorily answered a given question by providing its explanation at one or at most two levels of complexity down from the level at which the question was originally asked.

1.6 GENERAL CONCEPTS AND FRONTIERS

The primary orientation of this book is ecological, with emphasis on the organ, organ-system, and whole-animal levels. We try to view the natural world as it appears to the animals we discuss. As you read we hope it will become apparent that the world is a very different place to most other species than it is to us.

Just what appearance the world has to an animal is dependent on many factors. These include the animal's body size, its locomotor abilities, its dietary habits, and the details of the ranges and patterns of sensitivity of its sensory systems. For example, the deserts of the American Southwest present an appearance to animals as large as man that is very different from their appearance to animals such as lizards, which are small enough to take advantage of tiny patches of shade under bushes, small crevices in rocks, holes in the ground, and so on. The challenge of dryness in the desert, so important to man, is virtually no challenge at all to many species of birds that feed primarily on high-water-content insects. Navigation around obstacles, feeding, finding mates, and so forth are no problem to the so-called electric fishes, even though many species live in turbid tropic waters.

Visibility in these waters is often only a matter of millimetres. The difference is largely the result of the highly specialized lateral line systems and electric organs possessed by these fishes.

Where animals live and what they do to get along where they live is determined not only by physical appearances but also by innate physiological and behavioral capabilities and evolutionary histories. Their evolutionary histories, indeed, not only put most of them where they are geographically but also determine their physical properties and their differing capabilities, which are the subject matter of this book. Additional information on animal evolution may be found in some of the books listed at the end of this chapter.

The importance of behavior in determining how well animals get along cannot be overemphasized. Much of what is said in later chapters will not explicitly consider behavior. However, whenever the topic relates to the effects of environmental conditions on bodily functions (for example, ambient temperature versus body temperature), remember always that animals in nature usually have microenvironmental choices open to them—choices they put to positive advantage by varying their behavior. A great deal of physiological experimentation is concerned with determining the capacities of animals to respond to, or to tolerate, specified environmental circumstances. Natural behavior patterns are explicitly prevented from occurring. As a result of this, it is almost stating a truism to point out that most of what we know of such things as the capacities of organisms to control their body temperatures, body-fluid concentrations, even many aspects of heart and circulatory-system function, is in large part laboratory artifact. We are only beginning to design experiments and develop observational procedures that permit the subjects to be anything like their natural selves.

Much of current physiological knowledge is ecological artifact in still another fundamental way. Even allowing for behavioral adjustments by animals, the natural environment is almost never constant in time. In most places there are short-period and diurnal fluctuations in temperature, light intensity, salinity, barometric pressure, oxygen partial pressure, and other quantities. There are also longer-period environmental fluctuations ranging from lunar monthly periods (for example, changes in tidal range) to annual seasonal changes, to secular trends developing slowly over many years. Here again, most physiological experimentation has been based upon exposure of organisms for short (acute experiments) or long (chronic experiments) periods of time to *constant* conditions, or conditions in which only one quantity varies. The rationale has been that this is essential to permit analysis of the effects of specific influences. This is only partly legitimate. The physiological effects of two simultaneously varying environmental quantities are not always simple combinations of the effects of each variable changing alone. The effects of temporal variations of single quantities are also not simple and easily predictable. The amplitude, frequency, and nature of variations all can have major influences on the effects. We know very little in detail about these areas.

In recent years there has been a great increase in awareness among biologists of the fact that organisms also vary their properties with time. Studies of the effects of age and season have, of course, been carried out almost since the earliest days of physiology—although these areas of study continue to be relatively neglected. The new awareness relates mainly to shorter-term variations, primarily variations on an approximately diurnal or circadian scale. It now appears that a very large number of functions, if not all, in a wide range of species vary with circadian frequencies. This variation persists, in many cases, even after very long periods of maintenance in environments kept constant in as many ways as possible (for example, constant light intensity, constant temperature, constant humidity, and so on). Functions that vary range from body temperature through metabolic rate to rates of excretion of various organic compounds in the urine. The mechanisms producing these variations are not yet understood nor are the influences controlling them. In most cases their adaptive significance is similarly obscure. However, they occur, and most physiological research to date has taken no account of their occurrence.

Finally, there are two additional considerations that may be of great importance on occasion. Most of the time physiologists try to avoid the necessity for considering these by appropriate statistical design of their experiments and by careful choice of subject animals. However, even these precautions are sometimes not enough. The two considerations are genetic makeup of the experimental animals and the possibility that animals used may be pathological in some important way—in any case, not "normal."

The importance of genetic uniformity among experimental animals probably varies considerably with the nature of the problem under study. It is probably of greatest concern in situations in which the effect or phenomenon looked for is small and perhaps easily masked by the biological variability of the animals used. In a number of cases that have been investigated, a significant portion of this biological variability has been shown to be due to genetic variability.

The problem is sometimes on a larger scale. There are many documented cases of organisms that belong to a single morphological species, but to widely separated populations of that species, that are physiologically as different from each other (for example, in dependence of metabolic rate on environmental temperature) as one often finds two different species to be. Careful studies in some of these cases have demonstrated these differences to be true genetic differences, not simply phenotypic adaptational changes.

The question of possible pathology is even more difficult to handle. Even in our best-known species, man, there is great dispute about what constitutes normality in any given property. We know so much less about normality and deviations from it in other species that it is presently virtually hopeless to do anything more than pick subjects that do not show any noticeable unusual features in appearance, behavior, or function, as we understand it.

The development in recent years of bacteria-, virus-, fungus-, and parasite-free (so-called axenic) strains of a few species has further complicated this aspect of physiology. Studies of such animals indicate that, in many important regards, what we think of as the normal function of the species is really some complex interaction between the inherent properties of the form itself and its present and previous bacterial, viral, fungal, and parasitic tenants. Axenic organisms completely lacking in the usual history of encounters with these hangers-on generally have so little resistance to many ordinary insults of the normal world that they soon succumb following even brief exposures. The development of techniques that permit maintenance and study of axenic animals opens up another fundamental and unexplored field of study.

We hope that those of you who have persisted in reading this chapter to this point have not become totally dismayed and disillusioned by the complexities outlined. Physiology *is* a complex game. It has been termed by some the "queen of the biological sciences," since its proper study involves a synthesis of mathematics, physics, chemistry, and most of the other biological fields. It has also been called "an old person's game," because its complexity and technical challenge require long experience before they can be fully mastered. It is both these things. As a result its study is simultaneously broadly challenging, endlessly varied, and continually rewarding. For those having the interest, determination, persistence, and drive, all that is needed is to begin.

ADDITIONAL READING

HANDBOOKS
Altman, P. L., and D. S. Dittmer (eds.). *Biological Handbooks (Blood and Other Body Fluids; Environmental Biology; Metabolism; Respiration and Circulation)*, Federation of American Societies for Experimental Biology, Bethesda, MD, 1961–1971.

————. *Biology Data Book*, 3 vols., Federation of American Societies for Experimental Biology, Bethesda, MD, 1972–1974.

ANIMAL SYSTEMATICS AND EVOLUTION
Borradaile, L. A., F. A. Potts, L. E. S. Eastham, J. T. Saunders, and G. A. Kerkut. *The Invertebrata: A Manual for the Use of Students*, 4th ed., Cambridge University Press, New York, 1961.

Hallam, A. (ed.). *Atlas of Palaeobiogeography*, American Elsevier Publishing Co., New York, 1972.

Hanson, E. D. *Understanding Evolution*, Oxford University Press, New York, 1981.

Hyman, L. H. *The Invertebrates*, 6 vols., McGraw-Hill Book Company, New York, 1940–1967.

Mayr, E. *Animal Species and Evolution*, Harvard University Press, Cambridge, MA, 1963.

————. *Principles of Systematic Zoology*, McGraw-Hill, New York, 1969.

————. *Populations, Species, and Evolution*, Harvard University Press, Cambridge, MA, 1970.

Olson, E. C. *Vertebrate Paleozoology*, J. Wiley and Sons, New York, 1971.

Romer, A. S. *Vertebrate Paleontology*, 3rd ed., University of Chicago Press, Chicago, 1966.

Solbrig, O. T., and D. J. Solbrig. *Introduction to Population Biology and Evolution*, Addison-Wesley Publishing Company, Reading, MA, 1979.

Young, J. Z. *The Life of Vertebrates*, 2nd ed., Oxford University Press, Fair Lawn, NJ, 1962.

HISTORY OF PHYSIOLOGY
Florkin, M., and E. Stotz. *A History of Biochemistry*, 4 vols., American Elsevier Publishing Co., New York, 1972–1975.

Fulton, J. F., and L. G. Wilson. *Selected Readings in the History of Physiology*, 2nd ed., C. C. Thomas, Springfield, IL, 1966.

Hodgkin, A., A. F. Huxley, W. S. Feldberg, W. A. H. Rushton, R. A. Gregory, and R. A. McCance. *The Pursuit of Nature: Informal Essays on the History of Physiology*, Cambridge University Press, New York, 1977.

Leicester, H. M. *Development of Biochemical Concepts from Ancient to Modern Times*, Harvard University Press, Cambridge, MA, 1974.

Rothschuh, K. E. *History of Physiology*, R. E. Kreiger Publishing Co., Huntington, NY, 1972.

PHYSIOLOGY OF DEVELOPMENT
Adolph, E. F. *Origins of Physiological Regulations*, Academic Press, New York, 1968.

Assali, N. S. (ed.). *Biology of Gestation*, 2 vols., Academic Press, New York, 1968.

Weber, R. (ed.). *The Biochemistry of Animal Development*, 2 vols., Academic Press, New York, 1967.

PHYSIOLOGY OF GROUPS
Boolootian, R. A. (ed.). *Physiology of Echinodermata*, Interscience, New York, 1966.

Brand, T. von. *Biochemistry of Parasites*, Academic Press, New York, 1966.

Crompton, D. W. T. *An Ecological Approach to Acanthocephalan Physiology*, Cambridge University Press, Cambridge, England, 1970.

Fallis, A. M. (ed.). *Ecology and Physiology of Parasites*, University of Toronto Press, Toronto, 1971.

Farner, D. S., and J. R. King (eds.). *Avian Biology*, 4 vols., Academic Press, New York, 1971–1974.

Florkin, M., and B. T. Scheer (eds.). *Chemical Zoology*, 8 vols., Academic Press, New York, 1967–1974.

Gans, C., and W. R. Dawson (eds.). *Biology of the Reptilia*, vol. 5, Physiology A, Academic Press, New York, 1976.

Gans, C., and K. A. Gans (eds.). *Biology of the Reptilia*, vol. 8, Physiology B, Academic Press, New York, 1978.

Gans, C., and F. H. Pough (eds.). *Biology of the Reptilia*, vol. 11, Physiology C, Academic Press, New York, 1981.

Hoar, W. S., and D. J. Randall (eds.). *Fish Physiology*, 8 vols., Academic Press, New York, 1969–79.

Lee, D. L. *The Physiology of Nematodes*, W. H. Freeman, San Francisco, 1965.

Lockwood, A. P. M. *Aspects of the Physiology of Crustacea*, W. H. Freeman Co., San Francisco, 1968.

Lofts, B. (ed.). *Physiology of the Amphibia,* vols. 2, 3, Academic Press, New York, 1974–1976.

Moore, J. A. (ed.). *Physiology of the Amphibia,* vol. 1, Academic Press, New York, 1964.

Rockstein, M. (ed.). *The Physiology of Insecta,* 2nd ed., 3 vols., Academic Press, New York, 1973.

Smyth, J. D. *The Physiology of Trematodes,* W. H. Freeman Co., San Francisco, 1966.

———. *The Physiology of Cestodes,* W. H. Freeman Co., San Francisco, 1969.

Sturkie, P. D. *Avian Physiology,* 3rd ed., Springer-Verlag, New York, 1976.

Wigglesworth, V. B. *The Principles of Insect Physiology,* 7th ed., Wiley-Interscience, New York, 1972.

Wilbur, K. M., and C. M. Yonge. *Physiology of Mollusca,* Academic Press, New York, 1964.

EQUIPMENT AND TECHNIQUES

ANIMAL CARE AND STUDY

Domer, F. R. *Animal Experiments in Pharmacological Analysis,* Charles C. Thomas, Springfield, IL, 1970.

Gay, W. I. (ed.). *Methods of Animal Experimentation,* 5 vols., Academic Press, New York, 1965–1974.

Institute of Laboratory Animal Resources. *Standards and Guidelines for the Breeding, Care, and Management of Laboratory Animals (Amphibians; Coturnix; Dogs; Fishes; Gnotobiotes; Nonhuman Primates; Rodents; Ruminants; Swine),* National Academy of Sciences–National Research Council, Washington, D.C., 1969–1974.

Institute of Laboratory Animal Resources. *Nutrient Requirements of Laboratory Animals,* National Academy of Sciences–National Research Council, Washington, D.C., 1972.

Institute of Laboratory Animal Resources. *Animals for Research,* 10th ed., National Academy of Sciences–National Research Council, Washington, D.C., 1979.

Institute of Laboratory Animal Resources. *Research in Zoos and Aquariums,* National Academy of Sciences–National Research Council, Washington, D.C., 1975.

Kirk, R. W., and S. I. Bistner. *Handbook of Veterinary Procedures and Emergency Treatment,* 2nd ed., W. B. Saunders Co., Philadelphia, 1975.

Lane-Petter, W., A. N. Worden, B. F. Hill, J. S. Patterson, and H. G. Veners (eds.). *UFAW Handbook on the Care and Management of Laboratory Animals,* 3rd ed., Williams and Wilkins, Baltimore, 1967.

Leonard, E. P. *Fundamentals of Small Animal Surgery,* W. B. Saunders Co., Philadelphia, 1968.

Lumb, W. V. *Small Animal Anesthesia,* Lea and Febiger, Philadelphia, 1963.

National Institutes of Health. *Guide for the Care and Use of Laboratory Animals,* DHEW Publication No. (NIH) 78–23, Revised, Office of Science and Health Reports, DRR/NIH, Bethesda, MD, 1978.

New, D. A. T. *The Culture of Vertebrate Embryos,* Logos Press, London, 1966.

Rafferty, K. A. *Methods in Experimental Embryology of the Mouse,* Johns Hopkins Press, Baltimore, 1970.

Ritchie, H. D., and J. D. Hardcastle (eds.). *Isolated Organ Perfusion,* University Park Press, Baltimore, 1973.

Short, D. F., and D. P. Woodnott (eds.). *The I.A.T. Manual of Laboratory Animal Practice and Techniques,* 2nd ed., C. C. Thomas Co., Springfield, IL, 1971.

Spotte, S. *Marine Aquarium Keeping: The Science, Animals, and Art,* Wiley-Interscience, New York, 1973.

———. *Fish and Invertebrate Culture: Water Management in Closed Systems,* 2nd ed., Wiley-Interscience, New York, 1979.

Vago, C. (ed.). *Invertebrate Tissue Culture,* 2 vols., Academic Press, New York, 1971–1972.

Zarrow, M. X., J. M. Yochim, J. L. McCarthy, and R. C. Sanborn, *Experimental Endocrinology: A Sourcebook of Basic Techniques,* Academic Press, New York, 1964.

STATISTICS AND COMPUTERS

Box, G. E. P., W. G. Hunter, and J. S. Hunter. *Statistics for Experimenters,* Wiley-Interscience, Somerset, NJ, 1978.

Daniel, C., and F. S. Wood. *Fitting Equations to Data,* J. Wiley and Sons, New York, 1974.

Gruenberger, F., and D. Babcock. *Computing with Mini-Computers,* J. Wiley and Sons, New York, 1974.

Hollander, M., and D. A. Wolfe. *Nonparametric Statistical Methods,* J. Wiley and Sons, New York, 1974.

Orr, H. D., J. C. Marshall, T. H. Isenhour, and P. C. Jurs. *Introduction to Computer Programming for Biological Scientists,* Allyn and Bacon, Inc., Boston, 1973.

Pollard, J. H. *A Handbook of Numerical and Statistical Techniques,* Cambridge University Press, New York, 1977.

CHEMICAL METHODS

Bergmeyer, H. U. (ed.). *Methods of Enzymatic Analysis,* 2nd English ed., 4 vols., Academic Press, New York, 1974.

Bloomfield, D. A. (ed.). *Dye Curves: the Theory and Practice of Indicator Dilution,* University Park Press, Baltimore, 1974.

Glick, D. (ed.). *Methods of Biochemical Analysis,* 26 vols., Wiley-Interscience, New York, 1952–1980.

Kirchner, J. G. *Thin-layer Chromatography,* 2nd ed., Wiley-Interscience, Somerset, NJ, 1979.

Mattenheimer, H. *Micromethods for the Clinical and Biochemical Laboratory,* Ann Arbor-Humphrey Science Publishers, Ann Arbor, MI, 1970.

Morris, C. J. O. R., and P. Morris. *Separation Methods in Biochemistry,* 2nd ed., J. Wiley and Sons, Somerset, NJ, 1976.

Natelson, S. *Microtechniques of Clinical Chemistry,* 2nd ed., C. C. Thomas, Springfield, IL, 1961.

Nicolau, C. (ed.). *Experimental Methods in Biophysical Chemistry,* Wiley-Interscience, New York, 1973.

Welcher, F. J. (ed.). *Standard Methods of Chemical Analysis,* vols. III A and B, *Instrumental Analysis,* 6th ed., Van Nostrand Reinhold Co., New York, 1969.

Williams, B. L., and K. Wilson (eds.). *Principles and Techniques of Practical Biochemistry,* American Elsevier Publishing Co., New York, 1975.

MATHEMATICS AND PHYSICS

Glantz, S. A. *Mathematics for Biomedical Applications,* University of California Press, Berkeley, CA, 1979.

Goel, N. S., and N. Richter-Dyn. *Stochastic Models in Biology,* Academic Press, New York, 1974.

Morowitz, H. J. *Entropy for Biologists: an Introduction to Thermodynamics,* Academic Press, New York, 1970.

Schepartz, B. *Dimensional Analysis in the Biomedical Sciences,* C. C. Thomas Co., Springfield, IL, 1980.

White, D. C. S. *Biological Physics,* Halsted Press, New York, 1974.

INSTRUMENTAL METHODS

Bruck, S. D. *Properties of Biomaterials in the Physiological Environment,* CRC Press, Inc., Boca Raton, FL, 1980.

Cobbold, R. S. C. *Transducers for Biomedical Measurements: Principles and Applications,* Wiley-Interscience, New York, 1974.

Covington, A. K. (ed.). *Ion Selective Electrode Methodology,* 2 vols., CRC Press, Inc., Boca Raton, FL, 1979.

Glauert, A. M. (ed.). *Practical Methods in Electron Microscopy,* 5 vols., North Holland Publishing Company, Amsterdam, 1972–1977.

Goldstein, J. I., J. J. Hren, and D. C. Joy (eds.). *Introduction to Analytical Electron Microscopy,* Plenum Publishing Corporation, New York, 1979.

Goldstein, N. N., and M. J. Free. *Foundations of Physiological Instrumentation,* C. C. Thomas Co., Springfield, IL, 1979.

Gray, P. (ed.). *The Encyclopedia of Microscopy and Microtechnique,* Van Nostrand, Reinhold, New York, 1973.

Gross, J. F., R. Kaufmann, and E. Wetterer (eds.). *Modern Techniques in Physiological Sciences,* Academic Press, New York, 1973.

Hayat, M. A. *Introduction to Biological Scanning Electron Microscopy,* University Park Press, Baltimore, 1978.

Hitchman, M. L. *Measurement of Dissolved Oxygen,* J. Wiley and Sons, New York, 1978.

Mackay, R. S. *Biomedical Telemetry,* 2nd ed., J. Wiley and Sons, New York, 1970.

Nastuk, W. L. (ed.). *Physical Techniques in Biological Research,* 6 vols., Academic Press, New York, 1961–1962.

Neuman, M. R., D. G. Fleming, W. H. Ko, and P. W. Cheung (eds.). *Physical Sensors for Biomedical Applications,* CRC Press, Inc., Boca Raton, FL, 1980.

Newman, D. W. *Instrumental Methods of Experimental Biology,* Macmillan Co., New York, 1964.

Slayter, E. *Optical Methods of Biology,* J. Wiley and Sons, New York, 1970.

Wyard, S. J. *Solid State Biophysics,* McGraw-Hill Co., New York, 1970.

ELECTRONICS

Offner, F. *Electronics for Biologists,* McGraw-Hill Co., New York, 1967.

Olsen, G. H. *Electronics: A General Introduction for the Non-Specialist,* Plenum Press, New York, 1968.

Zucker, M. H. *Electronic Circuits for the Behavioral and Biomedical Sciences: A Reference Book of Useful Solid-State Circuits,* W. H. Freeman Co., San Francisco, 1969.

RADIOISOTOPES

Hendee, W. R. *Radioactive Isotopes in Biological Research,* Wiley-Interscience, New York, 1973.

Lassen, N. A., and W. Perl. *Tracer Kinetic Methods in Medical Physiology,* Raven Press, New York, 1979.

Wang, C. H., D. L. Willis, and W. D. Loveland. *Radiotracer Methodology in the Biological, Environmental, and Physical Sciences,* Prentice-Hall, Inc., Englewood Cliffs, NJ, 1975.

PHOTOGRAPHY

Blaker, A. A., *Photography for Scientific Publication: A Handbook,* W. H. Freeman Co., San Francisco, 1965.

Engel, C. E. (ed.). *Photography for the Scientist,* Academic Press, New York, 1968.

Ettlinger, D. M. T. (ed.). *Natural History Photography,* Academic Press, New York, 1975.

MISCELLANEOUS

Commission on Biochemical Nomenclature. *Enzyme Nomenclature,* American Elsevier Publishing Co., New York, 1973.

Day, R. A. *How to Write and Publish a Scientific Paper,* ISI Press, Philadelphia, 1979.

DeHart, W. D., and R. J. Siegel. *Audiovisual Aids Useful in the Teaching of Physiology,* American Physiological Society, Bethesda, MD, 1973.

Knudsen, J. W. *Biological Techniques: Collecting, Preserving and Illustrating Plants and Animals,* Harper & Row, New York, 1966.

Leader, R. W., and I. Leader. *Dictionary of Comparative Pathology and Experimental Biology,* W. B. Saunders Co., Philadelphia, 1971.

Lincoln, R. J., and J. G. Sheals. *Invertebrate Animals: Collection and Preservation,* Cambridge University Press, New York, 1980.

O'Connor, M., and F. P. Woodford. *Writing Scientific Papers in English,* American Elsevier Publishing Co., New York, 1975.

Smith, R. C., and R. H. Painter. *Guide to the Literature of the Zoological Sciences,* 7th ed., Burgess Publishing Co., Minneapolis, 1966.

Tichy, H. J. *Effective Writing for Engineers, Managers, Scientists,* J. Wiley and Sons, New York, 1974.

C. Barker Jørgensen

2

NUTRITION

2.1 INTRODUCTION

Nutrition means the ingestion, digestion, absorption, and assimilation of food materials. The food represents energy to be used in metabolism and growth. In an ecological context, animal nutrition is closely connected with the concept of energy flow between trophic levels in ecosystems. Energy captured from sunlight by the first trophic level, the primary producers, most often represented by green plants, is utilized by the herbivores, the first order of consumers. The herbivores are eaten by the primary carnivores, which in turn are eaten by the secondary carnivores, and so on. These flows of energy can be described in such general terms that the organisms that perform the actual energy transfers remain anonymous, being present only in the form of energy equivalents in the flow models. However, in order to obtain the basic data from which these energy equivalents are derived, it is necessary to study nutrition in specific animals under specified conditions.

The study of animal nutrition in terms of types of foods and feeding mechanisms probably represents the most diverse of all areas of animal physiology. It constitutes a rich field for exploration of adaptive mechanisms. Practically every type of organic matter can be utilized as food. Food is obtained by animals by a diversity of mechanisms, the nature of which being determined to some degree by the taxonomic position of an animal, but primarily by the nature of the food. Within single taxonomic groups of animals, which may otherwise show great similarities in structure and form, representatives of many different feeding types can often be found. In this chapter emphasis is on major strategies by which animals exploit the primary production, the quantitatively most important link in food-chain bioenergetics.

At the cellular and biochemical levels of nutrition, most diversity vanishes and unity prevails. The basic nutritional requirements of animals are almost universal in terms of both energy sources (that is, proteins, lipids, and carbohydrates) and specific substances (that is, essential amino acids, vitamins, essential fatty acids, and inorganic trace elements). Throughout the animal kingdom the same types of enzymes are responsible for the digestion of food, and the same types of mechanisms transport the products of digestion across the cell membranes lining the digestive tract.

This change from unity to diversity, when we move from the biochemical and biophysical levels of animal nutrition to the physiological level, which deals with feeding mechanisms and types of food, seems to reflect a fundamental trend in the nature of evolution. Apparently all basic biochemical machineries had already evolved during procaryote evolution. Eucaryote and metazoan evolution presumably proceeded with only modest innovations at the cellular and biochemical levels.

2.2 FEEDING TYPES, FOOD CHAINS, AND FOOD WEBS

Studies on animal nutrition long focused on elucidating structure and function of the digestive tract and other structures of feeding. Comparative physiologists were especially concerned with relationships between types of food and adaptive mechanisms of feeding and digestion. On the basis of the extensive knowledge that accumulated: several feeding types could be distinguished: herbivores, carnivores, omnivores, detritus feeders, deposit feeders, fluid feeders, ectoparasites, and endoparasites. These main groups could be further subdivided. Several principles have been adopted in trying to arrive at satisfactory classifications of the diversity of feeding types among animals. One widely used classification is shown in Table 2-1. The major groups are characterized by the size of the food elements; the subdivisions are based on the types of mechanisms that the animals use to acquire these elements.

The rising interest in ecological aspects of animal nutrition expanded this traditional basis for classification of feeding types with the concepts of food chains and food webs; thus, organisms are classified according to the trophic level they occupy. This is the approach used in the following sections. Feeding types and nutritional adaptations will be described within an ecological framework. Further details on feeding types and their adaptations, mostly treated in a more traditional context, may be found in the references cited at the end of this chapter.

TABLE 2-1. Classification of Feeding Mechanisms Used by Animals

Mechanisms for Dealing With Food	Examples
Microphages	
Pseudopods	Radiolarians, foraminiferans
Cilia	Ciliates, sponges, many worms, brachiopods, bivalves, anuran tadpoles (e.g., *Xenopus, Bufo,* and others)
Mucus, nets	Several gastropods, tunicates, *Amphioxus,* Ammocoetes
Setae and similar filtering structures	*Daphnia,* copepods, and other crustaceans, basking shark, some teleosts, flamingo, whalebone whales
Macrophages	
Mechanisms for swallowing the surrounding medium (mud, sand, earth, etc.)	Many burrowing and digging forms in sediments in the sea and fresh waters, earthworms
Mechanisms for seizing prey	Most coelenterates, many polychaetes, most non-mammalian vertebrates, some mammals
Mechanisms for seizing and masticating prey, and for biting, rasping, grazing, etc., often combined with mastication of food	Many gastropods, cephalopods, crustaceans, and insects; cyclostomes, some birds and most mammals
Feeders on Fluids and Dissolved Food	
Mechanisms for sucking fluids	Trematodes, nematodes, leeches, parasitic copepods, several insect groups, young of mammals
Mechanisms for absorbing dissolved food through external surfaces: parenteral food uptake	Many parasites, marine invertebrates

Source: C. M. Yonge, *Biol. Rev.,* **3,** 21–76 (1928); H. J. Jordan, and G. C. Hirsch, in A. Bethe et al. (eds.), *Handbuch der normalen und pathologischen Physiologie,* vol. 3, Springer-Verlag OHG. Berlin, 1927, part II, pp. 24–101.

2.2.1 Utilization of Primary Production

The primary production may enter the energy flow through the food webs in a diversity of ways, but two are more important than any others. In the first, the primary production, in the form of green plants, is eaten and digested by the consumers. In the second, the plants are utilized indirectly by means of microorganisms that cause decomposition.

The relative importance of the two pathways varies with the habitat, mainly depending upon the size and structure of the plants (Figure 2-1). In aquatic habitats, the sea and lakes, by far the greatest part of the primary production is carried on by microscopic plants, the phytoplankton. Usually phytoplankton cells are naked or have only thin walls, and constitute easily digestible food for their consumers. Terrestrial plants, seaweeds, and sea grasses are mostly large, and significant parts of their synthetic products are deposited as structural carbohydrates, especially cellulose, which usually cannot be broken down by animal digestive enzymes. Because of this difference, the main problem to be solved by aquatic primary consumers is how to acquire their highly dispersed food particles in sufficient amounts. In the terrestrial and near-shore aquatic habitats the main problem for the consumers is how to exploit the energy contained in the primary producers. As a result, various types of suspension feeding have come to characterize aquatic environments, whereas detritus pathways predominate in terrestrial and near-shore food webs (Figure 2-1).

2.2.2 Aquatic Environments

Figure 2-2 diagrams a food chain and a food web characteristic of the euphotic zone (the zone in which plants can photosynthesize) of the ocean or a lake. The total primary production of phytoplankton in the oceans is estimated at about 10^{13} kg of carbon per year; that is of the same order as the total primary production of terrestrial plants.

Figure 2-1

Main types of utilization of primary production.

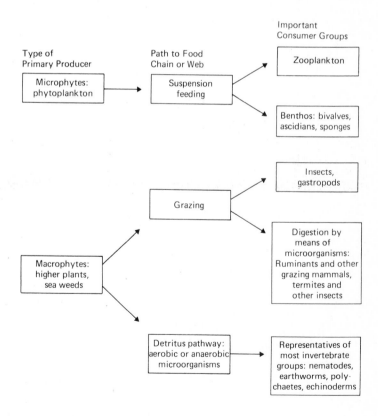

Food Chain

Figure 2-2
Diagrams of an aquatic food chain and a food web.

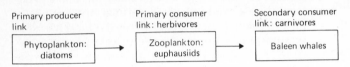

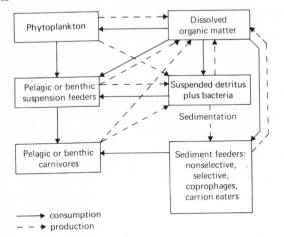

Aquatic and terrestrial primary producers are therefore about equally important in the economy of nature.

The phytoplankton production constitutes the basis for the web depicted in Figure 2-2. The algae are consumed by the grazing zooplankton and other planktonic or benthic suspension feeders, which are again eaten by carnivores. However, significant fractions of the primary production are only utilized indirectly by organisms representing higher trophic levels. The phytoplankton organisms seem to release about $\frac{1}{10}$ of the synthetic products directly to the surrounding water, where they contribute to the pool of dissolved organic matter. The part of the phytoplankton that is not eaten dies and becomes detritus; that is, dead decaying matter. On the diagram suspended detritus is shown as a complex with the bacteria that cause decomposition. The role of free-living bacteria is not fully understood. (See Additional Readings: Suspension feeding).

Also, animals release small organic molecules to the surrounding water, and when animals die they,

too, turn into detritus-bacteria complexes. These complexes, when present as small suspended particles, may be eaten by suspension feeders along with other suspended materials.

Other parts of the detritus particles sink to the bottom to be deposited with other sediments. The sediment feeders constitute a heterogeneous group of animals. At the one extreme, animals ingest the sediment unselectively, for instance several burrowing polychaetes. But many sediment eaters select material more rich in organic matter than the average. Probably, it is not the dead organic matter in the sediments that constitute the primary source of food energy to the sediment eaters but the bacteria that grow on the detritus and other particles.

In soft sediments, rich in organic matter, a large part of the detritus may consist of fecal pellets, which constitute the main diet of many sediment eaters. Again the fecal pellets mainly serve as a substratum for the bacteria on which the coprophages nourish. The special role that dissolved organic matter plays in the nutrition of marine animals will be dealt with in Section 2.2.4.

2.2.3 Suspension Feeding

The main characteristics of suspension feeding derive from the small sizes of the food particles compared with the size of the consumer. As a consequence the typical suspension feeder cannot sense and seize individual food particles or other particles suspended in the surrounding water. In order to obtain food the animals process the water through structures that retain suspended particles. Some forms use structures that may act as filters, for example the mucus nets of tunicates and several gastropods, and the setae of crustaceans (Table 2-1). Other suspension feeders use ciliary mechanisms for transporting water and retaining particles, for example many polychaetes, brachiopods, bivalves, and the pelagic larvae of numerous bottom invertebrates. The particle-retaining mechanisms in these latter forms are not yet fully understood (see Additional Reading).

The rates at which suspension feeders process the surrounding water and the efficiencies with which they retain suspended particles are both basic parameters in the physiology of feeding in these animals. The efficiency of particle retention is primarily determined by structural characteristics. Suspension feeders typically efficiently retain particles down to some few micrometers in diameter.

Suspension feeders process the surrounding water at rates that are largely independent of either the presence or the concentrations of food and other particles in the water. Undisturbed animals process water continuously and retain particles nonselectively, that is, independently of their value as food. The retained material is usually also ingested. The capacity for processing the surrounding water therefore is an innate, genetically determined behavioral and functional pattern, which has become established through evolution as an adaptation to the concentrations of suspended food in the habitat.

The typically automatic, stereotyped, and nonselective characteristics of their feeding behaviors do not, however, imply that rates of water processing by suspension feeders cannot be modified by factors in the environment. In fact, most suspension feeders are highly sensitive toward adverse environmental conditions, especially the presence of excessive amounts of suspended material, such as silt. They respond to adverse conditions by reducing or discontinuing water processing.

It is useful to compare the rates at which suspension feeders process the surrounding water with the energy needs of the animals, because this relationship provides information about the food levels that act as the ultimate factor in determining the rates of water processing. Table 2-2 provides some examples of the relationships existing between rates of water processing and metabolic requirements, as expressed by rates of oxygen consumption, in various types of suspension feeders that live in various habitats. Suspension feeders like salps, which live in the open ocean, process water at about 10 times higher rates than suspension feeders inhabiting coastal waters. Zooplankton crustaceans from freshwater and saline waters process the water at low rates.

The rates of water processing should be compared with the concentrations of food particles and other particles in the surrounding waters. Table 2-3 shows that concentrations of both inorganic and organic particulate matter vary greatly but tend to be an order of magnitude or more higher in coastal waters than in the open ocean. Concentra-

TABLE 2.2. Relationship Between Rates of Water Processing and Oxygen Consumption in Various Suspension Feeders

Habitat and Species	Liters of Water Processed For Each ml of O_2 Consumed
Oceanic waters	
Salp (Pegea confederata)	300
Coastal waters	
Sponges	15–30
Bryozoans	10–80
Bivalves	15–80
Ascidians	10–20
Copepods (Calanus spp, Acartia spp)	10–60
Freshwaters	
Daphnians	3–4
Saline waters	
Artemia salina	3

Source: C. B. Jørgensen. *Suspension Feeding. Handbook of Nutrition*, CRC Press, Boca Raton, Florida, 1982, in press.

TABLE 2.3. Particulate Inorganic and Organic Matter, Phytoplankton, and Bacteria Within the Euphotic Zone* of the Sea (mg dry weight l⁻¹)

Habitat	Inorganic Particulate Matter	Total Organic Particulate Matter (Detritus)	Phytoplankton	Bacteria
Oceanic waters	≃ 0–2	0.02–0.5	0.01–0.1	≃ 0.01
Coastal waters	≃ 0– > 100	0.2–10	0.2–2	0.01–0.1

* The water column within which the light intensity is high enough to support phytoplankton growth.

Source: C. B. Jørgensen. *Suspension Feeding. Handbook of Nutrition,* CRC Press, Boca Raton, Florida, 1982, in press.

tions vary even more in freshwaters, but the trend is toward even higher concentrations than in marine coastal waters.

The relative abundances of different types of particles are much the same in all aquatic environments. Inorganic particles usually constitute the largest fraction of suspended particles, followed by dead organic matter, detritus. Among living cells phytoplankton predominate, and during phytoplankton blooms the concentrations may reach those of detritus. Bacterial mass is small in unpolluted waters.

As shown in Table 2-2, coastal water representatives of widely differing taxonomic groups using widely differing feeding mechanisms process the water at about equal rates. This indicates that in all these forms the water-processing capacity has become adapted to about the same concentrations of food particles in the water. Phytoplankton cells are probably predominant among these food particles. From the data for the rates of water processing (Table 2-2) and for the phytoplankton concentrations (Table 2-3), we can estimate whether phytoplankton present in the surrounding waters meet the nutritional requirements of the suspension feeders. These estimates are based on the assumption that 1 ml of oxygen consumed corresponds to 1 mg of organic matter combusted. A typical suspension feeder from coastal waters, processing 20 liters of water for each milliliter of oxygen it consumes, can therefore obtain sufficient food to cover its maintenance requirements if the water contains only 0.05 mg of absorbable food per liter. During periods of growth or reproduction, food requirements are, of course, higher. Thus, at least during the productive seasons of the year, phytoplankton concentrations appear to be sufficient to cover energy requirements for both maintenance and growth. The same conclusion holds for salps that live in oceanic waters with low concentrations of phytoplankton.

The basic features of suspension feeding as just described thus seem to represent adaptations to feeding on minute particles, especially phytoplankton cells, present in highly dilute suspensions. These basic features apply to suspension feeders whose body size is large compared with phytoplankton cells.

This characterization of suspension feeding requires modification in suspension feeders of small size, such as the copepods which constitute the most important primary consumers in marine food chains. More complex relationships arise from overlaps in size between phytoplankton organisms and zooplankton organisms. Thus, chains of diatoms may be larger than both the smallest adult copepods and the young stages of large copepod species.

Until recently it was customary to distinguish between filter feeding and raptorial (carnivorous) copepods. However, this distinction can no longer be maintained. Filter-feeding copepods, of which the calanids are by far the most important, have also been found to be capable of actively seizing food particles, for example, large diatoms or animal prey. Whether the animals use filter feeding or direct seizing of individual food particles depends upon the nature of the food available, especially the sizes of the particles in relation to the sizes of the grazers. This versatility in feeding mechanisms also applies to euphausiid crustaceans and may be a main factor in the success of calanids and euphausiids, as expressed by their numerical predominance in marine zooplankton communities.

As indicated in Table 2-2, even within the euphotic zone of the sea, detritus usually constitutes a larger fraction of the total particulate organic matter than does phytoplankton. However, relatively little is known about the food value of detritus. It probably consists primarily of material that is resistant to the digestive enzymes of suspension feeders. Detritus seems most often to be broken down by bacteria attached to the detritus particles. It thus becomes available to suspension feeders as bacterial mass (see Section 2.2.6).

2.2.4 Dissolved Organic Matter as Food

In addition to particulate matter, natural waters also contain other fractions of organic matter that range in physical state from colloid organic complexes to small molecules in solution. This range of organic colloids and solutes is spoken of as dissolved organic matter, DOM. Its concentration is about 1 mg liter^{-1} in the sea and, in contrast to the particulate organic matter, it is rather uniformly distributed throughout the water masses down to the bottom of the oceans. The total mass of DOM present in the sea amounts to about 10^{15} kg, corresponding to about 100 years of primary production in the oceans.

The importance of the DOM as a food for aquatic organisms has been much debated ever since August Pütter, at the beginning of the twentieth century, proposed the theory that DOM constitutes the main source of food for aquatic animals, both invertebrates and fishes. However, it has only been recently that methods have become available that permit definite answers to the question.

Studies of the uptake of small organic molecules by aquatic organisms were greatly facilitated when radioactively labeled sugars and amino acids became available to measure influxes of molecules. Experiments using these showed that the integuments of marine invertebrates generally possess active transport systems for a variety of organic molecules, including glucose and amino acids. Such

systems are lacking or show only low activities in freshwater invertebrates. Euryhaline species, such as the polychaete *Nereis diversicolor*, take up organic molecules in seawater but not in freshwater. The integumentary transport of organic molecules thus depends upon the presence of salt in the surrounding medium. Integumentary transport systems have been analyzed in a large number of marine invertebrates. They have been characterized in terms of saturation kinetics similar to those of enzymatic systems.

Less is known about the significance of dissolved organic molecules in the energetic balance sheets of marine invertebrates. However, highly sensitive fluorescence techniques now permit measuring naturally occurring concentrations of small organic molecules and their net uptake rates by marine organisms. The major components of DOM are resistant macromolecules that are only slowly degraded microbially. Easily metabolizable DOM, including sugars and amino acids, seems mainly to derive from phytoplankton. It may represent the loss from phytoplankton cells of synthetic products through the cell membranes. Even within the euphotic zone, however, these molecules are present at such low concentrations, usually less than about 1 μM, that they probably play only minor or insignificant roles in the energy balance of the animals exposed to the water.

Much higher concentrations of small organic molecules have been measured in the interstitial waters of soft sediments rich in organic matter, where concentrations of amino acids may reach values up to 100 times those in the water above the sediments. Burrowing invertebrates, such as polychaete worms, may take up amino acids from such enriched interstitial waters in amounts sufficient to cover their metabolic needs.

The integumentary transport systems for small organic molecules are usually thought of as adaptations for the uptake and utilization of DOM in the ambient water. However, the transport systems may serve other functions as well. Intracellular concentrations of amino acids in the integumentary cells of marine invertebrates are many orders of magnitude higher than concentrations in the water outside, thus favoring a loss of amino acids by diffusion. It is probable that the integumentary transport systems for amino acids significantly reduce this loss by reabsorbing molecules that pass through the cell membrane before they reach the external bulk phase of water.

2.2.5 Symbiosis Between Aquatic Plants and Animals

Food is generally the most important biotic environmental factor to animals. An exception to this rule is provided by animals that have established symbioses with photosynthetic algae.

In the middle of the last century reports began to appear of chlorophyll-containing animals. Animal types involved were spongillids, turbellarians, hydra, and protozoans. The observations shook the then established belief in a sharp distinction between plants and animals. However, by the 1880s the "animal" chlorophyll was found to be contained in unicellular algae living intracellularly in the tissues of the host animals. Brandt introduced the term *Zoochlorella* for the green algae found in freshwater animals, such as *Hydra, Spongilla,* and *Paramecium,* and *Zooxanthella* for the yellowish algae in marine forms, such as radiolarians and actinians. The algae could be cultured outside the animal hosts. The animals obtained the algae with their food.

Brandt also observed that forms such as radiolarians and hydra survived normally in filtered water with no external food supply, if they were exposed to light. He concluded that symbiotic algae present in sufficient amounts could cover the nutritional requirements of their hosts. The host animals, however, under natural circumstances might also feed in the same ways as more usual animals.

In the time that has passed since Brandt's pioneering work, the number of aquatic invertebrates found to contain symbiotic algae has increased to about 150 genera, representing 8 phyla. Symbiosis has turned out to be remarkably common among aquatic Protozoa and Coelenterata.

The zoochlorellae of freshwater animals are species of green algae belonging to the order Chlorococcales. The zooxanthellae of marine invertebrates are almost all dinoflagellates, the most common and widely distributed species being *Gymnodinium microadriaticum*. This dinoflagellate species is found in all reef-building corals as well as several other groups of coral reef inhabitants, including the giant clams of the family Tridacnidae and most sea anemones.

The nature of the role played by the symbiotic algae in the physiology of the host animals has long been a subject of study. Within the last 15 years conclusive evidence has accumulated. Experiments with carbon-14 labeled carbon dioxide have shown that the photosynthesizing symbiotic algae release significant parts of their synthesized products to the host cells. Thus, in the coelenterate *Chlorohydra viridissima* about half the carbon fixed in photosynthesis moves into the animal's tissues. The high release rates seem to represent an adaptation that serves the needs of the host. Isolated algae kept in normal culture greatly reduce the release of photosynthates to the medium. This reduction in release could be correlated with changes in the fine structure of the cell walls of the algae.

The host animals readily utilize the photosynthates released by the symbiotic algae. There is little experimental support for the idea that hosts also, perhaps predominantly, profit from the symbiotic relationship by digesting the algae. Digestion of the algae by the animal host remains a potential, though generally unproven nutrient source.

Probably in some way the host animals control the growth, division, and numbers of symbiotic algae. The anemone *Anemonia sulcata* appears to be able to recognize aged or defective algae, which are then expelled by the normal mechanisms for excretion of undigestible materials. *Tridacna* also has been shown to recognize aged and degenerate symbiotic algae. Such algal cells are phagocytized by the amoebocytes.

A number of attempts have been made to evaluate quantitatively the nutritional roles of the symbionts by comparing the primary productivity of the algal populations with total respiration of the complexes of algae and hosts over 24-hr periods. Studies including freshwater sponges, hydra, and reef-building corals have shown that symbiont primary production exceeds respiratory requirements under the environmental conditions prevailing within the aquatic euphotic zones in nature. Conversely, it has also been found that at least some corals can subsist entirely on animal food. Thus, the reef coral *Porites porites* fed nauplii of brine shrimp *Artemia* ate several times more than needed to cover metabolic requirements.

The relative roles played by symbionts and external food sources in the nutrition of reef-building corals in nature has long been, and still is, a controversial subject. Some investigators have found zoo-plankton abundances over some reefs to be adequate to meet the food requirements of corals. In contrast, several other reefs have been found to require about 10 times as much food for their maintenance as the zooplankton available would seem able to provide. In the latter instances the main role of the zooplankton may be to provide the reef corals with essential food elements not produced by the symbionts, for example, essential amino acids.

The symbiotic algae of corals have been found to release alanine and carbohydrates, such as glycerol and glucose. The corals apparently obtain needed lipids from the cell membranes of the algae. In return, the symbionts get ammonia produced by the host, use it for synthesis of amino acids, including alanine, and thus establish a recycling system of nitrogen between coral and algae. The symbionts may also promote growth and calcification by the hosts.

Ecological, biochemical, and physiological data indicate that symbiotic algae may well be of major importance in the nutrition and growth of coral reefs. They are important to the reef-building corals themselves and also to such other important reef inhabitants as the majority of sea anemones and other coelenterates and the giant clams. The clams may owe their large size to the symbiosis with *G. microadriaticum.* Symbiosis between animals and algae in coral reef habitats appears to be a highly successful adaptation for solving nutritional problems in nutrient-poor tropical seas. Photosynthetic production by the zooanthellae of a coral reef may be three times higher than that of the phytoplankton production close to the reef.

2.2.6 Terrestrial Environments: The Detritus Pathway for Utilization of Primary Production

The large plants of most terrestrial and aquatic habitats are consumed by herbivores to only a minor extent (Table 2-4). This may be correlated with several factors. As was mentioned previously, most animals lack digestive enzymes that would permit them directly to utilize the cellulose and other structural polysaccharides that constitute a major part of the synthetic products of terrestrial plants and aquatic macrophytes. Further, higher plants

TABLE 2-4. Fraction of Primary Production Consumed by Herbivores

Plant community	Production Consumed (%)
Phytoplankton	60–90
Grass lands	12–45
Kelp beds	~10
Spartina marshes	~ 7
Mangroves	~ 5
Deciduous forests	1.5–5

Source: T. Fenchel and T. H. Blackburn. *Bacteria and Mineral Cycling,* Academic Press, London & New York, 1979.

have often developed protective means against predators, varying from aromatic resins and toxins to thorns. Since so few of the plants actually are eaten, the major parts of the vegetation, for instance leaves from trees or sea weeds or sea grasses from coastal or marsh habitats, simply die and begin to decompose. During this process the structural polysaccharides accumulate in the detritus, mostly because they represent the most refractory parts of the plants. However, simultaneously, the plant material is converted into microbial mass by the organisms of decomposition, bacteria, protozoans, and fungi, which can be eaten and digested by animals. The microbial activity thus converts parts of the primary production that is not accessible to most animals into matter of high nutritional value to the rich fauna that feeds on the soil of forests and fields or on the sediments of coastal waters, lakes, and other aquatic habitats. The microbial flora and fauna thus constitute the most important first links in most terrestrial habitats that direct the energy of the primary production into the food webs via the various types of detritus feeders.

The decomposing microbes do not convert all food energy into growth and multiplication. Part of the energy is lost to cover metabolic requirements. This loss varies from insignificant to very high, depending upon conditions for growth. Under aerobic conditions the rate at which bacteria respire seems to be independent of the rate of growth. At low growth rates, the greater part of the substratum may therefore be used as energy of oxidation; whereas at high growth rates, more than 90% of the energy of the substratum may be converted into bacterial mass.

Under anaerobic conditions the microbes will ferment their substrate and consequently convert a major fraction of structural polysaccharides into small organic molecules. Such anaerobic conditions prevail in aquatic habitats within sediments rich in organic matter. They also prevail in the intestinal tracts of larger animals, where the anaerobic microbial breakdown has resulted repeatedly in the establishment of symbiotic associations between herbivores and decomposing microorganisms.

2.2.7 Utilization of Primary Production by Means of Symbionts

Just as suspension feeding was the major nutritional adaptation to a herbivorous life in aquatic habitats, digestion by means of symbionts may be considered the major adaptation in the evolution of herbivorous habits in terrestrial environments. Digestion in herbivorous mammals illustrates the complex adaptations that may develop between host and microorganisms in the establishment of symbiont-aided digestion.

Microbial fermentation is probably characteristic of the large intestine of mammals and of large animals generally, not only of herbivores. Even in dogs volatile fatty acids reach high concentration in the colon content and in the blood leaving the colon wall. The importance of fermentation in the economy of the animal is therefore a question of the size and location of the fermentative segments of the digestive tract.

Fermentation may be of great importance in the utilization of food even in such unspecialized feeders as, for instance, the pig and the rat. In these forms one fourth to one third of ingested plant fibers may be digested by means of microbial fermentation, especially in the colon and cecum.

Many herbivores have specialized by enlargement of the colon and cecum, thus increasing the size of the fermentation chambers of the digestive tract. The green turtle, the ptarmigan, the dugong, and the horse and its relatives are examples. In other herbivores, such as rabbits and other lagomorphs, both the large intestine and the stomach are of equal importance as fermentation chambers. Rabbits thus represent a form intermediate between herbivores exhibiting postgastric microbial

digestion and ruminants with pregastric symbiont-aided digestion.

The rabbit has also attracted interest due to its habit of coprophagi. It produces two types of fecal pellets, one the well-known day type and the other formed in the cecum and excreted during the night. The latter are picked up by the animal directly from the anus and eaten. The pellet is enclosed by a gelatinous coating and consists mainly of bacteria. The eaten pellets are retained in the fundus of the stomach, in which a significant fermentation of carbohydrates and formation of lactic acid takes place. The colobid monkeys have adopted fermentation in the enlarged first part of the stomach, the saccus gastricus.

Symbiont-aided digestion has reached its highest development in the ruminants and other groups of ruminantlike animals. Groups that ruminate include the Tylopoda (camels and llamas), hippopotamus, sloths, and macropode marsupials (for instance, the quokka, *Setonix brachyurus*). This wide phylogenetic distribution gives rise to several examples of convergent evolution. Among recent herbivores, the true ruminants have been especially successful. Their upsurge in the Tertiary coincided with the radiation of grasses and expansion of grazing grounds.

Large-scale symbiont digestion of cellulose and other insoluble and resistant polysaccharides needs space and is time consuming. The digestive tracts of animals that have adopted this feeding habit are consequently equipped with greatly expanded sections in which the voluminous mixture of microorganisms and substrate are stored for the protracted periods required to break down the plant material, often several days. In ruminants, the reticulorumen serves as this storage organ (Figure 2-3). In cattle its capacity amounts to about 15% of the body mass.

The ruminants owe their name to their habit of returning some of the reticulorumen content to the oral cavity for further chewing. This chewing increases the surface of the food available to the action of the digestive enzymes in the rumen. In the mouth the repeated returns of the rumen contents further mix the cud with large quantities of saliva. Sheep and goats secrete 10–15 liters and cows 200 liters of saliva daily. Saliva production is continuous but increases in amount during feeding and rumination. The saliva is rich in bicarbo-

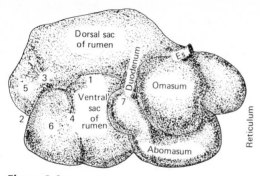

Figure 2-3
Stomach of ox, right view. Code: Es, Esophagus; 1, right longitudinal groove of rumen; 2, posterior groove of rumen; 3, 4, coronary grooves; 5, 6, posterior blind sacs of rumen; 7, pylorus. [From S. Sisson and J. D. Grossman. The Anatomy of Domestic Animals, *W. B. Saunders Company, Philadelphia, 1938.]*

nate, about 100 mM, which serves to neutralize the acids produced by fermentation in the reticulorumen. The pH is maintained at about 6.5.

In the reticulorumen, digestion occurs practically exclusively by means of populations of bacteria and ciliates, the bacteria being the most important. Ruminants can remain healthy even in the absence of the ciliate fauna. No digestive enzymes are secreted by the walls of the reticulorumen. These walls are covered by an epidermis consisting of a stratified squamous epithelium without glands. However, a great deal of absorption of the nutrients made available by the activity of the microorganisms takes place through the reticuloruminal epithelium.

The floras in the reticulorumen of the various ruminants include a great variety of types and species of bacteria, not all of which are considered genuine rumen bacteria. In order to qualify as a symbiont, a bacterium must fulfill certain criteria. It should be present at high concentration ($\geq 10^6$ per gram of rumen content), be anaerobic, possess a temperature optimum at 39°C, require growth factors present in the rumen fluid, and perform chemical processes known to take place in the rumen. Genuine rumen bacteria include both cocci and rods, Gram-positive as well as Gram-negative, and spirilla. They are often specialized to ferment certain substrates, and the composite flora can

break down almost all types of carbohydrates, including starch, cellulose, hemicelluloses, pentosanes, pectins, and so on.

Some structural components of plants are resistant to bacterial degradation. The most important are the lignins. Lignin constitutes up to 20% of the dry weight of higher plants. It is mainly degraded by fungi, which do not belong to the ruminant symbiont floras. Because lignin is not attacked at all, it has been used as a marker to estimate the efficiency with which food is assimilated in ruminants and other animals that digest by means of symbionts.

The initial stages in the breakdown of the food in the rumen are carried out by hydrolases secreted by the microorganisms. The enzymes secreted by the microorganisms include amylase and oligosaccharases in addition to glycosidases that split cellulose and other structural polysaccharides. It appears to be typical of digestion by means of symbionts that the host does not directly utilize the sugars that are set free by the activity of the microbial glycosidases. These sugars are largely absorbed and utilized by the microorganisms themselves. Owing to the anaerobic conditions of the environment, which has a redox potential of about −400 mV, sugar catabolism by the microorganisms is a fermentation that results in the formation of various volatile fatty acids. Acetic acid always appears to dominate, but large amounts of propionic and butyric acids, as well as some valeric acid, are generally also produced. It is these fatty acids that constitute the host's share in the carbohydrates of the food.

Very little glucose is left for absorption by the host animals. The glucose found in the blood of ruminants is synthesized from propionic acid. The fact that blood sugar concentrations are relatively low in ruminants, both domesticated and wild, may be a consequence of this situation. Values of about 3 mM appear to be typical of ruminants, compared to about 5 mM in most other mammals.

Another consequence of ruminant digestion is utilization of food proteins by the symbionts. The proteins of the food are split by proteolytic enzymes secreted by the microorganisms. Liberated amino acids may be absorbed through the walls of the reticulorumen, but the greatest amounts appear to go into the competing microorganisms to be used for their growth and multiplication. The protein, however, is not lost to the ruminant. The microorganisms themselves eventually pass from the reticulorumen to the abomasum, or the stomach proper, in which ordinary digestion takes place. The microorganisms thus represent a most important protein source to the host. Whether or not the transformation of food proteins into bacterial proteins is an advantage to the ruminant will depend upon the quality of the food protein.

The rumen microorganisms are able to utilize ammonia and urea for the synthesis of proteins. This capacity can be of great importance in the nitrogen economy of the ruminant when proteins are sparse in the diet. Urea produced in the liver during protein catabolism is returned to the reticulorumen in the saliva and by diffusion from the blood, to be reutilized in the protein synthesis of the microorganisms.

Digestion by means of symbionts thus endows the host with several advantages. In addition to those already mentioned, there is also a considerable saving in dietary vitamin requirements. The microorganisms synthesize many vitamins, especially those of the B group, which can be utilized by the host. Dietary vitamin needs are thus restricted almost entirely to vitamins A and D.

The energetic price of digesting food by means of symbionts depends both upon the cost of maintaining the microbial population and the efficiency with which this population utilizes the food. The rumen can be considered as an ecological system in dynamic equilibrium with respect to input (food and saliva) and output (absorption of free fatty acids through the wall of the reticulorumen, transfer of rumen content to the omasum and further down the digestive system, and loss of methane by belching). An extensive literature deals with this energy balance of the rumen and the ruminant because of the great economic implications of the subject for such things as meat and milk production.

The energy budget for a ruminant, like a sheep fed clover or lucerne hay, is shown in Figure 2-4. Input, that is, ingested food, is usually expressed in terms of organic matter or calories. Of the ingested organic matter, or calories, about 65% leaves the rumen after a residence time of 11–15 hr. Of this 65%, about 50% consists of nondigested plant materials, the remaining 15% of bacteria. Thus about half of the organic matter contained in the ingested food is digested in the rumen.

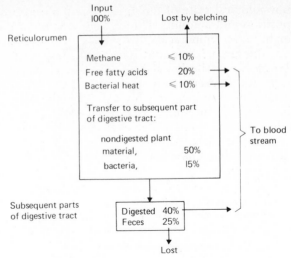

Figure 2-4
Energy budget for a ruminant (sheep).

About 20% of the ingested food becomes fermented to free fatty acids and absorbed. About 10% is converted to methane by *Methanobacterium ruminantium* and is lost by belching. Heat production amounts to another 10%.

The 50% of plant materials that were not digested in the reticulorumen are subject to ordinary digestion in the subsequent parts of the digestive tract. These processes include digestion by means of the enzymes secreted by the glands of the intestinal tract and also further fermentation by means of additional symbionts, especially in the colon. In this way about half of the plant materials that left the reticulorumen may become accessible to the host, leaving about one quarter ingested organic matter to be excreted as feces.

2.3 FOOD

Food supplies animals with energy to function, grow, and multiply. Food energy can be derived from proteins, carbohydrates, or lipids. These energy sources are to a large extent interchangeable in most animal functions. However, special functions, including cell repair and growth, require that certain amino acids are present in the food. The requirements depend upon the species, the age and physiological state of the animals, and upon composition and amounts of food. But it is the same eight to ten amino acids that are essential throughout the animal kingdom from protozoans *(Tetrahymena)* to mollusks, arthropods, and vertebrates (teleosts, chicken, rat, pig, man): namely, leucine, isoleucine, lysine, methionine, phenylalanine, threonine, tryptophan, valine, to which may be added histidine and arginine, especially for young, growing animals. Animals are able to synthesize the remaining amino acids according to their need for protein synthesis.

Besides sources of energy, the macronutrients, animals require small amounts of other organic molecules in the food. These micronutrients include vitamins and essential fatty acids. Some are listed in Table 2-5, and their functions are indicated.

The food further supplies animals with essential minerals, but minerals are also obtained from water. The amounts needed of the essential minerals vary from relatively large amounts to such minute traces that they may be difficult to detect. Interestingly, trace elements are still being added to the list of essential minerals, concurrently with improved chemical analytical techniques. The essential minerals and some of their main functions are listed in Table 2-6. The table also indicates the amounts of minerals that are typically present in the bodies of mammals.

TABLE 2-5. Vitamins and Essential Fatty Acids

Name	Biochemical Action
Fat-soluble vitamins:	
Vitamins A	Aldehyde of vitamin A_1, retinine, conjugates with the colorless protein opsin to form the visual purple, rhodopsin, of the retinal rods (see Chapter 10)
Vitamins D	Precursors to metabolites (hormones) that stimulate intestinal calcium absorption and bone calcium metabolism (antirachitic vitamin)
Vitamins E	Essential for normal gonadal function (e.g. in rats: antisterility vitamins)
Vitamins K	Essential for synthesis of prothrombin and thus for normal blood coagulation
Water-soluble vitamins:	
B vitamins	
Thiamine (vitamin B_1)	Constituent of the coenzyme cocarboxylase
Riboflavin	Constituent of the coenzymes flavin mononucleotide (FMN) and flavin adenine dinucleotide (FADN) that form the prostetic groups of a number of dehydrogenases (the flavoproteins)
Niacin (nicotine acid)	Nicotine acid amide (nicotinamide) constituent of the coenzymes nicotinamide adenine dinucleotide (NAD) and nicotinamide adenine dinucleotide phosphate (NADP)
Pyridoxine (vitamin B_6)	Constituent of the coenzyme pyridoxal phosphate (codecarboxylase)
Cyanocobalamin vitamin B_{12})	Essential for maturation of nucleated red cells in the bone marrow (anti pernicious anemia factor)
Ascorbic acid (vitamin C)	Essential for maintenance of normal functioning of mesenchymal structures, e.g. collagen formation (antiscorbutic factor)
Essential fatty acids:	
Linoleic acid	Precursors in biosynthesis of prostaglandins (see Chapter 12)
γ-Linolenic acid	
Arachidonic acid	

It is customary to distinguish between elements that are present in the body in relatively large amounts and the trace elements, but as may be seen from Table 2-6 there is no sharp delineation between the two groups. The quantities of essential minerals decrease gradually over many orders of magnitude. By definition, trace elements are present in concentrations smaller than 1:200,000. Iron thus occupies a transitional position between the groups.

During the last decade vanadium, silicon, tin, and arsenic have been recognized as trace elements needed for normal growth in chicks and rats, but their mechanism of action is little understood.

TABLE 2-6. Essential Minerals

Name	Concentrations Representative of Mammals (% of Body Mass)	Main Actions
Calcium	2.0	99% in bone and teeth, as apatite; 1% in soft tissues, for example blood, essential for normal membrane functions, transport of ions and molecules across cell membranes, contraction of muscles, excitability of nerves.
Phosphorus	1.0	80–85% in bone and teeth (see calcium), component of cell membrane macromolecules (lecithin, cephalin) nucleic acids (DNA and RNA), energy rich adenosintriphosphate (ATP, buffer system $HPO_4^{2-}/H_2PO_4^-$.
Potassium	0.35	Predominant intracellular cation, essential in membrane potentials and impulse conduction of excitable tissue.
Sulfur	0.25	Constituent of amino acids (threonine and cystein), chondroitin sulfate in cartilage, and some vitamins (for example thiamine).
Sodium	0.15	Predominant extracellular cation, essential in absorption of glucose and amino acids.
Chlorine	0.15	Predominant extracellular anion constituent of HCl secretion in stomach, activator of amylase in saliva and pancreatic juice.
Magnesium	0.05	60% in bone, essential for muscular contraction and normal excitability of muscles and nerves.
Iron	0.005	Constituent of oxygen transporting proteins (hemoglobin, myoglobin) and coenzymes (cytochromes, catalase), stored in ferritin and hemosiderin.
Zinc	0.002	Constituent of about 10 enzymes, including carboxypeptidases A and B, carbonic anhydrase, alkaline phosphatase (esterase); or required for activity of enzymes, for example, arginase and aminopeptidase.
Copper	0.00015	Constituent of many enzymes, such as cytochrome oxidase, tyrosinase, uricase, and the oxygen transporting protein hemocyanin (mollusks, arthropods).
Iodine	0.00004	Constituent of thyroid hormones.

TABLE 2-6. Essential Minerals (continued)

Name	Concentrations Representative of Mammals (% of Body Mass)	Main Actions
Manganese	0.00003	Activator of a series of enzymes involved in the energy metabolism.
Cobalt		Constituent of the vitamin cyanocobalamin (B_{12}).
Molybdenum		Constituent of enzymes, for example, xanthine oxidase and aldehyde oxidase.
Fluorine		Optimal doses important for strength of bone and teeth.
Selenium		Essential for normal growth and reproduction, acts synergistically with vitamin E, mechanisms of action little understood.
Chromium		Essential for action of insulin.

2.4 DIGESTION

The bulk of most foods is made up of macromolecules that animals cannot utilize directly as sources of energy or for synthesis of new cell constituents. The first step in the treatment of food therefore consists of splitting these macromolecules into their subunits: proteins into amino acids, carbohydrates into monosaccharides, and lipids into fatty acids and glycerol (or other alcohols, as in waxes). The splitting of the macromolecules is a hydrolysis performed by means of digestive enzymes.

Until recently animal digestive enzymes were characterized by the substrates they were able to hydrolyze. The enzymes were generally studied as mixtures in extracts of the digestive tract or glands, and no clear understanding could be reached concerning the exact number of enzymes, their specificity towards their substrates, or their chemical relationships. This situation is rapidly changing with the isolation of individual proteins, including enzymes, and the determination of their amino acid sequences. Such studies have shown that the enzymes, like most other proteins, constitute molecular families having long ancestries, their origins mostly preceding metazoan evolution. It is therefore not surprising that the same types of digestive enzymes may act either intracellularly or extracellularly according to the organization of the digestive system.

2.4.1 Intracellular and Extracellular Digestion

Most metazoans possess a digestive tract in which digestion may occur intracellularly in the cells lining the digestive tract, extracellularly in the lumen of the tract, or both. The actual differentiation of the digestive tract and the predominance of extracellular or intracellular digestion apparently depend upon both the phylogenetic position of the animal and the nature of the food.

Intracellular digestion is presumably the primitive condition taken over from the protozoans. It is still the only method of digestion in sponges, which lack a digestive tract. In animals with a digestive tract, intracellular digestion is more often observed in groups that are considered to represent phylogenetically primitive levels.

Coelenterates, many platyhelminthes, and *Limulus* are considered to be examples of animals in which the occurrence of intracellular digestion represents a primitive condition. In other groups, digestion is practically completed in the lumen of the digestive tract before absorption takes place. Examples are the crustaceans, insects, cephalopods, tunicates, and all vertebrates. The latter groups are generally considered to be highly developed.

The influence of type of food is suggested by the fact that feeding on suspended, particulate food in aquatic animals is often associated with intracellular digestion. Intracellular digestion is thus typical of microphagous groups such as brachiopods, rotifers, bivalves, and cephalochordates (Table 2-1).

Macrophagous feeding, that is, feeding on large food, does not, however, necessarily result in purely extracellular digestion. Among the bivalves, the carnivorous or carrion-eating septibranchs digest much of their food intracellularly. This process follows a predigestion and breakdown, in the gut lumen, of the food masses to particles of suitable size for intracellular uptake. The coelenterates, platyhelminthes, and *Limulus* are all macrophagous carnivores possessing intracellular digestion.

A major difference between extracellular and intracellular digestion is the great specialization of the digestive tract with which extracellular digestion is correlated. This specialization may result in regional differentiations for (1) food uptake; (2) storage; (3) digestion and perhaps mechanical disintegration of the food; (4) absorption, which often overlaps with the previous section; and (5) formation and evacuation of feces. This functional differentiation is most advanced in vertebrates.

2.4.2 Digestive Enzymes

The present treatment of animal digestive enzymes is based on the classification and nomenclature of enzymes adopted by the International Union of Biochemistry (Dixon and Webb, 1979). However, trivial names have been used with the systematic names when this was appropriate. In some groups of enzymes, such as most peptide hydrolases, only trivial names are available.

2.4.3 Carbohydrates and Carbohydrases

Some Important Carbohydrates. The most widely distributed carbohydrates are built up from hexoses. Hexoses, such as glucose in nectar, may occur freely or in disaccharides, especially sucrose and lactose. Polysaccharides, however, dominate among plant products and therefore constitute by far most important sources in the diets of herbivores. Polysaccharides, such as starch or inulin (which is stored in the roots of Compositae), may

serve as energy depots in plant cells and organs. Of even greater quantitative importance in the higher plants are the structural polysaccharides of the cell walls, especially cellulose, but also lignin, xylan, hemicelluloses, and pectic substances. Other structural polysaccharides, such as lichenine of lichens and laminarin of sea weeds, are of more restricted occurrence within the plant kingdom. Many of these substances, for example, starch, cellulose, lichenine, and laminarin, are built up from glucose units. Inulin is polymerized from fructose units; xylan from the pentose xylose; whereas hemicelluloses contain hexoses, pentoses, and also glucuronic acids. The pectic substances are typically found in fruits and are compounds of galacturonic acid, a smaller or larger proportion of which may be present as methyl esters. Many more polysaccharides are known, both widely distributed or of more limited occurrence in the plant kingdom. They are often badly defined and only partially known chemically. It is, however, apparent that most herbivores ingest a wide array of saccharides, especially polysaccharides, with their diet.

We should add the two major animal polysaccharides to the list of plant compounds. They are *glycogen,* a reserve polysaccharide of universal distribution in the animal kingdom, and *chitin,* the structural polysaccharide of the external skeleton in the arthropods. Glycogen is closely related to the starch component amylopectin, whereas chitin is built up from N-acetyl glucosamine units.

Carbohydrates are characterized by their glycosidic bonds and are largely classified according to the types of these bonds. As the types of glycosidic bonds determine to a large extent the specificity of the carbohydrate-splitting digestive enzymes, the classification of the enzymes quite naturally follows rather closely that of their substrates. Figure 2-5 is an example of the structure of a composite car-

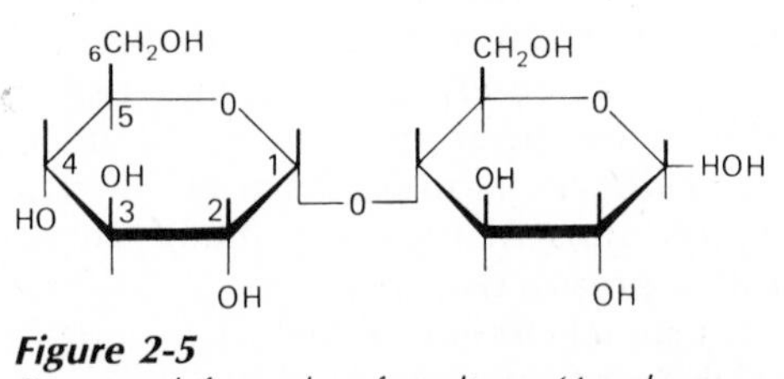

Figure 2-5
Structural formula of maltose (4α-glucopyranoside glucose).

bohydrate molecule drawn in the manner usually adopted for presenting structural characteristics. The example is the disaccharide maltose, which is produced from starch when it is hydrolyzed by amylase. Maltose is chemically a glucose, α-glucoside, possessing a pyranose type of ring isomerization (an oxygen bridge between carbon atoms 1 and 5 of the glucose unit).

Further information on types of carbohydrates and their structure and chemistry may be obtained in textbooks of biochemistry.

Carbohydrate-Splitting Enzymes. Glycosidases are characterized by their great specificity toward their substrates. The ability of a particular glycosidase to split carbohydrates appears to depend upon a number of properties of the carbohydrate. The most important of these properties are the types of glycosidic linkages, the type of ring isomerization, and the size and structure of the molecule.

The importance of molecular structure to glycosidase specificity is seen in the digestion of starch, which is composed of α-glucose units. In animals, the complete digestion of the long α-1,4-glucosidic chains of starch (amylose) to glucose is carried out in two stages by enzymes that are both α-glucosidases. The intact molecule and its larger fractions are digested by a specific glycosidase, α-1,4-glucan-4-glucanohydrase (trivial name, α-amylase). This enzyme produces primarily maltose and some oligosaccharides but only small amounts of glucose. The larger fragments are eventually broken down to free glucose by the second enzyme, the α-D-glucoside glucohydrolase, also known as maltase. A very similar two-stage digestion is seen in the digestion of cellulose. Cellulose is a β-glucoside.

The vertebrate digestive tract typically produces the polysaccharide-splitting enzyme α-amylase, and the disaccharide-splitting enzymes α-glucosidase (maltase), oligo-1,6-glucosidase, and β-galactosidase (lactase; lactase activity is especially high in suckling mammals). In various invertebrates many more glycosidases, especially for attacking polysaccharides, have been described. These enzymes should enable their producers to digest inulin and a smaller or greater number of the totally insoluble structural polysaccharides such as cellulose, lichenin, mannan, xylan, hemicellulose, and pectin, as well as chitin. Enzymes for digesting such materials have been described most often in herbivorous or wood-eating insects and mollusks and in earthworms.

The termites provide good examples of the adaptive diversity occurring among invertebrates in the utilization of cellulose and other structural carbohydrates. They also illustrate the shortcomings of attempts to classify animals into sharply delineated digestive types.

Lower (more primitive) termites digest cellulose by means of symbionts in an enlarged region of the hindgut, the paunch. They are thus comparable to vertebrates with postgastric microbial digestion of cellulose (see Section 2.2.7). But the higher (more advanced) termites (Termitidae), which constitute about 75% of termite species, lack the cellulose-digesting symbionts; and cellulose is exploited by a variety of mechanisms that apparently vary between groups. The fungus-growing termites (Macroterminae) obtain their assemblage of cellulose-digesting enzymes partly from the fungi they eat and partly by secreting the enzymes from the salivary glands and from the midgut epithelium.

2.4.4 Peptidases

Peptide hydrolases split protein and peptide molecules by hydrolysis of peptide bonds according to the general equation

$$
\begin{array}{c}
\underset{\substack{|\\H}}{\overset{R_1}{\underset{|}{NH_2-C}}}-\overset{O}{\overset{\|}{C}}-\underset{H}{\overset{H}{N}}-\underset{\substack{|\\H}}{\overset{R_2}{C}}-\overset{O}{\overset{\|}{C}}-OH + HOH = \\[3ex]
\underset{\substack{|\\H}}{\overset{R_1}{\underset{|}{NH_2-C}}}-\overset{O}{\overset{\|}{C}}-OH + \underset{\substack{|\\H}}{\overset{R_2}{\underset{|}{NH_2-C}}}-\overset{O}{\overset{\|}{C}}-OH
\end{array}
$$

Until recently it was assumed that the digestion of proteins was initiated by certain peptidases, known as proteinases or proteases, that specifically acted at various places along the peptide chain but did not attack terminal bonds. These enzymes were called endopeptidases. It is now realized that the specificities of peptidases generally are determined by the nature of the amino acids surrounding the peptide bond that is being hydrolyzed. Examples of such specificities in some vertebrate peptidases are shown in Figure 2-6. This figure also indicates that some peptidases require either a free amino terminus (aminopeptidases) or a free carboxyl terminus (carboxypeptidases). These latter two groups together are known as exopeptidases, whereas the remaining peptidases hydrolyze the appropriate bond wherever it is situated in the chain.

An important group of peptidases has been termed serine proteases because the amino acid serine constitutes part of the active site. This group includes the pancreatic digestive enzymes, trypsin and chymotrypsin, which have been found widely distributed throughout the animal kingdom, down to the coelenterates. Even bacterial trypsins are known. Serine proteases also include a great number of other enzymes, which are characterized by their functional diversity. Serine proteases are found in procaryotes as well as eucaryotes, and the family of enzymes constitute an excellent example of molecular evolution (see also Section 12-7).

The acid proteases constitute another important group of homologous proteases, with an ancestry comparable to that of the serine proteases. In vertebrate digestion they are represented by pepsin. Pepsins are optimally active at highly acid reactions, at pH about 1 to 2, in contrast to trypsin and chymotrypsin with optimum activity in neutral or slightly alkaline media. Pepsins are therefore restricted to animals with hydrochloric acid-producing sections along the digestive tracts, such as the stomachs of vertebrates, including most fishes. Also chymosin (= rennin), which is responsible for the clotting of milk within the abomasum of young ruminants, belongs to the family of acid proteases.

2.4.5 Lipids and Esterases

The lipids of the diet of most terrestrial animals consist mainly of esters of glycerol and long chain fatty acids, though some shorter chain fatty acids also occur (Figure 2-7). The triglycerides are known as fats. Until recently it was assumed that fats were also the major lipid types in the diets of aquatic animals. However, during recent years it has been realized that in marine food chains waxes are often of greater importance than fats. Waxes are esters of a long-chain fatty acid and a long-chain fatty alcohol.

Waxes have been found to constitute the major energy stores in nontropical copepods. The general pattern of development in such copepods seems to be that the copepods hatch and then feed in the euphotic surface waters during the season of high productivity. They go through a number of growth stages, reach adult size, and finally fill up their fat bodies with lipids, predominantly waxes. During autumn and winter they migrate to deeper and cooler waters, where they eventually mature sexually. Owing to the predominance of copepods as primary consumers, it has been estimated that about half the total primary production of the sea passes through the wax stores of the copepod populations. As a result, waxes form a major part of the food of the animals that eat copepods and also of subsequent links in the food chains. Waxes also seem to play important roles in the energy transfer in other marine ecosystems, for instance, in coral reef biotopes.

Lipids, waxes, and other esters are split by digestive enzymes that belong to the group of serine esterases. The serine esterases are related to the serine proteases just mentioned. Both groups of enzymes presumably share a common ancestral precursor. This relationship explains why serine

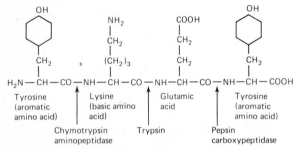

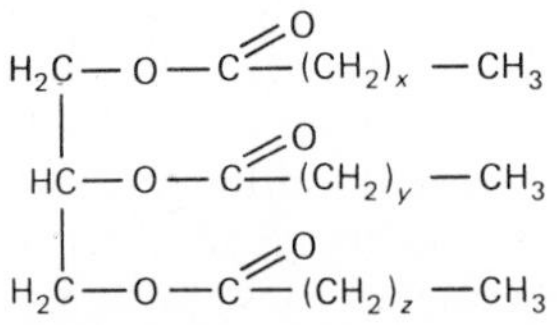

Figure 2-6
Specificities of some peptidases.

$$H_2C-O-C{\overset{O}{\diagup}}(CH_2)_x-CH_3$$

$$HC-O-C{\overset{O}{\diagup}}(CH_2)_y-CH_3$$

$$H_2C-O-C{\overset{O}{\diagup}}(CH_2)_z-CH_3$$

Figure 2-7
General structure of triglyceride fat molecule.

proteases, such as trypsin, chymotrypsin, and thrombin, can hydrolyze certain carboxyesters quite rapidly. The esterases have turned out to be more diverse than probably any other enzyme system. This extreme multiplicity of esterases combined with the overlapping substrate specificity of these multiple forms have made the identification of individual esterases difficult. Esterases belong to the standard equipment of animal digestive tracts. In vertebrates they are secreted in the pancreatic juice. The glycerol ester hydrolases are known as *lipases.* The pancreas of fishes, such as the anchovies, sardines, and herrings, which prey on copepods, secretes an esterase that hydrolyzes the waxes stored in the prey organisms.

2.4.6 Coordination of Secretion of Digestive Enzymes

Animals that feed continuously generally seem also to digest their food more or less continuously. In other animals a noncontinuous, coordinated secretion of the digestive juices occurs in response to feeding. Complex regulatory mechanisms evolved by the vertebrates ensure that the digestive secretions enter the lumen of the digestive tract in the proper amounts and at the proper times.

The following discussion relates mainly to mammals, which are the most thoroughly studied group in this regard.

The major breakthrough in the understanding of the function of the digestive tract occurred in the early part of the twentieth century. The rather simple functional pattern that emerged from these pioneer studies was considered reasonably conclusive and exhaustive until recent years, when the field of gastrointestinal research reopened. Today it is acknowledged that complex mechanisms control the function of the digestive tract, especially the gastric, pancreatic, and intestinal secretions. The secretion of hydrochloric acid in the stomach and releases of the various digestive enzymes are controlled by interactions between nerves and hormones. Some of the hormones act through the general circulation, others act locally by diffusion through intercellular spaces from the cells or nerves of origin to the target cells or nerves.

These intricate interactions are still only partly understood. Thus even an elementary treatment of gastrointestinal physiology must remain preliminary.

Saliva. Saliva is secreted from numerous glands in the oral cavity, especially from three pairs of large glands, the parotids, sublingual, and submaxillary glands. Saliva is a mucoid secretion that serves to lubricate the food and thus assists in its easy passage through the esophagus during swallowing. The saliva of some vertebrates contains α-amylase, which aids in the digestion of starch.

The secretion of saliva is generally coordinated with the intake of food. Olfactory and gustatory stimuli normally initiate the nervous reflex that results in the stimulation of salivary secretion. It is, however, well known that conditioned reflexes resulting in salivary secretion following visual or auditory sensory stimulation are easily established. Secretion from the salivary gland cells is predominantly controlled by parasympathetic nerve fibers.

Gastric Juice. The chief cells lining the long tubular glands of the fundus of the stomach secrete pepsin, whereas the more scattered parietal cells secrete hydrochloric acid. The two secretions together constitute the main components of the gastric juice. The secretion of gastric juice has been studied very thoroughly in dogs, largely by means of various surgical procedures that have allowed direct observation of the secretory activities of the stomach (Figure 2-8). The experimental arrangement often involved sectioning the esophagus of a dog, then ligating the cut ends into the surface of the skin. Such operated dogs can eat normally and the food exerts its normal stimulatory effect in the foremost parts of the digestive tract, but without reaching the stomach. Gastric secretions may be obtained by means of a gastric fistula that drains the stomach to the exterior, or a pouch may be isolated from the rest of the stomach to open on the surface. Such stomach pouches have been constructed either with normal innervation (the Pavlov pouch) or denervated and separated from the rest of the stomach (the Heidenhain pouch).

These and other preparations have been extensively used to study the functioning of the stomach during feeding and digestion. It has been found that much the same stimuli that result in nervous stimulation of the salivary glands also cause prolonged secretion of the gastric glands. Conditioned

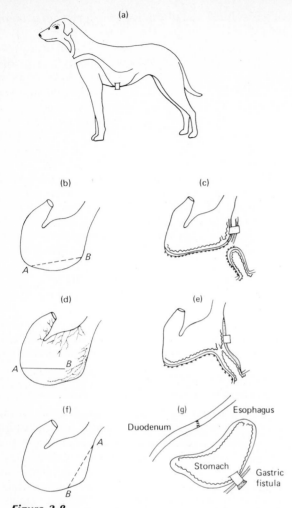

Figure 2-8
(a) Dog with esophagotomy and a gastric fistula. (b) Direction of incision for Heidenhain pouch. (c) Heidenhain pouch and gastric fistula. (d) Direction of incision for Pavlov pouch. (e) Pavlov pouch and gastric fistula. (f) Direction of incision for Brestkin and Savich pouch. (g) Frémont isolated stomach. [*From B. P. Babkin.* Secretory Mechanisms of the Digestive Glands, *Paul B. Hoeber, Inc., New York, 1950.*]

reflexes are also easily established for gastric secretion. The secretory innervation is cholinergic and the fibers run in the vagus nerve.

This cephalic phase of secretion of gastric juice is followed by a gastric phase which is initiated by the food that enters the stomach. The distention of the stomach acts as a mechanical stimulant, and peptides and certain amino acids act directly as chemical stimulants that cause secretion of the hormone gastrin. Gastrin is produced in specific cells located in the antral epithelium within the stomach. It is liberated into the blood to reach the other parts of the stomach by way of the general circulation. Gastrin stimulates the secretion of hydrochloric acid (Figure 2-9), apparently in complex interaction with acetylcholine liberated from nerve ends and with histamine. The cell membranes of the parietal cells carry specific receptors for each of these three chemical transmitters.

The secretion of hydrochloric acid is also under inhibitory control. An inhibitory factor has been localized in the small intestine. It is released into the blood stream in the presence of fat in the duodenum. This inhibitory factor has been called enterogastrone (Figure 2-9), but it has not yet been definitely chemically identified.

Pancreatic Secretion. The digestive enzymes of the intestine originate mainly from the pancreas but also from the intestinal epithelium. Ingestion of food leads to only moderate secretion from the pancreas. This secretion results from vagal stimulation. Prolonged and profuse secretion of pancreatic juice is first initiated when food passes from the pylorus into the duodenum. The mechanism of secretion is very similar to the chemical phase of secretion in the stomach (Figure 2-9). The elucidation of the chemical mechanism at work in the intestine antedated the elucidation of the gastric mechanism, and is considered to be the first true hormonal mechanism described. The crucial experiments were performed by Bayliss and Starling in 1902.

Bayliss and Starling observed that a crude extract of the duodenal mucosa stimulated the flow of secretion from the pancreatic duct when injected intravenously. The substance from the mucosa thought to be responsible for the stimulation of the pancreatic tissue was termed ***secretin***. Later they adopted the name ***hormone*** as a general term for substances that are produced in one tissue and control and coordinate the function of another tissue to which they are transported in the bloodstream (secretin is such a substance).

More recent work has shown that not only secretin but a second hormone produced by the intesti-

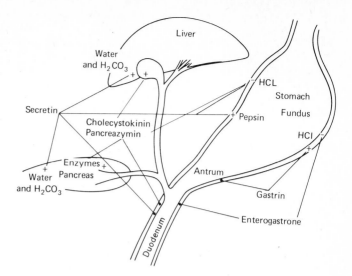

Figure 2-9
Hormonal control of digestive secretions. [*Modified from A. Gorbman and H. A. Bern.* A Textbook of Comparative Endocrinology, *John Wiley & Sons, Inc., New York, 1962.*]

nal mucosa is involved in the control of pancreatic secretions. Secretin, especially, stimulates a secretion that is rich in bicarbonate and poor in enzymes. The other hormone, pancreozymin, stimulates the secretion of enzymes from the pancreatic acinar cells. The two hormones are released by different mechanisms, which appear to be of obvious adaptive significance.

Secretin-producing cells in the epithelium of the duodenum release the hormone in response to the presence of acids in the gut content. The hormone release reaches a maximum at pH 3.0 and the hormone is secreted in amounts that are proportional to the amounts of acids entering the duodenum. The entrance of the acid stomach content into the duodenum will thus result in its rapid neutralization by the secretin-mediated secretion of bicarbonate. Pancreozymin release and the ensuing secretion of pancreatic enzymes into the duodenum are caused by the digestive products in the chyme that enter the duodenum from the pylorus, especially amino acids and long chain fatty acids.

The emptying of bile stored in the gall bladder into the duodenum is also controlled by a hormone known as cholecystokinin (CCK). This hormone has turned out to be identical with pancreozymin.

Two families of gastrointestinal hormones can be distinguished, gastrin and CCK belonging to one and secretin to the other. Gastrin and CCK, which both exist as larger and smaller molecules,

share a common carboxy-terminal pentapeptide. The secretin family of peptides includes the hormone glucagon, in addition to a number of members whose physiological significance is uncertain or unknown, for example, vasoactive intestinal polypeptide (VIP) and gastric inhibitory peptide (GIP) (see also Section 12-7).

The gastrointestinal hormones typically have a variety of functions (Figure 2-9). They stimulate or inhibit secretion and motility and also exert trophic effects on the gut mucosa and its glands. Moreover, the various hormones overlap in their activities, which may partly be explained by their close chemical relationships.

Table 2-7 summarizes mammalian digestive secretions, activities, and control mechanisms.

2.5 INTESTINAL ABSORPTION

The food of animals using extracellular digestion in the lumen of the gut is eventually degraded to products that can be absorbed through the epithelium of the intestinal wall. Absorption may be facilitated by the enlargement of the mucosal surface.

2.5.1 General Features

Such surface enlargement is especially conspicuous in the small intestine of mammals (Figure 2-10).

TABLE 2-7. Mammalian Digestive Glands: Control Mechanisms and Principal Enzymes

Section of Digestive Tract	Secretions and Their Origin	Mechanism of Control	Enzymes and Their Actions	Specificity
Oral cavity	Saliva, especially from parotid, sublingual, and submaxillary glands	Nervous control, especially through parasympathetic secretory nerves	α-Amylase, splits starch and glycogen to dextrins and maltose; enzyme of sporadic occurrence in mammals and other vertebrates	Hydrolyzes nonterminal α-1,4-glucoside bonds
Stomach	Gastric juice, mainly composed of HCl from parietal cells and pepsin from chief cells	Nervous: Vagus secretory nerve for HCl and pepsin secretion Hormonal: Gastrin from antral mucosa stimulates HCl secretion; enterogastrone(s) from duodenal mucosa inhibits gastric secretion and motility	Pepsin, splits proteins and polypeptides endopeptidase	Hydrolyzes peptide bonds adjacent to aromatic or dicarboxylic amino acids
			Rennin, in young animals	Specificity similar to pepsin
Small intestine	Pancreatic juice, composed of water and salts, especially bicarbonates, and enzymes	Nervous: Vagus, secretory nerve for enzymes Hormonal: Secretin, from intestinal mucosa, stimulates secretion of water and salts; CCK stimulates secretion of enzymes; control predominantly hormonal	Lipase, splits triglyceride fat molecules into free fatty acids, diglyceride, monoglyceride, and glycerol	Hydrolyzes glycerol esters
			α-Amylase	Similar to the saliva amylase
			Trypsin, splits proteins and polypeptides	Hydrolyzes peptide bonds adjacent to basic amino acids

TABLE 2-7. Mammalian Digestive Glands: Control Mechanisms and Principal Enzymes (continued)

Section of Digestive Tract	Secretions and Their Origin	Mechanism of Control	Enzymes and Their Actions	Specificity
Small intestine (cont'd)			Chymotrypsin, splits proteins and peptides	Hydrolyzes peptide bonds adjacent to aromatic amino acids
			Carboxypeptidases split free amino acids from dipeptides (exopeptidases)	Hydrolyze peptide bonds at carboxy-terminal end of peptide chain
			Ribonuclease and deoxyribonuclease	Hydrolyze RNA and DNA to nucleotides
	Intestinal secretion, "succus entericus," from intestinal crypts and autolyzed villus cells		α-Glucosidase, splits maltose and other α-glucoside sugars e.g., sucrose	Hydrolyzes α-D-glucosides to D-glucose and an alcohol
			Oligo-1,6-glucosidase, splits branching dextrins derived from amylopectin moiety of starch and from glycogen	Hydrolyzes α-1,6-glucoside bonds
			β-Galactosidase, splits lactose, in suckling young mainly	Hydrolyzes β-D-galactosides
			Aminopeptidases, split free amino acids from dipeptides (exopeptidases)	Hydrolyze peptide bonds at amino-terminal end of peptide chain
			Dipeptidases, split dipeptides into free amino acids	Hydrolyze specific dipeptides

TABLE 2-7. Mammalian Digestive Glands: Control Mechanisms and Principal Enzymes (continued)

Section of Digestive Tract	Secretions and Their Origin	Mechanism of Control	Enzymes and Their Action	Specificity
Small intestine (cont'd)	Bile, composed of water and solids, especially bile salts	Nervous: Vagus secretory nerve Hormonal: Secretin increases volume of bile, but not output of bile salts; CCK stimulates contraction of gall bladder wall and emptying of contents into small intestine Chemical: Bile salts themselves most important substances in increasing bile secretion	The bile does not contain enzymes, but the bile salts are important in the digestion and absorption of the lipids in the food	

Source: M. Dixon and E. C. Webb. *Enzymes,* 3rd ed., Longmans, Green & Co., London, 1979; R. A. Gregory. "Secretory Mechanisms of the Gastro-Intestinal Tract," *Physiol. Soc. Monog.,* **No. 11,** Intestinal Physiology," *MTP International Review of Science, Physiology Series One,* vol. 4, University Park Press, Baltimore, Md., 1974.

Mucosal folds increase the surface area by a factor of 3. The densely spaced 1-mm-long villi cause a further tenfold increase. Finally, the luminal sides of the epithelial cells are covered with microvilli, about 1 μm long, which also add considerably to the increase in area of the absorbing surface.

A single intestinal mucosal villus represents an absorbing unit. It is a finger-shaped structure (Figure 2-11) covered by a single layer of columnar epithelium under which lies a dense capillary network. A lymphatic vessel is centrally located. Longitudinally arranged smooth muscle fibers contract rhythmically, producing alternating shortening and (passive) lengthening of the villus.

A striking property of the mucosal epithelium is the speed with which it is renewed. Epithelial cells at the base of a villus divide continuously and daughter cells move up the villus. The cells at the top of the villus are simultaneously shed into the intestinal lumen. It only takes a few days for cells to migrate from their place of origin to the top of a villus and to become detached.

The estimated amount of the mass of epithelial cells thus transferred to the intestinal lumen is about 250 g daily in man. Autolysis of this cell mass releases enzymes, especially oligosaccharases (for example, maltase and lactase) and peptidases, which may carry out digestive functions in the lumen. However, it seems that the enzymes are exerting their principal action intracellularly (see Section 2.5.2). The rapid turnover of cells is considered to be primarily a mechanism to secure the maintenance of an intact intestinal epithelium.

Once transferred across the epithelial cells the nutrients leave the villus either by the capillaries or by the lymphatic vessel. Which route is most important depends upon the properties, especially the molecular size, of the nutrients entering the intercellular space of the villus. Small molecules enter both capillaries and lymph and thus are trans-

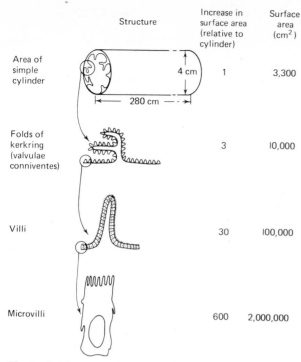

Structure		Increase in surface area (relative to cylinder)	Surface area (cm²)
Area of simple cylinder	4 cm / 280 cm	1	3,300
Folds of kerkring (valvulae conniventes)		3	10,000
Villi		30	100,000
Microvilli		600	2,000,000

Figure 2-10
Mechanisms for increasing surface area of the small intestine. [From T. H. Wilson. Intestinal Absorption, *W. B. Saunders Company, Philadelphia, 1962.]*

ported farther by both systems. However, blood flow has been found to be several hundred times greater than lymph flow in the villi. If the nutrients are distributed equally between blood and lymph space, only negligible amounts will actually leave the intestine with the lymph. Large molecules, such as fat synthesized in the mucosal epithelium, cannot pass the walls of the capillaries and thus be transported by the blood. Such molecules enter the lymphatic vessel through perforations in the wall.

2.5.2 Mechanisms

The transfer of substances across cell walls or membranes occurs either by passive diffusion or some type of energy-requiring transport process. We speak of *active transport* if the substance in question moves against an electrochemical potential gradient (when we are dealing with electrically charged particles) or against a chemical potential gradient (when we are dealing with nonelectrolytes). When these criteria do not apply we speak of *passive transport.*

Both active and passive transport processes are involved in the absorption of nutrients from the intestinal tract, but their relative importance varies strongly with the type of substance being transported. The mechanisms of movements of water and solutes will be discussed in greater detail in Sections 7.4.2 and 7.4.3.

Mechanisms of uptake of the various components of the digested food have been especially thoroughly studied in mammals. Enough is known about gastrointestinal absorption in other vertebrates, especially in frogs and several teleosts, to suggest that the mechanisms are strikingly similar throughout the vertebrates.

Carbohydrate Absorption. Carbohydrates were previously believed to be absorbed only in the form of monosaccharides. It is now widely accepted that significant amounts of disaccharides are absorbed by the intestinal cells before they are split by intracellular enzymes. The enzymes have been located in the brush border of the cell. The brush border thus functions as a cell organelle that unites both absorptive and digestive capacities. In man the cells of the intestinal epithelium contain at least eight disaccharide-splitting enzymes, including the lactose-splitting β-glycosidase and six α-glycosidases. The latter all split maltose to glucose, but one is especially adapted to split saccharose into glucose and fructose and isomaltose into glucose.

The monosaccharides presented for absorption include various hexoses and pentoses. They pass the membrane of the epithelial cell bound to a carrier molecule, the binding being determined by the pyranose ring of the monosaccharide. The nature of the carrier-mediated transport is not yet fully understood. It depends, however, upon the simultaneous transport of sodium ions. Absence of sodium on the mucosal side of the epithelium, or poisoning of the sodium transport by drugs such as ouabain, inhibits the transport of glucose or other monosaccharides. The inhibition is found not only in the mammalian intestine but also in frog and goldfish intestines, indicating the general nature of the sodium-dependent transport of sugars.

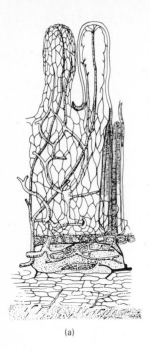

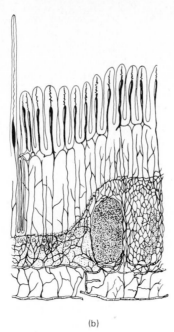

Figure 2-11
(a) Blood vessels and (b) lymphatic vessels of the intestinal mucosa of the dog. [*From T. H. Wilson. Intestinal Absorption, W. B. Saunders Company, Philadelphia, 1962.*]

(a)

(b)

Protein Absorption. Uptake of protein is negligible in the adult intestine, but dipeptides appear to be absorbed to some extent and to be hydrolyzed intracellularly in the intestinal epithelium, as was the case with disaccharides. The remaining part of ingested protein is degraded to amino acids before absorption.

Amino acids are absorbed by means of several pathways that are specific to groups of amino acids. Three of these pathways specifically transport neutral, basic, or acidic amino acids, respectively. A fourth pathway is specific for imino acids, such as proline and hydroxyproline.

As in sugar uptake, uptake of amino acids is coupled to transport of sodium ions. Moreover, the transport systems for sugars and amino acids may interact, but the details of this interaction are not finally established.

Fat Absorption. Small amounts of fat are taken up unhydrolyzed, apparently as micelles of sizes up to 100 Å in diameter. Significant amounts are absorbed as monoglycerides, and 25–60% of the fat appears to be fully hydrolyzed to free fatty acids before it is absorbed.

Water and Electrolytes. Absorption of these substances has been extensively studied by means of both stable and radioactive isotopes, especially of water and of sodium (Na^+) ions and chloride (Cl^-) ions. It has been found that all sections of the vertebrate digestive tract are highly permeable to water, somewhat permeable to monovalent ions, but rather impermeable to di- and trivalent ions. It further appears that water is only passively transported, and its movement is generally coupled to the transport of ions, especially Na^+. Even though Na^+ may diffuse passively through the intestinal wall, the greater part of normal uptake is by means of an active, energy-requiring process. Specific active uptake of Cl^- has also been demonstrated in the intestinal tracts of some animals, but passive transport of Cl^- is more important. The Cl^- is transferred across the intestinal epithelium by the positive electric potential arising across the epithelium as a result of the transport of Na^+.

As noted, the permeability of the intestine is generally low toward di- and trivalent ions. Many of these ions constitute vital elements of the food. Specific active mechanisms for their absorption from the intestinal contents appear to exist. Such transport mechanisms have been indicated for calcium (Ca^{2+}) ions and for iron (Fe^{2+}) ions.

2.6 PARTITIONING OF FOOD ENERGY

An important but long-neglected field of physiology involves the study of how animals utilize the energy derived from the food they eat for maintenance, growth, and reproduction, and which physiological mechanisms control this partitioning of food energy. The field first attracted the interest of veterinary physiologists because of its great economic implications for animal husbandry. Basic knowledge of the energetics of production by farm animals and poultry was essential for determining the most economical methods for producing meat, milk, and eggs. Impetus for studies like these was initially lacking in human physiology and still remains limited. The development of fish farming and other types of aquaculture expanded studies in animal bioenergetics to include species of teleosts and various invertebrates. Still more recently the growing concern of ecologists with the flow of energy between trophic levels in food webs and ecosystems has stimulated interest in understanding how various types of animals partition food energy. Research to answer questions in this area is rapidly expanding, but the literature is scattered and results are still fragmentary.

In any animal some of the food ingested is digested, and the products of digestion are absorbed. Undigested and undigestible parts leave the intestinal tract as feces, which thus represent energy lost to the organism. Some of the absorbed food energy is utilized in the metabolism to cover energy requirements of the organism for maintenance, work, temperature regulation, and so on (see Chapters 3 and 8); some of the energy is utilized for production of new cells and tissue, or for repair; and some is lost by excretion in the urine or through the epidermis.

These basic features of partitioning of the food energy may be expressed in the equation

$$C = P + R + U + F$$

where C is the amount of food energy consumed, P is the amount converted into production, R is the amount used to cover respiration (metabolic needs), U is the amount lost in the urine or other excreta, and F is the amount lost as feces.

Ideally, these entries in the energy budgets of an animal should be expressed in terms of energy units, such as calories or joules (see Chapter 3). In praxis, however, they are usually determined as dry organic matter or organic carbon and expressed in these terms.

The terms $P + R + U$ ($= C - F$) in the energy balance equation represent absorbed food; the terms $P + R$ represent the part of the food energy consumed that has been assimilated in the body. In some studies no clear distinction is made between the concepts of absorbed food and assimilated food, and the term U is frequently omitted from the equation.

There are two primary reasons for disregarding the energy lost by excretion in calculations of animal energy budgets. First, the concept is less well defined and more difficult to assess than the other items in the budget. Second, it often seems that the loss by excretion is small. However, this last may not be generally true, as can be seen from Figure 2-12. This figure shows the relationship that exists between absorbed and assimilated food energy in trout fed on the amphipod *Gammarus*. About 5–10% of the absorbed food energy was lost by excretion. Studies on other animals have yielded similar or lower, but also sometimes higher, values for energy lost by excretion. More work is needed to clarify the relationships existing between absorbed and assimilated food in various types of animals.

The efficiency with which food is assimilated determines how much of the food consumed enters the detritus pathway as feces and other excreta (see Section 2.2.6). The fraction of assimilated food that is utilized for production is termed the net production efficiency. Assimilation and production efficiencies determine how much of the food consumed becomes available to the next link in the food chain.

We will now discuss the major concepts involved in the energy balance equation.

Digestibility. The fraction of the consumed food that is digested and absorbed indicates the digestibility of the food and/or the capacity of the gut

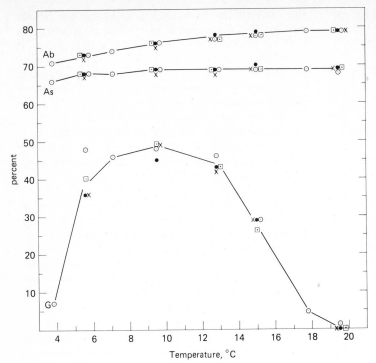

Figure 2-12
Effect of temperature on absorption and assimilation of food energy, and on net growth efficiency in brown trout, Salmo trutta, *fed maximum rations of amphipods (Gammarus). Sizes of fish were 12 g (●), 50 g (⊙), 80–90 g (□), and 250–280 g (x). Ab, absorption; As, assimilation; G, net growth efficiency. [Modified from J. M. Elliott. J. Anim. Ecol., **45**, 923–948 (1976).]*

to digest the amounts of food ingested. When food is plentiful many animals eat larger amounts of food than the gut is capable of digesting.

The digestibility of the food varies primarily with the feeding type of the consumer. Among the various feeding types, carnivores are generally the most efficient in utilizing their food. They usually digest and absorb about 90% of ingested food. On diets of pure meat, the percentage may go as high as 99%. Herbivores also utilize their food efficiently. Suspension feeders grazing on phytoplankton typically digest and absorb 80% or more of the ingested food. Terrestrial herbivores that digest their food by means of symbionts are slightly less efficient, only about two thirds of the ingested food energy being available to the host (see Section 2.2.7). In contrast, detritus feeders, especially nonselective sediment eaters, exhibit low overall utilization of the ingested food. Typically, only a few per cent of the material passing through the gut are digested and absorbed.

Production. Production refers primarily to growth of the body and reproduction. However, processes such as seasonal buildups of energy stores, fattening, or replacement of molted furs, plumages, or other exuviae (see Section 12.3.4) also constitute significant features of production in many animals.

Growth. Growth rates and the sizes of animals result from intricate interactions between a variety of factors, including the genetic constitution of the individuals, physiological processes requiring growth factors and hormones for their proper functioning (see Chapter 12), and environmental factors, especially food availability and temperature.

Growth varies with both the amount and quality of available food. When food is absent or in short supply, growth is negative, metabolic needs being covered by the body's own resources. The proportion of food energy used for growth, the growth efficiency, increases with food supply until it reaches a maximum, at which point food is no longer a limiting factor. This maximal growth efficiency expresses the (maximal) capacity for growth. Comparisons of actual growth efficiencies under various conditions demonstrate the adaptability of organisms to changing environmental conditions, especially changing availability of food. Some organisms exhibit wide variations in growth patterns, others narrow. Bivalves and copepods may serve as examples.

Mussels grow fast or slow depending upon feeding conditions. Individuals living under optimal conditions in the subtidal zone may reach sizes within a few months that individuals living at the upper limit of their range in the intertidal zone may not reach within 15 years. However, the old mussels maintain their capacity for growth.

Whereas bivalves are capable of growing continuously, copepods and other crustaceans grow in stages, with the stages separated from one another by moults. Moulting occurs at intervals that vary with the temperature, but are practically independent of food availability. At low food levels the growth attained during a developmental stage is small, and mortality rapidly increases with decreasing food levels.

Even under optimal conditions, rates of growth eventually decline in most animals. In mammals and birds the proportion of assimilated food energy used for growth, the net growth efficiency, starts to decline early in life. The pattern of growth in nestlings of the house sparrow is an example (Figure 2-13). During the first two days of life after hatching, about half the assimilated food energy was used for body growth. However, during the following several days, the growth efficiency rapidly declined to a level of 15–20%. Some few days before the young were ready to leave the nest, final body size was reached and growth efficiency became zero, or even slightly negative.

Until recently this pattern, of growth efficiencies starting to decline early in life, was thought to be characteristic of animal growth generally. It now seems to be a specialization of birds and mammals, that is, of endothermic vertebrates.

The early and steep declines in growth efficiencies that have been observed in the young of many mammals and birds are correlated with the high energetic costs of endothermy (see Chapter 8). This relationship is well illustrated in Figure 2-13). The period during which the fraction of assimilated food energy partitioned into growth declines from one half to one sixth coincides with the period in the life of passerine nestlings when they change from ectothermy to endothermy (see Section 8.11).

In poikilothermic vertebrates, as well as in invertebrates, high growth efficiencies can be maintained until sexual maturity is reached and the gonads start to compete with the rest of the organism for energy (see Reproduction, following). In addition,

Figure 2-13
Net growth efficiencies, expressed as percent of assimilated food energy, in nestlings of the house sparrow, Passer domesticus. *G, net growth efficiency (·); B. m., body mass in grams (x). [Modified from C. R. Blem,* Comp. Biochem. Physiol., **52A**, *305–312 (1975).]*

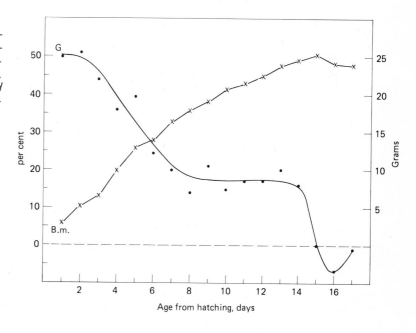

net growth efficiencies measured under optimal conditions are usually higher than those observed even in the early lives of birds and mammals. Net growth efficiencies around 70% or higher have been measured in representatives of the amphibians, teleosts, crustaceans, annelids, gastropods, nematodes, rotifers, and protozoa. Only in insects do most net growth efficiencies recorded range from 20 to 50%.

Published net growth efficiencies may not always represent maximal capacities for growth, because it is not always certain that the measurements were made under optimal conditions for growth, including optimal temperature. The effect of temperature on the efficiency of growth is apparent in the experiment on growth in the trout shown in Figure 2-12. The growth efficiency rapidly declined at temperatures below 6°C and above 13°C. The growth efficiency was not affected by increases in body mass over the large range from about 12 g to 250–280 g.

Reproduction. Chapter 13 deals in more detail with the diverse patterns of animal reproduction. At this time we will consider basic patterns in the partitioning of energy for reproduction and the efficiency of production of eggs, fetuses, and milk.

In young animals gonads typically grow at about the same relative rate as the other organs in the body. Sexual maturation results in the allocation of larger proportions of food energy to the gonads and other organs of reproduction. Somatic growth is usually simultaneously reduced. In some species of invertebrates the combined efficiency of somatic and reproductive growth remain more or less constant into adult life, with increasing proportions of assimilated food energy being diverted into reproduction in older individuals. We are not yet certain of how representative this pattern is of animal growth efficiency and partitioning of food energy in a wide range of kinds of animals.

Values for the proportion of assimilated energy that is diverted into reproduction are usually higher in invertebrates and ectothermic vertebrates than in endotherms, because of the high energy cost of endothermy. In invertebrates about half or more of the assimilated energy may be used in egg production, compared with only about 20% in chickens that have been selected for energetic efficiency in production of eggs.

In mammals energy for growth of the fetuses constitutes only a few percent of the assimilated energy. During lactation, however, production efficiency of milk may amount to about 40%. The period of lactation may represent such a heavy load on the energy budget that the energy in the food does not suffice to cover milk production. Energy then is mobilized from the storage depots of the mother. It is interesting to note that mammalian females typically deposit fat during pregnancy, apparently as an adaptation to meet the high energy costs of lactation.

Fattening. Animals from the temperate zone, and other environments with changing life conditions, usually are exposed to nutritionally favorable seasons alternating with less favorable periods. Growth is often restricted to the favorable season. During this season animals may also deposit energy reserves that can be used during periods when food is scarce or lacking. Animals may thus adapt to seasonally changing food supplies by eating a surplus when food is plentiful. The surplus is mostly deposited as fat, as in birds in the premigratory phase, or in animals preparing for hibernation. Sometimes the fat stores are deposited in special structures, such as the fat bodies in insects, amphibians, and reptiles. Glycogen may also serve as the main store of energy, as in bivalves.

Production efficiency, expressed as the proportion of assimilated food energy deposited as body mass, increases strongly during periods of fattening. Pigs are the best studied example. Figure 2-14 shows that during the period of growth from a body mass of about 20 kg to about 80 kg production efficiency increased from 16% to 50% of the assimilated food energy. During the same period true growth, as represented by protein retained, decreased from 16% to 8% of the assimilated food energy. Simultaneously, the proportion of energy deposited as fat increased from zero to more than 80% of the total energy deposited.

Energetic Costs of Production. The preceding examples indicate that production efficiency increases with the proportion of fat deposited, or decreases with increasing proportions of protein deposition, that is, growth in the strict sense of the word. This correlation between high production efficiencies and high fat contents in the products indicates that

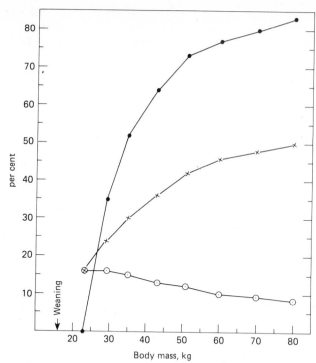

Figure 2-14
Relationship between fattening and growth in pigs. Production efficiency (x); protein retained, expressed as percent of assimilated food energy (○); fat retained, expressed as percent of total energy retained (●). [Modified from G. Thorbek. Nutr. Metab., 21, 105–118 (1977).]

the synthesis and deposition of fat in fat tissues, eggs, or milk is energetically cheaper than the synthesis of protein. However, production efficiencies do not directly tell us about the energetic costs of production. Attempts are therefore being made to distinguish between costs of maintenance and costs of production, including the costs of synthesis of protein, fat, and other constituents of growth, reproduction, and fattening. This central field of animal bioenergetics is rapidly expanding, with studies at the levels of thermodynamics, biochemistry, and physiology. At present final answers seem about to be reached only at the physiological level.

Priorities in Allocations of Food Energy. Assimilated food energy may be used to cover metabolic needs, or it may be deposited in growth, in energy stores, or in reproduction. Physiological processes control the proper allocation of food energy for these various purposes. Priorities in allocation when food energy becomes limiting vary among species and depend upon the developmental and functional states of the organisms. Usually, metabolic needs have first priority. In extreme instances this may result in negative growth, tissues of the body being broken down to serve as energy for maintenance of the organism. Growth and storage of energy are often assigned second priority in energy allocation; in such cases reproduction depends upon optimal food conditions. However, many variants exist in the pattern of allocation of food energy for reproduction. In one extreme, represented, for example, by lampreys and salmons, growth of gonads mainly occurs after feeding and bodily growth have ceased. In such instances reproduction depends upon the mobilization of energy from stores in the body and its incorporation into the growing gonads, especially the gametes.

The allocation of food energy for its various uses is controlled by hormonal mechanisms. In vertebrates, an intricate interplay between hormones, especially those secreted by the pancreatic islets, controls the partitioning of food energy during the phases of absorption, deposition, and mobilization that are associated with discontinuous feeding (see Section 12.4.2). Hormones that control growth operate at two levels. Hormones such as growth hormone and the thyroid hormones control growth of the whole organism, whereas local hormones and growth factors participate in regulating growth of individual organs and tissues (see Section 12.2.2). In comparison with the hormonal mechanisms that control partitioning of energy for metabolism and growth, the mechanism that controls allocation of energy for reproduction appears simple. Reproduction depends upon gonadotropin secreted by the pars distalis of the pituitary gland. Secretion of gonadotropin in amounts needed for reproduction depends upon central nervous stimulation, which may again depend upon the nutritional state of the organism (see also Section 13.5).

ADDITIONAL READING

GENERAL
Biedermann, W. "Die Aufnahme, Verarbeitung und Assimilation der Nahrung." In Hans Winterstein (ed.).

Handbuch der vergleichenden Physiologie, vol. 2, sec. 1, pp. 1–1563, Gustav Fischer, 1911.

Crawford, M. A. (ed.). "Comparative nutrition of wild animals," *Symposia of the Zoological Society of London*, **No. 21**, 1968.

Denison, R. H., S. Springer, B. Schaeffer, D. E. Rosen, E. C. Olson, C. Gans, and D. D. Davis. "Evolution and dynamics of vertebrate feeding mechanisms," *Am. Zool.*, **1**, 177–234 (1961).

Dixon, M., and E. C. Webb, *Enzymes*, 3rd ed. Longmans, Green & Co., London, 1979.

Florkin, M., and E. H. Stotz (eds.). *Metabolism of Vitamins and Trace Elements, Comprehensive Biochemistry*, vol. 21, 297 pp., Elsevier, Amsterdam, 1971.

Hoekstra, W. G., J. W. Suttie, H. E. Ganther, and W. Mertz. *Trace Element Metabolism in Animals*, University Park Press, Baltimore, 1974.

House, H. L. "Nutrition, digestion," in M. Rockstein (ed.), *The Physiology of Insects*, 2nd ed., vol. 5., Academic Press, New York, 1974.

Jennings, J. B. *Feeding, Digestion and Assimilation in Animals*, 2nd ed., Macmillan, London, 1972.

McDonald, P., R. A. Edwards, and J. F. D. Greenhalgh. *Animal Nutrition*, 2nd ed., Oliver and Boyd, Edinburgh, 1973.

Passmore, R., B. M. Nicol, M. N. Rao, G. H. Beaton and E. M. DeMayer. *Handbook of Human Nutritional Requirements*. WHO Monograph Series No 61, Geneva, 1974.

Steele, J. H. (ed.). *Marine Food Chains*, Oliver and Boyd, Edingburgh, 1970.

T-W-Fiennes, R. N. (ed.). *Biology of Nutrition*, Pergamon Press, Oxford, 1972.

Vonk, H. J. "Comparative biochemistry of digestive mechanisms." In M. Florkin and H. S. Mason (eds.). *Comparative Biochemistry*, vol. 6, pp. 347–401, Academic Press, New York, 1963.

SUSPENSION-FEEDING

Conover, R. J. "Zooplankton-life in a nutritionally dilute environment," *Am. Zool.*, **3**, 107–118 (1968).

Jørgensen, C. B. *Biology of Suspension Feeding*, Pergamon Press, New York, 1966.

Jørgensen, C. B. "Comparative physiology of suspension feeding," *Ann. Rev. Physiol.*, **37**, 57–79 (1975).

Jørgensen, C. B. "A hydromechanical principle for particle retention in *Mytilus edulis* and other ciliary suspension feeders," *Mar. Biol.*, **61**, 277–282 (1981).

King, K. R., J. T. Hollibaugh, and F. Azam. "Predator-prey interactions between the larvacean *Oikopleura dioica* and bacterioplankton in enclosed water columns," *Mar. Ecol.*, **56**, 49–57, 1980.

DISSOLVED ORGANIC MATTER AS FOOD

Jørgensen, C. B. "August Pütter, August Krogh, and modern concepts on role of dissolved organic matter as food for aquatic animals," *Biol. Rev.*, **51**, 291–328 (1976).

Stephens, G. (ed.). "The role of uptake of organic solutes in nutrition of marine organisms," *Am. Zool.*, 1981, in press.

SYMBIOSIS BETWEEN PLANTS AND ANIMALS

Goreau, T. F., N. I. Goreau, and C. M. Yonge. "On the utilization of photosynthetic products from zooxanthellae and of a dissolved amino acid in *Tridacna maxima* f. *elongata* (Mollusca: Bivalvia)," *J. Zool. London*, **169**, 417–454 (1973).

Jennings, D. H., and D. L. Lee (eds.). *Symbiosis*, Cambridge University Press, New York, 1975.

Lewis, J. B. "The importance of light and food upon the early growth of the reef coral *Favia fragum* (Esper)," *J. Exp. Mar. Biol. Ecol.*, **15**, 299–304 (1974).

Muscatine, L. "Nutrition of corals." In O. A. Jones and R. Endean (eds.). *Biology and Geology of Coral Reefs*, vol. II, Biology 1, pp. 77–115, Academic Press, New York, 1974.

Muscatine, L. "Chloroplasts and algae as symbionts in molluscs," *Int. Rev. Cytol.*, **36**, 137–169 (1973).

Taylor, D. L. "The cellular interactions of algal-invertebrate symbiosis," *Adv. Mar. Biol.*, **11**, 1–56 (1973).

DETRITUS PATHWAY

Edwards, C. A., and J. R. Lofty. *Biology of Earthworms*, Chapman and Hall, London, 1972.

Fenchel, T. M., and T. H. Blackburn. *Bacteria and Mineral Cycling*, Academic Press, New York, 1979.

Fenchel, T. M., and B. B. Jørgensen. "Detritus food chains of aquatic ecosystems: the role of bacteria," *Adv. Microbiol. Ecol.*, **1**, 1–58 (1977).

Melchiorri-Santolini, U., and J. W. Hopton (eds.). "Detritus and its role in aquatic ecosystems," *Mem. Istituto Ital. Idrobiol.*, vol. 29, Suppl., 1972.

DIGESTION BY MEANS OF SYMBIONTS

Blaxter, K. L. "The nutrition of ruminant animals in relation to intensive methods of agriculture," *Proc. R. Soc. (London)*, **B.183**, 321–336 (1973).

Clarke, R. T. J., and T. Bauchop (eds.). *Microbial Ecology of the Gut*, Academic Press, New York, 1977.

Dougherty, R. W. (ed.). *Physiology of Digestion in Ruminants*, Butterworth and Co. (Publishers), Ltd., London, 1965.

Howard, B. H. "Ruminants and other vertebrates with microbial symbiotes." In S. Mark Henry (ed.). *Symbiosis*, vol. 2, pp. 317–385, Academic Press, New York, 1967.

Hungate, R. E. *The Rumen and its Microbes*, Academic Press, New York, 1966.

Phillipson, A. T. (ed.). *Physiology of Digestion and Metabolism in the Ruminant*, Oriel Press, Newcastle-upon-Tyne, England, 1970.

DIGESTION AND ABSORPTION

Augustinsson, K. B. "The evolution of esterases in vertebrates." In N. Van Thoai and J. Roche (eds.). *Homologous Enzymes and Biochemical Evolution,* pp. 299–311, Gordon and Breach, New York, 1968.

Binder, H. J. (ed.). *Mechanisms of Intestinal Secretion,* Alan R. Liss, New York, 1979.

Glass, G. B. J. (ed.). *Gastrointestinal Hormones,* Raven Press, New York, 1980.

Jacobson, E. D., and L. L. Shanbour. *Gastrointestinal Physiology,* University Park Press, Baltimore, 1974.

Martin, M. M., and J. S. Martin. "The distribution and origins of the cellulolytic enzymes of the higher termite, *Macrotermes natalensis,*" *Physiol. Zool.,* **52,** 11–21 (1979).

Masters, C. J., and R. S. Holmes. "Isoenzymes, multiple enzyme forms, and phylogeny," *Adv. Comp. Physiol. Biochem.,* **5,** 109–195 (1974).

Ockner, R. K., and K. J. Isselbacker. "Recent concepts of intestinal fat absorption," *Rev. Physiol. Biochem. Pharmacol.,* **71,** 107–146 (1974).

Rose, R. C. "Water-soluble vitamin absorption in intestine," *Ann. Rev. Physiol.,* **42,** 157–171 (1980).

Sargent, J. R. "Marine wax esters," *Sci. Prog.,* **65,** 437–458 (1978).

Stroud, R. M. "A family of protein-cutting proteins," *Sci. Am.,* 74–88 (July 1974).

PARTITIONING OF FOOD ENERGY AND BIOENERGETICS

Atkinson, D. E. *Cellular Energy Metabolism and Its Regulation,* Academic Press, New York, 1977.

Calow, P. "The cost of reproduction—a physiological approach," *Biol. Rev.,* **54,** 23–40 (1979).

Cole, D. J. A., K. N. Boorman, P. J. Buttery, D. Lewis, R. J. Neele, and H. Swan (eds.). *Protein Metabolism and Nutrition,* Butterworth, London, 1976.

Hoar, W. S., D. J. Randall, and J. R. Brett (eds.). *Fish Physiology,* vol. VIII, *Bioenergetics and Growth,* Academic Press, New York, 1979.

Miller, P. J. (ed.). *Fish Phenology: Anabolic Adaptations in Teleosts,* Academic Press, London and New York, 1979.

Müller, H. L., and M. Kirchgessner. "Zur Energetik der Proteinsynthese beim Wachstum," *Z. Tierphysiol., Tierernährg. u. Futtermittelkde,* **42,** 161–172, 1979.

Waterlow, J. C., P. J. Garlick, and D. J. Millward. *Protein Turnover in Mammalian Tissues and in the Whole Body,* North-Holland, Amsterdam, 1978.

George A. Bartholomew

3

ENERGY METABOLISM

3.1 INTRODUCTION

Organisms are not "things" in the same sense that rocks or minerals are "things." They are transient concentrations of extremely complicated arrays of molecules, which, as long as they are alive, have virtually no fixed and permanent constituents. It is profitable to think of them as chemical systems in which energy transformations are continually taking place. These complicated energy relationships form the essential bonds that hold the transient chemical aggregations together as coherent, functioning systems.

If we adopt this kinetic approach, we are forced to look at organisms in a way that the layman would find unconventional. We can no longer consider them to be just the familiar animals and plants that we see living in nature or housed in the laboratory. We must treat them as intricate interactions between complex, self-sustaining, physicochemical systems and the substances and conditions that we usually think of as the environment. Looking at organisms in this manner, we are forced to the conclusion that Claude Bernard reached a century ago—organism and environment form an insepara-

ble pair; one can be defined only in terms of the other. Organisms are delicate but highly adaptable dynamic systems that, as long as they are alive, exist in a state of continuous exchange of energy and materials with the environment that surrounds them.

It is obvious that the interacting dynamic systems that constitute the organism can be maintained only by the continuous expenditure of energy. The energy that is at the base of the pyramid of chemical and physical relationships that constitutes an organism is released by the oxidation of carbon and hydrogen. The biochemical interactions involved in the transformation of the energy of oxidation into precisely regulated physiological work are enormously complex. The entire system of energetics is concealed with the single word *metabolism.*

3.2 CELLULAR ENERGY METABOLISM

This book is concerned primarily with the level of biological integration involving functionally intact multicellular organisms. We shall do no more

46

than identify the major patterns of the metabolic pathways of the energy transformations involved in the utilization of carbohydrates, fats, and proteins by cells. It must, however, be appreciated that intermediary (cellular) metabolism supplies an integrative theme that, like cellular organization, unites virtually all organisms. The principal features of energy metabolism are similar in most cells. Hence there is an essential unity in biological activity at the cellular level. We should, however, bear in mind that the steps leading to, and following, cellular respiration in complex organisms involve staggeringly complex patterns of activity. Among these are (1) the behavior involved in pursuit and capture of food; (2) feeding, digestion, and absorption into the body fluids (considered in Chapter 2); (3) transportation of nutrients and oxygen to the cell and metabolic wastes from the cell via the circulatory system (considered in Chapters 5 and 6); and (4) processing of the wastes and their elimination into the external environment (considered in Chapter 7).

In cellular metabolism, food materials are absorbed from the body fluids and fragmented into smaller molecules. These are eventually oxidized by precise and orderly chains of reactions that are regulated by complex systems of controls involving substrates, enzymes, cofactors, and hormones.

Cellular oxidation produces heat, but, except for the few animal groups that are endothermic (see Section 8.3), any energy that appears as heat in a cellular chemical reaction is energy that is biologically wasted. Much of the energy resulting from cellular respiration does not immediately appear as heat but is stored in chemical form as adenosine triphosphate (ATP; see Figure 4-7), which is a virtually universal source for immediate energy in cellular metabolism.

Glucose is one of the more frequently used of the fuels for cellular metabolism. If 1 mol (180 g) of glucose is ignited in an atmosphere of oxygen, it yields 673 kilocalories (kcal) (see Section 3.4) as a result of its uncontrolled combustion to form carbon dioxide (CO_2) and water (H_2O). Such an explosive release of energy could do useful work in a heat engine, but the cell is not a heat engine. It is a system that performs work by means of chemical energy. A basic principle of thermodynamics requires that a fixed amount of energy be liberated by the oxidation of a given amount of a substance

no matter what the process or mechanism of the oxidation. Consequently, the slow and regulated oxidation of glucose inside a cell must yield 673 kcal/mol, just as it does in rapid combustion. For the chemical machinery of the cell to use the energy freed by the oxidation of glucose, it must become available in small units, units of the order of magnitude of the energy-rich bonds in adenosine triphosphate (ATP). It is not surprising, therefore, that the process of oxidation goes forward in a series of small, precisely controlled steps, each one involving oxidizing agents that approach progressively closer to the oxidizing potential of the terminal agent, oxygen. At each successive stage, energy in utilizable quantities becomes available and is stored as ATP.

The unraveling of these relationships has been one of the most dramatic results of scientific investigation during the present century. Descriptions of the details of these processes are offered in virtually every contemporary text in general physiology, cell biology, and biochemistry, and need not be repeated here.

3.3 ENERGY METABOLISM OF WHOLE ANIMALS

The rate of energy metabolism probably integrates more aspects of animal performance than any other single physiological parameter. Indeed, from a simplistic point of view, the proverbial "struggle for existence" can profitably be thought of as a competition for physiologically utilizable energy.

A major preoccupation of students of the energy metabolism of whole animals has been the determination of the rate, or intensity, of the entire process. The rate of an animal's metabolism is a critical property of its total physiology, particularly if we know the pattern of variation of the rate through time. By integrating the temporal variations, we can estimate the total impact of a given animal on the energy resources of the environment. If adequate population data are available, we can even make a quantitative estimate of the success of a given type of organism in capturing some part of the limited amount of energy available for sustaining living systems under a given climatic regime. The temporal pattern of energy metabolism is also revealing because it is a quantitative record of the pattern of timing that an individual animal follows in its energy utilization. This in turn allows a realistic appraisal of the adjustments that it makes to the availability of energy both on a daily and a seasonal basis. Such data also give an insight into the complicated interplay involved in the energy relationships of the different organisms that are members of the same bioenergetic system.

3.4 SOME DEFINITIONS AND UNITS

The ultimate source of energy for metazoans is the oxidation of carbon and hydrogen. Since this oxidation yields heat, heat production is a measure of energy metabolism. The most familiar and widely employed unit of heat is the *calorie*, which was defined before the equivalence of work and heat was established. A calorie is the amount of energy required to raise 1 g of water 1°C. This quantity varies with temperature, but the variation can be eliminated when the calorie is redefined in terms of work; 1 cal = 4.1868 joules. For most physiological purposes the calorie is an inconveniently small unit and a unit 1000 times as large, the kilocalorie (kcal) is commonly used. The calorie is easy to visualize, but it is awkward to use when dealing with work, power, or energy flux. It is interesting to examine why its usage persists.

Science is one of the most deliberately and self-consciously historical of human activities, and many of the notations and units used in physiology persist simply because of long usage and familiarity. The use of calorie instead of joule, or calories per unit time instead of watts are cases in point. Every physiologist knows what a calorie is and its use in communication introduces a minimum of ambiguity. Thus it persists despite its awkwardness. Its awkwardness is compounded by the fact that it is often used in combination with body weight. One of the most commonly used expressions in animal energetics is heat production per unit weight (cal/g or kcal/kg), which presents many difficulties. Most persons reading this are familiar with the difference between mass and weight, but in most of the physiological literature mass and weight are expressed in the same units. This is because physiologists have been employing a system in which force was arbitrarily defined as the weight of a standard mass in pounds or kilograms. In the SI system of units (Systeme International), however, mass is the arbitrary unit and weight is defined as a force equal to the product of mass times the acceleration of gravity (9.806650 meter per second squared) and the unit of weight is the newton. This is not just an item of esoteric intelligence, it is critical to most calculations relating animal energetics to work and power.

Because the usage is unfamiliar, it is not yet practical to express animal weights in newtons, but in the present chapter, in all cases where it is computationally significant, the distinction made by the SI system between weight and mass will be observed.

The International System of Units, abbreviated SI, has been accepted by all the principal industrial nations including the United States, and since 1964 has been adopted by the U.S. National Bureau of Standards. There is one SI unit for each of the seven physical quantities, length, mass, time, electric current, thermodynamic temperature, luminous intensity, and amount of substance. These units are meter, kilogram, second, ampere, kelvin, candela, and mole, respectively. From these basic units a series of consistent derived units have been obtained by multiplication and division. Selected SI units relevant to energy metabolism are defined in Table 3-1, and conversion factors for the units commonly used are given in Table 3-2. A complete description of the SI System has been prepared by A. E. Mechtly [1973] and is available as NASA SP-7012 for less than a dollar from the U.S. Government Printing Office, Washington, D.C. 20402.

Communication between physiologists requires a commonly accepted system of notation as well as a system of units. Table 3-3 contains a few key rules for notation that are based on recommendations prepared for the American Physiological Society.

3.5 MEASUREMENT OF ENERGY METABOLISM: GENERAL

The point has previously been made that organisms are complex chemical arrays in which energy transformations are taking place. Another way of phrasing this is to state that organisms are chemical sys-

TABLE 3-1. Basic Units of the SI System for Animal Energetics

meter (m)	The meter is the length equal to 1,650,763.73 wavelengths in vacuum of the radiation corresponding to the transition between the levels $2p_{10}$ and $5d_5$ of the krypton-86 atom.
kilogram (kg)	The kilogram is the unit of mass; it is equal to the mass of the international prototype of the kilogram—a particular cylinder of platinum-iridium alloy which is preserved in a vault at Sèvres, France, by the International Bureau of Weights and Measures.
second (s)	The second is the duration of 9,192,631,770 periods of the radiation corresponding to the transition between the two hyperfine levels of the ground state of the cesium-133 atom.
kelvin (K)	The kelvin, unit of thermodynamic temperature, is the fraction 1/273.16 of the thermodynamic temperature of the triple-point of water.
newton (N)	The newton is that force which gives to a mass of 1 kilogram an acceleration of 1 meter per second per second.
joule (J)	The joule is the work done when the point of application of 1 newton is displaced a distance of 1 meter in the direction of the force.
watt (W)	The watt is the power which gives rise to the production of energy at the rate of 1 joule per second.
mole (mol)	The mole is the amount of substance of a system which contains as many elementary entities as there are carbon atoms in 0.012 kg of carbon-12. The elementary entities must be specified and may be atoms, molecules, ions, electrons, other particles, or specified groups of such particles.

Prefixes and Symbols for Decimal Multiples of Units

10^6	mega	M
10^3	kilo	k
10	deca	da
10^{-1}	deci	d
10^{-2}	centi	c
10^{-3}	milli	m
10^{-6}	micro	μ
10^{-9}	nano	n
10^{-12}	pica	p

tems that maintain their integrity by controlling the rates of storage and release of energy. Energy, of course, is convertible from one form to another, and, as every schoolboy knows, the energy available for biological systems comes ultimately in the form of radiation from the sun. Once the radiation is converted to chemical form by the process of photosynthesis, the energy metabolism of all organisms consists of the regulated release and storage of this chemical energy.

Although a complete description of the processes of intermediary metabolism may never be achieved, the total energy transactions involved in this complicated process can be measured in a variety of fairly simple ways.

3.5.1 Measurement of Energy Metabolism: Direct Calorimetry

Direct calorimetric measurement of metabolism involves placing an organism in a device that allows one to determine the amount of heat per unit time.

TABLE 3-2. Conversion Factors for Units Used in Energy Metabolism

The names of the units are spelled out on the left of the matrix and abbreviations are given at the top. The conversion factors are given as five-digit numbers followed in parenthesis by the power of 10 by which they must be multiplied. For example, to convert kilocalories to joules, multiply by 4.1868×10^3.

Work, Energy, Heat

To Into↓ convert→	cal	kcal	J	cm^3O_2*	$1O_2$*
calorie	—	1.0000(+3)	2.3885(−1)	4.8000(0)	4.8000(+3)
kilocalorie	1.0000(−3)	—	2.3885(−4)	4.8000(−3)	4.8000(0)
joule	4.1868(0)	4.1868(+3)	—	2.0097(+1)	2.0097(+4)
cubic centimeter oxygen*	2.0833(−1)	2.0833(+2)	5.0073(−2)	—	1.000(+3)
liter oxygen*	2.0833(−4)	2.0833(−1)	5.0073(−5)	1.0000(−3)	—

* Conversion factors for O_2 consumed vary with RQ (Section 3.3.5). Value of RQ used here is ~0.79.

Power, Energy Consumption

To Into↓ convert→	W	kW	kcal min⁻¹	kcal hr⁻¹	kcal day⁻¹
watt	—	1.000(+3)	6.9780(+1)	1.1630(0)	4.8458(−2)
kilowatt	1.0000(−3)	—	6.9780(−2)	1.1630(−3)	4.8458(−5)
kilocalorie per minute	1.4331(−2)	1.4331(+1)	—	1.6667(−2)	6.9444(−4)
kilocalorie per hour	8.5985(−1)	8.5985(+2)	6.0000(+1)	—	4.1667(−2)
kilocalorie per day	2.0636(+1)	2.0636(+4)	1.4400(+3)	2.4000(+1)	—

Specific Metabolic Rate, Specific Power

To Into↓ convert→	W kg⁻¹	kcal g⁻¹ hr⁻¹	kcal g⁻¹ day⁻¹	kcal kg⁻¹ hr⁻¹	kcal kg⁻¹ day⁻¹
watt per kilogram	—	1.1630(+4)	4.8458(+1)	1.1630(0)	1.6440(+3)
kilocalorie per gram hour	8.5985(−4)	—	4.1667(−2)	1.000(−3)	4.1667(−5)
kilocalorie per gram day	2.0636(−2)	2.4000(+1)	—	2.4000(−2)	1.0000(−3)
kilocalorie per kilogram hour	8.5985(−1)	1.000(+3)	4.1667(+1)	—	4.667(−2)
kilocalorie per kilogram day	2.0636(+1)	2.4000(+4)	1.0004(+3)	2.4000(+1)	—
meter per second kilometer	—	2.7778(−1)	4.4704(−1)		

TABLE 3-2. Conversion Factors for Units Used in Energy Metabolism (continued)

Into↓ convert→ To	Speed		
	$m\ s^{-1}$	$km\ hr^{-1}$	$mile\ hr^{-1}$
per hour	3.6000(0)	—	1.6093(0)
miles per hour	2.2369(0)	6.2137(−1)	—

Into↓ convert→ To	Force, Weight		
	N	dyn	$kg\ f$
newton	—	1.0000(−5)	9.8067(0)
dyne	1.0000(+5)	—	9.8067(+5)
kilogram-force	1.0197(−1)	1.0197(−6)	—

This direct measurement is possible because, as discussed in Section 3.2, the quantity of heat produced during a chemical reaction is independent of the number of intermediate steps involved in the reaction.

Direct calorimetry is relatively simple in concept but, like most physical measurements, it becomes complicated technically as one attempts to increase its accuracy. In its simplest form the method consists of enclosing an animal in an insulated chamber and then measuring the rise in temperature of the medium adjacent to the animal as heat is lost from the animal's body. If the specific heat of the substance acquiring the heat is known, the calories lost by the animal per unit time can be calculated from the increase in temperature. The archetypal direct calorimeter was designed by Lavoisier and Laplace almost 200 years ago. They determined metabolism by measuring the amount of ice melted by the heat from an animal rather than by measuring an increase in temperature caused by the heat. In their calorimeter a small mammal was placed in a chamber completely enclosed in ice and water. This chamber was put in a double-walled chamber also filled with a mixture of ice and water. Such a mixture maintains its temperature constant at 0°C. Because the chamber surrounding the small mammal was filled with a mixture of ice and water, the temperature there remained at 0°C. Therefore, no temperature difference existed between the

TABLE 3-3. Some Rules of Notation

1. A bar over a symbol denotes mean value $(\bar{x})$.
2. A dot over a symbol denotes a time rate ($\dot{V}_{O_2}$ indicates rate of oxygen consumption).
3. Capital T is used for temperatures; Kelvin, centigrade, or celsius. K is used when T is raised to a power. Otherwise °C is used.
4. Physiological subscripts usually have a lettered similarity to the property referred to:
   P_w = partial pressure of water vapor
   T_b = body temperature
5. The product of two units may be presented as:
   kcal hr   kcal · hr   kcal × hr
6. The quotient of two units may be presented as:
   $\dfrac{J}{s}$ J/s   J s^{-1}
7. No more than one solidus (/) should be used in an expression. cm³/(g hr) or cm³/(g hr) or cm³ g^{-1} hr^{-1}, but not cm³/g/hr.

outer and inner chambers and no heat flow could take place. The small mammal was continuously giving off heat by its metabolism, and since this heat could not produce any temperature change in the surrounding environment, its effect was to cause the ice to melt. The amount of ice melted was measured simply by collecting the meltwater.

In the many generations since Lavoisier and Laplace's crude pioneer investigations, numerous more sophisticated varieties of direct calorimeters have been designed. Historically, the principal importance of direct calorimetry on whole animals has been to establish that the heat produced by animals is equal to the heat of combustion of the compounds catabolized minus the heat of combustion of the various materials excreted. However, in recent years the availability of extremely sensitive and precise electric thermometric devices has allowed the development of methods of microcalorimetry that extend the units of measurement to include the entire range from megacalories to microcalories and to make virtually instantaneous measurements of heat loss from whole organisms or parts thereof. All types of direct calorimetry, however elaborate, suffer from the defect that the conditions for measurement of metabolism place stringent restrictions on the environment and behavioral condition of the animal being studied.

3.5.2 Measurement of Energy Metabolism: Indirect Calorimetry

Many of the biologically interesting parameters of aerobic energy metabolism can be conveniently measured by indirect methods that depend on determinations of the quantities of oxygen used, quantities of carbon dioxide given off, or the amounts of food utilized. Measurements of metabolic rate based on determinations of oxygen utilization, carbon dioxide production, or combinations of the two have been most frequently employed.

In the physiological literature oxygen consumption and carbon dioxide production are expressed both as volumes (V) or as quantities (M). These symbols can be converted to time rates by placing dots above them. Thus, $\dot{V}$ (pronounced "vee dot") is volume per unit time and $\dot{V}_{O_2}$ is volume of oxygen consumed per unit time, with units such as cubic centimeters per minute (cm^3/min) or liters per hour (liter/hr). Similarly $\dot{M}_{O_2}$ is the amount of oxygen consumed per unit time, with units such as moles per hour (mol/hr) or millimoles per second (mmol/sec). V and M are interconvertible; 1 mmol of a dry gas at standard temperature and pressure (STPD) is equal to 22.4 cm^3.

The rate of uptake of oxygen by an aquatic animal can be determined by serial measurements of the amount of oxygen dissolved in the water in which it floats or swims. The traditional method for such measurements has been titration, but gasometric methods of a high level of precision are now in common use. Polarographic systems using oxygen electrodes are also commercially available, and, when used in conjunction with recording potentiometers, allow continuous direct recording of amounts of dissolved oxygen.

Under circumstances in which it is not disadvantageous to maintain a constant environmental temperature, the amount of oxygen removed by a terrestrial animal from a given volume of air can conveniently be measured (1) by changes either in volume or pressure in a closed system, or (2) by adding measured amounts of oxygen to replace that removed, thus maintaining pressure or volume constant. Closed systems for measurement of metabolism of either aquatic or terrestrial animals are obviously inconvenient if one is dealing with very large animals or with very high rates of oxygen consumption—in the first case for obvious reasons, and in the second case because a great many precautions must be taken to keep environmental conditions at adequate levels for physiological processes to go forward normally.

Open systems for indirect calorimetry avoid some of these problems. In principle they operate by moving air or water through a chamber in which an animal is enclosed and then measuring the differences in quality of the air, or the gases dissolved in the water, between the input and output parts of the system. One of the simplest of such methods for air-breathing animals consists of putting the outflow air through known weights of various chemicals that combine with carbon dioxide and water, then weighing these chemicals to determine the quantities of water and carbon dioxide produced. From such weights and the changes in weight of the animal itself, one can calculate the amount of oxygen it has utilized.

In recent years the use of physical sensors to monitor continuously by electrical means the amount of oxygen or carbon dioxide in the outgoing and incoming air or water has made it possible to obtain continuous measurements of relatively undisturbed animals over very long periods of time. Such systems as the Pauling paramagnetic oxygen sensor, which directly measures the number of oxygen molecules per unit volume; infrared spectrophotometers, which can be set to measure the infrared absorption of carbon dioxide; or polarographic electrodes, which measure oxygen concentration either in water or gas, allow the continuous monitoring of rates of metabolism in a wide variety of organisms under a variety of environmental conditions.

All the methods of direct and indirect calorimetry described in the preceding paragraphs share a common disadvantage. They can be carried out only on animals that are enclosed or are wearing masks. This disadvantage can be avoided, under conditions where instantaneous measurements are not critical, by determining the amount of food consumed by captive animals. By standardizing the diet and determining its caloric content one can calculate the amount of energy available in the food ingested. One can determine by difference the metabolic rate of the animal by collecting its excretory and fecal products and determining their caloric content by bomb calorimetry. The difference between the calories ingested and the calories egested, when combined with a caloric correction for any changes in body weight, then gives a measure of the calories utilized by the animal during a unit of time. This method is rather tedious and does not take into account the fraction of the animal's food that is metabolized by the microorganisms in its gut (or, in the case of aquatic forms, in the medium). However, it yields data that cannot otherwise be obtained.

All the methods mentioned require that the animals being measured be kept under the unnatural conditions of confinement. Consequently, they yield data which can be related to energy exchange under natural conditions only by inference.

A technique is available that makes possible the measurement of the total energy metabolism of unconfined animals. The method uses water in which both the hydrogen and the oxygen are labeled isotopically. It depends on the fact that the oxygen in respiratory carbon dioxide is in isotopic equilibrium with the oxygen in the body water. The hydrogen of the body water is lost to the environment in water, but the oxygen is lost both in water and in carbon dioxide. The turnover rate for hydrogen in body water is, therefore, slower than that for oxygen. The difference in the turnover rates of hydrogen and oxygen is related to the carbon dioxide expired and can in principle be used as a measure of energy metabolism.

This isotopic method requires complex instrumentation, but it allows access to information that is available by no other means and is therefore an extremely powerful tool for the study of physiological energetics. Laboratory validation studies on a variety of species, including mammals, reptiles, birds, and insects, show that the doubly labeled water method is accurate to within 8%, but under field conditions, particularly if evaporative water loss is extensive, the error may be greater. A detailed discussion and an extensive bibliography is available in Nagy [1980].

It is a common observation that heart rate varies directly with general bodily activity. Numerous investigators have attempted to quantify this correlation in order to use heart rate, which is easily measured, to estimate total energy metabolism. These efforts have been quite successful with regard to human beings under controlled conditions, and recent advances in radiotelemetry have stimulated attempts to apply this method to free-ranging wild animals.

The method involves simultaneous laboratory measurement of heart rate and oxygen consumption as functions of some ecologically realistic factors such as ambient temperature and level of activity, and the subsequent monitoring of heart beat by radio telemetry of a free-ranging animal as it goes about its normal activities. Surgically implanted electrodes pick up the bioelectric potentials generated by the beating heart. By means of tiny battery-powered radio transmitters mounted on, or implanted in, the animal, these signals are relayed to a remotely located tape recorder for processing and computer analysis.

Radiotelemetered electrocardiograms have been obtained from a variety of animals including forms as small as songbirds, mice, and bats. These data are of interest in themselves and, in addition, have established a reasonably strong correlation between heart rate and energy metabolism in resting and moderately active animals. Nevertheless, they have not greatly extended knowledge of the total energy metabolism of free-ranging animals because of the variability in the relation between heart rate and high levels of activity, the marked differences between heart rate and activity in different taxa, between individuals in the same taxa, and in the same individuals on consecutive days. This variability is not surprising because heart rate is only one of the parameters involved in cardiac performance (Section 6.6.3). Oxygen consumption ($\dot{V}_{O_2}$) is equal to the product of heart rate (HR), stroke volume (SV), and the difference in oxygen content of the arterial and venous blood (A-V diff).

$$\dot{V}_{O_2} = (HR)(SV)(A\text{-}V \text{ diff})$$

Consequently, the total oxygen consumption of an animal can be determined from cardiovascular data only when all these parameters are measured simultaneously. See Gessamen [1980] for an evaluation of this method.

3.5.3 Respiratory Quotient and the Energy Equivalence of Oxygen

From the preceding discussion it is apparent that units of measurements that can be used to describe the rate of energy metabolism are various and depend on the method being employed. The units most commonly used are calories dissipated per unit time or volume of O_2 consumed per unit time. However, it is often useful to express energy metabolism in terms of some additional parameter that has particular biological relevance or is critical to the specific situation under consideration; for example, energy metabolism per unit time per unit body mass, per unit body surface, per unit fat-free mass, per unit protein-bound nitrogen, or in the case of populations, per unit area, per unit volume, per unit biomass, per individual, or per social unit. The choice depends on the biological context and the interests of the investigator, but whatever the units, all can in principle be converted to calories, joules, or watts per relevant unit.

Metabolic rate is most frequently measured by determining rate of oxygen uptake, and additional data—the relative amounts of carbon and hydrogen oxidized—are required for the conversion of O_2 consumption to calories. Such data are virtually unobtainable for aquatic forms. Even for terrestrial animals it is impractical to measure directly the amount of oxidative water produced per unit time, because water of undetermined origin is continuously being lost from the moist respiratory surfaces and by insensible perspiration through the integument. It is, however, relatively simple to measure the carbon dioxide (CO_2) expired, although measurements made over short periods may be seriously biased by the release of CO_2 formed at some previous time and then washed out of the animal's pool of carbonates during the period of measurement. If one simultaneously measures the O_2 consumed and the CO_2 produced, the amount of water produced can, in theory, be calculated.

The ratio of CO_2 produced to O_2 consumed (mol CO_2/mol O_2) is called the *respiratory quotient* (RQ). If the RQ is known, the relative amounts of carbohydrate and fat being oxidized can be estimated and the caloric yield of O_2 can be computed. The RQ has a value of 1.0 for carbohydrates, 0.8 for proteins, and 0.71 for fats. The oxidation of 1 g of mixed carbohydrate yields 4.0 kcal, 1 g of a typical fat yields about 9.5 kcal, and 1 g of protein yields about 4.5 kcal. As all three classes of compounds are being oxidized simultaneously, one cannot compute their proportional utilization from the RQ alone, for it is at best only a crude index of relative roles of the various pathways of intermediary metabolism. The amount of protein being used, however, can be calculated if one measures the amount of nitrogenous waste being produced (urea, uric acid, or ammonia, depending on the animal under consideration). If the protein contribution is known, one can more accurately calculate from the RQ the relative amounts of fat and carbohydrate being utilized. However, relatively little protein is oxidized by fasting animals in good condition; therefore, in practice when measurements of standard and basal metabolism are made, the RQ is usually not corrected for its protein component. Typical caloric equivalents of oxygen are shown in Table 3-4. The values used in calculating the heat production of a fasting aerobic animal from

TABLE 3-4. Caloric Equivalents of Oxygen, Protein Utilization Ignored

(For conversion factors for other units see Table 3-2)

RQ	Kcal/liter O_2 (STP)
0.71	4.73
0.75	4.77
0.80	4.86
0.85	4.91
0.90	4.95
0.95	5.04
1.0	5.09

its oxygen consumption are usually rounded off to either 4.7 or 4.8 kcal/liter of O_2.

3.6 STANDARD AND BASAL METABOLISM

The rate of energy metabolism of any given animal is highly variable. Its metabolic rate must match the energy cost of all its activities, and these activities may be extremely diverse. A measurement of gross energy metabolism is an umbrella that covers all the manifold activities within an organism. It is clear, however, that such a measurement cannot be taken at random or in a capricious manner but must be made under carefully chosen conditions. Otherwise, one has no basis from which to operate or from which to make comparisons. The list of conditions known to affect energy metabolism is long. Virtually any physiological activity observable at the organismic level can be measured in terms of energy cost. For example, metabolic rate varies with environmental temperature, muscular activity, digestion, quality of diet, lactation, pregnancy, estrus, time of day, time of year, sex, age, emotional state, posture, and so on ad infinitum.

To establish a point of reference it is necessary to have a convention accepted as the standard metabolic rate. Ideally the standard metabolic rate should be an animal's metabolism under the simplest and least physiologically demanding conditions. To be biologically realistic, the conditions for the determination of such a standard rate will vary with the kind of animal being measured. The environmental circumstances of minimum stress differ almost from species to species and certainly from higher taxon to higher taxon. Consequently, the precise circumstances appropriate for measuring the minimum resting metabolism of a given species can be determined only by the investigator. The biological appropriateness of his determination will depend on how well he knows the species with which he is working. In general, one attempts to set up conditions in which the animal is at ease psychologically, at ease physiologically and posturally, and at the time of day when it is normally at rest so that it will remain quiet and allow all its physiological activities to approach a minimal rate. Having set up the appropriate conditions, one then measures the animal's energy metabolism, either continuously or periodically, depending on the method being used, until it reaches a stable minimal level.

In the case of human beings and other mammals (and also birds), this stable minimal rate of energy metabolism, which represents an approximation of the rate of metabolism of a fasting adult animal at rest in its *thermal neutral zone* (that is, the range of temperatures in which energy metabolism is unaffected by changes in temperature, see Figure 8-19), is referred to as the *basal metabolic rate* (BMR). Basal metabolism is an unfortunate term because of the ambiguity of the word basal, but it is too firmly embedded in the physiological literature to change.

Because poikilotherms have no thermal neutral zone (see Chapter 8), they have no metabolic state to which the term basal can be applied. For animals other than mammals and birds, the minimum metabolism of fasting individuals at a given environmental temperature is referred to as the standard metabolic rate (SMR). Comparisons between taxa of poikilotherms are usually made in terms of the standard metabolism at some specified and biologically meaningful temperature. Although 95% or more of the research effort expended on energy metabolism has been devoted to the measurement of basal and standard rates, such rates are only a small segment of an animal's performance and should be considered only in the perspective of the total picture of the physiology of energy metabolism.

3.7 METABOLIC RATE AND BODY SIZE: THE SCALING OF A PHYSIOLOGICAL VARIABLE

There are at least 10 million kinds of organisms on earth, and most of these have not yet been named let alone studied physiologically. Even among large conspicuous animals such as the vertebrates, the number of kinds is impressive. There are over 8000 named species of living birds and almost as many species of mammals. There are over 20,000 species of fish and several thousand species of reptiles and amphibians. Not only are there many kinds of animals, they cover an enormous range of sizes. The mass of adult multicellular animals extend over many orders of magnitude, from rotifers and nematodes weighing fractions of a nanogram to whales, weighing more than 10^5 kg. Even within a single group of vertebrates, the range of body mass may span several log cycles. Mammals range in mass from shrews (Order Insectivora) weighing about 2.5 g to the blue whale (*Balaenoptera musculus*), which weighs as much as 120 metric tons (1.2×10^8 g). Birds range in mass from hummingbirds of the genus *Calypte*, weighing less than 2 g, to ostriches weighing 100 kg.

It is obviously impractical to study the physiology of all these kinds and sizes of animals; a selective sample must suffice. Fortunately within any given major taxon such as an order or a class, all species, despite differences in size and external appearance, share a fundamental similarity in morphology and a similarity in physiological functions. The lungs, kidneys, hearts, and other major organs are strikingly similar in hummingbirds and ostriches, or bats and elephants. The 1-g heart of a rodent looks like, and performs like, the 100-kg heart of a baleen whale, even though one of these organs is 100,000 times the weight of the other. Time and again it has been found that if physiological measurements from animals within a major taxon are ranked on the basis of the size of the individual from which they were obtained, an orderly and regular relation

exists that is some function of body size. Metabolic rate is the classic example.

3.7.1 Scaling, Power Functions, Exponents

A priori it would seem reasonable to assume that oxygen consumption should increase directly with increasing body size; each additional unit of mass adding an equal increment to the metabolic requirement of the whole organism. However, the situation is not that simple. Amazingly, the energy demand of a unit mass of tissue depends on the size of the animal of which it is a part. The energy metabolism of 1 g of tissue in a mouse is 20 times that of 1 g of the same kind of tissue in a bison. Thus, despite the fundamental uniformity of all animals at the cellular level and the close similarity at the organ level in structure and function of all the species in any given group, standard metabolism per unit mass decreases markedly as body mass increases.

Before examining this phenomenon further we must familiarize ourselves with some forms of notation and with some ways of interpreting relationships using body mass as a base of reference.

If the size of an object (including organisms) is changed without altering its shape, certain of its properties change in a predictable manner. For example, in any plain figure area (A) increases as the square of the length (l), $A \propto l^2$; in a three-dimensional object, volume (V) increases as the cube of the length, $V \propto l^3$. In any given group of animals mass (M) will be directly proportional to volume, $M \propto V$. Consequently, length and surface area can be expressed as functions of mass.

$$l \propto M^{1/3} \quad \text{or} \quad l \propto M^{0.33} \quad \text{or} \quad l \propto \sqrt[3]{M}$$

$$A \propto M^{2/3} \quad \text{or} \quad A \propto M^{0.67}$$

Thus, if the mass of an animal is known, its length and surface area can be predicted. Procedures of this type are called *scaling* and are applicable to both physiological and morphological variables as long as comparisons are based on geometrically similar organisms that differ only in size (scale), not in proportions. The consequences of size are expressed most accurately as power functions (exponents), but are most readily visualized graphically. For this reason the relations of physiological variables to body mass are often expressed in the scientific literature both as equations and as graphs. The relationships in scaling are usually expressed in terms of the allometric equation

$$Y = aX^b \qquad (3\text{-}1)$$

where Y is any physiological variable, a is a proportionality constant that characterizes that variable in a group of animals, X is body mass, and b describes the effect of size on the variable. The principal features of the exponent b in the present context are quite straightforward and are shown graphically in Figures 3-1 and 3-2.

1. Positive exponents indicate an increase with body size.
2. Negative exponents indicate a decrease with size.
3. An exponent of 0 indicates that the variable is unaffected by size.
4. Exponents of 1.0 and 0 are straight lines when plotted on an arithmetic grid.

5. Fractional positive exponents and all negative exponents are curves when plotted on an arithmetic grid.
6. All exponents yield straight lines when plotted on logarithmic coordinates.
7. On logarithmic coordinates the exponent b is equal to the slope of the linear regression of Y on X (Figure 3-2).

Since linear relations are much easier to manipulate than curvilinear ones, the allometric equation (3–1) is frequently converted to its logarithmic form

$$\log Y = \log a + b \log X \qquad (3\text{-}2)$$

It is in this form that the equation most often appears in the physiological literature and graphical presentations usually employ logarithmic coordinates. Because the functions described by allometric equations are linked, the equations can be combined and manipulated algebraically to predict physiological relationships.

Figure 3-1
Arithmetic plot of the allometric equation Y = aX^b with values from −1.0 to 1.5 assigned to b; Y is any physiological variable and X is body mass. Note that only the exponents 0 and 1.0 yield straight lines and that only a narrow range of values (usually only 1 or 2 orders of magnitude) can be clearly displayed. For convenience of plotting, a value of 1.0 has been assigned to a for the positive exponents and a value of 100 has been assigned to a for negative exponents.

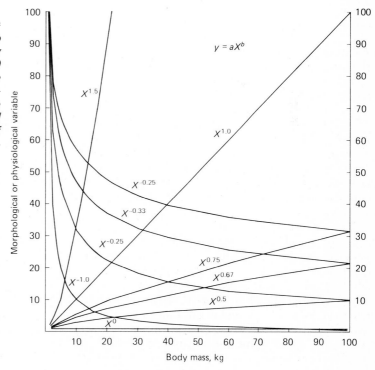

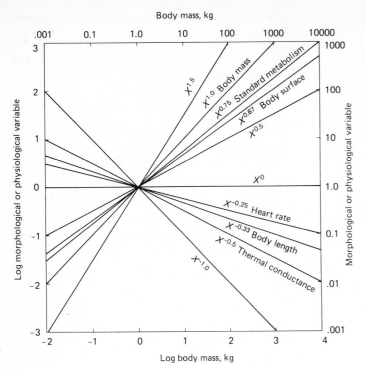

Body mass, kg

Figure 3-2
The allometric equation Y = aX^b plotted on logarithmic coordinates, with values for b as in Figure
3-1. A value of 1 has been assigned to a for both
positive and negative exponents. Note that all exponents now yield straight lines and each is equal
to the slope of its line. This graph encompasses
six orders of magnitude and can be expanded indefinitely on both axes. The scaling of a few familiar biological variables is indicated.

3.7.2 Metabolism and Body Mass

As pointed out in Section 3.5 it is important to
distinguish between an animal's mass and its
weight. This is particularly apparent when dealing
with energy metabolism because weight is a force,
the magnitude of which varies with the strength
of gravity. An astronaut traveling in space or orbit-
ing the earth is weightless. His metabolism calculated on the basis of weight would be infinite—
obviously an absurdity. The mass of a body is independent of gravity and mass is the appropriate
physical quantity against which to compare energy
metabolism—except in relation to the performance
of physical work (see Section 3.13).

Even though the relations being dealt with are
relatively simple, the variety of notations and types
of measurement used in describing metabolic rates
often make the situation confusing, particularly
when relating metabolism to body size. As indicated in Section 3.6, metabolic rate may be measured in a variety of units. The units in most common use are volume of oxygen converted to
standard pressure and temperature consumed by

an animal per unit time (cm^3 O_2/hr; liters O_2/day;
mol O_2/unit time), and kilocalories lost by an animal per unit time (kcal/hr; kcal/day).

The word "specific" before the name of an extensive physical quantity is restricted to the meaning
"divided by mass." *Specific energy metabolism* is usually expressed either as heat lost or as volume or
moles of oxygen consumed per unit time per unit
mass. In the S.I. system (Tables 3-1 and 3-2) the
units are watt per kilogram (W/kg). A variety of
equivalent notations for specific metabolism are
employed in the physiological literature. For example, the expression "kilocalories per kilogram per
hour" may appear in any of the following forms,
all except the last of which are acceptable (see Table 3-3):

kcal/kg · hr

kcal/kg hr

kcal kg^{-1} hr^{-1}

kcal (kg hr)$^{-1}$

kcal/kg/hr

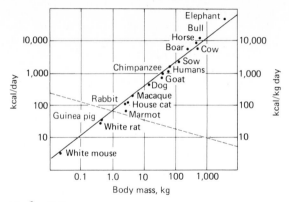

Figure 3-3
Linear relation of the logarithm of body size in mammals to the logarithms of energy metabolism (solid line) and mass specific metabolism (broken line). [Modified from M. Kleiber. Hilgardia, **6**, 315–353 (1932).]

In any one group of organisms BMR and SMR have a remarkably constant relation to body size. This is shown most familiarly in the "mouse to elephant curve" for mammals, but it is equally well demonstrated by data on fish, amphibians, reptiles, birds, and a variety of groups of invertebrates (Figures 3-3 to 3-7, and Table 3-5). From the graphs it is clear that the logarithms of SMR and BMR are linearly related to the logarithms of body mass; which is the same as saying that they are propor-

tional to some exponential function of body mass. This relationship is described by the allometric equation discussed in Section 3.7.1.

$$\dot{E}_m = aM^b \tag{3-3}$$

where $\dot{E}_m$ is the rate of energy metabolism, a (the proportionality constant) is the metabolic rate of an animal of unit mass, M is mass, and b is the exponent. Equation (3–3) can be converted to indicate specific metabolism by dividing by M.

$$\dot{E}_m M^{-1} = aM^{(b-1)} \tag{3-4}$$

As pointed out in Section 3.7.1 equation (3–3) may be transformed to logarithmic form and written

$$\log \dot{E}_m = \log a + b \log M \tag{3-5}$$

It is this logarithmic form that is usually used for computation. When an allometric equation is plotted on double logarithmic paper, it yields a straight line with a slope of b (Figures 3-3 to 3-6). This is convenient not only for curve fitting and statistical treatment but has the advantage that differences from the fitted line are in terms of percentage deviation rather than absolute deviation. As Brody pointed out many years ago, it is biologically reasonable to assume that a given percentage deviation has the same significance for animals of very

Figure 3-4
The relation between BMR and size in birds, excluding members of the order Passeriformes. [Modified from R. C. Lasiewski and W. R. Dawson. Condor, 69, 13–23 (1967).]

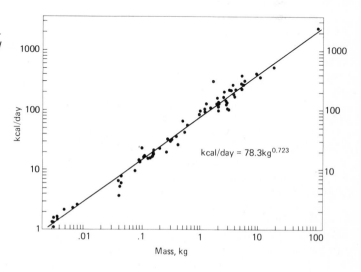

kcal/day = 78.3kg$^{0.723}$

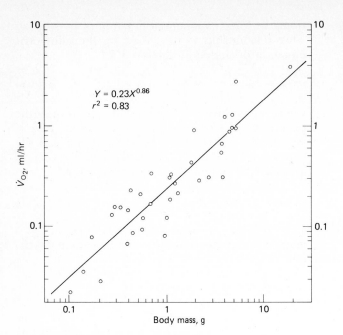

Figure 3-5
The relation of oxygen consumption of beetles at rest to body mass. [Modified from G. A. Bartholomew and T. M. Casey. J. Thermal Biol., **2,** 173–176 (1977).]

different sizes, but the same assumption cannot be made for a given absolute deviation—consider the shrew and the whale.

Historically there has been little controversy over $\dot{E}_m$, M, and a. These are matters either of experimental measurement or the units employed. However, b is something else again. Not only is there an element of personal judgement in selecting the data to be used in its calculation but it is the key element in determining the success of predicting metabolism from mass. At times there has even been controversy over the third decimal place in this fractional power!

Representative examples of the equations used to predict metabolic rate on the basis of body mass are given in Table 3-5. The average value for b in the equations for total metabolism lies between 0.7 and 0.8. The data under consideration here are based on only a few groups. However, after reviewing the relation of energy metabolism to body size in all organisms for which data were available, Hemmingsen concluded that a b value of 0.75 best fits the data for unicellular organisms and multicellular plants as well as metazoans. Thus, for organisms in general, for each doubling in mass, standard metabolism increases on the average by about three quarters. This relationship is empiri-

cally satisfying in that the exponent has predictive value but, despite its convenience as a rule of thumb, there is no present consensus concerning the precise theoretical significance of this value although an interesting series of ideas relating the three-quarter value to elasticity and buckling stresses has attracted attention [McMahon, 1973]. This lack of consensus may be related to the fact that the values of b are almost complete abstractions that obscure virtually all the metabolic variations existing within the species, genera, or families being considered. For example, species differ in fat content, development of protective or insulative structures, heaviness of skeleton, and other physical characteristics, as well as in temperament and behavior. Usually few, often none, and never all, of these parameters are taken into account in compiling the data on which the relationships are calculated.

Moreover, the data are highly variable and until recently confidence limits for the estimates of b have usually not been presented. The conditions of measurement differ from one instance to the next. The size of the samples represented by the means that are used vary from one or two for the largest species to a score or more for the smaller ones. The slopes of the various curves plotting me-

Figure 3-6
The relation of standard metabolism to body mass in five groups of vertebrates. The regressions are calculated from lines (2), (4), (6), (7), and (12) in Table 3-5.

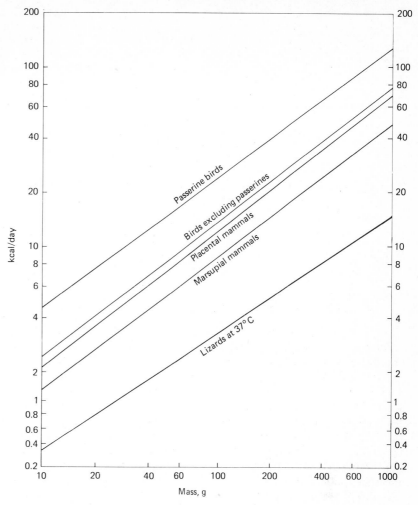

tabolism against mass for major taxa are virtually all based on pooled data. A given species, genus, or family is apt to be represented by only one, or at most a few, points. The various values of b are composites of many different averaging procedures, and the different data are not of uniform reliability. The value of b for a major taxon such as a class may, in fact, be the mean of many different values. If one had sufficient data to determine the b for each subgroup—say, a species, a genus, or a family within the total—different slopes might result. This is illustrated (1) by the standard metabolism of fishes, where, depending on the group considered, values varying from the controversial 0.67 of the "surface law" (see Section 3.7.3) to more than 0.9, which is close to direct proportionality, have been obtained; and (2) by the fact that for reptiles as a group the best estimate for b is 0.77, but for lizards it is 0.83 and for turtles, it is 0.86.

3.7.3 Basal Metabolism and the "Surface Law"

During the early years of the study of animal energetics mammals received more attention than other groups, and the resultant dearth of broadly comparative data led to misconceptions that are still

3.7 METABOLIC RATE AND BODY SIZE: THE SCALING OF A PHYSIOLOGICAL VARIABLE   **61**

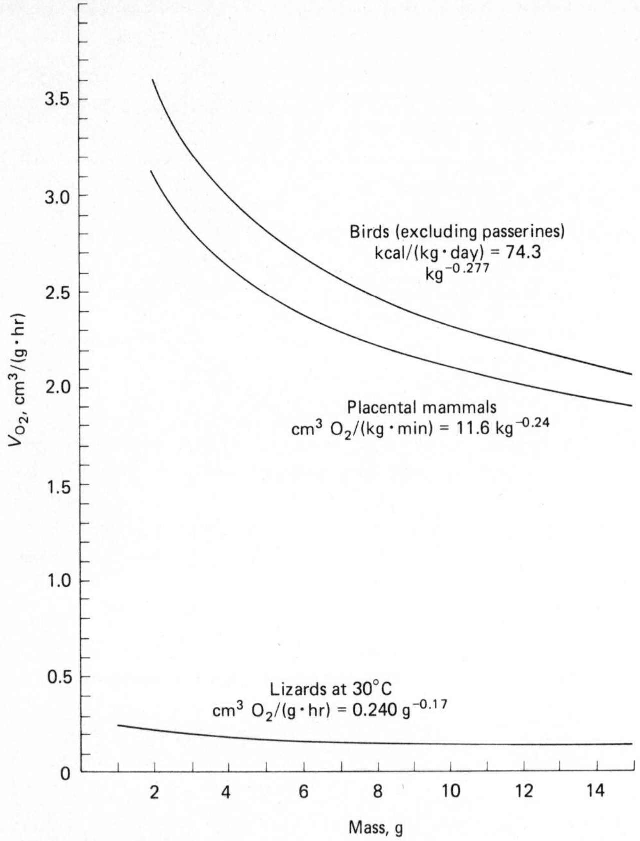

Figure 3-7
Mass-specific standard and basal metabolism in relation to body size in small members of three classes of vertebrates. The curves were calculated from the equations shown which are derived from lines (3), (6), and (11) in Table 3-5, p. 63.

The plot shows V_{O_2}, $cm^3/(g \cdot hr)$ on the vertical axis versus Mass, g on the horizontal axis, with three labelled curves:

Birds (excluding passerines)
$kcal/(kg \cdot day) = 74.3 \; kg^{-0.277}$

Placental mammals
$cm^3 \; O_2/(kg \cdot min) = 11.6 \; kg^{-0.24}$

Lizards at 30°C
$cm^3 \; O_2/(g \cdot hr) = 0.240 \; g^{-0.17}$

firmly entrenched in the secondary physiological literature, particularly in the clinical literature.

As discussed in the preceding section, the most striking generalization to appear in the early history of the study of mammalian energetics was that simple body mass was not an adequate base reference for metabolism. Other things being equal, the rate of cooling of a body is proportional to its surface area. A mammal in its "basal condition" has a constant body temperature. Therefore, it is producing heat at a rate equal to its heat loss (see Chapter 8 for a detailed discussion). If heat production equals heat loss and heat loss is proportional to body surface, then heat production must also be proportional to surface area. This series of relationships was first formulated in 1838 by Sarrus and Rameaux and independently proposed by Bergmann (see Section 8.9.1) who pointed out that in dogs of widely different sizes the rate of metabolic heat production divided by body surface area was essentially constant. In the years since then, the nature of the correlation between basal metabolism and body surface in mammals has generated an enormous outpouring of words, as well as much wheel-spinning and controversy, perhaps because the relationships can be arranged in an attractively (and probably speciously) symmetrical pattern.

1. The rate of heat loss is proportional to surface area.
2. Physical objects of similar geometry have surface areas that are proportional to the two-thirds power of their volumes (or their weights if specific gravity is constant).
3. Surface area can be calculated by an equation of the same form ($Y = aX^b$) that is used for predicting BMR; surface area $= aM^{0.67}$ (see Section 3.7.1).

TABLE 3-5. Formulas for Calculating Standard and Basal Metabolic Rates of Vertebrates from Body Mass

For convenience of comparison all the equations have been tabled in the form $\dot{E}_m = aM^b$. To calculate specific metabolism, b should be converted to the form $b - 1$. Thus, if $b = 0.72$, $b - 1 = -0.28$ etc. To convert to joules, watts, calories, or $\dot{V}_{O_2}$ use the appropriate conversion factors from Table 3-2.

		$\dot{E}_m$	a	M	b	Source
(1)	Placental	kcal/day	70.5	kg	0.734	Brody, 1945
(2)	mammals	kcal/day	70.0	kg	0.75	Kleiber, 1961
(3)		cm³ O₂/min	11.6	kg	0.76	Stahl, 1967
(4)	Marsupial mammals	kcal/day	48.6	kg	0.737	Dawson & Hulbert, 1970
(5)	Dasyurid marsupials	cm³ O₂/hr	2.45	g	0.735	MacMillan & Nelson, 1969
(6)	Birds, excluding Order Passeriformes	kcal/day	78.3	kg	0.723	Lasiewski & Dawson, 1967
(7)	Birds of Order Passeriformes	kcal/day	129.0	kg	0.724	Lasiewski & Dawson, 1967
(8)	Activity phase of daily cycle Passeriformes	kcal/day	140.9	kg	0.704	Aschoff & Pohl, 1970
(9)	Resting phase of daily cycle Passeriformes	kcal/day	114.8	kg	0.726	Aschoff & Pohl, 1970
(10)	Reptiles at 30°C	cm³ O₂/hr	0.278	g	0.77	Bennett & Dawson, 1976
(11)	Lizards at 30°C	cm³ O₂/hr	0.24	g	0.83	Bennett & Dawson, 1976
(12)	Lizards at 37°C	cm³ O₂/hr	0.424	g	0.82	Bennett & Dawson, 1976
(13)	Snakes at 30°C	cm³ O₂/hr	0.28	g	0.76	Bennett & Dawson, 1976
(14)	Turtles at 30°C	cm³ O₂/hr	0.66	g	0.86	Bennett & Dawson, 1976
(15)	Lunged salamanders at 25°C	ml O₂/hr	116.4	g	0.740	Feder, 1976
(16)	Plethodontid (lungless) salamanders at 25°C	ml O₂/hr	95.0	g	0.802	Feder, 1976
(17)	Frogs of family Ranidae at 25°C	cm³ O₂/hr	0.286	g	0.75	Modified from Hutchison, Whitford and Kohl, 1968
(18)	Salmon at 15°C	mg O₂/hr	0.156	g	0.846	Brett & Glass, 1973
(19)	Four species of freshwater fish at 25°C	cm³ O₂/hr	0.199	g	0.70	Adapted from Kayser & Heusner, 1964

4. In resting mammals heat production is equal to heat loss.
5. In mammals heat production is proportional to approximately the three-quarter power of body weight.
6. The exponents for calculating surface area ($\frac{2}{3}$) and heat production ($\frac{3}{4}$) are very similar when graphed and the confidence limits for estimating the latter often include the former.
7. Therefore, metabolic heat production is determined by surface area.

There can be no doubt that the BMR of mammals and birds varies in a way that roughly parallels surface area, but this does not mean that body surface area is an appropriate base reference for metabolic rate for the following reasons.

1. The effective surface area of a live animal changes continuously and is difficult and often impossible to measure accurately. (In practice, surface areas are calculated from empirical formulas which do not take into account postural adjustments).
2. Animals of different taxa are *not* geometrically similar, nor are their specific gravities the same.
3. Rates of heat loss through the body surface are neither passive nor constant and are under physiological control.
4. Heat loss per unit area differs in the various parts of an animal's body.
5. It is not physiologically possible for surface area to be the control mechanism for metabolic rate, since metabolic rate is under the control of a complex array of subcellular, cellular, endocrine, and neural factors.

From the preceding discussion, there appears to be little justification for using surface area as the base reference for metabolic rate, and it is clear that surface area *as such* does not stand in immediate causal relation to either standard or basal metabolism. Nevertheless, surface area cannot be dismissed as a factor in the energetics of homeotherms. This is true because, in dealing with functions of intact organisms, we must think not only in terms of the details of physiological mechanisms but also in terms of evolution and natural selection.

Thus, although surface area has no direct causal control over metabolic rate in birds and mammals, the strong positive correlation between surface area and heat loss should favor the selection of controls that would result in a rather close match between surface area and basal metabolism (see Chapter 8 for a more detailed discussion). Consequently, surface area could have been at least one of the important ultimate determinants of BMR in homeotherms because it would favor the evolution of control mechanisms for heat production that function in rough proportion to body surface.

However, other factors have also been involved in the evolution of standard and basal metabolism. The "surface law" was first proposed for mammals at a time when there were few data on other groups. From the perspective of present knowledge of comparative physiology it is clear that body surface cannot have been the primary causal factor that fixed the rates of standard metabolism for a simple but rather overwhelming reason. The slope of the regression of metabolic rate on mass is similar for mammals, birds, reptiles, fish, and most other groups of animals (Table 3-5 and Figure 3-6) whether endothermic, ectothermic, poikilothermic, or homeothermic (see Section 8.3 for definitions). Poikilotherms and homeotherms show essentially the same proportionality between metabolic rate and size, so either surface area has no causal relation to metabolic rate or the factors involved in this causal relationship are different in each group. Certainly it is safe to say that the surface law, which proposes that metabolic rate per unit area is the same for large and small animals, has little utility for animals in general and is only an empirical approximation, even for homeotherms (see Section 8.9.1).

3.8 BASAL METABOLISM AND CLIMATIC ADAPTATION

Birds and mammals under natural conditions are, of course, rarely, perhaps never, in a basal metabolic state, so *a priori* one might predict that the BMR should have no specifically adaptive relation to climatic conditions. Furthermore, from the fact that it is possible to construct the mouse-to-elephant and hummingbird-to-ostrich curves, it is apparent that the BMR is not fundamentally adaptive to climate. If it were, one would expect that tropical endotherms of a given size would have lower BMRs than arctic ones of the same size, which is not the case. Birds and mammals that live in the extremely low temperatures of the arctic region are sometimes exposed to air temperatures as much as 100°C below their core body temperatures. They live for months at a time in temperatures more than 50°C below their core temperatures, yet in most cases their mass specific BMRs do not differ significantly from those of most birds and mammals of similar size living under tropical conditions.

Although BMR does not participate importantly in the physiological adaptations of most mammals to cold terrestrial climates, the BMR of marine mammals is often higher than predicted on the basis of size. The northern sea otter, *Enhydra lutris,* which lives in the continuously cold waters of the Aleutian Islands and the coastal waters of mainland Alaska, is a dramatic example. Its BMR is 0.75 cm^3O_2 $g^{-1}hr^{-1}$ which is more than twice the expected value for a mammal of its size (18 kg).

A more complex situation exists with respect to the BMRs of birds and mammals living under conditions of sustained high temperatures and aridity (such as are found in the horse-latitude deserts), and with respect to birds in the less extreme but still continuously warm environment of the humid tropics. *A priori,* a lower-than-normal basal metabolism should be advantageous to an animal living under conditions of high environmental temperatures, because such a reduced basal rate would minimize the production of endogenous heat and so reduce the burden of losing heat to a hot environment (see Section 8.10 for a more extended discussion).

The data available, however, reveal several adaptive patterns with regard to this problem among birds and mammals living in the low-latitude deserts. Most desert passerine birds appear to have a metabolism appropriate to their size. The antelope ground squirrel (*Ammospermophilus leucurus*), a common rodent of the American Southwest, has a basal metabolism slightly higher than expected. Mice of the family Heteromyidae (kangaroo rats and pocket mice), which live in the same arid region, generally have metabolic rates lower than that predicted on the basis of size. A similar situation exists in the desert species of the mouse genus

Peromyscus. Because these mice are nocturnal and so avoid severe heat stress, there are difficulties in interpreting their low basal metabolism as an adaptation to high environmental temperature.

There is, however, at least one instance among birds in which a conspicuously reduced metabolic rate appears to be clearly adaptive. The poorwill (*Phalaenoptilus nuttallii*), an aerial-feeding, insectivorous bird of Western North America, has a basal metabolic rate less than one half that of other birds of its size (40 g). The poorwill is one of the very few birds that can undergo long periods of dormancy (see Section 8.12). Over much of its summer range it is exposed to severe conditions of heat. It is a crepuscular feeder but has the habit of spending the daylight hours, even in the desert, sitting quietly in the open, completely exposed to the sun. This behavior pattern imposes on it a much greater heat load than is faced by most other desert birds. The unusually low level of the BMR of this species minimizes the metabolic contribution to the total heat load and appears to be of critical importance in allowing the poorwill to exploit the desert environment. The poorwill is a member of the order Caprimulgiformes, all members of which appear to have much lower than predicted metabolic rates. Consequently this species is an unusually good example of physiological preadaptation to the difficult desert environment.

It is well established that BMR is strongly affected by the thyroid gland, which is regulated by the anterior pituitary which, in turn, has complex feedback relations with the hypothalamus. Consequently BMR can be modified by many different factors via the endocrine system. It is not surprising to find, therefore, that despite the regularity of its relationship to size within major taxa, BMR is not a fixed and immutable quantity. Not only does it vary between individuals but, in species that have been examined closely, it varies with season, local environmental conditions, and time of day.

For example, house sparrows, *Passer domesticus,* conspicuously reduce BMR when acclimated to high temperatures. The BMR of house sparrows in Houston, Texas, averages 20% lower than that of the same species in Boulder, Colorado, which lies 1100 km to the north, whereas in Iowa (40° N lat.) it is about 40% lower in summer than in winter.

Similar responses have been found in mammals;

in the prairie vole *(Microtus ochrogaster)* mass specific BMR is 24% higher in winter than in summer.

Conditions of temperature and humidity underneath the ground are much more constant and less extreme than those on the surface—a circumstance exploited at least some of the time by most small mammals. Those burrowing rodents which remain underground all of the time tend to have basal metabolic rates that are below the levels predicted on the basis of size.

A more detailed discussion of metabolic adaptations to temperature is included in Chapter 8.

3.9 CIRCADIAN RHYTHMS AND METABOLIC RATE

One of the most conspicuous attributes of organisms is the cyclic or rhythmic nature of most of their functions and activities. Many of these rhythms have a periodicity of approximately 24 hr. In an evolutionary sense this 24-hr periodicity is undoubtedly related to the period of the earth's rotation and hence the length of one light-dark cycle. In animals under natural conditions such a 24-hr periodicity is daily entrained or reinforced by the normal day-night cycle. However, the absence of daily reinforcement does not lead to extinction of biological rhythmicity. Nearly all organisms, and by far the majority of animals that have so far been tested, maintain a 24-hr periodicity even when kept in an unchanging environment in which they receive no clues about naturally occurring day-night cycles. One of the most striking features of these rhythms is their insensitivity to environmental temperature. Even in poikilotherms, wide variations in environmental temperature have only slight, or often no discernible, effects on the daily periodicity of behavior or function.

As might be expected from the all-pervading variability that characterizes biological systems, in the absence of an external timegiver *(Zeitgeber)*, these endogenous rhythms rarely have a period that is exactly 24 hr; it varies between 22 and 26 hr. Any given individual animal, however, keeps a constant periodicity to its daily rhythm that may be slightly longer or slightly shorter than the environmental cycles. These daily endogenous cycles are called **circadian rhythms** (L., *circum*, about; *dies*, day) be-

cause they are approximately, but not precisely, 1 day in length.

The importance of daily rhythms to physiological measurement can hardly be overemphasized, because nearly any measurable physiological quantity shows a striking daily cycle: blood sugar, testicular mitoses, sensitivity to radiation damage, sensitivity to photoperiodic stimulation, locomotor activity, bioluminescence, body temperature, and of course, oxygen consumption, to name but a few. Because all these and many other factors vary in a cyclic manner, time of day should be taken into account in virtually all physiological measurements.

Oxygen consumption, activity, and body temperature can be monitored over long periods of time without disturbance, so the effects of circadian rhythmicity are well documented with regard to these functions. Typically the body temperature of diurnal homeotherms is 1–3° higher during the daytime than at night. For nocturnal forms the same situation exists, but the cycle is reversed. The daily rhythms of the metabolic responses of vertebrates are more complex than those of body temperature, but they generally show the same fundamental pattern with rates being high during the hours of activity and low during the hours of rest and inactivity. This daily metabolic cycle is most clearly demonstrated for daytime active birds in which the BMR is about 20% higher during the day than at night—compare the values for *a* in lines (8) and (9) of Table 3-5.

One may reasonably ask why such circadian rhythms should have evolved, since as long as there has been life on earth there has been the normal day-night cycle to entrain the biological rhythm daily. The most probable answer is that the persistence of such endogenous rhythms allows the animal to be ready to function adequately before the environmental stimulus appears. That is, it is anticipatory and thus enhances the effectiveness of responses. Some of the endocrine mechanisms that may control these rhythms are discussed in Chapter 12.

3.10 THE COST OF LIVING

As discussed in Sections 3.7.2 and 3.7.3, most of the controversy in the physiological literature concerning the equation $\dot{E}_m = aM^b$ has centered on the exponent *b*. However, the constant *a* is obviously equally important biologically. If we know both *a* and *b* we can compare the standard metabolisms of animals of any given size in different taxa, which is not only of physiological interest but is also a matter of ecological and evolutionary interest. In Figure 3-6 the SMR and BMR for five groups of vertebrates have been plotted for body masses ranging from 10 to 1000 g. From such a plot several matters of biological interest are immediately obvious. (1) Passerine (song) birds have a much higher BMR than do other kinds of birds or either of the two major groups of mammals. (2) Birds (excluding members of the order Passeriformes) have essentially the same mass specific metabolism as placental mammals. In fact, the regression lines of metabolism for these two groups are not statistically distinguishable. (3) The BMR of marsupial mammals is approximately 30% lower than that of placentals—this despite the fact that the two groups maintain body temperatures at similar levels. (4) Even when their level of body temperature is the same as that maintained by most mammals, lizards have an SMR about one sixth to one tenth the BMR of birds and mammals.

By comparing points of equal metabolism on the different curves, one can assay the relative energy demands of animals of different sizes in different taxa. For example, the same number of calories are required to maintain a 10-g song bird and a 26-g mouse for a given time in a basal condition, or a 700-g varanid lizard with a body temperature of 37°C requires no more energy input per unit time than a 100-g carnivorous marsupial.

For computational and graphical convenience it is customary to display the relation between metabolism and size on a double log plot, but other sorts of presentations are helpful to emphasize particular relationships. For example, since most species of vertebrates weigh less than 100 g, it is a matter of particular interest to examine the mass specific metabolism of these small creatures so that we can have some appreciation of the energetic price that they must pay merely to stay alive in a resting condition. Using values from Table 3-5, we can compare the minimum energetic expenses that the smaller terrestrial vertebrates must meet in the absence of temperature stress and activity (Figure 3-7).

An arithmetic plot of mass specific oxygen consumption as a function of body size emphasizes the enormous energy expenditure per gram of very small birds and mammals, even when they are at rest and under no temperature stress. Despite this great energy expenditure the adults of many species of birds and mammals weigh less than 5 g—shrews, bats, mice, and hummingbirds are familiar examples. The smaller shrews, such as *Sorex cinereus*, which occurs in central and eastern United States, and the smaller hummingbirds, such as the bee hummingbirds of the genus *Calypte*, which occur in the West Indies, weigh only about 2 g. It takes large quantities of fuel to sustain such high levels of metabolism. Some shrews eat more than their body weight every day and will starve to death in a few hours if food is withheld. However, despite their prodigal rates of energy expenditure at rest, small homeotherms can increase their oxygen consumption greatly during activity (see Section 3.13).

Mass-specific oxygen consumption, of course, also shows an inverse relationship to body size in ectothermic vertebrates, but their levels of oxygen consumption are much lower. It is of interest that many species of reptiles, amphibians, and particularly fish are much smaller than the smallest birds and mammals. This at once leads one to wonder about the metabolism of infant birds and mammals. This matter is closely associated with temperature regulation and is examined in Chapter 8.

Both the BMR and the SMR are based on physiological states that are rarely, certainly reluctantly, and perhaps never, attained by animals living under natural conditions. To be in a basal, or standard, condition an animal must be nonfeeding, nondigesting, nongrowing, nonmoving, and nonbreeding. To put an animal in this conventionalized minimal state places it under so many constraints that it becomes difficult for the investigator to extrapolate from the BMR or SMR of an individual animal to its performance in nature. The most important things about metabolic rate relate to variability associated with performance and activity, not to predictability under conditions of minimal function. However, measurements of BMR and SMR offer essential baselines from which to evaluate the rates of energy metabolism that actually exist under conditions of stress or activity that occur under natural conditions.

Total Energy Metabolism. The total energy metabolism of an animal carrying out a normal pattern of activities in its natural habitat can be measured directly by the use of doubly labeled water (see Section 3.5.2), or it can be calculated by adding the energy requirements of all of its various function to its BMR or SMR. The tenuous nature of the relation of BMR and SMR to energy expenditure under natural conditions is illustrated by measurements of the daily energy metabolism of free-living western fence lizards (*Sceloporus occidentalis*) using doubly labeled water. As pointed out previously, a resting reptile at 37°C, a typical mammalian body temperature (see Chapter 8), has a metabolic rate that is 10–17% that of a mammal of the same size. Nevertheless, the daily energy expenditure of free-ranging *Sceloporus* (mass, 12 g) is only 1.67 kJ, which is 4% that of a mammal of the same size. This discrepancy is explained by the difference in daily temperature regimes between the reptile and the mammal. Although *Sceloporus* maintains its body temperature close to the mammalian level while it is active, this condition exists only during the middle hours of the day. At night its body temperature falls to ambient, and its metabolic rate is correspondingly depressed. In contrast the mammal's body temperature remains near 37°C night and day, and its metabolic rate never falls below basal.

Although in specific cases highly restricted categories of energy costs, such as the cost of hovering in hummingbirds, can be usefully employed, in general practice the sources of energy expenditure can be grouped into a few major components, each of which is additive with regard to BMR and SMR. These include (1) the energy cost of digestion and assimilation usually referred to as *specific dynamic action* or SDA, (2) the energy cost of activity, (3) the energy cost (for endotherms) of regulating body temperature, (4) the energy cost of production which in a broad sense includes the growth of cells and tissues, the storage of reserves of chemical energy as lipids, and the creation of new individuals during reproduction. These major components can be combined to give a variety of standardized measures of energy metabolism, such as the rates of (1) resting metabolism (RMR); (2) maintenance metabolism (MMR); (3) existence metabolism (EMR); (4) average daily metabolic rate (ADMR), each of which is definable in terms of BMR, or SMR, plus additives. Although these terms have been variously defined by different workers,

the following examples are illustrative of the way they have been used with regard to birds and mammals:

$$RMR = BMR + cost \ of \ thermoregulation$$

$$MMR = BMR + SDA + cost \ of \ thermoregulation + cost \ of \ activity$$

$$ADMR = BMR + SDA + thermoregulation + activity + production$$

The magnitude of these energy components will obviously vary with taxon, size, mode of locomotion, environmental temperature, and food habits. Some aspects of the roles of size and taxon have already been considered, temperature is dealt with in Chapter 8, and locomotion is discussed in Section 3.13.

Specific Dynamic Action. The processes of food assimilation are accompanied by substantial increases in oxygen consumption that are independent of other activities that an organism may be carrying out. The calorigenic effect, often called for historical reasons specific dynamic action (SDA), has been found in all animals so far studied. It is a function partly of digestion *per se* and partly of the subsequent molecular transformations of the digested food in the liver and elsewhere. SDA is of sufficient magnitude that it must be taken into account in all analyses of energy metabolism. When expressed in terms of the energy content of the food ingested, it is about 6% for sucrose, 13% for fats, and 30% for proteins. Clearly the percent by which SDA elevates the standard metabolism will depend on both the kind and amount of food eaten and the size and kind of animal being studied.

3.11 AEROBIC VERSUS ANAEROBIC METABOLISM

Biological oxidation proceeds in many small steps, and the actual release of energy for doing work does not directly involve the participation of oxygen (Section 3.2). However, oxygen is ultimately required for the complete oxidation of the fats, carbohydrates, or proteins that serve as fuel. In nearly all animals two main chemical pathways exist for the release of metabolic energy. Both pathways generate ATP as the immediate source of energy. One of these pathways is aerobic (requires oxygen), completely oxidizes its fuel, and produces water and carbon dioxide as end products. The other is anaerobic (does not require oxygen) and releases energy primarily by the fermentation of carbohydrates to form lactic acid. This anaerobic pathway is often called anaerobic glycolysis, or simply *glycolysis* because its principal immediate fuel is the starch, glycogen (see Section 4.5.4). The energy yield of anerobic glycolysis is low. The aerobic catabolism of glycogen yields more than 12 times as much ATP as is obtained anaerobically (Figure 3-8).

In complex animals aerobic catabolism can be sustained only with the support of the respiratory and cardiovascular systems, which must continuously supply oxygen and remove carbon dioxide. In most free-living organisms aerobiosis satisfies all continuous and long-term energy requirements. In addition, aerobiosis disposes of the lactic acid which is the principal end product of anaerobiosis, either by oxidizing it to CO_2 and H_2O, or by supplying the energy for its excretion or its conversion back to glycogen or glucose.

In contrast, anaerobiosis depends exclusively on material already present in the cells (in the present context, glycogen and the glycolytic enzymes in muscle fibers) and can provide extremely rapid mobilization of energy because it is free of the time lag inherent in dependence on external transport systems. In most vertebrates the primary role of glycolysis is the support of short-term bursts of muscular activity when the energy demands of such activity exceed the capacities of aerobic metabolism. Intense muscular activity routinely involves the expenditure of energy at rates which exceed the maximum that can be maintained by oxidation.

Activity involving aerobically supported muscle contractions can be sustained for long periods of time, allowing an animal in good condition to remain active almost indefinitely without muscle fatigue. In contrast, in most vertebrates activity supported by glycolysis can be sustained only briefly because of the rapid accumulation of metabolites, particularly lactic acid which has a relatively low coefficient of diffusion and accumulates not only in the blood but in the muscles where it is produced. Lactic acid is toxic in high concentration

and its build-up is one of the factors that may limit the duration of the period during which glycolysis can operate. In any event, intense anaerobically supported activity leads to rapid exhaustion and is followed by a variable period of increased oxygen consumption. Because this increase in oxygen consumption occurs after overt activity has ceased, and because it is accompanied by the gradual oxidation of the accumulated lactic acid, during a burst of anaerobic activity an animal is said to acquire an *oxygen debt*. The magnitude of this "debt" can be determined by measuring the increase of oxygen consumption above the resting level during the period of recovery following the interval of intense activity, but the precise quantitative relationship of oxygen debt repayment to the removal of lactic acid is not simple. The magnitude of the oxygen debt and the rate at which it is repaid differ widely between taxa. In both bony fish and reptiles, for example, the removal of a substantial oxygen debt requires many hours, whereas in a diving mammal it may be paid off in minutes. In vertebrates the central nervous system and the heart appear to be completely dependent on aerobic metabolism, and even under anaerobic conditions heart and brain are usually protected from anoxia by special circulatory adjustments (see Section 6.9.1).

Quantitative determinations of the relative roles of aerobic and anaerobic metabolism during muscular activity are difficult to obtain. The magnitude of the oxygen debt can serve only as a crude index to the extent of glycolysis because oxygen debt involves more than the elimination of lactic acid, and in most cases oxygen consumption returns to resting levels while lactic acid levels still show significant elevation. Lactic acid, the principal metabolite of anaerobic metabolism, is produced in the muscles, transported by the blood, and oxidized (or converted back to glycogen or glucose) in the liver. Because of the dynamic and compartmental nature of the physiology of the production and removal of lactate, no single assay of its abundance will suffice. The method that comes nearest to measuring it adequately is to grind up the entire animal and measure the lactate concentration in the resulting homogenate. This procedure, of course, gives only a single instantaneous measurement and is practical only when experimental animals are both abundant and small.

If one assumes that oxygen consumption and lac-

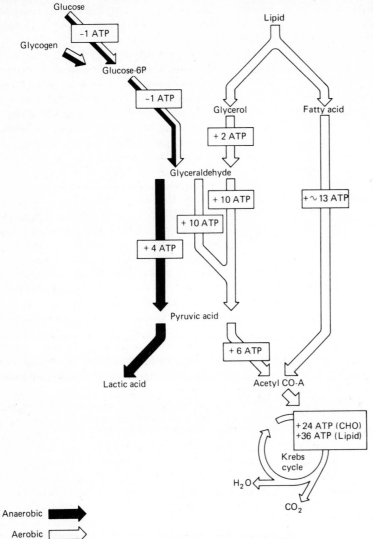

Figure 3-8
Schematic diagram of the main elements in aerobic and anaerobic energy metabolism.

Glucose

−1 ATP

Glycogen

Glucose-6P

−1 ATP

Glyceraldehyde

+ 4 ATP

Pyruvic acid

Lactic acid

Lipid

Glycerol   Fatty acid

+ 2 ATP   +∿13 ATP

+ 10 ATP

+ 10 ATP

+ 6 ATP

Acetyl CO-A

+24 ATP (CHO)
+36 ATP (Lipid)

Krebs
cycle

H_2O

CO_2

Anaerobic

Aerobic

Boxes and equations show approximate yield of ATP in moles per six carbon
segment of substrate

SUMMARY EQUATIONS

Glucose $\xrightarrow{\text{Aerobic}}$ CO_2 + H_2O + 38 ATP

Glucose $\xrightarrow{\text{Anaerobic}}$ Lactic acid + 2 ATP

Glycogen $\xrightarrow{\text{Aerobic}}$ CO_2 + H_2O + 39 ATP

Glycogen $\xrightarrow{\text{Anaerobic}}$ Lactic acid + 3 ATP

LIPIDS $\xrightarrow{\text{Aerobic}}$ CO_2 + H_2O + ∿48 ATP tic acid production account for all the functionally important energy release during activity, aerobic and anaerobic metabolism can be compared by expressing both oxygen consumption and lactate formation in terms of ATP production:

$$1.0 \text{ mg lactate formed} = 0.0167 \text{ mmol ATP}$$

$$1.0 \text{ ml } O_2(STP) \text{ consumed} = 0.290 \text{ mmol ATP}$$

From the discussion in Section 3.10 it is clear that birds and mammals can sustain much higher levels of aerobic metabolism than fish, reptiles, or amphibians. In fact, the standard metabolic rates of birds and mammals at rest are as high as the maximum rates reached during activity in other vertebrates. It is of interest that although the activities of aerobic enzymes in lizards show far lower activities than in mammals, on a protein-specific basis the glycolytic enzymes of lizards match and sometimes exceed those of the white rat. Dependence on glycolysis results in the accumulation of large amounts of lactic acid which, in high concentrations, may cause disruption of blood equilibria and is accompanied by rapid and severe exhaustion. Despite these deleterious effects and its relative low energy yield, glycolysis usually accounts for a substantial fraction of the ATP production during intense activity in fish, amphibians, and reptiles.

Data on oxygen consumption and whole-body lactate content are available for some small amphibians. In these animals the situation is obscured by the fact that estimates of the relative magnitudes of aerobic and anaerobic metabolism vary with the methods used to induce activity. However, during 3-min periods of mechanically induced sustained activity, anaerobiosis can account for 20–60% of the energy that these small amphibians utilize (Table 3-6). The anaerobic contributions during intense activity are similarly high in reptiles (Table 3-7). Field measurements of blood lactic acid are available for a few reptiles. Among these is the Galapagos marine iguana, *Amblyrhynchus cristatus*, a large lizard that feeds underwater on marine algae but spends most of its time on shore. In this species all routine activities including cruise-swimming and diving for several minutes to forage under water on algae are aerobically supported; only bursts of rapid running or swimming, and unusually prolonged dives (20–60 min) depend on anaerobiosis (Figure 3-9).

In ectotherms as in endotherms, routine activities such as foraging and walking are sustained aerobically; ordinarily only bursts of intense effort, such as running at maximum speed, depend on anaerobiosis. However, in a few cases anaerobiosis appears to be used during long intervals of moderate to low activity. The goldfish (*Carassius auratus*), a member of the carp family (Cyprinidae), can operate under sustained anaerobic environmental conditions. In cold areas many fish, frogs, and turtles overwinter under the ice in lakes in what appear to be anaerobic conditions, and in deep tropical lakes a number of species of fish appear to be able to spend long periods in the bottom water, which is anoxic. Among air breathers, anaerobic capacity

TABLE 3-6. Aerobic and Anaerobic ATP Production in Some Amphibians During 3-min Periods of Mechanically Induced Activity

(See text for method of calculation.)

Species	Mass (g)	ATP (μmol/g)		
		Aerobic	Anaerobic	Total
Frogs				
Hyla cadaverina	3.1	17.2	4.5	21.7
Hyla regilla	3.6	16.0	10.2	26.2
Lungless salamander				
Batrochoseps attenuatus	0.6	11.0	19.5	30.5

Data from S. S. Hillman, V. H. Shoemaker, R. Putnam, and P. C. Withers, *J. Comp. Physiol.*, **129**, 309–313 (1979).

TABLE 3-7. Relative Contributions of Aerobic and Anaerobic Energy Metabolism During Intense Activity in Some Reptiles

Species	T_b (°C)	Length of Activity (min)	Aerobic $\left(\dfrac{\mu mol}{ATP/g}\right)$	Anaerobic $\left(\dfrac{\mu mol}{ATP/g}\right)$	Anaerobic contribution (%)
Lizards					
Amblyrhynchus cristatus	40	2	6.3	22.1	78
Dipsosaurus dorsalis	40	2	21.9	30.3	58
Eumeces obsoletus	40	2	20.2	19.7	49
Iguana iguana	35	5	9.1	16.2	64
Sceloporus occidentalis	30	2	10.8	28.8	73
Turtles					
Pseudemys scripta	30	2	5.9	9.2	61
Terrapene ornata	40	2	6.2	7.5	55

Data from A. F. Bennett. In C. Gans and F. H. Pough (eds.), *Biology of the Reptilia,* vol. 6, Academic Press, New York, 1982.

is particularly impressive in turtles. Not only do they depend on glycolysis to support short bursts of activity, but they also use it as an energy source during prolonged periods of diving. The Green sea turtle *(Chelonia mydas)* is capable of dives lasting for hours, and the fresh water red-eared turtle, *Pseudemys scripta,* can stay under water for as long as 2 weeks at temperatures of 16 to 18°C despite the complete absence of electron-transport mediated oxygen consumption.

3.12 METABOLIC SCOPE

Metabolic scope is defined as the difference between the minimum and maximum rates of energy metabolism under a standard set of conditions. The concept was first proposed by Fry [1947] with regard to aerobic metabolism. A few years later a similar idea, the index of metabolic expansibility, was proposed and defined as the ratio of peak metabolic activity during sustained muscular work to basal metabolism. For comparative purposes metabolic scope is the more useful term because it is based on SMR rather than BMR and consequently is applicable to poikilotherms as well as to homeotherms. When comparing animals with very different rates of standard metabolism, instead of expressing metabolic scope as the difference between active and resting rates of oxygen consumption,

it is often informative to use the ratio of active to resting metabolism. This dimensionless number is called *factorial scope.*

Metabolic scope is important because it measures the amount of energy that an animal can release over and above the amount needed for the maintenance of its physiological machinery. However, as originally defined scope serves as an indicator only to the capacity for aerobically supported activity. Consequently, in the absence of determination of the magnitude of oxygen debt, the extent of the lactate build up, and knowledge of mechanical efficiency under given conditions of temperature and physiological state, direct translation to behavioral performance is not possible. Restricting the idea of scope to aerobiosis clearly reduces its general applicability and minimizes its functional significance in animals such as fish, amphibians, and reptiles, which routinely depend on glycolysis.

It is instructive, therefore, to subdivide metabolic scope into aerobic and anaerobic components, the two summing to yield the total energy that an animal can make available for activity or external work. In parallel with the original concept of aerobic scope, *anaerobic scope* may be defined as the maximum rate of lactic acid formation by glycolysis during activity. This must be distinguished from *anaerobic capacity,* which is equal to the total amount of lactic acid produced when activity is continued to exhaustion, and measures the total amount of

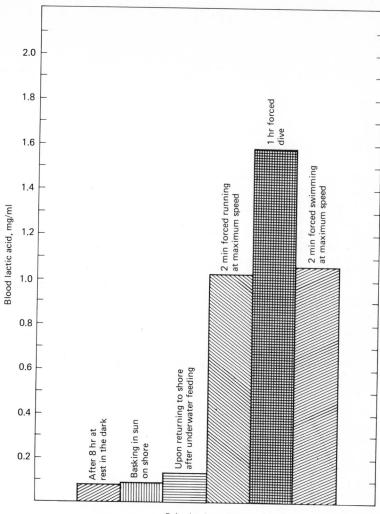

Figure 3-9
The relation of the concentration of lactic acid in the blood to activity in the marine iguana. Blood lactate is an index to the extent of anaerobic metabolism. [*Data from T. D. Gleeson.* Physiol. Zool., **53,** *157–162 (1980); and from G. A. Bartholomew, A. F. Bennett, and W. R. Dawson,* Copeia, **1976,** *709–720.*]

anaerobically derived energy that an animal can mobilize during a sustained bout of activity. It must be recognized that these definitions, although operationally useful, are not entirely satisfactory because they ignore pyruvic acid with which lactic acid exists in a complex equilibrium and which occupies a key position at the junction between the aerobic and anaerobic pathways (Figure 3-8).

The ability of mammals and birds to increase aerobic metabolism is impressive, particularly in view of their high rates of resting oxygen consumption. However, their maximal energy expenditure is even higher than indicated by the increase in oxygen consumption because of the increment of energy made available by their capacity to acquire an oxygen debt as a result of anaerobic metabolism. For mammals, the best data are available for man and domestic species. A man in good physical condition can increase oxygen consumption 15–20 times the basal level. Similar values have been obtained for draft horses and dogs. Heavy work sustained for a period of several hours by a man causes a four- to eightfold increase in oxygen consumption above the basal level. However, during a few seconds of maximal exertion, energy expenditure can be as much as 100 times the resting level; under these circumstances the immediately mobilized energy is derived from anaerobic sources and an oxygen debt is incurred. The fact that muscular work and oxygen consumption can be temporarily uncoupled is obviously of enormous adaptive significance—consider, for example, diving birds and mammals which, despite their high metabolic rates, breathe only at long intervals and engage in intense activity beneath the surface of the water (see Chapter 6).

The data on the relative roles of aerobic and anaerobic metabolism during periods of intense activity in small mammals are not adequate to support any broad generalizations. However, the available data (Table 3-8) from rodents running on treadmills inside respirometers suggest that their metabolic scope is much smaller than that of large mammals. This may be a function of the high mass specific oxygen consumption of small mammals at rest, or it may be correlated with the fact that small mammals, in contrast to large ones, rarely travel at high speeds for long distances—their evasive maneuvers typically consist of quick short dashes to cover rather than straight-away running at high speeds for long distances. Hence, natural selection may have favored the evolution of a capacity to incur an oxygen debt rather than the acquisition of large metabolic scope.

Among poikilotherms, the most complete data on aerobic metabolic scope are those on trout (including salmon). The fishes are placed in a special respirometer built in the form of a water tunnel in which their oxygen consumption is measured by analysis of the oxygen content of the water while they are forced to swim against water of controlled temperature and rate of flow. If the fish fails to swim fast enough to hold its position in the stream of water, it drifts back against an electrified grid from which it receives a mild shock. Thus it can be forced to swim at a sustained speed during the period of measurement.

When resting (standard) and active aerobic metabolic rates of sockeye salmon (*Oncorhynchus nerka*) are compared at 15°C, a pattern of relationships between size and metabolism is revealed that emphasizes the futility of looking for a single answer to the relation of metabolism to body size such as that envisioned in the surface law previously discussed (Section 3.7.3) for example.

1. The slope of metabolism on size increases regularly with increasing activity, starting with 0.78 for resting metabolism and reaching 0.97 during maximal sustained swimming.
2. The ratio of active to standard aerobic metabolism increases regularly with size in salmon, from less than 5 for fishes with a mass of 5 g to more than 16 for fishes with a mass of 2.5 kg.

Although all fishes that have been measured show essentially complete dependence on glycolysis during bursts of swimming at maximum speed, there is little information on which to assess the relative contributions of aerobiosis and anaerobiosis during periods of less intense activity. There is, however, a considerable body of data on the total amount of glycolysis occurring in salmonids that have been forced to swim until they were ex-

---

TABLE 3-8. Relation of Standard to Maximum Oxygen Consumption in Rodents

	Mass (g)	Max. $\dot{V}_{O_2}$/std. $\dot{V}_{O_2}$
Redbacked mouse (*Clethrionomys glareolus*)	18	4.0
White footed mouse (*Peromyscus leucopus*)	25	5.7
House mouse (*Mus musculus*)	33	7.2
Lemming (*Dicrostonyx groenlandicus*)	61	3.0
Chipmunk (*Eutamias merriami*)	75	7.0
Hamster (*Cricetus auratus*)	100	6.5

Data from: B. Wunder, *Comp. Biochem Physiol.,* **33,** 821–836 (1970); P. Pasquis, A. Lacaisse, and P. Dejours, *Resp. Physiol.,* **9,** 298–309 (1970).

hausted. By back-calculating from the lactic acid produced it can be estimated that about 2 mg of glycogen per gram of muscle are used. This is equal to the production of about 0.0334 mmol of ATP per gram of muscle.

Data on the activity metabolism of amphibians and reptiles are of particular interest because they allow a relatively precise evaluation of the aerobic and anaerobic components of metabolism in these poikilotherms (Tables 3-6 and 3-7).

The activity patterns of most lizards consist of short bouts of activity supported in part by glycolysis, which are separated by much longer aerobic periods of relative quiet. Nevertheless, all species so far studied show substantial aerobic scope. In many forms aerobic scope is maximal near the preferred body temperature (see Section 8.7.2). The responses summarized in Figure 3-10 are typical of the families Iguanidae, Agamidae, and Scincidae. However, in the large active and predatory monitor lizards (Family Varanidae) of Africa, Asia, and Australia, aerobic scope continues to increase even after body temperature exceeds the preferred level. It is noteworthy that monitor lizards show much less behavioral dependence on oxygen debt than other reptiles and are capable of longer periods of sustained activity than other kinds of lizards.

In those reptiles in which both aerobic and anaerobic scope have been measured, the energy yield from glycolysis during sustained, rapid locomotor activity exceeds that from oxidation. Because glycolysis in lizards appears to be less affected by temperature than oxidation, its relative importance in these animals is greater at low temperatures than at high temperatures. At 20°C about 90% of energy is anaerobic during the first 30 sec of activity in iguanid lizards, whereas at 37°C about 80% is anaerobic.

In poikilotherms the relative importance of aerobic and anaerobic scopes varies from group to group. However, caution must be exercised in making adaptive interpretations of these differences be-

Figure 3-10
The relation of aerobic metabolic scope to body temperature in representatives of three families of lizards. The arrows indicate the preferred body temperature. [Data from K. J. Wilson. Copeia, No. 4, 920–934 (1974); and A. F. Bennett, W. R. Dawson, and G. A. Bartholomew. J. Comp. Physiol., 100, 317–319 (1975).]

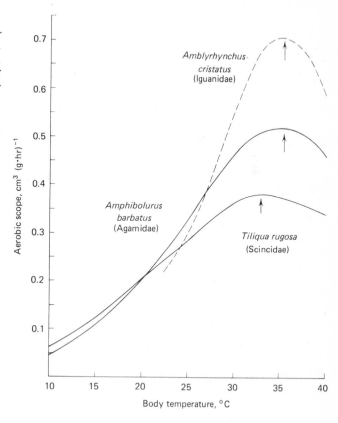

cause rates of active metabolism have the same general dependence on size as do resting metabolic rates, with small animals achieving higher rates of oxygen consumption per unit mass than larger ones during both activity and rest. Nevertheless, despite the limited comparative data available, there appears to be strong correlation with ecology and behavior. Species that are continuously active show larger aerobic scopes than species in which activity occurs in short bursts. Conversely, the latter show greater dependence on anaerobic metabolism than do the former.

The data for fishes are complicated by differences in size and also by differences in experimental procedures, but the general picture is clear. Active animals have higher aerobic scopes than sedentary ones. For example, cave fishes such as *Amblyopsis rosea* and *Typhlichthyes subterraneous*, which are slow-moving and sluggish, have aerobic scopes at 13°C of only 0.05 and 0.07 cm³ $O_2g^{-1}hr^{-1}$, whereas the strong swimming sockeye salmon *Oncorhynchus*

nerka at 15°C has an aerobic scope of 0.85 cm³ $O_2g^{-1}hr^{-1}$.

The differences between varanids and other lizards with regard to metabolic scope and dependence on glycolysis have already been mentioned. Instructive examples can also be drawn from the amphibian order Anura. Spadefooted toads *(Scaphiopus hammondii)* bury themselves deep in the ground, and digging requires a large expenditure of energy. Leopard frogs *(Rana pipiens)* spend most of the time sitting quietly and escape from predators and also capture their food by a single powerful leap or a short series of leaps and a rapid burst of swimming. These behavioral and ecological differences are clearly paralleled by their aerobic scopes (Figure 3-11). A similar contrast exists between toads of the genus *Bufo,* which carry out sustained activity, and tree frogs of the genus *Hyla,* which are active only briefly and intermittently.

In any event, natural selection has favored extensive dependence on anaerobiosis in most poikilo-

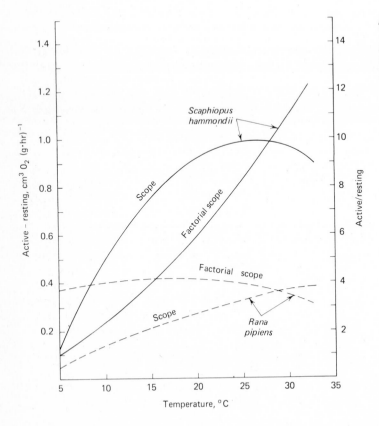

Figure 3-11
The relation of absolute and factorial aerobic metabolic scopes to body temperature in a spadefoot toad and the leopard frog. [Data from R. S. Seymour. Copeia, No. 1, 103–115 (1973).]

therms which, of course, experience a much wider range of body temperatures than do homeotherms. Some of the factors that have contributed to this situation seem clear. As previously pointed out, aerobiosis involves a transport system so complex that it cannot be instantly mobilized, particularly at low temperatures. Homeotherms minimize this constraint by maintaining a continuously high metabolic rate, whereas poikilotherms, for the most part, have simply by-passed it by depending on anaerobiosis. This has been possible because anaerobiosis is much less temperature-dependent than aerobiosis and thus allows mobilization of energy at all body temperatures normally encountered by active animals. The advantages of glycolysis are counteracted by obvious disadvantages—low energy yield per unit of fuel and the disruptive effects of accumulation of large quantities of lactate, which makes it impossible to sustain high levels of muscular performance for any prolonged period.

3.13 ENERGY METABOLISM, WORK, AND POWER

In the preceding sections it has been implicitly assumed that, by measuring the energy metabolism of an animal one can estimate the total energy transactions going on within it and between it and the environment. In the succeeding sections we shall make this assumption explicit by examining the metabolic costs of locomotion, which involves work and power. Before doing so we should briefly consider some of the relations between the chemical and physical aspects of energy metabolism.

From elementary physics we know that (1) *energy* is the capacity for doing work; (2) *work* is force times distance; (3) *power,* the rate of energy transfer, is the product of force times velocity; and (4) *velocity* is distance divided by time. If we measure metabolism in terms of oxygen consumption and we know the kind of fuel being oxidized, we can express metabolic rate either as calories per unit time, or joules per unit time (Section 3.4). The unit of power is the watt (W) and 1 W is equal to 1 joule per second (J/s). Consequently one can convert oxygen consumption to power (watts) and vice versa (Table 3-2). Similarly, if one knows an

animal's mass and the distance and velocity it has traveled, one can compute the metabolic energy it must have expended, because a joule is the work done when 1 newton (N) is moved 1 m, and a newton is the force required to give a mass of 1 kg an acceleration of 1 m/sec (Table 3-1).

The interconvertability of physiologically measured metabolic rate and physically determined work and power has proved to be of great importance to understanding energy metabolism in relation to locomotion because it allows one to evaluate metabolism by means of two completely different but compatible approaches; one is based on chemical and physiological considerations, the other is based on the kinetics of activity and concerned with mass, distance, power, and velocity.

Both metabolic rate and power are expressed as energy per unit time. Metabolic rate measures the total energy release in an animal, hence is the functional equivalent of the total *power input* (P_i) required for the animal to operate. Similarly, the work done by physiological processes can be treated as *power output* (P_o). No machines, including physiological ones, are completely efficient ($P_o/P_i < 1$) and, since energy is always conserved, the difference between P_i and P_o has to be accounted for by heat production.

It is instructive to view energy metabolism during locomotion in terms of power inputs and power outputs. From Figure 3-12 it is apparent that part of the P_i released by oxidation of fuel goes to locomotor muscles and that part of it goes to support systems. Either way, most of it is converted to heat rather than mechanical work. Only a fraction of the power input to muscle appears as power output. The ratio of P_o to P_i depends on the kind of animal (see Chapter 4), the type of muscle, and the conditions of contraction. The efficiency (P_o/P_i) of vertebrate skeletal muscle was first adequately measured by the British physiologist A. V. Hill in 1939, and it varies between 0.2 and 0.3. (An efficiency of 0.2 or 20% is customarily used for computational purposes). Thus, about four fifths of the power input to locomotor muscle ends up as heat and only about one fifth ends up as power output ($P_o/P_i = 0.2$).

The compartmentalization of P_o and the magnitude of its principal components depends on the type of locomotion—swimming, flying, or walking. (1) Many aquatic animals are neutrally buoyant and

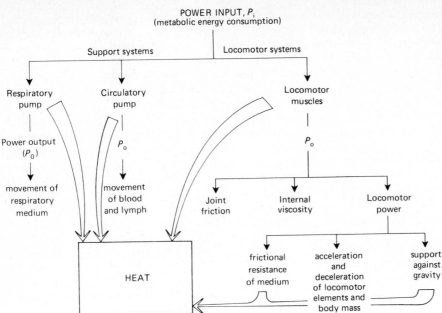

POWER INPUT, P_i
(metabolic energy consumption)

Support systems   Locomotor systems

Respiratory   Circulatory   Locomotor
pump   pump   muscles

Power output P_o P_o
(P_0)

movement of   movement   Joint   Internal   Locomotor
respiratory   of blood   friction   viscosity   power
medium   and lymph

HEAT   frictional   acceleration   support
                                  resistance   and   against
                                  of medium   deceleration   gravity
                                                  of locomotor
                                                  elements and
                                                  body mass

Figure 3-12
Diagram of the distribution of the power input made available by an animal's energy metabolism to the various elements in its locomotor system. About one fifth of the power input to the locomotor muscles is eventually available for propulsion; the rest is lost in heat.

spend little or no energy to support themselves against the force of gravity, but they move through a dense medium of high viscosity in which resistance, or drag, is large and varies in a complex manner with speed and body length. (2) Flying animals move in a gaseous medium of low density and low viscosity. They must support their bodies against gravity, since they move rapidly, they must overcome substantial drag, and during flapping flight they must also deal with the inertial forces involved in rapid and repeated reversal of direction of wing movement. (3) Walking animals face a still different situation. When moving horizontally at a moderate speed, there is no frictional energy loss between feet and ground; air resistance is negligible; no useful external work is performed; and the power required for support against gravity is no greater than when standing still. Nevertheless a running animal consumes much more energy than one standing motionless. All this increase in energy is used to overcome frictional resistance in the joints (a relatively small quantity) and the repeated accelerations and decelerations of mass in the limbs and other parts of the body.

In a runner, most of the power output is used accelerating and decelerating locomotor elements. In a swimmer most of the power output is used for overcoming drag. In a flier, however, power output has three primary components which function (1) to support body weight—induced power, (2) to overcome the drag of body—parasite power, and (3) to move the wings up and down and through the air—profile power. [See Pennycuick, 1972a, for a clear elementary exposition of the power requirements of flight, and Webb, 1975, for an excellent discussion of the power requirements of swimming.]

With adequate estimates of velocity, mass, bodily dimensions, frequency of stride, or wing beat, or tail undulation, it has proved possible to predict with remarkable accuracy the energy expenditure required for locomotion. This figure together with estimates of the power inputs required by the support systems (respiration and circulation) can then be added to the basal metabolic rate to give the total energetic cost of locomotion. The accuracy of such computations can be checked by direct measurement of the oxygen consumption of running, swimming, and flying animals.

3.13.1 Energy Metabolism of Flight

The physical and aerodynamic constraints under which natural selection has necessarily operated with regard to flight and the paucity of major taxa that depend on powered flight give to this mode of locomotion a homogeneity that is found neither in swimming nor terrestrial movement. In part for this reason, the physical and physiological analysis of the energetics of flight have produced remarkably congruent and compatible results despite differences in methodology. We shall, therefore, begin our examination of locomotion with a consideration of flight.

The capacity for sustained metabolically powered flight has evolved in only two phyla, Chordata and Arthropoda. Among living animals powered flight is used only by birds, bats, and insects. Most flying animals, whether large or small, fly at high speeds for periods of time so long that anaerobiosis with its attendant rapid fatigue and oxygen debt offers no energetic advantages. In fact, except in a few cases exemplified by quail and pheasants of the avian order Galliformes, which employ powered flight only for brief spurts, anaerobiosis seems incompatible with locomotor dependence on flight. In fliers, natural selection has favored the capacity to achieve extremely high levels of sustained aerobic metabolism. In animals that depend on powered flight, the magnitude of aerobic scope and the coordination of effective support systems for transport of materials to and from the muscles is of much greater importance than the enhancement of anaerobic capacities.

The aerobic energy cost of flight has been measured in still air in closed respirometry systems on flying insects and hovering hummingbirds, and it has been measured on a variety of animals trained to fly in wind tunnels. The latter technique has proved to be a particularly powerful and versatile analytical tool. When flight speed precisely matches the speed of the air stream in a wind tunnel, ground speed is zero and the flying animal remains stationary with reference to the observer (Figure 3-13). Under such circumstances various parameters of the physiological performance of a flying animal can be measured while it is airborne. If the tunnel is tilted so that the front end is higher or lower than the rear end, energy expenditure can be measured during ascent and descent as well as during level flight. The relative power requirements of gliding and flapping flight at various speeds can also be measured by appropriate combinations of wind speed and angle of flight.

The first direct measurements of oxygen consumption during flight in a wind tunnel were obtained in 1968 from a parakeet, the budgerigar, *Melopsittacus undulatus*, a small (35–40 g), strong-flying parrot widespread in the arid parts of Australia. Since then, similar measurements have been made on a variety of birds (hummingbirds, crows, starlings, falcons, gulls), one insectivorous bat (suborder Microchiroptera) and two fruit bats of the genus *Pteropus* (suborder Megachiroptera). These physiological measurements have been paralleled by estimates of the power requirements of natural flight based on measurements of weight, wingspread, velocity, angle of flight, and drag (wind resistance) together with relevant aspects of aerodynamic theory developed with regard to both fixed-wing aircraft and helicopters. These two approaches, one empirical and physiological, the other theoretical and physical, have merged to form a coherent picture of animal flight that encompasses both small insects and large vertebrates [see the references to Pennycuick, Rayner, Tucker, and Weis-Fogh in Additional Reading]. A key step in meshing these two different approaches is the equating of the physiologist's metabolic rate ($\dot{E}_m$) and the engineer's power input (P_i) as discussed in the preceding section.

The physiological performance of the budgerigar conforms nicely to aerodynamic theory and may be treated as typical for flapping flight in medium sized birds. In smooth air during level flight its oxygen consumption is minimal (21.9 ml O_2/hr) at a velocity of (35 km/hr) and increases at both lower and higher velocities. Descending flight is less energy-demanding than either level or ascending flight (Figure 3-14). Flight in turbulent air at a given velocity requires 2 or more times the power input of flight in smooth air. The factorial metabolic scope of budgerigars is at least 25, and even at their most economical flight speed they use oxygen at much higher rates than the maxima reported for terrestrial mammals of similar size. Using the double-labeled water technique (Section 3.5.2), it has been calculated that during a 500-km flight racing pigeons weighing 384 g use 57 cal$g^{-1}hr^{-1}$. This figure is 13.4 times the BMR calculated from line (6) in Table 3-5. A 3-g Allen hummingbird, *Selasphorus alleni* (see Figure 8-40), uses 40 cm^3 O_2 $g^{-1}hr^{-1}$ while hovering, which is 12 times its SMR.

There are conspicuous interspecific differences

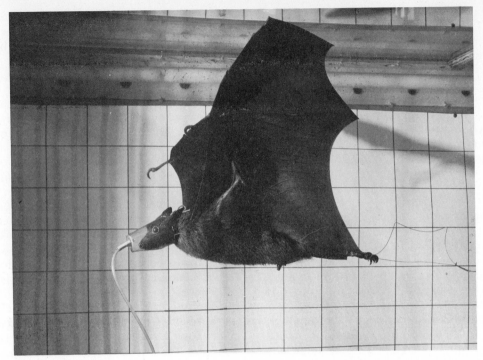

Figure 3-13
An 800-g flying fox (Pteropus poliocephalus) *flying at 28 km/hr while wearing a plastic mask and carrying thermocouples for measuring wing-surface temperature and deep body temperature. The thermocouple leads are secured to the left hind foot. Expired gases collect inside the mask and are drawn off through the flexible tube. [Photograph courtesy of Roger E. Carpenter.]*

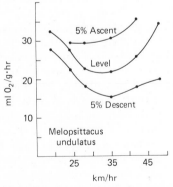

Figure 3-14
The relation of oxygen consumption to flight speed in smooth air in the budgerigar. [Modified from V. A. Tucker. J. Exp. Biol., **48**, *67–87 (1968).]*

in patterns of flight and wing loading, and the volumes of oxygen used during flight are 10 or more times the BMR. Nevertheless, the data now available strongly suggest that the slope of the regression of oxygen consumption on mass of birds and bats engaged in flapping flight is strikingly similar to that of their basal metabolic rates. A similar relationship exists between the slopes of the regressions of resting metabolism and flight metabolism on body mass in moths, such as sphinx moths, that generate high body temperature during flight (Figure 3-15).

The measured oxygen consumption of three birds and two mammals during flapping flight in a wind tunnel averages 10 to 15 times basal (Table 3-9) and 6 or 7 times the resting level determined under the same physical conditions as the flight metabolism. The highest mass-specific rates of oxy-

Figure 3-15

*The relation of oxygen consumption to body mass during flapping flight and at rest in insects, birds, and mammals. [Data on flying birds from M. J. Berger, S. Hart, and O. Z. Roy. Z. vergl. Physiol., **66,** 201–214 (1970). Data on flying bats and birds from S. P. Thomas. J. Exp. Biol., **63,** 273–293 (1975). Data on moths from G. A. Bartholomew and T. M. Casey. J. Exp. Biol., **76,** 11–25 (1978). BMR's from Table 3-5.]*

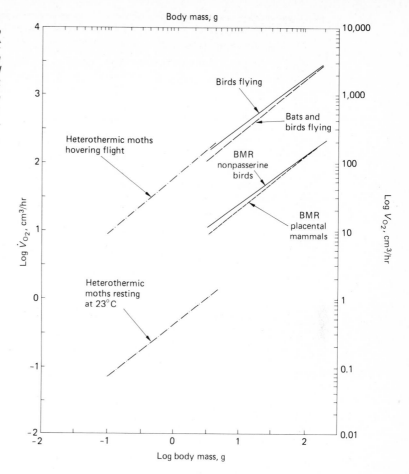

gen consumption in vertebrates are 40–45 cm³ O_2 g^{-1}hr^{-1} measured during hovering flight in several species of hummingbirds ranging in mass from 2 to 6 g. These rates are 12 to 15 times basal depending on body mass.

Flying insects in the same general range of weight (1–3 g) have metabolic rates as high or higher than those of hovering hummingbirds. For example, sphinx moths of the genera *Manduca* and *Hyles* consume approximately 50 cm³ O_2 g^{-1}hr^{-1} during flight at 30°C. This value is more than 100 times their resting metabolism at the same temperature. This enormous factorial scope is a function of the fact that, unlike birds, moths at rest expend no energy on maintenance of body temperature (see Chapter 8).

Many birds, particularly large ones such as hawks, vultures, gulls, pelicans, storks, and frigate birds, glide for long periods of time without flapping their wings. Soaring flight on fixed wings exploits an interaction between ascending air and the force of gravity. The bird continuously glides downward, but as long as its sinking rate is equal to or less than the rate at which the air is rising, it maintains or gains altitude. By seeking out and remaining in areas of upward moving air, a bird may fly for hours without resorting to wing-flapping as a power source.

Oxygen consumption of the herring gull *(Larus argentatus)* during gliding flight has been measured in a wind tunnel. The metabolic rate during gliding in this species is 3 to 4 times basal and 2 to 2.5 times the resting level measured in the experimental set up. From this data it appears that, in the

TABLE 3-9. Energy Requirements for Flapping Flight Based on Wind Tunnel Measurements of Oxygen Consumption

(S.I. units are used in this table. See Table 3-2 for conversion factors.)

Species	Mass (kg)	Flight Velocity (m/s)	Metabolic Rate (W) Basal	Metabolic Rate (W) Flight	Flight $\dot{E}_m$ / Basal $\dot{E}_m$
Budgerigar (Melopsittacus undulatus)	0.035	10.7	0.33	3.7	11.2
Laughing gull (Larus atricilla)	0.322	13.0	1.64	24.5	14.9
Fish crow (Corvus ossifragus)	0.275	11.0	2.42	24.8	10.2
Starling (Sturnus vulgaris)	0.073	13.5	0.94	8.9	9.5
Spearnosed bat (Phytostomus hastatus)	0.0927	8.0	0.67	9.1	13.6
Gould's fruit bat (Pteropus gouldii)	0.780	9.9	3.12	47.1	15.1

Data from: V. A. Tucker, *J. Exp. Biol.*, **58**, 689–709 (1973); and J. R. Torre-Bueno and J. Larochelle, *J. Exp. Biol.*, **75**, 223–229 (1978).

confines of a wind tunnel, gliding requires only about a third the expenditure of energy of flapping flight. Under natural conditions where a gliding bird can freely vary its velocity and glide angle, the relative cost of gliding is probably even less, particularly in the case of very large birds especially adapted for soaring [see Pennycuick, 1972b, for a detailed and informative treatment of this topic].

Direct measurements of the metabolic rate during flight are difficult to make, and often require exacting training of the experimental animals as well as elaborate and expensive instrumentation. Therefore, a method that allows accurate prediction of the power requirements for flight from a few easily measured parameters has practical as well as theoretical value. Such methods based on body dimensions, wing-beat characteristics, and helicopter aerodynamic theory have been developed for both forward flight and hovering [Pennycuick, 1968, 1969; Tucker, 1973, 1975; Rayner, 1979; Weis-Fogh, 1972, 1973].

The details are beyond the scope of this treatment but, for illustrative purposes, let us consider the main features of this theoretical approach for powered forward flight. If the mass and wingspread of a bird are known, it is possible to calculate, usually to within 8% and sometimes exactly, the power required for it to fly. The power output (P_o) is computed on the basis of aerodynamic considerations such as body weight, drag, viscosity of the air, velocity, and the area swept out by the wings during each beat. The power input (P_i) required to obtain the needed output is calculated by adding to the BMR the energy required by the respiratory and circulatory support systems and the required P_o corrected for the efficiency of converting muscle contraction to mechanical work. The final equation, which consists of a summation of the power inputs to all the major processes that consume power during flight, is impractical to solve without automatic facilities, but some of its solutions can be closely approximated by simple estimating equations. Two examples are given.

For the power input in watts at the flight speed where the cost of transport (see Section 3.14.4) is a minimum and m is mass in kilograms

$$P_i = 84.7 \ m$$

For the minimum cost of transport by flying

$$P_i/wv = 0.927\ w^{-0.2}$$

where w is weight in newtons and v is velocity in meters per second.

3.13.2 Energetics of Terrestrial Locomotion

As one would expect from the physical complexity of the terrestrial environment and the large number of major taxa that have adapted to it, there are many patterns of terrestrial locomotion. Even among vertebrates the variety of locomotor patterns is impressive. Frogs hop. Salamanders walk. Snakes and limbless lizards crawl. Lizards run quadrupedally and bipedally. Mammals walk, run, hop, and gallop quadrupedally and walk, run, and brachiate bipedally. In addition some mammals use the tail as a third locomotor appendage (spider monkeys) or a fifth locomotor appendage (kangaroos).

Despite this diversity of locomotor styles, all vertebrates depend on the same general morphology of bone and muscle, the same physiological support systems, and the same biochemical processes for the release of energy. It should be noted that in birds and mammals the energetics of locomotion are closely linked to temperature regulation because, as previously mentioned, about 80% of the power input for locomotion appears as heat. However, this aspect of the topic will be deferred to Chapter 8.

Direct measurements of oxygen consumption have been made on a variety of vertebrates running on treadmills or in exercise wheels. When the power inputs required for terrestrial locomotion are scaled against size, the relationships are impressively regular. Presumably this regularity depends on the following: (1) Most of the energy for terrestrial locomotion regardless of type is used for accelerating and decelerating the limbs and for cyclically accelerating and decelerating the center of mass of the body. (2) Skeletal muscles of all vertebrates generate nearly the same maximum force for a given cross-sectional area. (3) The diameter of muscles (hence their cross-sectional area) changes directly with muscle length and muscle length is a function of animal size.

Quadrupedal Locomotion. When running at speeds that can be sustained without acquiring a significant oxygen debt, the power input (metabolic rate) of both lizards and mammals increases linearly with increasing velocity. Aside from some very detailed recent work on ponies, the published data show no breaks in the linearity of the increase of P_i with velocity even though most quadrupeds employ different gaits at different velocities, typically progressing from walk, to trot, to gallop. The rate of increase of P_i with velocity is greater in small animals than in large ones, but both energy expenditures and velocities are much lower in lizards than in mammals (Figures 3-16 and 3-17). The slopes of the linear regressions shown in these figures can be used as an index to the energy requirements of locomotion in relation to body size. When plotted on logarithmic coordinates against body mass these slopes can be fitted by a straight line with a negative exponent (-0.4), that is, the power input required to move a unit of mass is less in large animals than in small ones. For example, it costs a 10-g mouse 15 times as much as it does a 10-kg dog to move 1-g a distance of 1 km. Although such a comparison is of physiological and theoretical interest, it is somewhat less than satisfying ecologically because for small mammals biologically meaningful distances are more apt to be measured in body lengths than in kilometers. One can, however, readily examine the energetic cost of locomotion on a length-specific basis by scaling; the exponent relating the length of geometrically similar objects to their mass is one third (Section 3.7.1). The body length of most mammals is approximated by $l = M^{0.33}$ where l is body length in centimeters and M is body mass in grams. Since the masses of all the mammals in Figure 3-17 are given, we can convert kilometers to body lengths. However, even on a length-specific basis the slope is still negative (-0.2), and the cost of running is still higher for small mammals than for large ones.

Bipedal Locomotion. In their terrestrial locomotion all birds are bipedal and a number of them are primarily, or exclusively, runners. For example, most members of the order Galliformes (quail, pheasants, chickens, turkeys) are effective runners that fly only occasionally, and the ratites (ostrich, emu, cassowary, and rhea) are all fast-running flightless forms. There are also a number of bipedal

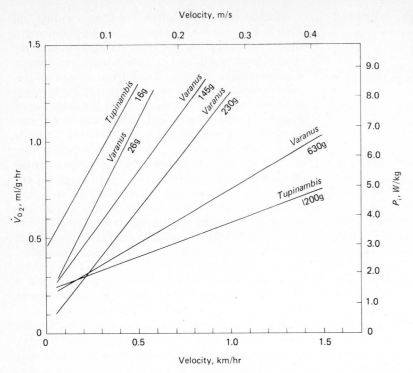

Figure 3-16
Oxygen consumption and power input in relation to velocity of running in lizards of the families Varanidae and Teiidae. [Data from R. T. Bakker. Physiologist, **15,** 76 (1972).]

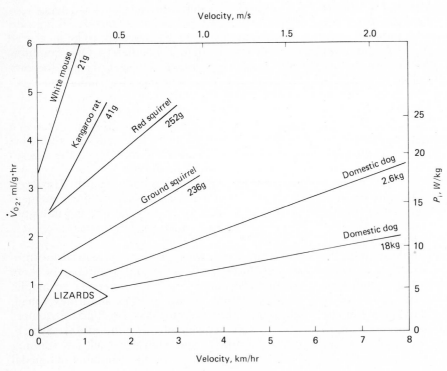

Figure 3-17
Oxygen consumption and power input in relation to velocity of running in mammals: (top to bottom) Mus musculus, Dipodomys merriami, Tamiasciurus hudsonicus, Citellus tereticaudus, and Canis familiaris. The polygon in the lower left of the graph encloses all the data on the lizards shown in Figure 3-16. [Data from C. R. Taylor, K. Schmidt-Nielsen, and J. L. Raab. Am. J. Physiol., **219,** 1104–1107 (1970); and B. A. Wunder and P. R. Morrison. Comp. Biochem. Physiol., **48a,** 153–161 (1974).]

lizards and mammals, but, with very few exceptions, living members of these groups are quadrupedal when moving slowly and become bipedal only at moderate to high speeds.

In cursorial birds as in mammals and lizards, power input increases linearly with velocity and the slopes of the increases are steepest in the smaller species (Figure 3-18). The energy expenditure of a mammal running quadrupedally and a bird running bipedally are about the same for animals of the same size, but there is a tendency for small birds to require less energy and large ones to require more energy to run a given distance than mammals of the same size.

Small to medium sized bipedal mammals (mass 3 kg) that hop show the same relation of energy expenditure to velocity as mammals that run quadrupedally. However, in large bipedal mammals the situation is less clear.

Man is not an efficient runner. The power input of a running man is about twice that of a quadrupedal mammal of the same size, but the performance of kangaroos shows that bipedal locomotion in mammals is not necessarily inefficient. Kangaroos

have an unusual combination of locomotor patterns. When traveling rapidly they hop bipedally on their hind legs with front legs held against the chest and the heavy tail used only for balancing. When moving slowly, however, the weight of the body is supported by the front legs and tail while the hind legs are being swung forward to initiate the next stride. The power requirements for the two modes of locomotion differ markedly. When moving slowly with the "five-leg" gait, a kangaroo uses more energy than a quadrupedal placental mammal of the same size traveling at the same speed. However, when a kangaroo reaches a speed of 1½–2 m/sec (5–7 km/hr) it starts to hop bipedally and its energy expenditure diminishes. This diminution of energy expenditure with increased velocity contrasts sharply with the situation in other terrestrial vertebrates.

The hypothesis that has been proposed to account for this paradoxical situation can be visualized in terms of the relationships shown in Figure 3-19. The kangaroo's rate of hopping is essentially independent of velocity, but the lengths of its hops increase directly with velocity. Thus, when a kanga-

Figure 3-18

Oxygen consumption and power input in relation to velocity of bipedal running in cursorial birds, five species from order Galliformes and one ratite bird from the order Rheiformes. [Data from M. A. Fedak, B. Pinshow, and K. Schmidt-Nielsen. Am. J. Physiol., 227, 1038–1044 (1974).]

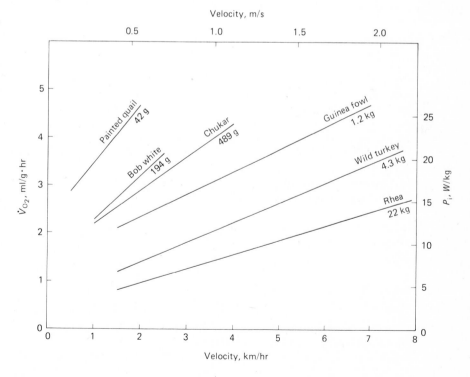

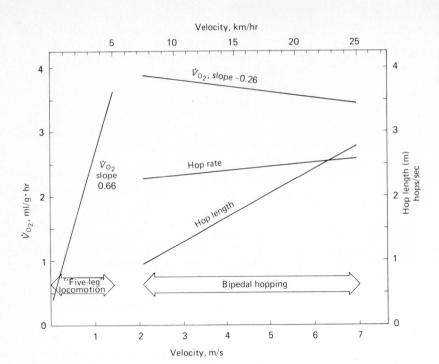

Figure 3-19
The energetics of locomotion in the red kangaroo, Megaleia rufa. [*Data from T. J. Dawson and C. R. Taylor. Nature, **246**, 313–314 (1973).*]

roo speeds up it does not change its rate of hopping; it merely increases the distance traveled per hop. This requires that the kangaroo increase the acceleration imparted to its mass at the start of each hop—paradoxically it is able to do this while decreasing the energy input per hop. The resolution of this paradox lies in the mechanical storage of energy. A hopping kangaroo is analogous to a man on a pogo stick. At the end of each hop its energy of momentum is temporarily stored in the elastic elements of the limbs (ligaments and tendons) and at the start of the next hop this stored energy is released. The efficiency of the kangaroo's elastic recoil system increases with velocity, so the faster it goes the less muscular energy it needs to spend accelerating and decelerating its limbs and center of mass, and the more efficient its locomotion becomes.

3.13.3 Energy Cost of Swimming

All the phyla of animals are fundamentally aquatic, and the modes of aquatic locomotion that have evolved are so numerous and diverse that no integrated treatment of the sort applied to powered flight is possible. Even if one considers only the kinds of fishes that swim by lateral undulations of the body, the situation is far from simple. There is a very large body of data on fish propulsion, particularly on its mechanics and hydrodynamics. This topic is outside the scope of the present text; interested readers should consult Webb [1975] for a coherent introductory treatment.

As pointed out in the introduction to this section (3.13) swimmers are supported by the water in which they float, but to swim they must overcome the substantial resistance (drag) imposed by the density and viscosity of the medium. At any given speed the *thrust* generated by the propulsive machinery of a swimmer must equal the drag generated by its body as it moves through the water (drag = thrust). Because drag is easier to measure than thrust, most studies of swimming have dealt with drag. However, the physiology of the fishes' locomotor apparatus is primarily concerned with the generation of thrust. The best data are available for salmonids, and we shall confine our discussion to this family.

The distribution of power associated with the conversion of metabolic energy to thrust can be

Figure 3-20

Oxygen consumption in relation to velocity of swimming in adult and immature sockeye salmon. Data can be converted to $\dot{V}_{O_2}$ or watts by use of Tables 3-1 and 3-2. [Data from J. R. Brett. J. Fish. Board Can., 22, 1491–1501 (1965).]

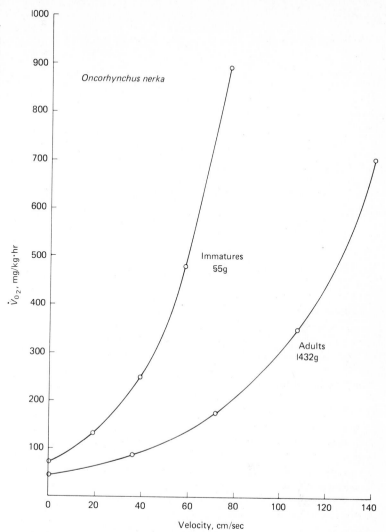

fitted into the diagram shown in Figure 3-12 by adding a support system for osmoregulation. The percentage of total power input required by the physiological support systems increases with swimming velocity. In salmonid fishes maximum values are calculated to be about 17% for osmoregulation, 10% for the respiratory pump, and 11% for the cardiac pump. In swimmers, metabolism increases exponentially with velocity not linearly with velocity as it does for runners, but in swimmers as in runners the rate of increase of oxygen consumption with increasing velocity is greater in small animals than big ones (Figure 3-20). However, relative to their length, small fishes can swim more rapidly than large ones. Length-specific velocity decreases with length, whereas velocity in centimeters per second increases with length. The data relating size to velocity are described by the familiar allometric equation (3-1) with b negative for body lengths per second and positive for centimeters per second (Figure 3-21).

3.13.4 Metabolic Cost of Transport

Data on the metabolic energy expenditure during locomotion, although available for relatively few

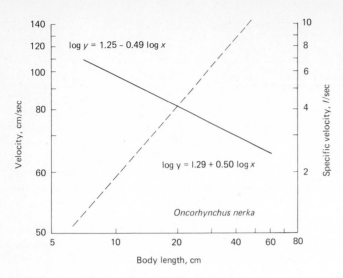

Figure 3-21
The relation between maximum sustained swimming speed and body length. Solid line is length-specific velocity. Broken line is velocity in centimeters per second. The equations are the logarithmic form of the allometric equation (3-1) Y = aX^b. [Data from J. R. Brett. J. Fish. Res. Board. Can., **22,** 1491–1501 (1965).]

species, has now become extensive enough to contribute to the understanding of behavioral and ecological problems as well as matters of strictly physiological interest. One can evaluate the economics of locomotion, examine the energetics of long distance migration, compare the efficiencies of swimming, flying and walking, and examine the effectiveness of the same locomotor mode in distantly related taxa and in closely related animals that differ in size.

A number of methods have been used for estimating the energy cost of locomotion with the criteria employed depending on the questions under consideration. A few examples will suffice. The *optimal speed* has been defined as the speed at which energy expended per unit distance traveled is minimal. In terrestrial animals this optimum speed can be related to the *minimum cost of running,* which has been defined as the slope of the curve relating oxygen consumption to velocity. Using this convention, the cost of running 1 km at a given velocity is divided by that velocity. Hence the cost of running diminishes with velocity and is minimal at or near the maximum sustainable speed, which thus is the optimal speed (Figure 3-17). For swimming and flying, however, the optimum speed is in the middle range of sustainable velocities (see Figure 3-14). The *net cost of locomotion* has several definitions, one obvious one being the total metabolism during locomotion minus resting metabolism. In the case of the data presented in Figures 3-15, 3-

16, and 3-17, the net cost of transport could also be defined as the slope minus the y intercept, or the slope minus the difference between resting metabolism and the y intercept. Instead of being in terms of meters or kilometers, any of the preceding could be in terms of body lengths, distance per tail undulation (swimmer), distance per wing beat (flier), or distance per stride (runner).

One of the most instructive ways for expressing the energy requirements of locomotion is the *cost of transport,* which is defined as the total metabolic energy expended while moving one unit of body mass (grams or kilograms) over one unit of distance. In physiological terms the cost of transport is the specific metabolic rate divided by velocity:

$$\frac{\text{specific metabolic rate}}{\text{velocity}} = \frac{\text{kcal kg}^{-1}\text{ hr}^{-1}}{\text{km hr}^{-1}}$$

$$= \frac{\text{kcal kg}^{-1}}{\text{km}} = \text{kcal kg}^{-1}\text{ km}^{-1} \quad (3\text{-}6)$$

Since time cancels out, we can compare animals moving at whatever speed their cost of transport is minimal, whether the animal is a high-speed flier or a low-speed crawler. If the SI system of units is used, the cost of transport as defined in equation (3–6) is the equivalent of P_i/wv, where P_i is the power input in watts, w is weight (the product of mass and the acceleration of gravity) and v is velocity in meters per second. Using the SI system, the cost of transport, P_i/wv becomes a dimensionless number because the units all cancel. This formulation was used in the preparation of Figure 3-22, which compares the minimum metabolic cost of transport in a variety of animals and a few man-made machines. To supply a little more immediacy, a few representative values of cost of transport are also presented as kilocalories per kilogram per kilometer in Table 3-10.

It is impressive that, regardless of their taxonomic affiliations, swimming, flying, and walking animals fall into separate groups and in each group the cost of transport diminishes as size increases. On the size scale of body mass employed (12 orders of magnitude), single lines can be used to describe the log-transformed values of the minimum cost of transport for swimmers, for fliers, and for walkers (if small mice are excluded).

It is intuitively obvious that the cost of transport for swimmers should be lowest. After all, they are passively supported by the medium they move through, but why should it require less energy for an animal of a given size to fly than to walk? Part of the answer is velocity. A small bird and a mouse have the same mass and roughly similar metabolic rates but the bird flies an order of magnitude faster

than the mouse can run, and hence has a cost of transport that is about an order of magnitude lower than that of the mouse.

A variety of factors other than speed are involved of course. For example, only part of the power output of the locomotor muscles of a running animal is used to move the body in a forward direction. During each step cycle a substantial fraction of the power output results only in vertical movements of the center of mass of the body. The muscles of one set of limbs contract and lift the animal upward and forward, but almost immediately the muscles of the other set of limbs must begin to slow down the rate of descent of the body as it is accelerated downward by gravity. In so doing these active muscles are stretched by the weight of the body they are decelerating. About half the work done by these stretching muscles as they decelerate the downward movement of the center of mass of the body is stored in the series elastic elements of the muscle system and released to do work in the next stride, but about half is converted to heat and eventually lost to the environment. In contrast to the limbs of runners, the wings of fliers move at right angles to the direction of locomotion and

Figure 3-22
The minimum cost of transport in relation to size. The geometric figures enclose the values for animals using the same general types of locomotion. The dotted lines are the linear regressions of the log-transformed values of X and Y. See text for explanation of P_i/wv. [Modified from V. A. Tucker. Am. Sci., **63**, *413–419 (1975).]*

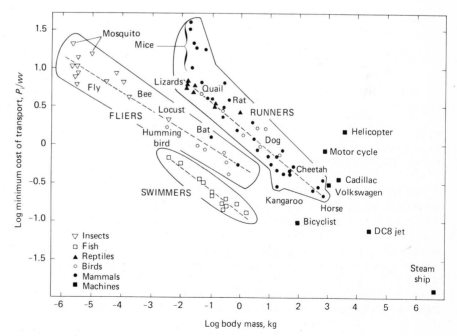

TABLE 3-10. Cost of Transport for Animals and Machines

Animal or Vehicle	Mass (kg)	Speed (km/hr)	Metabolic Rate (kcal/kg hr)	Cost of Transport (kcal/kg km)
Flyers				
Fruit fly (Drosophila)	2.0×10^{-6}	7.4	104	14.05
Blow fly (Lucilia)	3.0×10^{-5}	30.0	450	15.0
Honey bee (Apis)	1.0×10^{-4}	30.0	450	15.0
Desert locust (Schistocerca)	2.0×10^{-3}	15.0	75	5.0
Hummingbird (Calypte)	0.003	49.0	197	4.02
Budgerigar (Melopsittacus)	0.035	35.0	102	2.91
Laughing gull (Larus)	0.310	31.0	44.9	1.45
Pigeon (Columba)	0.384	58.0	55.1	0.95
Walkers and Runners				
White footed mouse (Peromyscus)	0.023	1.08	37.8	35.0
Lemming (Dicrostonyx)	0.061	1.54	63.0	40.9
Ground squirrel (Citellus)	0.198	2.80	15.2	5.43
Lizard (Iguana)	0.90	0.370	2.31	6.24
Dog (Canis)	10.7	5.84	7.94	1.36
Sheep (Ovis)	30.0	2.88	2.88	1.0
Man (Homo)	70.0	5.58	4.06	0.73
Horse (Equus)	707.0	5.00	2.61	0.52
Swimmer				
Salmon (Oncorhynchus)	0.19	2.8	1.1	0.39
Vehicles				
Cherokee airplane	978	200	240	1.2
Sikorski S62 helicopter	3,590	148	524	3.54
F104F jet fighter	17,200	1,150	1,750	1.52
DC-8 jet	107,000	965	550	0.57
Volkswagen	1,010	81	59.6	0.74
Cadillac	2,210	119	98.3	0.83

Sources: T. Weis-Fogh. *Trans. Ninth Int. Congr. Ent.,* **1,** 341–347 (1952); V. A. Tucker. *Comp. Biochem. Physiol.,* **34,** 841–846 (1970).

by virtue of their capacity to generate aerodynamic lift can convert the downward acceleration of the body mass due to gravity into forward motion without doing large amounts of muscular work. Merely by extending its wings a bird can change downward fall into forward flight.

The cost of transport of machines can be directly compared with that of animals if one knows the energy released by the oxidation of their fuel, the amount of fuel burned, the mass of the machine and its velocity. As can be seen from Figure 3-22 on the basis of mass the cost of transport of machines is of the same order as that of animals.

3.13.5 Migration and Cruising Range

The energetics of locomotion can be used to enhance our understanding of ecology as well as physiology. Consider the spectacular phenomenon of long distance migration. Fliers and swimmers with masses of only a few grams migrate, but no running or walking animals below the 10–100 kg range do so. The correlation of migration with cost of transport is striking. It is apparently not energetically feasible for an animal with a cost of transport of more than 2.0 (0.3 on the log-transformed y axis of Figure 3-22) to be a long-distance migrant.

It is instructive to consider long distance migration of birds. If one knows a bird's metabolic rate during flight, its flight speed, and the quantity of fat it has stored, one can calculate how far it can fly without feeding. The budgerigar can be used as an example. The RQ of a budgerigar in flight is 0.78, which indicates that it is oxidizing mostly fat. The oxidation of 1 g of fat yields about 9.5 kcal or 39.77 kJ (Table 3-2). The metabolic rate of a 40-g budgerigar flying at a velocity of 10.7 m/sec is 3.7 W or 13.32 kJ/hr (Table 3-9). If 10% (4 g) of the budgerigar's mass is fat, one can calculate how long it can remain airborne on this amount of fuel. Knowing its flight speed, one can calculate how far it could fly without having to refuel (feed).

$$\text{flight time} = \frac{4 \times 39.77}{13.32} = 11.94 \text{ hr}$$

$$\text{flight distance} = 11.94 \times (10.7 \times 3600)$$
$$= 459.7 \text{ km}$$

This type of analysis is of particular interest with regard to long-distance aerial migration—for example, transoceanic or trans-Saharan migratory flights by small land birds. In addition, cruising range is more than just a matter of interest to biologists; it has played a major role in the evolution of the capacity for annual migration in birds. The biological importance of cruising range is documented by the astounding capacity that birds, particularly those weighing less than about 50 g, have evolved to deposit fat stores during the weeks immediately prior to spring and fall migrations. At the onset of migration the smaller migratory birds accumulate fat deposits, which may constitute 50% or more of their body weight. Such fat deposition does not occur in nonmigratory species.

During migration some ruby-throated hummingbirds (*Archilochus colubris*) travel from Florida to Yucatan. Is their cruising range great enough to allow them to fly directly across the Gulf of Mexico, or must they fly in many stages along the coasts of the southern United States and eastern Mexico? The fat-free weight of a male ruby-throated hummingbird is 2.5 g. Prior to migration it accumulates 2.0 g of body fat. Since the fat is consumed during flight we shall base our calculations on an RQ of 0.71 and a body weight of 3.5 g, the weight it would have with half of its fuel consumed. Oxygen consumption during hovering flight is about 42 $cm^3g^{-1}hr^{-1}$, or 0.69 kcal/hr, and its flight speed

Figure 3-23

*The flight range per unit fuel as a function of flight speed in (a) a hummingbird and (b) a domestic pigeon. The arrows indicate the most economical cruising speeds. [Modified from C. J. Pennycuick. Ibis, **111**, 525–556 (1969).]*

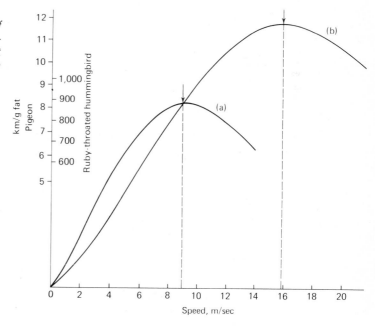

is 25 mi/hr. On the basis of these figures, a ruby-throat should be able to fly 650 mi without refuelling. The longest over-water flight it would have to make in crossing the Gulf of Mexico is 500 miles; this is well within the bird's capacity unless it runs into headwinds. This computation is extremely conservative because it assumes that the energetic cost of flight at 25 mi/hr is the same as that of hovering, although in fact it is much less. The estimate can be refined by use of other methods.

The sustained flight of birds represents a state of steady expenditure of energy, and the energy required for long flights can be approached from the point of view of aerodynamic analysis rather than indirect calorimetry. One can fly a bird in a wind tunnel and calculate the mechanical power required for flight at different speeds. The work done can then be converted to terms of fuel and oxygen consumption. Large birds can carry relatively less extra weight than small ones and so cannot devote such a large proportion of their weight to fuel (fat) as small ones. Small birds fly more slowly than large ones and therefore are more affected by headwinds than large ones. Figure 3-23 shows some of the consequences of these relationships. The hummingbird gets more distance from a given quantity of fat than does a pigeon, but it flies much slower and, of course, the absolute quantity of fat that the hummingbird can carry is only a small fraction of that carried by the pigeon. With regard to the transgulf migration of the ruby-throated hummingbird, Figure 3-23 indicates that in the absence of headwinds this tiny bird should be able to fly 900 km on a single gram of fat if it stayed at its most economical cruising speed, 9 m/sec (32.4 km/hr, or 20.1 mi/hr). Its range falls off sharply at both lower and higher speeds.

ADDITIONAL READING

Aschoff, J. (ed.). *Circadian Clocks,* North Holland, Amsterdam, 1965.

Aschoff, J., and H. Pohl. "Der Ruheumsatz von Vögeln als Funktion der Tagezeit und der Körpergrösse," *J. Ornith.,* **111,** 38–47 (1970).

Bartholomew, G. A. and T. M. Casey. "Oxygen consumption of moths during rest, pre-flight warm-up, and flight in relation to body size and wing morphology," *J. Exp. Biol.,* **76,** 11–25 (1978).

Benedict, F. G. "Vital energetics," *Carnegie Institution of Washington Publications,* **No. 503,** Washington, D.C., 1938.

Bennett, A. F., and W. R. Dawson. "Metabolism." In C. Gans and W. R. Dawson (eds.), *Biology of the Reptilia,* vol. 5, *Physiology,* Academic Press, New York, 1976.

Bennett, A. F. and K. A. Nagy. "Energy expenditure in free-ranging lizards," *Ecology,* **58,** 697–700 (1977).

Berger, M., and J. S. Hart. "Physiology and energetics of flight." In D. S. Farner and J. R. King (eds.). *Avian Biology,* Chap. 5, pp. 415–477, Academic Press, New York, 1974.

Brett, J. R., and N. R. Glass. "Metabolic rates and critical swimming speeds of sockeye salmon (*Orcorhynchus nerka*) in relation to size and temperature," *J. Fish. Res. Board Can.,* **30,** 379–387 (1973).

Brody, S. *Bioenergetics and Growth,* Reinhold Publishing Corporation, New York, 1945.

Calder, W. A., and J. R. King. "Thermal and caloric relations of birds." In D. S. Farner and J. R. King (eds.), *Avian Biology,* Chap. 4, pp. 259–412, Academic Press, New York, 1974.

Carpenter, R. E. "Flight metabolism in flying foxes." In T. W. T. Wu et al. (eds.), *Swimming and Flying in Nature,* vol. 2, pp. 883–890, Plenum Publ. Corp., New York, 1975.

Dawson, T. J., and A. J. Hulbert. "Standard metabolism, body temperature, and surface areas of Australian marsupials," *Am. J. Physiol.,* **218,** 1233–1238 (1970).

Dawson, W. R. "Avian physiology," *Ann. Rev. Physiol.,* **37,** 441–465 (1975).

Feder, M. E. "Lunglessness, body size, and metabolic rate in salamanders," *Physiol. Zool.,* **49,** 398–406 (1976).

Fry, F. E. J. "Effects of the environment on animal activity," *Publications of the Ontario Fisheries Research Laboratory,* **No. 68,** 1947.

Gessamen, J. A. "An evaluation of heart rate as an indirect measure of daily energy metabolism of the American kestrel," *Comp. Biochem. Physiol.,* **65A,** 273–289 (1980).

Hazel, J. R., and C. L. Prosser. "Molecular mechanisms of temperature compensation in poikilotherms," *Physiol. Rev.,* **54,** 620–667 (1974).

Hemmingsen, A. M. "Energy metabolism as related to body size and respiratory surfaces and its evolution," *Rep. Steno Mem. Hosp. Nord. Insulin Lab.,* **9,** 7–110 (1960).

Hochachka, P. W., and G. N. Somero. *Strategies of Biochemical Adaptation,* W. B. Saunders Co., Philadelphia, 1973.

Hutchison, V. H., W. G. Whitford, and M. Kohl. "Relation of body size to surface area and gas exchange in anurans," *Physiol. Zool.,* **41,** 68–85 (1968).

Kayser, Ch., and A. Heusner. "Etude comparative du metabolisme energetique dans la serie animale," *J. Physiol. (Paris),* **56,** 489–524 (1964).

Kendeigh, S. C. "Energy requirements for existence in relation to size of bird," *Condor,* **72,** 60–65 (1970).

Jolicoeur, P., and A. A. Heusner. "The allometry equation in the analysis of the standard oxygen consumption and body weight of the white rat," *Biometrica,* **27,** 841–855 (1971).

Kleiber, M. *The Fire of Life,* John Wiley & Sons, New York, 1961.

Lasiewski, R. C., and W. R. Dawson. "A reexamination of the relation between standard metabolic rate and body weight in birds," *Condor,* **69,** 13–23 (1967).

Lefebvre, E. A. "The Use of D_2O^{18} for measuring energy metabolism in *Columba livia* at rest and in flight," *Auk,* **81,** 403–416 (1964).

McMahon, T. "Size and shape in biology," *Science,* **179,** 1201–1204 (1973).

MacMillen, R. E., and J. E. Nelson. "Bioenergetics and body size in dasyurid marsupials," *Am. J. Physiol.,* **217,** 1246–1251 (1969).

Mechtly, E. A. *The International System of Units: Physical Constants and Conversion Factors* (2nd rev.), **NASA-7012.** National Aeronautics and Space Administration, Washington, D.C. (1973).

Morrison, P. R., M. Rosenmann, and J. A. Estes. "Metabolism and thermoregulation in the sea otter," *Physiol. Zool.,* **47,** 218–229 (1974).

Nagy, K. A. "CO_2 production in animals: analysis of potential errors in the doubly labeled water method," *Am. J. Physiol.,* **238,** R466–R473 (1980).

Paynter, R. A., Jr. (ed.). *Avian Energetics,* Publication Nuttall Ornithology Club, Cambridge, Mass., 1974.

Pennycuick, C. J. *Animal Flight,* Edward Arnold Ltd., London, 1972a.

———. "Soaring behavior and performance of some East African birds observed from a motor-glider," *Ibis,* **114,** 178–218 (1972b).

Rayner, J. M. V. "A new approach to animal flight mechanics," *J. Exp. Biol.,* **80,** 17–54 (1979).

Somero, G. N. "Molecular mechanisms of temperature compensation in aquatic poikilotherms." In F. E. South et al. (eds.), *Hibernation and Hypothermia, Perspectives and Challenges,* pp. 55–80, Elsevier Press, New York, 1972.

Stahl, W. R. "Scaling of respiratory variables in mammals." *J. Appl. Physiol.,* **22,** 453–460 (1967).

Thompson, S. D., R. E. MacMillen, E. M. Burke, and C. R. Taylor. "The energetic cost of bipedal hopping in small mammals," *Nature,* **287,** 223–224 (1980).

Tucker, V. A. "Bird metabolism during flight: evaluation of a theory," *J. Exp. Biol.,* **58,** 689–709 (1973).

Webb, G., and H. Heatwole. "Patterns of heat distribution within the bodies of some Australian pythons," *Copeia,* **No. 2,** 209–220 (1971).

Webb, P. W. *Hydrodynamics and Energetics of Fish Propulsion,* **Bull. 180,** pp. 1–158, Fisheries Research Board of Canada, 1975.

Weis-Fogh, T. "Quick estimates of flight fitness in hovering animals, including novel mechanisms for lift production," *J. Exp. Biol.,* **59,** 169–230 (1973).

———. "Energetics of hovering flight in hummingbirds and Drosophila," *J. Exp. Biol.,* **56,** 79–104 (1972).

Malcolm S. Gordon

MOVEMENT: THE PHYSIOLOGY OF MUSCLE

4.1 INTRODUCTION

Physical movements of many kinds use up the greatest part of the energy most organisms obtain from the food they eat. The spectrum of energy-requiring movements extends from the molecular level (for example, active transport of inorganic ions and organic molecules across membranes) to grand-scale phenomena such as migrations of whole populations. Considering the diversity of the phenomena, one might expect a similar diversity in the underlying mechanisms producing them. There is indeed diversity at the mechanistic level but apparently considerably less than at the level of the movements themselves.

The last statement must be qualified by specifying the organizational level of the mechanism we are discussing. Subcellular and cellular movements such as active transport, cytoplasmic streaming, movements of cellular organelles, or locomotion of unicellular microorganisms are clearly outside our frame of reference. Movements resulting from

ciliary and flagellar activity will also be omitted. These exclusions narrow the field almost completely to movements resulting from some type of muscular activity.

Even muscle physiology, however, must be approached at different levels. At the whole-animal, organ-system, and organ levels, it is basically functional morphology, a broad and important area of overlap between physiology and anatomy. The functional aspects of this field concern themselves primarily with the physical, engineering kinds of analyses concerning such matters as patterns of whole animal locomotion, the mechanics of how individual muscles and groups of muscles interact to produce limb and body movements, and so on. These are interesting and valuable subjects, and, with regard to structural level and direct ecological significance at least, they fit in logically with most of the rest of this book. However, there is not enough space available to permit adequate discussion of these areas in addition to the discussion that must be included of the subcellular-to-tissue levels, which tell us the most about how muscles produce movement. Several books and papers cited in the Additional Reading list provide introductions to important aspects of functional morphology.

It is at the subcellular-to-tissue levels that the greatest uniformities in muscle physiology appear. The diversity that exists, at least in the forms studied so far, appears to represent variations on only a few major structural and biochemical themes.

4.2 HISTORY

The history of the development of our understanding of how muscles work includes a series of almost classical illustrations of the generalization that scientific progress, especially in experimental fields, is fundamentally tied to technology. Two examples will illustrate the point.

The process that triggers muscle contraction was a subject for long and diverse speculations up until the end of the eighteenth century. It was not until that time, however, that enough had been established about the nature of electricity to permit scientists to have in their laboratories machines that let them produce sparks at will. The Italian scientist Galvani had such machines in 1792 and frequently used them. Galvani, among other things, was also a professor of anatomy and often had frogs dissected in his laboratory. One day Galvani had a freshly dissected pair of frog legs lying on a table near one of the spark machines. An associate apparently touched a nerve on one of the legs with the tip of his metal scalpel at the same time another associate was operating the machine. The muscles of the legs immediately contracted violently. Galvani and his associates repeated the process many times, with the same result. Galvani's interpretation of this was that animal cells contain the same kind of electricity as produced by his machine and that perturbation of this electricity triggered activity in the muscle. We know now that this is essentially what does happen in the initial stages of muscle activation. Thus a technological advance in the form of the spark machine combined with chance, in the presence of a good observer, to open up the whole field of muscle excitation and, eventually, electrophysiology.

A second fundamental issue that produced discussion and debate for many years before it was resolved was the structural basis for the pattern of cross banding so conspicuous in skeletal muscle. The resolution of this problem depended heavily on two technological developments. These were the perfection of the electron microscope, with its great resolving power [down to the range of 5 angstroms (Å) or less], and the development of techniques for preparing very thin sections of biological materials embedded in plastics or other hard matrices. The electron microscope was necessary because the dimensions of many of the essential structures are on the order of only 40 Å. Thin-sectioning procedures were essential because the geometric arrangements of the various parts are such that there is superposition of closely underlying layers of structure in sections thicker than about 150–250 Å. This superposition makes unambiguous interpretation of observations virtually impossible. Both these developments reached the necessary levels of perfection during the middle 1950s.

4.3 METHODS AND DEFINITIONS

The thin-sectioning, electron microscope, x-ray diffraction, and related techniques needed for ultrastructural studies of muscle or other tissues cannot be done justice in a brief space. We suggest that students wishing to learn more about these matters confer with appropriate faculty members at their schools.

There is also great variety and complexity in studies of the physical and chemical properties of muscles. However, there are in these areas a few basic procedures and terms that are so widely used that it will be worthwhile to review them.

Muscle can be studied in either the resting, unstimulated state, or in the active, stimulated state. One of the major physical parameters of resting muscle (at least for those types of muscle in which it is possible to define the quantity) is its rest length. For skeletal muscle this is its length when it is fully extended but not under any tension. Rest length is the reference point for many types of experiments studying the dependence of various properties on muscle length.

Active muscle is most often studied under one

of two sets of circumstances. The first, called *isotonic recording,* has the muscle attached firmly at one end, with the other end free to move as different loads are applied and other experimental conditions are varied. Isotonic experiments monitor the changes that occur in muscle length.

The second arrangement is called *isometric recording* and measures force development. Here the muscle is firmly attached at both ends, so that only very small changes in length can occur (changes in length are usually only fractions of millimeters). The attachment at one end of the muscle is to some type of device that can monitor changes in tension. The starting point is often, but not always, rest length. Figure 4-1 shows an experimental setup usable for both isotonic and isometric experiments.

The procedures used for stimulating muscles also vary. Electric stimuli are most widely used,

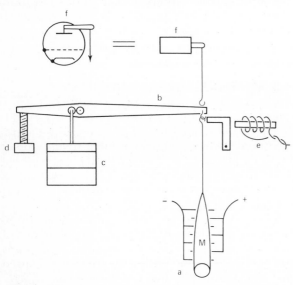

Figure 4-1
Generalized experimental setup for the study of muscle contraction. Code: M, muscle; a, multiple stimulating electrodes; b, light level; c, load (placed near fulcrum to reduce inertia), for use in isotonic experiments; d, mechanical stop, permitting adjustment of initial length of muscle; e, electromagnetic stop, permitting controlled release of muscle at preset intervals following stimulation: f, electronic strain gauge, for use in isometric experiments. [*From D. R. Wilkie.* Brit. Med. Bull., **12,** *177–182 (1956).*]

both because they are most nearly physiological and because they can be so readily varied and controlled. There are two basic routes for application of stimuli, direct and indirect. *Direct stimulation* means that electrodes are applied directly to the muscle fibers. For many purposes arrays of multiple electrodes are used to ensure that all parts of the muscle are stimulated simultaneously. *Indirect stimulation* means stimulation via the nerves innervating the muscle. The Instrumental Methods and Electronics sections of the Additional Reading list for Chapter 1 include several references giving many technical details of the electronic and other physical methods used to stimulate and record from muscles.

The chemical study of muscle constitutes a complex and specialized field of biochemistry. Many of the methods used in this study are discussed in the references in the Additional Reading list for this chapter. We shall mention here only two of the more important of the specialized muscle procedures.

The full cycle of chemical changes associated with a single contraction cycle in many mammalian skeletal muscles operating at normal body temperatures (near 37°C) is over in less than 100 milliseconds (msec). Even so-called slow muscles in other kinds of animals, when operated at temperatures near 0°C, generally complete their contraction cycles within maximum times of no more than a few seconds. Conventional biochemical procedures are far too slow to follow the details of these events.

Increasing use has been made in recent years of optical methods for following what occurs. Many of the important compounds active in muscle metabolism have characteristic absorption spectra or, in some cases, fluorescence spectra. By observing appropriately thin sheets of muscle during contraction cycles at the wavelengths of light of peak absorption or (under proper illumination) fluorescence, it is possible to follow changes in the concentrations of the compounds of greatest concern. High-speed recording and computing methods are usually used in association with the light detectors.

More conventional biochemical isolation and analysis procedures can also be used if it is possible to stop the contraction of a muscle almost instantaneously at a given stage in the cycle and to preserve it in that state. The technical problems are considerable, and it is often not possible to produce the stoppage as fast as is desirable. However, important contributions to understanding of muscle chemistry have come from procedures in which fairly thin muscles are stimulated and then, after preset delays, plunged rapidly into isopentane at liquid-nitrogen temperature. Desired analyses are then carried out on the frozen muscle, usually after breaking it up into fine powder.

4.4 BASIC STRUCTURES AND PROCESSES

The structural basis for muscle contraction in essentially all types of muscles so far studied is the movement of intracellular filamentous protein structures of some kind. These filaments are widely variable in numbers, lengths, diameters, and orientations. Both light and electron microscopic observations of many different muscles in species belonging to most of the animal phyla have shown that there are three broad categories of fiber structures which result from the very varied arrangements of filaments. These are (1) striated muscles, (2) intermediate muscles, and (3) smooth muscles.

Striated muscles contain contractile machinery composed of two kinds of protein filaments (see Section 4.5.2) aligned in such a way that the fibers are regularly cross banded along their length. The bands are alternating optically dense A bands and less dense I bands. The letters assigned to the bands describe their transmission properties for polarized light. The letter *A* means *anisotropic* and refers to bands that transmit preferentially light polarized in only certain planes. The letter *I* means *isotropic* and refers to bands that transmit light polarized in all planes equally well. Striated muscles are found in virtually all kinds of animals, from coelenterates to vertebrates.

Intermediate muscles generally also contain two kinds of filaments in their contractile machinery. The arrangements of these filaments are much more variable and are often much less regular than the filament arrangements in striated muscles. Some intermediate muscles are striated along only one side of a fiber (the noncontractile parts of the cell contents being concentrated along the other side), others are obliquely striated, and still others contain helically arranged bundles of filaments that produce a diamond-lattice appearance when viewed under the light microscope. Intermediate muscles are known from many invertebrate phyla but, so far, have not been described from the vertebrates.

Smooth muscles show no cross striations and may contain one or two kinds of filaments. Whatever types of filaments are present, they are irregularly, often virtually randomly arranged. Smooth muscles have been found in all types of multicellular animals investigated.

Table 4-1 provides additional, more detailed structural information about some of the more important subcategories of each of these three major categories of muscle types. The table concentrates on three of the best known phyla in these contexts: molluscs, arthropods, and vertebrates.

The remainder of this chapter will be devoted primarily to the muscles of vertebrates. Some limited information on invertebrate muscles is scattered throughout and Section 4.9 is devoted specifically to invertebrate muscles.

Among the vertebrates the majority of muscle fibers go through a characteristic sequence of events in a contraction cycle. The resting, relaxed fiber maintains an electric potential (resting potential) across its external membrane that is very similar in magnitude and properties to membrane potentials in nerve cells (see Chapter 9 for details). Following stimulation (normally via action potentials carried on associated nerves), graded depolarization of the areas of fiber membrane adjacent to the nerve endings begins. In most cases there is a critical degree of depolarization (threshold) that triggers generation of a self-propagating action potential in the muscle membrane (see Chapter 9 for a discussion of neuromuscular transmission and action potentials). This action potential sweeps over the fiber surface inaugurating a series of processes known collectively as *excitation-contraction coupling.* This coupling stimulates the actual shortening of the contractile machinery in the fiber. Once the fiber has contracted, a new series of internal changes occurs that produces relaxation and restoration of the fiber to its resting condition. This

TABLE 4-1. Some Major Structural Characteristics of Important Subtypes of Muscle Fibers*

Fiber Types and Subtypes	Fiber Structure
Striated	
Vertebrate skeletal fast (Type II fibers)	Numerous regularly arranged intracellular fibrils. Sarcomeres uniform length, short. T and SR tubules abundant. Innervation usually single point, by one nerve.
Vertebrate skeletal slow (Type I fibers)	Fewer, less regularly arranged fibrils. Sarcomeres uniform length, sometimes longer. T and SR tubules less abundant. Innervation usually multiple points, by one nerve (multiterminal).
Vertebrate cardiac	Fibers often branched, anastomosing with adjacent fibers. Sarcomeres uniform length, short. T and SR tubules variably abundant. No direct innervation (only via other muscle cells).
Crustacean phasic (fast)	Sarcomeres somewhat variable length, primarily short. T and SR tubules more abundant. Innervation variable, often single point by one nerve in fastest fibers.
Crustacean tonic (slow)	Sarcomeres somewhat variable length, primarily long. T and SR tubules less abundant. Innervation variable, usually multiple point by more than one nerve (polyneuronal).
Insect resonant flight (fibrillar)	Extremely regularly arranged fibrils. Sarcomeres uniform length, short. Abundant T, but few SR tubules. Innervation multiterminal by one nerve.
Intermediate	
Diagonally striated (nematodes, molluscs, annelids)	Fibrils regularly arranged, in staggered arrays. Sarcomeres variable length, sometimes long. T and SR tubules variably abundant. Innervation variable, often both multiterminal, polyneuronal and via adjacent muscle cells
Smooth	
Vertebrate classical	Short, often spindle-shaped fibers. No sarcomeres. No T, few SR tubules. Innervation variable, often indirect (via other muscle fibers). If direct, multiterminal and polyneuronal.
Molluscan catch	Fibers containing large diameter fibrils of special chemical composition. No sarcomeres. T and SR tubules rare or absent. Innervation multiterminal and polyneuronal.

* For definitions of technical terms see text.

sequence of events will provide the framework for the discussion that follows.

4.5 STRIATED MUSCLE: SKELETAL MUSCLE

Skeletal muscles comprise the greatest part of the body mass in most vertebrates, in many cases amounting to 80% or more of the soft tissues. They occur virtually everywhere in vertebrate bodies and are involved in the functioning of practically every organ system. They are used for a vast range of functions, from locomotion of the entire organism to minute movements of the eyes.

4.5.1 General Structure

At the level of gross morphology, skeletal muscles vary considerably. Any text on comparative anatomy will document this variation. All the varieties of movement that result from the shortening of the fibers of these muscles following nervous stimulation are due to different fiber lengths, different geometric arrangements of fibers, different groupings of fibers within connective tissue sheaths, and different placements of origins and insertions of entire muscles on skeletal or other structures. The details of the mechanical effects of shortening of the fibers are also influenced by differences in the relative amounts and physical properties of the connective tissue sheaths and tendons.

As would be expected, many of the gross morphological variations between muscles are the result of structural differences requiring light- or even electron-microscopic observations for their clarification. There are great differences between muscles in terms of number and diameters of fibers. Muscles in positions requiring the generation of great tension usually have more fibers per unit cross-sectional area than muscles in places lacking such stringent demands. This kind of variation provides more contractile machinery in the places where it is needed most. Reasons for variations in fiber diameter are more complex, relating not only to mechanical demands for strength but also to such things as speed of shortening (see Chapter 9 for discussion of the relationship between conducting-cell diameter and action-potential conduction velocity). Most vertebrate skeletal muscle fibers have diameters between 10 and 100 μm. In some invertebrate muscles, notably some muscles of the barnacle *Balanus nubilus* from the north-eastern Pacific Ocean, fiber diameters are as large as 0.5–3.0 mm.

There are also important differences between muscles in terms of overall appearance. The colors of muscles vary a great deal, both within individual animals and between different species. In some cases, such as the occurrence of red, orange, or white muscle in trout and salmon, the differences seem to be related primarily to diet and not to anything having to do with contraction. In most vertebrates, however, variations in shades from whitish through pink to dark red are related to the relative abundances of two histologically and histochemically fairly distinct fiber types. "Type I" fibers are relatively uncommon in most skeletal muscles; they are usually red in color and contain many mitochondria (sarcosomes) and some quan-

tity of the respiratory pigment myoglobin (see Chapter 6). "Type II" fibers are abundant in most skeletal muscles; they are usually whitish in color and contain few sarcosomes and little myoglobin. The contractile material is arranged differently in the two fiber types; it occurs in large, ribbonlike areas in Type I fibers, whereas Type II fibers characteristically contain large numbers of well-defined fibrils. The fiber types differ also in patterns of innervation (see Section 4.5.3).

Not surprisingly, the structural differences between the fiber types have a range of biochemical and biophysical consequences. Type I fibers obtain most of their metabolic energy from aerobic processes, Type II fibers (especially during periods of contractile activity) primarily from anaerobic processes. Type I fibers generally contract fairly slowly, hence are called slow fibers, whereas Type II fibers usually contract fairly rapidly. Unless stated otherwise, the discussion of skeletal muscle physiology in this chapter describes Type II, fast fibers. Figure 4-2 shows many of the important features of fast skeletal muscle structure as seen at moderate magnifications in the electron microscope.

Slow fibers have been found in a variety of muscles in all five of the major vertebrate groups. They are usually mixed in with ordinary fibers.

Muscle color by itself is an unreliable indicator of muscle function. Exceptions have been described to the distinctions between fiber types just discussed. In these exceptions the features of the two types are somewhat mixed. The clearly differentiated and separated red and white muscle fibers found in the trunk musculature of many fishes belong in the exceptional category. In the silver carp, both types of fibers have myofibrillar structures similar to those of fast fibers in other vertebrates. Their contractile properties most closely resemble those of the opposite fiber type, though both types are fairly unique in these properties.

There also are a number of special situations in which red muscle color results from stringent metabolic demands produced by special environmental conditions. An example of this is the very dark red muscle of diving birds and mammals. The color of these muscles results from high concentrations of myoglobin, which forms part of a set of respiratory adaptations for diving (see Chapter 6). The red muscle of these diving animals is not slow.

Figure 4-2
Electron micrographs of (a) a longisection (35,000×) and (b) cross section (60,000×) of skeletal-muscle fibrils from semitendinosus muscle of frog (Rana pipiens). Fresh muscle fibers fixed, in resting condition, in glutaraldehyde; post-fixed in osmium tetroxide; embedded in Vestopal W; sectioned at 100–200 Å; section stained in uranyl acetate and lead citrate. Code: A, A band; I, I band; H, H band; Z, Z band; Sr, sarcoplasmic reticulum. [Courtesy of Prof. Fritiof S. Sjöstrand.]

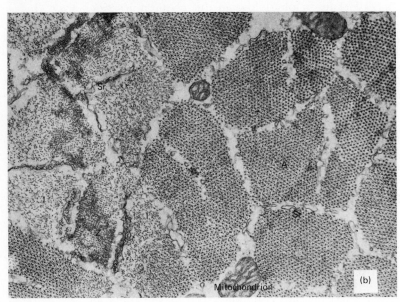

The contractile machinery in fast skeletal muscle is contained within the intracellular myofibrils. The myofibrils are grouped, usually in numbers varying from 4 to 20. These groups of fibrils have been called *muscle columns.* The individual fibrils are very close to each other (only 200–350 Å apart), the spacing varying primarily with contractile state.

The muscle columns are separated from one another by 0.2–0.5 μm. The sarcosomes are located within the sarcoplasm, frequently in close proximity to those parts of the sarcoplasmic reticulum (the endoplasmic reticulum of muscle cells) called *triads.* The triads appear to play a major role in the process of excitation-contraction coupling, as do associated transverse tubules (called the **T system**) joining the triads to the outer sarcolemmal membrane. Figure 4-3 diagrams the general appearance of the sarcoplasmic reticulum and its relationship to the myofibrils. A somewhat different, but basically similar, reticulum exists in the striated muscle fibers of many invertebrates.

The amount and complexity of development of the sarcoplasmic reticulum are generally closely correlated with the fiber diameter and speed of contraction of muscles. Thin-fibered, or slow, muscles have poorly developed reticula with few or no triads. Thick-fibered, or fast, muscles have an extensively developed reticulum with many triads.

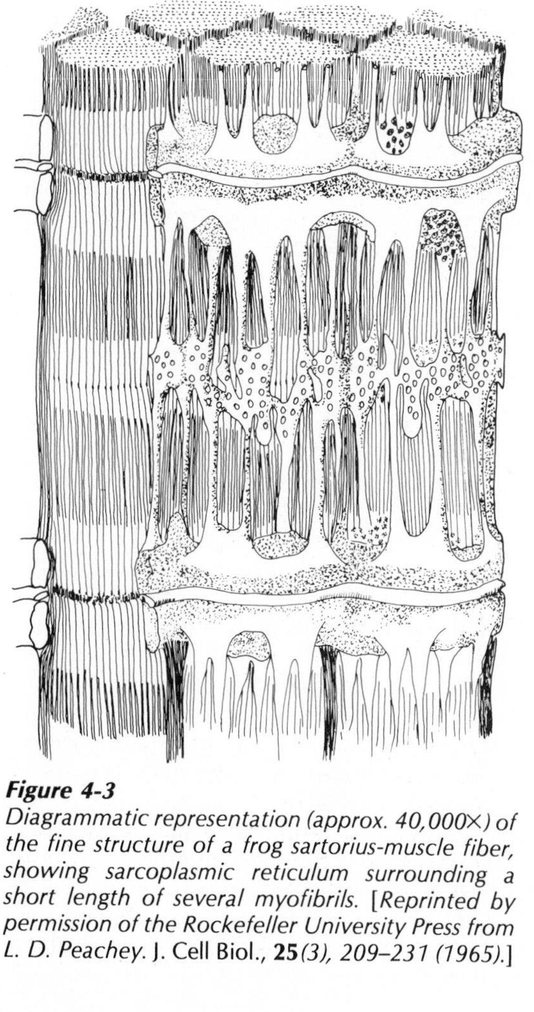

Figure 4-3
*Diagrammatic representation (approx. 40,000×) of the fine structure of a frog sartorius-muscle fiber, showing sarcoplasmic reticulum surrounding a short length of several myofibrils. [Reprinted by permission of the Rockefeller University Press from L. D. Peachey. J. Cell Biol., **25**(3), 209–231 (1965).]*

There are many exceptions to these statements. Tortoise muscles, for example, are often very slow in their mechanical responses, despite the fact that they contain a well-developed reticulum with many triads.

Skeletal-muscle fibers are syncytic, forming from the fusion of up to several hundred embryonic mesodermal myoblasts. As a result of this origin, fully developed fibers are multinucleate. The nuclei are usually distributed rather irregularly in the sarcoplasm, just beneath the outer sarcolemmal membrane. There does not seem to be a special metabolic modification of these nuclei that specifically serves the contractile function of the fibers.

4.5.2 Structure of the Contractile Machinery

The long, now historic, controversy over the structural basis for the cross banding of skeletal muscle was mentioned in Section 4.2. The resolution of the controversy came in the late 1950s when it became technically possible to make sections of muscle thin enough to avoid superposition effects, to take electron microscopic pictures of very high resolution, and to analyze properly a vast amount of x-ray diffraction data.

The *sliding-filament* model is now firmly established as the basic structure of the contractile elements of essentially all striated muscles. The cross striations are a consequence of this structure.

The cellular organelle, which is itself striated, is the myofibril. The entire fiber appears striated because the striations of the individual myofibrils are in very close register with each other. The appearance of one of the repeating units of striation, a sarcomere, as seen under the light microscope, is diagrammed in Figure 4-4. The reasons for the names of the A (anisotropic) and I (isotropic) bands were given previously (Section 4.4). The Z lines (also called Krause's membrane) in earlier times were sometimes thought to represent cell membranes (name from German *Zwischenscheibe*). The central lighter zone is called H for *helles*, meaning light in German.

Figure 4-5 illustrates the fine structure of a skeletal muscle in the relaxed state. Diagrams of the essential components of the contractile mechanism, based on analysis of large numbers of electron micrographs such as those in Figure 4-5, are included

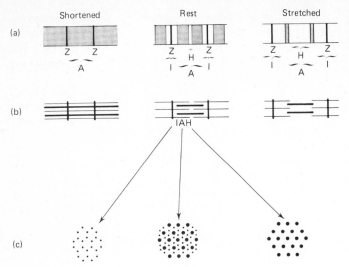

Figure 4-4
Diagrammatic representation of the cross-banding pattern of a single vertebrate striated-muscle fibril at rest, partially stretched and greatly contracted. (a) As would be seen under the light microscope. (b) A small part of the width of the fibril as would be seen under very high magnification in the electron microscope. The parts of the sarcomere are in register with the labeled bands in (a). Symbols as in Figure 4-2. Note that the I bands correspond to areas containing only thin filaments, the A bands to areas of overlap between thick and thin filaments and the H bands to areas of only thick filaments. (c) Representations of very high magnification electron microscopic views of cross sections through the I, A, and H bands, respectively. Figures 4-2b and 4-5b illustrate the actual appearance of such sections of medium and high magnifications. Note the hexogonal symmetry in fiber arrangements, especially in the A bands.

in Figure 4-4. The major features of these structures are as follows.

The cross section of a myofibril is occupied by a large number of highly ordered, parallel, longitudinal filaments. One group of these filaments arises from the equally highly ordered, closely spaced, pyramidal (under very high magnification) units making up the Z lines. These relatively thin filaments (40–80 Å in diameter) usually are from one third to one half the length of the sarcomere. In about the central half of the resting sarcomere there is a second highly ordered array of thick filaments (100–140 Å diameter) that does not attach to the Z lines but is connected, by areas of short cross bridges near the ends, to the thin filaments. Both types of filaments are remarkably uniform in length in vertebrate skeletal muscles, the thin at approximately 2.0 μm and the thick at approximately 1.6 μm.

The I bands of the sarcomere are occupied only by thin filaments, the A bands by both thin and thick filaments, and the H zone by only thick filaments. The hexagonal symmetry shown in the arrangement of both thin and thick filaments viewed in cross sections cut at certain levels in sarcomeres is due to some basic features of the molecular structures of the proteins of which they are composed (see Section 4.5.4).

The pattern just described seems to be virtually universal in striated muscles. This does not mean,

however, that there is no variation in ultrastructure of contractile machinery. There is considerable variety, mostly in dimensions, but also, in many invertebrate muscles, in numbers of thin filaments per thick filament and in the regularity of thin-filament arrangement (see Section 4.9).

4.5.3 Innervation

Vertebrate skeletal muscles are also frequently called *voluntary muscles*. This is because skeletal muscle fibers will only contract when stimulated from outside in some way. The usual method of stimulation is via nerves that are themselves under the control of the nonautomatic parts of the central nervous system. The vast majority of motor nerves terminating in skeletal muscles do so in accordance with a single basic pattern.

A single skeletal muscle will usually contain several hundreds, if not thousands, of fibers. Each of these fibers is innervated by a single nerve fiber terminating at one or two specialized locations on its surface. These spots are called the *motor end plates*. Chapter 9 discusses the structure and function of motor end plates.

The axon terminating at a motor end plate is usually only one of possibly hundreds of branches from the main axon stem of a single motor nerve cell. This being the case, production of an action potential by the motor nerve cell will result in es-

Figure 4-5

Electron micrographs of (a) longisection (37,500×) and (b) cross section (100,000×) of sliding filaments and other fine structures of skeletal-muscle fibrils from semitendinosus muscle of the frog (Rana pipiens). Tissue preparation as in Figure 4-2. Code (in addition to those used in Figure 4-2): features of fine structure of triads of Sr are T, cross section of transverse tubule; L, sections of lateral tubules (portions indicated are terminal cisternae). [Courtesy of Prof. Fritiof S. Sjöstrand.]

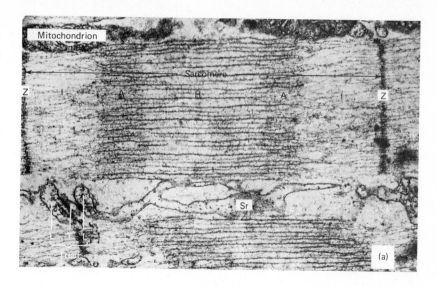

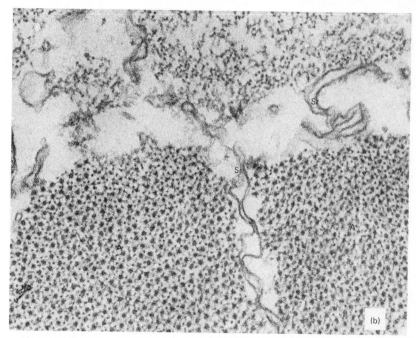

sentially simultaneous activation of all the muscle fibers innervated by the branches of that motor cell. This group of simultaneously acting muscle fibers is called a *motor unit.* Depending on the size of the muscle, there may be only a few or even a hundred or more such motor units present. The number of motor units is equal to the number of motor nerve cells innervating the muscle. The significance of this structural pattern will be discussed later (see Section 4.5.10).

The motor-unit pattern of innervation is almost exclusively a vertebrate arrangement. Most inver-

TABLE 4-2. Correlations Between Types of Sarcomeres and Some Important Physiological Properties in Crustacean Striated Muscles

Type of Sarcomere	Rate of Isometric Tension Development [kg/(cm² sec)]	Maximum Isometric Tension (kg/cm²)	Probability of Action Potential or Large Graded Potential
Short (4 μm)	Fast (~40)	Small (~0.8)	High
Intermediate (4–8 μm)	Intermediate (~8)	Intermediate (~4.5)	Intermediate
Long (8 μm)	Slow (~1)	Large (~5)	Low

Source: H. L. Atwood. *Am. Zool.,* **13,** 357–378 (1973).

tebrates differ in several ways. Invertebrate striated muscles usually lack well-defined motor end plates, the nerve endings terminating on normal-appearing sarcolemmal membranes. There are also usually multiple nerve endings per fiber, rarely only one.

Crustaceans are probably the best studied group of invertebrates in these respects. As indicated in Table 4-1, crustacean striated muscle fibers may be placed in two broad groups, phasic or fast fibers, and tonic or slow fibers. The distinction is, however, an artificial one; actually there is a continuum of fiber structures and properties ranging from fibers having short sarcomeres, single motor end plates, and properties generally similar to those of vertebrate fast fibers, to fibers having long sarcomeres, innervated by multiple endings (multiterminal innervation) from up to five different motor nerves (polyneuronal innervation), and having variably slow, graded, and local contractions. Table 4-2 and Figure 4-6 summarize major features of this variability.

Patterns of innervation similar to some of those found in invertebrates occur in a number of places in vertebrate skeletal muscle. The most widespread are the various types of muscle *proprioceptors* (internal sense organs relaying information to the central nervous system about the state of contraction of the muscles in which they are located). The structure and function of these organs are discussed in Chapter 10. In addition to muscle spindles and other proprioceptors, many slow muscle fibers and some fast fibers in fish muscles lack motor end plates and have multiple nerve endings on their surfaces.

One of the most intriguing developments in muscle physiology has been the unequivocal demonstration that, in several vertebrate groups, the identity of the nerve innervating a muscle plays a major role in determining whether the fibers of the muscle are fast or slow. The nature of this neuronal regulation is still in dispute and may involve both the pattern of electrical activity in the nerve and some kinds of "trophic" substances produced by the nerve. Whatever the mechanism may be, it is possible to cross innervations between previously fast and slow muscles, which results in virtually complete conversion of the fast muscle into a slow one and significant speeding up of the slow muscle (see Chapter 11).

A recent surprising finding is that it is also possible to experimentally modify the contractile properties of some crustacean skeletal muscle fibers, primarily by controlling the amounts of use the muscles are given during larval development. Crustaceans are arthropods, and arthropods are protostome animals in which essentially final genetic determination of the developmental fates of virtually all cells takes place quite early in embryonic (prelarval) development. However, Atlantic lobsters (*Homarus*), given no opportunity to manipulate objects with their claws during larval development, develop two similar, slender, "cutter" claws; the normal heavy "crusher" claw on one side does not develop under these conditions. The closer muscles of crusher claws are composed entirely of slow fibers; cutter claw closer muscles contain 65–75% fast fibers and only 25–35% slow fibers. Thus, levels of muscle use during larval stages can change patterns of differentiation of closer muscle fibers,

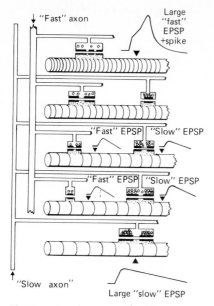

Figure 4-6
*Diagram illustrating the matching between types of innervation and types of muscle fiber in a crustacean (crab) muscle. Phasic ("fast") nerves form the primary innervation of fibers having short sarcomeres and formed motor end plates. Phasic fibers also show fast excitatory postsynaptic potentials and action potentials following stimulation. Tonic ("slow") nerves form the primary innervation of fibers having long sarcomeres. Tonic fibers also usually lack formed motor end plates, and show slow graded, nonpropagated excitatory post-synaptic potentials following stimulation. Intermediate fibers have varying amounts of innervation of both types. (EPSP, excitatory post-synaptic potential) [Modified from H. L. Atwood. Am. Zool., **13**, 357–378 (1973).]*

patterns which supposedly were fully genetically fixed at much earlier stages.

4.5.4 Chemical Composition

The chemical environment within skeletal muscle cells is, in broad terms, similar to that within most other cells in an organism. By this we mean that the average osmotic concentration of the sarcoplasm is equal to the osmotic concentration of the extracellular fluids, that there is a high concentration of intracellular potassium, and that there is a low concentration of intracellular sodium and chloride. The concentrations of assorted low molecular weight organic compounds, such as sugars and amino acids, are also not unusual. More detailed discussion of these and related aspects of intracellular ionic and osmotic concentrations in tissues of various animals may be found in Chapter 7. Resting potentials and various membrane phenomena related to internal chemical composition are discussed in Chapter 9.

The internal chemical environment in skeletal muscle cells is, however, unusual in several specific ways. Two of the most important of these are the pattern of distribution and movements of inorganic calcium ions (Ca^{2+}) and the presence of significant concentrations of an organic phosphagen—universally creatine phosphate (CP) in the vertebrates (see Figure 4-7). Details of the functional significance of these two substances are discussed in Sections 4.5.7 to 4.5.9.

The contractile machinery contained in the myofibrils is composed primarily of two proteins: actin in the thin filaments and myosin in the thick filaments. Additional proteins are also present: α-actinin, M-protein, troponin, and tropomyosin. α-Actinin is found in the Z bands, where it may be concerned with the steric arrangement of the thin filaments. M-protein performs the equivalent function for the thick filaments at the M line (the M line is a series of cross bridges between thin and thick filaments near the center of the sarcomere; it can be seen, though it is not identified, in both Figures 4-2a and 4-3). Troponin is found along the full length of the thin filaments; it is the sole calcium-receptive protein of the contractile system. Tropomyosin also occurs along the full length of the thin filaments; it acts morphologically to provide a binding site for troponin and functionally as the mediator of the signal for contraction which is transmitted from troponin to actin (see Section 4.5.8).

The isolation, purification, identification, and cytological localization of all of these molecules constitutes a distinguished and significant chapter in the history of biochemistry (see Additional Reading

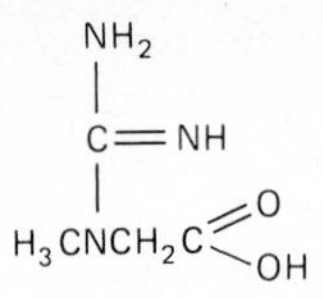

NH_2

$C = NH$

$H_3CNCH_2C\begin{smallmatrix}O\\OH\end{smallmatrix}$

Creatine (C)

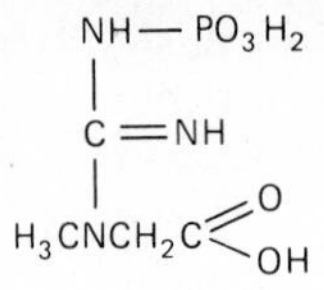

$NH — PO_3H_2$

$C = NH$

$H_3CNCH_2C\begin{smallmatrix}O\\OH\end{smallmatrix}$

Creatine phosphate (CP)

$NH — PO_3H_2$

$C = NH$

NH

$(CH_2)_3$

$H_2NCHC\begin{smallmatrix}O\\OH\end{smallmatrix}$

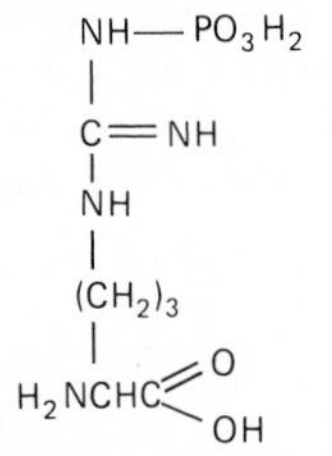

Arginine phosphate (AP)

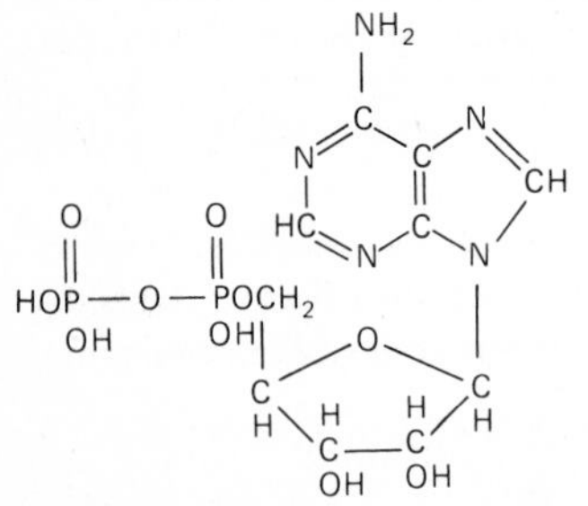

Adenosine diphosphate (ADP)

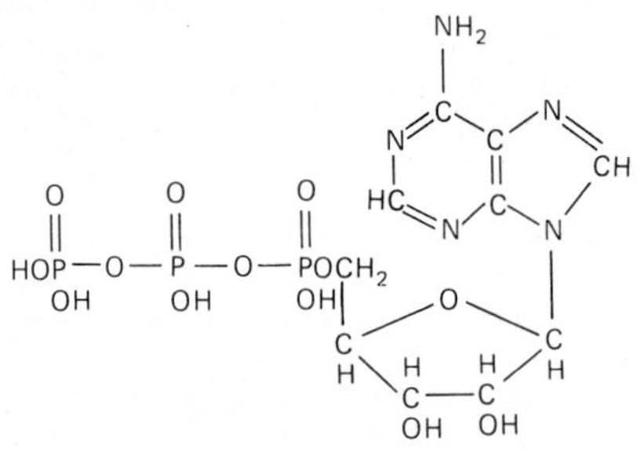

Adenosine triphosphate (ATP)

Figure 4-7

Structural formulas of five of the most important low molecular weight compounds involved in supplying energy for contraction of muscle. Creatine phosphate is the universal phosphagen in vertebrate muscles and also occurs in the muscles of several groups of invertebrates. Arginine phosphate is a widely distributed invertebrate phosphagen.

list). Additional work on their properties continues today.

Actin and myosin have both turned out to be extremely complex molecules. Undenatured actin isolated from skeletal muscle occurs in one of two forms: a globular (G) form and a filament (F) form. F-actin appears to be a polymer of large numbers of G-actin subunits.

Pure preparations of G-actin have a molecular weight of 43,000. The molecule consists of a single polypeptide chain containing 374 amino acid residues and having a higher order structure such that the overall shape of the molecule is ovoid, with an axial ratio of 3 or less. The amino acid composition has been completely determined, as has the amino acid sequence. Each molecule binds to itself one molecule of adenosine triphosphate (ATP; see Figure 4-7). The structure and properties of actin are very uniform in many different kinds of animals.

The formation of F-actin from G-actin is a complex process. The polymer is a linear double-stranded helix of G-actin monomers. There is no characteristic molecular weight, but fairly large, long-filament polymers predominate in solutions. These polymers will themselves aggregate, producing larger filaments composed of two helically wound strands made up of identical, spherical subunits (under the electron microscope). These larger filaments are the backbone of the thin filaments of myofibrils.

F-actin binds to itself adenosine diphosphate (ADP; see Figure 4-7). In the presence of free magnesium ion it has the property of activating the ATP-splitting (ATPase) enzymatic activity of myosin (see later in this section).

Myosin is a much larger and more complex molecule than actin. Myosins from comparable muscle types in different species, and from different muscle types in single species, differ from each other in a number of small but functionally significant ways. There is, however, a central molecular structure that is quite uniform in all myosins studied so far (Figure 4-8).

The rod portion of the molecule is about 1450 Å long and 20 Å in diameter, has an almost completely α-helical secondary structure, and has a molecular weight near 110,000. The LMM (light meromyosin) portion of the rod is about 900 Å long and has a molecular weight of about 72,000. Each S-1 fragment of HMM (heavy meromyosin) has a molecular weight of about 115,000, appears to be shaped somewhat like a blunt-ended banana, is 100–150 Å long and 30–40 Å in diameter.

The light chain fractions of HMM form one of the most variable parts of the structure. Myosins from different sources have different numbers of light chains, with different molecular weights and electrical charges. Vertebrate skeletal muscle myosin has four light chains associated with it, two with molecular weights of about 18,000, one with a molecular weight of about 21,000, and one with a molecular weight of about 16,000.

Adding the figures just given, the total molecular weight of vertebrate skeletal muscle myosin is near 413,000 (other myosins have weights to 460,000).

Myosin molecules, like G-actin molecules, will spontaneously and reversibly aggregate when in solution under conditions similar to those existing within muscle cells. As seen under the electron microscope, this aggregation begins with formation of dimers bound together at the ends of the rod-shaped "tails." The dimers themselves aggregate in orderly fashion to form filaments very similar in appearance and dimension to the thick filaments of myofibrils. The molecules arrange themselves so that their "heads" are always directed away from the midpoints of the filaments, as is also found to be the case in the A bands of sarcomeres. The heads correspond to the cross bridges observed between thick and thin filaments in sarcomeres (Section 4.5.2). Pure myosin filaments lack at least the M-protein found in "native" thick filaments.

As mentioned earlier, myosin is not only a basic part of the structure of the contractile machinery but also acts as a powerful ATPase. This enzymatic activity is activated by the presence of actin and magnesium ion. The characteristics of this catalysis form another lively field of biochemical study. The activity is located in the HMM part of the molecule. Two of the light chain peptides (18,000 molecular weight) are needed for this activity in vertebrate skeletal muscle. It constitutes an important part of the energy-supply mechanism for contraction (see Section 4.5.8). Myosins from different sources have different levels of ATPase activity.

The basis for contraction is an interaction between F-actin and myosin that will be discussed further in Section 4.5.6. Chemically this interaction appears to involve one G-actin per myosin molecule.

Considerable attention has also been paid to the chemical characterization of the four other proteins involved in the contractile machinery. Compared with actin and myosin, all are present in muscle in only small amounts (Table 4-3). Two (α-actinin and M-protein) are present in very small amounts. As a result, these molecules have been more difficult to study than actin or myosin.

α-Actinin is a complex protein with an apparent monomeric molecular weight of about 160,000. The monomers of vertebrate α-actinin aggregate to varying extents depending upon the ionic strengths of the solutions surrounding them. Aggregation occurs most strongly at ionic strengths close to those found in living vertebrate cells. The aggregation process acts to produce cross connections between thin filaments in the sections of these

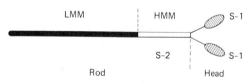

Figure 4-8
Schematic diagram of central molecular structure of a single myosin molecule. The structure can be fragmented by proteolytic enzyme treatment into two parts: light meromyosin (LMM), made up of about 60% of the rod portion, and heavy meromyosin (HMM), made up of the rest of the rod and the head. Proteolytic enzyme treatment of HMM further fractionates it into four parts: two S-1 portions of the head, several relatively low molecular weight peptide chains (light chains), and the S-2 portion of the rod. [Modified from J. M. Squire. Ann. Rev. Biophys. Bioeng., **4,** *137– 163 (1975).]*

TABLE 4-3. Abundance and Selected Chemical Properties of Major Proteins in the Contractile Machinery of Vertebrate Skeletal Muscle*

Protein	Estimated Weight in Myofibrils,† (%)	Molecular Weight	Estimated Bound Ca in Myofibrils,‡ (%)
Myosin	60	413,000	6
F-actin	20	variable	2
Tropomyosin	4	70,000	0
Troponin	3	78,000	87
α-Actinin	1.5	160,000	0

* Rabbit white muscle

† Other proteins, both identified (e.g. M-protein) and unknown, may amount to 10–15% of total myofibrillar protein by weight.

‡ Assumes that the amount of Ca bound to these five proteins is 95% of total exchangeable Ca.

Source: S. Ebashi, M. Endo, and I. Ohtsuki, *Quart. Rev. Biophys.,* **2,** 351–384 (1969).

filaments in and adjacent to the Z-lines. The chemical properties of α-actinin are otherwise poorly known.

M-protein is even less well defined. It has a molecular weight of about 155,000.

Tropomyosin is a double-stranded, α-helical, coiled-coil, rod-shaped molecule composed of two almost identical polypeptide chains. Each chain has a weight of 35,000, producing a molecular weight close to 70,000. The molecular rod is about 410 Å long and 20 Å in diameter. Each thin filament in vertebrate skeletal muscle includes two helical filaments of tropomyosin, each filament composed of molecules bonded end to end. Each tropomyosin molecule contacts seven G-actins. The filaments are wound into the grooves of the actin cores of the thin filaments (see Figure 4-9).

Troponin is a complex molecule composed of three globular subunits. These are, respectively, troponin-T (TN-T, molecular weight 38,000), troponin-C (TN-C, molecular weight 18,000), and troponin-I (TN-I, molecular weight 22,000).

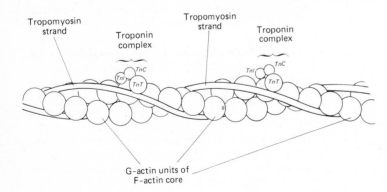

Tropomyosin strand · Troponin complex · Tropomyosin strand · Troponin complex

TnI · TnC · TnT · TnI · TnC · TnT

G-actin units of F-actin core

Figure 4-9

Diagram of major features of the molecular structure of a small portion of a thin filament in vertebrate striated muscle, showing the spatial relationships of the three types of proteins involved as they are in the absence of free Ca^{2+}. The central actin core is composed of a double helical array of G-actin molecules all oriented in the same direction. The rod-shaped tropomyosin molecules are bonded end to end in continuous strands that lie along the sides of the grooves in the actin core. Each tropomyosin molecule contacts seven G-actin molecules. One troponin complex molecule is bonded to each tropomyosin, about one third of the distance from one end of the tropomyosin. For clarity the troponin complexes are shown on only one side of the actin helix. [Sources: J. M. Squire. Ann. Rev. Biophys. Bioeng., **4,** *137–163 (1975); C. Cohen. Sci. Am.,* **233,** *36–45 (1975).]*

TN-T is the component that bonds the complex to tropomyosin (there is one troponin complex bound to each tropomyosin molecule in the thin filaments, see Figure 4-9). TN-C is the component binding calcium ions, the property of troponin most important in its regulatory role in contraction (see Section 4.5.8). TN-C bonds to both TN-T and TN-I, but does not bond to tropomyosin or to actin. TN-I acts to inhibit bonding between actin and myosin. It bonds to both TN-C and actin, but not to tropomyosin. The bonding of TN-I to actin is weakened by the bonding of calcium ions to TN-C.

The fundamental biochemical energy sources for skeletal muscle cells are the same as those for most other cells. We shall mention here only a few pertinent details.

Skeletal muscle cells, like other cells, produce usable energy (primarily in the form of ATP) most efficiently from aerobic glycolysis. On many occasions, however, muscles have to function effectively over long periods of time in the absence of oxygen [for example, in diving animals during dives (see Chapter 6), and during periods of sustained activity (see Chapter 3)]. Accordingly, the machinery for anaerobic glycolysis is well developed. Resting muscle cells accumulate considerable quantities of glycogen, which serve as reserves for use under these conditions.

The regeneration of muscle glycogen reserves following activity occurs in different ways in different groups of animals. In frogs the major route appears to be a simple reversal of the machinery to reuse the lactic acid that has been produced. In mammals the process occurs via what is called the *Cori cycle* (named for the biochemists C. F. Cori and G. T. Cori, who described it in 1941). This cycle involves diffusion into the blood of lactic acid produced in the muscles. The acid is then transported to the liver. In the liver it is converted to liver glycogen and stored. The liver glycogen is continuously turning over, producing glucose that diffuses into the blood. This blood glucose is taken up by the muscles and is used to synthesize muscle glycogen.

4.5.5 Contraction: Excitation

Physiological stimulation of vertebrate skeletal muscle contraction involves arrival of a nerve action potential at the motor end plate. The nerve poten-

tial in turn produces an end-plate potential (EPP). The EPP ordinarily involves a depolarization of the end-plate membrane (see Chapter 9). This depolarization produces longitudinal electric-potential gradients between the end-plate membrane and the surrounding, normally polarized sarcolemmal membrane. Inorganic ions within these potential gradients begin moving in directions appropriate to their charges, producing local electric currents. The movements of charge in these currents produce depolarization in the adjacent sarcolemmal membrane. In fast fibers, when this depolarization reaches threshold levels, a propagated, all-or-none action potential is generated, which then sweeps over all parts of the sarcolemma of the fiber. This potential results primarily from movements of potassium and sodium ions. A latent period of up to a few milliseconds follows the passage of the action potential over the sarcolemma of any given sarcomere. Contraction then begins.

The details of the electric, ionic, and other biochemical processes occurring in the membranes of ordinary skeletal muscle fibers at rest and during production of EPPs and action potentials are discussed in Sections 9.4 to 9.6. Only two points will be mentioned here. First, there are several small but probably physiologically significant differences between muscle- and nerve-action potentials. For example, under similar environmental conditions muscle potentials often last longer than nerve potentials, and muscle potentials generally show a much larger after-potential than do nerve potentials. Second, the ionic currents responsible for action-potential generation in many invertebrate striated muscles (especially crustacean muscles) derive from movements of calcium and potassium ions rather than sodium and potassium ions.

Excitation events in slow fibers are somewhat different. Arrival of nerve potentials at the multiple endings on the surfaces of these fibers produces local depolarizations that spread outward from the endings, decreasing in amplitude with distance. No all-or-none action potentials are generated. Contraction occurs only in the sarcomeres influenced by these potentials. Contraction amplitude in a sarcomere is proportional to the amount of depolarization that occurs.

The slow-fiber pattern of excitation resembles the situations found in many invertebrate muscles. Considering the animal kingdom as a whole, the ordinary striated muscles of vertebrates are, in this regard, a quite unusual special case. It is usual for invertebrate muscle fibers to be polyneuronally innervated, with multiple endings for each neuron on each fiber. It is usual for there not to be motor end plates. It is usual for there to be inhibitory motor fibers in addition to the usual stimulatory fibers. And it is usual for there to be a lack of all-or-none pattern of contraction, associated with a lack of all-or-none action potentials (see Table 4-1 and Figure 4-6). A more detailed and extensive discussion of patterns of invertebrate muscle innervation and activation may be found in the monograph by Bullock and Horridge [1966].

4.5.6 Contraction: Structural Events

The large-scale result of stimulation of skeletal muscles is their shortening (under isotonic conditions) or their generation of tension (under isometric conditions). Under most circumstances in the bodies of animals some combination of these two phenomena occur. The underlying process in both cases is the shortening of stimulated sarcomeres. It is the presence of elasticity within the contractile machinery itself, in the sarcolemmas of the fibers, in the connective tissue sheaths of the fibers and the muscle as a whole, and in the ligaments and tendons that permits conversion of this underlying shortening process into all the combinations of shortening and tension that occur.

Under the light microscope one sees a characteristic sequence of changes in the banding pattern of individual sarcomeres (Figure 4-4). The changes are readily explained on the basis of the sliding-filament model (Figure 4-4). The basic sequence of events in vertebrate muscle, both qualitatively and quantitatively, is fully accounted for on this basis.

Once again, the situation among the invertebrates is more complex. The several types of invertebrate striated muscle referred to in Section 4.4 show a variety of sequences of banding-pattern changes during contraction. For example, the giant muscle fibers from the barnacle *Balanus nubilus* can shorten to small fractions of their rest length—a degree of shortening not found in vertebrate muscle. This extreme shortening is associated with an unusual change in the banding pattern. This change has been shown (by electron microscopic studies) to be caused by the passage of the thick filaments through perforations in the Z bands, the thick filaments continuing even farther to overlap with the thick filaments of adjacent sarcomeres. The fundamental situation in all types of striated muscles appears to be as it is in these barnacle fibers. No matter what the observed sequence of banding-pattern changes, it is possible to account for these changes on the basis of two types of filaments that slide relative to each other.

The existence of sliding filaments immediately raises the question of how they slide. A complete description of the structural events producing sliding at the molecular level is not yet available, but intensive work in recent years has established many of the main features of the process. The cross bridges linking thick to thin filaments are responsible for the sliding. The bridges are the globular ends of the HMM parts of the myosin molecules, which comprise the backbones of the thick filaments (Figure 4-8). As was mentioned earlier these myosin molecules have a definite structural polarity, their heads being directed away from the midpoints of the filaments. As a result all the cross bridges in one half of an A band have the same polarity, and this polarity is reversed in the opposite half of the same A band. A corresponding reversal of polarity of the structure of the thin filaments is found on either side of a Z line. There is, therefore, an electron microscopically visible structural basis for sliding of the filaments.

Beyond this point a variety of complexities and inconsistencies are found. A number of rather complicated theories have been proposed, none of which adequately accounts for all available observations.

The major structural events that are firmly established are basically two: (1) During the course of contraction cycles, the cross bridges (the myosin heads) change their orientations with respect to the long axes of the thick filaments in ways that produce filament sliding. (2) The orientation changes of the cross bridges are not synchronized. Thus, only a small fraction of the total number of bridges present in a fiber are actually attached at any given moment in a contraction.

Whatever the detailed mechanism of sliding turns out to be, it must satisfy two requirements:

First, the force of contraction must be produced as a result of a precisely determined set of structural changes associated with the splitting of a molecule of ATP (see Section 4.5.8). Second, the force-generating mechanism must work equally well over a wide range of side spacings between the actin and myosin filaments (see Section 4.5.1).

There is one specialized type of invertebrate striated muscle that represents a somewhat simpler and perhaps more tractable situation in this regard. This is the *fibrillar flight muscle* in insects (see Section 4.9). Fibrillar muscles are unusual in many respects, an important one of these being that they contract extremely rapidly and only for short distances. Sarcomere shortening in these muscles is so slight that electron microscopic studies indicate that it can be fully accounted for by changes in the angles between the interfilamentary cross bridges and the long axes of the filaments. The mechanism producing the changes in bridge angles is unknown.

4.5.7 Contraction: Excitation-Contraction Coupling

One of the areas of muscle physiology that has recently progressed with almost explosive rapidity is that of excitation-contraction coupling. The basic outline of the process in fast vertebrate muscle now seems to be firm. Vertebrate slow muscle fibers, also invertebrate muscles, are as diverse and complicated in this regard as they are in other ways.

The several milliseconds of latent period that begins with the passage of an action potential over the sarcolemmal membrane of a given sarcomere and ends with the beginning of contraction in that sarcomere is deceptive in its general appearance of inactivity. In reality a complex chain of events occurs during this brief time, the main function of which is the conversion of the external, primarily electrical action potential signal into a chemical signal that triggers contraction throughout the mass of the sarcomere (refer to Figures 4-2 and 4-3 for the structures participating in these events).

The transverse tubules (T system) of the sarcoplasmic reticulum form direct connections between all parts of the cross section of a sarcomere and its outer surface. The tubules terminate in pores opening through the sarcolemma. The pores and the T system as a whole are closely associated with either the Z bands or the A-I junctions (Figure 4-5).

Several types of experiments have shown that the pores and the T system seen under the electron microscope are indeed open in living muscle fibers. The most convincing experiments show that there is diffusional exchange of dissolved substances between the contents of the tubules even deep inside muscle fibers and the extracellular fluid bathing the outside. One series of experiments made use of an iron-containing protein, ferritin, which is easily seen under the electron microscope. Frog muscles were soaked in ferritin-containing Ringer solution before being fixed and sectioned for electron microscopic observation. Ferritin was found throughout the T systems of the muscle fibers but nowhere in the sarcoplasmic reticulum.

Exactly what happens along the walls of, and in the lumina of, the T-system tubules following passage of action potentials over the sarcolemma is an intriguing area of current research. The longitudinal electric *currents* associated with action potentials cannot account for what occurs. One piece of evidence supporting this statement is that it is possible, using microelectrodes placed inside single muscle fibers, to pass longitudinal electric currents of the same duration as normal action-potential currents (but of up to several times greater magnitude) through sarcomeres and produce no contractile activity at all.

The membrane *potential* changes occurring in action potentials are most significant. The essential role played by the T tubules in transmitting these changes to the contractile machinery has been demonstrated by experiments in which the T-tubule system is selectively virtually completely disrupted. This disruption is accomplished by brief treatment of muscles in a glycerol-Ringer (physiological saline) solution, followed by return of the muscles to normal Ringer. Electron microscope studies of muscles treated in this way show that only about 2% of normal T tubules remain, but that the sarcolemma, the sarcoplasmic reticulum, and the myofibrils are all intact. Treated muscles do not respond mechanically to electrical stimulation, though they produce nearly normal action potentials and their contractile systems respond normally to various chemical stimuli.

How the T tubules transmit electrical potential changes inward is presently a subject for active study. The weight of evidence favors a regenerative process similar to an action potential, at least in fast fibers.

As was mentioned previously, slow fibers appear not to be so completely dependent upon their T tubules. Slow fibers from frog muscles which have had their T tubules destroyed produce quite normal contractions in response to electrical stimulation. The basis for this difference from fast fibers is uncertain, but it may support the idea that slow fibers resemble cardiac muscle fibers in their mechanism for handling calcium ion (see Sections 4.5.9 and 4.7.3).

When a suitable electrical stimulus reaches the inner parts of the T tubules of a muscle fiber, it triggers a series of events that ultimately results in the initiation of filamentary movements. Calcium ions (Ca^{2+}) play the key role in these events, which can be divided into two parts: (1) electro-calcium coupling; and (2) calcium-mechanical coupling. Electro-calcium coupling is the coupling of an electrical membrane change to the release of free Ca^{2+} in the myoplasm. Calcium-mechanical coupling is the tying together of changes in the concentration of free Ca^{2+} in the myoplasm and contraction.

The mechanism of electro-calcium coupling is unknown. The process, however, is divisible into two parts: (1) The transmission of either electrical influences, probably potential changes, or chemical influences, perhaps Ca^{2+}, from the walls of the T tubules to the walls of the terminal cisternae of the SR. (2) The release of free Ca^{2+} into the myoplasm by the terminal cisternae. Recent work indicates that process (2) can be triggered by free Ca^{2+}. Detailed electron microscope studies of the structures of the membranes in striated muscle triads have shown that there may be physical connections ("feet") between T tubules and the SR. The structural results appear to make possible a direct electrical coupling of the two systems.

Calcium-mechanical coupling has been the subject of tremendous interest. As a result a great deal is known about its mechanism.

The terminal cisternae appear to be the principal cellular organelles in skeletal muscle that accumulate calcium. This accumulating ability is powerful enough for isolated cisternae, in the presence of ATP, to reduce the free Ca^{2+} concentration in muscle homogenates to as low a level as 0.02 micromolar (2×10^{-8} M). This concentration level is below the threshold Ca^{2+} concentration (10^{-7} to 10^{-6} M) that stimulates contraction in model muscle systems such as glycerol-extracted fibers. The fact that it is indeed the cisternae that produce the lowering of calcium concentration has been demonstrated in elegant experiments using living muscle fibers from which the sarcolemma had been removed. When such stripped fibers are soaked in solutions containing suitable concentrations of oxalate, the terminal cisternae within the fibers accumulate sufficient calcium within themselves so that the solubility product of calcium oxalate is exceeded. The internal precipitates of calcium oxalate that form are observable using an electron microscope.

Many experiments, including some very difficult ones involving microinjections of calcium-containing solutions into specific locations within single sarcomeres, have shown that the intracellular free Ca^{2+} concentration is the major determinant of whether or not a sarcomere will contract. It is possible, in such injection experiments, to produce contraction in the I band on only one side of a single Z band. This contraction can be maintained for at least as long as one full second—probably as long as it is possible to maintain a high-enough calcium concentration in the area of overlap between the thick and thin filaments.

The final step in excitation-contraction coupling thus appears to be the release from the terminal cisternae of enough calcium to raise the concentration of free Ca^{2+} in the area of overlap between thin and thick filaments above threshold for the initiation of filamentary sliding. The detailed nature of the mechanisms controlling calcium uptake and release from the terminal cisternae is an area of continuing research interest.

4.5.8 Contraction: Chemical Events

Resting skeletal muscle can be stretched quite easily. This fact implies that, at rest, there is little interaction between thick and thin filaments in the sarcomeres. Chemically this means that actin and myosin are dissociated from each other.

Almost immediately following activation it becomes very difficult to stretch muscle (see Section 4.5.10). This phenomenon presumably results from the formation of tight cross bridges between the two groups of filaments. These cross bridges then begin the process that results in sliding of the filaments relative to each other and contraction. Chemically this means the formation of a complex of actin and myosin called *actomyosin.*

We have previously mentioned that actin plus myosin (actually in the combined actomyosin form), in the presence of free Mg^{2+}, form a powerful ATPase. The action of this *actomyosin ATPase* releases energy from ATP by splitting off the terminal pyrophosphate group according to equation (4-1).

$$ATP \underset{Mg^{2+}}{\overset{\substack{actomyosin \\ ATPase}}{\rightleftharpoons}} ADP + inorganic\ P + energy$$

(4-1)

This reaction provides much of the energy needed to produce sliding of the filaments. Actomyosin ATPase is located in the S-1 portions of the myosin molecule.

Equation (4-1) focuses attention on an apparently contradictory fact. Section 4.5.7 indicated that increases in intracellular free Ca^{2+} concentration, not Mg^{2+} concentration, are the principal trigger for contraction.

The resolution of this contradiction was not easy. A summary of the current view follows.

Most of the magnesium present in muscle fibers is in the form of a magnesium-ATP (MgATP) complex. It is this complex that is the true substrate in equation (4-1) and appears to be primarily responsible for most other actions of ATP in the contractile process. Free Mg^{2+} and ATP are both present in small amounts, but their roles in contraction cycles are not well understood (see the previous discussion).

The process of actomyosin formation is in some way directly influenced by MgATP, over a relatively narrow range of MgATP concentrations. Higher concentrations lead to dissociation, lower concentrations to association. In the presence of troponin and tropomyosin the exact range of concentrations over which this control operates can be shifted by changes in the Ca^{2+} concentration in the medium. Higher Ca^{2+} levels shift the range to higher MgATP concentrations. Since intracellular MgATP levels usually remain quite constant, appropriate changes in the concentration of Ca^{2+} can cause either acto-

myosin formation or dissociation by shifting the sensitive range of the system from one side of the existing MgATP level to the other. Figure 4-10 diagrams the situation.

Troponin and tropomyosin act to confer Ca^{2+} sensitivity on the actin-myosin system in a rather complex way. The mechanism appears to be an example of what molecular biologists and biochemists call allosteric regulation. Allosteric regulatory mechanisms are control mechanisms on enzymes (in this case magnesium-activated actomyosin ATPase) or other functionally active macromolecules (such as hemoglobin), which are based upon molecular conformational changes resulting from the interaction of the macromolecules with other smaller molecules that are not themselves the substrate for the reaction involved. These smaller molecules interact with the macromolecules at sites different from the sites active in the reaction. The sequence for the actin-myosin system is as follows:

In the absence of Ca^{2+} the troponin-tropomyosin

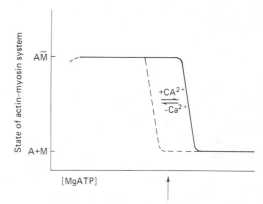

Figure 4-10
*Diagram illustrating the effects of MgATP and Ca^{2+} concentrations on the response of the contractile system. AM and A + M represent the associated and dissociated states of the actin-myosin system. The solid line represents the behavior of highly purified systems that lack calcium sensitivity. The dashed line indicates the nature of the calcium sensitivity of systems including troponin and tropomyosin. The arrow indicates the normal intracellular concentrations of MgATP. [Figure modified from S. Ebashi, M. Endo, and I. Ohtsuki. Quart. Rev. Biophys., **2**, 351–384 (1969).]*

complex inhibits actomyosin ATPase activity by blocking the activating sites for the enzyme on the actin filament. As indicated in Figure 4-9, under these conditions the tropomyosin strands extending along the lengths of the thin filaments are located somewhat to the sides of the grooves in the helical actin cores. Each tropomyosin strand is aligned with one of the two actin strands and probably controls the activity of that actin strand.

In the presence of Ca^{2+}, two individual Ca^{2+} bind to each troponin complex. The first ion binds to TN-C (see Section 4.5.4.), the second possibly to a site on another component of the complex (the second site is apparently activated by the binding of the first ion). Whatever the details, Ca^{2+} binding to the troponin complex is a function of still obscure cooperative interactions between the components of the complex.

The molecular consequences of this Ca^{2+} binding are still obscure. The major result, however, is that the activated troponin complex causes a shift in the position of the tropomyosin molecule to which it is bound, resulting overall in a movement of the tropomyosin strands toward the centers of the grooves in the actin cores. The movement appears to be an actual rolling to the side of the strand. This movement of tropomyosin derepresses actin-myosin binding (seven actins are activated for each tropomyosin molecule moving) and activates the actomyosin ATPase. ATP splitting and filament sliding follow.

The primary role of ATP (in the form of MgATP) as a supply of energy for contraction is apparent. We should note, however, that total ATP concentrations in muscles at rest are rarely higher than about 10^{-3} M. If contractile activity is maintained for any length of time, this amount of ATP is soon completely used up. How, then, do muscles continue to function?

A truly complete answer to this question is not yet possible. Detailed studies of aspects of muscle intermediary metabolism during and shortly after periods of contraction indicate the occurrence of a variety of biochemical changes associated with energy-supply processes. However, the most rapid and apparently the most important process is a fairly simple one. This process involves the resynthesis of ATP from ADP by transfer of a phosphate group from creatine phosphate (CP) according to equation (4-2).

$$ADP + CP \underset{\substack{\text{creatine} \\ \text{phosphokinase}}}{\rightleftharpoons} ATP + C \qquad (4\text{-}2)$$

(where C represents free creatine). This reaction is very rapid. The concentration of CP in resting muscle is usually 4 to 5 times higher than the ATP concentration.

The reserves of CP are themselves replenished by running the reaction backward. New ATP synthesized in both anaerobic and aerobic glycolysis reacts with free creatine to produce CP and ADP.

An important question concerning the rate at which energy can be supplied during contraction arises with respect to equation (4-2). This is the question of whether all the molecules involved are more or less free in solution (although apparently localized near the Z band and in the A band), hence are in thermodynamic equilibrium with each other according to the equilibrium constant for the reaction, or if they are usually separated from each other in different compartments. The latter case would probably mean slower reaction rates than the former. At present there is no unequivocal evidence on this issue.

An important area of research in muscle biochemistry deals with the general problem of metabolic controls. Some of the basic questions are the following: What are the chemical mechanisms involved in switching on the metabolic machinery that supplies the energy for contraction? How are the rates of the various reactions kept in their proper relationships with each other? How is the rate of energy supply kept closely correlated with the actual power output in performing work?

The discussions earlier in this chapter of the roles of Ca^{2+} and Mg^{2+} in activating contraction are illustrations of specific situations in muscle. We shall mention one additional and striking case. This is the activation of the complex process of anaerobic glycolysis.

Muscle produces lactic acid even under fully aerobic conditions. Under anaerobic conditions the rate of production is larger. In anaerobic activity the rate of lactic acid production increases to as much as 150–200 times the basal rate. The amount of increase appears to be proportional to the rate of stimulation of the muscle, not to the amount of work being performed. This and other facts make it seem possible that the control mechanisms for anaerobic glycolysis, like that for actomyosin

ATPase activity, has something to do with excitation-contraction coupling. Perhaps free Ca^{2+} is involved here as well as in contraction.

Another important research area that continues to contribute basic information to our understanding of muscle energetics is that of the time relations of heat production by muscles stimulated in various ways. This is a complex area based on its own specialized and sophisticated technology. Detailed summaries of the major facts may be found in the Additional Reading list.

4.5.9 Relaxation

Everything we have said so far about skeletal muscle function has related to contraction. The next question is how a contracted sarcomere gets back to its resting condition.

The mechanism of relaxation was a major stumbling block for many years. The major phenomena now have been worked out. This has been done with the help of sophisticated experimental procedures, many of which were developed in the later 1960s.

The mechanism is the reversal of the sequence of events occurring during excitation-contraction coupling. Once the stimulus that caused the terminal cisternae of the sarcoplasmic reticulum to release free Ca^{2+} has passed, the calcium-uptake mechanism in the cisternae rapidly proceeds to reabsorb the previously released ions. Contraction continues as long as the free Ca^{2+} concentration around the sliding filaments is above about 10^{-6} M. Once cisternal reabsorption lowers the concentration below this level, the ATPase activity of the actomyosin complex is no longer activated, cutting off the major portion of the energy supply. The actomyosin complex also begins to redissociate into F-actin and myosin. The sarcomere again becomes easily mechanically stretchable and the elastic forces generated during contraction in tendons, ligaments, and associated connective tissues pull the thin filaments back to their original positions. It is possible that internal elasticity in the contractile machinery also plays a part in this. The cisternal calcium-uptake mechanism apparently maintains a free Ca^{2+} concentration in resting fibers of about 10^{-7}M.

A great deal of evidence has accumulated in the past few years that demonstrates that calcium uptake and release by the sarcoplasmic reticulum is a widely distributed property, not restricted only to the terminal cisternae. The physiological significance of this fact is not yet understood.

In Section 4.5.1 we pointed out that considerable variations exist among types of striated muscles in terms of the degree of development of their sarcoplasmic reticula and the abundance of triad or similar structures. These variations seem to be correlated primarily with speed of contraction in the muscles (faster muscles have more reticulum). It is possible that the thinner, slower muscles that lack reticulum and triads use a different mechanism for controlling the free Ca^{2+} concentration around their contractile machinery. These fibers appear to have a better developed calcium-uptake mechanism in their sarcolemmal membranes than do the faster and larger fibers. This uptake mechanism apparently begins to operate at the onset of activity. The fibers do not accumulate calcium internally during periods of activity, so the inference is that however much is taken up (presumably during contraction) is soon lost again (presumably during relaxation). The basic control mechanism producing contraction-relaxation cycles is thus thought to be the same in these fibers as in the "usual" ones, but the sources and destinations of the Ca^{2+} ions involved are different. The relative importance of the sarcolemmal pump probably varies inversely with the degree of development of the sarcoplasmic reticulum.

4.5.10 Physical Properties

Much of the earliest work in muscle physiology involved the physical properties of skeletal muscles. Many phenomena described in this experimental work provided the basis for a large part of the structural and biochemical effort that produced the results described previously. However, the entire field has progressed sufficiently far that it seems better to reverse the actual historical sequence and consider muscle biophysics last. The structural and biochemical background now make the physical results more readily intelligible. We shall consider here only some of the major features.

Figure 4-11a is an analog model of muscle that provides the framework for virtually all discussions of its physical properties. Any muscle can be considered to be made up of three basic components:

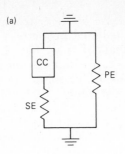

Figure 4-11
(a) Analog model of the structure of skeletal muscle. Code: CC, contratile component; SE, series elasticity; PE, parallel elasticity. (b) Tension/length diagram for resting vertebrate skeletal muscle. (c) Tension/length diagram for tetanically activated vertebrate skeletal muscle. Tensions expressed as ratios to full tetanic tension P_0 at rest length L_0. Lengths expressed as ratios to L_0.

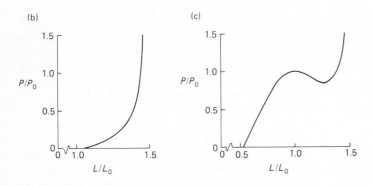

a *contractile component* (CC); some elastic elements in series with the CC, called collectively the *series elasticity* (SE); and other elastic elements in parallel with the CC, called collectively the *parallel elasticity* (PE). The specific identities of all the components making up the SE and PE are still matters of some dispute. There is also argument about the extent to which there is elasticity in the CC itself. Problems such as these, however, do not prevent us from discussing what happens in muscle in terms of a black box containing only these three components.

Before proceeding, some comment on definitions and experimental procedures is needed. Different muscles vary widely in size and strength of contraction, so it is usually most convenient to express experimental results in terms of ratios to some uniform reference points usable for all muscles rather than in absolute units. Accordingly, in this discussion, muscle lengths are expressed as ratios of actual length L to rest length L_0. Tensions generated *(P)* are expressed as their ratios to the maximum tetanic tension (P_0; see following paragraphs for definition) generated by the same muscle when it is stimulated isometrically.

The types of experiments one performs with muscles will depend on the particular components one wishes to study. If the CC is to be the subject, it is necessary to eliminate the effects of the SE and PE. This can be done by recording length changes in isotonic experiments, care being taken to avoid transient inertial effects when loads are applied to the muscle. Because loads are constant in isotonic experiments, the elastic components are kept at constant length, and length changes occurring must be in the CC.

Alternatively, if the SE is to be the subject, one would eliminate the PE by carrying out isometric experiments at L_0 or other shorter lengths. Tension changes occurring under these circumstances could only be due to stretching of the SE as a result of CC shortening.

With relatively few exceptions (for example, extensor and retractor muscles of the tongue), most skeletal muscles *in vivo* contract under isometric or nearly isometric conditions. Thus the results of isometric experiments most nearly represent physiological conditions. The tensions that fully activated muscles from various sources can generate

116

under isometric conditions are quite large, ranging from 1 to 10 kg/cm² of cross section.

Modern high-speed methods of monitoring physical events make it possible to record and resolve adequately the time courses of physical phenomena in muscles even when they are operating at room or normal avian or mammalian body temperatures. It is frequently desirable and convenient, however, to slow things somewhat to permit more complex manipulations during periods of activity. Thus many physical studies of muscle have been carried out at temperatures near 0°C.

Changes in the pattern of stimulation of muscles also permit study of different physical aspects. The most basic situation is the single twitch, which follows application of a single adequate stimulus. As stimulus frequency increases, there is progressively greater overlap between successive twitches, resulting in greater and greater partial summation of the tensions produced. As stimulus frequency is increased still more, a frequency is reached above which individual responses can no longer be resolved. The stimulus frequency at which fusion occurs is primarily a property of the contractile machinery, not the sarcolemma. The tension generated under these circumstances is the maximum that can be generated by the muscle at its particular length. This situation is called *full,* or *fused, tetanus.* It is the end point of this particular summation process, called *temporal summation.* It is also the most usual stimulus situation under which muscles function in place in animals. Muscles *in situ,* however, rarely tetanize fully—stimulus frequencies are tetanic but durations are short. The time courses of tensions generated under isometric conditions by muscles in single twitches and full tetanus are diagrammed in Figure 4-12.

Figures 4-11b and c illustrate two of the most important physical relationships existing in muscles: the resting and fully tetanized tension/length curves. The nonlinearity of the resting curve indicates that the elastic elements in muscle are not simple springs obeying Hooke's law but are more or less rubberlike in their properties. This similarity to rubber permits full recovery from stretches considerably greater than would exceed the yield points of substances acting like springs. Most of this resting-muscle elasticity probably resides in the connective and supporting tissues.

The tension/length curve for fully tetanized mus-

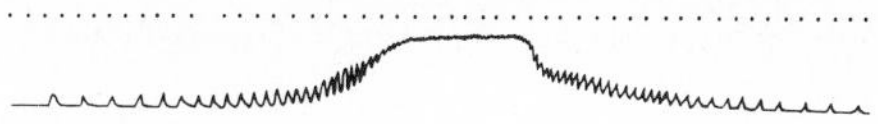

Figure 4-12
Isometric tension of single vertebrate skeletal-muscle fiber during continuously increasing and decreasing stimulation frequency (2–50 per sec). Time intervals at top of record, 0.2 sec. [*From F. Buchthal.* Dansk. Biol. Medd., **17,** 1–140 (1942).]

cle represents the static-tension/length curve for the CC. This is so because, in full tetanus, the steady tension developed is the result of equalization of stresses in both the SE and CC, the CC being maintained at constant length. The most important feature of this curve is that, in the physiological range of lengths, the CC is capable of generating its greatest tension at or near rest length. Most muscles in place in animals usually operate at lengths near rest length. There is considerable variation between muscles from different sources in both the maximum tensions they can generate and the range of lengths over which they can function.

A connection between the physical properties of intact muscles and the ultrastructure of single fibers has been indicated by studies of the variations in isometric tension with sarcomere length in tetanically stimulated isolated single fibers of frog fast muscle. Figure 4-13 summarizes the major findings. Only those parts of the single fiber preparations were studied in which sarcomere lengths were uniform.

The results are interpretable as supporting the sliding-filament theory. Tension is close to zero at point A, which corresponds with the sarcomere length at which the thick and thin filaments just do not overlap (see Section 4.5.2 for filament lengths). There is a linear rise in tension as the sarcomeres shorten to point B, which coincides with complete overlapping by the thin filaments of all of the cross bridges protruding from the sides of the thick filaments. This result supports the idea that the tension developed per cross bridge is uniform, the total tension being a simple product of the single-bridge tension times the number of bridges making interfilamentary contact. The curve is horizontal between points B and C. Point C is close to the length at which the ends of thin filaments from opposite Z bands begin overlapping each other. A slow, linear decrease in tension occurs between points C and D. Point D coincides with the sarcomere length at which the ends of the thick filaments make first contact with the Z bands. Point E, where the tension is once again zero, does not coincide with any particular filamentary event. It lies about two thirds of the way between point D and the length at which the ends of the thin filaments make first contact with the Z bands of the opposite ends of their sarcomeres.

The curve in Figure 4-13 is modified in intact muscles by two major influences. First, there is substantial nonuniformity of contraction states, even in fully tetanized muscles, in sarcomeres at different positions along the length of single fibers and also between sarcomeres in different fibers. This circumstance rounds the corners present in the single-fiber curve. Second, in muscles stretched significantly beyond L_0, the elastic properties of the other parts of the muscle (connective tissue, tendons, and so on) contribute increasingly to the total tension measured. At great lengths, these other parts produce the entire tension. Taking these two influences into account, one can readily see how Figure 4-13 can be integrated into, and made to account for a major portion of, Figure 4-11c.

Mention should be made of the fact that sarcomere-length/tension curves generated in experiments using single-twitch tension rather than tetanic tension do not agree so closely with theoretical predictions. Other lines of evidence, not summarized here, make it seem possible that variations in amounts of Ca^{2+} released from the sarcoplasmic reticulum may also play an important role in determining the shapes of these curves.

The curves just discussed describe static properties of muscle. It is also necessary to understand their dynamics. One of the most important dynamic properties is the velocity of shortening under different conditions of load. Figure 4-14 illustrates such a force/velocity curve.

The quantity of main concern is the shortening velocity of the CC when loaded to different extents at constant length. To make such measurements, the muscle must be mounted isotonically (to provide constant stretch to the PE and the SE) and it must be afterloaded. *Afterloading* means that the experimental setup is arranged so that the muscle does not actually begin shortening from the desired, preset length until it has generated tension equal to the weight of the load (see Figure 4-1 for a diagram of a setup permitting such experiments). The shortening velocity in the first instant after the load is lifted is then measured, usually in single twitches. Many repetitions using different loads result in curves such as that in Figure 4-14.

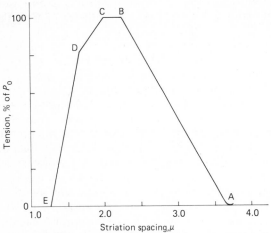

Figure 4-13
Length/tension curve for sarcomeres of frog fast muscle. Measurements made isometrically on fully tetanized, isolated single fibers. See text for discussion. [*Modified from A. M. Gordon, A. F. Huxley, and F. J. Julian. J. Physiol.,* **184**, *170–192 (1966).*]

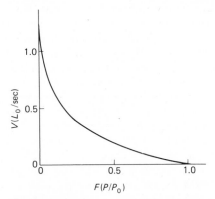

Figure 4-14
Force/velocity curve for vertebrate skeletal muscle. Ordinate, velocity of shortening in units of L_0 per second. Abscissa, isotonic load (force) relative to P_0. Muscle stimulated tetanically.

4 MOVEMENT: THE PHYSIOLOGY OF MUSCLE

Force/velocity curves are segments of hyperbolas that fit general equation (4-3)

$$(P + a)(V + b) = k \qquad (4\text{-}3)$$

where a, b, and k are constants. The term k can be evaluated from a measurement of tension in the muscle while it is in isometric tetanus at a specified length l. This measurement is indicated symbolically as $P_{0,l}$. Under these conditions $V = 0$ and

$$k = (P_{0,l} + a)b \qquad (4\text{-}4)$$

Equation (4-3) thus becomes

$$(P + a)(V + b) = (P_{0,l} + a)b \qquad (4\text{-}5)$$

This is called **Hill's characteristic equation.** It is named for its discoverer, A. V. Hill, a British biophysicist and Nobel Prize winner. Hill derived it from purely mechanical results such as those just described and, independently, from careful measurements of the rate of heat production by muscles during contraction. More recent work has shown this equivalency is not exact.

The physiological significance of these curves is manifold. We shall mention only two aspects. First, the mechanical curves are quantitative demonstrations that the faster a muscle shortens, the less force it can develop. Second, the near identity of the mechanical and caloric curves indicates that the energy-supply mechanism is largely arranged so that increased velocity of sliding of the filaments of the contractile machinery results in a decrease in the amount of energy available for doing work.

Another major point regarding physical properties relates to how rapidly a muscle switches over from the resting to the active state. The active state is defined as the tension the CC can exert when it is at constant length following stimulation. From this definition it follows that full development of the active state is represented by the isometric tetanic tension at any given length (see Figure 4-12).

The processes of excitation-contraction coupling described previously lead to the development of the active state. Ingenious mechanical measurements show that this state develops quite rapidly following stimulation, although by no means instantaneously. The basic experiments leading to this conclusion are termed **quick-stretch experiments.**

These involve the sudden stretching of previously stimulated muscles by an amount (previously determined) just sufficient to extend the SE fully and thus eliminate the need for the CC to shorten further to exert its full tension. Figure 4-15 illustrates the results of such experiments and compares them with the shape of the tension/time curve for an unstretched muscle in an isometric twitch.

As indicated in Figure 4-15, the CC of a skeletal muscle is activated very quickly, even in single twitches. The maximum single-twitch tension, however, does not even closely approach the tetanic value. The reason for this relates partly to the force/velocity curve of the CC. In the unstretched muscle the CC must stretch the SE to exert tension. It has to shorten to do this, its shortening occurring with the velocity profile of the force/velocity curve. It therefore takes time to stretch the SE. Other types of experiments have shown that the active state is fairly short lived. The dashed line in Figure 4-15 indicates its time course. Single-twitch tension thus never reaches tetanic levels because shortening the CC takes so long that the active state has begun to decay before the CC has fully shortened. There is also evidence that the active state in a single twitch is never as fully developed as the active state in a complete tetanus.

Many efforts are being made by research workers in different laboratories to tie in the results of studies of the physical properties of muscles with the results of ultrastructural and biochemical studies. We have mentioned the length/tension curve and its correlation with the sliding-filament model. A second example is the correlation of the active state with intracellular Ca^{2+} release.

Experiments with giant barnacle and frog muscle fibers support the idea that the time course for the active state coincides closely with the time course for the release and reabsorption of Ca^{2+} by the sarcoplasmic reticulum (Section 4.5.7). In these experiments a luminescent protein, aequorin (isolated from and named for a jellyfish), is injected into an isolated, single fiber. Aequorin emits a bluish light as a specific response to the presence of Ca^{2+}. The intensity of light emission is proportional to Ca^{2+} concentration over the range of concentrations normally occurring in contracting muscle fibers. The photoelectronically measured time course of light emission from injected aequorin in contracting fibers closely parallels the indepen-

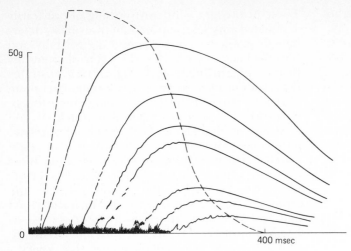

Figure 4-15

Time course of the active state during single twitches in frog skeletal muscle at 0° C. Solid lines are experimental tension/time curves obtained from delayed-release experiments. Dashed line, which passes through peaks of experimental curves, shows time course of intensity of active state. [*From D. R. Wilkie,* Brit. Med. Bull., **12,** *177–182 (1956).*]

50g

0

400 msec

dently physically-measured time course for the active state in those fibers.

Considerable attention has been given to the relationships existing between predictions of the sliding-filament model and force/velocity curves. According to the model, the generation of force is a function of the relative numbers of actomyosin cross bridges formed. The velocity of shortening against a given load should depend upon the rate of cycling of cross-bridge formation and release. The rate of cycling in turn should be related to the rate of ATP splitting.

Experimental studies have demonstrated a clear relationship between the speeds of contraction of different muscles and the activity levels of the actomyosin ATPases in those muscles. There also is a comparable relationship between speeds of contraction and the energetic efficiencies of contraction.

There is also evidence that intracellular Ca^{2+} concentrations control shortening velocity as well as force development. The mechanism of this control is still uncertain, but Ca^{2+} binding to troponin appears to be involved.

The increased precision of thought and experimental design which integrative studies like these require is producing two important results. First, they are generating an elegant, internally consistent view of how muscles operate at several levels of complexity. Second, they are also unearthing a wealth of new phenomena which are forcing contin-

uing reevaluations of many points previously considered firmly established.

One final point must be made with respect to the physical properties of skeletal muscle. This relates to the ways in which vertebrates can control the amplitude or strength of contraction of their muscles. The force/velocity curve indicates that this is partially done within the fibers themselves. On a larger scale it is done by using some combination of two types of summation processes. Temporal summation, relating to variations in stimulus frequency, has already been discussed. The second process is called *spatial summation.* Spatial summation refers to variations in amplitude or strength of contraction resulting from variations in the number of motor units simultaneously activated within a muscle. The more active units there are, the larger or stronger the contraction. As was indicated in Section 4.5.3, invertebrates achieve similar control by quite different methods.

4.5.11 Environmental Correlations

This discussion of vertebrate skeletal muscle has thus far treated the subject largely in a general physiological framework. It is apparent that this approach is a successful one. However, evidence has been cited which indicates that all skeletal muscle fibers, even within single muscles, are not the same, either ultrastructurally or functionally. There is significant probability that some of the more puzzling problems remaining may be resolved by giving more explicit recognition to this fact. Further, the greatest part of what we have said is based upon experiments carried out on muscles from (or in) a very few types of vertebrates—with frogs, rats, and rabbits constituting the vast majority. As has been the case in other fields of physiology, and as is already the case in invertebrate muscle physiology, we should expect a multitude of surprises as comparative vertebrate muscle physiology develops.

One of the potentially most significant areas for comparative studies is that of environmental effects on muscle structure and function. Some concrete evidence already exists indicating that the combination of muscle physiology with natural history will lead to the discovery of new phenomena and the generation of new insights into contractile mechanisms. We will illustrate the point with some examples of temperature effects on skeletal muscle function.

Many species of ectothermous vertebrates are physically active over a wide range of diurnally, seasonally, or geographically varied temperatures. These animals adapt to environmental temperatures in several ways, all of which permit them to remain more independent of their physical environment, in terms of activity, than biologists used to think was possible (see Sections 8.6 and 8.7). This is particularly true with respect to chronic exposures to different temperatures. Both individual (phenotypic) and populational (possibly genotypic) adaptations occur.

One investigation studied phenotypic thermal adaptations in toad muscles. The results indicate that adjustments in contractility within individuals may be due to changes in the duration of the active state. Sartorius muscles of California toads (*Bufo boreas*) adapted to 10°C appear to have a significantly shorter duration of the active state, at given test temperatures, than is shown by similar muscles taken from toads adapted to 30°C.

In sartorius muscles from temperate zone *Rana pipiens,* the curve relating P_0 (therefore full development of the active state) to temperature shifts significantly toward low temperatures in frogs adapted to 6°C as compared with frogs adapted to 20°C. This effect occurs in both summer and winter.

Another investigation compared the contractilities of isolated gastrocnemius muscles taken from two species of frogs from two climatically different regions: *Rana sylvatica* from arctic central Alaska and *Rana pipiens* from subtropical coastal west Mexico. All frogs were kept at 23°C for from several days to 1 week before being used. Previous studies indicated that these periods were long enough for complete phenotypic thermal adaptation to occur. When directly stimulated to contract isometrically at different temperatures from 0 to 25°C, the muscles from the two species behaved in opposite ways. *R. pipiens* muscles showed single twitch amplitudes that declined as temperature decreased, the muscles becoming inexcitable at 0°C. *R. sylvatica* muscles, in contrast, developed 150–200% more tension at 0°C than they did at 25°C, and they did not become inexcitable until they physically froze, at about −3°C. The indications thus are that the contractile properties of the muscles of these frogs are strongly determined, perhaps genetically, by the environmental temperatures of their home regions.

It is worth noting that much of the literature on temperature effects on muscle contraction is based upon studies of frog muscles from temperate-zone populations of either *R. pipiens* or two common species of English and European frogs, *R. esculenta* and *R. temporaria.* These other studies all show patterns similar to that for the Alaskan *R. sylvatica*—which is not surprising when one considers the cold winters of the temperate zone. The muscles of subtropical and tropical frogs may well all be substantially different.

A complex series of thermal adaptations occurs in the muscles of lizards. The major determinant of the temperature response of lizard skeletal muscle is the preferred body temperature of the species (see Section 8.7.2). Clear correlations exist between preferred body temperature and the upper thermal limits for full normal responses in both myosin ATPase activity and isometric twitch-tension development in isolated limb muscles (Figure 8-12). These correlations appear to be independent of individual temperature adaptations, phylogenetic position of the species studied, their geographic origins, or activity patterns.

The mechanisms producing these varied effects of temperature on the contractile machinery of ectotherm skeletal muscle are unknown.

No comparably clear thermal adaptation phenomena have been described from the muscles of endothermous vertebrates. This appears to be true even for the skeletal muscles of mammalian hibernators (see Section 8.12), some of which have been intensively studied.

Perhaps the most widely occurring environmentally related aspect of muscle function is that of heat production. The heat produced as a by-product of contractile activity provides a substantial fraction of the calories used by endothermic animals to maintain body temperatures higher than ambient temperatures at ambient temperatures below the lower limits of their thermoneutral zones (see Chapter 8 for extensive discussions of temperature regulation). In mammals, shivering is a major mechanism for heat production. Among invertebrate endotherms at least three groups of flying insects (Lepidoptera, Hymenoptera, Diptera) include species that are termed "myogenic flyers." These forms warm their thoraxes in preparation for flight by rapid vibrations of their flight muscles (see Section 8.13). The flight muscles themselves will function effectively only after they have reached an elevated temperature.

Space limitations prevent discussion of many other environmentally significant adaptations of muscle function. However, two additional related aspects are of sufficient importance that they must at least be mentioned. These are adaptations for endurance and the phenomena collectively termed fatigue.

Activity patterns of different kinds of nonsessile animals are tremendously varied, ranging from almost sedentary to continuously, furiously active. The striated muscles of animals having different activity patterns, especially those muscles used in locomotor movements, are almost equally varied in their performance characteristics. Among the more active animals two major adaptive patterns occur widely (see Section 3.13).

Muscles of animals that are on the move frequently, rapidly, and for long periods generally have much higher aerobic metabolic capacities and can remain active for longer periods than muscles from less motile forms. They are adapted for endurance. In contrast, muscles in animals that, although active, are less motile and instead do large amounts of heavy physical work (such as digging or burrowing in hard soil) often have hypertrophied fibers with unusually great strength. Their endurance is likely to be less than that of the mus-

cles of the motile animals, and they may show greater capacities for anaerobic metabolism (see Section 3.11). At the level of individuals, similar adaptations, within limits, are possible with suitable training regimes. Much of human sports physiology is based upon this fact.

The ultimate result of activity is, of course, fatigue. Despite its near universal occurrence, and great research interest in its causes, the metabolic bases for fatigue still remain surprisingly unresolved. In any given situation fatigue of skeletal muscles almost certainly results from a combination of a variety of different conditions. These conditions may include, but are by no means limited to, depletion of metabolic reserves, accumulation of metabolic wastes, shortage of oxygen, changes in properties of neuromuscular transmission, changes in mechanisms of excitation-contraction coupling, and so on.

4.6 CARDIAC AND SMOOTH MUSCLE: GENERAL COMMENTS

Space will not permit as full a discussion of cardiac and smooth muscle as we have had of skeletal muscle. And it is true that understanding of these two types of muscle has, in many ways, not yet progressed to the level of sophistication found in the subject of skeletal muscle. This situation is partially due to a smaller number of active workers in these areas and partially to an inherently greater intractability on the part of the tissues themselves.

The discussions that follow will once again concentrate on the subcellular and cellular levels. Discussions of the heart as an organ are included in Chapter 6. We shall not consider here such "unusual" organs as the caudal hearts of hagfish or the lymph hearts of amphibia.

4.7 STRIATED MUSCLE: CARDIAC MUSCLE

4.7.1 Structure

Cardiac muscle cells are not as uniform in general appearance as skeletal muscle cells. They are divisible into two major categories, the primarily contractile cells and the modified cells of the conducting system *(Purkinje fibers)*.

The cells of the sinoatrial and atrioventricular nodes are very similar to contractile cells. We shall consider here only the contractile cells. Figure 4-16 is an electron micrograph of parts of several contractile cells.

Contractile cardiac cells usually are not simply straight fibers but are often branched, forming anastomoses with adjacent fibers. The anastomoses usually contain an intercalated disc. Electron microscopic studies have demonstrated that the classical notion that heart muscle is a syncytium is not correct. The intercalated discs contain morphological cell boundaries as well as several other things.

The single cells bounded by intercalated discs differ from most skeletal muscle fibers in a number of ways (Table 4-1). They develop from single myoblasts rather than by fusion of a number of such cells. As a result they usually contain only a single nucleus. They contain large numbers of mitochondria, usually many more than are found in most skeletal muscle fibers. Sarcoplasmic reticulum may or may not be present and, where present, is developed to varying degrees. Unlike most skeletal muscle, cardiac muscle sarcoplasmic reticulum is continuous from sarcomere to sarcomere. T-system tubules are also variably present (they are missing from most of the atrial muscle of mammalian hearts and from frog ventricle). Where present, the T system is not only a series of radial tubules opening onto the sarcolemmal surface but also includes widespread lateral ramifications throughout the fibers. As in skeletal muscle, there is no open communication between the T system and the sarcoplasmic reticulum. There is no direct innervation, so there are no motor end plates. A part of the intercalated disc, called the **nexus,** serves to transmit electric excitation between cells.

The contractile machinery appears to have the same structure as that in skeletal muscle.

4.7.2 Chemical Composition

The inorganic composition of heart muscle is generally similar to that of skeletal muscles in the same animal. It is difficult to generalize about the content of low molecular weight organic substances. This is because cardiac muscle varies widely in its metabolic requirements in different organisms. For example, the hearts of birds and mammals generally have very little capacity for functioning anaerobically, whereas the hearts of many ectothermous

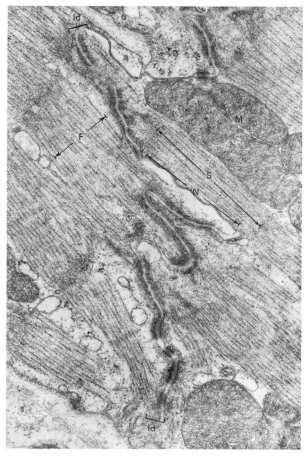

Figure 4-16
Electron micrograph of a longisection (30,000✕) of parts of two adjacent cardiac-muscle cells from the heart of a guinea pig (Cavia porcellus). *Fresh muscle fixed, in strongly contracted state (denser contraction bands indicate locations of Z lines at ends of sarcomeres), in glutaraldehyde phosphate; postfixed in osmium tetroxide and uranyl acetate; embedded in Vestopal W; sectioned at 300–400 Å; section stained in uranyl acetate and lead citrate. Code: F, width of single fibril; S, length of sarcosome; M, mitochondria; Id–Id, approximate limits of intercellular boundary that would appear as intercalated disk under light microscope; N, nexus, area of closely apposed cell membranes within intercalated disk.* [Courtesy of Prof. Fritiof S. Sjöstrand.]

vertebrates have considerable capacity for such function. These differences are probably due to differences in the degree of development of the machinery for anaerobic glycolysis.

The contractile machinery of heart muscle is similar to that in skeletal muscle. Cardiac myosin apparently differs somewhat from skeletal muscle myosin in that it is somewhat larger, breaks up in a somewhat different way under the attack of proteolytic enzymes, and shows a lower ATPase activity under certain conditions. These differences do not appear to cause appreciable differences in the function of the myosin.

4.7.3 Contraction

Sarcolemmal action potentials are responsible for excitation in cardiac muscle just as they are in skeletal muscle. This is true even in isolated pieces of heart muscle that show spontaneous, self-generated activity. Chapter 6 discusses the roles of pacemakers and the conducting system in controlling this spontaneous activity.

The mechanisms resulting in spontaneous contraction in heart muscle, the electrical and ionic details of the action potentials, and related electrophysiological questions are discussed in several of the papers cited in the Additional Reading list. Here we shall point out only that cardiac action potentials are extraordinarily long (over 1-sec duration in frog heart at room temperature) and that intercellular transmission via the intercalated discs is by electrical means, rather than the chemical mechanisms operating in most nerve synapses. Intercellular transmission occurs very easily, permitting much of the heart to function as a syncytium even though it is physically not one.

The contraction cycle of heart muscle is much longer than the durations occurring in most skeletal muscles. In frog ventricle the duration of contraction is proportional to the duration of the action potential. Frog heart is similar in this respect to invertebrate muscles such as barnacle muscle. Mammalian myocardium is more nearly like fast skeletal muscle.

The processes of excitation-contraction coupling and of contraction itself appear to be similar in cardiac muscle and in skeletal muscle. There are however, a number of quantitative differences. For example, the Ca^{2+} sensitivity of intact cardiac muscle is much greater than that of skeletal muscle. The Ca^{2+} affinity of cardiac troponin is only one third that of skeletal muscle troponin. The ATPase activity of cardiac myosin is appreciably lower than that found in equivalent skeletal muscle myosin preparations.

The variable presence and degree of development of sarcoplasmic reticulum also produces some differences in details of mechanisms. The virtual absence of sarcoplasmic reticulum in frog myocardium inevitably means that the source of Ca^{2+} activating contraction must be different. The extracellular fluid, via the sarcolemma, is the source; a significant fraction of the total amount of Ca^{2+} needed for a single contraction enters during the action potential, primarily during the plateau phase. The remaining Ca^{2+} needed is apparently released from binding sites on intracellular membranes, probably primarily the inner surface of the sarcolemma.

Cardiac cells with better developed sarcoplasmic reticula (such as many mammalian cells) are like skeletal muscle in deriving activating Ca^{2+} almost entirely from intracellular sources. There is, however, debate as to whether the reticulum is the sole or even the primary source of this Ca^{2+}. Both the abundant mitochondria and the sarcolemma may also be involved. This question is presently not resolved.

4.7.4 Relaxation

The underlying mechanism of cardiac muscle relaxation appears to be the same as in skeletal muscle. There are, however, again differences in details. For example, Ca^{2+} uptake by preparations of cardiac sarcoplasmic reticulum requires two groups of biochemical cofactors, whereas the same phenomenon in preparations from skeletal muscle requires only one of these groups.

4.7.5 Physical Properties

The relative simplicity and clarity with which it is possible to describe the physical properties of skeletal muscle derive in large part from two major features. First, there is the availability of simple geometries. It is easy to obtain muscles in the form of thin sheets made up of parallel fibers covered by minimal amounts of connective tissue. Second, there are a number of basic properties of the fibers themselves. Three of the most important of these properties are (1) the abrupt beginning of contraction, (2) the possibility of development of full tetanus, and (3) the small contribution of resting tension to total tension at lengths near L_0.

Heart muscle structurally is very complex, and its cells possess none of these properties. Accordingly, it is very difficult to determine force/velocity curves or the existence, nature, and duration of the active state. The results of efforts to carry out the necessary experiments are complex and not easily susceptible to the types of analysis that have worked so well with skeletal muscle. The field is, therefore, one of considerable challenge. An important result of work on cardiac muscle mechanics is the demonstration that the active state is slow to develop, prolonged in duration, and variable in maximum intensity.

Present understanding of cardiac muscle physiology is based largely upon experiments with tissues from four types of animals: frogs, rabbits, cats, and dogs. Just as in skeletal muscle, we should not be surprised if new studies on other forms produce surprising new results.

4.8 SMOOTH MUSCLE

Vertebrate smooth muscles belong to the category of "classic" smooth muscles defined in Section 4.4. They are widely distributed in the bodies of animals, constituting important parts of the walls of such organ systems as the alimentary tract, reproductive system, and vascular system. In many places smooth muscle forms a layer around a hollow space of some kind. Its major function usually is to exert pressure on the contents of the space, most often as a way of emptying the space.

Smooth muscles have long been used as one of the favorite tissues for assay of drug or other chemical activities by pharmacologists and others. As a result, a few aspects of their properties have been extensively studied, but primarily as means to other ends rather than as intrinsically valuable studies in themselves. This fact, combined with considerable variability in properties of muscles obtained from different sources, makes it difficult to provide a coherent, consistent description of its structure and properties.

4.8.1 Structure

Smooth muscle fibers apparently are always single cells containing only one nucleus. They are frequently spindle-shaped and usually quite small, most vertebrate cells being 50–300 μm long and 5–50 μm in diameter in their central portions. Some invertebrate fibers are much larger. This is especially true in nematodes such as *Ascaris*, which has cells as large as 1 mm in diameter. Figure 4-17 illustrates the structure of some visceral smooth muscle fibers.

Electron microscope studies have resolved a long-standing debate as to whether cytoplasmic

Figure 4-17
Electron micrograph of a longisection (6000×) of parts of several classical smooth muscle fibers from the taenia coli of a guinea pig (Cavia porcellus). Fresh muscle fixed in glutaraldehyde; postfixed in osmium tetroxide; embedded in araldite; sectioned at 300–400 Å; section stained with uranyl acetate and lead citrate. Code: Pm, plasma membrane; Mf, myofilaments; M, mitochondria; Db, dense bodies (possibly correspond to Z bands of striated muscle). [Courtesy of Dr. Catherine F. Schoenberg.]

bridges exist between adjacent smooth muscle cells. There apparently are a few bridges in some muscles. More often, however, there are areas of intimate contact (nexus) and perhaps even partial fusion of sarcolemmal membranes of neighboring cells.

Many smooth muscle cells lack direct innervation. Those cells that are directly innervated may have one or many nerve endings terminating on their surfaces. Cells with multiple endings probably are innervated by more than one nerve cell. Neuromuscular junctions in smooth muscle are much simpler in structure than skeletal muscle motor end plates.

A sarcoplasmic reticulum is present in vertebrate and other classic smooth muscles. Details of its organization, variations in degree of development in muscles with different properties, and other details have not yet been worked out. T tubules are absent. Mitochondria are present, but variations in numbers, sizes, and distribution have not yet been studied adequately (see Table 4-1).

The contractile apparatus in smooth muscle is made up of ultramicroscopic fibrils. There is, however, no regularity in their pattern and distribution, at least in classic smooth muscles. Thin and thick filaments both occur, thin filaments generally being much more abundant. The thin filaments are composed of actin and tropomyosin. The thick filaments are made up of myosin. Different types of structures occur in the other types of smooth muscle (helical and paramyosin) (see Table 4-1 and Section 4.9).

Electron microscope studies have shown that several types of invertebrate muscle, traditionally considered to be structurally close to classic smooth muscle, actually possess a regular, though geometrically complex sliding-filament type of structure. The fibers, such as the trunk muscles of *Ascaris,* turn out to be obliquely striated. This means that the arrangement of thick and thin filaments is staggered so strongly that the structural equivalents of the Z bands (called Z bundles, and closely associated with dense bodies similar to those shown in Figure 4-17) form angles of only about 6° with the long axis of the fibrils. In addition, the apparent direction of the striations changes with the plane of sectioning of the fibers and may vary all the way from radial to longitudinal. The result of this geometrical complexity is the appearance, especially at lower magnifications, of a disorderly, smooth-muscle type structure.

4.8.2 Chemical Composition

The inorganic and low molecular weight organic compositions of vertebrate smooth muscles appear to be generally similar to those of skeletal muscle cells in the same animals. There appear, however, to be important variations in organic composition in at least some smooth muscles. These and other aspects of smooth muscle composition are discussed in the reviews in the Additional Reading list.

The contractile machinery of vertebrate smooth muscles appears to be composed of actin, myosin, and tropomyosin of essentially the same composition and properties as in skeletal muscle. Actin and tropomyosin are both proportionately more abundant than they are in skeletal muscle. Invertebrate helical and paramyosin muscles contain other proteins as well.

4.8.3 Contraction and Relaxation

Vertebrate smooth muscles can be separated into two categories on the basis of their contractile properties. Muscles such as those of the iris, ciliary body, and nictitating membranes of the eyes, and many muscles in the walls of blood vessels, are called *multiunit muscles.* These fibers depend on external nervous or hormonal stimulation to initiate contraction. Their contraction cycles are relatively short, and fiber innervation appears to be arranged so that activation occurs in motor units similar to those in skeletal muscle. Contraction in these muscles follows excitation by all-or-none action potentials in their sarcolemmas. The action potentials are complex and variable in form, depending on the nature of the stimulus situation.

Most other vertebrate muscles, especially those in the walls of visceral organs, are called *visceral muscles.* These fibers exhibit spontaneous, often rhythmic, contractility and are easily stimulated mechanically. Their contraction cycles are usually long. Large masses of a visceral muscle will contract as a unit, acting as a functional syncytium. Electrical activity in the sarcolemmas of these fibers is complex and variable, sometimes, but not always, including action potentials similar to those occurring in skeletal muscle. Intercellular conduction apparently occurs by electrical means at the nexus contacts between adjacent cells. Overall patterns of electrical activity, hence of contraction, are controlled by autonomic nerves or, as in the mammalian uterus, by hormones (see Chapter 12).

Despite the lack of structural organization of the intracellular fibrils, smooth muscle contraction results from a sliding-filament mechanism. Free Ca^{2+} plays a role in smooth muscle contraction similar to that which it plays in skeletal muscle. At least in vascular smooth muscle the near absence of troponin from the thin filaments means that activation of contraction results from binding of Ca^{2+} directly to myosin. The 20,000 molecular weight light chains of the myosin are the sites of this binding, which produces activation after phosphorylation of the light chains also occurs. Once again relatively poor development of sarcoplasmic reticulum appears to be correlated with use of extracellular fluid as a source of needed Ca^{2+}, although intracellular sources are also used.

The energy-supply mechanism in smooth muscle is imperfectly known. The usual components appear to be present (ATP and CP), but the concentration of CP is often much lower than in skeletal muscles.

4.8.4 Physical Properties

Despite the obvious differences between vertebrate smooth and skeletal muscle in terms of structure and internal organization, the careful studies of visceral smooth muscle physical properties that have been carried out have demonstrated a surprising degree of similarity. Specifically, the shapes of the resting and fully active (tetanic) length/tension curves are very similar, as are the force/velocity curves. Hill's characteristic equation, in fact, seems to apply to visceral smooth muscle just as well as it does to skeletal muscle.

These facts lend support to the position that the uniformities that apparently exist in terms of chemical composition of the contractile machinery are more important than the differences in patterns of excitation, durations of contraction cycles, and so on, that separate visceral smooth from skeletal muscles. The only significant physical property that appears to separate the two types of muscle is the rate of development of the active state. This apparently develops only slowly in visceral smooth muscle (full development requires several seconds, rather than only a few milliseconds). Such slow development results in a dependence of active-state development in smooth muscle on muscle tension as well as time. The basis for this difference is uncertain, but it may be related to the relative lack of sarcoplasmic reticulum, the disarray of the intracellular fibrils, or a combination. Whatever the cause of the difference, the basic fact remains that the active state does develop and, once developed, shows the same properties as are found in skeletal muscle.

In closing this section we must once again sound a cautionary note. Smooth muscles from different organs within single animals, as well as those from equivalent organs in different animals, vary tremendously in their behavior and properties. Only a few of the many variations have been studied in any detail, and then usually only with respect to a few properties. Thus we know little about visceral smooth muscles and even less about multiunit smooth muscles. We should not be surprised if future research requires some fundamental reorientations.

4.9 INVERTEBRATE MUSCLES

A reasonably complete description and discussion of present knowledge of the patterns, mechanisms, and controls of muscle function in the various phyla of invertebrates would require at least as much space as that we have devoted to vertebrate muscles. That not being available, we will restrict our consideration to this one section, plus the various references to invertebrate muscles made previously (see especially Sections 4.4, 4.5.1, 4.5.3, 4.5.5, and 4.5.6). We will emphasize here only important aspects of invertebrate muscle physiology not adequately discussed previously. As was the case in Section 4.4, we will concentrate upon the two best known invertebrate phyla, the molluscs and the arthropods.

Diversity is probably the best single word to use in describing invertebrate muscles. Table 4–1 lists some of the major structural characteristics of several of the principal types of invertebrate muscles. Figure 4–18 provides additional visual documentation of the diversity.

4.9.1 Catch Muscles

Molluscan muscles are among the types of invertebrate muscles most different from vertebrate muscles. This is particularly true for the so-called "catch" muscles. Catch muscle fibers are most often found in the adductor muscles of bivalve molluscs—the muscles that close, and hold closed, the shells of clams, oysters, and so on. Catch muscles are characterized by their ability to maintain tension and remain in a contracted state for extended periods of time. Oyster adductor muscles can remain in the catch state for 20–30 days.

In the catch state, molluscan muscles are physically changed. A relaxed muscle is plastic and stretchable, a muscle in catch is rigid and resistant to stretch. Development and termination of the catch state in live bivalves is dependent upon nervous stimulation. There are separate nerve fibers inducing contraction and relaxation. Maintenance of catch is usually associated with continuing low frequency nerve stimulation, but, in many forms, cutting the stimulatory nerves produces no change in state.

Catch fibers are generally thin (~5 μm) and nonstriated. The fibers contain both thick and thin filaments, usually in ratios of 12 or more thin filaments per thick filament. The thin filaments are composed primarily of actin; troponin is absent, though tropomyosin is present. The thin filaments are about 11 μm long and 55–60 Å in diameter. Their basic structure is the same as that of striated muscle thin filaments, except for the absence of troponin (see Figure 4-9).

The thick filaments are much larger and more variable in size than vertebrate thick filaments. They average about 30 μm in length, and range from 200–1200 Å in diameter. The thick filaments are composed of an outer layer of myosin (similar in structure and properties to vertebrate myosins) surrounding a central core of another protein, paramyosin. The paramyosin is thought to serve two functions: (1) to give the filament structural rigidity, and (2) to modify the properties of myosin ATPase so that the muscle can go into the catch state. How paramyosin produces these effects and also the question of whether or not it has other functions are not known. Paramyosin occurs in many invertebrate muscles, not all of which show the catch phenomenon.

Paramyosin itself is a rod-shaped molecule similar to the rod portion of myosin (see Figure 4-8). It has a molecular weight of about 200,000. It is packed in the thick filaments in a regular array, which gives paramyosin-containing filaments a characteristic latticelike appearance in the electron microscope.

The molecular mechanism for contraction in catch muscles is thought to be similar to that in vertebrate muscles. The myosin molecules on the surfaces of the thick filaments are oriented in bipolar fashion, the ends of the filaments being polarized in opposite directions. Cross-bridge formation and filament sliding probably occur. The role of Ca^{2+} in these processes is uncertain. The action potentials in the sarcolemmal membranes are primarily calcium spikes. There are no T tubules and little, if any, SR. Given the absence of troponin, it is probable that the controls for contraction in these muscles reside in the thick, not the thin, filaments.

The energy supply for catch-type contractions is ATP. The phosphagen serving as the source of phosphate needed to regenerate ATP from ADP is arginine phosphate (see Figure 4-7), rather than the creatine phosphate used by most vertebrate muscles.

4.9.2 Fibrillar Flight Muscles

A number of the differences that distinguish arthropod muscles, especially crustacean muscles, from vertebrate muscles have already been adequately described. A great deal of attention has also been given to the muscles of insects, which possess several types of muscles that are at least as different from vertebrate striated muscles as are molluscan catch muscles (see especially the book by Usherwood in the Additional Reading list). The most spectacular of these muscle types we have only referred to in passing—fibrillar (or asynchronous) flight muscle.

Fibrillar muscles are one of three structurally distinguishable types of insect flight muscles. They are found principally in the orders Diptera (two-winged flies), Coleoptera (beetles), Hymenoptera (bees, wasps), and Hemiptera (true bugs), usually in locations in the flight musculature that require high frequencies but small amplitudes of contraction. They are called fibrillar muscles because their

Figure 4-18
Electron micrographs of transverse sections through six different types of muscles from both vertebrates and invertebrates, illustrating the diversity existing in terms of both arrangements and relative numbers of thin and thick filaments: (1) garter snake (Thamnophis) *striated muscle (thin: thick ration 2:1); (2) insect* (Benacus) *fibrillar flight muscles (3:1 ratio; note double ring structure of thick filaments); (3) rabbit* (Sylvilagus) *striated (psoas) muscle (2:1 ratio); (4) unidentified copepod larva, antennulary muscle (3:1 ratio); (5) locust* (Schistocerca) *jumping leg muscle (4:1 ratio); (6) crab* (Podophthalmus) *pink eye-raiser muscle (7:1 ratio). All photographs at 100,000× magnification.* [*Source: G. Hoyle.* Am. Zool., **7,** *435–449 (1967).*]

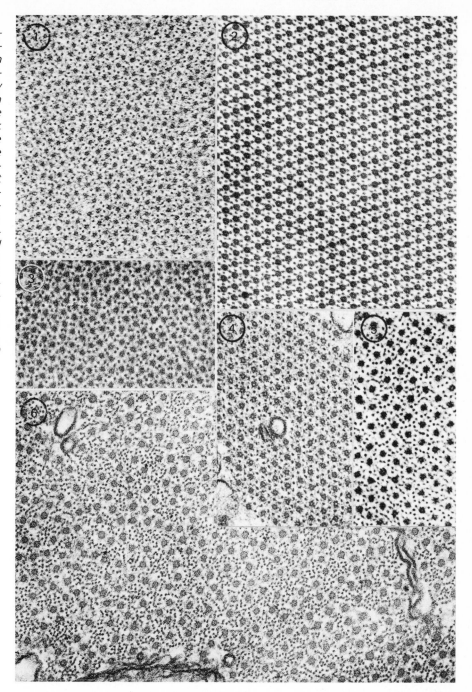

fibers, which usually are large in diameter (0.05 to more than 1.00 mm), contain many large diameter (2–4 μm) fibrils. These muscles are functionally virtually unique in the animal kingdom in that their contraction frequency, once contractile activity has begun, is often far higher than, and relatively independent of, motor nerve impulse frequencies. Contractile frequencies in excess of 1000 Hz (cycles per second) have been measured. These frequencies are at least several times higher than maximal frequencies for action potentials in the motor nerves to the muscles.

Other important structural features of fibrillar muscles include: extremely regular internal arrangements of fibrils, and of filaments in them (Figure 4-18); large numbers of mitochondria arranged in longitudinal columns interspersed among the fibrils; well developed T-tubule systems; SR systems reduced to isolated vesicles; and a rich supply of respiratory tracheoles. The sarcomeres in the fibrils have the normal striated muscle arrangements of thick and thin filaments, but the I bands are very short and the ratio of numbers of thin to thick filaments is higher than in vertebrate striated muscles (3:1 rather than 2:1).

The thick filaments are thicker than vertebrate thick filaments, are composed primarily of myosin (other components are not yet identified), and contain about twice the amount of myosin per unit length as do vertebrate thick filaments. The myosin is similar in molecular structure and properties to mammalian myosin. The detailed arrangement of the myosin molecules is still uncertain, but at least the outermost layer of molecules is oriented in bipolar fashion, polarizing the ends of the filaments in opposite directions.

The thin filaments have the same protein composition as vertebrate thin filaments; a double-helically arranged actin core, plus tropomyosin and troponin. The details of the positions of the latter two molecules have not yet been established.

The mechanism of contraction appears to be similar to that in vertebrate striated muscles. The details of the role of Ca^{2+} are not yet clear, however. It is established that both the thin and thick filaments bind Ca^{2+} and participate in the control of contraction. The processes of excitation-contraction coupling are clearly somewhat different from those in vertebrate muscles since the fibrillar muscle SR is greatly reduced, and there is no correla-tion between contraction cycling frequencies and motor action potentials. The amplitudes of contraction are small (usually only 1–2% of rest length), possibly based upon some difference in structural details of actomyosin interactions.

The basis for the independence of contractile cycles in fibrillar muscles relates to physical responses of the muscle fibers to mechanical stretches. The mechanism of the stretch response is complex and well worked out. Interested students are referred to Usherwood's book listed in Additional Reading.

There is much more of interest and value that can be said about invertebrate muscles. The few examples described here have been chosen to illustrate several points: (1) the diversity of muscle structures; (2) the diversity of muscle functions; (3) the diversity of muscle controls. They also illustrate strikingly that, despite the diversities just mentioned, the underlying contractile mechanisms all appear to be amazingly uniform. The Additional Reading list includes several reviews and books providing further information.

ADDITIONAL READING

FUNCTIONAL MORPHOLOGY

Alexander, R. M. *Animal Mechanics,* University of Washington Press, Seattle, 1968.

————. *The Chordates,* Cambridge University Press, New York, 1975.

Elder, H. Y., and E. R. Trueman (eds.). *Aspects of Animal Movement,* Cambridge University Press, New York, 1980.

Gans, C. *Biomechanics: an Approach to Vertebrate Biology,* J. B. Lippincott Co., Philadelphia, 1974.

————, and W. Bock. "The functional significance of muscle architecture: a theoretical analysis," *Ergeb. Anat. u. Entwicklungsgesch.,* **38,** 115–142 (1965).

Gray, J. *Animal Locomotion,* W. W. Norton, New York, 1968.

COMPARATIVE PHYSIOLOGY

Bolis, L., S. H. P. Maddrell, and K. Schmidt-Nielsen (eds.). *Comparative Physiology: Functional Aspects of Structural Materials,* American Elsevier Publishing Co., New York, 1975.

Bourne, G. H. (ed.). *The Structure and Function of Muscle,* 2nd ed., 4 vols., Academic Press, New York, 1972.

Bullock, T. H., and G. A. Horridge, *The Nervous System of Invertebrates,* W. H. Freeman and Company, San Francisco, 1966.

Clarke, M., and J. A. Spudich. "Nonmuscle contractile proteins: the role of actin and myosin in cell motility and shape determination," *Ann. Rev. Biochem.*, **46**, 797–822 (1977).

Hatano, S., H. Tshikawa, and H. Sato (eds.). *Cell Motility: Molecules and Organization*, University Park Press, Baltimore, 1979.

Heinrich, B. *Bumblebee Economics*, Harvard University Press, Cambridge, MA, 1980.

Hoyle, G. "Comparative aspects of muscle," *Ann. Rev. Physiol.*, **31**, 43–84 (1969).

———. "Neural control of skeletal muscle." In M. Rockstein (ed.). *The Physiology of Insecta*, 2nd ed., vol. 4, Academic Press, New York, 1974.

Huddart, H. *The Comparative Structure and Function of Muscle*, Pergamon Press, Elmsford, NY, 1975.

Needham, D. M. *Machina Carnis: the Biochemistry of Muscular Contraction in its Historical Development*, Cambridge University Press, New York, 1971.

Perry, S. V., A. Margreth, and R. S. Adelstein. *Contractile Systems in Non-muscle Tissues*, Elsevier/North Holland, New York, 1977.

Sherman, R. G. (ed.). "Symposium on invertebrate neuromuscular systems," *Am. Zool.*, **13**, 235–446 (1973).

Tregear, R. T. (ed.). *Insect Flight Muscle*, Elsevier/North Holland, New York, 1977.

Usherwood, P. N. R. (ed.). *Insect Muscle*, Academic Press, New York, 1975.

SKELETAL MUSCLE

Adelstein, R. S. (ed.). "Phosphorylation of muscle contractile proteins," *Fed. Proc.*, **39**, 1544–1573 (1980).

Adrian, R. H. "Charge movement in the membrane of striated muscle," *Ann. Rev. Biophys. Bioeng.*, **7**, 85–112 (1978).

Bárány, M., and K. Bárány. "Phosphorylation of the myofibrillar proteins," *Ann. Rev. Physiol.*, **42**, 275–292 (1980).

Bessman, S. P., and P. J. Geiger. "Transport of energy in muscle: the phosphorylcreatine shuttle," *Science*, **211**, 448–452 (1981).

Caputo, C. "Excitation and contraction processes in muscle," *Ann. Rev. Biophys. Bioeng.*, **7**, 63–83 (1978).

Curtin, N. A., and R. C. Woledge. "Energy changes and muscle contraction," *Physiol. Rev.*, **58**, 690–761 (1978).

deMeis, L., and A. L. Vianna. "Energy interconversion by the Ca^{2+}-dependent ATPase of the sarcoplasmic reticulum," *Ann. Rev. Biochem.*, **48**, 275–292 (1979).

Ebashi, S. "Excitation-contraction coupling," *Ann. Rev. Physiol.*, **38**, 293–313 (1976).

Eisenberg, E., and L. E. Greene. "The relation of muscle biochemistry to muscle physiology," *Ann. Rev. Physiol.*, **42**, 293–309 (1980).

Elbrink, J., and I. Bihler. "Membrane transport: its relation to cellular metabolic rates," *Science*, **188**, 1177–1184 (1975).

Endo, M. "Calcium release from the sarcoplasmic reticulum," *Physiol. Rev.*, **57**, 71–108 (1977).

Goslow, G. E., Jr. (ed.). "Skeletal muscle tissue," *Am. Zool.*, **18**, 97–166 (1978).

Hill, A. V. *First and Last Experiments in Muscle Mechanics*. Cambridge University Press, Cambridge, England, 1970.

Holloszy, J. O., and F. W. Booth. "Biochemical adaptations to endurance exercise in muscle," *Ann. Rev. Physiol.*, **38**, 273–291 (1976).

Homsher, E., and C. J. Kean. "Skeletal muscle energetics and metabolism," *Ann. Rev. Physiol.*, **40**, 93–131 (1978).

Josephson, R. K. "Extensive and intensive factors determining the performance of striated muscle," *J. Exp. Zool.*, special issue, 1975.

Kotani, M. (ed.). *Some New Approaches to Muscle Contraction*, University Park Press, Baltimore, 1980.

Martonosi, A. N. (ed.). "Calcium pumps," *Fed. Proc.*, **39**, 2401–2441 (1980).

Mauro, A. (ed.). *Muscle Regeneration*, Raven Press, New York, 1979.

Morgan, H. E., and K. Wildenthal (eds.). "Protein turnover in heart and skeletal muscle," *Fed. Proc.*, **39**, 7–52 (1980).

Sugi, H., and G. H. Pollack (eds.). *Cross-bridge mechanism in muscle contraction*, University Park Press, Baltimore, 1979.

Tregear, R. T., and S. B. Marston. "The crossbridge theory," *Ann. Rev. Physiol.*, **41**, 723–736 (1979).

CARDIAC MUSCLE

Alpert, N. R., B. B. Hamrell, and L. A. Mulieri. "Heart muscle mechanics," *Ann. Rev. Physiol.*, **41**, 521–537 (1979).

Berne, R. M., N. Sperelakis, and S. R. Geiger (eds.). *The Heart. Handbook of Physiology*, sec. 2, vol. 1. American Physiological Society, Washington, D.C., 1979.

Fabiato, A., and F. Fabiato. "Calcium and cardiac excitation-contraction coupling," *Ann. Rev. Physiol.*, **41**, 473–484 (1979).

Fozzard, H. A. "Heart: excitation-contraction coupling," *Ann. Rev. Physiol.*, **39**, 201–220 (1977).

Lieberman, M., and T. Sano (eds.). *Developmental and Physiological Correlates of Cardiac Muscle*, Raven Press, New York, 1975.

Morad, M. (ed.). *Biophysical Aspects of Cardiac Muscle*, Academic Press, New York, 1978.

Sanger, J. W. "Cardiac fine structure in selected arthropods and molluscs," *Am. Zool.*, **19**, 9–27 (1979).

SMOOTH MUSCLE

Bohr, D. F., A. P. Somlyo, and H. V. Sparks, Jr. (eds.). *Vascular Smooth Muscle. Handbook of Physiology*, sec. 2, vol. 2. American Physiological Society, Washington, D.C., 1980.

Bülbring, E., and M. F. Shuba (eds.). *Physiology of Smooth Muscles,* Raven Press, New York, 1975.

Hellstrand, P. "Mechanical and metabolic properties related to contraction in smooth muscle," *Acta Physiol. Scand.,* **Suppl. 464,** 1–54 (1979).

Murphy, R. A. "Filament organization and contractile function in vertebrate smooth muscle," *Ann. Rev. Physiol.,* **41,** 737–748 (1979).

Prosser, C. L. "Smooth muscle," *Ann. Rev. Physiol.,* **36,** 503–537 (1974).

Stephens, N. L. (ed.). *The Biochemistry of Smooth Muscle,* University Park Press, Baltimore, 1977.

Fred N. White

5

RESPIRATION

5.1 INTRODUCTION

Lavoisier and Laplace (1780) said that "life is a combustion." Essentials for that combustion are the acquisition of oxygen, elimination of carbon dioxide, and the transport of these gases to and from the combustion sites in the tissues. Demands for oxygen vary widely among animals and with their state of activity (Table 5-1). In this chapter we shall be concerned with the manner in which these demands of metabolism are met by internal transport and exchanges with the environment.

Because oxygen diffuses slowly in aqueous media, dependence on diffusion alone can support an actively metabolizing organism no larger than about 0.5 mm in radius. The evolution of larger animals has been accompanied by the development of specialized systems for extracting oxygen from, and eliminating carbon dioxide to, the environment. Indeed, metabolism is often expressed in terms of oxygen consumption or carbon dioxide production (see Chapter 3). This exchange with the environment by animals stands in a remarkable balance with the botanical mass of the world, for plants utilize carbon dioxide, yielding up oxygen

which can then enter once again into animal metabolism.

Diverse means, such as gills, integument, and lungs, are utilized for gaseous exchange. These generally have in common the presentation to the external environment of a large, well-vascularized surface area across which diffusion of gases occurs. But this is not enough to meet the problem, for the tissues consuming oxygen may be distant from the respiratory exchange surfaces. Transport systems must move the acquired gas to its destination. This is accomplished by the circulatory system under the control of delicate mechanisms that regulate local blood flows in balance with oxygen and other nutritive requirements of the tissues. Because the solubility of oxygen is low in aqueous media, the plasma holds only a small fraction of the oxygen circulated. Specialized pigments, hemoglobins in vertebrates (hemoglobins and other pigments in many invertebrates), capture the bulk of the oxygen at the respiratory exchange surfaces and unload the precious cargo at the tissue. Hemoglobins also enter into the transport of carbon dioxide on the return trip to the respiratory surfaces.

Varying demands associated with activity and dif-

TABLE 5-1. Utilization of Oxygen by Various Animals under Conditions of Rest and Activity

	Weight	Oxygen Consumption [ml/(kg hr)]
Paramecium	0.001 mg	500
Mussel (Mytilus)	25 g	22
Crayfish (Astacus)	32 g	47
Butterfly (Vanessa)	0.3 g	
resting		600
flying		100,000
Carp (Cyprinus)	200 g	100
Pike (Esox)	200 g	350
Mouse	20 g	
resting		2,500
running		20,000
Man	70 kg	
resting		200
maximal work		4,000

Source: A. Krogh. *The Comparative Physiology of Respiratory Mechanisms.* University of Pennsylvania Press, Philadelphia, 1959.

ferential oxygen consumption by the tissues are met by a wonderful system of controls on respiratory exchange. By neural elements, this system senses changes in the gas or hydrogen ion composition of the blood. The central nervous system then institutes appropriate modifications in respiratory activity. Anyone who has done strenuous work has been aware of the increases in respiratory rate and depth that result; conversely, quiet sleep or rest are associated with reduction in respiratory activity.

Various environmental situations offer special respiratory problems. The solution of some of these problems will form the subject matter of a good deal of this chapter.

5.2 GASES IN THE EXTERNAL ENVIRONMENT

Comparison of the gaseous composition and physical characteristics of air and water emphasizes the range of respiratory problems faced by terrestrial and aquatic vertebrates. The oxygen content of air is about 20 times that of water saturated with air. It is thus apparent that, to extract a given quantity of oxygen, aquatic vertebrates must pass over their

respiratory surfaces a far greater volume of the milieu than air breathers. This situation is further complicated by the fact that the diffusion rate for oxygen in water is very low compared to that in air. The density of water (1000 times that of air at normal pressure) and its viscosity (100 times that of air) impose a larger work load on respiratory pumping, which compounds the problem of extraction from an aquatic environment. From these physical properties it is not surprising to find that fishes expend about 20% of their oxygen consumption in support of external respiration at rest as compared to 1–2% in mammals.

The carbon dioxide (CO_2) content of natural waters is low and often almost nil. CO_2 is extremely soluble in water and diffusion is rapid compared to oxygen. Most of the CO_2 enters into H_2CO_3 (carbonic acid) or is found in the form of carbonates or bicarbonates. This buffering capacity maintains a reduced partial pressure of free CO_2 and aids in maintaining a favorable gradient for exchange by water dwellers.

Rising temperature reduces the solubility of gases in fresh water. A change from 5°C to 35°C reduces the oxygen content from 9 ml/liter to 5 ml/liter. A rise in body temperature produces an increase in metabolism, so a fish experiencing a body-temperature increment is under a double handicap. More water must be pumped across the gill surfaces to extract the same amount of oxygen as was needed at the lower temperature. In addition, the increased metabolism must be supported. Table 5-2 illustrates the influence of temperature on oxygen content of water, metabolism, and ventilation in a resting and an active fish.

Another limitation on oxygen availability in natural waters is imposed by the amount of dissolved salts. Whereas 3.6% NaCl (approximate salinity of the ocean), contains 38 ml O_2/liter when equilibrated with 1 atm of O_2 at 0°C, 2.9% NaCl contains 40 ml O_2/liter, and pure water has 49 ml O_2/liter at the same temperature and partial pressure. Thus elevated temperature and increasing salinity combine to limit oxygen availability.

Finally, bodies of water may have oxygen-poor zones (swamps, below the thermocline of deep lakes), owing to poor circulation and oxygen consumption of bottom detritus.

Extraction of oxygen from different media thus presents special problems as a result of the physical

TABLE 5-2. Influence of Temperature on Oxygen Content of Water, Metabolism, and Ventilation in the Goldfish

Temp. (°C)	Oxygen Content in Water (ml/l)	Oxygen Consumption [ml/(kg hr)] Inactive	Oxygen Consumption [ml/(kg hr)] Active	Ventilation in Liters/(kg hr) at 75% Utilization
5	9.0	8	30	1.3
15	7.0	50	110	9.0
25	5.8	140	255	32.0
35	5.0	225	285	60.0

Source: F. E. J. Fry and J. S. Hart. *Biol. Bull.,* **94,** 66–67 (1948).

characteristics of the environment. Table 5-3 summarizes the major contrasts of aquatic and aerial conditions.

5.2.1 Gas Pressures and Diffusion

The fact that gases have mass means that the gases of our atmosphere are attracted by the earth's gravitation field. Thus the atmosphere exerts a pressure proportional to the weight of a column of air above the surface of the earth which extends to the limit of our atmosphere. Normal atmospheric pressure at sea level is sufficient to support a mercury column 760 mm in height. We thus refer to sea-level atmospheric pressure as 760 mm Hg. As air is composed chiefly of nitrogen (79.02%), oxygen (20.94%), and carbon dioxide (0.04%), each contributes to the total pressure in direct proportion to the percent composition. Thus the pressure exerted by nitrogen at sea level is 79.02% of 760 mm Hg, or 600.55 mm Hg; that of oxygen is 159.16 mm Hg; and for carbon dioxide the pressure exerted is 0.30 mm Hg. This is in accord with **Dalton's law of partial pressures,** which states that in a gas mixture each gas will exert a pressure as if it alone occupied the entire volume.

TABLE 5-3. Major Contrasts between Aquatic and Aerial Respiration

	Aquatic (Fish)	Aerial (Mammals)
Viscosity of medium	H_2O 100 times air	
Density of medium	H_2O 1000 times air	
Diffusion rate, O_2	Low	High
O_2 content of medium on inspiration	Nil to 10 ml/liter	100–130 ml/liter: lower than outside air due to dead space
CO_2 content of medium on expiration	Low (0–13 ml/liter)	>100 ml/liter
Respiration by:	Gills—exchange across secondary lamellae	Lungs—exchange across alveolus
Ventilation	Continuous	Tidal
O_2 utilization, %	Up to 80	25
O_2 consumption used for respiratory pump, %	20	1–2

Modified from: G. M. Hughes. *Comparative Physiology of Vertebrate Respiration,* Harvard University Press, Cambridge, Mass., 1963.

The presence of water vapor in a gas mixture will, of course, contribute to the total pressure of the mixture. For this reason, atmospheric gases are usually dried before measurements are made and expressed in terms of dry air. On inspiration and at a temperature of 37°C, the atmospheric air becomes saturated with water vapor, the water-vapor pressure being equivalent to 47 mm Hg. To calculate the partial pressures of the respiratory gases, this value must be subtracted from the total atmospheric pressure. For oxygen, $760 - 47 = 713$ mm Hg $\times 0.209 = 149$ mm Hg. This is the partial pressure of oxygen on inspiration and saturation with water vapor at 37°C. The water-vapor pressure for a range of physiologically important temperatures is given in Table 5-4.

The pressure exerted by a gas depends also on the volume of the gas. If we compress a given number of molecules of a gas to half their original volume, the pressure will increase. Conversely, as we ascend to high altitudes, the gases are compressed less and their partial pressures fall. At 25,000 ft the atmospheric pressure is 282 mm Hg and the P_{O_2} about 59 mm Hg (see Figure 6-47).

An elevation of temperature increases the rate of molecular movement and thus the pressure exerted by the gas. So long as the volume remains constant, the pressure of a gas is directly proportional to the absolute temperature.

In calculating the partial pressures of gases, then, it is necessary to consider temperature, volume, vapor tension, and total pressure. Table 5-5 summarizes the partial pressures of the respiratory gases in mammals. (Note, when looking at the table, that the partial pressure of oxygen in inspired and expired gas is less than in the atmosphere because of the saturation with water vapor and diffusion of some of the oxygen into the blood stream.)

Because a large volume of gas may be dissolved in a liquid does not necessarily mean that its partial pressure is high. If we drive off all gas from water by boiling or chemical absorption and place the gas in a chamber over the water, the gas molecules will diffuse into the water until the gas pressures are equal in both phases when an equilibrium is attained. If the gas were extremely soluble in water but exerting a low partial pressure in the gas phase above the water, a large volume of gas, but at a low partial pressure, would exist in the aqueous phase. Thus, very soluble gases, such as acetone, may be dissolved in large quantities at low partial pressures. Conversely, gases of low solubility may be present in small quantity but at very high partial pressures.

The effect of increasing the partial pressure (tension) of a gas that will come to equilibrium with an aqueous medium is illustrated by the observation that the oxygen content of water in equilibrium with air is about 5 ml/liter. In equilibrium with pure oxygen at atmospheric pressure, the oxygen content of water would be about 5 times higher. This is because the partial pressure of pure oxygen is approximately 5 times that of oxygen in ordinary air. Diffusion is toward the lower partial pressure.

TABLE 5-4. Water-Vapor Pressures (P_{H_2O}) at Various Temperatures

Temp. (°C)	P_{H_2O} (mm Hg)	Temp. (°C)	P_{H_2O} (mm Hg)
20	17.5	29	30.0
21	18.7	30	31.8
22	19.8	31	33.7
23	21.1	32	35.7
24	22.4	33	37.7
25	23.8	34	39.9
26	25.2	35	42.2
27	26.7	36	44.6
28	28.3	37	47.0

Source: J. H. Comroe, Jr. *Physiology of Respiration,* The Year Book Medical Publishers, Inc., Chicago, 1966.

TABLE 5-5. Partial Pressures of Respiratory Gases (mm Hg)—Man

	P_{O_2}	P_{CO_2}	P_{N_2}	P_{H_2O}	Total
Atmosphere (dry)	159	0	601	0	760
Inspired gas (saturated)	149	0	564	47	760
Expired gas (saturated)	116	28	569	47	760
Alveolar gas (saturated)	100	40	573	47	755
Arterial blood	95	40	573	47	755
Venous blood	40	46	573	47	706
Tissues	<30	>50	573	47	700

Source: P. C. Johnson. "Respiratory Gas Exchange and Transport." In E. E. Selkurt (ed.). *Physiology,* pp. 417–443, Little, Brown and Company, Boston, 1963.

Fick's law describes the diffusion of a substance from a region of higher to that of owner concentration through a cross-sectional area A. The law takes the form

$$\frac{ds}{dt} = - DA \frac{dc}{dx}$$

where ds/dt is the differential expression for rate of transport (moles of s in time dt), dc/dx the concentration gradient between the points being considered, and D the diffusion coefficient is dependent on the character of the medium. The *diffusion coefficient* of a gas is the number of milliliters diffusing through 1 cm²/min when the concentration gradient is 1. Its units are thus milliliters per area per time per concentration gradient. Note that Fick's law utilizes *concentration* gradients. A consideration of a simple experiment illustrates that the law, when applied to gases, is more correctly expressed in terms of activity gradients or partial pressure. In the experiment saline was equilibrated at a P_{O_2} of 100 mm Hg and olive oil was equilibrated at 40 mm Hg. Both liquids were held at 37°C. The concentration of oxygen in olive oil was about twice that in saline when the two solutions were placed on either side of a membrane. Oxygen, however, diffused into the olive oil, not toward a region of lower concentration but toward a lower partial pressure. It is best, therefore, to deal with the diffusion coefficient in terms of pressure, and the units of D become milliliters per area per time per pressure gradient. Substituting p for c in Fick's equation we have the expression as currently utilized:

$$\frac{ds}{dt} = - DA \frac{dp}{dx}$$

The actual calculation of D for a biological membrane is most difficult, especially where more than one membrane is involved and determination of true pressure gradients between environmental medium and capillaries represents a difficult technical problem. The diffusion coefficient of oxygen for the human lung $[D_{L_{O_2}}]$, ml of dry oxygen (STP), diffusing across the pulmonary membrane per minute per mm Hg of partial pressure between alveolar air and the pulmonary capillary blood, has been estimated at about 15 ml/min per mm Hg at rest.

5.2.2 Determinants of Gas Content in a Medium

One of the primary properties that determines the amount of O_2 or CO_2 carried by air or water is the *capacitance coefficient*, β, of the medium for the particular gas. The capacitance coefficient is defined as the increment in concentration of the gas (C_g) per increment in partial pressure: $\Delta C_g/\Delta P$ (mmol · liter⁻¹ · torr⁻¹). Note that "torr" is used interchangeably with mm Hg. In liquids where

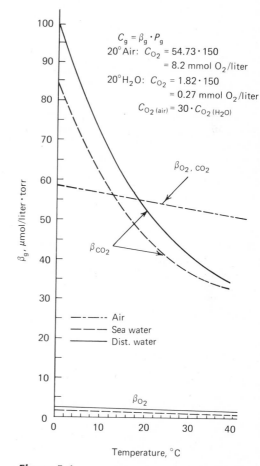

Figure 5-1
The capacitance coefficient for O_2 CO_2 (β) as a function of temperature for air, distilled and sea water. C_g = concentration of the gas; β_g = capacitance coefficient of the gas; P_g = partial pressure of the gas.

no chemical binding of the gas occurs, β is equal to the physical *solubility coefficient,* α. Since chemical binding for CO_2 occurs in blood and water and O_2 binds with hemoglobin in blood, the solubility coefficient is not the appropriate characteristic to utilize where *quantities* of the gas are of concern.

For a gas phase, one can show from the ideal gas law: $P \cdot V = M \cdot R \cdot T$, that $\beta = M/V \cdot P = 1/R \cdot T$, where P = partial pressure, V = volume; M = amount of the substance; R = gas constant; and, T = absolute temperature. The capacitance coefficient is equal for all gases in an "ideal" gas mixture. For O_2 in water, as indicated above, β is equal to physical solubility.

When we consider CO_2 in water containing buffer substances, such as bicarbonate or borates, β will depend on the influence of P_{CO_2} on the formation of the CO_2 containing buffer substances. Thus, for sea water: $CO_2 + CO_3^{2-} + H_2O \rightleftharpoons 2\ HCO_3^-$; and, $CO_2 + H_2BO_3^- + H_2O \rightleftharpoons HCO_3^- + H_3BO_3$. These equations state that due to formation of HCO_3^- as P_{CO_2} increases, the CO_2 concentration will be higher than the physical solubility would predict.

Temperature exerts an important influence on β, which, at any given temperature, is expressed to varing degrees for O_2 and CO_2 depending on the medium involved. At 20°C, β_{CO_2} is much higher than β_{O_2} in both seawater and distilled water (Figure 5-1). It can also be seen that the concentration of O_2 in water, at an atmospheric P_{O_2} of 150 mm Hg, is about $\frac{1}{30}$ of that in air ($C_g = \beta_g \cdot P_g$). This characteristic of water forces water breathing animals to convect relatively large quantities of the medium across the respiratory surfaces in order to achieve an O_2 uptake equivalent to that of an air breather. Since β_{CO_2} exceeds β_{O_2} in water, the high ventilation requirement results in low expired and inspired partial pressure differences for CO_2, that is $(P_e - P_i)_{CO_2}$ (Figure 5-2). Since in air, $\beta_{CO_2} = \beta_{O_2}$, $(P_e - P_i)_{CO_2}$ will be similar to $(P_i - P_e)_{O_2}$, differences only reflecting respiratory quotient deviations from 1 (RQ = CO_2 production/O_2 consumption). The R lines, for animals that rely on a combination of lungs, gills, and skin (bi- and trimodal breathers), occupy positions intermediate between the extremes of water and air breathers.

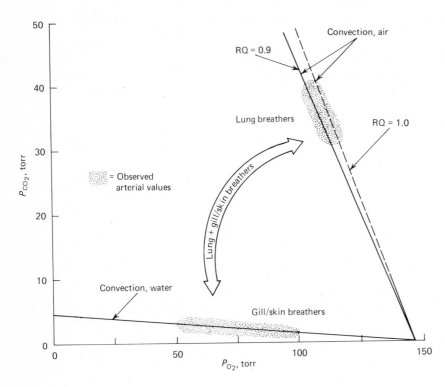

Figure 5-2
The O_2–CO_2 diagram for respiratory transport of gases by convection in air and water. Note that bimodal and trimodal breathers are in an intermediate zone indicated by the arrow. [*Based on J. Piiper and P. Schied. In J. G. Widdicombe (ed).* Internat. Rev. Physiol., Respir. Physiol. 2, *Vol. 14, University Park Press, Baltimore, 1977.*]

5.3 NATURE OF VERTEBRATE RESPIRATORY ORGANS

The organs of external respiration of all vertebrates present to the environment large, usually thin, and richly vascularized surfaces. This basic structure produces a close association of the blood with the external medium across distances small enough that gas exchange can take place rapidly by diffusion. Three principal organs are used by vertebrates to accomplish exchange. Gills represent either totally external appendages (as in the neotenic salamander *Necturus*) or filamentous leaflets within a protected opercular chamber (as in fishes). Some fishes and a number of amphibians also use the integument. Dependence on cutaneous respiration is seldom total and is usually associated with either gills or lungs (the terrestrial lungless salamanders are among the exceptions). Both gills and lungs are embryologically derived as outpouchings of the gut; most have the advantage of a protected internal location connected to the exterior by tubes. Circulation of gas is usually accomplished by rhythmic pumping mechanisms; however, water may be lost in this exchange, especially in endothermic air-breathing animals. Such water loss is not a problem for aquatic forms, a situation that aids water homeostasis. Gills, however, generally provide a poor solution to air breathing, because the finely divided and thin filaments collapse against each other, exposing a smaller surface area for exchange.

Respiratory exchange by lungs involves an aqueous interface because lungs are kept moistened by a fine film of fluid through which gases must diffuse. This film is necessary to prevent dehydration of the exchange surface.

5.3.1 General Model for Gas Exchange

Because diffusion of gases is a slow process, organisms beyond a certain small, critical size must depend on convective systems to assure rapid turnover of gases in blood and external medium. The result is the establishment of more favorable partial pressure gradients to drive diffusion, which is the intermediate process between both gas acquisition and gas elimination. A generalized model of external gas exchange (Figure 5-3) emphasizes these two processes, convection and diffusion, and the prop-

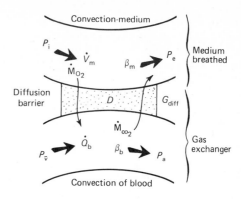

For Convective transport by medium:
$$\dot{M}_g = \dot{V}_m \cdot \beta_m \cdot (P_i - P_e)$$

For Convective transport by blood:
$$\dot{M}_g = \dot{Q}_b \cdot \beta_b \cdot (P_a - P_{\bar{v}})$$

Figure 5-3
Basic components of a gas exchange model. Both convective and diffusive transport of gases are involved. The major diffusive barrier consists of the tissue and fluids that separate the medium from blood. Since diffusion is slow in aqueous media, it is important that this barrier is thin if effective gas transfer is to occur. Convection of medium breathed and blood are necessary in order to assure renewal of gas partial pressure gradients that drive gas exchange across the diffusion barrier.

erties that determine gas transfer between organism and environment. In the model, the following symbols are used:

$\dot{M}_g$ *(transfer rate)*: the amount of gas exchanged over unit time, that is, O_2 consumption and CO_2 output (mmol · min^{-1}).

$\dot{Q}_b$ *(blood flow)*: flow rate of blood (ml · min^{-1}).

$\dot{V}_m$ *(flow of medium)*: flow rate of air or water (ml · min^{-1})

β *(capacitance coefficient)*: the increment in concentration per increment in partial pressure, $\Delta C/\Delta P$ (mmol · liter^{-1} · torr^{-1}). When no chemical combination is involved β = the solubility coefficient, α.

P *(partial pressure)*: (torr = mm Hg).

D *(diffusing capacity)*: depends on characteristic of the barrier (mmol · min^{-1} · torr^{-1}).

G_{diff} *(diffusive conductance)*: equals D from Fick's law of diffusion: $G_{diff} = D = D \cdot \alpha \cdot A/x$, where A = area of barrier, x = thickness, and α = the physical solubility of the gas in the barrier material.

In the model both blood and external medium are brought into close approximation but are of necessity separated by a barrier where diffusion of the gases links the transfer between the convective streams. If the barrier were thick and characterized by low diffusion coefficient, a limitation on the gas transfer process would occur, that is, G_{diff} may become small and limiting, especially in hypoxic circumstances.

The transfer rate of a gas is the product of several physical parameters: for convective transport by the medium, $\dot{M}_g = \dot{V}_m \cdot \beta_m \cdot (P_i - P_e)$; for blood, $\dot{M}_g = \dot{Q}_b \cdot \beta_b \cdot (P_a - P_{\bar{v}})$. Organisms are highly dependent on altering convection and partial pressure gradients in order to meet changing gas transport needs. Thus, in exercise, increases in $\dot{V}_m$ and $\dot{Q}_b$ are commonly involved. It should be apparent that generally equivalent increases in the convection of *both* blood and medium must be made, that is, $\dot{V}_m/\dot{Q}_b$ should remain rather constant for efficient transfer of gases. Reduction in this ratio would create more unfavorable partial pressure gradients for transfer of both O_2 and CO_2 across the gas exchanger.

Values for both β_{CO_2} and β_{O_2} in blood may not remain constant. This is especially the case of O_2 since a plot of blood O_2 content against P_{O_2} is nonlinear over the physiological range of P_{O_2}'s (see Figure 5-4). At both high and low P_{O_2}'s, β_b declines, but, importantly, at normal venous P_{O_2}'s the value is large, a factor augmenting unloading of O_2 to the tissues and promoting rapid uptake by hemoglobin when venous blood approaches partial pressure equilibrium in gas exchange organs.

Specific features of the general model of gas exchange differ among the vertebrate groups. The manner in which these arrangements influence gas exchange will be discussed in Section 5-6.

5.3.2 Ventilation and Respiratory Volumes

Exchange of gases requires constant renewal of the environmental medium in contact with the respiratory surfaces. The respiratory pump, alternating between inspiratory and expiratory activity, performs the work of ventilation. There passes through the gills or lungs within 1 min a quantity of environmental medium known as the *ventilation volume* (or *minute volume*). It is the product of the respiratory rate and the volume of air or water drawn across respiratory structures during each respiratory cycle. Alterations in rate or depth of respiratory activity bring about adjustments in ventilation volume. This variable-flow feature of ventilation provides one means of adjusting gas exchange in accordance with the metabolic needs of animals.

The greatest volume of gas contained by the lungs is present after a forced inspiration. In man, the maximum lung capacity is about 6 liters, and this total has been divided into several components by physiologists. During quiet breathing, about 500 ml, the *tidal volume,* is inspired and expired with each respiratory cycle. A certain quantity remains within the lungs at the close of expiration, and a greater volume can be taken in with a more forceful inspiration.

The respiratory volumes may be measured by the use of a recording spirometer (Figure 5-5) into which a subject breathes. Air within the lungs is

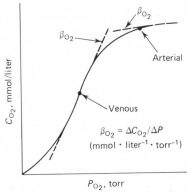

Figure 5-4
Typical sigmoid hemoglobin dissociation curve for vertebrate blood. The capacitance coefficient, β_{O_2}, is not constant, but varies as a function of the partial pressure. The dashed lines represent the slopes of the curve, which are equivalent to β_{O_2}, at the arterial and venous points.

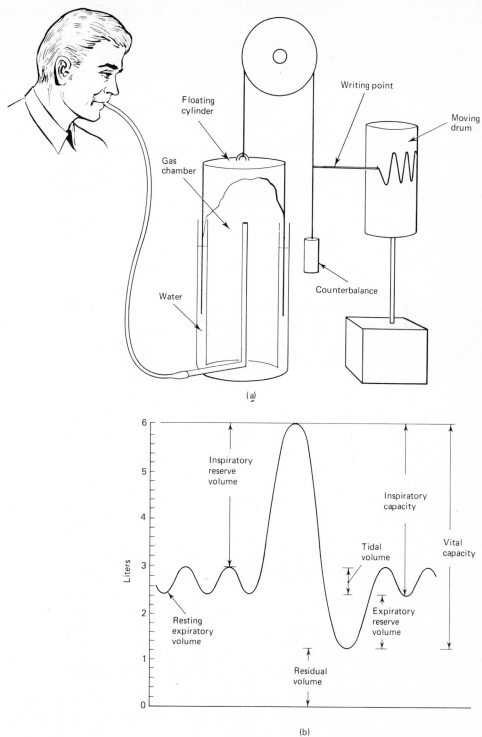

Figure 5-5
Lung volumes and capacities of a human subject and their measurement.

Floating cylinder

Writing point

Moving drum

Gas chamber

Counterbalance

Water

(a)

Inspiratory reserve volume

Inspiratory capacity

Tidal volume

Vital capacity

Liters

Resting expiratory volume

Expiratory reserve volume

Residual volume

(b)

transferred through the tube in the subject's mouth into the gas chamber of the spirometer during expiration, forcing the floating cylinder to rise and thus causing the writing point to inscribe a downward deflection on the moving drum of a kymograph. During quiet respiration, rhythmic fluctuations are recorded and the volume displaced is the tidal volume. If, at the end of a normal expiration an additional forceful expiratory effort is made, about 1200 ml of gas, the *expiratory-reserve volume,* may be expelled. Following a normal inspiration, an additional 3000 ml may be drawn into the lungs, a quantity termed the *inspiratory-reserve volume.* The sum of the expiratory-reserve, tidal, and inspiratory-reserve volumes represents the *vital capacity* (about 4800 ml), a measure of the maximum ventilation volume that can be attained. During exercise, the tidal volume increases and inspiratory and expiratory reserve volumes decrease.

The lungs cannot be collapsed by voluntary thoracic movements. Thus, even after the most forceful expiratory effort, a *residual volume* of approximately 1200 ml remains in the lungs.

Because portions of the airways, trachea, bronchi, and bronchioles do not participate as respiratory exchange surfaces, the inspired gas that fills these structures is not subject to gas exchange with the blood. This *anatomical dead space* is about 150 ml in volume. Thus, of a 500-ml tidal volume, only 350 ml ventilates the alveoli, a relationship of which physiologists must be aware when using tubes (dead space) to connect an experimental animal to a respiratory pump. In addition, nonfunctional alveoli may be ventilated, adding a *physiological dead space.*

The ventilation volume of a man during rest (at a tidal volume of 500 ml and a respiratory frequency of 12 per min) is 6 liters/min. This is a fair figure in resting man, but during heavy exercise both rate and tidal volume increase and ventilation volume may reach values in excess of 100 liters/min.

The anatomic analogs of the alveoli in fishes are the thin-walled secondary lamellae of the gills (see Section 5.3.3). Different portions of the respiratory water may be distinguished during gill ventilation. The axial stream of water passing adjacent lamellae loses little if any of its oxygen and may be regarded as physiological dead space. The gill filaments normally touch at their tips and most of the water passes between the secondary lamellae. However, with excessive pressure gradients across the gills induced by increased respiratory pumping, some water may be shunted between the filament tips. This volume, which fails to pass across the secondary lamellae, is equivalent to the anatomical dead space of the lung. The anatomical dead space is normally small, because the movements of the branchial arches and the gill filaments tend to maintain the entire gill as a sieve of secondary lamellae across the path of the water current. No data exist concerning the relative magnitudes of physiological and anatomical dead spaces in fishes. G. M. Hughes regards the anatomical dead space as a protective shunt that operates when the pressure differential across the gills becomes excessive.

Ventilation is increased in fishes by a combination of increasing respiratory rate and *respiratory stroke volume* (a preferable term to tidal volume for a flow-through system). The pattern is variable, however; some fishes (eel) use a combination of increased respiratory stroke volume and rate, whereas others (trout) rely predominantly on an increase in respiratory stroke volume.

5.3.3 Gills and Skin

The gills of most fishes are the major gas-exchange surfaces. Gas exchange is not the exclusive function of gills, however; they also function as secretory and excretory organs (see Section 7.4.1).

A series of gill arches form the major support of the gills. From the arches project gill filaments, each of which bears on the dorsal and ventral sides a group of leaflike secondary lamellae (Figure 5-6). These secondary lamellae are the respiratory exchange surfaces. Blood vessels passing through the gill arches send branches into the filaments from which smaller vessels of the secondary lamellae arise. As indicated in Figure 5-6c, the flow of blood in the vessels of secondary lamellae is in a direction counter to water flow.

Secondary lamellae consist of two epithelia that are continuous at the lamellar margins. The epithelia are separated by numerous pillar cells, the flanges of which enclose the blood spaces (Figure 5-7). Fibrillar material is present in the pillar-cell cytoplasm. The similarity of the pillar fibrils with that of smooth muscle has led Hughes and Grimstone to speculate that the pillar cells may be con-

Figure 5-6
(a) Position of the four gill arches beneath the operculum on the left side of a fish. (b) Part of two of these gill arches are shown with the filaments of adjacent rows touching at their tips. The blood vessels that carry the blood before and after its passage over the gills are shown. (c) Part of a single filament with three secondary folds on each side. The flow of blood is in the opposite direction to the water. [*From G. M. Hughes.* New Sci., **11**, 346–348 (1961).]

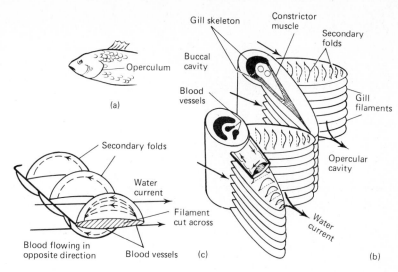

tractile. If true, shortening of the pillar-cell bodies and subsequent constricting of the blood spaces would be an important determinant of blood flow in the secondary lamellae. Secretion of the basement membrane, which is interposed between epithelia and pillar cells, is also a function of pillar cells. Observations on a cichlid (*Haplochromis multicolor*) and the pollack (*Gadus pollachius*) reveal that the water-to-blood pathway is from 0.31 to 2.0 μm and 1 to 3 μm respectively. These figures are remarkably close to the air-to-blood distance (0.36–2.5 μm) observed in the mammalian lung. Hughes and Grimstone break down the total distance in the pollack as follows: external epithelia layer, 0.4–2.5 μm; basement membrane, 0.3 μm; pillar-cell flange, 0.1–0.3 μm. These components represent the diffusion pathway across which gases move into and out of the blood.

Hughes, who measured gill area in a large series of marine teleosts, found that more active fish not only have a larger gill area for gas exchange but that the area is increased in such a way as to keep resistance to flow to a low value.

The arrangement of blood flow in secondary lamellae in a direction counter to water flow has considerable bearing on the efficiency of gas exchange across the respiratory surfaces. Consider the two arrangements of tubes shown in Figure 5-8. It can be appreciated that, if the objective were to transfer heat from the tube containing warm water to that containing cold water, the counterflow system

would be the most efficient arrangement. Likewise, the transfer of oxygen to blood is greatly facilitated by countercurrent exchange. The importance of countercurrent flow is emphasized by studies in which oxygen extraction from water was measured before and after experimentally reversing the direction of water flow across the gills of teleosts. About 80% extraction was observed in the normal countercurrent situation. When the direction of water flow was reversed, the extraction fell to 10%. Countercurrent exchange is a feature of the gills of cyclostomes and elasmobranchs as well as teleosts. Engineers, utilizing this principle to conserve valuable calories in industrial situations, were anticipated by millions of years by primitive vertebrates.

External gills are utilized by a number of vertebrates. They are present in a number of larval fishes and are lost with the appearance of the adult branchial structures. The development of external gills of larval amphibians shows a close correlation with oxygen content of the water. Low environmental-oxygen tensions are associated with increases in length and total surface area of the gills in both *Rana* and *Salamandra*. Among the curious examples of external gills are the filamentous respiratory structures of the male lungfish, *Lepidosiren*. A mass of vascular filaments grows on the pelvic fins during the period when the male cares for the nest. The lungs are recharged with air during periodic excursions to the surface. On returning to the nest, the pelvic gill filaments, supplied with well-oxygenated

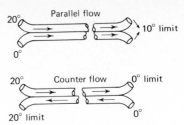

Figure 5-8
Exchange through parallel flow and counterflow. [*From P. F. Scholander.* Hvalradets Skrifter Norske Videnskaps-Akad. Oslo, **44**, *1–24 (1958).*]

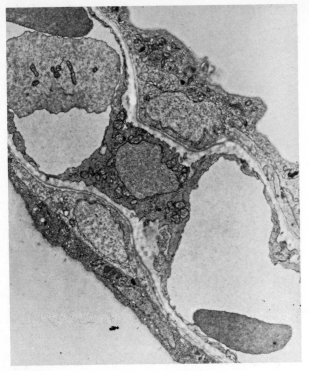

Figure 5-7
An approximately median section (5000×) of a pillar-cell body and its flanges. Note the large nucleus and the abundant mitochondria and tubules set in a dense cytoplasmic matrix. A column is present, enclosed in an infolding of the plasma membrane. Note the junction of one of the pillar-cell flanges with that of an adjacent pillar cell. In the epithelia, note the location of the nuclei opposite the pillar-cell body and the presence of more than one cell layer. Both nucleus and cytoplasm in the outermost cell of the lower epithelium are markedly denser than those of the other cells. [*From G. M. Hughes and A. V. Grimstone.* Quart. J. Microsc. Sci., **106**, *343–353 (1965).*]

blood, apparently serve as an oxygen supply for the eggs.

Cutaneous respiratory exchange is associated, in fishes and amphibians, with a vascular, thin, and moist skin. As metabolism, in some groups, is closely related to surface area (see Section 3.7.3), it is theoretically possible for the skin to serve as the sole respiratory surface. An obvious limitation is imposed by the vulnerability of a thin, highly vascular, integument. Furthermore, in terrestrial situations, a moist integument presents a major av-enue of water and electrolyte loss. A number of fishes and amphibia, however, rely on the skin for a significant portion of their respiratory exchange. Eels can exchange a large fraction of the respiratory gases over the skin (see Figure 5-18), as can frogs. During hibernation, practically all gas exchange in amphibians is cutaneous.

5.3.4 Lungs

The development of internal nares, lungs, and the buccal force pump by Devonian fishes, similar to present-day lungfishes, paved the way for the amphibian migration to land. These respiratory developments were associated with modification in limb structure, making terrestrial locomotion possible.

The lungs of vertebrates vary tremendously in complexity, ranging from the simple saclike lungs of the Dipnoi to the complex divided lungs of mammals and birds. The general evolutionary trend is toward increasing subdivision of the airways and greater surface area at the exchange surfaces.

There are exceptions to the trend, however, for some amphibia (Plethodontidae) have secondarily lost their lungs and depend primarily on cutaneous respiration.

In amphibians the trachea is not subdivided into secondary tubes or bronchi but ends abruptly at the anterior pole of the lungs, where cartilage plates maintain the connection patent. The lungs are subdivided by incomplete septa in frogs, and between these larger septa are secondary septa that bound the alveoli or terminal air spaces where exchange occurs. The alveolar diameter of lower vertebrates is larger than in mammals (frog alveolus is about 10 times the diameter of human alveolus). This is associated with the much larger relative surface area of the mammalian lung. The respiratory surface of the frog, *Rana*, is approximately 20 cm² for each cubic centimeter of air contained, whereas that of man is about 300 cm². The nature of the gas-conducting and -exchange area of the human lung is shown in Figure 5-9.

Air and blood are brought into closest contact in the alveoli, but are separated by a barrier composed of the epithelial lining of the alveolus, interstitial elements (basement membrane and connective tissue), and the capillary endothelial cells. Furthermore, there is a thin surface film covering the alveolar epithelium. Gases exchanged between air and blood must cross this barrier, which, in the rat lung, is approximately 0.5 μm thick.

The lungs are elastic bodies. (Elasticity is the property of materials to return to their initial state following the removal of a deforming force. Elastic tissues act, then, like springs.) Inflation of the lungs is accompanied by an increase in potential energy. The conversion of this potential energy into kinetic energy during deflation supplies a portion of the motive force needed for expulsion of gases. In other words, the lungs exhibit elastic recoil; a part of the energy put into expansion is recovered on deflation.

The elasticity of the lungs can be studied by inflating them with air or saline and measuring the resulting pressure under static conditions. This procedure equates force and stretch with pressure and volume. When several points are determined, the slope of the plot gives a measure of the stiffness (Figure 5-10). This measure is referred to as *compliance*, the change in volume per unit change in pressure, measured in liters per centimeter of water.

In 1929 von Neergaard made a startling discov-

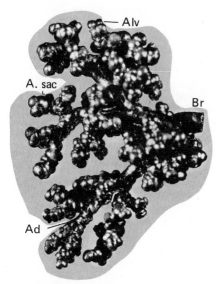

Figure 5-9
Cast of a small secondary lobule from the periphery of a newborn infant's lung. The principal bronchiole (Br) gives rise to two further generations of smooth-walled (nonrespiratory) bronchioles, which, in turn, produce two ranks of alveolar ducts (Ad), tipped by alveolar sacs (A sac) bearing terminal alveoli (Alv). Prepared by C. W. Mueller, Curator, Department of Anatomy, University of Maryland School of Medicine. [From V. E. Krahl. In W. O. Fenn and H. Rahn (eds.), Handbook of Physiology, *sec. 3, vol. 1, pp. 213–284, American Physiological Society, Washington, D.C., 1964.]*

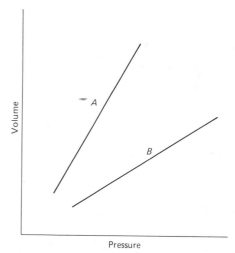

Figure 5-10
Plot of pressure and volume of two closed tubes. Line A *represents a body of lesser stiffness (greater distensibility) that line* B. A *is the body of greater compliance.*

ery in his studies of pressure/volume curves generated by lungs filled with saline or air. He found that the pressure necessary to enlarge the lungs to a given volume was less when the lungs were filled with liquid than when they were charged with air. The liquid-filled lung did not recoil with the same force as the air-filled lung! It was deduced that the differences were caused by the nature of the environment—alveolar interface, the interface being liquid-liquid, on the one hand, and gas-liquid, on the other. In the saline-filled (liquid-liquid interface) lung, surface tension must be reduced to nearly zero, so the curves obtained from these lungs must result from the elastic properties of the lung wall; in the air-filled situation the result must represent the cumulative effects of wall properties *plus* surface tension of the film lining the lungs.

The attractive forces of molecules maintain the continuity or cohesiveness of matter. These forces are equally distributed in direction except at an interface such as water with air. Here the surface molecules of water are attracted more strongly from below and laterally than from above. As a result, the surface molecules tend to "sink" (literally are pulled downward), producing the characteristic meniscus seen in a liquid-filled burette. The magnitude of this surface force perpendicular to a unit length is the surface tension. For pure water it is about 70 dynes/cm; solutions of detergents have lower surface tensions, about 25–45 dynes/cm.

Favorite objects for the study of surface phenomena are soap bubbles. The elastic films of bubbles obey the *law of Laplace* (a proportionality exists between tension T and the product of pressure and the radius). For soap bubbles the expression takes the form $4T = Pr$, or, rearranged, $P = 4T/r$. Now, if one considers the circumstance of two bubbles of the same soap but of differing radius (Figure 5-11a), the law of Laplace predicts that the pressure in the larger bubble will be less than in the smaller. Upon connecting the tubes (Figure 5-11b) the small bubble will empty into the large one. The alveoli of the lungs are not uniform in size and it can be seen that, given a uniform surface tension, small alveoli would tend to empty into large ones; this would result in the unstable condition of collapse in some populations of alveoli and overexpansion in others. The very fact that this does not normally occur suggests that the surfaces of lungs do not behave like soap bubbles, and other factors are operative.

Extracts of lung-surface coating exhibit properties that offer an explanation for the stability of alveoli of differing sizes. The surface material, called *surfactant*, is a complex of dipalmityl lecithin and protein. The remarkable property that characterizes this material is that the surface tension changes directly with surface area, unlike the soap bubbles described. Thus, when the lungs are filled, those alveoli that are most inflated have surface tensions greater than those which are relatively underdistended. This serves to stabilize alveoli of differing sizes, preventing small alveoli from emptying into larger ones, as would be the case for soap bubbles.

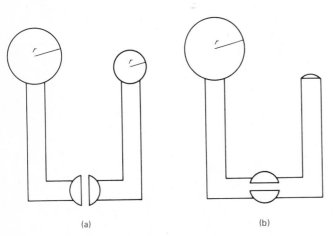

Figure 5-11
When two soap bubbles of differing radius (a) are connected through a tube (b), the bubble of smaller radius will empty into the larger bubble.

(a) (b)

Because of the low surface tension provided by surfactant, the lungs are more compliant than would be the case with a simple lining of water. Thus, the work of expanding the lungs is reduced by the presence of the material.

Surface forces that tend to collapse the alveoli tend to suck fluid into the alveoli from the capillaries. Surfactant, by reducing these forces, aids in preventing transudation of fluid, and this aids in maintaining a short diffusion path between air and blood.

Some newborn infants exhibit "respiratory distress syndrome," a condition in which there is a deficiency of surfactant. The condition is characterized by stiff lungs in which many alveoli are either collapsed or fluid filled.

Whether surfactant is important in preventing collapse of the relatively large connective tissue supported alveoli of amphibians and reptiles or the fixed-volume avian lung is unknown. However, its roles in reducing the work of respiration by making the lungs more compliant and in preventing transudation of fluid into the respiratory air spaces are probably of fundamental significance for all lunged vertebrates.

The airways of lungs offer resistance to the flow of gas by virtue of their geometry. This relationship follows the form derived by Poiseuille:

$$\dot{Q} = \Delta P\left(\frac{\pi r^4}{8L}\frac{L}{\eta}\right)$$

where $\dot{Q}$ is the flow of a fluid or gas, ΔP the pressure gradient along the tube, r the radius of the tube, L the length of the tube, and η the viscosity of the fluid or gas. This expression is often simplified to $\dot{Q} = \Delta P/R$, where R is resistance. In this form the relationship resembles Ohm's law ($I = E/R$). The resistance is determined then by tube dimensions and viscosity. The relationship predicts that flow is directly proportional to the pressure gradient and the fourth power of the radius and inversely proportional to the tube length and viscosity. The striking feature here is the marked effect of the radius of tubes. If the radius decreases by one-half at a given pressure, the flow will decrease to one sixteenth of its original value. A system of multibranching tubes of smaller and smaller radius toward the terminals offers an inherently higher resistance to flow. This factor then may be expected to play a larger role in mammalian than in amphibian lungs, where branching of tubes is minimal.

Figure 5-12

Simplified model representing the lungs, body wall, and airways. P_{inf} = the pressure necessary to inflate the lungs. P_{inf} must overcome the elastic elements (with compliance), and the resistive elements. M_b, active element developing pressures as a result of forces exerted by the respiratory muscles; C_b, compliance of body wall; R_b, body-wall resistive element; R_l, lung resistive element; C_l, compliance of lungs; R_a, airway resistive element.

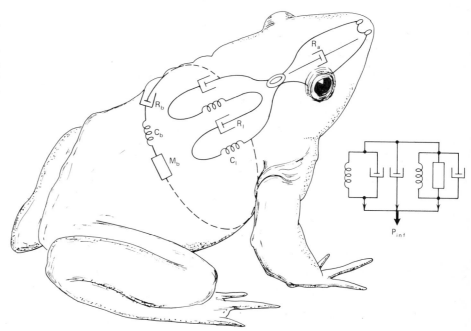

One fact that tends to compensate to some extent for the progressive decrease in tube radius is the fact that as the tubes branch they become shorter.

The motive power for lung inflation is derived from muscular effort; this is translated into pressure, and the pressure required to inflate the lungs must be great enough to overcome (1) the elasticity of the lung and its surface lining, (2) frictional resistance of the lungs, (3) elasticity of the thorax or thoracoabdominal cavity, (4) frictional resistance of the body-wall structures, (5) resistance inherent in muscle, and (6) airway resistance. These relationships are depicted schematically in Figure 5-12.

5.4 RESPIRATORY MECHANISMS OF AQUATIC VERTEBRATES

The gill structures of cyclostomes, both lampreys and hagfishes, are in the form of pouches (marsipobranchs) which are connected internally with the pharynx and open to the exterior, either by a fusion of the excurrent gill ducts into a single tube (*Myxine glutinosa*) or individually by separate gill slits (*Petromyzon* and *Eptatretus*).

5.4.1 Cyclostomes

In all these forms the gill lamellae form a ring around the margins of the gill sac, and the series of sacs is supported by flexible branchial skeletal elements. The number of paired pouches varies (7 in *Petromyzon* and 6–14 in hagfishes). In lampreys, the pharynx divides into a dorsally situated esophagus and a ventral blind tube from which the gill pouches arise. There is a nostril penetrating the dorsal pharynx in hagfishes but none in lampreys. Thus, in lampreys, the only anterior access is via the mouth. While embedded in the flesh of its host, the lamprey maintains a flow of water through the gills by alternate contractions of the gill pouches. Filling occurs as the gill-pouch muscles relax; thus water is sucked in as the pouches expand. A set of valves directs the flow of water over the gill lamellae during the expiratory phase of the cycle only. Contraction of the branchial region apparently prevents reflux of water into the pharynx while expiration is occurring. This form of ventilation of the respiratory surfaces is well adapted to periods of feeding when the head of the lamprey is embedded in the flesh of its prey.

Respiration in cyclostomes had been most thoroughly studied in the hagfish, *Myxine glutinosa*. Dependence on cutaneous respiration has been suggested, but quantitative data are not available. The major oxygen supply is derived from the water that is drawn in through the nostril, which opens into the pharynx anterior to the gill pouches. Just posterior to the nasopharyngeal aperture is a peculair structure called the **velum**. The velum is suspended from the dorsal midline of the pharynx by a membrane, the velar frenulum, from which a horizontal bar arises in the configuration of an inverted T. Bilateral velar scrolls arise from the horizontal bar and are supported by velar skeletal structures. Through the action of a series of muscles, the velar scrolls extend ventrad and then roll upward after the manner of a window shade. The scrolls literally scoop up water and direct it toward the frenulum and posteriorly toward the gill pouches. This action has been confirmed by direct observation and by high-speed cineradiographic studies in which opaque media were introduced through the nostril. Observations on the frequency of velar pulsations range from 11–15 pulses/min to 50–100 pulses/min. Such a wide range may reflect differences in metabolic state and activity of the animals under observation.

Johansen's radiographic studies have also demonstrated the presence of sphincters on both afferent and efferent gill ducts. In addition to acting as filters removing particles in the water, the sphincters take an active part in directional flow through the gill pouches. Contraction of the afferent sphincters occurs during the exhalant phase and the gill-pouch volume decreases, suggesting that contraction of muscle elements within the pouches may contribute to expulsion of water. If this is true, contraction of efferent and relaxation of afferent sphincters coupled with elastic rebound of the gill pouches following expiratory contraction may produce a negative pressure within the gill pouch; this would create a favorable pressure gradient for filling. Whether velar contractions are coordinated with gill-pouch contractions is not known.

The introduction of suspended particles into the nostril produces a violent "sneeze" and expulsion of the foreign material from mouth and nostril.

Figure 5-13
Blood and water flow through the gill body of the hag-fish. Note that water flow is counter to that of blood, an arrangement facilitating gas exchange.

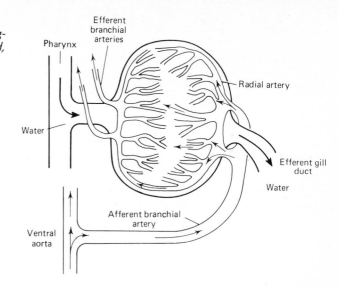

The sneeze is accomplished by contraction of the pharyngeal constrictor and velar chamber muscles accompanied by closing of the gill-duct sphincters. This reaction may have physiological significance in avoidance of smothering of the respiratory surfaces in an animal having common respiratory and alimentary ducts.

Blood flow in the gills of cyclostomes is in a direction counter to that of water flow (Figure 5-13). This arrangement increases the efficiency of exchange of gases across the respiratory surfaces (see Section 5.6).

5.4.2 Elasmobranchs and Teleosts

Cartilaginous and bony fish employ a double-pump mechanism for maintaining an almost continuous flow of water across the gill surfaces. For water flow to occur, a favorable pressure gradient must be established across the sieve formed by the secondary lamellae. The flow of water is directly proportional to the pressure difference across the sieve and inversely proportional to the resistance: $\dot{Q} = \Delta P/R$ (see also Section 5.3.4). The major variables in the system are the pressures generated by the pump and the resistance. Resistance depends on the configuration of the sieve, which may change from time to time depending on the degree of interdigitation and approximation of the secondary lamellae of adjacent gill filaments. The control of

this variable has thus far defied objective analysis.

The buccal cavity serves as a positive-pressure pump, whereas the opercular cavity, in teleosts, or the parabranchial cavity, in sharks and skates, functions as a suction pump. Flow into the oral cavity occurs following depression of the buccal floor. This depression produces a pressure below that in the surrounding water (Figure 5-14). The opercular or gill flaps are closed during this phase. As the opercular muscles relax, the opercular cavity expands, creating a pressure on the back side of the gill filaments that is lower than the pressure in the buccal cavity. The pressure differential produces flows across the gill surfaces. This initial inspiratory phase is followed by closure of the mouth and active elevation of the buccal floor, creating a positive pressure in the mouth cavity. This produces additional water flow to the still-negative-pressure opercular or parabranchial chambers. As the pressure head builds up in the anterior pump, the pressure in the post-gill chambers, caused by the buccal force pump, rises above that in the external medium. This results in forcing open the opercular or gill-flap valves and in the discharge of water to the outside. A slight reversal of flow may occur just prior to the following filling phase. Flow is, for practical purposes, continuous. This pattern of events has been worked out from studies of the pressures and movements of the chambers and gill flaps (Figure 5-15).

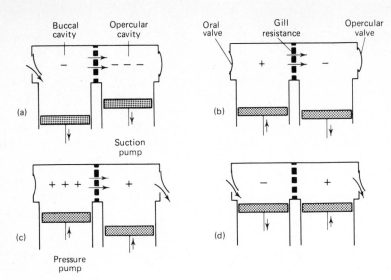

Buccal cavity   Opercular cavity

Oral valve   Gill resistance   Opercular valve

(a)

(b)

Suction pump

(c)

(d)

Pressure pump

Figure 5-14

Diagrammatic representation of the double-pumping mechanism, which maintains an almost continuous flow of water across the gills. The two major phases of the cycle are (a) in which the opercular suction pumps are active and (c) when the buccal pressure pump forces water across the gills. The two transition phases (b) and (d) each take up only one tenth of the whole cycle. The pressures in the cavities are given with respect to the water pressure outside. [From G. M. Hughes, New. Sci., 11, 346–348 (1961).]

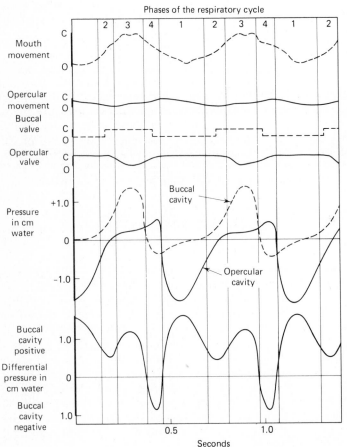

Phases of the respiratory cycle

Mouth movement

Opercular movement

Buccal valve

Opercular valve

Pressure in cm water

Buccal cavity

Opercular cavity

Buccal cavity positive

Differential pressure in cm water

Buccal cavity negative

Seconds

Figure 5-15

Breathing movements of the mouth and operculum of a 70-g trout, together with associated pressure changes in the buccal and opercular cavities. The differential pressure between these cavities is shown below. O and C indicate the opened and closed positions of the mouth, operculum, and their associated valves. Temperature, 17°C. [From G. M. Hughes and G. Shelton. J. Exp. Biol., 35, 807–823 (1958).]

The spiracle of elasmobranchs, an anterior reduced gill slit, also provides a channel of water flow into the orobranchial chamber. The spiracle is valved, and dependence on spiracular water flow is correlated with habitat. Bottom-dwelling forms such as rays and skates have relatively larger spiracles than their pelagic relatives, and the preponderance of water flow passes through the spiracle rather than the ventrally oriented mouth, apparently assuring a flow of water containing little particulate matter.

The pumping mechanisms previously described is not the invariable method of ventilation, for sharks have been observed, while swimming, to keep both mouth and gill flaps open, assuring a constant flow across the gill surfaces. Upon slowing, or settling to the bottom, pumping activity is resumed. The mackerels and tunas are teleosts incapable of maintaining sufficient blood oxygen saturation when prevented from swimming. They make no active respiratory movements, in fact, almost lack branchial musculature. Respiration in these fishes depends on the current resulting from their forward motion through the water. They can never stop swimming.

5.4.3 Air Breathing in Fishes

The limitation of fishes to a purely aquatic situation is implied by the common phrase "a fish out of water," used to describe a tenuous or hopeless situation. However, many fishes depend to varying degrees on aerial respiration and some, like the mudskipper, *Periophthalmus*, spend much of their time on land near the water's edge. Eels and some catfish (*Clarias*) make extensive overland migrations, during which they rely heavily on cutaneous respiration.

Air breathing is commonly seen in ecological situations in which the O_2 content of water is low or zero. Stagnant waters containing considerable organic matter are often anaerobic. High temperatures and CO_2 levels also combine to make such situations something less than ideal. Two general means of acquiring O_2 are employed in these extreme circumstances. Some fishes keep close to the air-water interface, where O_2 tension, owing to surface diffusion, is highest. This restriction on vertical location is avoided by most stagnant-water forms in which accessory respiratory structures have appeared. These structures are usually elaborations within the alimentary tract and represent highly vascularized portions of the branchial cavity, pharynx, or stomach. Air gulping at the surface serves to charge the respiratory surface (pharyngeal epithelium in *Electrophorus*, stomach in *Plecostomus*), and the frequency of rising to the surface to gulp air corresponds to the activity and oxidative metabolism of the fish.

The South American electric eel (*Electrophorus electricus*) is an inhabitant of muddy and O_2-deficient streams. It is an obligate air breather with reduced gills and depends on periodic charging with air of the highly vascular and diverticulated oropharyngeal space. Gas, from which O_2 has been extracted, is expelled through the opercular slits. The respiratory organ represents about 15% of the total surface area. The efferent vascular drainage of the oral mucosa is via the jugular veins and hence mixes with the general systematic venous return (see Figure 6-3). The O_2 capacity of the blood is high, of the order of 20 vol% (similar to mammals), and correlates with high hemoglobin and hematocrit (percent of blood as red blood cells) values. The high O_2 capacity may be related to the mixed condition of the systemic arterial blood (about 60% saturated). *Electrophorus* also exhibits elevated arterial CO_2 tensions; bicarbonate levels are high, assuring greater buffering capacity, and the effect of CO_2 on the O_2-dissociation curve is relatively slight (see Section 5.7). The exchange of CO_2 is principally across the skin and to a smaller extent through the vestigial gills. Blood flow to the respiratory organ is highest immediately following a breath and declines during extraction. Breathing hypoxic gas is associated with vasoconstriction and reduction of blood flow to the exchange site. These adjustments bring about a more favorable relationship between ventilation and perfusion. The frequency of air gulping may be driven by O_2-sensitive chemoreceptors, since ventilation volume doubles as O_2 tension of air is experimentally reduced to around 70 mm Hg.

Amia calva (a Holostean; the bowfin) possesses both gills and an air bladder that can be utilized for respiration. The mode of respiration utilized depends upon both temperature and O_2 content of the water. *Amia* is almost exclusively a water breather at 10°C, a temperature associated with a low level of physical activity. Air-breathing rate increases with temperature and activity, and at 30°C about 3 times more O_2 is extracted from air than water. The elimination of CO_2 is principally across the gills. The frequency of air breathing stands in inverse relationship to O_2 content of the water and, as O_2 tensions of water fall below 40–50 mm Hg at 20°C, air breathing largely replaces water breathing. The cardiac output is distributed, via local vasomotor responses, in such a manner that blood largely bypasses an exchange site (gill or air bladder) that is not being utilized as a principal site of gas exchange.

Erythrinus unitaeniatus (the yarrow), a swamp-dwelling teleost of British Guiana, utilizes both aquatic and aerial respiration. Use of one or the other depends on the gas content of the water. Willmer, who studied this fish in waters of varied O_2 and CO_2 content, obtained the results in Figure 5-16. Below an O_2 content of about 1.3 ml/liter, purely gill respiration does not occur. Levels of CO_2 above 35 ml/liter or below about 3 ml/liter

are associated with aerial respiration only. Utilization of both modes of respiration characterizes the intermediate conditions. The use of aerial respiration when CO_2 content is low but O_2 content is high is the most puzzling feature of the fish's behavior. Willmer has postulated that the respiratory areas of the brain that control gill movements are not sufficiently stimulated under conditions of low CO_2. Closure of the gill chambers when the O_2 content of water is low may prevent loss of the O_2 acquired by air gulping to the water across the gill surfaces. In *Erythrinus* it is clear that both CO_2 and O_2 levels are determinants of respiratory activity. The capacity to utilize the appropriate modes of respiration apparently extends the conditions under which survival can occur.

Eels out of water use the skin as a major respiratory surface. Although total O_2 consumption falls when eels *(Anguilla vulgaris)* are experimentally removed from water, dependence on cutaneous respiration increases while O_2 uptake across the gills

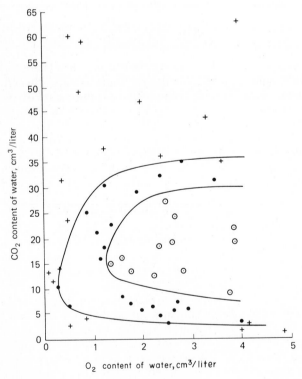

Figure 5-16
*Relationship between the type of respiration of the yarrow and the CO_2 and O_2 content of the water. [From E. N. Willmer, J. Exp. Biol., **11**, 283–306 (1934).]*

○ Aquatic respiration • Intermediate respiration + Aerial respiration

5 RESPIRATION

falls. An increase in cutaneous blood flow probably occurs, because it has been observed that, when transferred to air, the large dorsal and anal fins become conspicuously reddened, owing to vasodilatation. Gill respiration is continued, however, by periodic inflation with air. A smaller utilization of swim-bladder gas also occurs. Figure 5-17 illustrates the shifts in dependence on these respiratory structures when eels are removed from water. Fishes when out of water exhibit a reduced heart rate and a shift toward anaerobic metabolism. Following return to the aquatic situation, the heart rate increases and O_2 consumption and blood lactic acid rises, a circumstance inducing a fall in blood pH. These changes depend on length of exposure to air and the temperature. Such adjustments have been observed in grunion during their sojourn on sandy beaches while breeding and in flying fishes during their brief aerial excursions (see Section 6.9.1).

Present-day lungfishes, which have fossil antecedents as far back as Devonian times, are thought to be close to the evolutionary origin of amphibians. They depend, to varying extents, on extraction of O_2 from ventral diverticula of the enteron, which are homologous to higher vertebrate lungs. The arterial circulation of these structures is derived

from the sixth aortic arch, and the venous drainage is directly to the heart as in terrestrial vertebrates. The South American *Lepidosiren* suffocates if denied access to air, the gill circulation being poorly developed. *Protopterus* of Africa appears to be intermediate in its dependence on air, whereas the Australian form, *Neoceratodus,* possesses well-developed and highly vascular gills and ventilates the gills in a typical piscine manner. The latter form cannot survive out of water for long periods.

Inflation of the lungs is brought about by rising to the surface and gulping a large bubble of air, which, on compression of the buccopharyngeal region, is forced into the lung. Little is known about the mechanics of lung inflation in this important group, but it appears to be similar to the buccal force pump of many amphibians.

Introduction of O_2-deficient gas into the lungs of *Protopterus* stimulates the rate of air breathing. Cardiac output (and lung perfusion) increases with each breath, declining in the interval prior to the next breath. The distribution of the cardiac output shifts in such a manner that the percentage of blood flow perfusing the lungs may rise to 70% of the cardiac output, gradually falling to about 20% prior to the next breath. Such shifts are due to regional changes in vascular resistance, the mediators for which are unclear. Although aerial respiration is of predominant importance for O_2 acquisition, aquatic breathing remains important for CO_2 elimination and the gills are about 2.5 times more effective for the removal of CO_2 than the lungs. Radiographic and blood-gas-distribution studies suggest that a relatively efficient separation of pulmonary and systemic venous returns is achieved in this form. The capacity to maintain a relatively efficient double circulation is an important transitional step toward assumption of a lung-breathing terrestrial life.

The external and internal nostrils of these forms apparently are not involved in respiration and are thought to be related to chemoreception. Their presence is a possible example of "preadaptation" to the respiratory function of the nares of higher vertebrates.

Both *Protopterus* and *Lepidosiren* withstand prolonged periods of drought and enter into a state of estivation during which metabolism and respiration are reduced. *Protopterus* secretes about itself a protective waterproof lipoprotein mucous tube

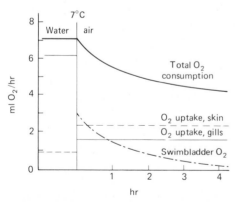

Figure 5-17
Development in total O_2 uptake and O_2 uptake from gills, skin, and swim bladder during an air exposure (eel). Note that the swim-bladder O_2 store is of importance only during the first hours. [*From T. Berg and J. B. Steen.* Comp. Biochem. Physiol., **15,** 469–484 (1965).]

that is connected to the air only through an aperture entering the oral cavity. On entering the cocoon, O_2 consumption falls by about 50% during the first week and to 10% of the preestivation level during the following weeks. Respiratory frequency falls to 1–2 per min and heart rate of 3 per min. In this state the animals may survive 3–5 years, until sufficient water is available to sustain active life (see Section 7.5.2).

5.5 RESPIRATORY MECHANISMS OF AMPHIBIANS AND TERRESTRIAL VERTEBRATES

The respiratory surfaces utilized by amphibians include the skin or derivatives of the skin, the buccal cavity, and the lungs. Dependence on these surfaces varies with subgroups and with habitat. The salamanders belonging to the Plethodontidae, for instance, are lungless and respire across the skin and to a lesser extent across buccal membranes. Frogs utilize buccal, cutaneous, and lung respiration simultaneously, but to varying degrees. In the lunged amphibia, as in the Dipnoi, the sixth aortic arch is the pulmonary arterial supply.

5.5.1 Amphibians

In the Anura (frogs and toads) the sixth arch bifurcates into pulmocutaneous arteries supplying the lungs and a portion of skin. The pulmonary vein returns blood from the lungs directly to the left atrium; however, the cutaneous venous drainage is into the systemic venous return.

All the respiratory surfaces utilized in other forms are seen in frogs. As tadpoles, frogs respire via the skin and gills. The gills enlarge in oxygen-poor waters and become smaller in water that is well aerated. The large tail fins are highly vascular and are a significant respiratory surface. Like the external gills, they are lost during metamorphosis.

Inflation of the lungs in the adult frog depends on a positive-pressure buccopharyngeal pump reminiscent of lungfish inflation. The sequence of respiratory and air flow patterns is illustrated in Figure 5-18. Immediately preceding expiration, the floor of the mouth is depressed and air is drawn in through the nostrils and into the depression in the oral floor (stage 1). The expiratory phase (stage 2) is powered by the body wall muscles and the elastic recoil of the previously inflated lungs. At this stage the glottis is held patent and a high-velocity jet of expired air passes through the nostrils. Gans, et al. have shown that mixture of the exhaled stream with the air previously drawn into the buccal floor is minimal, owing to a jet stream effect across the upper portion of the oropharyngeal airway. At the close of expiration the buccal floor is elevated while the nostrils are closed (stage 3). The positive pressure generated by buccal floor elevation results in lung inflation as air is forced through the open glottis. After inflation there ensues a sequence of buccopharyngeal pumping actions with glottis closed and nostrils open (stage 4). Although it was previously thought that this pumping phase was primarily related to olfaction or "sniffing," it has been shown that this phase serves the respiratory function of flushing out any residual expired gas from the previous expiration and results in charging the buccopharyngeal cavity with ambient air.

Szarski, who evaluated the vascularity of the respiratory surfaces in a number of amphibians, found that the skin of the back and thighs (areas exposed to air) of frogs contains significantly more capillary meshes than the skin of the underparts. The capillary surface area of the buccal membranes is relatively small compared to that of other respiratory surfaces. Buccopharyngeal capillarity is greatest in lungless forms, where it represents 5–10% of the total respiratory capillary area. In the aquatic newt (*Triton*), which utilizes both cutaneous and lung respiration, the skin contains about 75% of the respiratory capillaries. This is in contrast to more aerial forms such as the tree frog, *Hyla arborea*, where the lungs contain 75% of the respiratory capillary surface area.

Gas exchange across the skin and lungs of frogs was measured by Krogh. He found CO_2 to be mainly excreted through the skin in both *Rana temporaria* and *Rana esculenta*. Uptake of O_2 through the lungs is about 3 times greater than through skin in the more terrestrial *Rana temporaria*. In *Rana esculenta*, which is restricted more to water, the lungs and skin play about equal roles in O_2 uptake.

In general the skin appears to be the major avenue for CO_2 exchange in amphibians, whereas the lungs are the prominent surface for O_2 uptake in anurans. The skin is the dominant surface for both

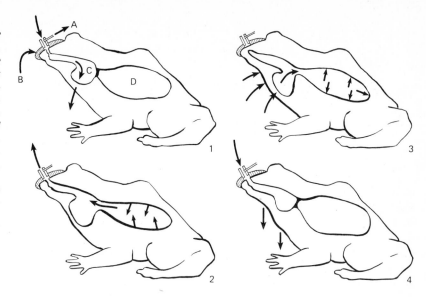

Figure 5-18
Successive stages of flow during the ventilatory cycle in Rana catesbeiana. *The gas sampling probe leading to the respiratory mass spectrometer (A) is supported by a plastic collar in the rubber mask (B). The arrows indicate the approximate sequence of gas flow into and out of the buccal cavity (C) and lung (D). Stage 2 shows how the pulmonary efflux apparently bypasses the inhaled gases, which rest in the posterior portion of the buccal cavity.* [*From: C. Gans, H. J. De Jongh, and J. Farber.* Science, **163,** *1223–1225 (1969).*]

CO_2 and O_2 exchange among urodelians. In lungless forms the skin contains 90–95% of the respiratory capillary area. The broad impression is one of increasing dependence on pulmonary ventilation with greater aerial habits.

5.5.2 Reptiles

The reptilian migration to land was associated with a greater dependence on lung respiration than seen in amphibians. The principal mechanism of inflation is a **suction pump** rather than the buccal force pump. In most reptiles, the motive force for inflation is derived from muscular expansion of the ribs and body wall, which creates a sub-atmospheric pressure within the lungs during inflation.

The lungs of reptiles have greater internal surface area than amphibians and the conducting airways are more complex. There is a distinct trachea, which subdivides into bronchi in most. Elastic and smooth muscle fibers are present in the lung wall. In general, the caudal portion of the lungs may be poorly vascularized, the principal exchange surfaces being anterior. In some snakes, fully two thirds of the lung is a simple membranous reservoir.

The respiratory cycle of reptiles is typically triphasic in character, as can be seen from observations of rib-cage movements in lizards (Figure 5-19). Inspiration is brought about by expansion of the rib cage as the glottis opens. Intraabdominal pressure falls below atmospheric pressure at this point, as does the intrapulmonary pressure. That the rib muscles are active during this phase is indicated by an increase in their electrical activity as contraction occurs. The inspiratory phase ends with closure of the glottis and is followed by an apnea or respiratory pause during which the lungs are inflated. The muscles of respiration relax during this period. As a result of their rebound, intraabdominal pressure rises, as does intrapulmonary pressure. The gas is held in the lungs under 2–4 mm H_2O pressure until the expiratory phase occurs. Expiration is achieved by passive elastic recoil of the lungs and active contraction of body-wall musculature.

Studies by Templeton on the respiratory mechanics of the chuckwalla lizard *(Sauromalus)* revealed that both intraabdominal and thoracic pressures remain subatmospheric during the respiratory cycle. Intrathoracic pressure remained consistently below intraabdominal. Although no true diaphragm is present in reptiles, this observation suggests anatomic partition of thorax and abdomen.

Pharyngeal pumping is commonly seen in lizards but occurs only in the apneic part of the respiratory cycle, presumably associated with olfaction. When disturbed, however, the pharyngeal pump may be employed to inflate the animal. Inflation in *Sauro-*

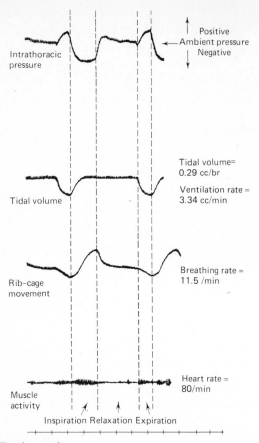

Intrathoracic pressure

Positive
←— Ambient pressure
Negative

Tidal volume

Tidal volume=
0.29 cc/br

Ventilation rate =
3.34 cc/min

Rib-cage movement

Breathing rate =
11.5 /min

Muscle activity

Heart rate =
80/min

Inspiration Relaxation Expiration

Time in seconds

Figure 5-19

Respiratory cycle of collared lizard at 25° C as revealed by simultaneous measurements of intrathoracic pressure, tidal volume, movement of the rib cage, and electrical activity of the thoracic muscles. Horizontal arrow right of the record for intrathoracic pressure indicates ambient pressure level; vertical arrows indicate directions of departure from this level representing increased (positive) and decreased (negative) pressures. Increases in tidal volume and expansion of the rib cage are represented by upward deflections on the appropriate records. The values for tidal volume, ventilation rate, and breathing rate during these measurements are given at the right of records of tidal volume, movement of rib cage, and muscle activity, respectively. Period of relaxation in respiratory cycle (respiratory pause), represented by the broadest interval between the vertical dotted lines, is bordered left and right, respectively, by intervals in which inspiration and expiration occur. [From J. Templeton and W. R. Dawson. Physiol. Zool., 36, 104–121 (1963).]

malus appears to serve as a defense mechanism, wedging the animal between rocks and preventing removal by predators. Inflation is accomplished by elevating the filled and depressed pharyngeal floor against closed nares and an open glottis. A series of pumping cycles may elevate intratracheal pressure to 75–250 mm H_2O. Intraabdominal pressure is elevated to above atmospheric (10–60 mm H_2O) but intrathoracic pressure remains subatmospheric. It has been suggested that the cranial portion of the lung contracts during inflation.

The rigid shell of turtles prevents general expansion and contraction of the body wall as the ribs are fused with the shell. Typically several respiratory cycles are followed by a period of apnea. Pharyngeal pumping is restricted to the apneic phase as in lizards and frogs. The lungs of turtles are situated in a relatively rigid frame and are located between the viscera and the domed carapace. Ventilation in the tortoise, *Testudo graeca*, is accomplished by changing the volume of the thoracoperitoneal cavity by altering the position of the limb flanks. The pressure necessary to power lung deflation is generated by the activity of muscles that draw the shoulder girdle back into the shell and compress the abdominal viscera in the rear. These muscle actions increase peritoneal-cavity pressure with a consequent transmission of pressure to the lungs. During inspiration antagonistic muscles produce the opposite effect, that is, bring about a decrease in peritoneal-cavity pressure with a consequent reduction of intrapulmonary pressure below atmospheric, thus favoring lung inflation. The cycle of events is shown in Figure 5-20.

Because of the rigidity of the shell in *Testudo* it is not possible to recapture the potential energy stored in expanded abdominal-wall structures as is the case in most terrestrial vertebrates, and both expiration and inspiration are active, energy-consuming events. The situation is somewhat different in the snapping turtle (*Chelydra serpentina*), which possesses a reduced and deformable plastron; some energy of inflation may be recovered during deflation due to rebound of the deformed undersurface structures. Furthermore, this form spends a large fraction of the time submerged and breathes by snorkeling (extending the head to the air-water interface). In this situation, expiration is largely passive owing to the hydrostatic compression effect of the water against the relaxed body wall.

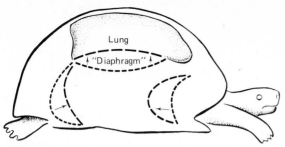

Figure 5-20
Diagram of changes in volume of the visceral cavity produced by changes in shape of its anterior and posterior borders during a ventilating cycle. These result in movement of the "diaphragm" and alterations in lung volume. Dashed lines, expiration; dotted lines, inspiration. Major muscles of expiration: Transversus abdominis, pectoralis. Inspiration: Serratus major, obliquus abdominis. [From C. Gans and G. M. Hughes. J. Exp. Biol., 47, 1–20 (1967).]

It was long believed that turtles utilize the buccopharyngeal pump to inflate the lungs. Measurements of intratracheal pressure show that this is not the case. Furthermore, pharyngeal pumping only occurs during periods of apnea. Sampling of air or water by this mechanism is probably related to olfaction, because the rate of pumping does not change when a strip of rubber membrane is held near the nostrils but increases severalfold when the membrane is first dipped in crabmeat extract.

The structure of reptilian lungs is variable, the varanid lizards having the most complexly divided subunits with greatest respiratory surface area. In turtles and snakes the ventilated gas is not homogeneously distributed to the various subdivisions of the lung in contrast to mammals. This has been demonstrated by observing the distribution of a radioactively tagged inert gas, ^{133}Xe, introduced into the inspired gas (Figure 5-21a). The time to reach a 50% equilibrium value is measured in regions of interest and used as an index of effective ventilation. As can be seen in the figure, the anterior regions are more effectively ventilated with fresh gas than is the posterior saclike portion of the lung. This pattern of inhomogeneous distribution is the result of the *in series* relationship of the compartments (Figure 5-21b). Those regions of more intense ventilation are also of greater vascularity. The posterior region has little functional respiratory surface and very little blood supply. This region is utilized as a "bellows," which is deflated and inflated by the posterior respiratory muscles.

Figure 5-21
(a) Activity levels of ^{133}Xe in various segments of a turtle lung (Pseudemys scripta). The animal breathes air in which the radioactive gas is introduced at a steady concentration. In time, all segments will reach an equilibrium concentration. The number of breaths required to reach 50% of the equilibrium value increases from the anterior to posterior region of the lung. (b) Model of the turtle lung as a series of n compartments, each having a resting volume of V_i and which may expire a volume, ΔV_i. This is in contrast to the mammalian lung, the subunits of which are interconnected in parallel. [Based on R. Spragg, R. Ackerman, and F. White. Respir. Physiol., 42, 73–86 (1980).]

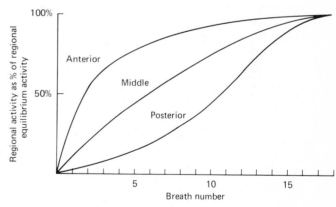

(a)

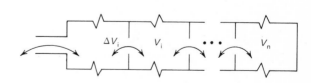

(b)

5.5.3 Birds

The relatively high rate of metabolism of resting birds and the added cost of flight requires a ventilation system capable of meeting high rates of gas exchange. Given such requirements, it initially appears surprising that the gas volume of bird lungs is small compared to that of mammals. However, the density of the gas exchange units, the air capillaries, is larger than alveolar density in mammals. The relatively small-volumed lung cavity of birds is connected with membranous air sacs by a series of tubes (Figures 5-22 and 5-23). The complex system of airways interconnecting the exchange paths of the lungs with the air sacs has posed great difficulty in understanding the pathway of gas flow within the lungs.

The essential features of the avian airways are shown in Figure 5-23. The trachea subdivides into two primary bronchi that course through each lung. The primary bronchi pass directly to the large, paired, abdominal air sacs while secondary bronchi supply the other air sacs and give rise to tertiary bronchi that penetrate the lung mass. Air capillaries, the walls of which are richly vascularized, arise from the tertiary bronchi (parabronchi), and gas exchange occurs across these capillary surfaces. The air capillaries are analogous to the alveolar exchange surfaces of mammals.

Most modern birds possess two distinct bilateral lung masses. The major mass (main lung) is the paleopulmo. A smaller more ventrally located mass is called the neopulmo. The latter lung region has been little studied. It is absent in penguins.

Inflation in birds is accomplished by enlargement of the thoracoabdominal cavity by the action of the inspiratory muscles. The sternum swings forward and downward, enlarging the thoracoabdominal space in a dorsoventral axis. At the same time the ribs move laterally, increasing the diameter of the cavity transversely. Antagonistic expiratory muscles increase the pressure within the airways when the cycle is reversed. In contrast to mammals, the compliance of the avian lung is slight, and the lung is relatively stiff. The compliant portions of the respiratory system are mainly the air sacs and surrounding body wall.

The total respiratory-system volume of birds is 2–3 times that of mammals of comparable size. This is owing primarily to the large volume of the air sacs. The total respiratory volume of a 2-kg chicken is around 200 ml, whereas tidal volume is approximately equal to lung air volume, about 25 ml.

Recent work, utilizing paired thermistor catheters (sensitive to differences due to direction of air flow) placed in various portions of the avian airway, allows the construction of a probable airflow scheme (Figure 5-23). It seems clear that air

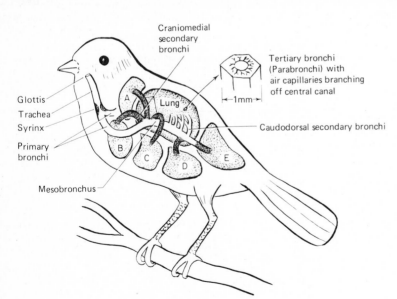

Craniomedial secondary bronchi

Tertiary bronchi (Parabronchi) with air capillaries branching off central canal

├—1mm—┤

Glottis
Trachea
Syrinx
Primary bronchi
Mesobronchus
Lung
Caudodorsal secondary bronchi

Figure 5-22

Connections of avian respiratory pathways. A, cervical air sac; B, interclavicular air sac; C, anterior thoracid air sac; D, posterior thoracic air sac; E, abdominal air sac. The air sacs are connected to the mesobronchus by direct connecting tubes, the secondary bronchi. Secondary bronchi also penetrate the lung where they further subdivide into tertiary bronchi (parabronchi) from which the air capillaries arise. Tertiary bronchi are a series of parallel tubes which are connected to both the caudodorsal and craniomedial secondary bronchi.

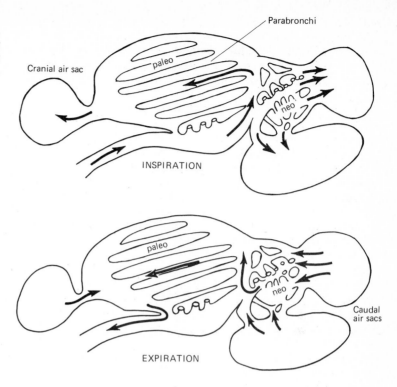

Figure 5-23
Pathways of gas flow in the lungs and air-sac system of birds during inspiration and expiration. Note that airflow is unidirectional through the parabronchi of the paleopulmo during both inspiration and expiration.

Parabronchi

Cranial air sac

paleo

neo

INSPIRATION

paleo

neo

Caudal air sacs

EXPIRATION

flow through the parabronchi and the respiratory exchange surfaces (air capillaries) occurs during both inspiration and expiration. Furthermore, expiratory air flow through the lung is largely from the posterior air sacs. Experiments in which a bolus of pure oxygen was introduced into the trachea of the ostrich, revealed that the oxygen was detected in the posterior air sacs with the initial inspiration, but appeared in the anterior sacs only during subsequent cycles. The anterior air sacs contain gas of lower P_{O_2} and higher P_{CO_2} than the posterior sacs. It appears that the anterior air sacs are the recipient of lung-gas efflux during the inflation phase, whereas fresh gas is derived through the primary and caudal secondary bronchi. During the expiratory phase the predominant gas flow across the exchange areas is derived from the posterior air sacs, where the gas is characterized by a relatively high P_{O_2} and low P_{CO_2}. The mechanisms by which resistances in the various tubing components are controlled is not understood.

The P_{CO_2} at the exchange surfaces of birds is lower than for mammals due to the fact that the avian lung undergoes a more complete ventilation by virtue to having the lungs in series between the primary bronchi and the large-volume air sacs. This is in sharp contrast with the cul de sac type of lungs of other vertebrates, which contain a large residual volume. The arterial blood reflects this situation, because the resting arterial P_{CO_2} is only around 28 mm Hg compared to 40 mm Hg for most mammals. Pigeons subjected to heat stress pant vigorously and become alkalotic due to a greater CO_2 washout during increased ventilation. This is in contrast to the ostrich, which does not develop hypocapnia (reduction in arterial P_{CO_2}) and alkalosis during heat-induced panting. It seems likely that the panting ostrich can shunt a portion of the increased minute volume to nonexchange pathways and thus augment evaporative cooling without hyperventilating the parabronchi and air capillaries.

During flight pigeons increase ventilation about 20 times above resting level, but with little change in tidal volume (in contrast to exercising man). The respiratory patterns are precisely synchronized with wing motion; peak expiratory flow rates correspond to the maximum downstroke of the wing beat. The ventilation during flight has been estimated at 2.5 times that required for metabolic needs. The "excessive" ventilation contributes to thermoregulation, because about 17% of the heat production during flight was lost through evaporative cooling of the expired gas. In contrast to the pigeon, both evening grosbeaks and ring-billed gulls increased ventilation during flight in proportion to metabolism. In these birds the ventilation increment was accounted for by increases in both respiratory frequency and tidal volume.

The sparkling violetear, a 6–9-g hummingbird, achieves a respiratory frequency of 330 breaths per minute while hovering at sea level. When at high altitude, the frequency rose to 380. Oxygen extraction in resting birds (sparrows, pigeons) increases during breathing of hypoxic gas as it does during flight at high altitude in hummingbirds. It is not clear whether increased effective ventilation of the lung or/and blood flow alterations in the lung are responsible. The explanation would go far toward explaining the presence of bar-headed geese at 9200 m during their semiannual migration across the Himalayas. This altitude is lethal for resting man. Clearly, birds have the advantage at high altitude.

Although the avian respiratory system primarily serves a metabolic function and is important as an adjunct to thermoregulation, it is also important in communication. The syrinx is the membranous structure responsible for sound production and is located at the lower extremity of the trachea near its bifurcation into the primary bronchi. Some small birds produce sustained song of long duration, especially in view of the high breathing frequencies. Furthermore, artificially imposed air flow produces sound only when the flow across the syrinx is in the expiratory direction. W. A. Calder studied the thoracic movements of canaries while simultaneously recording their songs. He found that song notes or pulses were in a 1:1 relationship to thoracic movements; however, trilled notes (up to 25 per sec) were formed by individual shallow breaths. The dorsoventral dimension could be maintained or even increased to augment air stores as trills were produced. Song was followed by an expiratory movement. Calder suggests that song duration in canaries would not necessarily be limited by the air volume taken in before song production, but more likely by the O_2 and CO_2 content of gas in the respiratory system after a period of shallow and rapid breathing.

5.5.4 Mammals

Although amphibians, and to a lesser extent, reptiles, utilize a positive-force pump for lung inflation, mammals rely almost entirely on a negative-pressure pump. The lung itself plays a passive but important role in both inflation and deflation. Inflation of the lungs is accomplished by contraction of the inspiratory muscles, which enlarges the thorax and reduces the pressure between the lungs and the thoracic wall (intrathoracic pressure) as well as the pressure within the lungs (intrapleural pressure). There is a consequent enlargement of the alveoli, their ducts, and the bronchioles. Air at atmospheric pressure then flows toward this lower pressure through the upper respiratory tract.

The major muscles of inspiration are the diaphragm and the external intercostals (Figure 5-24). The diaphragm, innervated by the phrenic nerve, is a domelike sheet of muscle separating abdominal from thoracic cavities. On contraction, the medial

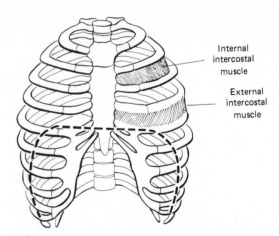

Figure 5-24
Muscles of respiration. The major dashed line represents the inspiratory position of the diaphragm. The internal and external intercostal muscles are drawn in only one interspace. [From J. H. Comroe, Jr. The Physiology of Respiration, The Year Book Medical Publishers, Inc., Chicago, 1965.]

portion of the diaphragm moves downward, acting after the manner of a piston. As a result of its downward movement the thoracic cavity is enlarged in a cephalic-caudad plane. Some compression of the abdominal contents results, with a consequent rise in intraabdominal pressure.

Rotation of the ribs upward and laterad, a result of the action of the external intercostal muscles, increases the circumference of the thorax on inspiration. In addition to the diaphragm and the external intercostals, accessory muscles, such as the scalene and sternomastoids, may be brought into action during severe exercise. Maximum contraction of inspiratory muscles may reduce intrapleural pressure by more than 75 mm Hg below atmospheric pressure, although in quiet breathing less extreme lowering of pressure is achieved. Inspiration is terminated with glottic closure and contraction of the abdominal muscles.

During quiet, unlabored respiration (eupnea), expiration may be achieved without an active contribution of the muscles of expiration. The energy stored in the lungs and thorax wall as a result of their expansion contributes an elastic recoil. This stored energy generates the motive force for expiration as the glottis opens and the inspiratory muscles relax. With larger ventilation volumes or with labored respiration (dyspnea), muscles of expiration, notably the internal intercostals and the abdominal muscles, are activated. Contraction of the internal intercostals, acting antagonistically to the external intercostals, results in depression of the ribs and a decrease in circumference of the thorax. Maximum activity of expiratory muscles can produce pressures in excess of 100 mm Hg. Both sets of intercostals, on contraction, contribute to the rigidity of the thorax wall, preventing bulging of the intercostal spaces.

In the newborn human infant, prior to the first breath, the pleural peritoneum is in close approximation to the parietal peritoneum, separated only by a thin film of fluid that occupies the intrapleural space. As the chest walls moves out with the first inspiration, these membranous layers move outward, held together by the adhesive force of the fluid between the two layers. As the intrapulmonary pressure falls below atmospheric, air enters the lungs. The elastic forces of the stretched lung, however, tend to recoil, and this force results in a subatmospheric pressure in the intrapleural space. In quiet inspiration the pressure recorded by a needle inserted between parietal and pleural peritoneal membranes is about −6 mm Hg with respect to atmospheric pressure. In the expiratory state the pressure rises to about −4 mm Hg. Intrapulmonary pressure (pressure measured in the trachea) is slightly subatmospheric (−3 mm Hg) at the start of inspiration as the lungs expand but is atmospheric at the close of inspiration. Expiration is characterized by a positive intrapulmonary pressure of about 3 mm Hg, returning to atmospheric at the close of expiration. Greater pressure fluctuations occur with deep respiration, and the intrapleural inspiratory pressure may be as low as −30 mm Hg, whereas expiratory intrapulmonary pressure may rise to 100 mm Hg when forceful expiratory efforts are made against a closed glottis. The fluctuations of pressure in the chest cavity may have profound influence on the flow of blood in the great veins returning blood to the heart, as will be discussed in Chapter 6.

5.5.5 Convection Requirements

In a previous section (5.2) we contrasted aerial and aquatic media as regards such factors as gas content, viscosity, and diffusion rate of gases. It is of interest to discover how the respiratory parameters of ventilation and perfusion vary in aquatic and lung breathers in regard to gas utilization and the character of the medium being breathed. Here we ask how much ventilation is associated with a given oxygen consumption and what are the perfusion requirements involved? The ratio of ventilation ($\dot{V}E$) to oxidative metabolism ($\dot{V}_{O_2}$), that is, ($\dot{V}E/\dot{V}_{O_2}$) is a useful way of comparing animals. The ratio is referred to as the air or water convection requirement, and the units are liters per millimole oxygen (liter/mmol O_2). A ratio of 0.9 for a rabbit would indicate that the animal must maintain a ventilation of 0.91 air for each millimole of oxygen consumed. The blood convection requirement, $\dot{V}_b/\dot{V}_{O_2}$ has the same utility and the same units. A rabbit having a ratio of 0.4 exhibits a pulmonary blood flow of 0.4 liter/mmol O_2 consumed. The ratio of the amount of oxygen used to the amount available is the extraction coefficient, E.

Figure 5-25 compares known values for convection requirements and extraction for a variety of air and water breathers as a function of the concen-

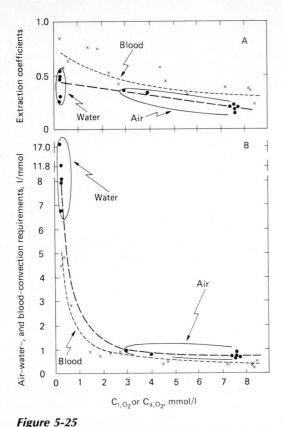

Figure 5-25

Oxygen extraction coefficients and convection requirements as functions of the oxygen concentrations in the ingoing medium, arterial blood (C_{a,O_2}), inspired water or air (C_{I,O_2}) in water breathers (Octopus, two crustaceans, various fishes), and in air breathers (iguana, hen and various mammals): for details concerning species, and ambient conditions, see Dejours, Garey, and Rahn (1970) from which this figure is redrawn.

*The extraction coefficient and the convection requirement decrease when the oxygen concentration in the ingoing medium increases. For a given oxygen concentration of the ingoing medium, the extraction coefficient is higher and the convection requirement is lower for blood than for the air and water. [Based on P. Dejours, W. F. Garey, and H. Rahn. Resp. Physiol., **9**, 108–117 (1970).]*

tration of oxygen in either the external medium or the blood perfusing the exchange organs. The data show us that the convection requirements for both the medium breathed and for blood markedly decrease with oxygen availability in the medium.

Water breathers are faced with much heavier relative demands for ventilation, perfusion, and mobilization of energy to support these functions. The countercurrent arrangement in the gills of fishes blunts these requirements and accounts in some measure for higher extraction from the medium.

A role for hemoglobin in ameliorating perfusion requirements is indicated. Most fish exhibit a blood convection requirement around 0.8 liter/mmol O_2. The corresponding figure for the antarctic icefish (no hemoglobin) is 8.36 and for mammals and birds 0.33–0.56. These figures suggest that the relative decrease in the blood convection requirement among air breathers has been partially accounted for by the generally higher hemoglobin concentration and oxygen-carrying capacity that characterizes the blood. Reinforcing this concept is the fact that both blood and air convection requirements of anemic man are increased above normal. It would be an interesting exercise to correlate, in a quantitative way, the relationship between oxygen-carrying capacity of various vertebrates and the blood convection requirement. Unfortunately, data on blood convection requirements are not abundant.

The general trend, from fishes to mammals, of increasing metabolic requirements, makes it of interest to compare oxygen flow rates for medium breathed and blood. The O_2 flow rate, for either ventilation or perfusion, is the product of mass flow and concentration of oxygen. Results of such calculations are shown in Figure 5-26. Although ventilatory and circulatory flow rates increase in the sequence, owing to increased O_2 requirements, the *ratio* of ventilatory to circulatory flow of O_2 decreases from 2.5 in fishes to 1 in mammals. This suggests that a generally more efficient extraction of O_2 from the environment has accompanied the transition from water to air breathing and endothermy.

The air convection requirements may be strongly influenced, in ectothermic lung breathers, by temperature. In the turtle, *Pseudemys scripta*, each body temperature experienced is characterized by a specific requirement, which was seen to shift progressively from 1.2 liters/mmol O_2 at 15°C to 0.5 liter/mmol O_2 at 37°C. Blood convection requirement remained relatively constant (1 liter/mmol O_2) over the temperature range while O_2 extraction increased. This thermally related change in ventila-

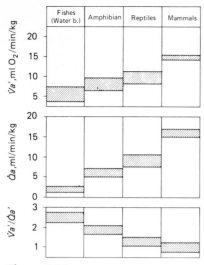

Figure 5-26
Ventilatory flow of oxygen (upper panel), circulatory flow of oxygen (middle panel), and ratio of both flow rates (lower panel) in several classes of vertebrates. [*From C. Lenfant and K. Johansen.* Resp. Physiol. **14**, *211–218 (1972).*]

tion intensity has an important influence on acid-base regulation (see Section 5.7.6).

5.6 MODELS OF GAS EXCHANGE ORGANS

In addition to the quantities represented in the general model (Section 5.3.1) we now add two conductances: convective conductance by ventilation, G_{vent} and convective conductance by perfusion, G_{perf}. These have dimensions as follows: $G_{vent} = \dot{V}_m \cdot \beta_m$ and $G_{perf} = \dot{Q}_b \cdot \beta_b$ (Figure 5-27). From the previously developed equation of the general form: $\dot{M}_g = \dot{V}_g \cdot \beta_g \cdot (P_i - P_e)_g$, it is clear that, by substituting conductances into this equation, we derive: $\dot{M}_g = G_g \cdot (P_i - P_e)_g$. We now may consider three conductances: G_{vent}, G_{perf}, and G_{diff}. It is seen from these relationships that $\dot{M}_{O_2}$, at any set of partial pressures, is dependent on the conductances (which, for the convective processes, incor-

porate both convective flow and capacitance coefficient of the gas for the medium involved).

Since the conductances are the product of convection and capacitance coefficients, it is evident that shifts in *either* capacitances or flows will alter the efficacy of gas transport. Because such alterations may occur in the external *or* internal convective systems, it is useful to express the conductances for the convective processes as ratios: G_{vent}/G_{perf}. For any of the gas exchange arrangements that will subsequently be examined, this ratio will allow an assessment of the relative efficiencies of the gas transfer process.

Vertebrates exhibit considerable variation in the structural and functional arrangements of gas exchangers (Section 5.3). These differences are adaptations to varied respiratory requirements. Four general types of systems will be considered: countercurrent system (fish gills); crosscurrent system (parabronchial lungs of birds); ventilated pool (mammal lungs); and, infinite pool (skin of amphibians). These are schematically presented in Figure 5-28, after the fashion of the previously discussed general model.

For the countercurrent system, it is seen that the first exchange of gases between the two convective streams is between venous blood and expired water. This arrangement allows arterialized blood to approach equilibrium with the P_{O_2} of inspired water. Should the arrangement be concurrent, the arterialized blood would approach equilibrium with expired water at a lower P_{O_2}. The crosscurrent model is characterized by blood flow that is effectively at right angles to the unidirectional flow of gas through the parabronchi. Arterialized blood is an admixture of efferent streams whose partial pressures vary along the length of the parabronchi. Note that there is an increase in arterial P_{O_2} from point to point along the length of the gas exchanger. The result is similar to that seen for countercurrent exchange, that is, the P_{O_2} of the arterial blood may exceed that of expired gas. This is in sharp contrast to the ventilated pool model where arterial P_{O_2} may approach equilibrium with the expired gas but never exceed the expired value. The infinite pool model allows the arterialized blood to approach P_{O_2} equilibrium with the ambient water P_{O_2}. G_{diff} for amphibian skin tends to be low by comparison with the other systems, owing to skin thickness and relatively low surface area. The sys-

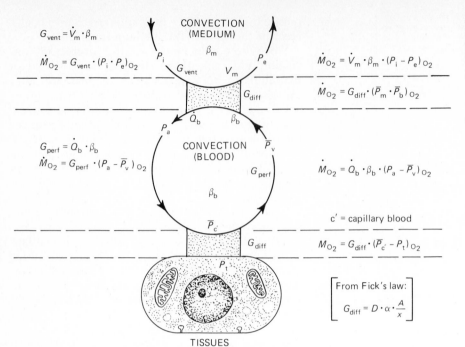

Equations in figure:

$$G_{vent} = \dot{V}_m \cdot \beta_m$$

$$\dot{M}_{O_2} = G_{vent} \cdot (P_i \cdot P_e)_{O_2}$$

CONVECTION (MEDIUM)

$$\dot{M}_{O_2} = \dot{V}_m \cdot \beta_m \cdot (P_i - P_e)_{O_2}$$

$$\dot{M}_{O_2} = G_{diff} \cdot (\bar{P}_m \cdot \bar{P}_b)_{O_2}$$

$$G_{perf} = \dot{Q}_b \cdot \beta_b$$

$$\dot{M}_{O_2} = G_{perf} \cdot (P_a - \bar{P}_v)_{O_2}$$

CONVECTION (BLOOD)

$$\dot{M}_{O_2} = \dot{Q}_b \cdot \beta_b \cdot (P_a - \bar{P}_v)_{O_2}$$

c' = capillary blood

$$M_{O_2} = G_{diff} \cdot (\bar{P}_{c'} - P_t)_{O_2}$$

From Fick's law:

$$G_{diff} = D \cdot \alpha \cdot \frac{A}{x}$$

TISSUES

tem is thus more limited by the nature of the diffusion barrier.

The relative efficiencies of the countercurrent, crosscurrent and ventilated pool models have been evaluated by Piiper and Scheid (Figure 5-29). These workers calculated the relative partial pressure differences for both medium and blood under differing ventilation-perfusion conductance ratios when G_{diff}/G_{perf} was set at infinity (that is, the diffusion barrier is not limiting of gas transfer) or when there was a significant limitation ($G_{diff}/G_{perf} = 0.5$). As indicated in the figure, the gas transfer rate, $\dot{M}$, is proportional to the length of the partial pressure bars for either medium breathed or blood. The sequence of decreasing efficiency is countercurrent, cross current, ventilated pool. This is especially manifested at $G_{vent}/G_{perf} = 1$. A strong diffusion limitation, at $G_{diff}/G_{perf} = 0.5$, has the effect of reducing efficiency greatly in all models. Under any conditions of partial pressures and diffusion barrier characteristics existing in nature, it is the ratio of the conductances for the medium breathed and for blood which is the primary determinant of gas transfer.

Much of the difference in gas transfer efficiencies between the various arrangements is due to differences in direction of convective flows and anatomical design. For mammals and birds, the dead space of the large conducting tubes of the airways is filled with gas equivalent to end-expired composition at the close of expiration. During inspiration, the dead-space gas is convected back into the gas exchange areas as fresh air enters the system. This has the effect of diluting the fresh gas. In short, there is no way by which the lungs can be ventilated with truly fresh air at external gas pressures. The fish gill demonstrates the advantage of unidirectional water flow, for there is a functional absence of dead-space water, which would have to be rebreathed.

In animals that have a complete separation of venous and arterial blood, the composition of venous and arterial blood is well defined (mammals, birds, and gill-breathing fishes). Among reptiles, amphibians, and bi- or trimodal breathing fishes, there are differing admixtures of venous and arterial blood within the heart or due to vascular shunts or drainage patterns from accessory gas exchange

Figure 5-28

*Models of vertebrate respiratory exchange organs. (a) Countercurrent (fish gills); (b) crosscurrent (birds); (c) ventilated pool (mammalian lung); (c) infinite pool (amphibian skin). Below each model is represented the equilibrium of the partial pressure of O_2. [Based on J. Piiper and P. Schied. Respir. Physiol., **23**, 209 (1975).]*

(a) Countercurrent

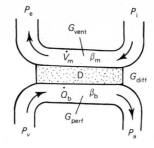

(b) Crosscurrent

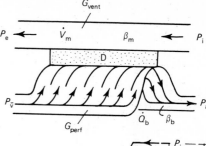

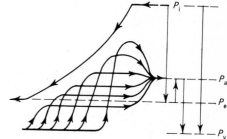

(c) Ventilated pool

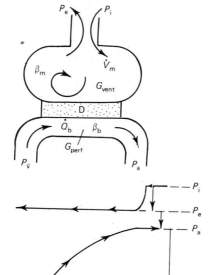

(d) Infinite pool

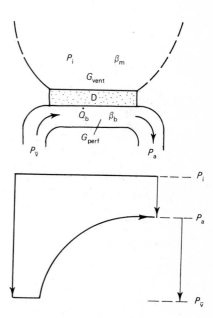

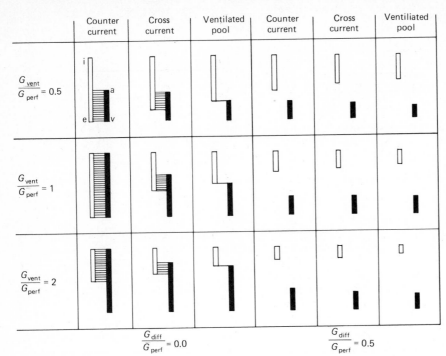

Figure 5-29
Partial pressure relations of medium breathed and blood for various conductance ratios for countercurrent, crosscurrent, and ventilated pool models. The height of the bars is proportional to M, the total gas transfer rate. M is a measure of the efficiency of the system under the conditions of conductance ratios indicated. For all conditions specified the sequence of decreasing efficiency is: countercurrent > crosscurrent > ventilated pool. [Based on J. Piiper and P. Scheid. In J. G. Whiddicombe (ed). Internat. Rev. Physiol., Respiratory Physiology, 2, vol. 14, p. 235, University Park Press, Baltimore, 1977].

organs. Such complications make the analysis of such systems extremely complex and remain an intriguing, if frustrating, challenge.

5.7 TRANSPORT OF RESPIRATORY GASES

The movement of gases between the environmental medium and the blood is in accord with the gas laws. The idea that gases are secreted by the lung or gill surfaces has been shown to be incorrect. The purely physical phenomenon of diffusion can account for gas transport across the respiratory membranes.

The complexity of the diffusion pathway for a gas across the environmental-blood barrier is illustrated in Figure 5-30. As long as the diffusion coefficient, the area of membrane exposure, and interstitial-fluid volume remain relatively constant, it can be appreciated that the major determinant of oxygen transport will be the difference in P_{O_2} between capillary blood and the gas at the respiratory surface. It is easy, therefore, to appreciate the impact of a fall in environmental P_{O_2} on the rate of oxygen diffusion across respiratory surfaces. Given the circumstances of adequate ventilation, however, and air or water of usual composition, favorable gradients for O_2 diffusion inward and CO_2 diffusion outward exist, because the P_{O_2} of the capillary blood of the respiratory membranes is lower and P_{CO_2} higher than in the environmental medium.

5.7.1 Morphometric Relationships

An interesting approach to gas exchange, morphometric analysis, seeks to understand how such physical parameters as number and size of alveoli, diffusion capacity, thickness of the exchange surfaces, lung volume, respiratory surface area, and so on, are related to oxygen consumption, body mass, and activity. The object is to derive general mathematical relationships that relate structure to function. This field is now yielding considerable information with improvements in fixation techniques, histological methodology, and accuracy of measurements.

The major morphometric findings for the lungs of mammals are summarized in Table 5-6 and Figure 5-31. For mammals of all masses, examined

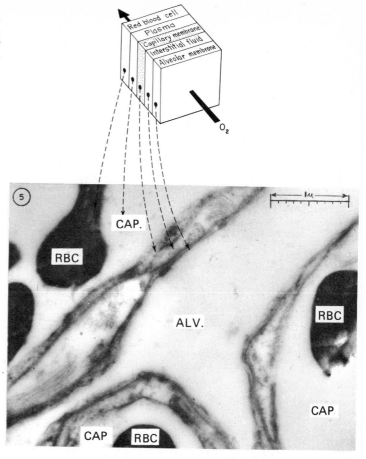

Figure 5-30

Tissues through which oxygen must pass from the gas phase in the alveolus until it combines with the hemoglobin within the red blood cell (10,000×). No attempt has been made to portray relative tissue thickness. [Photograph originally in F. N. Low. Anat. Rec., **117**, 241 (1953); reproduced from J. H. Comroe, Jr., R. E. Forster, A. B. Dubois, W. A. Briscoe, and E. Carlsen. The Lung, The Year Book Medical Publishers, Inc., Chicago, 1962.]

TABLE 5-6. Allometric Factors Describing Morphometrics of Mammalian Lungs

Y	$=$	a	$\cdot$	W^b	$r=$
V_L	$=$	0.032	$\cdot$	$W^{1.05}$	0.995
S_a	$=$	0.0038	$\cdot$	$W^{0.98}$	0.993
τ_{ht}	$=$	0.300	$\cdot$	$W^{0.05}$	0.843
D_L	$=$	0.0083	$\cdot$	$W^{0.96}$	0.991
$\dot{V}_{O_2}$	$=$	0.122	$\cdot$	$W^{0.72}$	0.973
D_L	$=$	0.167	$\cdot$	$\dot{V}_{O_2}^{1.28}$	0.978

Symbols: V_L = lung volume, ml; S_a = respiratory (alveolar) surface area, M^2; τ_{ht} = harmonic mean barrier thickness, μm; D_L = diffusion capacity for O_2, ml O_2/(min · torr); $\dot{V}_{O_2}$ = oxygen consumption, ml · min^{-1}; W = body mass, kg. [Based on: E. R. Weibel. Respir. Physiol., **14**, 26–43 (1972).]

lung volume and alveolar surface area (Figure 5-31b) are closely proportional to body mass. The diffusion barrier thickness, τ_{ht}, increases slightly, but significantly with W (Figure 5-31b, top). Thus, small and metabolically intense animals are less prone to diffusion limitations than are large forms. In Figure 5-31c, we observe the D_L is also proportional to W. So far, the equations are linear on log-log plots; however, when S_a is plotted against $\dot{V}_{O_2}$, the results are curvilinear (dashed line, Figure 5-31a). This is somewhat disappointing—we might expect a slope of 1, and indeed this appears to hold until mass declines to the mouse-bat-shrew range. The problem may reside in difficulties in estimating equivalent $\dot{V}_{O_2}$'s for small and large animals. But the morphometrist is looking for general rules. The composite data suggest that D_L is pro-

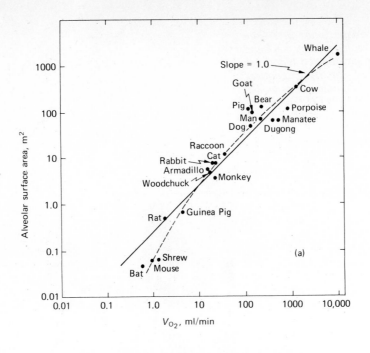

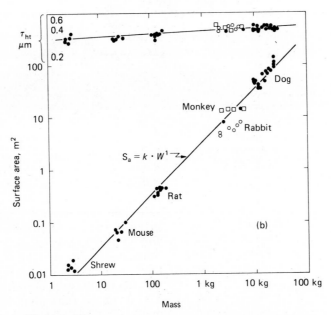

Figure 5-31
(a) Relationship between alveolar surface area and oxygen consumption in mammals of various sizes. Broken line stresses a nonlinear relationship. (b) Plot of alveolar surface area and mean respiratory barrier thickness (air to blood) against body mass.

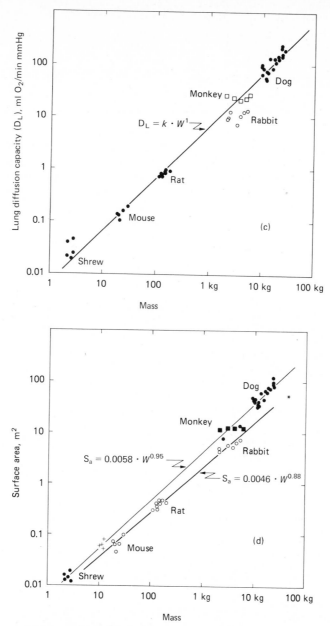

Figure 5-31 (continued)

(c) Diffusion capacity (D_L) plotted against body mass. (d) Separation of "free-living" (upper thin line) and "captive" species in plot of alveolar surface area against body mass. Asterisk relates value for man and crosses indicate waltzing mice, a hyperactive genetic strain of the "captive" category which falls on the "free-living" line. [*From (a) S. M. Tenney and J. E. Remmers.* Nature, **197,** *54–56 (1963); (b)–(d) from E. R. Weibel.* Physiol. Rev., **53** *(2), 419–495 (1973).*]

portional to $\dot{V}_{O_2}^{1.28}$; however, when dealing with animals larger than mice, a direct proportionality to W is probably more appropriate.

When log-log plots are before us they tend to visually "hide" the scatter of the data. At times, one can find further value in separating out data, based on objective criteria. When cage-kept animals are evaluated separately from wild or "athletic" animals, an interesting result is found (Figure 5-31d). Rabbit, rat, and mouse, all cage-reared, occupy a curve below the "athletic" group; that is, the more sedentary animals exhibit lower alveolar surface areas than active forms of similar mass. Note that the asterisk, representing man, finds us in the "captive" group. The small crosses on the graph represent a genetic strain of mice, called waltzing mice, which are characterized by hyperactivity. The results strongly suggest that a behavioral factor, physical activity, could be determinant in the control of the size of the gas exchange organ. An additional factor in determining diffusion capacity appears to be the prevailing partial pressure of O_2 in the atmosphere. Rats reared to adulthood under hypoxic conditions, resembling the high altitude mountain situation, exhibited larger D_L values than their "low altitude" counterparts. Interestingly, adult rats are not capable of showing this adjustment. A similar lack of adjustment characterizes lowland man sojourning at high altitude.

Morphometric data are also accumulating for nonmammalian vertebrates. The relatively small lungs of birds have been shown to contain a gas exchange area larger than for mammals on a per gram body weight basis. This is achieved through a very dense arrangement of blood capillaries in association with air spaces, the "air capillaries," which are only 3–10 μm in diameter. The surface density of air capillaries was found, by Duncker, to be from 190–300 mm^2/mm^3. Mammals of the same size range as the birds studied had an alveolar surface density of 50–60 mm^2/mm^3.

Amphibian lung volume appears to be proportional to metabolic rate, that is, $W^{0.75}$. However, when both pulmonary and skin respiratory surfaces were considered, total respiratory surface area was found to increase in direct proportion to body mass. It appears that pulmonary surface area of reptiles is proportional to metabolic rate.

Data on the lamellar surface area of fishes indicate that such active fishes as tuna exhibit a greater respiratory exchange area per unit body mass than more sluggish species. Thus, similar factors may determine respiratory surface area from fish to mammals.

A significant morphometric question has been asked concerning gas exchange in the avian egg: since egg mass varies widely among birds, in what manner does the conductance of water and gases vary with mass and with the porosity of the shell? It was found that water conductance varied with egg mass in a manner similar to the relationship of oxygen consumption and egg mass, that is, $W^{0.78}$ (Figure 5-32). The transport of all gases (CO_2, O_2, H_2O) across the egg shell is by diffusion, and it was found that the conductance for any gas is proportional to the product of the diffusion coefficient of the gas and the ratio of pore area to shell thickness, that is, $W^{0.78}$. It was concluded that since water loss is independent of metabolic rate, it is likely that the particular pore geometry that predetermines gas conductance evolved in response to the metabolic needs of the embryo. The fact that water conductance varies as $W^{0.78}$ merely reflects

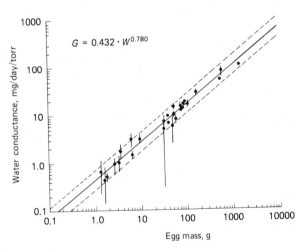

Figure 5-32
Regression of egg shell water vapor conductance (G) on the initial mass (W) of egg. Points represent grand means. Vertical bars indicate ± standard deviations. The dotted lines enclose ±2 standard errors of estimate, and represent the 95% confidence limits for the log of conductance. [From A. Ar, C. V. Paganelli, R. B. Reeves, D. G. Greene, and H. Rahn. The Condor, 76, 153–158 (1974).]

its dependence on metabolic requirements for gas exchange. That the avian shell gland can manufacture an encasement that subscribes so precisely to the upcoming metabolic requirements of the embryo is a remarkable revelation of morphometric studies which can only deepen our appreciation of the evolutionary process while leaving us deeper in mystery as to the nature of the mechanisms involved in the accomplishment. There is more implied when we break our breakfast eggs than coagulated protein!

5.7.2 Structure and Properties of Hemoglobin

Vertebrate blood, after traversing the respiratory exchange surface, contains about 5–25 ml $O_2/100$ ml. The plasma, however, contains only about 0.3 ml $O_2/100$ ml; thus the major portion of the oxygen is associated with the red blood cells in the form of a complex with hemoglobin. This pigment is common to all vertebrates except some antarctic fishes that do not possess a respiratory pigment.

It has been commonly held that the presence of hemoglobin in red cells has the advantage of reducing the blood viscosity and colloid osmotic pressure below that which would exist if the large molecule were in solution in the plasma. Work by K. Schmidt-Nielsen and C. Taylor has shown that, when the red cells of dog or goat blood were disrupted with ultrasound, giving a solution with the same hemoglobin concentration, the relative viscosity was reduced. They suggested that the existence of red cells does not contribute to a reduced blood viscosity. It is probable that a major advantage is achieved, however, by the close association of hemoglobin with localized intracellular enzymes that accelerate the reactions of hemoglobin loading (that is, glutathion reductase, which reduces methemoglobin to the oxygen-carrying form) or participate in the acceleration of the buffering function of hemoglobin (carbonic anhydrase, see Section 5.7.4).

Hemoglobins of vertebrates consist of four iron-containing (heme) groups attached to a larger protein "carrier" portion of the molecule, the globin. The molecular weight is 68,000 and each molecule contains four heme groups (Figure 5-33), each of molecular weight 572. An oxygen molecule may unite reversibly by combining with one of the four

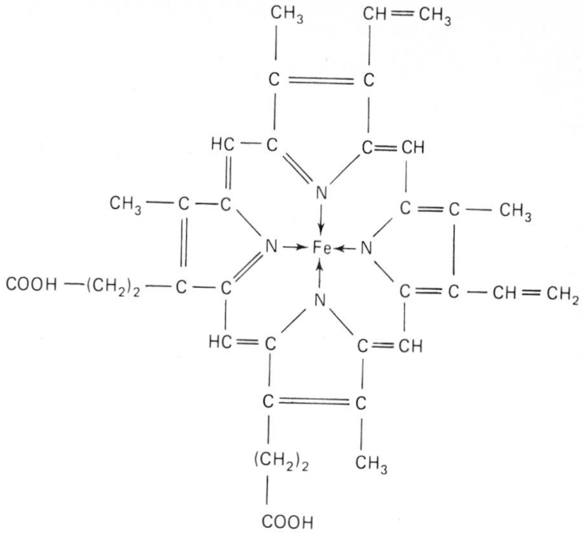

Figure 5-33
Basic heme structure of the hemoglobin molecule.

iron atoms that is attached by valency bonding to four pyrrole groups that make up the heme molecule. A remaining sixth valency bond of iron apparently attaches with the globin. A fully saturated molecule of hemoglobin (HbO_2) actually contains four molecules of oxygen. In this form the hemoglobin is called *oxyhemoglobin* and is said to be fully saturated. The degree of saturation of hemoglobin is dependent, among other factors, on the partial pressure of oxygen in the plasma. By exposing blood at a constant pH to different partial pressures of oxygen and, following equilibration, determining the oxygen content of the red cells, a curve representing the combining capacity at varying partial pressures of oxygen is obtained. Oxygen content may be expressed as percent saturation. If, after exposure to pure oxygen, the oxygen content of hemoglobin was found to be 19 ml/100 ml of blood (100% saturated), a similar sample exposed to a lower P_{O_2} and containing 9.5 ml $O_2/100$ ml would be 50% saturated. Such plots are called *oxygen dissociation curves* and serve to illustrate the importance of oxygen tension on both loading and unloading (Figure 5-34a).

The pH and P_{CO_2} of the blood influence the shape of the oxygen dissociation curve in an important manner. High P_{CO_2} and low pH both shift the curve to the right (*Bohr effect*) (Figure 5-34b and c). Be-

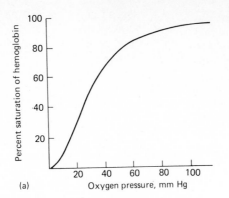

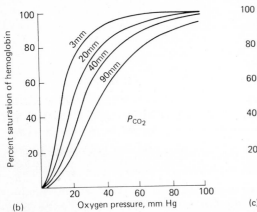

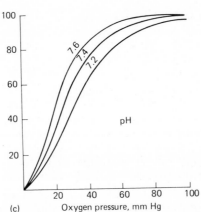

Figure 5-34
(a) Oxygen dissociation curve. (b and c) Effects of variation in P_{CO_2} and pH on the dissociation curve.

cause the P_{CO_2} in the capillaries of an active tissue is elevated and the blood slightly more acid, it can be seen that the shift to the right facilitates unloading and thus oxygen diffusion. Although elevations in P_{CO_2} decrease the pH of the blood, there is evidence that the mechanisms by which P_{CO_2} and pH influence the dissociation curve are different.

The portion of the dissociation curve of greatest interest is the steep zone in the range around 40 mm Hg P_{O_2}, a P_{O_2} close to that of tissues. In this zone a small fall in P_{O_2} is associated with a fairly large unloading of oxygen. At the upper end of the curve a fairly large decrease in the oxygen tension of alveolar oxygen has a much smaller effect on oxygen saturation.

A rise in temperature, like a fall in pH or an increase in P_{CO_2}, also causes a shift to the right in the hemoglobin dissociation curve. At 20°C, human hemoglobin is well saturated at low P_{O_2}'s, mak-

ing it of little value as an oxygen carrier at such a low temperature. The effect of temperature may be of considerable importance for lower vertebrates. Because body temperature, and hence metabolism, is more closely correlated to ambient temperature in ectotherms, shifting of the dissociation curve to the right facilitates unloading at higher temperatures, an effect that may be of great importance in forms experiencing wide ranges of ambient temperature.

While an elevation in P_{CO_2} is generally associated with a shift in the oxygen saturation curve to the right, exceptional behavior of hemoglobin in response to elevated P_{CO_2} has been observed. Root found that, for many fish hemoglobins, an elevation in P_{CO_2}, or reduction in pH, resulted in lower oxygen saturation even at very high P_{O_2}. This behavior is known as the ***Root effect.*** The blood of fishes exhibiting the Root effect also exhibit a marked

Bohr effect. Since the Root effect has been generally associated with fishes that have gas bladders, it is thought of as a mechanism for oxygen secretion into the bladders (see Section 6.9.5). The Root effect has been reported for the hemocyanin of *Octopus.*

The hemoglobin of muscle, *myoglobin,* has a much greater oxygen affinity at low partial pressures of oxygen than does circulating hemoglobin. The pigment contains a single heme group. When the dissociation curves of mammalian myoglobin and hemoglobin are compared, it is observed that at the venous P_{O_2} of 40 mm Hg, hemoglobin is 60% saturated whereas myoglobin is over 90% saturated. The curve for myoglobin is far to the left of hemoglobin and is hyperbolic, in contrast to the sigmoid hemoglobin curve. Myoglobin is about 50% saturated at the low partial pressures at which the cytochrome oxidase system is just saturated, and it has been suggested that myoglobin facilitates the transport of oxygen from hemoglobin to cellular enzyme systems. This idea is based on the observation that diffusion of oxygen in water is at a more rapid rate in the presence of myoglobin. The effect is thought to be due to the diffusion of the pigment itself. In muscle cells, it is argued that since the pigment is mobile in the fibers, it is likely that it facilitates oxygen diffusion to mitochondria. The resolution of the functional role of myoglobin *in vivo* must await measurements of the oxygen equilibrium curve under intracellular conditions. If facilitated diffusion is a major function of myoglobin, it could be imagined that this was a primary evolutionary function of the pigment that would augment oxygen transport beyond the limitations of simple diffusion. Incorporation of this monomer of hemoglobin into circulating cells and subsequent aggregation into tetrameric hemoglobin would then occur as a further supplement to oxygen transport as well as to the buffering capacity of the blood. At present, all this is a matter of conjecture.

Another way of looking at the hemoglobin equilibrium was devised in 1910 by A. V. Hill. His approximate equation

$$\frac{y}{100} = \frac{KP^n}{1 + KP^n}$$

gives the equilibrium curve, where y is the percent saturation with oxygen, P the partial pressure of oxygen, K the equilibrium constant, and n a measure of the interaction between heme groups. The equation can be rearranged in the form

$$\frac{y}{100 - y} = KP^n$$

Taking the logarithms of the two sides of the equation yields

$$\log\left(\frac{y}{100 - y}\right) = \log K + n \log P$$

A plot of log $[y/(100 - y)]$ against log P yields a straight line, the slope of which gives n. The intercept on the log $[y/(100 - y)]$ axis is equal to the equilibrium constant K. In the case of myoglobin, which contains only one atom of iron, Hill's equation yields $n = 1$ over the entire equilibrium curve. For hemoglobin, which contains four heme groups, the portion of the dissociation curve of greatest physiological interest (between 20 and 98% saturation) is characterized by $n = 2.7$. Below and above these levels n increasingly approaches 1.

Myoglobin ($n = 1$), containing a single iron atom, exhibits iron-oxygen interaction independent of the heme groups of other myoglobin molecules. Hemoglobin, on the other hand, behaves over a wide range of saturations as if, instead of containing single units of iron-oxygen complexes, the iron atoms were members of a complex of 2.7 atoms of iron. Because the complex behaves as a unit in combining with oxygen, the phenomenon is referred to as *heme-heme interaction,* and is presumably caused by the fact that the oxygen affinity of a heme group is influenced by the state of the other heme groups of the molecule. Oxygenation of one heme group facilitates oxygenation of the others. Heme-heme interaction is thought to be due to changes in the protein rather than in electrostatic forces between adjacent hemes. Elevations of P_{CO_2} or increasing acidity produces the characteristic Bohr effect, a circumstance in which the facilitatory heme-heme interaction (n value) is reduced.

The hemoglobin of lampreys is peculiar among chordates in that it, like myoglobin, contains only one basic unit. This hemoglobin shows no heme-heme interaction ($n = 1$) and the dissociation curve is hyperbolic. *Myxine,* an Atlantic hagfish, is intermediate, having hemoglobin of one and two basic units, a mixture suggestive of a state in the evolution of heme-heme interaction.

Shulman, et al. (see Additional Reading, Respiratory Pigments) have presented evidence that challenges the more classical concept just presented. Using nuclear magnetic resonance and electron paramagnetic resonance techniques, they fail to detect changes in the heme groups themselves and have suggested that the "cooperativity" of oxygen loading is associated with the free-energy changes derived from interaction in the protein moiety of the subunits. They suggest the term **subunit interaction** rather than heme-heme interaction to describe the facilitatory effect of oxygenation on oxygen affinity of hemoglobin. The mechanisms by which subunit interactions produce facilitation is as yet unclear.

5.7.3 Environmental Correlations and Hemoglobin Behavior

The oxygen dissociation curves of various animals exhibit different relative positions and shapes. This diversity is probably related to different structural relationships in hemoglobin from one species to another. A dissociation curve to the extreme right of the range would indicate a hemoglobin requiring relatively high oxygen tensions for loading. An animal possessing such a pigment would have an oxygen-transport capability of limited utility when moving into zones of low oxygen tension, which allow only partial saturation of the hemoglobin. One can thus appreciate how a biochemical characteristic may limit distribution of animals. Among vertebrates (and invertebrates as well) there exist rather striking correlations between the form of the oxygen dissociation curve and ecological distribution.

Among fishes, those that inhabit waters of high P_{O_2} (trout) generally exhibit oxygen dissociation curves to the right of forms living in stagnant waters (for example, carp and catfish). The behavior of these hemoglobins has an important influence on the environmental restriction placed on animals. At 15°C and a P_{CO_2} of 1–2 mm Hg, the P_{O_2} required to bring about 50% saturation of the hemoglobin (P_{50}) for carp blood is 5 mm Hg, whereas that for rainbow trout is 18 mm Hg. It is apparent then that carp can range into water of low P_{O_2} forbidden

to trout by virtue of the loading characteristics of their hemoglobin. In addition, the effect of elevated P_{CO_2} (Bohr effect) on the blood of stagnant-water forms is less than for those fishes inhabiting well-oxygenated waters (Table 5-7). A shift to the right induced by a change of P_{CO_2} from 1 to 10 mm Hg may shift the dissociation curve so far to the right in a trout that only partial saturation may be attained, and the fish may suffocate in an abundance of oxygen.

It was pointed out previously that elevations in temperature, like increasing P_{CO_2}, shift the oxygen dissociation curve to the right (elevate the P_{50} or half-saturation oxygen pressure). This effect, as is true for the Bohr effect, varies among different hemoglobins. Figure 5-35 gives some idea of the range of variation seen in several fishes. It can be appreciated from these data that *Anguilla*, for example, is at an advantage compared to *Raja* when the temperature is elevated from 5 to 30°C, because at the higher temperature almost full saturation can be attained at partial pressures of oxygen far below that required by *Raja*.

Extensive comparisons of animals inhabiting zones of varying oxygen availability have revealed that aquatic forms load the hemoglobin at lower oxygen pressures (lower P_{50} values) than their terrestrial relatives. This left-handedness of the oxygen dissociation curve seen in aquatic forms allows extraction of oxygen from an environment poor in oxygen relative to air.

Mammals that are native to high altitudes (for example, the llama and vicuña) possess hemoglobin of greater oxygen affinity than their low-altitude relations. These residents of high altitude have **hematocrits** (percent of blood as red cells) and red

TABLE 5-7. Effect on P_{50} (Half-Saturation O_2 Pressure) Induced by Elevating P_{CO_2} by 10 mm Hg

Forms	Habitat	Change in P_{50} (mm Hg)
Catfish, carp, bowfin	Stagnant waters	<10
Trout (three species)	Well-oxygenated streams and lakes	>35

Source: C. L. Prosser, *Comparative Animal Physiology* W. B. Saunders Company, Philadelphia, 1961, p. 218.

Figure 5-35

Oxygen dissociation curves for the bloods of three species of fishes in relation to temperature and P_{O_2} together with the positions of the half-saturation points of the bloods of various other species determined at single temperatures. [From F. E. J. Fry. In M. E. Brown (ed.). The Physiology of Fishes, *vol. 1, pp. 1–63, Academic Press, New York, 1957*).]

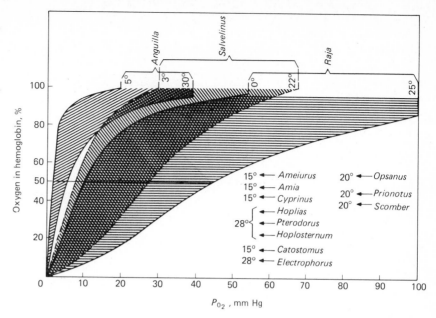

cell counts similar to their sea-level relatives. When low-altitude animals (for example, dog, sheep, and man) are acclimated to high altitude, both hematocrit and red cell count rise, as does total hemoglobin. The effect is to increase oxygen capacity. This response is not exclusively mammalian, as goldfish exposed to a reduced P_{O_2} respond with an increase in hemoglobin.

The dissociation curves of mammals ranging in size from mouse to elephant exhibit an array in which body mass correlates directly with steepness of the curve. Smallness correlates with higher P_{50} values; that is, the blood of small mammals will unload oxygen under higher P_{O_2} than larger mammals (Figure 5-36c). Small animals also exhibit a greater Bohr effect, that is, the acidification of blood at the tissue level has a greater effect on unloading in the mouse than in the elephant (Figure 5-36b). These characteristics of hemoglobin correlate with the greater relative need for oxygen by smaller animals (see Section 3.7).

There is increasing evidence that hypoxia induces, at least in mammals, the release of a hormone, erythropoietin, which accelerates red cell production and release from the bone marrow. A search for the source of erythropoietin indicates that the kidney is a likely site of production.

Variation in the behavior of hemoglobin has been observed at different stages of the life cycle in a number of vertebrates. For example, the oxygen dissociation curve of the fetus of both sheep and man is displaced to the left of that seen in the adult. Oxygen can thus be more efficiently transferred from maternal to fetal blood across the placenta. After parturition, the fetal pigment is replaced by adult hemoglobin. Fetal hemoglobins have been observed in all major vertebrate groups. Among fishes, viviparous forms such as the dogfish exhibit fetal hemoglobins of greater oxygen affinity than in adults, and the same relationship has been observed in viviparous snakes. Frog tadpoles' hemoglobin dissociation curves lie to the left of the adult and the Bohr effect is quite small. In this relation the tadpole is more piscine that the adult, a situation that appears quite appropriate for the transition from an aquatic to a more terrestrial, air-breathing life. A similar transition occurs in birds. Fetal hemoglobins appear to be well suited for oxygen procurement and are an important supporting feature of viviparity and the transition from aquatic to terrestrial life during the developmental cycle.

The position of the hemoglobin dissociation curve is influenced by the concentration of organic

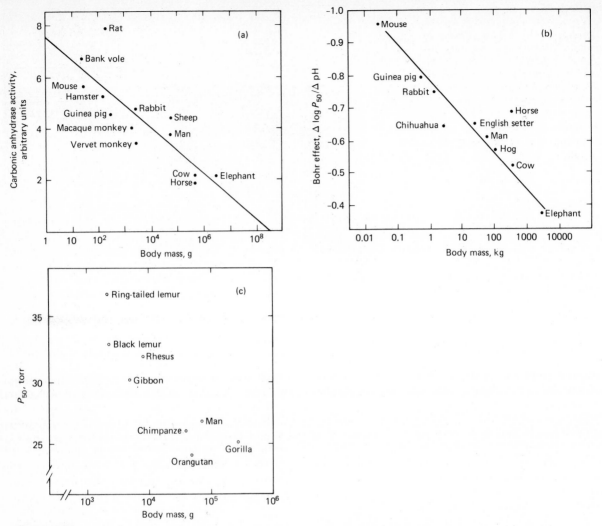

Figure 5-36
Relationships between body mass and (a) carbonic anhydrase activity of red blood cells; (b) Bohr effect; and (c) partial pressure of oxygen at 50% hemoglobin saturation, P_{50}. [From (a) E. Maqid, Comp. Biochem. Physiol., **21,** *357–360 (1967); (b) A. Riggs,* Physiol. Rev., **45,** *619–673 (1965); (c) from D. Dhindsa, J. Metcalfe, and A. Hoversland,* Respir. Physiol., **15,** *331–342 (1972).]*

phosphates within the red cells and the particular phosphates involved differ among animal groups. In the mammalian red cell, the major organic phosphate is 2,3-diphosphoglycerate (2,3-DPG); birds and some turtles have inositol hexaphosphate (IHP); most reptiles, including many turtles have mostly adenosine triphosphate (ATP) and, fish erythrocytes contain either ATP or guanosine tri-

phosphate (GTP). Organic phosphate compounds are also found in embryos or larval red cells in concentrations often differing from adult levels. At high erythrocyte concentrations, these compounds interact with hemoglobin to produce an allosteric effect on oxygen affinity by which oxygen affinity is reduced (right shift of the dissociation curve). Decreases in organic phosphate concentrations have the reverse effect, that is, increasing oxygen affinity.

In addition to the allosteric effect on oxygen affinity, the concentration of the unbound fraction of these phosphate compounds, which do not escape across the red cell membrane, has an influence on the Donnan distribution of protons in such a manner that lowering their concentration increases the pH of the cell. In the case of the carp (studies by Greaney and Powers), it may be concluded that the effects may be complementary. The concentration of red cell ATP decreases during exposure of the fishes to hypoxia, owing to a decrease in erythrocyte oxidative phosphorylation. The effect is to increase cell pH via the altered Donnan distribution of protons across the membrane. Additionally, ATP binding is preferentially to deoxyhemoglobin. Thus, increased ATP binding serves to amplify the increase in pH since it reduces the unbound ATP concentration.

This combination of allosteric and pH effect increases O_2 affinity and allows loading to occur more effectively in the gills in the presence of a reduction in the P_{O_2} of the water being breathed. As loading occurs, ATP is released from its bound form to *reduce* the cell pH via the Donnan effect. Further acidification occurs across the capillaries, allowing right shift (reduced hemoglobin affinity for O_2) and unloading of oxygen to the tissues. Without these changes in pH, *unloading* during hypoxia would become a limiting factor.

A number of fishes have been demonstrated to modulate hemoglobin-O_2 affinity in response to changes in acclimation temperature as well as to oxygen availability, the ATP levels being inverse to temperature. Such modulation counteracts the temperature-induced right shift in the O_2 dissociation curve seen at constant ATP concentration. Additionally, specific phenotypes, exhibiting different ATP erythrocyte levels, which correlate with both thermal and oxygen availability aspects of their environments, have been identified for the killifish, *Fundulus heteroclitus*. Thus, genetic selection for capacity to regulate organic phosphate levels in regard to specific features of the environment may partially explain the wide ecological and geographic distribution of this fish.

The response of mammals, including man, when transported to the hypoxic circumstances of high altitude, is contrary to that of fishes exposed to hypoxic water, that is, erythrocyte 2,3-DPG *increases*. The initial effect of hypoxia is hyperventilation with consequent reduction in blood P_{CO_2} and alkalinization of the blood. This effect is also expressed in a elevated red cell pH. It has been shown that this pH effect accelerates the enzyme pathways leading to 2,3-DPG synthesis. The acidification effect of the unbound phosphate compound, as discussed, tends to counter the alkalinization effect of hyperventilation and, it is thought by some, that the *in vivo* O_2 dissociation curve during acute exposure to altitude may be unchanged from the low-altitude curve. One may speculate that during longer term altitude exposure, as renal mechanisms that lower the bicarbonate concentration of the blood compensate for the respiratory alkalosis, the levels of 2,3-DPG of early exposure may be reduced, with a consequent modulation of the O_2 dissociation curve. This phase of altitude adjustment is in need of study.

Although specific fetal hemoglobins, with high O_2 affinity, have been identified in all vertebrate groups, there is also evidence that their oxygen affinity is modulated by phosphate compounds. Human maternal blood (HbA) exhibits lower O_2 affinity than the HbF of the fetus. When purified, HbA exhibits *greater* affinity than HbF. It is the fact that HbA is more sensitive to DPG that accounts for the higher O_2 affinity of fetal relative to maternal blood. Such an adaptation, favoring transfer of O_2 from the maternal to the fetal circulation, is reminiscent of the evolutionary trends for Hb of greater O_2 affinity in aquatic as compared with air-breathing forms.

5.7.4 Carbon Dioxide Transport

The CO_2 produced in the tissues is transported to the lungs or gills, where it is exchanged by diffusion. Little of the CO_2 is in physical solution in the plasma but rather is largely found in reversible chemical combinations, both in plasma and in the erythrocytes. Much of the plasma CO_2 combines with water to yield carbonic acid ($CO_2 + H_2O \rightleftharpoons H_2CO_3$). Carbonic acid dissociates readily into hydrogen and bicarbonate ions ($H_2CO_3 \rightleftharpoons H^+ + HCO_3^-$). This reaction would acidify the plasma (free H^+) except for the presence of protein (A^-), which acts as a buffer, sopping up free hydrogen ($H^+ + A^- \rightleftharpoons HA$). Bicarbonate ion forms an association with the plasma cation, chiefly Na^+, to yield the reversible complex, $NaHCO_3$. This buffering capacity of plasma components drives the reaction to the right, further facilitating formation of HCO_3^- and H^+ from carbonic acid, and "sucking in" more CO_2. These reactions are summarized:

$$CO_2 + H_2O \rightleftharpoons H_2CO_3 \qquad (5\text{-}1)$$

$$H_2CO_3 \rightleftharpoons H^+ + HCO_3^- \qquad (5\text{-}2)$$

$$H^+ + A^- \rightleftharpoons HA \qquad (5\text{-}3)$$

$$HCO_3^- + Na^+ \rightleftharpoons NaHCO_3 \qquad (5\text{-}4)$$

or, in more abbreviated form:

$$CO_2 + H_2O + NaA \rightleftharpoons HA + NaHCO_3$$

The formation, in plasma, of H_2CO_3 is much too slow to account for the rate at which CO_2 is taken up into the plasma. The observation by Christiansen and coworkers in 1914 that oxygenated blood takes up less CO_2 at any given P_{CO_2} than does deoxygenated blood (presence of reduced hemoglobin) provided the clue that hemoglobin was acting as a buffer after O_2 dissociation in the tissues (Figure 5-37). Some 50% of the increase in CO_2 content of the blood after passing the capillaries can be accounted for by the buffering capacity of hemoglobin. The CO_2 involved in this reaction ends up in the form of bicarbonate, and the role of hemoglobin buffering in this set of reactions can be summarized as follows:

1. CO_2 enters red cell from plasma.
2. In the red cell $CO_2 + H_2O \xrightarrow{\text{CA}} H_2CO_3$ (this reaction is catalyzed by the enzyme carbonic anhydrase, and is very fast).
3. $H_2CO_3 \rightleftharpoons H^+ + HCO_3^-$.
4. Hemoglobin, after giving up O_2 (Hb^-), enters into union with H^+ : $Hb^-H^+ \rightleftharpoons HbH$.

This reaction is of great importance in stabilizing the pH of the cell.

The addition of CO_2 to the blood is accompanied by a shift in Cl^- from plasma to the interior of the cell. As HCO_3^- is formed within the erythrocyte at a high rate, HCO_3^- diffuses to the plasma. Electrical balance could be maintained across the membrane if the HCO_3^- were accompanied by cation

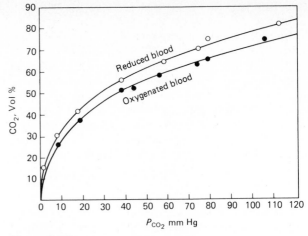

Figure 5-37
CO_2 dissociation curve of oxygenated and reduced blood. [From F. J. W. Roughton. In W. O. Fenn and H. Rahn (eds.). Handbook of Physiology, *sec. 3, vol. 1, pp. 767–825, American Physiological Society, Washington, D.C., 1964.]*

(such as K^+ or Na^+), but the cell membrane resists passive cation movement. The principal anion of the plasma is Cl^-, and, because the membrane is quite permeable to Cl^-, this anion enters the cell as HCO_3^- leaves, maintaining the electrical balance across the membrane. There is, then, an exchange of HCO_3^- for Cl^-. The phenomenon is often referred to as the *chloride*, or *Hamburger, shift*.

This chain of events moves along very rapidly because of the presence of carbonic anhydrase. The enzyme is found in high concentration in erythrocytes, and its importance can be appreciated when the handling of CO_2 is studied in the presence of drugs (such as diamox) that inhibit carbonic anhydrase. Under these conditions, the reaction $CO_2 + H_2O \rightleftharpoons H_2CO_3$ occurs at a slow rate and is quite incomplete by the time the blood traverses the capillaries and reaches the pulmonary circulation. As a result, the P_{CO_2} in the venous blood decreases as it reaches the pulmonary capillaries although some CO_2 is excreted. The P_{CO_2} rises, however, as it traverses the arteries as CO_2 unloading continues toward completion. In consequence of this continued unloading, the P_{CO_2} of the tissues rises.

Small mammals exhibit higher mass-specific cardiac outputs than larger members of the group.

Figure 5-38
Major reactions involved in the transport of CO_2.

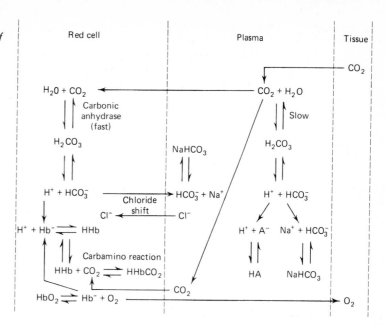

High blood flow through the lungs is presumably associated with reduced red cell transit time, potentially limiting the processes by which lung and blood gas equilibria occur. The inverse relationship between red blood cell carbonic anhydrase activity and body mass, among a series of mammals (vole to elephant), provides a biochemical means by which CO_2 equilibria may be optimized among animals of differing metabolic intensity and size (Figure 5-36a).

Another reaction, involving hemoglobin, accounts for a smaller portion of the total CO_2 than does HCO_3^- and is independent of carbonic anhydrase. Amino groups of the hemoglobin molecule interact with CO_2 in physical solution to yield a carbamino compound.

$$HbNH_2 + CO_2 \rightleftharpoons HbNHCOOH \rightleftharpoons HbNHCOO^- + H^+$$

Some 20% of the CO_2 is transported in the form of carbaminohemoglobin, 5% in physical solution, and 75% in the form of bicarbonate ion.

The principal reactions involved in CO_2 transport are summarized in Figure 5-38. With the addition of O_2 in the lungs the sequence is rapidly reversed, again under the influence of carbonic

anhydrase, and CO_2 is unloaded into alveolar gas.

From the previous considerations it can be appreciated that the unloading of O_2 and loading with CO_2 in the tissues are mutually related phenomena, because an increase in P_{CO_2} shifts the O_2 dissociation curve to the right, thus facilitating unloading (Bohr effect). Unloading of O_2 on the other hand, facilitates CO_2 uptake, owing to the role of Hb^- in buffering the H^+ that results from the dissociation of carbonic acid.

5.7.5. Carbon Dioxide and Acid-Base Regulation

In their classic monograph of 1932, Peters and Van Slyke emphasized that an important condition for tissue survival was the maintenance in the arterial blood of a reaction slightly to the alkaline side of the neutral point. They were speaking of mammals where pH 7.4 is usual for arterial blood at 37°C body temperature. The prevailing $[H^+]$ (concentration of hydrogen ion) has a profound effect on the net charge of proteins and thus the level of dissociation of ions bound to protein. Alterations in net charge thus affect the distribution of ions and water between intracellular and extracellular compartments. Additionally, enzymatic activities are sensitive to [H⁺]. Alterations in extracellular [H⁺] are associated with parallel alterations in intracellular [H⁺], although the latter is typically to the acid side of the extracellular [H⁺]. These influences of [H⁺] on cellular environment and water and electrolyte distribution are the bases for the seriousness with which physicians treat disturbances in acid-base status of their patients.

Carbon dioxide transport is a primary determinant of the blood acid-base status. Because the respiratory organs are a major avenue for CO_2 excretion, the level of ventilation controls, to a major degree, the prevailing pH of the blood. Essential to understanding the role of CO_2 in these processes is the reaction $H_2CO_3 \rightleftharpoons H^+ + HCO_3^-$. The dissociation constant for this reaction is

$$K_A = \frac{[H^+][HCO_3^-]}{[H_2CO_3]}$$

From the fact that $[H_2CO_3]$ is proportional to the product of P_{CO_2} and the solubility coefficient for CO_2 (α_{CO_2}; also designated S), it can be written

$$K_A = \frac{[H^+][HCO_3^-]}{\alpha CO_2 \cdot P_{CO_2}}$$

Expressed in logarithmic form

$$\log K_A = \log [H^+] + \log \frac{[HCO_3^-]}{\alpha CO_2 \cdot P_{CO_2}}$$

from which can be written

$$-\log [H^+] = -\log K_A + \log \frac{[HCO_3^-]}{\alpha CO_2 \cdot P_{CO_2}}$$

Since pH is the negative log of [H⁺] and the negative log of K_A has been designated pK, one can write

$$pH = pK + \log \frac{[HCO_3^-]}{\alpha CO_2 \cdot P_{CO_2}}$$

This expression is known as the **Henderson-Hasselbalch equation.** The $[H_2CO_3]$, in millimoles per liter, is the product of the prevailing P_{CO_2}, torr, and the appropriate solubility factor. Since we must deal with animals at different temperatures, it is impor-

tant to realize that both pK and the solubility coefficient are functions of temperature (Tables 5-8 and 5-9).

Taking normal values for man and other mammals, we find for arterial blood:

$$pH = 6.1 + \log \frac{24}{0.03 \times 40} = 7.4$$

We can alter this relationship by holding our breath or hyperventilating. In the latter case, by "blowing off" CO_2, the P_{CO_2} of blood leaving the lungs will fall, resulting in a higher pH for arterial blood—a condition known as respiratory alkalosis. Although alterations in P_{CO_2} can be rapidly achieved by changing ventilation, the $[HCO_3^-]$ is regulated more slowly by renal mechanisms or by transport across the gills and/or kidneys in fishes. Compensations achieved by such mechanisms are referred to as metabolic compensations, and acid-base disturbances caused by production of acid (lactic acid) or failure in renal excretory function produce metabolic acidosis or alkalosis.

5.7.6 Acid-Base Regulation and Temperature

Blood, held in a gas tight syringe so that the CO_2 content cannot change by equilibration with air, exhibits a striking change in pH as temperature is changed ($\Delta pH/°C = -0.014$). First described by Rosenthal [1948], this behavior allows one to calculate the pH of a blood sample of an animal at, say 22°C, even though the pH electrode is at a higher temperature (37°C). By applying the Rosenthal correction factor, $pH_{22°C} = pH_{37°C} + (37 - 22)(0.014)$, we can estimate that pH 7.4, measured at 37°C, is translated to pH 7.6 at 22°C. The Rosenthal factor finds extensive use during hypothermic surgical procedures and allows reports from the blood gas laboratory, where measures are routinely done at 37°C, to be easily corrected to the low temperature of a patient's blood. We normally think of the arterial blood of man (37°C) to be characterized by pH 7.4. In reality, endotherms represent a heterogeneity of arterial pH's. Once blood has traversed the lungs, the arterial blood is at constant CO_2 content and, on entering a tissue that is cool or warm, will follow the

TABLE 5-8. Solubility Coefficents, S, for CO_2 in Plasma at Various Temperatures

Temp. (°C)	S	Temp. (°C)	S
40	.0288	32	.0345
39	.0294	30	.0362
38	.0301	28	.0381
37.5	.0304	26	.0402
37	.0308	24	.0425
36	.0315	22	.0450
35	.0322	20	.0478
34	.0329	15	.0554
33	.0337	10	.0575

$$S = H_2CO_3 \ (mmol/L/P_{CO_2}, \ mm \ Hg)$$

Source: J. W. Severinghaus, M. Stupfel, and A. F. Bradley, *J. Appl. Physiol.*, **9**, 189–196 (1956).

TABLE 5-9. Nomogram for Calculation of Serum pK

(A straight line between the known pH and temperature intersects the center line at the appropriate pK.)

Temp. (°C)	pK	pH
	6.00	
45	6.02	8.0
	6.04	
40	6.06	7.8
	6.08	
35	6.10	7.6
	6.12	
30	6.14	7.4
	6.16	
25	6.18	7.2
	6.20	
20	6.22	7.0
	6.24	
15	6.26	6.8
	6.28	
10	6.30	6.6

Source: J. W. Severinghaus, M. Stupfel, and A. F. Bradley. *J. Appl. Physiol.*, **9**, 197–200 (1956).

Rosenthal curve as the precapillary blood equilibrates with the tissue temperature (Figure 5-39).

It is important to realize that, when the total CO_2 of blood is held constant, the P_{CO_2} will vary directly with temperature because of the thermal dependence of the solubility coefficient of CO_2.

The temperature-pH behavior of blood, at constant CO_2 content, has been ascribed to the buffering behavior of imidazole, a moiety of histidine that is present in proteins of plasma, tissue, and red blood cells. There is a sufficient concentration of titratable imidazole groups to account for the $\Delta pH/°C$ observed at constant CO_2 content. Thus, two buffer systems influence pH; imidazole of histidine and the carbonic acid–bicarbonate system. The *slope* of pH against temperature is largely governed by imidazole buffering.

Change of body temperature as a function of environmental temperature, is the *sine qua non* of ectotherms. Many species of invertebrates and vertebrates are known to exhibit a $\Delta pH/°C$ slope for arterial blood that is quite similar to the Rosenthal curve (Figure 5-40). Hermann Rahn and his associates have pointed out that the temperature-pH curve is parallel to the neutral pH of water (pN_{H_2O}) as well as the pOH. The pN_{H_2O} changes with temperature because the dissociation constant of water is temperature dependent. Thus, there is a constant $[OH^-]/[H^+]$ and a constant alkalinity *relative* to the pN_{H_2O}. These ectotherms behave *as if* they were closed systems. Since they exchange respiratory gases with the environmental medium, they are clearly *open* systems that, as temperature changes, must regulate either $[HCO_3^-]$ of the P_{CO_2} in order to achieve $\Delta pH/°C \cong -0.014$.

Table 5-10 exhibits data from a number of lung-breathing vertebrates in which can be found a clear trend—the arterial P_{CO_2} is a direct function of temperature whereas $[HCO_3^-]$ remains rather constant. For each species in the table, the ratio of ventilation to oxygen consumption (or CO_2 production) is *inverse* to temperature. Thus, the ventilatory poise is specific for each temperature, and the relative underventilation at higher temperatures is responsible for the elevation in P_{CO_2} required to maintain the arterial blood at constant CO_2 content. This allows the pH to slide up and down a curve, the slope of which is determined by the ubiquitous α-imidazole of histidine. Typical of this behavior is the Galapagos marine iguana, an animal that expe-

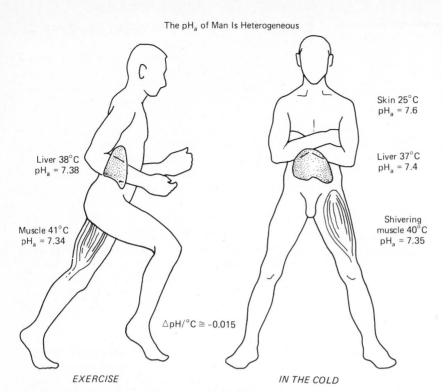

The pH$_a$ of Man Is Heterogeneous

Skin 25°C
pH$_a$ = 7.6

Liver 38°C
pH$_a$ = 7.38

Liver 37°C
pH$_a$ = 7.4

Muscle 41°C
pH$_a$ = 7.34

Shivering
muscle 40°C
pH$_a$ = 7.35

ΔpH/°C $\cong$ -0.015

EXERCISE

IN THE COLD

Figure 5-39
Arterialized blood in man may represent a spectrum of pH's depending on the tissue temperature with which the precapillary blood equilibrates.

riences a diurnal body temperature range of around 20°C (Figure 5-41).

The high β_{CO_2} of water and the flow-through nature of water across the gills is responsible for the low arterial P_{CO_2}'s of fishes. Fishes are thus left with the alternative of regulating arterial CO_2 content via HCO_3^- transport, an active process that requires greater time than the process lung breathers use, in order to achieve an acid-base equilibrium appropriate to a new thermal state.

This elaborate system of controlling the acid-base status has an important biochemical meaning. The achievement appears to be the maintenance of a constant charge state of histidine α-imidazole. The fractional dissociation of the α-imidazole groups, will remain constant along the pH slope achieved by organismal regulation of CO_2 content because the pK of imidazole (pK = 7 at 25°C) changes with pN$_{H_2O}$. So long as pH changes with temperature in parallel to pN$_{N_2O}$, the charge state of α-imidazole will be conserved.

The intracellular pH (pH$_i$) has been examined for a number of ectotherms at different tempera-

tures. In general pH$_i$ appears to be maintained near pN$_{H_2O}$. This means that α-imidazole charge state will remain constant. Thus, enzymatic reactions requiring protonated histidine for ligand binding to facilitate the reaction in one direction may occur in the reverse direction when binding requires uncharged imidazole. An enzyme known to involve a histidine residue in substrate binding is lactate dehydrogenase (LDH). LDH functions as both a pyruvate reductase and a lactate oxidase. Yancey and Somero examined the pyruvate reductase behavior of this enzyme in a variety of ectothermic muscles as well as in an endotherm, the rabbit. Optimum enzyme function is maintained when the binding constant (Michaelis constant, K_m) is held in a narrow range. K_m is quite sensitive to [H$^+$]. In the case of the LDH's examined the K_m of pyruvate must be stabilized at values allowing the enzyme to increase activity as pyruvate rises, as during muscular work. These workers found that the K_m of pyruvate was conserved quite narrowly when ΔpH = -0.017 (this is the slope of pN$_{H_2O}$). When the pH of the reaction medium was maintained

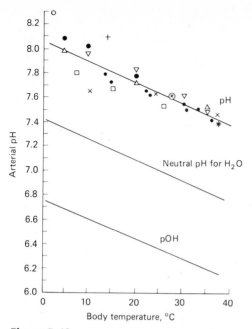

Figure 5-40
*Arterial pH values for various vertebrates (fish, amphibia, reptiles) and an invertebrate (crab) against body temperature. The asterisk represents man. Note that the pH curve is parallel to the neutral pH of H_2O, which shifts with temperature. The calculated pOH values also parallel the neutral pH of water. [Based on B. J. Howell, et al., Am. J. Physiol., **218**, 600 (1970) with added points for additional species.]*

at 7.4 while temperature was altered, K_m varied widely at differing temperatures.

A number of enzyme systems have been shown to behave in this fashion: Na-K-ATPase, acetyl CoA carboxylase, fatty acid synthetase, NADH-cytochrome c reductase and, succinate-cytochrome c reductase.

An additional feature of regulation of pH along the curve of pN_{H_2O} is the stabilization of the Donnan distribution of ions between intra- and extracellular fluids. Cell volume is also maintained constant.

Thus, the respiratory poise of lunged vertebrates and the regulatory processes governing $[HCO_3^-]$ are set in such a manner that stabilization of important cellular mechanisms occurs. Endotherms appear to be isothermal cases of the same phenomenon, excepting mammalian hibernators, which maintain the arterial pH relatively constant, around 7.4, when cold. The acidosis exhibited is due to respiratory underventilation. It has been argued that the destabilization of enzyme functions at the "wrong" pH, by reducing rates of processes, is a mechanism of lowering overall energy requirements beyond that expected from the effects of temperature alone. Such a mechanism would allow the energy stores of the animal to be conserved during the winter season when food supply is too low to maintain metabolism at normothermic levels.

The fact that vertebrates of various classes subscribe to the same pH-temperature slope indicates common requirements for arterial $[OH^-]/[H^+]$.

TABLE 5-10. Arterial Acid-Base Parameters in Lung Breathing Vertebrates

Species	Temperature Range (°C)	P_{CO_2} Range (torr) (low-high temperature)	$[HCO_3^-]mEq/L$ Mean ± SD	$\Delta pH/°C$
Rana catesbeiana (Bullfrog)	5–34	7.0–23.1	22.6 ± 1.27	−0.016
A. mississippiensis (Alligator)	15–35	11.8–23.2	15.1 ± 0.95	−0.013
Pseudemys scripta (Turtle)	10–30	14.1–31.9	33.3 ± 1.46	−0.010
P. floridana (Turtle)	22–37	20.8–34.8	29.6 ± 1.43	−0.013
Sauromalus obesus (Lizard)	16–40	11.0–40.0	14.1 ± 1.63	−0.017
Dipsosaurus dorsalis (Lizard)	17–42	9.9–32.0	19.5 ± 1.20	−0.014

[Literature sources: *Rana:* R. B. Reeves, *Respir. Physiol.,* **14,** 219 (1972); *Alligator:* D. G. Davies and M. T. Kopetzky. *Fed. Proc.,* **35,** 840 (1976); *P. scripta:* D. C. Jackson and R. D. Kagen. *Am. J. Physiol.,* **230,** 1389 (1976); *P. floridana:* J. L. Kinney, D. T. Matsuura, and F. N. White. *Respir. Physiol.,* **31:** 309 (1977); *Sauromalus:* E. C. Crawford and R. N. Gatz. *Comp. Biochem. Physiol.,* **47A,** 529 (1974); *Diposaurus:* P. Bickler, unpublished observations.]

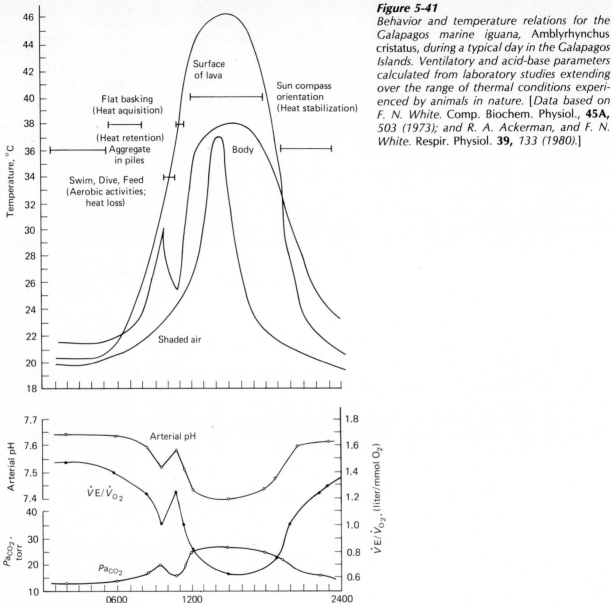

Figure 5-41
Behavior and temperature relations for the Galapagos marine iguana, Amblyrhynchus cristatus, *during a typical day in the Galapagos Islands. Ventilatory and acid-base parameters calculated from laboratory studies extending over the range of thermal conditions experienced by animals in nature. [Data based on F. N. White. Comp. Biochem. Physiol.,* **45A,** *503 (1973); and R. A. Ackerman, and F. N. White. Respir. Physiol.* **39,** *133 (1980).]*

Since man shares a common position with ectothermic vertebrates at 37°C, certain medical implications are suggested: Chilling the body (hypothermia) is widely used by surgeons as they repair crippling cardiac defects in children. The purpose of reducing the temperature is to achieve lower tissue metabolism and requirements for blood flow as well as additional time for performing the corrective procedures. Because man is endothermic, it has become usual to think of pH 7.4 as normal. However, the question arises whether an arterial pH of 7.4 is appropriate at 25°C? A major problem experienced at pH 7.4 at reduced temperature is the progressive loss of the heart's capacity for doing work and the tendency to develop ineffective contractions (ventricular fibrillation). Recent experiments on hypothermic dogs revealed that altering the arterial pH to 7.7 at 28°C (as exhibited by ectothermic vertebrates) resulted in marked improvement in cardiac function, blood flow, and oxidative metabolism. These experiments suggest that consideration of acid-base regulation by lower vertebrates may be valuable in establishing a firmer rationale for acid-base requirements for man at temperatures other than 37°C. There are obvious implications for the field of organ storage as well as for physiologists who study isolated tissues bathed in artificial solutions. If we are to play turtle games we might do well to study turtle rules!

5.8 CONTROL OF RESPIRATION

The late eighteenth-century description by Lavoisier of oxygen and carbon dioxide exchange in respiration gave credence to the suspicions of de Saussure (1796) that the increase in respiratory activity at high altitude was a physiological response to low partial pressures of oxygen.

5.8.1 Basic Structures and Phenomena

Some 80 years later the German physiologist Pflüger demonstrated stimulation of respiration both by low oxygen and elevated carbon dioxide pressures in inspired air. It was soon observed that acids were also respiratory stimulants. Much of the subsequent work on respiratory control has centered on these observations in attempting to elucidate the mechanisms by which appropriate respiratory adjustments are brought about in response to chemical alterations in respiratory gases and body fluids. Most of our knowledge is derived from studies on mammals, and our insight into control of lower vertebrate respiration is fragmentary at best.

That the brain plays a cardinal role in respiratory control was demonstrated by Galen, who found that cutting across the spinal cord near the brain led to a cessation of respiration. Cesar Legallois (1811) more precisely localized the medulla as the major

central nervous system area in respiratory control. He did so by transsection of the brain just posterior to the pons, following which he progressively sectioned the medulla, a few millimeters at a slice, until respiration ceased. Reflexly mediated responses by the vagus nerve were suggested in 1850 by Marshall Hall, and that the "venosity" of the blood of the cerebral circulation influenced respiration was postulated by Kussmaul and Tenner (1857), who observed respiratory stimulation following occlusion of the blood supply to the brain.

These early observations do not do justice to the history of the control of respiration but are mentioned because they form a framework upon which subsequent work has been based. Collectively, these observations imply that (1) appropriate respiratory adjustments may be made in response to alterations in environmental gas composition or chemical alterations in body fluids; (2) the central nervous system, especially the medulla, is essential for respiratory activity; and (3) alterations in chemical composition may directly or reflexly modify the activity of the centrally mediated motor acts of respiration.

The medulla is the site of neurons that activate inspiratory as well as expiratory activity in all vertebrates. Coordination of these two groups of neurons is not obligatorily linked with higher regions of the brain. Separation of the medulla from the spinal cord and anterior brain regions was followed in *Raja*, for example, by a coordinated opercular rhythm. Sectioning of the medulla in the midline did not destroy coordinated respiratory movements, but the rhythm on the two sides could vary independently.

Similar experiments in all major vertebrate groups have confirmed a central role of the medulla in respiratory control, and two groups of neurons, one expiratory and the other inspiratory, seems a universal vertebrate feature. These two groups of neurons constitute the "respiratory centers," but it is clear that no discrete loci are devoted to expiration and inspiration, the two types of neurons overlapping anatomically. Experiments in which the peripheral nerves, which bring sensory impulses to the respiratory areas, have been cut demonstrate that the isolated medulla exhibits spontaneous discharging, expiratory alternating with inspiratory bursts. The system behaves as if an oscillating relationship involving inhibitory and excitatory nerves

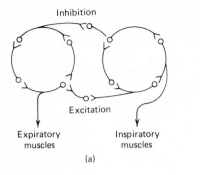

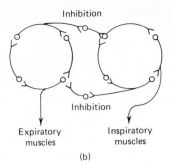

(a)

(b)

Figure 5-42

Two hypothetical systems for the rhythmicity of the medullary respiratory "center." In (a) activity of the inspiratory circuit is associated with excitatory connections to the expiratory circuit, which must reach a threshold before activating the expiratory neurons. Once activated, inhibitory connections reach the inspiratory circuit and cause withdrawal of the inspiratory to expiratory excitatory circuit activity. In (b) mutual inhibitions between the two oscillators is proposed. In this model, it is necessary to postulate some sort of "fatigue" or periodic refractory period to stop activity in one circuit while withdrawing inhibition from the other. In either case, the basic rhythm of the system is modified by neural connections from higher centers and mechano- and chemoreceptor systems.

link the two sets of neurons. Two hypothetical rhythmicity systems are seen in Figure 5-42.

Although the medulla appears to be the site of rhythmic respiratory activity, the medullary site is not completely independent of higher centers be-

cause visual or olfactory stimuli alter respiratory rhythm in fishes as do pain, heat, and cold stimuli applied to various peripheral areas in higher vertebrates. Furthermore, at least two additional centers in the pontine area have been located in mam-

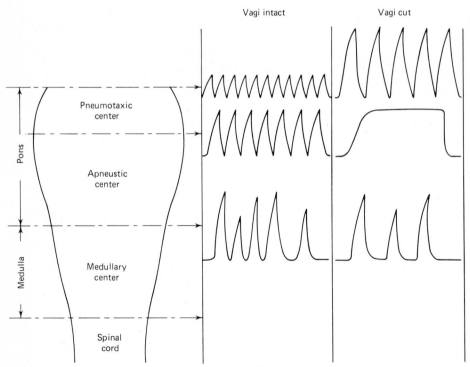

Figure 5-43

Patterns of respiration after brain-stem sections. The four patterns are representative of those that follow complete sections at each level. Section below the medulla results in complete apnea. [From J. H. Comroe, Jr. The Physiology of Respiration. The Year Book Medical Publishers, Inc., Chicago, 1965.]

mals (Figure 5-43). These centers have been found by surgical section of the brain stem at various levels. Section of the brain between the medulla and the spinal cord results in a cessation of respiration because the motoneurons that drive the respiratory muscles no longer retain their connections with the medullary respiratory neurons. Section between the medulla and lower pons is characterized by an abnormal pattern of respiration. Thus, connections between the medulla and higher centers are necessary for the normal pattern. A higher section, leaving the lower pons connected to the medulla establishes a more normal pattern; however, the ventilation volume is abnormally large and rate reduced below normal. In such a state, cutting of the vagus nerve produces an inspiratory spasm.

This is due to the absence of pulmonary stretch receptor influence that signals a cessation of respiration as the lung tissues are stretched during inspiration (see Section 5.8.3). When the brain is sectioned only at a higher pontine level, leaving everything below the pons intact, a relatively normal pattern of respiration is established. This higher pontine area includes the pneumotaxic center. Since the tidal volume is lower when the pneumotaxic center is intact than when its lower connections are sectioned, it appears that the pneumotaxic center has restraining influence on depth of respiration. A diagramatic scheme of the major interactions of peripheral and central neural components that control respiration is shown in Figure 5-44. These will be discussed in the following sections.

Figure 5-44
Some major pathways involved in the control of ventilation. A medullary oscillator consisting of expiratory and inspiratory components is the "heart" of the system. The outputs of the oscillator's components are modulated by influences of receptors sensing alterations in blood gas composition (central and peripheral chemoreceptors), alterations in the mechanics of the lung and chest wall (mechanoreceptors), as well as cortical and pontine higher influences. Shown also is the muscle spindle system which, at the spinal level, aids in stabilizing ventilation as mechanical conditions change. As, annulospiral (stretch) receptors; If, intrafusal respiratory muscle; Ef, extrafusal (major) respiratory muscle fibers.

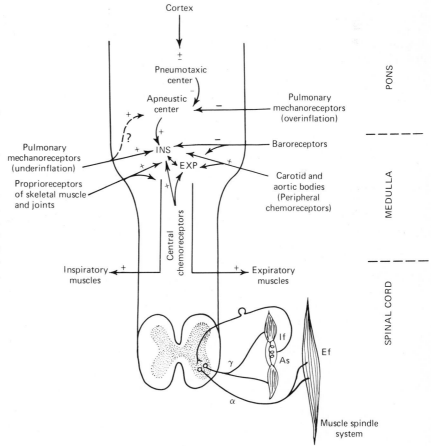

5.8.2 Chemoreceptors and Respiration

Since the middle of the eighteenth century, anatomists have known of the **carotid bodies** (Figure 5-45), small vascular spherules located at the bifurcation of the internal and external carotids. In 1936 Heymans received the Nobel Prize for his demonstration that the carotid bodies, and similar structures on the aorta, the **aortic bodies,** where chemosensitive. He elicited increases in tidal volume and frequency of respiration when the bodies were perfused with blood of low P_{O_2}. Nerve fibers from the carotid bodies (nerve IX) and the aortic bodies (nerve X) form the afferent limb of the reflex in mammals. The carotid body of the fowl is situated, not at the divergence of internal and external carotids but on the common carotid just rostral to the origin of the carotid artery. Nerve X supplies the innervation in birds rather than nerve IX. Both carotid and aortic bodies receive a profuse arterial supply, and blood flow is of the order of 20 ml/g of tissue per minute, the largest regional blood flow known. It has been theorized that the aortic and carotid bodies of mammals represent branchial arch derivatives, which are important in gilled forms for the detection of oxygen content in the water. Decreased P_{O_2} of the water is a positive respiratory stimulant in fishes, which are more sensitive to low P_{O_2} than to high P_{CO_2}. Little is known, however, about the sensitivity of carotid or aortic bodies in nonmammalian forms.

Present evidence, derived from mammalian studies, suggests that the chemosensitive cells of these bodies are active at the normal P_{O_2} of arterial blood. A great increase in firing rate occurs, however, when the P_{O_2} is reduced to 50 mm Hg. The P_{CO_2} plays a role in the sensitivity of these cells to P_{O_2} because, when P_{CO_2} is elevated, the response to decreasing P_{O_2} is exaggerated (Figure 5-46). Thus asphyxia, which reduces P_{O_2} and elevates P_{CO_2}, is a circumstance in which the receptors are greatly activated. In addition to the influence of carotid and aortic body stimulation on respiration, connections with central cardiovascular control areas are indicated. Perfusion of the carotid and aortic bodies of the dog with blood of low P_{O_2} results in increased heart rate and peripheral vasoconstriction with a consequent elevation in arterial pressure.

Hypotension (low blood pressure) is another circumstance that may elicit chemoreceptor activity. The reduced carotid and aortic body blood flow that results may lead to anoxia of the bodies even in the presence of a normal P_{O_2}. Thus the receptors are, indirectly, flow sensitive. A combination of reduced blood flow and low arterial P_{O_2} (anoxemia) is a more effective stimulus than either alone, and it is probable that in this circumstance locally produced CO_2 or H^+ plays a role in stimulation.

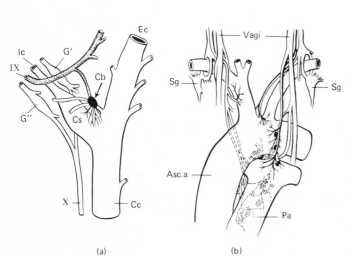

(a)   (b)

Figure 5-45

(a) Carotid body in a dog. The carotid body (Cb) receives its arterial blood from the occipital artery, which is the first branch of the external carotid (Ec). Its afferent fibers run in the carotid branch of the IXth cranial nerve. Code: X, vagus nerve; Cc, common carotid artery; Cs, carotid sinus; Ic, internal carotid; G', superior cervical sympathetic ganglion; and G", nodose ganglion of the vagus. (b) Aortic bodies in a dog. Aortic chemoreceptor cells lie on the anterior (black ovals) and posterior (hatched ovals) surfaces of the aortic arch and pulmonary artery (Pa). They receive their blood from a small branch of the aortic arch. Code: Sg, stellate ganglion: Asc a, ascending aorta [(a) redrawn from W. E. Adams. The Comparative Morphology of the Carotid Body and Carotid Sinus. Charles C. Thomas, Publisher, Springfield, Ill., 1958; (b) redrawn from J. F. Nonidex. Anat. Rec., 69, 299 (1937). From J. H. Comroe, Jr., The Physiology of Respiration. The Year Book Medical Publishers, Inc., Chicago, 1965.]

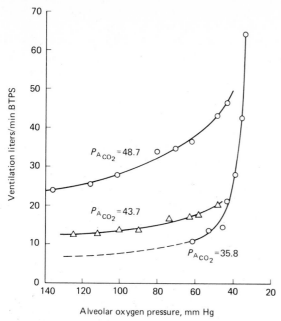

Figure 5-46
*Increase in ventilation in a human due to hypoxia associated with low and high levels of CO_2. Alveolar P_{CO_2} ($P_{A_{CO_2}}$) was maintained close to 35.8 mm Hg (bottom curve), 43.7 mm Hg (middle curve) and 48.7 mm Hg (top curve) in a subject given low O_2 mixtures to breathe. Each curve starts (left) at a higher level because of the effect of higher CO_2 tensions. The slope of the upper curve becomes steeper than that of the lower ones because of interaction between low O_2 and high CO_2 stimuli. Volume of ventilation corrected to BTPS (=gas at body temperature, ambient pressure, saturated with water vapor. [Redrawn from H. H. Loeschke and K. H. Gertz, Arch. ges Physiol., **267**, 460 (1958). (From J. H. Comroe, Jr. The Physiology of Respiration. The Year Book Medical Publishers, Inc., Chicago, 1965).]*

Aside from the effects of CO_2 on peripheral chemoreceptors, there is abundant evidence that increases in arterial CO_2 of only 2–4 mm Hg (levels that do not stimulate carotid or aortic bodies) elicit increased ventilation by acting on central chemoreceptor areas in the brain. It has been supposed that the respiratory centers, as such, were sensitive to small changes in P_{CO_2}; however, perfusion of the cerebrospinal fluid with solutions of high P_{CO_2} or low pH produces increased ventilation before

CO_2 or H^+ can reach the respiratory centers by diffusion. Areas on the ventral surface of the medulla have been located that are sensitive to application of solutions of high P_{CO_2} or low pH, and these receptors seem to have connections with the respiratory centers. The receptors are probably sensitive, not to CO_2 but to H^+. Because the protein content and thus the buffering capacity of the cerebrospinal fluid is low and CO_2 diffuses rapidly from blood to cerebrospinal fluid, the receptors are ideally situated for detection of slight changes in H^+. Without hemoglobin buffering, the reaction

$$CO_2 + H_2O \rightleftharpoons H_2CO_3 \rightleftharpoons H^+ + HCO_3^-$$

will decrease the pH of cerebrospinal fluid far more than for blood. The critical location of the receptors is illustrated in Figure 5-47, and it can be appreciated that through H^+ sensitivity, a critical monitor of blood P_{CO_2} and cerebrospinal fluid pH is attained. An elevation of arterial P_{CO_2}, through its effect on cerebrospinal fluid pH, will initiate increased ventilation, thus blowing off CO_2 and bringing arterial P_{CO_2} and cerebrospinal fluid pH back to normal ranges.

A relative insensitivity to increases in alveolar

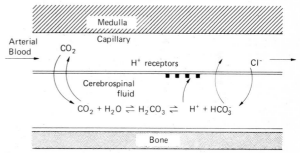

Figure 5-47
Central chemoreceptors are bathed in cerebrospinal fluid which is separated from blood by the blood-brain barrier, a barrier that is relatively impermeable to H^+ and HCO_3^-; CO_2 freely crosses the barrier. The buffering capacity of cerebrospinal fluid is small by contrast with blood. When P_{CO_2} is elevated in arterial blood (as with underventilation) the CO_2 which diffuses into cerebrospinal fluid promotes formation of H^+ and HCO_3^-. The H^+ receptors stimulate ventilation and thus correct arterial P_{CO_2} toward normal. When HCO_3^- is present in excess, it is thought to be actively transported to blood in exchange for Cl^-.

P_{CO_2} has been reported for such mammalian divers as the harbor seal, beaver, and muskrat, and increased ventilation does not occur until the CO_2 level reaches 5%. Only a doubling of ventilation was attained in the harbor seal, by elevating to 10% the CO_2 content of inspired air. This is in sharp contrast to man, because elevating the CO_2 in man by only 0.2% elicits a doubling of pulmonary ventilation.

The fact that representatives from all vertebrate groups respond with increased ventilation to elevated P_{CO_2} or low P_{O_2} in the respiratory medium suggests that chemoreceptors are widespread. Peripheral chemoreceptor locations, their behavior and afferent neural connections are known for only a few submammalian species.

Fishes are sensitive to changes in O_2 levels, whereas their responses to CO_2 appear to be less pronounced. Some evidence suggests that branchial arterial receptors, responsive to arterial P_{O_2}, may be present. The precise location of these putative receptors is unclear. The tench exhibits a decreased response to water of low P_{O_2} following sectioning of nerves IX and X, which innervate the gill epithelium.

Amphibians appear to utilize aortic chemoreceptors since the increase in ventilation to hypoxia by *Rana esculenta* is reduced after denervation of the carotid labyrinth. Additional evidence for a chemoreceptor role in control of respiration is the observation that electrical activity in the carotid nerve of *Bufo vulgaris* increases when the carotid perfusate is made anoxic, whereas it diminishes if hyperoxic fluid is used. Carbon dioxide acts to stimulate respiration at some site other than the carotid labyrinth. It has been suggested by Foxon that, under natural conditions, in contrast to higher vertebrates, frogs should be able to eliminate the mass of their CO_2 cutaneously (see Section 5-5-1) and that it should be expected that pulmonary ventilation is regulated more by O_2 requirements than in response to the CO_2 produced.

Current evidence suggests that both birds and reptiles possess receptors in the airways that are sensitive to CO_2. In both cases the sensory fibers reach the central respiratory areas of the brain via the vagus nerves. Records of the avian receptor activity have been obtained by recording discharge frequency from single fibers in the vagus nerves as a function of intrapulmonary P_{CO_2} (Figure 5-48).

It is important to distinguish whether such receptors respond to mechanical alterations (pressure = stretch). The figure illustrates that discharge frequency for a single unit was virtually the same at high or low intrapulmonary pressure. The avian receptor, unlike known chemoreceptors of mammals, decreases firing rate as P_{CO_2} is elevated. It is thought that the receptor activity is determined by both the P_{CO_2} of pulmonary arterial blood and alveolar gas. The receptors yield information concerning the mean CO_2 concentrations and the rate of change in the lung. Figure 5-49 is a system diagram illustrating how these receptors may interact with the central respiratory pacemaker elements to control the regulated variable (CO_2 concentration) at the gas exchange site.

Arterial chemoreceptors within the carotid body

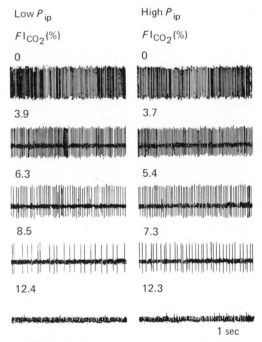

Figure 5-48
*Static discharge frequency of an intrapulmonary CO_2-sensitive receptor at various $F_{I_{CO_2}}$ at low and high intrapulmonary pressures. $F_{I_{CO_2}}$ is the fractional concentration of CO_2 in the inspired gas. [From M.R. Fedde, R. N. Gatz, H. Slama, and P. Scheid. Resp. Physiol., **22**:99–114 (1974).]*

Figure 5-49

Avian F_{CO_2} regulator as an error actuated feedback system. Ref. is the set point, ϵ is the error signal generated by the comparator, $\dot{I}$ is the rate of firing over the respiratory motor nerves, and $\dot{V}$ is ventilation ($\dot{V} = f \cdot V_t$). The receptor, comparator and controller process feedback information about the regulated variable, F_{CO_2}, to control the plant, which does the physical work. [From A. L. Kunz and D. A. Miller. Resp. Physiol., **22**, 167–177 (1974).]

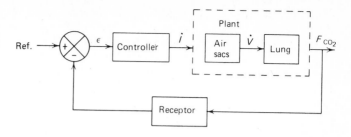

are also involved in determining ventilation level in birds. These vagal receptors are thought to be responsive to the P_{O_2} of the arterial blood, causing increased ventilation under hypoxic conditions. How these receptors interact centrally with the CO_2 receptors of the airways is not established.

The evidence for CO_2-sensitive receptors in the lungs of reptiles is based on neural recording and ventilatory responses to fast-time changes in alveolar gas and alterations in ventilation in response to vagal denervation of the lungs. A rapid elevation in alveolar P_{CO_2} has been observed to cause interruption of apnea and breathing within 4–7 sec—a time course too short for central neural detection of the effects of CO_2 in arterial blood. Furthermore, section of the vagus nerves to the lungs is followed by hyperventilation and alkalosis, indicating a key role for the pulmonary chemoreceptors in the control of acid-base status. Hypoxia is much less effective in altering the ventilatory pattern of reptiles than is CO_2. Recordings from single vagal fibers in turtles and in a lizard demonstrate that, CO_2-sensitive receptors associated with the airways behave in a fashion similar to the avian receptors, that is, firing rate is inverse to alveolar P_{CO_2}. Although stretch sensitive elements are also present, the hyperventilation following vagal section is probably partially explained by denervation of the CO_2-sensitive elements. The absence of impulses from these receptors apparently indicates to the central respiratory areas that an infinitely high P_{CO_2} exists in the lungs.

5.8.3 Nonchemically Initiated Respiratory Reflexes

Sectioning the vagus nerves of mammals produces an increase in inspiratory depth and a decrease in rate of respiration. This and other observations led Hering and Breuer (1868) to conclude that the vagus nerves contained fibers coursing to the respiratory centers and that these fibers initiated inhibition of inspiratory activity when the lungs were inflated sufficiently to stimulate their peripheral receptor endings. By interrupting inspiration, these receptors play a role in determining the duration of the respiratory cycle. Their inhibition of inspiration permits expiration to occur earlier in the cycle. Evidence from experiments in which the brain stem has been sectioned at various levels suggests that these vagal afferents enter at no higher level than the apneustic center.

Destruction of the bronchiolar membranes by steam, with no damage to the alveolar membranes, abolishes the Hering-Breuer reflex, indicating that the receptors are located in the bronchioles or bronchi. Electrical activity recorded from the vagus and corresponding to the discharge rate of the Hering-Breuer receptors shows a better correlation with the pressure in the airways than with lung volume and alveolar distention.

The Hering-Breuer reflex may play a role in limiting inflation to the most efficient level of inspiration. As inspiratory volume increases above a certain level, the muscular effort to produce a given increase in volume increases disproportionately. By limiting inspiratory volume, respiration is maintained at a more efficient level. The increase in respiratory depth accompanying exercise, acidosis, or anoxemia probably represents an overriding of the Hering-Breuer reflex by the powerful effect on the respiratory centers of elevated chemoreceptor activity.

The inflation reflex is evidently of long phylogenetic standing. Inflation of the pharynx of the dogfish inhibits inspiratory activity, an effect that vanishes after sectioning of the spiracular branches of nerves VII, IX, and X. Receptors, whose fibers are found in these cranial nerves, have been located at the junction of the pharynx and gill pouches. A Hering-Breuer reflex is present in birds, and it is probably a general vertebrate feature that aids in maintaining an efficient balance between respiratory effort and ventilation volume. Mechanoreceptors sensitive to underinflation have also been identified. They promote inspiration, but their influence is apparently limited at a critical inspiratory volume by the overinflation (Hering-Breuer) reflex.

Muscle spindles are skeletal muscle fibers containing sensory endings sensitive to stretch. These elements, often called *intrafusal fibers,* are numerous in intercostal muscles and have been observed in the diaphragm. The sensory element, called the *annulospiral ending,* is in series with the intrafusal fibers and in parallel with the major mass of extrafusal fibers (Figure 5-44). Activation of γ neurons, which innervate the intrafusal fibers, or relaxation of the extrafusal fibers are effective in initiating impulses from the annulospiral endings. There is some evidence to suggest that the γ neurons may be activated by fibers having their origin in the respiratory centers. The functional significance of the muscle spindle in respiration is a much debated question, and several alternative hypotheses have been suggested. A plausible scheme, called the *follow-up length servo hypothesis,* pictures the central respiratory efferents in synaptic association with γ neurons and thus activating the intrafusal fibers. As a result of annulospiral ending discharge, the α fibers are reflexly activated. In this scheme the shortening of the intrafusal fiber would indicate the degree of extrafusal fiber activity necessary to attain a desired tidal volume.

Proprioreceptors, sensitive to distortion, are also present in tendons and joints. Their activation, during exercise, stimulates respiration via spinal pathways that reach the central respiratory areas.

Proprioreceptors may be important in inhibiting ventilatory activity during the transition from rhythmic to ram ventilation as fishes go from rest to active swimming. Here the idea is that proprioreceptors, associated with the muscles of locomotion, inhibit central respiratory activity as speed and frequency of tail beat increase. However, this does not seem to be the case for the hitchhiker fish, the remora, which attaches to its host by a sucker disk. Although they show a transition from rhyth-mic ventilation to ram ventilation as the host starts to cruise, their bodies do not undulate.

A host of other peripheral receptors reflexly modify respiratory activity. The pathways, and indeed the physiological significance of most of these reflexes, are poorly understood. Pain, temperature, and proprioceptive stimuli all influence respiration. The stretch receptors located in the carotid sinus and aortic arch (baroreceptors), aside from their influence on the arterial pressure, have been shown to modify respiration. Increased stretch (elevated arterial pressure) induces an increase in the firing rate of the receptors. Nerves IX and X transmit the impulses to the brain stem, where inhibition of centrally mediated vasoconstrictor and sympathetic cardiac activity occurs. The result is slowing of the heart and peripheral vasodilation and a consequent fall in arterial pressure toward normal levels; respiration is also inhibited. A decrease in carotid sinus pressure augments respiration. The reflex may play a role in improvement of arterial P_{O_2} and P_{CO_2} under circumstances, such as hemorrhage, in which arterial pressure is reduced. A similar stretch-receptor mechanism is present in the branchial arteries of fishes; thus as perfusion pressure decreases, a mechanism for improvement of ventilation is brought into play.

(*Note:* Additional ecological correlations of the respiratory and cardiovascular systems are discussed in Section 6.9.)

ADDITIONAL READING

JOURNALS
Journals that frequently publish research papers on the comparative physiology of respiration: *Respiration Physiology; Comparative Biochemistry and Physiology; American Journal of Physiology; Journal of Experimental Biology; Journal of Physiology; Journal of Comparative Physiology; American Zoologist.*

MONOGRAPHS, SYMPOSIA AND COMPENDIA
Davies, D. G., and C. D. Barnes (eds.). *Regulation of Ventilation and Gas Exchange,* Academic Press, New York and London, 1978.

Dejours, P. "Comparative physiology of respiration in vertebrates," *Resp. Physiol.,* **14,** 1–236 (1972).

———. *Principles of Comparative Respiratory Physiology,* North Holland/American Elsevier, New York, 1975.

Fenn, W. O., and H. Rahn (eds.). *Handbook of Physiology,* sec. 3, vols. 1 and 2 (*Respiration*), American Physiologial Society, Washington, D.C., 1964.

Hughes, G. M. *Comparative Physiology of Vertebrate Respiration,* Harvard University Press, Cambridge, MA, 1963.

Krogh, A. *The Comparative Physiology of Respiratory Mechanisms,* University of Pennsylvania Press, Philadelphia, 1941.

Steen, J. B. *Comparative Physiology of Respiratory Mechanisms,* Academic Press, London, 1971.

Wood, S. C., and C. Lenfant (eds.). *Evolution of Respiratory Processes, a Comparative Approach,* Marcel Dekker, New York and Basel, 1979.

GLOSSARY

Bartels, H., P. Dejours, R. H. Kellogg, and J. Mead. "Glossary on respiration and gas exchange," *J. Appl. Physiol.,* **34,** 549–558 (1973).

GASES

Cherniak, N. S., and G. S. Longobardo. "Oxygen and carbon dioxide gas stores of the body," *Physiol. Rev.,* **50,** 196–243 (1970).

Forster, R. E. "Diffusion of gases." In W. O. Fenn and H. Rahn (eds.), *Handbook of Physiology,* sec. 3, vol. 1, pp. 839–872, American Physiological Society, Washington, D.C.

Livingstone, D. A. "Chemical composition of rivers and lakes." In *Data of Geochemistry, Geological Survey Professional Paper* **440G,** chap. G, pp. 1–64, Washington, D.C., U.S. Government Printing Office.

Nahas, G., and K. E. Schaefer (eds.). *Carbon Dioxide and Metabolic Regulations,* pp. 1–372, Springer-Verlag, New York.

RESPIRATORY PIGMENTS

Bartlett, G. R. "Phosphate compounds in vertebrate red blood cells." *Am. Zool.,* **20,** 103–114 (1980).

Bauer, C. "On the respiratory function of haemoglobin," *Rev. Physiol. Biochem. Pharmacol.,* **70,** 1–31 (1974).

Ghiretti, F. *Physiology and Biochemistry of Haemocyanins,* Academic Press, New York, 1968.

Greeney, G. S., and D. A. Powers. "Allosteric modifiers of fish hemoglobins," *J. Exp. Zool.,* **203,** 339–349 (1978).

Grigg, G. C. "Respiratory function of blood in fishes." In *Chemical Zoology,* vol. 1, pp. 331–368, Academic Press, Inc., 1974.

Hemingsen, E. A., E. L. Douglas, K. Johansen, and R. W. Millard. "Aortic blood flow and cardiac output in the hemoglobin-free fish *(Chaenocephalus aceratus).*" *Comp. Biochem. Physiol.,* **43A,** 1045–1051 (1972).

Kilmartin, J. V., and L. Rossi-Bernardi. "Interaction of hemoglobin with hydrogen ions, carbon dioxide, and organic phosphates," *Physiol. Rev.,* **53,** 836–890 (1973).

Kreuzer, F. "Facilitated diffusion of oxygen and its possible significance; a review," *Resp. Physiol.,* **9,** 1–30 (1970).

Mangum, C. P. "Evaluation of the functional properties of invertebrate hemoglobin," *Neth. J. Sea Res.,* **7,** 303–315 (1973).

Manwell, C. "Comparative physiology: blood pigments," *Ann. Rev. Phys.,* **22,** 191–244 (1960).

Schmidt-Nielsen, K., and C. R. Taylor. "Red blood cells: why or why not?" *Science,* **162,** 274–275 (1968).

Shulman, R. G., S. Ogawa, K. Wüthrich, T. Yamene, J. Peisach, and W. E. Blumberg. "The absence of 'heme-heme' interactions in hemoglobin," *Science,* **165,** 231–257 (1969).

Tazawa, H., T. Mikami, and C. Yoshimoto. "Respiratory properties of chicken embryonic blood during development," *Resp. Physiol.,* **13,** 160–170 (1971).

Wittenberg, J. G. "Myoglobin-facilitated oxygen diffusion: role of myoglobin in oxygen entry into muscle," *Physiol. Rev.,* **50,** 559–636 (1970).

MORPHOMETRICS

Burri, P. H., and E. R. Weibel. "Morphometric estimation of pulmonary diffusion capacity. II. Effect of P_{O_2} on the growing lung. Adaptation of the growing rat lung to hypoxia and hyperoxia," *Resp. Physiol.,* **11,** 247–264 (1971).

Czopek, J. "Quantitative studies on the morphology of respiratory surfaces in amphibians," *Acta Anat.,* **62,** 296–323 (1965).

Geelhaar, A., and E. R. Weibel. "Morphometric estimation of pulmonary diffusion capacity. III. The effect of increased oxygen consumption in Japanese waltzing mice," *Resp. Physiol.,* **11,** 354–366 (1971).

Hemmingsen, A. M. "Energy metabolism as related to body size and respiratory surfaces, and its evolution," *Reports Steno Memorial Hospital,* 110 p., 1960.

Piiper, J., and P. Scheid. "Gas transport efficacy of gills, lungs, and skin: theory and experimental data," *Resp. Physiol.,* **23,** 209–221 (1975).

Rahn, H. "Aquatic gas exchange: theory," *Resp. Physiol.,* **1,** 1–12 (1966).

Stahl, W. R., "Scaling of respiratory variables in mammals," *J. Appl. Physiol.,* **22,** 453–460 (1967).

Szarski, H. "The structure of respiratory organs in relation to body size in amphibia," *Evolution,* **18,** 118–126 (1964).

Tenney, S. M., and J. E. Remmers. "Comparative quantitative morphology of the mammalian lung: diffusing area," *Nature,* **197,** 54–56 (1963).

Tenney, S. M., and J. B. Tenney. "Quantitative morphology of cold-blooded lungs: amphibia and reptilia," *Resp. Physiol.,* **9,** 197–215 (1970).

Weibel, E. R. "Morphological basis of alveolar-capillary gas exchange," *Physiol. Rev.,* **53,** 419–495 (1973).

Weibel, E. R. "Oxygen demand and the size of respiratory structures in mammals." In S. C. Wood and C. Lenfant (eds.) *Evolution of Respiratory Processes, a Compar-*

ative Approach, pp. 289–346, Marcel Dekker, New York and Basel, 1979.

ACID-BASE REGULATION

Filley, G. F. *Acid Base and Blood Gas Regulation,* Lea and Febiger, Philadelphia, 1971.

Howell, B. J., F. W. Baumgardner, K. Bondi, and H. Rahn. "Acid-base balance in cold-blooded vertebrates as a function of body temperature," *Am. J. Physiol.,* **218,** 600–606 (1970).

Howell, B. J., H. Rahn, D. Goodfellow, and C. Herreid. "Acid-base regulation and temperature in selected invertebrates as a function of temperature," *Am. Zool.,* **13,** 557–563 (1973).

Hurtado, A., and H. Aste-Salazar. "Arterial blood gases and acid-base balance at sea level and at high altitudes," *J. Appl. Physiol.,* **1,** 304–325 (1948).

Jackson, D. C., S. E. Palmer, and W. L. Meadow. "The effects of temperature and carbon dioxide breathing on ventilation and acid-base status of turtles," *Resp. Physiol.,* **20,** 131–146 (1974).

Leusen, I. "Regulation of cerebrospinal fluid composition with reference to breathing," *Physiol. Rev.,* **52,** 1–56 (1972).

Randall, D. J., and J. N. Cameron. "Respiratory control of arterial pH as temperature changes in rainbow trout *Salmo gairdneri,*" *Am. J. Physiol.,* **225,** 997–1002 (1973).

Reeves, R. B. "An imidazole alphastat hypothesis for vertebrate acid-base regulation: tissue carbon dioxide content and body temperature in bullfrogs," *Resp. Physiol.,* **14,** 219–236 (1972).

Reeves, R. B. "The interaction of body temperature and acid-base balance in ectothermic vertebrates," *Ann. Rev. Physiol.,* **39,** 559–586 (1977).

Robin, E. D. "Relationship between temperature and plasma pH and carbon dioxide tension in the turtle," *Nature,* **195,** 249–251 (1962).

Yancey, P. H., and G. N. Somero. "Temperature dependence of intracellular pH: its role in the conservation of pyruvate apparent K_m values of vertebrate lactate dehydrogenase," *J. Comp. Physiol.,* **125,** 129 (1978).

VENTILATION-PERFUSION RELATIONSHIPS

Dejours, P., W. F. Garey, and H. Rahn. "Comparison of ventilatory and circulatory flow rates between animals in various physiological conditions," *Resp. Physiol.,* **9,** 108–117 (1970).

Dejours, P. "Comparison of gas transport by convection among animals," *Resp. Physiol.,* **14,** 96–104 (1972).

Hemmingsen, E. A., E. L. Douglas, K. Johansen, and R. W. Millard. "Aortic blood flow and cardiac output in the hemoglobin-free fish *(Chaenocephalus aceratus),*" *Comp. Biochem. Physiol.,* **43a,** 1045–1051 (1972).

Hughes, G. M. (ed.). *Respiration in Amphibious Vertebrates,* Academic Press, New York and London, 1976.

Piiper, J. and P. Scheid. "Comparative physiology of respiration: functional analysis of gas exchange organs in vertebrates." In J. G. Widdicombe (ed.). *International Review of Physiology,* vol. 14, University Park Press, Baltimore, 1977.

Rahn, H., and L. E. Farhi. "Ventilation, perfusion, and gas exchange—the $\dot{V}_A/\dot{Q}$ concept." In W. O. Fenn and H. Rahn (eds.). *Handbook of Physiology,* vol. 1, pp. 735–766, The American Physiological Society, Washington, D.C., 1964.

West, J. B. *Ventilation/Blood Flow and Gas Exchange,* Blackwell Scientific Publications, Oxford, 117 p.

White, F. N. "Comparative aspects of vertebrate cardiorespiratory physiology," *Ann. Rev. Physiol.,* **40,** 471–499 (1978).

FISHES

Ballintijn, C. M., and G. M. Hughes. "The muscular basis of the respiratory pumps in the trout," *J. Exp. Biol.,* **43,** 349–362 (1965).

Ballintijn, C. M. "Efficiency, mechanics and motor control of fish respiration," *Resp. Physiol.,* **14,** 125–141, 1972.

Brown, C. E., and B. S. Muir, "Analysis of ram ventilation of fish gills with application to skipjack tuna," *J. Fish Res. Board Can.,* **27,** 1637–1652 (1970).

Carter, G. S. "Air breathing." In M. E. Brown (ed.). *The Physiology of Fishes,* vol. 1, pp. 65–79, Academic Press, New York.

Dejours, P. "Problems of control of breathing in fishes." In L. Bolis, K. Schmidt-Nielsen, and S. H. P. Maddrell (eds.). *Comparative Physiology, Locomotion, Respiration, Transport and Blood,* pp. 117–133, North-Holland/American Elsevier, 1973.

Guimond, R. W., and V. H. Hutchinson. "Aerial and aquatic respiration in the congo eel *Amphiuma means means* (Garden)," *Resp. Physiol.,* **20,** 147–159 (1974).

Hoar, W. S., and D. J. Randall. "The nervous system, circulation, and respiration," *Fish Physiology,* vol. IV, Academic Press, New York, 1970, 531 p.

Johansen, K. "Comparative physiology: gas exchange and circulation in fishes," *Ann. Rev. Physiol.,* **33,** 569–612 (1971).

Johansen, K., C. Lenfant, and D. Hanson. "Gas exchange in the lamprey, *Entosphenus tridentatus,*" *Comp. Biochem. Physiol.,* **44a,** 107–119 (1973).

Hughes, G. M. "Respiration of amphibious vertebrates," a symposium held in Bhagalpur, India (Oct. 28–1st Nov., 1974), Academic Press, London, 1976.

Hughes, G. M., and M. Morgan. "The structure of fish gills in relation to their respiratory function," *Biol. Rev.,* **48,** 419–475 (1973).

Lenfant, C., and K. Johansen. "Respiration in the African lung fish *Protopterus aethiopicus.* I. Respiratory properties of blood and normal patterns of breathing and gas exchange," *J. Exp. Biol.,* **49,** 437–452 (1968).

Muir, B. S., and G. M. Hughes. "Gill dimensions for three species of tunny," *J. Exp. Biol.,* **51,** 271–285 (1969).

Piiper, J., and D. Schumann. "Efficiency of O_2 exchange in the gills of the dogfish, *Scyliorhinus stellaris,*" *Resp. Physiol.,* **2,** 135–148 (1967).

Randall, D. J., and J. N. Cameron. "Respiratory control of arterial pH as temperature changes in rainbow trout *Salmo gairdneri,*" *Am. J. Physiol.,* **225,** 997–1002 (1973).

AMPHIBIA

De Jongh, H. J., and C. Gans. "On the mechanism of respiration in the bullfrog, *Rana catesbeiana:* a reassessment," *J. Morphol.,* **127,** 259–290 (1969).

Foxon, G. E. H. "Blood and respiration." In J. A. Moore (ed.). *Physiology of the Amphibia,* pp. 151–209, Academic Press, New York, 1964.

Guimond, R. W., and V. H. Hutchinson, "Aquatic respiration: an unusual strategy in the hellbender (*Cryptobranchus alleganiensis alleganiensis* (Daudin)," *Science,* **182,** 1263–1265 (1973).

Hughes, G. M. "Respiration of amphibious vertebrates," a symposium held in Bhagalpur, India (Oct. 28–1st Nov., 1974), Academic Press, London.

Ishii, K., K. Honda, and K. Ishii. "The function of the carotid labyrinth in the toad," *Tohoku J. Exp. Med.,* **88,** 103–106 (1966).

Lenfant, C., and K. Johansen. "Respiratory adaptations in selected amphibians," *Resp. Physiol.,* **2,** 247–260 (1967).

Vinegar, A., and V. H. Hutchison. "Pulmonary and cutaneous gas exchange in the green frog. *Rana calamitans,*" *Zoologica,* **50,** 47–53 (1965).

REPTILES

Fedde, M. R., W. D. Kuhlmann, and P. Scheid. "Intrapulmonary receptors in the tegu lizard: I. Sensitivity to CO_2," *Respir. Physiol.,* **29,** 35–48 (1977).

Frankel, H. M., A. Spitzer, J. Blaine, and E. P. Schoener. "Respiratory response of turtles *(Pseudemys scripta)* to changes in arterial blood gas composition" *Comp. Biochem. Physiol.,* **31,** 535–546 (1969).

Gans, C., and G. M. Hughes. "The mechanism of lung ventilation in the tortoise *Testudo graeca* Linné," *J. Exp. Biol.,* **47,** 1–20 (1967).

Jackson, D. C. "Ventilatory response to hypoxia in turtles at various temperatures," *Resp. Physiol.,* **18,** 178–187 (1973).

Jackson, D. C., S. E. Palmer, and W. L. Meadow. "The effects of temperature and carbon dioxide breathing on ventilation and acid-base status of turtles," *Resp. Physiol.,* **20,** 131–146 (1974).

Kinney, J. L., D. T. Matsuura, and F. N. White. "Cardio-respiratory effects of temperature in the turtle, *Pseudemys floridana.*" *Respir. Physiol.,* **31,** 309–325 (1977).

Nielsen, B. "On the regulation of respiration in reptiles. II. The effects of hypoxia with and without moderate hypercapnia on the respiration and metabolism of lizards," *J. Exp. Biol.,* **39,** 107–117 (1962).

Spragg, R. G., R. Ackerman, and F. N. White. "Distribution of ventilation in the turtle, *Pseudemys scripta,*" *Respir. Physiol.,* **42,** 73–86 (1980).

BIRDS

Bouverot, P., and L.-M. Leitner. "Arterial chemoreceptors in the domestic fowl," *Resp. Physiol.,* **15,** 310–320 (1972).

Bretz, W. L., and K. Schmidt-Nielsen. "Bird respiration: flow patterns in the duck lung," *J. Exp. Biol.,* **54,** 103–118, 1971.

Dejours, P. "Receptors and control of respiration in birds," *Resp. Physiol.,* **22,** 1–216 (1974).

Duncker, H.-R. "Structure of avian lungs," *Resp. Physiol.,* **14,** 44–63 (1972).

Jones, D. R., and M. J. Purves. "The effect of carotid body denervation upon the respiratory response to hypoxia and hypercapnia in the duck," *J. Physiol.,* **211,** 295–309 (1970).

Lawiewski, R. C. "Respiration function in birds." In D. S. Farner, J. R. King, and K. C. Parkes (eds.). *Avian Biology,* pp. 271–342, Academic Press, New York, 1972.

Rahn, H., and A. Ar. "The avian egg: incubation time and water loss," *Condor,* **76,** 147–152 (1974).

Scheid, P. "Mechanisms of gas exchange in bird lungs," *Rev. Physiol. Biochem. Pharmacol.,* **86,** 137–186 (1979).

Scheid, P. "Respiration and control of breathing in birds," *The Physiologist,* **22,** 60–64 (1979).

Scheid, P., and J. Piiper. "Direct measurement of the pathway of respired gas in duck lungs," *Resp. Physiol.,* **11,** 308–314 (1971).

Tucker, V. A. "Respiration during flight in birds," *Resp. Physiol.,* **14,** 75–82, 1972.

MAMMALS

Bouverot, P., V. Candas, and J. P. Libert. "Role of the arterial chemoreceptors in ventilatory adaptation to hypoxia of awake dogs and rabbits," *Resp. Physiol.,* **17,** 209–219 (1973).

Comroe, J. H., Jr. *The Physiology of Respiration,* The Year Book Medical Publishers, Chicago, 1966.

Davies, D. G., and C. D. Barnes (eds.). *Regulation of Ventilation and Gas Exchange,* Academic Press, New York and London, 1978.

Fenn, W. O., and H. Rahn (eds.). *Handbook of Physiology,* sec. 3, vols 1 and 2 *(Respiration),* American Physiological Society, Washington, D.C., 1964.

Kooyman, G. L. "Respiratory Adaptations in Marine Mammals," *Am. Zool.,* **13,** 457–468 (1973).

Leusen, I. "Regulation of cerebrospinal fluid composition with reference to breathing," *Physiol. Rev.,* **52,** 1–56 (1972).

McCutcheon, F. H. "Organ systems in adaptation: The respiratory system." In D. B. Dill (ed.). *Handbook of Physiology,* sec. 4, pp. 167–191, American Physiological Society, Washington, D.C., 1964.

Otis, A. B., W. O. Fenn, and H. Rahn. "Mechanics of breathing in man," *J. Appl. Physiol.,* **2,** 592–607 (1950).

Rahn, H. "A concept of mean alveolar air and ventilation-blood flow relationships during pulmonary gas exchange," *Am. J. Physiol.,* **158,** 21–30 (1949).

West, J. B. *Respiratory Physiology—The Essentials,* The Williams and Wilkins Company, Baltimore, 1974.

Widdicombe, J. G. "Respiratory reflexes in man and other mammalian species," *Clin. Sci.,* **21,** 163–170 (1961).

Widdicombe, J. G. (ed.). "Respiratory physiology II." In *International Review of Physiology,* vol. 14, University Park Press, Baltimore, 1977.

Fred N. White

6

CIRCULATION

6.1 INTRODUCTION

August Krogh, in his famous monograph *The Comparative Physiology of Respiratory Mechanisms,* made the following calculation: (1) Assume a homogeneous spherical body in which O_2 is used at a constant rate throughout the sphere and that the O_2 tension in the center is zero. (2) Consider E. N. Harvey's equation: $Co = Ar^2/6D$ where Co is the concentration at the surface, in atmospheres, A is the oxygen consumption of the sphere (ml g^{-1} min^{-1}), r the radius of the sphere in centimeters, and D the diffusion coefficient (ml O_2/min diffusing through 1 cm^2 area and a path length of 1 cm at a pressure difference of 1 atm). At constant metabolism the required O_2 tension difference is proportional to the square of the radius. Krogh then gives an example of an organism of 1 cm radius consuming O_2 at 100 ml kg^{-1} hr^{-1} (1/600 ml g^{-1} min^{-1}). D, the diffusion coefficient was set at 0.000011, based on his previous measures for connective tissue. The necessary O_2 pressure outside will be: $Co = 1/600 \cdot 6 \cdot 0.000011 = 25$ atm. Clearly, such an organism cannot live by diffusion alone. For an organism of 1 mm radius, the requirement is just met in air-saturated water. Krogh concluded that diffusion can provide the oxygen demands, at reasonable rates of metabolism, of organisms of 1 mm or less in diameter; or, that larger forms dependent on diffusion alone must have very low rates of metabolism. Further limitation on diffusional support of metabolism occurs when cells are organized into tissues in which the pathway over which gases must be transported becomes lengthened.

Although the final step in gas exchange remains diffusion, it is apparent from Krogh's example that some convective process is essential in the support of metabolically intense organisms with complex tissues and ranging in mass to many tons. Internal convection of gases, to and from the specialized respiratory structures and remote tissues, is a primary role of the cardiovascular system. Only one group of metabolically intense organisms seem to have evolved a design for gas exchange that does not appear to rely on cardiovascular convection of gases—the insects. Their tracheoles penetrate so profusely within the tissues that the limitations of diffusional exchange are largely overcome. The fact that diffusion in the gas-filled tracheal system is much more rapid than in liquids allows a butterfly of 0.3-g mass to support the metabolism of flight at 165 times the resting level. The blood vascular system of insects plays a minimal role in the gas transport required for this amazing aerobic range. However, for organisms that develop specialized respiratory surfaces, such as gills or lungs, circulatory transport of the gases is a requirement. Indeed, limitations of circulatory transport may be partially responsible for setting the upper levels of sustained oxygen consumption. Sustained elevation in oxygen consumption of exercising man is limited to around 4–5 liters/min (20 times resting), a limit largely due to maximum cardiovascular performance.

In addition to the cardiovascular link between gas exchange with the environment and gas exchange at the cellular level, many other functions are served. In birds and mammals, the regulation of body temperature depends on distribution of blood either toward the core or peripherally, where heat may be dissipated. The balance between renal blood flow and related kidney functions is essential to electrolyte and nitrogen regulation. Hormonal agents are secreted into the blood and find access to target sites via the blood. These diverse activities, including facultative transport of gases, require propulsion of the blood by a well-regulated variable-output pump and control of resistances in the peripheral tissues to assure appropriate regional distribution of blood flow.

Although the vertebrate circulatory system is closed in the sense that it is a system of continuous tubes, it is in another sense open: exchange of water, electrolytes, and other materials across the capillaries is continuously taking place. An appropriate balance in fluid exchange is necessary to maintain volume distribution between vascular and interstitial compartments. This "cooperation" is governed by complex interactions between neural, hormonal, physical, and local tissue factors. When considering the complexities of such a system we can empathize with William Harvey who stated in the first chapter of *Exercitatio Anatomica de Motu Cordis et Sanguinis in Animalibus* (1628) that he feared that the motion of the heart and blood was to be comprehended only by God.

In what follows we will explore the basic features of circulatory systems, the nature of their construction, how they meet the varying demands of the specialized tissues and organs, the manner in which the system is held under control, and, in general, how the blood is propelled round and round a circle to provide renewal of its properties to support the next moments of life.

6.2 THE NATURE OF THE CIRCUITRY

"Coepi egomet mecum cogitare, an motionem quandam quasi in circulo haberet" ("I began to think within myself whether it [the blood] might have a sort of motion, as it were, in a circle."). This thought, verified by Harvey's experiments, was man's first step toward understanding the meaning of the vertebrate circulation. Yet, there are many organisms that do not propel the blood around in a circle. In the shallow California tidal bays may be found *Urechis caupo*, an echiuroid tube worm that propels a flow of water through its tube by peristaltic waves of the muscular body wall. There is no organized set of blood vessels nor a heart, yet the large coelomic cavity contains a red blood cell rich fluid that is convected by the body wall motions (Figure 6-1a). The red blood cells contain hemoglobin, which releases its bound oxygen during low tides when oxygen availability is low.

Many invertebrates utilize specialized pumps and series of vessels or sinuses to convey the blood. One finds in mollusks a cyclical flow of oxygen-depleted venous blood passing through the gas exchange organ and hence to the heart, where propulsive energy is supplied to keep the system going (Figure 6-1). Although invertebrates vary widely in the details of their circulatory patterns, this principle of specialized muscular pumps, often driven by a rhythmic pacemaker, is the basic plan of vertebrate circulatory design. A comparison of the cephalopod system with that of mammals (Figure 6-2) illustrates a marked convergence in general design in which a dual pump system is utilized. In both groups the venous blood is propelled through the gas exchanger through a special set of vessels. The function of the cephalopod branchial hearts is equivalent to the mammalian right ventricle. In both cases, the oxygen-rich venous return from the gas exchange organ is energized by an additional pump that propels the blood to the tissues.

A survey of the circulation of the various vertebrate groups reveals several significant trends.

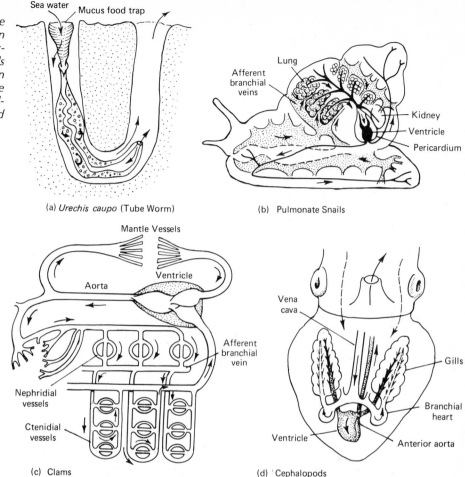

Figure 6-1
Circulatory patterns in some invertebrates. (a) An open circulation in which convection of the red blood cells and dissolved O_2 depends on peristaltic movements of the body wall. (b–d) Various mollusks in which specialized pumps energize blood.

(a) *Urechis caupo* (Tube Worm)

(b) Pulmonate Snails

(c) Clams

(d) Cephalopods

Among these is the adoption of a "closed" circulation of continuous vessels. The development of progressively higher pressure circuits characterized by high flow rates and greater separation of arterial and venous blood is evident; this development reaches its highest expression in mammals and birds.

The open, or lacunar, type of peripheral circulation is only evident in cyclostomes to any marked extent. Such low-pressure circuits depend on body-wall contractions for propulsion of the blood, a situation common in many invertebrates. Accessory hearts are present in hagfishes, apparently to supply an additional energy input favoring venous return to the heart.

The circulation passes through a system of capillaries in all vertebrates, and propulsion through these small vessels requires the development of considerable pressure by the action of the heart. The gill circulation of cyclostomes and fishes imposes a capillary circuit between the heart and the major distributing vessel, the aorta. There is thus a pressure drop across the gills. Such a circulatory arrangement does not require a complexly divided heart to circulate arterialized blood to the tissues.

With the development of lungs and dependence on aerial respiration, a problem in circuitry arises. The elimination of gills between the heart and aorta removed an obstacle to the development of a higher pressure circulation. However, how is oxy-

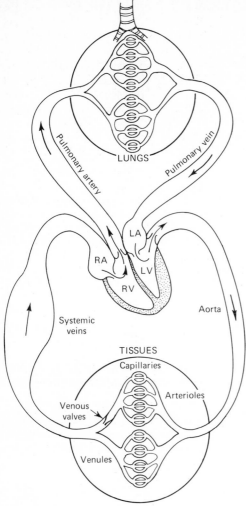

Figure 6-2
The circulatory pattern of mammals in which the circulation to the gas exchange organ (pulmonary circulation) and other organs (systemic circulation) is maintained separate and is energized by a dual pump. LA, left atrium; LV, left ventricle; RA, right atrium; RV, right ventricle.

ventricle exist. The bulbus arteriosus is also partially divided by endocardial ridges, providing separate channels for the ventricular output. The pulmonary veins return blood to the left atrium, and it has been shown by radiographic studies that left atrial blood is selectively distributed to the systemic circuit, whereas the venous blood of the right atrium is shunted, apparently with the aid of the ridges of the bulbus, to the lungs via the pulmonary artery.

Amphibians and reptiles exhibit, in variable degree, the capacity to maintain a double circulation. The major portion of the venous return is distributed to the lungs, whereas the pulmonary venous blood is largely directed to the systemic circuit. Birds and mammals have a complete double circulation. This situation is achieved by the development of the complete ventricular septum, with the pulmonary artery originating from the right ventricle and a single aorta originating from the left. Both groups, being endothermic, have a high metabolic rate and require relatively higher tissue perfusion rates than the ectotherms. The high-pressure double circulation of birds and mammals was a necessary cofactor of endothermism and the diversity of distribution of these forms.

The general circulatory arrangements seen in the major vertebrate groups are presented in Figure 6-3.

The closed circulatory system of vertebrates requires that a system of branching tubes, ever narrowing to microscopic dimensions in the tissues, must then reconverge into large venous conduits to return the blood to the heart. William Harvey was left mystified by this geometry, for he could not discern the nature of the tiny vessels of the tissues. It took Marcello Malpighi, who was born the year of Harvey's great tract (1628), to discover what his predecessors could not see. The invention of the microscope allowed him to observe the smallest blood vessels, which he named *capilus*, from which we derive *capillary*.

The geometry of the blood vessels gives some hint of the functions of the various types of vessels making up the circuit (Table 6-1). The progressive subdivision of large into smaller and yet smaller vessels provides the means of achieving an enormous number of capillaries and a large surface area across which nutrient, fluid, and gas exchange, takes place within the metabolizing tissues (Figure

genated blood to be distributed from the lungs to the body and the venous return selectively propelled to the pulmonary arterial circuit? One stage in the early attempts to solve this problem may be illustrated by the Dipnoi (lungfishes), in which completely divided atria and an extensively divided

6-4). As subdivision on the arterial side takes place, the vessels become shorter. On the venous side of the circuit the process is repeated in mirror fashion; however, the total cross-sectional area on the venous side is large, resulting in a greater volume of blood within the low-pressure side of the circuit. From the figures in Table 6-1 it can be seen that about 70% of the blood volume resides in the venous system. This volume is not static and can be mobilized by the muscular compression of exercise or by nerve-mediated constriction of the veins un-der conditions requiring greater cardiac output.

Because energy is lost in the system of branching tubes, the arterial circuit must be perfused under sufficient pressure to maintain the system in motion. This is a functional requirement that demands pressures within the arterial system higher than those in the venous side. Reference to Figure 6-5 depicts the transitions in hydrostatic pressure characterizing the mammalian circulation and the disposition of vessels of differing character. The pressure profile reveals wide swings in pressure within

Figure 6-3

Circulatory arrangement in various vertebrates. Code: Acv, anterior cardinal vein; Da, dorsal aorta; Dc, ductus Cuvier; Eb, efferent branchial arteries; Ej, external jugular vein; Fp, foramen Panizzae; Ha, hepatic artery; Hpv, hepatic portal vein; Hv, hepatic vein; L.at, left atrium; Lsa, left systemic arch; Lv, left ventricle; Pa, pulmonary artery; Pcv, posterior cardinal vein; Pv, pulmonary vein; Pvc, posterior vena cava; Ra, renal artery; R.at, right atrium; Rsa, right systemic arch; Rpv, renal portal vein; Rv, right ventricle; V, ventricle; Va, ventral aorta. Arteries open; veins black. (Continued on next page.)

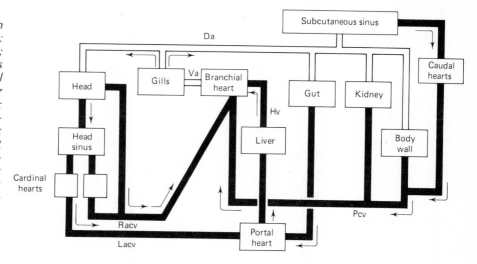

(a) Hagfish

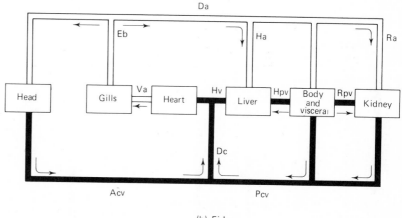

(b) Fish

Figure 6-3 (continued)

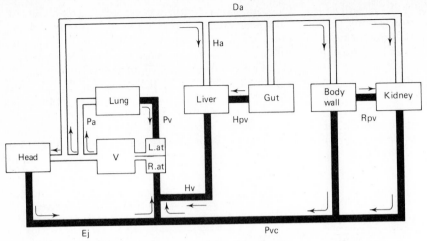

(c) Urodele Amphibian

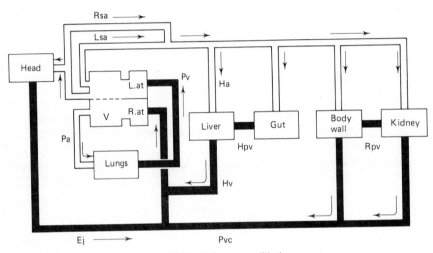

(d) Reptile (non-crocodilian)

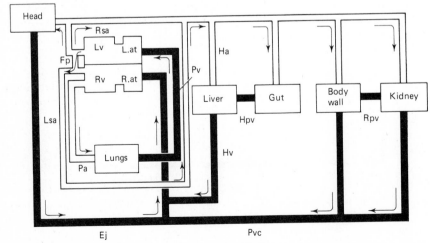

(e) Reptile (crocodilian)

Figure 6-3 (continued)

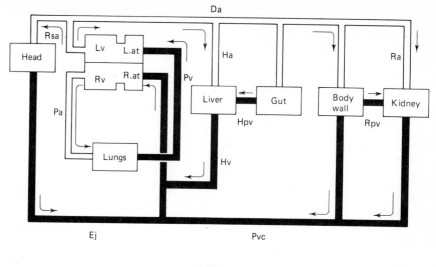

(f) Bird

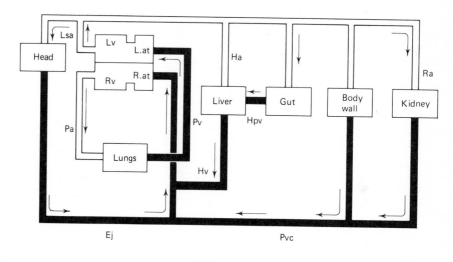

(g) Mammal

the left heart as it alternately fills and ejects its contents into the arterial system (the Windkessel vessels). In these vessels pressure oscillates between peak (systolic) and valley (diastolic) pressures around a mean value of 90–100 mm Hg. As the cross-sectional area of the system widens with branching, the pressure fluctuations become progressively damped and pressure declines to very low levels as blood is returned to the right heart.

Again, the blood is energized, but not to such high levels, as the right ventricle pumps the venous return from the tissues to the lungs.

On average, the same quantity of blood flows through the systemic and pulmonary circuits. Each of these circuits may be considered to contain series-coupled segments: Windkessel vessels, precapillary resistance vessels, precapillary sphincters, capillary exchange vessels, postcapillary resistance

TABLE 6-1. Geometry of Mesenteric Vascular Bed of the Dog

Kind of Vessel	Diameter (mm)	Number	Total Cross-sectional Area (cm²)	Length (cm)	Total Volume (cm³)
Aorta	10	1	0.8	40	30
Large arteries	3	40	3.0	20	60
Main artery branches	1	600	5.0	10	50
Terminal branches	0.6	1800	5.0	1	25
Arterioles	0.02	40,000,000	125	0.2	25
Capillaries	0.008	1,200,000,000	600	0.1	60 ⎤
Venules	0.03	80,000,000	570	0.2	110 ⎥
Terminal veins	1.5	1800	30	1	30 ⎥ 740
Main venous branches	2.4	600	27	10	270 ⎥
Large veins	6.0	40	11	20	220 ⎦
Vena cava	12.5	1	1.2	40	50 ⎦
					930

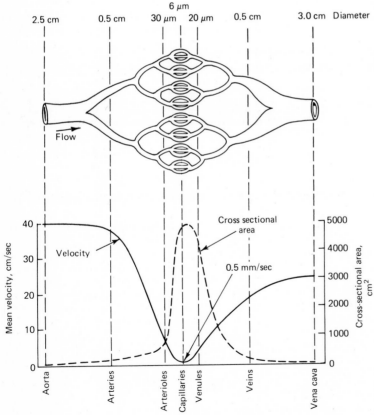

Figure 6-4

Cross-sectional area and velocity profile of the major components of the systemic circulation of a dog. The velocity of blood is inversely proportional to the cross-sectional area and is around 0.5 mm/sec in the capillaries where gas and nutrient exchanges must take place.

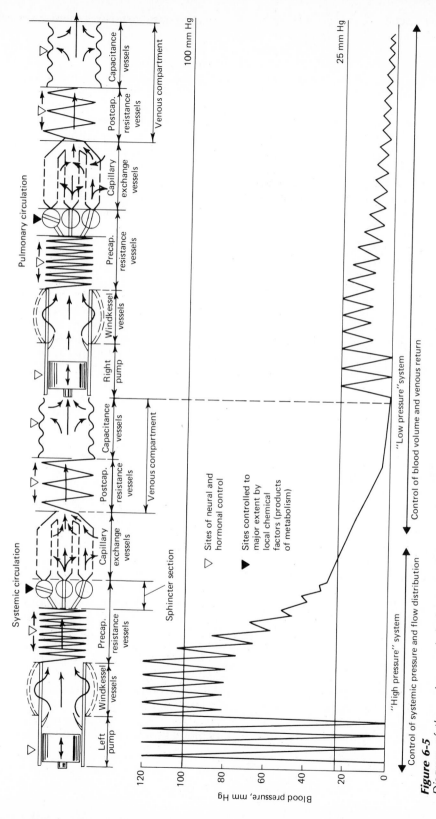

Figure 6-5

Diagram of the series-coupled elements of the mammalian or avian circulatory system and their relation to vascular pressures (pressures are those characteristic of mammals). The system is influenced by variable pump output, resistance, and capacitance. [Modified from B. Folkow and E. Neil. Circulation, Oxford University Press, London, 1971.]

Systemic circulation

Pulmonary circulation

Left pump

Windkessel vessels

Precap. resistance vessels

Sphincter section

Capillary exchange vessels

Postcap. resistance vessels

Capacitance vessels

Venous compartment

Right pump

Windkessel vessels

Precap. resistance vessels

Capillary exchange vessels

Postcap. resistance vessels

Capacitance vessels

Venous compartment

▽ Sites of neural and hormonal control

▶ Sites controlled to major extent by local chemical factors (products of metabolism)

100 mm Hg

25 mm Hg

"High pressure" system

"Low pressure" system

Control of systemic pressure and flow distribution

Control of blood volume and venous return

Blood pressure, mm Hg

120 100 80 60 40 20 0

vessels, and capacitance vessels. Each segment displays characteristic structural properties of functional significance:

Windkessel vessels are thick-walled and distensible (less so than veins) but contribute little resistance to flow. Their thick walls withstand the higher pressures within this segment, yet they distend during the ejection phase of the left ventricle. This "damps" the pressure during the ejection phase; however, during the filling phase of the ventricle, the arterial vessels recoil and the stored energy in the wall sustains the pressure and blood flow to more distal segments of the system until recharged again by the following ventricular systole.

Precapillary resistance vessels are the site of the greatest resistance to flow. These small arteries and arterioles, by contraction or relaxation of their muscular walls, have a profound effect on regional blood flow and the pressure within capillary beds. The radius of these vessels is influenced by intrinsic myogenic tone, which may be altered by local chemical factors, circulating agents, and the influence of sympathetic nerves that innervate them. They represent a major site of reflex control of the arterial pressure and distribution of the cardiac output.

Precapillary sphincters contribute also to the arterial resistance but are largely under the control of local chemical factors, the levels of which reflect the metabolic state of a particular tissue. These vessels determine the exchange area within tissues, and through their control of the number of perfused capillaries, they also control the length of the diffusion pathway across which gases, nutrients, and metabolites are exchanged.

Capillaries are thin-walled, consisting of a single layer of endothelial cells supported by a ground substance. Through the pores and across the walls of their endothelia pass electrolytes, water, lipid and water soluble substances, and the gases that are consumed or produced by tissue metabolism. The capillaries are the true business end of the circulation.

Postcapillary resistance vessels are small veins and venules that may, due to their state of contraction, influence the pressure gradient across capillary beds. This gradient determines the prevailing capillary hydrostatic pressure and, hence, the level of fluid transfer occurring between the vascular space and the tissue fluids (interstitial fluid).

Capacitance vessels contain the largest fraction of the blood volume. The wall stiffness of these vessels may be altered by the sympathetic nerves, which control the state of the smooth muscle in the vessel wall. These vessels are also subject to compressive forces of exercising skeletal muscle. Alterations in capacitance of the venous system importantly influence cardiac output, for it is the venous volume that must prime the pump.

On average, the equivalent of the cardiac output passes through each of the above series-related segments of the circulation. Since the velocity of the blood is inverse to the cross-sectional area, those segments of large areas are characterized by relatively low velocity. The low velocity of the capillaries (Figure 6-4) is of functional significance, since effective gas and nutrient exchanges with the tissues require a finite residence time of blood within the capillaries. The low velocity allows the blood and interstitial fluid to approach equilibrium and effect the necessary exchanges.

In our discussion of the functionally distinct segments of the circulation, we have considered these segments to be organized in series, one with the other. It is important to note that the arterial supplies of the organs are arranged *in parallel,* an arrangement that insures each organ or tissue will be subject to an equivalent arterial pressure. Although each organ is perfused by blood under a similar pressure, this does not imply that each equivalent mass of different organs receives a similar blood flow. Indeed, the blood flows of the organs differ greatly, both at rest and during stress or exercise (Table 6-2).

Inspection of the figures in Table 6-2 reveals that the distribution of the increased cardiac output is based on a new set of priorities during exercise. The skeletal muscles receive the king's share—up to a 20-fold increment over the resting state. The elevation in skin flow reflects the importance of circulatory transport of excess heat to the periphery. The 5-fold increase in cardiac output is supported by a roughly proportional increase in coronary blood flow while some tissues (gastrointestinal

TABLE 6-2. Alterations in Cardiac Output and Distribution in Rest and Exercise

Organs and tissues	Percent cardiac output (liters/min)	
	Rest	Exercise
Cardiac output	100 (5.25)	100 (25.00)
Brain	13–15 (0.78)	3–4 (0.85)
Heart	4–5 (0.25)	4–5 (1.13)
Liver and gastroin-testinal tract	20–25 (1.24)	3–5 (1.00)
Kidneys	20 (1.10)	2–4 (0.75)
Muscle	15–20 (0.96)	75 (18.75)
Skin	3–6 (0.25)	10 (2.50)
Skeleton, marrow, fat, etc.	10–15 (0.69)	1–2 (0.38)

Based on figures of B. Folkow, and E. Neil, *Circulation,* Oxford University Press, London, 1971.

tract, kidneys) exhibit an absolute decrease in blood flow during exercise. This remarkable redistribution is based on a combination of local, hormonal, and neurogenic controls that support the organs of locomotion and aid in the dissipation of the excess heat generated by the working muscles.

6.3 SOME HEMODYNAMIC PRINCIPLES

Poiseuille's Law. J. L. M. Poiseuille published a brief note in 1846 that still stands as the core idea and analytical tool of circulation physics. His experiments were elegant in simplicity. They consisted of comparing the flow of water through glass capillary tubes, ranging from 30 to 14 μm diameter, when driven by compressed air. His findings may be expressed by the equation $\dot{Q} = K(Pr^4/L)$ which reads: the flow is directly proportional to the product of pressure head (P) and the tube radius to the fourth power (r^4) and inversely proportional to tube length (L), where K is a temperature dependent constant characteristic of the fluid. Poiseuille's constant, K, is currently symbolized by η (eta) which denotes viscosity. Later derivations have fur-

ther refined the relationship to its currently used form:

$$\dot{Q} = \frac{\pi P r^4}{8 L \eta}$$

where $\pi/8$ is a numerical factor arising in the course of integrations of the calculation. One can view $1/\eta$ as a "viscosity factor," and r^4/L as a "geometric factor." The single most powerful component of the equation is tube radius, for it enters to the fourth power. All other factors held constant, this means that halving the radius will reduce flow to a diminutive one sixteenth of its former value! Small changes in the diameters of the muscular arterioles thus profoundly effect regional blood flow.

The law of Poiseuille is strictly applicable only for streamline flow in nonpulsitle, constant viscosity systems. Although these conditions are not met by the circulatory system, the law still provides a close and useful approximation for evaluating hemodynamics. In practice, Poiseuille's law is used by analogy to Ohm's law, which describes the relationship between current, voltage, and resistance: $I = E/R$. Let blood flow equate with current (I), pressure head with voltage (E), and resistance to blood flow with R (electrical resistance). The geometric factors that contribute to resistance may be lumped in the form: $R = 8L\eta/\pi r^2$. We can write:

$$\dot{Q} = \frac{P_1 - P_2}{R}$$

$P_1 - P_2$ is the pressure gradient across the system, that is, the effective pressure. The pressures used are *mean* pressures, not systolic or diastolic pressures. Since, for most ordinary situations we can consider both the length of the tubes and viscosity as relatively invariable, R will be primarily influenced by the radii of the peripheral resistance vessels.

Rearranging $\dot{Q} = (P_1 - P_2)/R$ we may calculate the resistance from $R = (P_1 - P_2)/\dot{Q}$ in peripheral resistance units, where P is expressed in mm Hg and $\dot{Q}$ in ml/min. Note that this is a dimensionless ratio. For comparative purposes the resistances of various tissues are best expressed on the basis of tissue mass:

$$PRU_{100} = \frac{P_1 - P_2}{\dot{Q}(ml\ min^{-1}\ 100\ g^{-1})}$$

Total peripheral resistance (TPR) for either the systemic or pulmonary circuits is usually calculated in terms of dyne sec/cm⁵ where cardiac output is expressed in ml/sec and pressure head in mm Hg. For example

$$TPR = \frac{100\ mm\ Hg \times 13.6 \times 980}{90\ ml/sec}$$
$$= 14,810\ dynes\ sec/cm^5$$

From the data of Table 6-2 and using some realistic pressure values, one can readily gain some idea of the magnitude of regional resistance changes that characterize rest and exercise in man. Let resting and exercising mean arterial pressures be 90 and 110 mm Hg respectively. For simplicity, we assume the venous pressures are 4 mm Hg for all vascular beds. Using $PRU = (P_1 - P_2)/\dot{Q}$, we find for muscle (90 − 4 mm Hg)/960 ml/min = 9×10^{-2} PRU; exercising muscle = 5.7×10^{-3} PRU, a sixteenfold difference. Reflecting on the fact that, from Poiseuille's law, a doubling of radius of the resistance vessels at constant pressure head will increase flow by 16-fold, we can conclude that the effective radius of the resistance vessels of muscle approached a doubling (not quite, because pressure increased by 20 mm Hg during exercise) to achieve a 19.5-fold increase in flow. You can also show that in other vascular beds the effective radius of the resistance vessels was reduced (kidney:

rest = 7.8×10^{-2} PRU; exercise = 0.14 PRU). Such regional vasoconstrictor responses are important in the maintenance of the arterial pressure and redistribution of the cardiac output to support the metabolic requirements of active tissues, such as muscle.

Viscosity of Blood. Plasma is about 1.8 times the viscosity of water. Whole blood is much more variable, depending on the hematocrit (percentage of whole blood represented by red cells). Viscosity increases with hematocrit or at lower temperatures. Plasma viscosity rises about 2% per degree Celsius decrease in temperature, a fact that may be of importance in hibernators, in forms in which body temperature fluctuates with ambient temperature, or in areas of the body subjected to heat or cold.

A marvelous behavior of blood, flowing through tubes of various dimensions, is illustrated in Figure 6-6a. At tube radii above around 1 mm in radius, the viscosity measurements are relatively constant; however, a powerful effect of radius occurs below this value. The arterioles, the site of the major resistance to flow, are of the order of 30–50 μm in diameter, and capillaries are of even smaller bore ($\cong 6\mu m$). This *anomalous viscosity* of blood, also called the *Fahraeus-Lindquist effect*, has not been satisfactorily explained.

A plausible theory is based on the fact that the measured hematocrit of blood within tissues is lower than that of arterial or venous blood. Yet, the same volume of red cells and plasma that flows through supply arteries must pass through the mi-

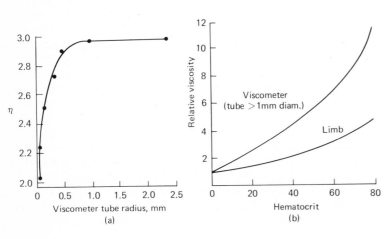

(a)

(b)

Figure 6-6
(a) Viscosity of blood determined in glass tubes of differing radius. Note the sharp decline in viscosity in tubes below 1-mm radius. (b) Contrast between the relative viscosity of blood flowing through a glass viscometer and the limb of an animal. [(a) based on R. Haynes, Am. J. Physiol., **198**, 1193–1200 (1960); (b) based on S. Whittaker and F. Winton, J. Physiol., **78**, 339–369, (1933).]

crovessels. The measured transient time of red cells is shorter than for plasma; that is, the red cells pass through the microcirculation at higher velocity than does plasma. The cells then "slow up" once more when entering the venous system. The probable reason for the faster transient time for red cells is their tendency to move into the faster axial stream, whereas the plasma zones near the wall of the vessel move more slowly owing to frictional factors. The reduction in hematocrit in the small bore resistance vessels is the probable cause of the reduction in effective viscosity.

This decrease in effective viscosity is of real physiological significance as can be shown by the apparent viscosities of blood in a glass viscometer (of greater than 1 mm diameter bore) compared to blood flowing through the hind limb of a dog when used as a biological viscometer (Figure 6-6b). At a normal hematocrit of 40% the relative viscosity in the dog limb preparation is 2, compared with 4 in the glass viscometer. Resistance (in Poiseuille's law) $= 8L\eta/\pi r^4$. Thus doubling η will produce a twofold increase in R, an event that would require a similar increment in pressure head if flow is to be maintained. In the absence of the anomalous viscosity of blood passing through the major resistance vessels, we would require a mean arterial pressure of around 200 mm Hg (instead of 90–100); or, a complete overhaul of vascular design. As Goethe's Mephistopheles remarked to Faust, "Blood is a truly remarkable juice."

Pressure and the Energy of Perfusion. It may seem intuitively correct that fluids will flow from points of higher to lower pressure. Although often the case, this is not strictly true. Rather, fluids flow down *energy* gradients and pressure measurements do not always reflect the true energy of a flowing system. This distinction is often overlooked, and confusion on this point may lead to the paradoxical and erroneous conclusion that blood flows "uphill," that is, against a pressure gradient. The confusion can be resolved by a consideration of Bernoulli's principle.

Before stating this principle in an equation, it might be enjoyable to make a simple device of common materials that illustrates the importance of Bernoulli's work. An ordinary sewing thread spool, a straight pin, a piece of cardboard cut to the circumference of the spool, and a bit of sticky tape is all that is required. The pin pierces the cardboard disc at its center and the pinhead is secured to the bottom of the disc with tape. The pin is then guided into the hole in the center of the spool until the disc is flush with one end of the spool. The trick is to try and blow the disc away from the spool. One must support the disc until a hefty air stream is established by blowing down the hole in the spool. The more forcefully you expire through the spool, the more tightly the disc "adheres" to the spool with no apparent support.

Bernoulli's principle states that the total fluid energy of a system is the sum of three components: (1) P, the potential energy of pressure; (2) potential energy due to gravity acting on a column of fluid, and (3) the kinetic energy of fluid motion, that is, a mass of fluid having a velocity. The equation is written

$$E = P + \rho g h + \tfrac{1}{2}\rho V^2$$

where E is in ergs/cm² = dynes/cm²; $P =$ pressure (dynes/cm²; 1 mm Hg = 1330 dynes/cm²; $\rho =$ fluid density (g/cm²); $g =$ acceleration due to gravity (980 cm/sec²); $h =$ height of the fluid column above a point of interest; and $V =$ velocity of the fluid at that point (cm/sec).

It is important to note that the forms of energy, kinetic and potential, are interconvertible. This point is made in Figure 6-7a, where a main conduit narrows and then widens again. Flow through all segments is at a constant level, so it is clear that velocity in the narrow segment must be relatively high. Side pressure (P in the equation) is lower in the constriction than in the downstream tubing; yet, the total fluid energy of the narrow section exceeds the downstream energy and flow is maintained. During exercise, pressure gradients of this type have been measured in the pulmonary artery as catheters with side or downstream facing orifices (Figure 6-7b) were advanced from the base to more distal segments of the pulmonary artery where the total cross-sectional area becomes larger. The difference between side pressure and end pressure, measured at a common point, is a valid estimate of the kinetic energy of the system. End pressures yield an estimate of total fluid energy.

Table 6-3 presents calculations of the relative importance of the potential and kinetic energy for various types of blood vessels during rest and at

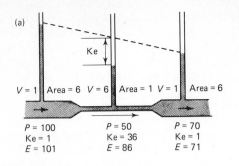

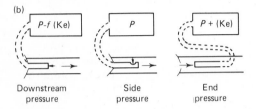

| Downstream | Side | End |
| pressure | pressure | pressure |

Figure 6-7

(a) Bernoulli's principle (where the tube narrows), demonstrating how flow occurs down a gradient of total fluid pressure E but not down a gradient of pressure P, (b) Kinetic-energy artifact in making measurements of pressures in blood vessels with catheters. [From A. C. Burton. Physiology and Biophysics of the Circulation. *The Year Book Medical Publishers, Inc., Chicago, 1965.]*

elevated cardiac output. The data indicates that the kinetic energy factor in the aorta is small at rest but rises significantly at higher cardiac output, as in exercise. In small vessels (arteries, capillaries, small veins), kinetic energy is negligible. The kinetic energy is important in the vena cava during exercise, not at rest. It is of a significant magnitude in the pulmonary artery at rest and of great importance at high cardiac output.

Peak side pressure and velocity occur during systole. From Table 6-3 it can be seen that during exercise 52% of the fluid energy is kinetic in the pulmonary artery during systole. The potential energy accounts for the distending forces that cause vessel wall stress and expansion. Imagine that one could instantly stop flow in the pulmonary artery during systole. This would result in more than a doubling of the distending pressure. For major vessels, especially during high cardiac output states, the increase in the ratio of kinetic to total fluid energy is of real significance in minimizing vascular wall stresses.

Turbulence may develop in flowing streams at critical velocities and tube dimensions. When it occurs, eddies and whirls develop and the flow profile becomes irregular. Such disorganization consumes energy and requires a greater pressure gradient to sustain flow (Figure 6-8). The critical velocity

TABLE 6-3. Amount and Relative Importance of Kinetic Energy in Different Parts of the Circulation*

| Vessel | Resting cardiac output | | | | Cardiac output increased 3 times | | |
	Velocity (cm/sec)	Kinetic energy (mm Hg)	Pressure (mm Hg)	Kinetic energy as % of total	Kinetic energy (mm Hg)	Pressure (mm Hg)	Kinetic energy as % of total
Aorta, systolic	100	4	120	3	36	180	**17**
Mean	30	0.4	100	0.4	3.8	140	2.6
Arteries, systolic	30	0.35	110	0.3	3.8	120	3
Mean	10	0.04	95	Neg.		100	Neg.
Capillaries	0.1	0.000004	25	Neg.	Neg.	25	Neg.
Venae cavae							
and atria	30	0.35	2	**12**	3.2	3	**52**
systolic	90	3	20	**13**	27	25	**52**
Mean	25	0.23	12	2	2.1	14	**13**

* The cases where kinetic energy should not be neglected—where it is more than 5% of the total fluid energy—are indicated by bold figures. When an artery is narrowed by disease processes, the kinetic energy becomes very important.
Source: A. C. Burton. *Physiology and Biophysics of the Circulation,* The Year Book Medical Publishers, Inc., Chicago. 1965.

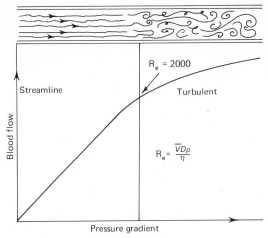

Figure 6-8
Relationship between flow and pressure gradient in a rigid tube. Above a critical pressure the flow becomes increasingly turbulent (Reynolds numbers exceeding around 2000). [*Based on N. Coulter and J. Pappenheimer. Am. J. Physiol.,* **159,** *401–408 (1949).*]

at which turbulence occurs is determined by calculating a nondimensional value, the Reynold's number: $\mathrm{Re} = \overline{V}D\rho/\eta$, where $\overline{V}$ is average flow velocity (cm/sec), D is the internal diameter of the tube (cm), ρ is fluid density and, η the viscosity in poise (η of water at $20.3°\mathrm{C} = 1$ centipoise). Turbulence appears when the Reynolds number exceeds 2000. It is probable that, in heavy exercise, this number is exceeded for brief moments during systole in the proximal aorta. Due to successive branching and increase in cross-sectional area of the vascular tree, linear velocity is reduced to the extent that Reynolds numbers become progressively lower and turbulent flow is not a normal feature of the peripheral arterial or venous circulation. Narrowing (stenosis) of supply arteries or of the valvular orifices of the heart may occur pathologically, with turbulence distal to the stenosis.

Perhaps the most commonly used diastolic tool of medicine is the sphygmomanometer. Its use depends on the purposeful creation of turbulence and its associated sounds to signal systolic and diagnostic pressures. The sounds heard via a stethoscope distal to a blood presure cuff (Korotkow sounds) occur as blood is forced through the partially col-

lapsed artery during systole, and disappear when cuff pressure falls below the diastolic pressure as streamlined flow is once again established at normal arterial diameter.

6.4 PERIPHERAL CIRCULATION

Blood vessels are elastic bodies and are constantly changing their dimensions. Both passive (connective tissue) and active (vascular smooth muscle) components are involved in these dimensional changes.

6.4.1 Walls of Blood Vessels

The abundance of different vessel components varies in different blood vessels (Figure 6-9). Elastin is a distensible material and provides an elastic tension, resisting stretch, that does not require expenditure of biochemical energy. Collagenous fibers, on the other hand, resist stretch to a greater extent than elastin, but are loosely arranged and do not reach their elastic limit until considerable distension of the vessel has occurred. They provide a stiffer buffer to expansion at larger diameters. Smooth muscle provides the variable in determining wall stiffness. This variability is energy-consuming, because development of muscle tension requires expenditure of biochemical energy. The arterioles, because of their large content of smooth muscle and relatively small lumen size, contribute much more to the total resistance than the larger vessels upstream.

Blood vessel walls are elastic bodies, so one might expect that their behavior, when subjected to distending forces, would be described by *Hooke's law:* The force per unit area developed by an elastic body when stretched is proportional to the length to which the material is stretched. Obviously there is a limit at which the body will come apart—the breaking stress.

The walls of blood vessels are not homogeneous and, instead of following Hooke's law, they resist stretch increasingly as greater stretching occurs. Their behavior can be studied from pressure/volume curves (Figure 6-10), which can be analogized to tension/length diagrams. Veins, being more distensible than arteries, develop a lower tension per increment of volume than do arteries. The veins

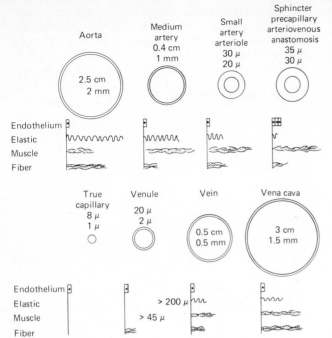

Figure 6-9

Variety of admixture of the four tissues in the wall of different blood vessels. The figures under the name of the vessel are the diameter of the lumen, below it the thickness of the wall. [*From A. C. Burton.* Physiol. Rev., **34**, *619–642 (1954).*]

are thus better suited to the role of blood volume reservoirs than the arteries, a fact correlated with the large proportion of the blood volume residing in the venous system. Norepinephrine produces contraction of vascular smooth muscle; with increasing smooth muscle contraction, both veins and arteries become "stiffer," developing greater tension with successive volume increments. Such smooth muscle activity may be induced in the intact animal by a variety of nervous and hormonal events.

The elastic behavior of cylindrical vessels is described by the law of Laplace: $T = Pr$, where T is the wall tension (dynes/cm), P the pressure (mm Hg), and r the tube radius. The law predicts that, at equilibrium, the wall tension is proportional to the product of pressure and radius. It is also apparent that in a larger vessel the wall tension at equilibrium must be greater than for small vessels at the same pressure. The smaller wall thickness and mass of small vessels is therefore compensated for, and a smaller wall tension is required to reach equilibrium than is true for the aorta.

The variable character of the smooth muscle component of tension makes it possible for several Laplacian curves to be derived from a single vessel. In Figure 6-11 the straight line from the origin designated P_1 is the equilibrium curve predicted at pressure P_1 from the law of Laplace. The curved lines represent the behavior of a hypothetical vessel under different degrees of smooth muscle activity. The point at which the curves cross Laplacian lines (P_1, P_2, and P_3) represent the radius and wall tension at which equilibrium will occur as the vessel is filled at the different pressures. For the relaxed vessel there is only one point where equilibrium will occur at pressure P_1. Note that at the larger radii and higher pressures (P_2 and P_3), a much higher wall tension is required to establish equilibrium. Note also that, with increasing smooth muscle activity, the equilibrium radius at any given pressure is smaller. In the case of the greatly contracted state, equilibrium will be established only at the higher pressures (P_2 and P_3), the vessel being closed at P_1 because the forces attributable to muscle and connective tissue exceed the distending force of pressure.

The pressure at which this collapse occurs is called the ***critical closing pressure.*** It obviously varies with smooth muscle activity, and its determination

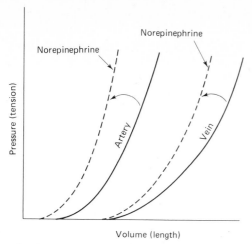

Figure 6-10
Pressure/volume diagrams of an artery and a vein. The artery is the least distensible of the two. Norepinephrine, either circulating or released from sympathetic nerve endings, produces contraction of vascular smooth muscle and increases vessel wall stiffness.

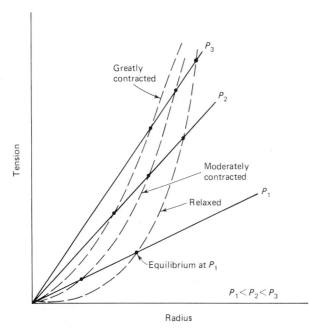

Figure 6-11
Tension/radius relationships of blood vessels at various distending pressures.

has been suggested as a means of assessing vasomotor tone. Furthermore, we can appreciate that some minimum critical pressure must be maintained in blood vessels if tissues are to be perfused. These factors may be important in considerations of the form of tissue perfusion utilized by animals generating very low pressures and possessing open or lacunar peripheral circulations, as seen in many invertebrates. The use of small branching tubes, simply from the consideration of the law of Laplace, sets certain minimum pressure requirements on the system if perfusion is to take place at all. Estimates of critical closing pressures in mammals (closure probably occuring in the arterioles) indicate that pressures of at least 20 mm Hg are required to maintain patency of the vessels.

6.4.2 Arterial Circulation

The arteries form the major conduits for distributing blood to the microcirculation, receiving blood intermittently from the heart. It is essential that sufficient energy be injected into the arterial system if perfusion of the tissues is to result. This is because of the resistance offered, chiefly, by the arterioles. The pressure measured in arteries is largely dependent on the volume injected into the system and the resistance, following Poiseuille's law ($P = R\dot{Q}$). There are thus two principal determinants of pressure: the amount of blood injected as a result of heart action and the number and diameters of the resistance vessels patent at a given time.

The time course of pressure recorded from a major vessel exhibits a pulsatile form rising to a maximum (systolic) and falling to a minimum (diastolic) pressure during each cardiac cycle. The ***mean pressure*** represents the average pressure during a cycle, and the difference between systolic and diastolic pressures is the ***pulse pressure.*** There is little fall in mean pressure in major arteries, but a decline occurs rather sharply as the blood passes through the arterioles. Because the vessel wall material is elastic, a portion of the energy put into the expansion of the walls by the action of the heart is recovered as the walls rebound against the fluid core. This effect promotes the maintenance of pressure and flow even after systolic ejection has ceased.

The form and magnitude of the pulse of arterial pressure accompanying each cardiac cycle is influenced by the volume output of the heart, the resistance offered by the arterioles, and the distensibility of the arteries. Distensibility of the arteries characteristically decreases with age in man. Furthermore, the arteries become less distensible as they are stretched; that is, they are not *linearly* distensible. This is the principal reason why many older people exhibit a large pulse pressure with the systolic level elevated more than the diastolic. Hypertension, due to increased peripheral resistance, elevates the mean arterial pressure and has a generally small effect on pulse pressure if the arteries are young and compliant. Elevation of the cardiac output alone increases pulse pressure as well as mean pressure; however, circumstances that produce increases in cardiac output (exercise) are usually associated with reduction in total peripheral resistance. The latter adjustment often compensates for the influence of cardic output and prevents excessive mean and pulse pressures from developing.

6.4.3 Microcirculation

A prominent physiologist is fond of stating that the heart, arteries, and veins may be considered an elaborate plumbing system that services the microcirculation. There is something to this, because the capillaries represent the true "business end" of the system—the area where exchange with interstitial fluid and tissues occurs.

The arrangement of vessels in the microcirculation varies considerably in different vascular beds. Generalizing, many authors present the simplified picture of arteries branching into arterioles, which in turn subdivide into capillaries that ramify in a network to reunite as small veins. The description is only true in a broad way, because the arrangement varies considerably and in a manner often correlated with the functional characteristics of the tissue. Figure 6-12 illustrates the pattern observed in mesentery.

One is immediately impressed by the alternative pathways available for blood flow. Shunting from arteriole to venule may occur, especially if the metarteriole constricts as the arteriolar-venular anastamosis relaxes. Blood may flow principally in the preferential channels if the smooth muscle cell at the entrance of true capillaries (precapillary sphincters) are constricted; a combination of preferential channel and capillary flow may occur, and so on. There is some difficulty in making judgments about nutritive flow when the flow to an entire organ is measured.

The glomerular circulation of the kidney and the vasa recta of the renal medulla (see Section 6.7.2) are special arrangements of the microcirculation subserving special functions, that is, filtration and countercurrent exchange, respectively. Such varied microcirculatory arrangements should make us wary of oversimplification of the situation.

6.4.4 Capillary Function

We shall now consider the functional aspects of microcirculatory exchange, a process occuring across the walls of capillaries. Two major exchange processes occur: bulk filtration and diffusion. Bulk filtration depends on transcapillary pressure gradients; diffusion is related principally to concentration differentials.

The most satisfactory explanation of the factors governing filtration was generated by the work of

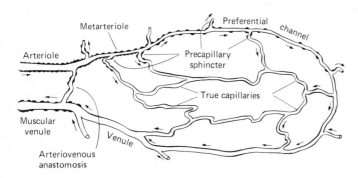

Figure 6-12
Schematic representation of the structural pattern of the capillary bed. The distribution of smooth muscle is indicated in the vessel wall. [*From B. Zweifach. In* Transactions of the 3rd Conference on Factors Regulating Blood Pressure, *pp. 13–52, Josiah Macy, Jr., Foundation, New York, 1950.*]

E. H. Starling. He realized the importance of both hydrostatic and colloid-osmotic (oncotic) pressures as determinants in governing the transfer of fluids across the capillaries. The presence of protein in the plasma in concentrations considerably above that in the interstitial fluid imparts to plasma an oncotic pressure tending to promote the movement of fluid from the interstitial fluid, across the capillary endothelium, and into the plasma. Opposing the influence of plasma protein, the hydrostatic pressure generated by the heart tends to promote filtration of fluid into the interstitial fluid compartment. The volume distribution of fluid between vascular and interstitial compartments depends on a delicate balance of these factors. The presence of lymphatics offers a safety factor because excess fluid entering the interstitial space finds its way into the lymphatics and ultimately back into the venous system. The lymphatic system is, however, limited in its capacity to remove excess fluid because, with increased interstitial fluid volume, the hydrostatic pressure of the tissues may rise to the point that the lymphatics collapse.

Figure 6-5 depicts the pressure drop between the arterial and venous systems in a mammal. The major pressure drop is across the arterioles (the principal resistance vessels). At the entrance to an idealized capillary, the pressure is approximately 35 mm Hg, falling to about 15 mm Hg at the venous end of the capillary. The plasma oncotic pressure is about 25 mm Hg in mammals, corresponding to a total serum protein of about 6.5%; most of the oncotic pressure is due to albumin. Figure 6-13 illustrates the relationships of the forces tending to promote filtration and absorption along the length of a capillary. At the arterial end of the capil-

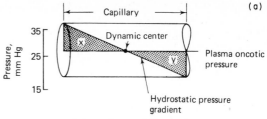

Area x = Forces of filtration (plasma hydrostatic + tissue oncotic pressure)

Area y = Forces of absorption (tissue hydrostatic + plasma oncotic pressure)

When area x = area y, there is no net movement of fluid.

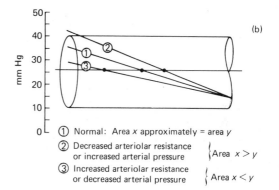

① Normal: Area x approximately = area y

② Decreased arteriolar resistance or increased arterial pressure } Area x > y

③ Increased arteriolar resistance or decreased arterial pressure } Area x < y

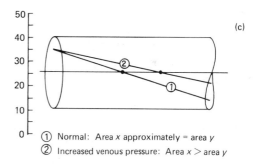

① Normal: Area x approximately = area y
② Increased venous pressure: Area x > area y

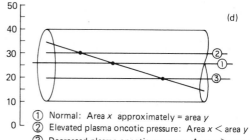

① Normal: Area x approximately = area y
② Elevated plasma oncotic pressure: Area x < area y
③ Decreased plasma oncotic pressure: Area x > area y

Figure 6-13
Factors influencing bulk filtration across capillaries (a). Forces promoting filtration are the plasma hydrostatic pressure and the tissue oncotic pressure (usually low, of the order of 2–10 mm Hg); forces favoring absorption are the tissue hydrostatic pressure (1–2 mm Hg) and the plasma oncotic pressure (25 mm Hg). When the forces favoring filtration and absorption are equal (dynamic center), no net filtration occurs. Effects of changes in arterial pressure or arteriolar resistance are shown in (b), alterations in venous pressure in (c), and changes in plasma oncotic pressure in (d).

lary the sum of the capillary hydrostatic and tissue oncotic pressures exceeds that of tissue hydrostatic and plasma oncotic pressure. The dominant forces result in filtration at the inflow segment of the capillary. A steady drop in hydrostatic pressure occurs and, as a result of upstream filtration of protein-poor fluid, the plasma oncotic pressure rises slightly. A point is reached (dynamic center) where the forces promoting filtration are exactly matched by the forces favoring absorption. No net movement in or out of the capillary occurs at this point. As we move downstream, absorptive forces dominate, principally because of the fall in hydrostatic pressure along the length of the capillary. Apparently the forces promoting filtration slightly exceed those of absorption. The excess interstitial fluid that results is returned to the vascular system by the lymph.

The description presented in the preceding paragraph is somewhat of an abstraction—an idealization of "average" capillary filtration and absorption. Capillaries may absorb throughout their length. This would be the case for capillaries with precapillary sphincters completely contracted and a very low hydrostatic pressure. A number of situations may shift the "dynamic center," resulting in either a predominance of filtration or absorption. Alterations of the forces promoting filtration and absorption may have rather profound physiological consequences. For instance, hypovolemia (low blood volume) may produce hypotension and a consequent fall in capillary hydrostatic pressure; the predominant net transfer of fluid across the capillaries would be from the interstitial compartment to the blood plasma. This tends to correct the plasma volume loss at the expense of interstitial fluid.

Diffusion is the major process involved in transport of materials across the capillary walls. Thus, carbon dioxide and oxygen move across the capillaries in a large quantity in accordance with the Fick diffusion equation, a transfer process that is independent of bulk filtration. For a discussion of diffusion see Section 5.2.1.

Capillary blood flow is closely correlated with the metabolic needs of tissues. This relationship was beautifully demonstrated by Krogh, who observed the capillaries, following india ink injection, in both resting and contracting skeletal muscle of the frog. He observed a 20–50-fold increase in the number of open capillaries in the contracting muscle and concluded that local blood flow is adjusted to meet metabolic demand. These adjustments occur even when the nerves are sectioned or when the tissue is isolated and perfused. Thus a local control mechanism exists that governs the number of open capillaries and the contractile state of precapillary sphincters.

Various workers have suggested that metabolic vasodilation is due to oxygen deficit or to local increase in carbon dioxide or the concentration of potassium ion. The nature of the dilator material is not yet known and we shall simply refer to such material as **metabolite.** Recent work supports the concept that the precapillary sphincters respond to local changes in osmolality of the interstitial fluid. According to this idea, locally produced metabolites, of whatever nature, contribute to osmolality elevations, the precapillary sphincter responding to such elevations with relaxation and consequent increase in blood flow.

A probable scheme linking local blood flow to metabolism appears in Figure 6-14. The precapillary sphincters are largely independent of the nervous system, and although the sympathetic nerves may produce massive arteriolar constriction, local metabolic dilator materials that accumulate during

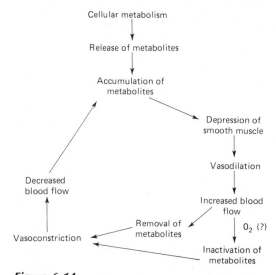

Figure 6-14
Metabolic feedback control of the microcirculation.

low-flow states may diffuse to the vicinity of arterioles, producing an *ascending vasodilatation,* thus overriding, to some extent, nerve-mediated vasoconstriction. Ascending vasodilatation may provide a protective mechanism for the tissue, even in states of massive sympathetic vasoconstriction. The phenomenon has been demonstrated most clearly in skeletal muscle.

6.4.5 Veins

The same structural elements are found in veins as in arteries. There is, however, an important quantitative difference, the veins being invested with a much thinner layer of elastic fibers. The thin-walled character of veins and their much greater distensibility constitute another important contrast to arteries. The smooth muscle component is absent in the small venules into which the capillaries drain; a muscular media occurs in veins greater than about 0.5 mm in diameter. The venules, then, lack the capacity to influence capillary exchange in the same manner as the muscular arterioles.

Veins are not passive tubes. There is rich sympathetic innervation in mammals, and there is good evidence that normally a sympathetically mediated venous tone exists. Sympathectomy abolishes venous tone, producing venodilation and a consequent increase in venous capacity. No dilator nerves have been demonstrated, and changes in venous tone apparently represent changes in degree of sympathetic venomotor activity. The veins also respond to circulating vasoactive agents, such as norepinephrine, an effect observed in reptiles as well as mammals. Valves are present in many veins, especially in the veins of extremities. The valves are important in determining the direction of venous flow, especially in veins orientated in a vertical plane, because the internal pressure may rise as a result of the height of the hydrostatic fluid column. Contraction of skeletal muscles creates a high intraluminal pressure, squeezing the veins and forcing fluid toward the heart and past venous valves. Backflow is prevented because the valves are unidirectional. Valves greatly increase the efficiency of this peripheral pumping, because with closure of the valves retrograde resistance is infinitely high.

The venous sytem, constructed of highly distensible tubes with a variable capacity, is ideally suited as a dynamic blood-volume reservoir. The largest proportion of the blood volume resides on the venous side of the circulation. Furthermore, the high compliance of the reservoir is emphasized by the fact that there is a minimal change in venous pressure when blood is injected rapidly into the venous system. A much greater pressure change is induced by the injection of an equal volume into the arterial system. The functional role of venous smooth muscle is related to the change in capacity induced by venoconstriction or dilation and the subsequent influence on venous pressure. Such capacity changes influence the pressure gradient from the periphery to the heart and, thus, the venous flow to the heart. As pointed out previously, changes in venous pressure also produce shifts in capillary filtration.

6.5 BLOOD VOLUME

Simple dilution techniques are utilized in the determination of circulating blood volume. A nontoxic substance that remains in circulation because of large molecular size or plasma-protein complexing is administered. Mixing time is allowed and the dilution of the material determined. Evans' blue (T–1824) or ^{131}I-tagged serum albumen is utilized for this purpose. The indicators are diluted in plasma, so the results include an estimate of plasma volume. The time of sampling plasma is important because the material does leave the circulation, although at a rather steady rate. Normally, 30 min after administering T-1824 the rate of fall in plasma concentration of the dye is steady because mixing is complete. Because a load was administered at zero time and the dye is obviously "leaking" from the circulation, the pertinent question is: What would be the concentration of the dye if complete mixing had occurred at zero time? An estimate of the proper concentration is obtained by extrapolating a line from the constant slope component back to zero. The volume of the plasma is thus obtained from the relation

$$\text{volume (ml)} = \frac{\text{load (mg)}}{\text{plasma concentration (mg/ml)}}$$

An estimate of total blood volume (TBV) may be obtained from a knowledge of plasma volume (PV) and hematocrit (Hct). Thus

$$TBV = \frac{PV}{1 - Hct}$$

For Example,

PV = 540 ml
Hct = 40%
$$TBV = \frac{540}{1 - 0.4} = \frac{540}{0.6} = 900 \text{ ml}$$

Total blood volume, determined in this manner, is subject to error because a single sample of venous or arterial blood does not yield a truly representative value for the hematocrit. Red blood cells may, however, be tagged with isotopes (^{51}Cr), and using the dilution technique, a good estimate of circulating red cell volume can be obtained.

Mammals, birds, and reptiles are characterized by blood volumes ranging between 6 and 10% of their body weight. Diving birds and mammals may have larger blood volumes than nondivers. Among fishes, elasmobranchs and cyclostomes fall within the same general range as higher vertebrates, but teleosts are characterized by lower volumes (1.5–3% of body weight). For representative blood volumes of various vertebrates, see Tables 7-5 and 7-12.

Variations of blood volume occur under a number of conditions. The increase in blood volume that occurs at high altitude is due primarily to an increase in circulating red cell mass. Loss of blood volume through hemorrhage is made up partially from interstitial fluid. Volume receptors in the mammalian left atrium contribute to blood- and interstitial-volume regulation. The receptors promote secretion of antidiuretic hormone under conditions of depleted volume. Volume receptors may also be important in making up fluid loss by promoting the sensation of thirst.

6.6 VERTEBRATE HEART

The role of the heart is to service tissues by propelling a continuous stream of blood through the capillaries.

6.6.1 General Features

The propulsion of blood is accomplished by the conversion of potential (chemical) energy into kinetic energy as the heart imparts movement to the blood (see Chapter 4 for a discussion of muscle energetics). The heart is a reciprocating type of pump, characterized by intermittent pulsing with valves governing direction of flow into and out of the heart chambers. Many invertebrates utilize a rotary type of pumping mechanism, whereby body-wall musculature literally squeezes blood through the tubes or spaces of the vascular system. Rotary pumping is also utilized by vertebrates, chiefly to impart energy to the veins during muscular exercise.

The importance of the cardiac output can be appreciated when one considers that the blood supply to all the tissues is dependent on its magnitude. Cardiac output is variable, a necessary characteristic, because the flow demands of organs vary under different circumstances. The factors that determine cardiac output include venous pressure at the atria, pulmonary or gill resistance and pressure, the systemic arterial pressure against which the heart must pump, heart rate, nutritional (coronary) blood flow, and the state of the myocardium. The pumping characteristics are controlled by complex feedback systems involving neural, hormonal, and physical components.

The venous return to the heart, over an extended period, must be equal to the cardiac output. The elastic properties of cardiac muscle allow a fairly wide range of diastolic filling volumes with little change in intracardiac pressures. Factors decreasing venous flow to the heart will ultimately be expressed as a decrease in cardiac output. The central venous pressure is normally low, and in the atria is only about 0.5 mm Hg. The peripheral venous pressure must be of higher magnitude to maintain a favorable gradient of flow toward the heart. Elevations in venous pressure will increase venous return to the heart and, secondarily, cardiac output. Because the force generated by contraction increases, within limits, with increasing myocardial-fiber length, the stroke output may be varied with little change in rate.

Speeding (*tachycardia*) and slowing (*bradycardia*) of the heart are effective means of altering the cardiac output as demand changes. The diastolic phase of the cycle is typically of longer duration than systole, and rate changes are primarily associated with lengthening or shortening of diastole (or filling time). Although filling time is reduced during tachycardia, this potential limitation is somewhat compensated for since most factors inducing cardiac acceleration (sympathetic stimulation, circulating epinephrine or norepinephrine) produce, in addition to their positive *chronotropic* (rate) effect, an augmentation of myocardial contractile force (positive *inotropic* effect) with more complete ejection of the contents of the heart.

6.6.2 Electrical Activity

Activity Sequence Activation of the vertebrate heart is initiated by pacemaker cells within the sinus venosus of lower vertebrates or by the *sinoatrial node* (S-A node), in mammalian and avian hearts (Figure 6-15). These specialized cells exhibit the property of *automaticity* and *rhythmicity* and continue to activate the heart beat even in the absence of nervous innervation. Such specialized regions are the *pacemakers* of the heart. The intrinsic rate of sinoatrial node cells is greater than that of *atrioventricular nodal* (A-V node) cells, which are capable of initiating ventricular contraction when the

S-A node is experimentally depressed, albeit at a slower rate.

The first electrical event precedent to a cardiac contraction cycle is the depolarization of the S-A node cells. This is followed by a spread of an excitatory impulse to the atrial cardiac fibers and to the A-V node. Atrial myocardial conduction is around 1 m/sec in man; however, there is a considerable delay of this excitatory process at the junctional region of the A-V node where conduction velocity is only around 0.5 m/sec. Once the A-V node is activated, the subsequent velocity of excitatory spread is 1 m/sec or greater. The A-V nodal delay is important functionally since it allows contraction of the atria to occur while the ventricles are being charged with blood prior to ventricular systole. From the A-V node the excitatory processes passes to the ventricles via a specialized conducting system consisting of the *atrioventricular bundle* (bundle of His) which subsequently branches, perforating the interventricular septum as so called *bundle branches.* These branches subdivide into a network of conducting fibers, the *Purkinje fibers.* These are the fastest-conducting fibers within the heart (1–4 m/sec), a feature allowing smooth activation of all regions of the endocardium of the ventricle.

Contraction is initiated slightly earlier in the in-

Figure 6-15
Conduction pathways of the mammalian heart.

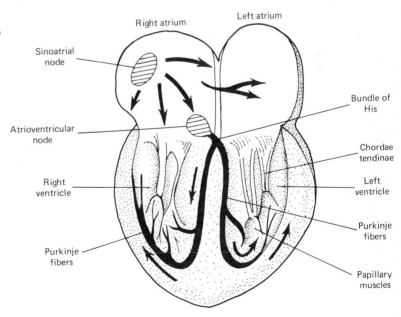

terventricular septum and papillary muscles. Early septal contraction provides a rigid anchor for the subsequently contracting ventricular wall. Since the papillary muscles provide the primary attachment between the fibrous *chordae tendinae* and the atrioventricular valve margins, early contraction of these muscles aids in preventing eversion of these valves during ventricular systole. The contraction wave spreads rapidly to the endocardium and hence, more slowly to the epicardium. The last portions of the heart to be excited are the epicardial regions of the base of the ventricle. This sequence ensures efficient ejection with contraction proceeding from apex toward the base.

Transmembrane Potentials. The activated state of the myocardium is preceded by the generation of an action potential that may be recorded by measuring the potential difference between two electrodes, one in the external fluid medium and the other (a microelectrode) inserted into a myocardial cell. In Figure 6-16a, point A shows the two electrodes in the external medium (no potential difference), whereas at point B, a ventricular fiber is pierced by the microelectrode. A resting potential difference of around −90 millivolts (mV) indicates electronegativity of the interior with respect to the exterior. At point C, a sharp depolarization occurs as a propagated action potential is transmitted to the cell. Typically, there is a slight positive overshoot of around 20 mV, followed rapidly by a fast (1), then slow (2), repolarization trend. The *plateau*

phase (2) is then followed by more rapid repolarization (3) toward the former resting potential level (4). The general relationship between these transmembrane alterations and development of muscle tension is shown in Figure 6-16b.

The generation of the action potential is accompanied by alterations in the permeability of the cell membrane to both K^+ and Na^+. The $[K^+]$ (concentration is indicated by square brackets) of the interior of myocardial cells, as with other tissues, exceeds that of extracellular fluid, that is, $[K^+]_i >$ $[K^+]_o$. The influence of a given ion species on the potential difference across a membrane depends on the ratio of the intracellular to extracellular concentration of the ion and may be calculated from the Nernst equation (see Section 9.4.1): for the potassium *equilibrium potential*, $E_K = -61.5$ log $[K^+]_i/[K^+]_o = -90$ mV. This value is approximately that measured for the resting potential (Figure 6-16a, point A). Na^+ equilibrium potential is around +60 mV. From the study of isotope exchanges across the resting myocardium, it has been demonstrated that conductance of K^+ far exceeds that of Na^+ and the resting membrane potential more nearly approximates E_K than E_{Na}. This potential is also very sensitive to variations in $[K^+]_o$ and follows closely the predictions of the Nernst equation. The $[Na^+]_o$ has little influence on the resting potential.

With the onset of depolarization, the conductance of these ions is dramatically changed, that for Na^+ exceeding K^+. The membrane potential

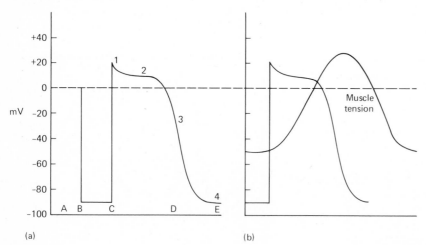

(a)

(b)

Figure 6-16
(a) Action potential from a mammalian ventricular fiber. (b) Relation of the action potential to muscle tension development. (See text for details.)

thus becomes Na^+ dominated and becomes positive. During this phase, $[Na^+]_o$ has a marked effect on the level of the action potential, just as $[K^+]_o$ does during the resting phase. Sodium conductance continues to exceed K^+ conductance through the plateau phase but then steadily declines as the resting potential is reestablished. Another ion species, Ca^{2+}, exhibits an influx during the plateau. This is especially important since Ca^{2+} is involved in the contraction process (see Section 4.5.8). Epinephrine dramatically increases contraction force and also augments the influx of Ca^{2+} during the plateau of the action potential. This may be a partial explanation for the action of epinephrine on the heart.

The firing of pacemaker cells is associated with an action potential profile that differs from that of cardiac muscle cells (Figure 6-17). The decrease in potential following depolarization grades into the subsequent action potential. Following a minimal level of depolarization, there is a steady slow rise in potential *(prepotential)* until a **threshold** is reached, followed by the action potential. The slope of the prepotential as well as the level of depolarization preceding this phase, are determinants of pacemaker rate—and, hence, heart rate. In Figure 6-17, the rate of firing decreases from A to C. Rate is normally determined by nervous influences on the pacemaker arising from the parasympathetic and sympathetic divisions of the autonomic innervation of the heart. Sympathetic activity increases heart rate through release of norepinephrine. This agent changes the slope of the prepotential (B to A), whereas acetylcholine, released by the parasympathetic neurons, reduces firing rate by lowering the resting potential *and* reducing the slope of the prepotential preceeding the attainment of threshold (C).

Both the velocity of conduction of the action potential and the level of the resting membrane potential are reduced with decreasing temperature. Animals facing osmotic stresses or experiencing changes in body temperature may exhibit shifts in resting membrane potential and conduction velocity. Depolarization is propagated at relatively slow velocities in ectotherms at lower body temperatures; frog ventricle strips exhibit conduction velocities of 16 cm/sec at 27°C, whereas at 18°C the velocity is 10 cm/sec.

The body surface electrocardiogram (ECG) takes advantage of the fact that the body fluids acts as an electrical conductor, allowing the recording of the electrical events associated with the cardiac cycle. The form of a typical ECG is shown in Figure 6-18a. The principal waves of the tracing have been correlated with specific electrical events by comparing intracellular microelectrode potentials with surface electrocardiograms. For example, in Figure 6-18b, it can be seen that the QRS complex is clearly associated with the depolarization of a ventricular fiber. The P wave of the ECG is associated with atrial depolarization, and the interval P–R is a measure of the time from atrial activation to the initiation of ventricular activation (in man around 0.15 sec). This period includes the conduction delay at the A-V node and conduction system. The QRS wave correlates with the ventricular depolarization process. During the S–T interval (about 0.4

Figure 6-17
Cardiac pacemaker potentials. Parasympathetic influences of the vagus nerve promote transformation from A toward C (slower rates). Sympathetic influences tend toward A (faster rates).

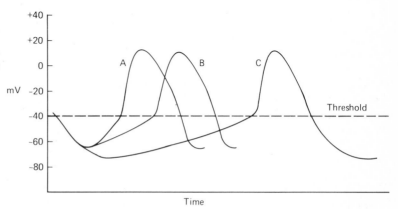

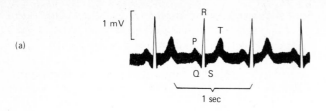

(a)

1 mV

R

P

T

Q S

1 sec

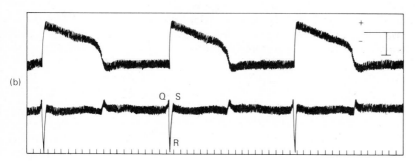

(b)

Q S

R

+

−

Figure 6-18

(a) Electrocardiogram (man). (b) Trans-membrane potential of a frog ventricular fiber in situ. Below is a simultaneous surface electrocardiogram. Note correspondence of QRS with the depolarization event recorded by the microelectrode. Calibration: upper right, 50 mV; horizontal bar, zero line; time marks, 0.1 sec. [From H. H. Hecht, Ann. N.Y. Acad Sci., **65**, 700 (1957).]

sec), the entire ventricle is depolarized, hence its isoelectric characteristic. Repolarization occurs during the T interval. The Q–T segment is often referred to as *electrical systole.* The exact form of the ECG depends on the position and condition of the heart and on the placement of the recording electrodes. Electrode placements are standardized in medical practice and much useful information can be gained about the condition of the heart with expert evaluation of the ECG.

The general form of the ECG is similar among the vertebrates, all forms exhibiting P, QRS, and T components. In those forms (fishes, amphibians, and reptiles) possessing a distinct sinus venosus, a V wave, preceding P, may be recorded. Fishes and amphibians often exhibit a small but distinct B wave, which slightly precedes T. This B wave represents activation of the conus arteriosus.

Although the geometric relationships of the cardiac chambers and flow patterns within the heart are quite different among various vertebrates, they share in common a similar sequencing of electrical and mechanical events, the rates being controlled by pacemakers, which may be modified by neural reflexes.

6.6.3 Cardiac Output and Heart Rate

Cardiac output and rate vary widely among animals and depend on such factors as animal group, pres-

TABLE 6-4. Approximate Values for Heart Rate and Cardiac Output in Various Vertebrates

	Cardiac output (ml/kg body wt per min)	Heart rate (beats/min)
Codfish (18°C)[1]	9–10	30
Amphiuma[2]	30	
Iguana		
(20°C)[3]	40	20
(38°C)	85	90
Turtle (22°C)[4]	60	40
Hummingbird (4 g)[5]		615
Ostrich (80 kg)[6]		60–70
Duck (3 kg)[7]	500	240
White rat[8]	200	350
Dog (sitting)[9]	150	100
Man (resting)	85	70
Man (heavy exercise)	510	180
Elephant (4000 kg)[10]		30

Source: (1) K. Johansen. *Comp. Biochem. Physiol.,* **7,** 169 (1962); (2) K. Johansen. *Acta Physiol. Scand. Suppl.,* **217,** 1–82 (1964); (3) V. A. Tucker. *J. Exp. Biol.,* **44,** 77 (1966); (4) F. N. White and G. Ross. *Am. J. Physiol.,* **211,** 15 (1966); (5) E. P. Odum. *Science,* **101,** 153 (1945); (6) A. J. Clark. *Comparative Physiology of the Heart,* The Macmillan Company, New York, 1927; (7) B. Folkow, N. J. Nilsson, and L. R. Yonce. *Acta Physiol. Scand. Suppl.,* **277,** 51 (1964); (8) G. Ross, F. N. White, A. W. Brown, and A. Kolin. *J. Appl. Physiol.,* **21,** 1273 (1966); (9) D. S. Dittmer and R. M. Grebe. "Handbook of Circulation," *WADC Technical Report 59–593,* 1959, p. 80; (10) K. L. Blaxter. *Vet. J. London,* **99,** 2 (1943).

ence or absence of endothermism, rest and activity, body temperature, and mass (Table 6-4).

We have seen in Section 3.7.2 that the weight specific oxygen consumption of various vertebrate groups is inverse to body mass and follows a regular allometric relationship described by equations of the form $\dot{E}_m = aM^b$. It follows that for smaller species of a class, a given tissue mass must require greater oxygen delivery per unit time to support the higher metabolism. The oxygen capacity of the blood of small and larger species is similar. Thus, the relative cardiac output of small species must be greater than for larger species. This could be accomplished by an inverse relationship between body and heart mass; however, it does not seem to be the answer for mammals. Rather, it has been shown by Stahl that, among mammals, heart mass is described by the equation $M_{\text{heart, kg}} = 0.0059 M_{\text{body,kg}}^{0.98}$.

The slope of 0.98 is sufficiently close to 1.0 to allow us to say that heart mass is directly proportional to body mass (Figure 6-19). The equation predicts that a 60-kg human should have a heart weighing 326 g, whereas the heart of a 5-kg dog will weigh only 28.6. In both cases, the heart mass lies between 0.54 and 0.57% of body mass. Thus, the increased blood flow requirement of smallness must be accomplished by rate adjustments. Stahl's findings, $f_{(\text{min}^{-1})} = 241 M_{\text{kg}}^{-0.25}$, tell us that this is so, the negative slope indicating that a 2500-kg elephant's heart rate is around 34 beats/min, whereas a 10-g rodent is depolarizing and repolarizing the cardiac muscle at a rate of 762 times/min (= 12.7 times/sec). These figures are for nonexercising animals.

Cardiac output increases during exercise, and we are left with the astounding figure of 1200–1500 beats/min in the humming bird hovering at the window feeder. Actually, birds exhibit a dual solution to the mass-blood convection problem. The equation $M_{\text{heart}} = 0.0082 M^{0.91}$ found by Lasiewski and Calder states that larger birds have a relatively smaller heart than their more diminutive relatives. The avian solution is thus a combination of both mass and frequency adjustments. In general, avian hearts are relatively larger than mammalian hearts,

Figure 6-19
Regression relationships of heart rate, heart mass, and mass specific metabolism as functions of body mass.

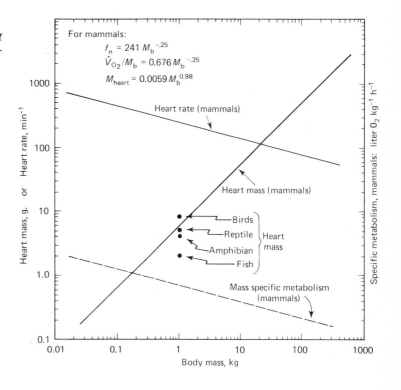

For mammals:
$$f_n = 241 M_b^{-.25}$$
$$\dot{V}_{O_2}/M_b = 0.676 M_b^{-.25}$$
$$M_{\text{heart}} = 0.0059 M_b^{0.98}$$

Heart rate (mammals)

Heart mass (mammals)

Birds
Reptile } Heart mass
Amphibian
Fish

Mass specific metabolism (mammals)

Heart mass, g.   or   Heart rate, min⁻¹

Specific metabolism, mammals: liter O₂ kg⁻¹ h⁻¹

Body mass, kg

and the differences in mass may be striking. For example, the heart of a 500-g bird may be expected to weigh 4.4 g; that of a mammal of comparable mass weighs only 3 g.

For mammals, the relationship between body mass and mass specific metabolism (oxygen consumption/unit body mass) is characterized by a negative slope of 0.25, the same as for heart frequency and body mass. Thus, for mammals, it appears that the percentage of the body mass represented by the heart is constant, and the inverse relationship between body mass and frequency rather perfectly correlates with the increase in mass specific oxygen consumption that accompanies smallness.

It is clear from these relationships that the cardiac output per unit body mass of small mammals is high compared with large members of the group. However, other studies have shown that the blood convection requirement (liters pumped/mmol O_2 consumed) of mammals is of the general order of 0.4–0.5 liter$\cdot$(mmol O_2)$^{-1}$, irrespective of body mass. Given this fact, one can approximate from the allometric relations previously discussed the differences in weight specific and absolute cardiac output and stroke volumes of mammals of differing mass.

From the specific metabolism of 0.1- and 10-kg mammals, we can calculate that 53 mmol O_2 and 16.96 mmol $O_2\cdot kg^{-1}\cdot hr^{-1}$ respectively are consumed (1 mmol O_2 = 22.4 ml O_2, STPD). At a blood convection requirement of 0.5 liter/mmol O_2, the mass specific cardiac outputs are 26.8 and 8.45 liters$\cdot kg^{-1}\cdot hr^{-1}$, a 3.2-fold difference. The absolute cardiac outputs are: 0.1-kg mammal = 45 ml$\cdot min^{-1}$; 10-kg mammal = 1410 ml$\cdot min^{-1}$. From the regression of heart rate and body mass, the expected frequencies are 428 and 135 beats/min; thus, the stroke volume of the 0.1-kg animals is around 0.11 ml/beat, that for the 10-kg animal about 10.4 ml/beat. The difference in stroke volume is about 100-fold and is similar to the difference in body mass. These are, of course, approximations based on regression analyses. Cardiac output requirements vary widely depending on metabolic requirements and reach maximum values during heavy exercise. We now turn to methods of measurement of the output of the heart and the factors determining its variability.

Measurement of the Cardiac Output. Cardiac output, usually expressed as milliliters per minute per unit of body weight or unit of surface area, may be measured by both indirect chemical or by physical methods. One of the most widely used indirect techniques is the Fick method. Assume we have an indicator material that is partially consumed (cleared) by an organ and not produced by the organ. With a knowledge of the amount of the material consumed per unit of time (X/t) and the difference in arterial and venous concentrations of the material ($C_a - C_v$), the blood flow to the organ may be calculated from the relationship

$$\dot{Q} = \frac{X/t}{C_a - C_v}$$

This is the basic Fick equation for blood flow. Oxygen is consumed by the organism; so, if one knows the oxygen consumption, arterial oxygen concentration, and the oxygen concentration of venous blood entering the lungs, the cardiac output may be calculated.

The venous blood sample must be taken in the pulmonary outflow tract by placing a catheter in the right ventricle via an appropriate vein. This is necessary because peripheral venous oxygen concentration varies from one locale to another; thus a well-mixed sample, representative of average oxygen concentration, is taken. Oxygen consumption may be measured in air-breathing forms by observing the volume decrement of a spirometer in which carbon dioxide is removed by chemical absorption.

The Fick equation for cardiac output (CO) becomes

$$\text{CO (ml/min)} = \frac{O_2 \text{ consumption (ml/min)}}{A_{O_2} \text{ (ml/ml)} - V_{O_2} \text{ (ml/ml)}}$$

Example

$$O_2 \text{ consumption} = 200 \text{ ml/min}$$
$$A_{O_2} = 0.2 \text{ ml/ml}$$
$$V_{O_2} = 0.15 \text{ ml/ml}$$
$$\text{CO (ml/min)} = \frac{200}{0.2 - 0.15}$$
$$= 4000 \text{ ml/min}$$

The Fick method is also used to measure blood flow in specific organs.

An indirect method of determining the cardiac output is the *indicator dilution technique.* The indicator (dye or isotope) is injected as a "slug" into the venous system as close as possible to the heart (preferably into the right heart). The indicator material is then measured in an artery immediately following its passage through the heart. A curve representing the appearance and decay of the material is constructed; that is, concentration is plotted against time. To calculate flow one needs a knowledge of the amount of indicator injected, the time for passage past the measurement point, and the mean indicator concentration. The equation for cardiac output is $CO = I/\Sigma C \, \Delta t$, where I is the amount of indicator injected and $\Sigma C \, \Delta t$ is the concentration time factor (it is equivalent to the $C_a - C_v$ oxygen difference in the Fick method). The value for $\Sigma C \, \Delta t$ is obtained by adding the concentrations observed during the first, second, third, . . . seconds over 1 min and dividing by 60 to obtain the minute concentration. The method is complicated by recirculation of indicator material past the point of measurement during the decay portion of the curve. It is therefore necessary to make an appropriate extrapolation of the decay curve to zero concentration at the point where recirculation occurs.

Cardiac output or organ blood flow may also be measured by placing an appropriate transducer on the root of the aorta or the artery to an organ. Two types of flowmeters are currently in use: electromagnetic and ultrasonic. The probes of such instruments may be surgically implanted so that blood flow may be measured in nonanaesthetized animals. Both instruments yield instantaneous flow patterns, or by appropriate integration circuits the mean flow may be obtained.

The electromagnetic flowmeter utilizes the Faraday effect and is based on the fact that, as a conductor flows through a magnetic field, an electric potential is developed proportional to the length and velocity of the conductor. The probe heads contain a small electromagnet designed around a partially open lumen into which the blood vessel may be fitted. Recording electrodes, housed in the probe head, contact the vessel wall and pick up the potential created as the blood stream (the conductor)

moves through the field of the magnet. By appropriate amplification, the instantaneous flow profile may be recorded.

Ultrasonic flowmeters are designed around the observation that the velocity of sound in a flowing stream and emanating from a single point is higher traveling downstream than upstream (Doppler effect). The velocity of the stream is one half the difference between the up- and downstream velocities. A pulse of high-frequency sound is directed upstream, and its transit time is compared with a downstream pulse. A major advantage of this method is that the signals may be telemetered to a recorder from considerable distance. This has enabled physiologists to study blood flow in a number of free-ranging animals, including baboons, horses, seals, and crocodiles.

6.6.4 Mechanics of the Cardiac Cycle

Although there are important differences in the functional morphology of the hearts of various vertebrate groups, the contraction sequences of the principal chambers (atria and ventricles) and the mechanics of filling and emptying are similar. Because the cardiodynamics of mammals have been more extensively studied we will concentrate on the mechanical aspects of this group. The relationship of the chambers is essentially the same in both mammals and birds.

The presence of the diaphragm and thoracic cavity has an important influence on venous return to the heart. Inthrathoraic pressue is negative, so the veins of the thorax tend to distend, because their intraluminal pressure is positive. The more negative the intrathoraic pressure, the greater the flow into the thoracic veins from the higher pressure extrathoracic veins. Thus, in inspiration, with intrathoracic pressures of about -6 mm Hg, venous flow to the heart is augmented over that of expiration, during which intrathoracic pressure is about -4 mm Hg. Fluctuations in mean arterial pressure at the same frequency as respiration are due to this phasic influence of intrathoracic pressure and the changes in inflow to the heart that affect cardiac output. As inspiration proceeds, with the lowering of the diaphragm, the abdominal pressure may rise. This is especially evident in heavy breathing. Compression of the intraabdominal veins elevates venous pressure, another factor favoring inflow to the lower pressure intrathoracic veins. Because both intrathoracic and abdominal pressures are involved, the respiratory effect on venous return is referred to as the ***thoracoabdominal pump***. As depth and frequency of respiration increase, as in exercise, the contribution of the thoracoabdominal pump may be quite great.

Some 70% of the blood volume is found in the veins, so any change in capacity of the venous system will influence venous return to the heart. Sympathetic innervation to the veins is abundant and under certain conditions these sympathetic nerves, releasing norepinephrine, induce venoconstriction. The net effect is an increase in venous pressure favoring a higher blood flow to the low-pressure right atrium and thus an increase in cardiac output.

An additional factor is involved in filling the heart. After systole, a residual volume of blood is left in the ventricular chamber. The magnitude of the residual volume limits the amount of additional blood that can be added through venous return. The residual volume depends on myocardial shortening during systole; the more forceful the contraction, the greater the degree of emptying.

The sequence of events of filling and emptying the left ventricle is shown in Figure 6-20. The general relations are also valid for the right ventricle. However, the maximum pressures generated by the right ventricle are much lower than for the left, because the pulmonary vascular resistance is low compared to systemic resistance. With the cessation of systole, the ventricular muscle relaxes very rapidly, owing to the release of elastic forces produced during contraction. During this early phase of diastole, the intraventricular pressure exceeds atrial pressure, so no filling occurs. For this reason, this phase is called ***isometric relaxation.*** As the elastic recoil progresses, pressure rapidly declines, both in the aorta and in the ventricular cavity. Backflow in the aorta carries the semilunar valves together as the second heart sound occurs. The fall in intraventricular pressure proceeds and becomes equal to atrial pressure within a short time. As soon as ventricular pressure falls below atrial pressure, filling begins as the mitral valves open. At this point a large fraction of left-atrial blood enters the ventricle (rapid-inflow phase). During this phase 50% or more of the inflow occurs, as a result of the higher left-atrial pressure and continued release

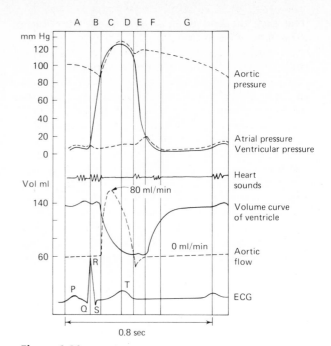

Figure 6-20
Major events of the cardiac cycle (man). Phases of cardiac cycle: A, atrial contraction (diastasis); B, isovolumetric ventricular contraction; C, maximum ejection; D, reduced ejection; E, isovolumetric relaxation of ventricle; F, rapid filling; G, diastasis, merging into A. Aortic flow measured at root of aorta.

of elastic forces in the ventricular wall. Thereafter, filling proceeds more slowly (period of diastasis), owing to the decline in atrial pressure and rebound forces of relaxation. The slower the heart rate, the longer the slow-filling phase, and a considerable volume may be added in this period. Note that atrial contraction, represented by a small positive terminal pressure and volume deflection, adds little to filling. This is in contrast to lower vertebrates, in which atrial contraction makes a greater contribution.

The total ventricular-filling volume *(V)* may be described by the expression

$$V = \int_a^b P_{tm}D \; dt$$

where *a* is the opening of the atrioventricular valves, *b* the end of diastole, P_{tm} the transmural pressure (intraventricular P— intrathoracic P), and D the distensibility characteristic, a dynamic property of the myocardium that changes through time. The determinants of transmural pressure are (1) atrial and venous pressure, (2) intrathoracic pressure (more negative in inspiration, thus elevating the transmural pressure), and (3) the pericardial fluid and membrane. With a large filling volume, the percardium offers a restraint to further filling because it is relatively nondistensible. The limitation on filling by the pericardium protects against pulmonary vascular congestion by limiting right-ventricular filling. This may be important, especially at slow heart rates, where filling time is so prolonged.

The time of filling depends closely on heart rate, because as heart rate increases there is a disproportionally larger reduction in diastolic time than for systolic. Thus, at very high rates the available filling time may become a limiting factor in the determination of stroke volume.

Filling ceases rapidly as the myocardium develops tension in early systole because, as soon as the intraventricular pressure slightly exceeds atrial pressure, the atrioventricular valves close. Intraventricular pressure rises rapidly in early systole, but until the pressure exceeds that in the aorta no blood can be ejected across the aortic orifice, still closed by the semilunar valves. The volume cannot change in this phase, so the period is referred to as the *isovolumetric* (or isometric) *phase of systole*. As soon as the pressure generated within the ventricle slightly exceeds that of the aorta, the semilunar valves are forced and the rapid ejection phase of systole commences. After reaching peak pressure, the ventricular pressure curve, almost coincident with that in the aorta, falls until it reaches a level below aortic. During this phase of reduced ejection, the myocardium diminishes its rate of shortening. Aortic outflow, at this time, exceeds uptake from the heart, which results in the decline of arterial pressure with ventricular pressure until valve closure. Aortic backflow occurs at this time, and because of pressure difference across the aortic valves, closure of the orifice of the aorta occurs. The small inflection (dicrotic notch, or incisura) on the aortic pressure trace is due to a reflected pressure wave off the semilunar valves.

Another way of looking at the ventricular cycle is shown in Figure 6-21. In this case left-

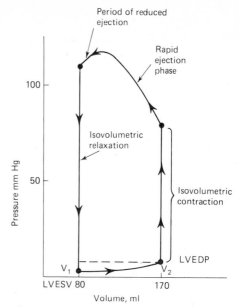

Figure 6-21
Pressure/volume diagram of the left ventricle of man.

intraventricular pressure is plotted against left-ventricular volume. V_1 is the residual volume (left-ventricular end systolic volume, LVESV). As diastole proceeds, filling occurs until the left-ventricular end diastolic volume (LVEDV) is reached. During isovolumetric (isometric) contraction, no volume change occurs until the aortic valves open. Volume rapidly declines as pressure rises to peak, followed by the period of reduced ejection toward the end of systole. The LVESV has been reached at this point and the cycle commences again. Note that work is performed by the ventricle during systole. During diastole, work is being done, chiefly by extraventricular forces. The dimensions of this diagram are pressure (1 mm Hg = 1.35 g/cm²) and volume (cm³). (Thus g/cm² × cm³ = g cm.) We have constructed a work diagram. The work of the ventricle can be expressed by pressure-volume curves and the appropriate representation of the major work of the left ventricle takes the form

$$W = \int_{V_1}^{V_2} P\, dV$$

In Figure 6-21, the work performed by the ventricle is represented by the area above the dashed line and enclosed by the pressure/volume curve. Although the right heart puts out a quantity of blood equal to the left, it works against a lower arterial pressure (about 22 mm Hg systolic and 8 mm Hg diastolic). Thus the work of right-ventricular ejection is less than that of the left side of the heart. Comparisons of the two sides yield work outputs for the left ventricle of about 7 times that of the right. This is correlated with the larger muscle mass of the left ventricle.

It is obvious that, to propel a greater volume of blood through the aorta against a higher pressure (as in an exercising animal), more work must be done. This increase in work is accomplished by increasing rate and force of ventricular contraction, yielding work diagrams similar to that in Figure 6-22. This shift to the left, accompanying increased rate, is directed by two important influences of the sympathetic nervous system. Increased cardiac sympathetic activity, through the liberation from nerve terminals of norepinephrine, increases rate as well as force of contraction and rate of relaxation. Owing to the increase in force of contraction, the heart "digs" into the reserve volume (LVESV) and, although filling time is shortened, the stroke volume may be maintained. The maintenance of

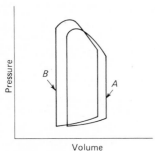

Figure 6-22
Pressure/volume relationships of the left ventricle before, A, and during, B, sympathetic stimulation of the heart. Stroke volume is changed little in B; however, cardiac output is increased due to the increase in rate of contraction.

stroke volume with increased rate yields a higher cardiac output and work per minute increases. The increased cardiac output must be maintained by increasing venous return. Elevated respiratory activity (thoracoabdominal pump) and sympathetic venoconstriction are two important factors augmenting the greater venous return.

Some feeling for the potential influence of the sympathetic nerves on the contractility of the heart can be gained from the experiment depicted in Figure 6-23. By clamping the aorta, the myocardium is forced to contract essentially in an isometric fashion. Electrical stimulation of the sympathetic nerve suppy to the heart under these conditions demonstrates their effect on the contractile force of the myocardium in the absence of peripheral vascular effects. Increased sympathetic activity is an important factor in allowing the heart to meet the increased work load of exercise.

Ernest Henry Starling in 1914 announced **Starling's law of the heart,** stating that the mechanical energy released on passing from the relaxed to the contracted state depends on the initial length of the muscle fibers. The relationship is certainly true for the isolated denervated mammalian heart. Starling showed in the isolated heart that, as venous pressure is elevated and ventricular end diastolic volume rises, the amount of blood ejected in the subsequent systole is increased; that is, stroke volume increases. This led to the general belief that, during exercise, for example, the increased cardiac output was due to increased filling and stretching of myocardial fibers, followed by a more forceful contraction and greater stroke output. Studies, performed principally on intact dogs, revealed that the heart behaves more like the case presented in Figure 6-22, and that the heart may actually become slightly smaller in exercise.

We are not ready to abandon Starling's law, however, even though it does not appear to explain the increased cardiac output of exercise. It does not seem likely that left- and right-ventricular stroke volume can be maintained at exactly equal levels on a stroke-to-stroke basis, although the volume passing through the two ventricles must balance in the longer run. The pulmonary vascular resistance decreases during inspiration; thus the outflow from the right heart increases for a few beats. If we postulate that this added pulmonary blood flow amounted to 50 ml/min and holding

Figure 6-23

Effect of arotic occlusion on left ventricular pressure. When the cardiac sympathetic nerves are stimulated, the systolic pressure is approximately doubled owing to the inotropic effect of these nerves. Stippled areas are bordered by systolic and diastolic pressures. Pressure in the aorta was recorded distal to the occlusion. [Based on R. F. Rushmer. Cardiovascular Dynamics, W. B. Saunders Co., Philadelphia, 1970.]

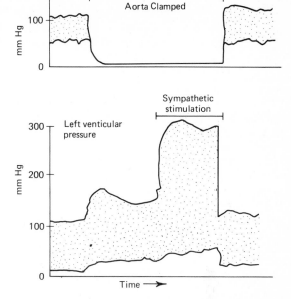

left-ventricular output constant, within 10 min, 500 ml of blood would accumulate within the lungs. This does not normally happen, however, and it is probable that the imbalances that occur between left and right stroke volumes are rapidly corrected by the mechanism of Starling's law. In the preceding instance correction would occur because of the increased pulmonary venous inflow to the left ventricle, the fibers being longer in diastole and followed by a slightly more forceful contraction at systole. In this manner the "excess" right-ventricular output may be corrected moments later by an increase in left-heart output.

6.6.5 Functional Anatomy—Hearts of Lower Vertebrates

Cyclostomes. It has been known for many years that the embryonic vertebrate heart commences to beat before nerves reach the heart. This absence of direct neural control of the heart is retained

in the adult only among the myxinoids. The myxinoid heart is without a nervous innervation and the sinus venosus is the pacemaker.

It is clear from the few studies on cyclostome hearts that the pressures generated are small compared to other vertebrates. In the California hagfish, *Eptatretus stouti*, ventricular diastolic pressure is about 0 mm Hg and systolic pressure is about 7 mm Hg. When the sinus and atrium are overdistended by injection of seawater into the posterior cardial vein, a definite atrial contribution to ventricular filling is observed because ventricular pressure rises with atrial contraction (Figure 6-24). These pressure studies also indicate that the heart can adjust its output to a higher level with an increase

Figure 6-24

Hagfish. Pressure drop from branchial ventricle to ventral aorta. Ventricular calibration is at left, aortic at right. At A, 0.30 ml of seawater was injected into posterior cardinal vein. Note the effect on atrial pressure (small rise at the beginning of the ventricular upstroke) as well as on ventricular response. [From C. B. Chapman, D. Jensen, and K. Wildenthal, Circ. Res., 12, 427–440 (1963).]

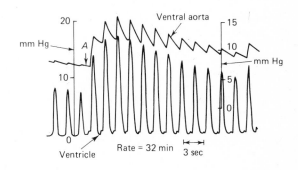

in venous pressure and return of blood to the heart. Regulation of the stroke volume may follow Starling's law of the heart [the force of contraction generated by heart muscle is a function of fiber length, the force increasing with resting fiber length up to a critical point, after which force declines with further increase in length (see also Section 6.6.4]. Increased filling in the hagfish heart may, by distending the myocardium (increasing resting fiber length), result in the ejection of a greater stroke volume.

The branchial heart is not the sole pump energizing the cardiovascular system in cyclostomes. Dorsal aorta pressure is greatly augmented by contraction of the branchial musculature, a situation analogous to the respiratory or thoracoabdominal pump of mammals (Section 6.6.4).

Additional auxiliary pumps provide a pressure booster for blood circulating through the sinuses to the veins. There is thus some decentralization of pumping from the branchial or main heart to auxiliary pumping stations on the low-pressure side of the circuit (Figure 6-3).

The accessory hearts of cyclostomes are unique among animal pumps. The caudal heart of myxinoids (Figure 6-25) is supported along the central axis by a cartilage rod that separates the heart into two valved chambers. Alternate contractions of skeletal muscle, attached to the two ends of the rod, flex the rod. Contraction of one set of muscles flexes the rod in such a manner that the rod compresses the chamber opposite the contracted muscle, forcing blood out of the compressed chamber. During this phase the contralateral chamber fills as a result of expansion.

It has been reported that the frequency of contractions of the caudal heart of *Eptatretus stouti* increases when the tail is lower than the head. Increased filling of the heart due to gravitational effects may represent a sensory input for reflex acceleration. Green has located a "caudal heart center" in the spinal cord and suggests that contraction of the heart is due to rhythmic discharges of motor impulses from the spinal cord.

The caudal hearts have no muscles of their own, compression apparently being caused by contractions of the skeletal muscle adjacent to the heart.

The portal heart, found only in myxinoids, is aneural and inherently rhythmic. Blood is propelled by means of this pump through the common portal vein to both lobes of the liver. Increases in frequency have been observed when inflow to the heart was increased, and systolic pressures of 3–6 mm Hg have been measured in the cavity of the heart. Pacemaking is under the control of a sinuslike area in the veins entering the heart. There is no synchrony between portal and branchial hearts, the rates being independent and apparently related to the rate of venous inflow or to the pressure in the afferent chambers.

Although both portal and branchial hearts are aneural, masses of chromaffin cells containing epinephrine and norepinephrine (adrenal medullary hormones) have been observed in the myocardium. There is no adrenal gland in the hagfish and the significance of catecholamine-containing cells in the noninnervated hearts of these animals is not clear. Catecholamines increase the rate (positive chronotropic effect) and force (positive inotropic effect) of vertebrate myocardium, so it seems possible that these agents may be released under conditions of increased venous return.

Fishes. The hearts of fishes are typically four-chambered, each chamber being valved on its outflow side to prevent retrograde flow. The course

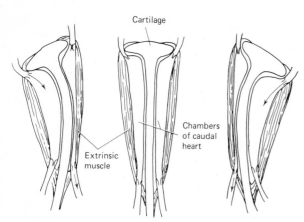

Figure 6-25
Caudal heart of the Pacific hagfish. The heart consists of a central cartilage rod, two chambers, and extrinsic muscles, which flex the rod and provide the power for filling and emptying the chambers. Unidirectional valves are found at the inflow and outflow points of the chambers. [*Based on the description of C. W. Green. Am. J. Physiol.,* **3**, *366–382 (1900).*]

of blood is from sinus venosus to atrium, then to the ventricle, and finally to the bulbus (teleosts) or conus arteriosus (elasmobranchs). In elasmobranchs, several pairs of valves are present in the conus. The coronary circulation is derived from the efferent branchial arteries or the dorsal aorta; hence the heart is provided with an arterialized nutrient flow.

Studies in which ligatures have been tied between chambers, or in which specific areas of the heart are cooled, reveal that the sinus venosus, or closely associated tissue, is the pacemaker zone. Pacemaker zones with intrinsic rates lower than the sinus have been located in the atrioventricular junction and the bulbus.

The hearts of both elasmobranchs and teleosts are innervated by the vagus. Numerous studies have demonstrated that sectioning of the vagi leads to cardioacceleration and that stimulation of the cardiac end of a cut vagus produces cardioinhibition, an effect caused by liberation of acetylcholine from the nerve terminals of the heart. Several workers have maintained that there is no sympathetic innervation to the fish heart and that rate is controlled by the parasympathetic vagal nerve alone, speeding being caused by release of vagal tone. The possible role of sympathetic augmentor nerves to the heart is controversial.

Teleost fish hearts generate greater pressure than cyclostome hearts. Pressures in the ventricle of unanesthetized codfish were from near 0 cm H_2O at diastole to above 60 cm H_2O at peak systolic pressure (Figure 6-26a). The resting cardiac output was 9–10 ml kg^{-1} min^{-1} as reported for other teleosts. Difficulties in comparison among species arise because the magnitude of the cardiac output is influenced by temperature and metabolic intensity among other variables. The most useful comparative value is the *blood convection requirement,* the number of liters pumped by the heart per millimole of O_2 consumed (liter/mmol O_2). From the limited data available, both sharks and teleosts fall in the general range of 0.75–0.85 liter/mmol O_2 (compare with 0.35–0.5 liter/mmol O_2 for mammals). The O_2 capacity of blood shows a strong correlation with blood convection requirement, the fishes having, in general, lower capacity for O_2 because of comparatively low hemoglobin concentrations. This correlation finds its extreme expression in the antarctic, where the hemoglobin-free ice-

fishes exhibit a blood convection requirement of around 8.5 liters/mmol O_2 consumed—the highest value of any vertebrate.

Simultaneous pressure recordings in the ventricle and bulbus of codfish, along with stroke volume (Figure 6-26b and c), reveal that the rapidly climbing ventricular pressure is converted to an extended, smoother pressure in the bulbus, the peak pressure of the bulbus being considerably smaller than that of the ventricle. Bulbus pressure is extended far into diastole as is positive aortic flow. The bulbus, being an elastic chamber, smooths the pressure pulse, preventing excessive pressure variations in the ventral aorta and protecting the gill capillaries. It is clear that the pressure-reservoir character of the bulbus affects the maintenance of blood flow far into ventricular diastole, a relationship that may be of considerable importance for the maintenance of gill flow, especially at low heart rates.

The piscine heart responds to an increased venous return, induced by infusion of blood into the intestinal veins, by increasing stroke volume (Figure 6-26c). It is interesting that, in spite of the increased stroke volume, bulbus pressure falls, indicating a decrease in gill-vessel resistance (dilation). The mechanism of this gill vascular response is not known. Studies of the blood pressure of swimming teleosts reveal that cardioacceleration is a minor component of the increase of cardiac output. Increased venous return, augmented by skeletal muscular pumping of the veins, may occur in exercise; thus, elevated cardiac output would be achieved by increasing stroke volume. This view corresponds with the idea that innervation of the teleost heart is chiefly parasympathetic and that sympathetic acceleration, as seen in mammals, may not occur in fishes. Increased stroke volume, as suggested for myxinoids, may be achieved by the Starling mechanism.

The heart weights of fishes are smaller in relation to body weight than for the frog (4% of body weight). There is an apparent correlation of heart size and activity, the heart weight of *Ophichthys imberbis,* an inactive species, being 0.15% of body weight compared to 2.5% for flying fishes.

Temperature variations, both in isolated hearts and in intact fishes, influence heart rate. In the isolated heart, pulse rate varies directly with temperature, increasing up to a critical temperature

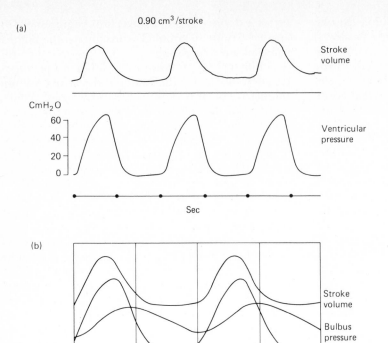

(a)

0.90 cm³/stroke

Stroke volume

CmH₂O

60
40
20
0

Ventricular pressure

Sec

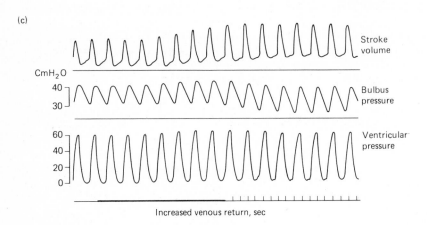

(b)

Stroke volume

Bulbus pressure

Ventricular pressure

Sec

(c)

Stroke volume

CmH₂O

40
30

Bulbus pressure

60
40
20
0

Ventricular pressure

Increased venous return, sec

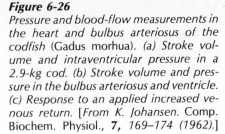

Figure 6-26

Pressure and blood-flow measurements in the heart and bulbus arteriosus of the codfish (Gadus morhua). *(a) Stroke volume and intraventricular pressure in a 2.9-kg cod. (b) Stroke volume and pressure in the bulbus arteriosus and ventricle. (c) Response to an applied increased venous return.* [*From K. Johansen.* Comp. Biochem. Physiol., **7,** *169–174 (1962).*]

characteristic for the species, after which the rate declines with further temperature elevation. Larger fish have slower heart rates than small species, but a good correlation between temperature of the water and heart rate has been observed. Because temperature increases oxygen consumption, one would expect a rise in cardiac output (gill blood flow) if

increased oxygen transport across the gills is to occur. Elevations in heart rate may be an effective means of increasing cardiac output in response to high temperature, but there are no published studies on this question.

As pointed out previously (see Section 6-2), the development of a pulmonary circulation by the Dipnoi was accompanied by subdividing the atrium by a complete septum. The return of blood from the lungs to the left atrium is characteristic of the majority of higher vertebrates. A major problem facing lungfish, amphibians, and most reptiles is distribution of oxygenated blood to the tissues. It has been assumed by most comparative anatomists that a rather thorough mixing of systemic venous (unoxygenated) and pulmonary (oxygenated) blood occurs in the ventricle of these forms. That this is not the case for the African lungfish has been demonstrated by studies in which radiographic, direct blood-flow measurement, and oxygen-distribution techniques have been utilized. A large fraction of the pulmonary venous return may be preferentially shunted past gill exchange areas and to the systemic arterial system, especially with the onset of pulmonary ventilation (see Figure 6-27).

Amphibians. The amphibians present an impressive array of forms, exhibiting variable dependence on lungs and cutaneous respiration in transition from aquatic to terrestrial habitats. Some aquatic salamanders are lungless, and the atrial septum may be absent. The degree to which a double circulation is achieved is a cotnroversial subject, and claims ranging from complete separation of the venous and pulmonary return to complete mixing within the ventricle have been made. Most attention has been given frogs, and excellent studies on the salamander *Amphiuma* have appeared.

The sinus venosus is a conspicuous structure in the frog heart. The atria are completely divided, the systemic venous return traversing the sinus venosus to the right atrium. The left atrium receives the pulmonary veins (Figure 6-28). Atrioventricular valves prevent retrograde flow from the ventricle, which is spongy and trabeculate. A peculiar spiral "valve" is attached along the dorsal edge of the bulbus. From the truncus arteriosus the major systemic and pulmonary arteries arise.

One requisite for providing a selective distribution of blood is met in most amphibians—separate return of pulmonary and systemic venous blood to the left and right atria, respectively. If a complete double circulation is to exist, however, the ventricular and arterial outflow tracts of the heart must maintain an equal flow of systemic venous blood to the pulmonary artery and pulmonary venous blood to the systemic circuit. To what degree these criteria are met in the frog is not clear, because studies indicating both complete mixing and complete separation exist in the literature. The major evidence, derived from watching dyes traverse the heart, suggests that a selective distribution exists but that separation is not absolutely complete. How any separation of right and left artial blood occurs is a complex physical problem that has thus far defied analysis, but the spiral valve may play an important role. The pulmonary artery resistance is somewhat lower than the systemic resistance, and it has been claimed that the blood returning from the right atrium occupies the trabecular spaces of the right portion of the ventricle. Ejection of this blood early in systole is thought to favor distribution around the spiral valve and into the lower resistance pulmonary circuit.

Johansen's study of the cardiac dynamics of *Amphiuma* has revealed that at least this amphibian

Figure 6-27
Simplified diagram of the perfusion pattern through the heart and major outflow channels in the lungfish. Protopterus aethiopicus. Immediately following a breath, as much as 95% of the pulmonary venous blood passes into the anterior branchial arteries, most of which are gill-less. [Based on K. Johansen, C. Lenfant, and D. Hanson. Z. vergl. Physiol., 59, 157–186 (1968).]

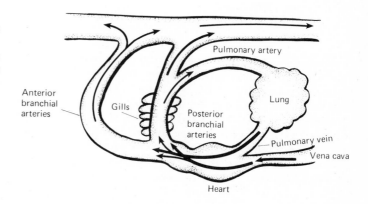

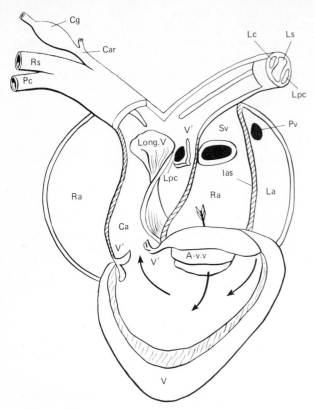

Figure 6-28
Ventral view of internal structure of a frog heart. Code:
A-v.v, atrioventricular orifice, guarded by two valves; Ca,
conus arteriosus; Car, carotid artery; Cg, carotid gland;
Ias, interatrial septum; La, left atrium; Lc, left carotid; Lpc,
left pulmocutaneous artery; Ls, left subclavian; Long.v,
longitudinal valve (spiral); Pc, right pulmocutaneous artery;
Rs, right subclavian; Sv, sinus venosus; V', valves at exit
to conus and to left pulmocutaneous artery; V, ventricle.
[*From J. S. Robb.* Comparative Basic Cardiology, *Grune*
& Stratton, Inc., New York, 1965.]

is capable of maintaining a fairly efficient double circulation. By injecting radiopaque medium in the posterior vena cava and following the course of the contrast material with high-speed cineradiography, he demonstrated that right-atrial blood is distributed to the trabecular spaces of the right ventricle. On ventricular systole, the systemic venous return is selectively distributed to the pulmonary arteries (Figure 6-29a), whereas contrast medium

injected into the pulmonary vein finds it way into the aortic arches (Figure 6-29b). Further evidence for separation came from oxygen analysis of blood in the pulmonary arch, aortic arch, sinus venosus, and pulmonary vein. Many animals showed an almost perfect matching of pulmonary vein and aortic arch oxygen content, indicating good selective distribution. However, the selective passage could be easily disturbed, especially if hemorrhage occurred. These data strongly support the idea that the amphibian heart, represented by *Amphiuma tridactylum*, can accommodate two separate blood streams with only slight mixing.

The cardiac output of *Amphiuma*, measured with electromagnetic flow meters, is of the order of 30 ml kg^{-1} min^{-1}. The heart adapts to artificially increased venous return by increasing stroke output. Tachycardia occasionally occurs in response to increased venous return, but overall cardiac output does not rise with increased rate, in contrast to mammals. The atria actively contribute to ventricular filling and the capacities of sinus venosus and atria are large and represent a significant cardiac reserve. In general, increased volume in any chamber is followed by more forceful contraction. The time relations of filling and emptying of the various chambers are shown in Figure 6-30.

The bulbus arteriosus actively contracts in *Amphiuma*, adding a last energizing boost to the blood. During early ventricular systole the bulbus is distended by the outflowing blood. Late in systole the bulbus contracts and displays a reduction in diameter. As in fishes, the bulbus smooths the pressure pulse and prolongs flow during its contraction phase.

Both vagal parasympathetic and sympathetic nerves supply the heart of the frog, and vagal slowing as well as sympathetic augmentation of rate have been observed in stimulation experiments. In *Amphiuma*, norepinephrine augmented stroke flow, indicating that it exerts a positive inotropic effect.

Johansen has summarized the major factors influencing cardiac output in *Amphiuma* (Figure 6-31). These influences are probably applicable to most amphibian hearts.

Although a tendency to develop a double circulation is evident among amphibians, because of the extreme specializations present in the class, one cannot describe a typical case. Futhermore, there

Figure 6-29

Distribution of radiopaque contrast medium in the heart of the amphibian, Amphiuma. (a) Contrast injection in the posterior vena cava. Note the clear demarcation between the right and left portion of the ventricle. The selective distribution is largely maintained in the bulbus cordis with a preference for filling of the pulmonary arteries. Animal in dorsal position. Rate of exposures: 2 frames/sec at 50 kV. Exposure time: 0.03 sec. Frames 1, 2, 3, 4, and 5 are consecutive. (b) Contrast injection in the pulmonary vein. Frame 1 shows the left atrium with venoatrial and atrioventricular junctions. Ventricular filling is restricted to the upper left portion of the ventricle (frames 1 and 2). Frames 3, 4, and 5 demonstrate a clear preference for filling of the systemic arteries. Animal in dorsal position. Rate of exposures, 2 frames/sec at 50 kV. Exposure time, 0.03 sec. 1, 2, 3, 4, 5 are consecutive frames. [From K. Johansen. Acta Physiol. Scand. Suppl., **217**, 1–82 (1964).]

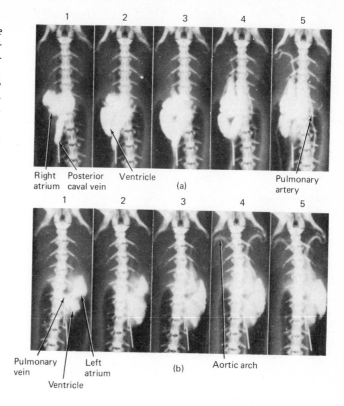

Figure 6-30

Demonstration of the interrelationship between the various cardiac compartments measured as projected areas on the roentgen films throughout a cardiac cycle of the amphibian Amphiuma. At the bottom of the figure is indicated the time sequence of the filling and contraction of the various compartments. The films on which the figure is based were exposed at a rate of 4 frames/sec at 52 kV. Exposure time, 0.04 sec. [From K. Johansen. Acta Physiol. Scand. Suppl., **217**, 1–82 (1964).]

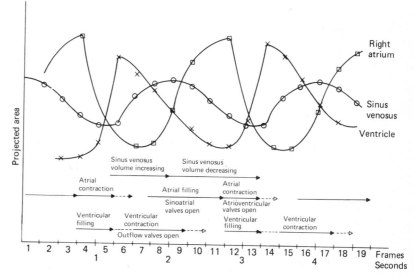

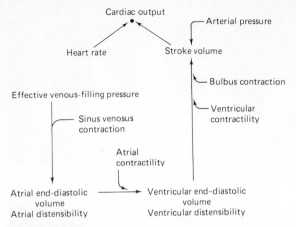

Figure 6-31
Simplified scheme of the various factors influencing the cardiac output in Amphiuma tridactylum. [*From K. Johansen. Acta Physiol. Scand. Suppl.,* **217,** *1–82 (1964).*]

is a variable dependence on cutaneous respiration, and the cutaneous venous drainage is to the right atrium along with the rest of the systemic venous blood.

Reptiles. From a structural point of view, two general architectural patterns appear in extant reptiles. The crocodilians possess a complete ventricular septum, whereas turtles, lizards, and snakes have a complexly chambered but confluent ventricle. All reptiles possess a distinct sinus venosus and a complete interatrial septum.

The ventricle of noncrocodilians is divided into three distinct subchambers, all in anatomic continuity with one another (Figure 6-32). A canal, the interventricular canal, connects cavum arteriosum with cavum venosum. In the absence of this canal the arterialized blood from the left atrium would enter a dead end. A conspicuous muscular ridge projects ventrad to partially subdivide the cavum venosum from the cavum pulmonare, from which the pulmonary artery originates. The free edge of the muscular ridge closely approximates the ventral ventricular wall. The atrioventricular valves are single flaps, somewhat swollen on their free ends. Both left and right atrioventricular valves originate from the connective tissue of the anterior portion of the interventricular canal. These valves may be easily depressed medially to occlude partially the interventricular canal. Their lateral excursions bring them into the atrioventricular orifices. Successive sections through the heart (Figure 6-33) serve to illustrate the anatomic continuity of the subchambers as well as the proximity of the muscular ridge to the outer ventricular myocardium. The bulbus arteriosus has been replaced by distinct arteries with separate origins, although these are bound together by connective tissue.

In crocodilians a complete ventricular septum divides the ventricle into distinct right and left chambers. There is a peculiarity in the origin of the great vessels. The right aortic arch originates from the left ventricle, and the left aortic arch takes its origin from the right ventricle, medial to the pulmonary arterial orifice. Furthermore, both left and right aortae are in communication near their

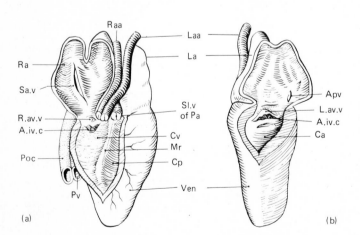

(a)   (b)

Figure 6-32
Snake heart, (a) Ventral view. (b) Left lateral view. Code: A.iv.c, aperature of interventricular canal; Apv, aperture pulmonary vein; Ca, cavum arteriosum; Cp, cavum pulmonare; Cv, cavum venosum; La, left atrium; L.av.v, left atrioventricular valve; Laa, left aortic arch; Mr, muscular ridge; Pa, pulmonary artery; Poc, post-cava; Pv, pulmonary vein; Ra, right atrium; R.av.v, right atrioventricular valave; Raa, right aortic arch; Sa.v, sinoatrial valves; Sl.v, semilunar valves; Ven, ventricle. [From F. N. White. Anat. Rec. **135,** *129–134 (1959).*]

Figure 6-33

Camera lucida drawings of successive sections through the heart of the snake Diadophis punctatus. Code: Ca, cavum arteriosum; Cp, cavum pulmonare; Cv, cavum venosum; La, left atrium; Laa, left aortic arch; Mr, muscular ridge; Pa, pulmonary artery; Ra, right atrium; Raa, right aortic arch; Sa.v, sinoatrial valve, Sv, sinus venosus. [From F. N. White. Anat. Rec., **135,** 129–134 (1959).]

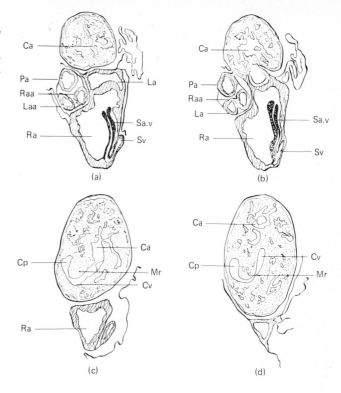

bases via an aperture, the **foramen Panizzae** (see Figure 6-3e).

Classical anatomical descriptions led to the physiological deduction that reptilian hearts were "inefficient" inasmuch as double circulation was impossible because of continuity of the interventricular chambers (noncrocodilians) or origin of the left arch from the right ventricle (crocodilians). More recent studies have shown, however, that representatives of all major reptilian groups exhibit selective distribution of the venous and pulmonary return to the heart.

Among snakes and lizards, selective distribution has been demonstrated by oxygen analysis of blood from critical locations. In the snake, *Coluber constrictor,* supported by artificial ventilation, the oxygen content of the right and left aortic arches and left atrium averaged 7.18, 7.22, and 6.9 ml/100 ml blood respectively. Contents for pulmonary artery and right atrium were 1.72 and 1.45 ml/100 ml blood. These figures indicate a high degree of separation of the pulmonary and systemic flows as they traverse the heart. Surgical damage to the muscular ridge interfered with the normal distributional pattern.

Turtles exhibit no significant difference in oxygen content of the two aortic arches, and the left atrial blood is similar in oxygen content to that of the aortic arches. There is, however, a recirculation of a portion of the pulmonary venous blood to the pulmonary artery (left-to-right shunt), which is especially pronounced during the ventilation phase of the respiratory cycle. This has been revealed by oxygen analysis, direct measurement utilizing electromagnetic flowmeters, and indicator-dilution analysis. When dye was introduced in the right atrium it rapidly appeared in the pulmonary artery (Figure 6-34). The appearance of dye in the systemic arteries occurred later, after dye had traversed the lungs. When injected into the left atrium, dye appeared almost simultaneously in pulmonary and systemic arteries, demonstrating left-to-right shunt.

Although the left aorta of crocodilians has its anatomic origin in the right ventricle, a double circulation is present, as revealed by oxygen analysis

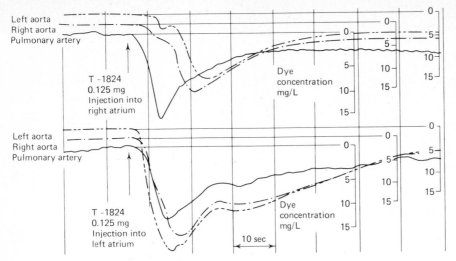

Figure 6-34
Demonstration by indicator-dye dilution curves of bidirectional shunting in the ventricle of the turtle's heart. [From R. R. Steggerda and N. E. Essex. Am. J. Physiol., **190**, 320–326 (1957).]

experiments similar to those discussed previously. Thus in spite of the apparent anatomic difficulties present in reptilian hearts, functional features of the heart appear to prevent the inefficiency of mixing postulated by earlier workers.

The most probable sequence of events compatible with physiological observations is shown in Figure 6-35 for the various reptiles. In all these forms the pulmonary arterial pressure and flow rise earlier in the cardiac cycle than for the systemic arches. This is because pulmonary diastolic pressure is much lower than in the aortic arches; blood is ejected earlier "downhill," toward the lower resistance. Changes in the pulmonary or systemic vascular resistance alter the distribution of blood and, as will be seen in Section 6.9.1, the left-to-right shunt of turtles may be completely reversed under diving conditions where pulmonary vascular resistance is high. The separation of pulmonary and systemic venous return demonstrated for crocodilians is due to the maintenance of a lower right-ventricular pressure than exists in the base of the aorta. Only in the event of a high pulmonary resistance, with a consequent rise in right-ventricular pressure above left-aortic pressure would venous blood enter the left arch. This circumstance occurs during the prolonged apnea associated with diving (see Figure 6-35).

The role of neural influences on the reptilian heart has been little studied. Vagal inhibition of heart rate has been reported for all species studied.

However, few studies of sympathetic nervous influences have appeared. Heating of the skin of *Iguana iguana* with heat lamps is followed within a few seconds by tachycardia and elevation of the arterial blood pressure. This response is not prevented by prior blockade with atropine of the vagal parasympathetics but is prevented with sympathetic ganglionic blocking agents and thus most probably represents a reflex activation of the cardiac sympathetic nerves. Both turtles and alligators respond to intravenous catecholamines with increased force of contraction. Cardiac acceleration has been reported for the alligator.

It is apparent that in reptiles significant advances have been made toward the development of a double circulation, correlated most probably with greater dependence on the lung as the major respiratory organ. Although some shunting of blood exists in turtles, and possibly to a lesser extent in squamates, this feature may be adaptive in aquatic situations (see Section 6.9.1).

6.7 CIRCULATION TO VARIOUS ORGANS

The distribution of the cardiac output to the various organs subserves diverse specific tissue functions that necessitate delivery of an appropriate blood flow. The supply of blood may be chiefly

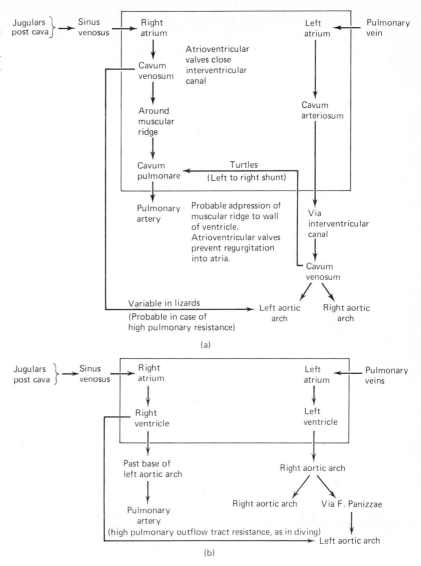

Figure 6-35
Probable pathways of blood in reptilian hearts. (a) Noncrocodilian forms. (b) Crocodilians. Ventricular diastolic events enclosed within the continuous lines.

related to local tissue metabolism or may reflect some broader aspect of metabolism such as body-temperature regulation. Local metabolic, hormonal, and nervous factors are involved in varying degrees in the regulation of local blood flow.

6.7.1 Blood Flow to Tissues Involved in External Respiration

Gill blood flow in fishes (except Dipnoi) is, of necessity, equal to cardiac output, because the entire cardiac output passes through the ventral aorta and efferent branchial vessels. The gill circulation imposes a resistance; thus a pressure drop occurs across the gills. The magnitude of the pressure drop varies in different fishes and probably under different conditions in the same fish. Pressure drops between ventral and dorsal aorta ranging from 20 to 45% have been reported for various species. Figure 6-36 illustrates the pressures recorded across the gills of the dogfish, *Squalus*. The protracted pressure plateau following peak afferent

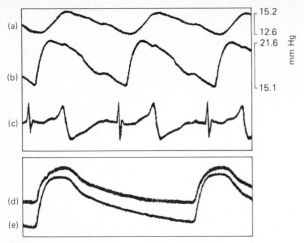

Figure 6-36
Oscilloscope tracings all from one specimen of S. lebruni, *14° C. (a) and (b) recorded simultaneously, (c) Placed in appropriate time relation, (d) and (e) recorded simultaneously, (a) and (b) blood pressure from celiac artery and first afferent branchial artery, respectively; calibration on right. (c) ECG. (d) Gill opacity (an index of blood flow) recorded with phototransistor. (e) Afferent branchial artery blood pressure.* [*From G. N. Satchell. J. Exp. Biol.,* **37,** *719–731 (1960).*]

branchial pressure coincides with contraction of the bulbus arteriosus. The pressure differential across the circuit is greatest in the early part of the cardiac cycle. Gill blood flow follows a form similar to the contour of the pressure gradient. A wide range of pressures have been recorded in various species (Table 6-5). Conditions of the various experiments must account for considerable variation in the values. The high ventral aortic pressure coupled with the relatively small pressure drop in the Chinook salmon *(Oncorhynchus tschawytscha)* indicates a relatively small hindrance to flow in this active species. Both epinephrine and norepinephrine produce brachial dilatation and increase gill flow in the perfused gill apparatus. Acetylcholine reduces gill flow (blenny). Gill resistance is reduced during stress or hyperactivity, perhaps secondary to catecholamine release. How such changes are mediated is unknown.

Reflex coordination of the heart beat with respiration has been reported for the dogfish. Vagal inhibition of cardiac activity occurs during the pha-

ryngeal inflation phase of respiration. The afferent limb of the reflex consists of pharyngeal receptors, sensitive to dilatation of the pharynx, which reach the medullary cardioinhibitor area via the branchial nerves. As water is expelled from the pharynx, the reflex inhibition is reduced and cardiac contraction occurs at an appropriate phase of water flow. Gill blood flow is probably maximum during maximum water flow across the gill filaments. Such phasic timing of blood and water flow favor maximum oxygen extraction, as does the countercurrent mechanism (Section 5.3.3). It is difficult to know how general this coordination of respiratory and cardiac activity is; however, a number of workers have reported that the intervals between heartbeats are multiples of the respiratory frequency with a general tendency for cardiac contraction to occur soon after the intake phase of respiration. Such feedback as described for the dogfish affords a beautiful control mechanism, linking gill ventilation to its essential counterpart, blood flow.

Accessory respiratory organs (see Section 5.4.3) of fishes (except the lung of Dipnoi) are generally perfused by arteries originating from the dorsal aorta, thus placing the accessory organs in parallel with the systemic circuit. No additional oxygen can be added to the blood in such areas when full saturation has occurred across the gills. Under conditions in which post-gill blood is only partially saturated, the amount of oxygen absorbed across the accessory respiratory surface will depend on inflow saturation, gas composition across the organ, and blood flow. Such organs could never supply a fully saturated arterial blood, because their drainage is combined with the systemic venous return to the heart. There is the additional complication, in habitats of low P_{O_2}, of loss of oxygen across the gill surfaces after its acquisition by accessory respiratory mechanisms. Diffusion from blood to the opercular chamber is reported to be minimized in some forms by cessation of gill water flow with the opercular flaps closed. How well the Dipnoi achieve high arterial saturation on the basis of lung perfusion depends on the distributional fate of the pulmonary and venous returns, and our information here is scanty.

In amphibians, reptiles, mammals, and birds, the pressures reported in the pulmonary artery are lower than systemic arterial pressure. Because the degree of separation of pulmonary and systemic

TABLE 6-5. Blood Pressure Drop across the Gills of Various Fishes

Species	Mean blood pressure (mm Hg) Proximal to gills	Mean blood pressure (mm Hg) Distal to gills	Blood pressure drop across gills (mm Hg)	Reference
Torpedo sp	16–18	6	10–12	Schoenlein and Willem (1894)
Scyliorhinus catulus	30–33	8–9	22–24	Schoenlein and Willem (1894)
Carcharias sp.	32	23.3	8.7	Lyon (1926)
*Squalus acanthias**	28.2	15.4	16.1/4.3	Lutz and Wyman (1932a)
Squalus acanthias	39/28	30/23	9/5	Burger and Bradley (1951)
Oncorhynchus tschawytscha	74.6†	53.3†	21.3	Greene (1904)
Anguilla anguilla‡	37.5	20	17.5	Mott (1950)

* Pithed posterior to sixth vertebra.
† These are apparent systolic pressures on a mercury manometer; if allowance is made for the apparent pulse pressure, the mean figure for the ventral aortic pressure is about 69 mm Hg.
‡ Anesthetized.
Source: J. C. Mott. "The Cardiovascular System." In M. E. Brown (ed.). *The Physiology of Fishes,* vol. 1, pp. 81–108, Academic Press, New York, 1957.

blood depends to a large degree, in amphibians and reptiles, on the pressures in the outflow tracts of the heart (see Section 6.6.5, Reptiles), the lower pulmonary resistance would appear to be an essential feature in selective distribution. Elevation of pulmonary resistance in the turtle (during diving) is responsible for shunting systemic venous blood into the systemic arteries.

The pulmonary circulation in all vertebrates is characterized by a large capillary area at the exchange surface. This is a major factor increasing the efficiency of gas exchange. High hydrostatis pressures in these capillaries, however, would promote excessive bulk filtration, leading to pulmonary edema and impaired exchange. In birds and mammals, where the anatomic aspects of vascular and cardiac circuitry ensure separation of pulmonary and systemic venous blood, the pulmonary arterial pressure is low by systemic arterial standards. Abnormal elevations of pulmonary venous or arterial pressure are associated with excessive pulmonary capillary filtration and impaired gas exchange, with consequent deleterious acid-base alterations and poor arterial saturation.

Our knowledge of the factors influencing pulmonary blood flow are derived chiefly from mammalian studies and, of what follows, one can only speculate as to the applicability to lower vertebrates.

The pulmonary capillaries are larger than those of the systemic circuits; in man they present a surface area of 50–60 m², accounting, in part, for the large diffusion capacity. The circulation time through the lungs is very rapid (within 1 sec), during which gas exchange occurs. Oxygen supply to lung tissue is derived from systemic arteries (the bronchial circulation), which supply about 1% of the pulmonary circulation. The pulmonary flow is equal to the cardiac output (left-ventricular output) because of its position between the left and right heart. The flow is thus determined by right-heart output and the pressure differential between right heart and left atrium.

Because of the relatively thin, less-muscular character of the pulmonary artery, distensibility of the inflow circuit is greater than for the aorta. This uptake character of the pulmonary artery is a factor, together with short arterioles and large capillaries, in promoting the lower pressure of the circuit. Pulmonary arterial distensibility also accounts for the rather large blood volume of the precapillary lung circulation. Of the 15–20% of the total circulating blood volume found in the pulmonary vessels, about 40% is arterial, 50% venous, and 10% capillary. There are not, however, static volumes and, should the right ventricle fall behind a few beats, left-heart output may be maintained because of the reservoir represented in the pulmonary circuit. Capillary volume increases as cardiac output rises, providing an increased exchange surface at elevated flows.

The pressure gradient across the lungs is from a mean arterial pressure of about 25 mm Hg to a venous pressure of some 4 mm Hg. Capillary pressure is of the order of 6 mm Hg. This is in sharp contrast to 25 mm Hg in systemic capillaries. Because of the plasma oncotic pressure, such a low capillary hydrostatic pressure favors the movement of fluid into, rather than out of, the capillaries. This is an important factor contributing to the "dryness" of the lungs and more effective gas exchange.

The arterioles of the lungs are sensitive to pH and gaseous environment. Low P_{O_2}, elevated P_{CO_2}, and low pH all promote arteriolar vasoconstriction. These local responses are important because they result in shunting the blood away from poorly ventilated alveoli toward well-ventilated alveoli.

Norepinephrine and epinephrine are pulmonary arteriolar constrictors, in contrast to their gill-dilator action in fishes. Knowledge of the role of the sympathetic nervous system as a determinant of pulmonary vascular resistance is fragmentary, but it would appear that neurogenic participation is not great in mammals.

Little is known of the mechanisms that control pulmonary vascular resistance in lower vertebrates. The pulmonary circulation of reptiles is chiefly under the control of parasympathetic vasoconstrictor fibers. Stimulation of the pulmonary vagus nerves produces pulmonary vasoconstriction in pulmonary arteries within the lung. This reaction is blocked by administration of atropine, a cholinergic antagonist. Recent direct observations of the pulmonary capillaries of turtles indicates that, unlike mammals, the precapillary resistance is unresponsive to alveolar hypoxia. It seems probable that the control of pulmonary blood flow is under reflex control; however, virtually nothing is known concerning the sensory limb or central neural components of this reflex. Pulmonary vascular resistance is reduced during the ventilatory phase of the respiratory cycle in turtles, whereas it increases as apnea progresses. Elevated heart rate is also associated with the ventilatory phase. This results in a maximization of pulmonary blood flow during the period of greatest gas turnover. The mean pulmonary blood flow shows a direct correlation with the turtle's oxygen consumption over a wide range of body temperatures. Thus, the ratio of pulmonary blood flow to oxygen consumption is maintained constant as temperature changes, unlike the ratio of ventilation to oxygen consumption (see Section 5.7.5) which is inversely related to temperature.

The control of pulmonary circulation in amphibians has been little studied; however, stimulation of the vagus nerves is reported to increase pulmonary vascular resistance in frogs. This observation hints at a control system similar to that observed in turtles. It seems likely that amphibians and reptiles regulate lung perfusion by mechanisms that are quite dissimilar to those employed by mammals. The situation in birds is completely obscure.

6.7.2 Blood Flows to Other Regions

All vertebrates possess, at least during embryonic life, a renal-portal circulation. The mammals are the only group that do not possess this system as adults. When present, the renal-portal veins drain the posterior portions of the body and enter the kidney, ramifying around the renal tubules as capillaries. The efferent drainage converges with the renal veins. The renal tubules of the mesonephros receive, in addition, a supply of capillaries from the efferent glomerular circulation. Among amphibians, reptiles, and birds, shunts connecting the renal-portal to the renal vein are present, making diversion of the renal-portal circulation to the renal vein possible. In aglomerular fishes, the renal-portal is the major blood supply to the tubules. Thus secretion by the renal tubules is dependent on this low-pressure circuit. Other forms possessing the renal-portal circulation receive an arterial supply to the glomeruli and are perfused by high- and low-pressure circuits. The preponderance of tubular excretion over glomerular filtration in the glomerular fishes, amphibians, and birds is evidence of considerable dependence on the portal circulation. The arterial pressure, and hence the glomerular filtration pressure, of fishes, amphibians, and reptiles is relatively low. Although glomerular filtration occurs, it is of a low order of magnitude, and it would appear that the renal-portal system provides the renal tubules with the secretory load without imposition of a large glomerular filtrate and the necessity of the high rate of renal reabsorption of solute and water that characterizes mammals (*see* Section 7.8.3).

With the appearance of the mammalian metanephros, glomerular filtration is the dominant feature, the renal-portal system being represented only in embryology. An elaborate reabsorptive mechanism providing the capacity to form urine hypertonic to the blood was an essential accompaniment in the shift toward dependence on glomerular filtration. Tubular excretion became less important, and this shift in emphasis paralleled the disappearance of the renal-portal system. Like so many evolutionary problems, it is difficult to say which of several related factors was the primary change, whether the disappearance of the renal-portal system fostered the emphasis on filtration or vice versa.

In mammals, the renal blood flow represents about 20% of the cardiac output. This large blood flow passes through organs weighing about 0.5% of the body weight. There is an abundant sympathetic vasoconstrictor supply, and the kidney is an active participant in cardiovascular reflexes. During exercise, excitement, or postural changes, the renal circulation contributes to the peripheral resistance adjustments. In general, increases in sympathetic outflow associated with declining arterial pressure or chemoreceptor stimulation (anoxemia of carotid bodies) elicit increased renal vascular resistance involving arteriolar vasoconstriction. This variable resistance, through which such a large fraction of the cardiac output passes, represents a very significant contribution of peripheral resistance adjustments.

The major work of the mammalian kidney is the reabsorption of constituents, chiefly Na^+, which are filtered into the tubules. It is not surprising to find, therefore, that O_2 consumption is closely related to Na^+ reabsorption. The arterial-venous O_2 difference is low (1.7 ml of O_2/100 ml of blood) compared to other organs (4–6 ml of O_2/100 ml of blood), yet O_2 consumption may represent some

8% of the total resting O_2 utilization. Reduction in blood flow is generally accompanied by a reduction in O_2 consumption with little change in the $(A - V)$ O_2 difference. This is in contrast to most other organs, in which reduction in blood flow over a fairly wide range is accompanied by an increase in the $(A - V)$ O_2 difference and thus little change in organ metabolism. The significant variable in the kidney is a reduction in glomerular filtration rate and hence in filtered Na^+ load.

Renin, an enzyme, is found associated with granules in the wall of the afferent artiole of the kidney. These elements are the juxtaglomerular cells, and the enzyme is released when perfusion pressure is low. The precise signal for release is not known, but current theories indicate that it is associated either with pressure sensitivity of the juxtaglomerular cells or is secondary to a reduction in glomerular filtration. Renin acts on plasma substrates to produce an octapeptide, angiotensin II, the most powerful arteriolar constrictor known. During hypodynamic states (hypotension and hemorrhage) renin is released and angiotensin may participate in the general increase in peripheral resistance. Angiotensin II is a powerful releaser of aldosterone, an adrenal corticoid. Aldosterone facilitates Na^+ reabsorption in the distal renal tubule. Aldosterone also augments salt excretion by the nasal glands of birds. Renin is reported to be present in mammals, birds, amphibia, and in bony fishes; however, it appears to be absent in Chondrichthyes and Agnatha. For further discussion on the role of some of these hormones see Section 7.10.

The coronary blood flow is closely related to the metabolism of the heart. Increasing work load of the heart is accompanied by elevated coronary flow. The driving force for coronary blood flow is the pressure at the coronary artery orifices at the base of the aorta. The capacity of the heart to perform work is closely linked to adequate nutrient flow, and from this point of view perfusion of the peripheral tissues is dependent on the coronary blood flow. This statement can be made for mammals and birds, but the lower vertebrates exhibit a marked capacity to utilize anaerobic sources of energy for running the cardiac contractile machinery and are thus less dependent over a protracted period on coronary O_2 supply.

In mammals the coronary $(A - V)$ O_2 difference is high, indicating high extraction. Thus elevation in cardiac metabolism cannot be met to any large degree by increased extraction. The coronary microcirculation exhibits vasodilatation in hypoxic states, indicating that coronary resistance is, to a large degree, controlled by local metabolic environment. Sympathetic amines, both circulating and locally released from cardiac sympathetic nerves, augment rate and strength of myocardial contraction, both leading to increased work. It is difficult to differentiate the direct effect of catecholamines on the coronary vessels because increases in metabolism are induced by these agents, and dilator effects may be secondary to metabolic changes. Dogs can increase coronary flow about 9 times in exercise, and the blood flow to the myocardium represents about 3–4% of the cardiac output. Cardiac O_2 utilization is about 10% of the total O_2 consumption.

The coronary vessels penetrate the myocardium, so the development of tension by cardiac contraction decreases transmural pressure and compresses the vessels. Negative flow may occur for short periods in early systole, especially during sympathetic stimulation when myocardial compressive forces are augmented (Figure 6-37). Flow is greatest during diastole, when tension of the myocardium has declined. Diastolic exceeds systolic flow by about 2.5 times. The heart is the only organ in which blood flow is maximal during the declining phase of the arterial pressure pulse.

Cerebral blood flow is not much influenced by reflexogenic factors and is remarkably constant over wide ranges of arterial pressures. Local adjustments of resistance are related to local metabolism, and the cerebral vasculature of mammals is exquisitely sensitive to the P_{O_2} and P_{CO_2} of blood. The dilator effect of elevated CO_2 is apparently a consequence of the decreases in pH caused by CO_2. The influence of O_2 is less striking than for CO_2. However, in most instances depressed P_{O_2} is accompanied by elevated P_{CO_2} and the two effects are reinforcing. Although circulating vasoactive materials are capable of modifying the cerebrovascular resistance, their effects are usually quickly overshadowed by local metabolic influences.

The magnitude of cerebral blood flow is impressive, representing about 15% of the cardiac output in man. Oxygen consumption is 3–5 ml/100 g min (roughly 20% of the total body O_2 consumption). Glucose utilizations of 5–7 mg/100 g min are cor-

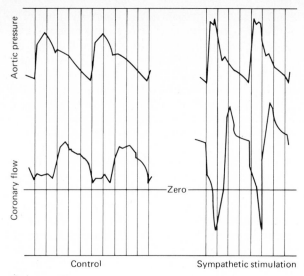

Figure 6-37
*Effect of cardiac sympathetic nerve stimulation on phasic coronary blood flow. The compressive forces of ventricular contraction during nerve stimulation produce negative flow during systole. Coronary flow is greater during diastole in both cases. In other vascular beds arterial blood flow follows a contour resembling the arterial pressure pulse. [Based on L. Granata et al., Circ. Res., **16**, 114 (1965).]*

related to a respiratory quotient of 0.99, indicating that glucose is a major energy source. Because glucose stores in the brain are limited, metabolism is dependent on substrate delivery as well as O_2 and CO_2 transport by the blood.

Greater dependence on anaerobic metabolism of neural tissue is indicated for lower vertebrates than for mammals and birds. Many reptiles and amphibians tolerate prolonged hypoxia without irreversible damage to neural tissues.

Perfusion of the liver is dual in all vertebrates. In addition to the hepatic arterial inflow, the hepatic portal vein is imposed between the major visceral organs (for example, the gut and spleen) and the liver (Figure 6-38). The entire venous outflow of the abdominal gastrointestinal organs passes through the liver before reaching the hepatic veins. The hepatic artery supplies oxygenated blood to the liver, and anastomotic connections with the portal vein are present. The capillaries derived from interlobular arterioles enter the sinusoids. Al-

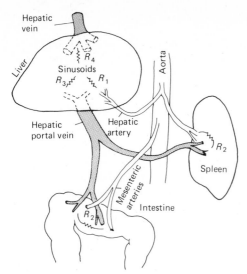

Figure 6-38
Resistance components of the splanchnic circulation. Code: R_1, hepatic arteriolar; R_2, splenic and intestinal arteriolar; R_3, portal venular; R_4, postsinusoidal.

though the portal vein functions as a transport medium for foodstuff, it also provides a significant portion of the O_2 supply. The terminal capillaries of the portal vein drain into the sinusoids, which, in turn, empty into central veins of the liver lobules, then converge as sublobular veins with the hepatic veins.

Arterioles, sinusoids, and intrahepatic veins are innervated by sympathetic vasoconstrictor fibers, and it is doubtful that vasodilator fibers are present. The precise role of sympathetic nerves in intrahepatic resistance changes is uncertain because the complexity of the vasculature makes evaluation difficult.

The resistance components in the splanchnic circuitry are all variable, and changes in intestinal resistance will be accompanied by secondary changes in hepatic portal flow. Gastrointestinal blood flow is elevated after ingestion of food, and studies indicate a greater shift toward mucosal than toward nonreabsorptive and nonsecretory tissues of the gut.

The splanchnic circulation participates actively in cardiovascular reflexes. The reflexogenic sympathic vasoconstrictor response accompanying a decline in aortic or carotid pressure is prominently represented in the gut. Splanchnic venoconstriction also occurs. The blood volume of the splanchnic circulation is approximately 20% of the total blood volume, so such responses lead to a shifting of volume toward the heart and arterial system. Thus this volume reservoir may be utilized in mobilization of blood reserves during emergency states. More than 20% of the cardiac output is devoted to splanchnic flow in mammals. Sympathetic vasoconstriction of splanchnic arterioles thus may contribute significantly to resistance adjustments leading to stabilization of the arterial pressure. During states of low perfusion an increased $(A - V)$ O_2 difference (increased extraction) favors a reasonably constant O_2 utilization.

The spleens of dogs, cats, and bats are significant reservoirs of red blood cells. Contraction of the splenic capsule, together with arteriolar vasoconstriction resulting from sympathetic stimulation, promotes the ejection of blood of elevated hematocrit. Discharge is initiated by reflex mechanisms, probably as a result of chemoreceptor stimulation during hypoxic situations. This may be an important accompaniment of arousal in hibernating bats (Section 6.9.2). During hypoxia, the splenic O_2 reserve may amount to 30–60 ml of O_2, released with the store of about 150 ml of blood stored in the dog spleen. The increase in circulating red cell mass also contributes to a larger O_2 capacity of the blood. Splenic discharge is apparently of small magnitude in man and rat.

The vasculature of the skin, in addition to supplying the nutrient needs of the tissue, participates very actively in the regulation of body temperature (Section 8.9.1). The capacity of the skin vessels is large, owing to the presence of an extensive venous plexus. This large capacity allows for the displacement of considerable circulating blood volume to the surface for heat dissipation. In some areas (distal extremities) arterial-to-venous anastomoses are present, allowing for the shunting of blood into or away from the venous plexus. These shunts are under sympathetic control.

Cutaneous blood flow exhibits enormous variability, ranging in magnitude from essentially nil to around 70 ml/100 g min. This variability is largely associated with the thermoregulatory state. Reduction in skin blood flow is due to sympathetic vasoconstriction and accomplishes an increase in the functional insulation of the body which buffers heat loss to the environment. Withdrawal of cutaneous sympathetic vasoconstrictor activity accomplishes the converse effect.

Some mammals, including man and horse, sweat when under heat stress. When the sweat glands are activated by cholinergic nerves, they release, in addition to sweat, a vasoactive polypeptide, bradykinin. This agent augments the sympathetically controlled vasodilator state of hyperthermia. Other circulating agents (angiotensin, epinephrine, and norepinephrine) produce cutaneous vasoconstriction.

Arteriovenous anastomoses are not extensive in the skin of man. Although they may give some hint of our emotional state, their role in thermoregulation is not great. These low resistance shunts bypass the capillaries and allow a relatively enormous flow of blood near the surface of the skin. Polar birds have rich fields of arteriovenous anastomoses in featherless areas of the feet or face. It has been found that under heat stress, the Adelie penguin can dissipate about 50% of the excess heat across the arteriovenous anastomoses of the flippers and feet, which represent only about 14% of the bird's body surface. Vasoconstriction of these areas and feathers of great insulative value prevent excessive heat loss when exposed to the cold. The importance of the vasoconstricted state of these vessels is illustrated in Figure 6-39. In the experiment shown, the bird was standing in sea water brine of −1.5°C. Pharmacological blockade of the nerves to one foot caused a marked rise in the foot web-

bing temperature as the arteriovenous anastomoses relaxed due to the nerve block. The body temperature could not be sustained at a normal level until the effect of the nerve block began to subside. Interestingly, even though the feet are exposed to subfreezing water, the vascular state allows sufficient blood flow to prevent freezing of the tissues.

The metabolism of muscle varies widely with muscle activity; elevated blood flow is also associated with activity. Resting blood flow of mammalian skeletal muscle (2–8 ml/100 g min) may increase severalfold (to about 40 ml/100 g min) during exercise. The increase in flow is correlated with an increase in the number of open capillaries. Metabolic vasodilator products of metabolism (or decreased P_{O_2}) promote relaxation of the precapillary sphincters, and something like 30–40 times the resting number of open capillaries has been observed in contracting muscle. That this mechanism is of local origin is suggested by the observation that the increase in flow of exercise occurs in denervated as well as innervated preparations.

Although the local influence of metabolism is independent of neural control, there is a rich supply of sympathetic nerves to the arterioles. There is normally a sympathetic adrenergic vasoconstrictor tone present and sectioning of the nerves leads to vasodilatation. Experiments in which the sympathetic vasoconstrictor fibers to the gastrocnemius of the cat are stimulated maximally reveal that metabolites generated by a working muscle are capable of overriding the vasoconstrictor effects of the sympathetic arteriolar constriction. This phenome-

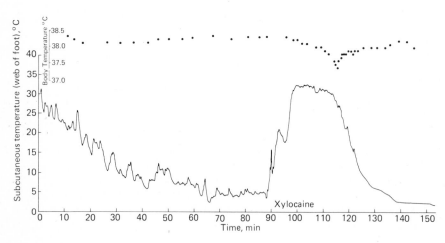

Figure 6-39
Adelie penguin standing in icy sea water (−1.5°C) from time zero. The tendency for body temperature to fall was reversed as the arteriovenous anastomosis in the foot webbing constricted. Blockade of the nerves to these vessels (one foot only) by a local anesthetic caused a large increment in blood flow and a fall in body temperature until the anesthetic effect subsided. [From F. N. White and K. Morgareidge, personal observations.]

non is called *ascending dilatation* and illustates the primacy of local factors, which in this case diffuse to the arterioles, as muscle flow determinants.

In cat, man, and the dog, cholinergic sympathetic vasodilator fibers also supply the arterioles of muscle. With the initiation of exercise, these fibers are reflexly activated and, coupled with metabolic determinants, augment muscle flow. Even the thought of exercise or situations in which exercise is likely to follow (fear) elicits vasodilatation on the basis of these cholinergic fibers, illustrating the presence of cortical levels of integretion. Thus the athlete, prior to a great moment "anticipates" exercise through muscle vasodilatation.

The fact that such responses occur in carnivores suggests a functional role for this neurogenic vasodilator system. Imagine a lion just prior to charging an antelope. Skeletal muscular activity is about to become maximal, a state that will, by virtue of local metabolic influences on precapillary vascular resistance, produce an enormous blood flow to the muscles that go into action during the attack. Such a large reduction in peripheral resistance must be accompanied by reflex controlled elevations in cardiac output and resistance in other vascular beds (skin, gut, kidney), else the arterial pressure would fall too low to sustain tissue perfusion to vital tissues including the heart and brain. These adjustments must occur during exercise; however, they require a few seconds of time. By "anticipating" the large muscle blood flow required during the upcoming burst of exercise, the reflex adjustments necessary to sustain perfusion pressure have been made *prior to* the act, and neurogenic vasodilatation of relative inactivity will be replaced by the metabolic vasodilatation of the chase. No reflex time is lost and no sudden fall in arterial pressure will occur during the critical phase of going from rest to activity. The presence of this vasodilator system in man may say something about how our ancestors made a living.

6.8 CONTROL OF CARDIOVASCULAR FUNCTIONS

The relative constancy of the mean arterial pressure in the face of fluctuations in local metabolism and blood flow is one of the remarkable features of the circulation. If flow to an organ is to increase, either the cardiac output or the resistance in other organs must rise if arterial pressure is to be kept from fluctuating widely. Both alternatives are utilized in arterial pressure homeostasis, and it is obvious that rather delicate control mechanisms must be involved in the integration of the variables. These controls are neural, hormonal, and physical in nature.

Departures from the normal (or reference) blood pressure are detected by sensory receptors within blood vessels. The firing rate of these receptors determines the input into central nervous control areas, where the sympathetic and parasympathetic output is adjusted to bring about appropriate corrective cardiac and vasomotor responses. Although we are far from a complete knowledge of these mechanisms, a good deal has been learned about such systems, especially in the common mammalian laboratory animals.

Sensory nerve endings responsive to stretch are located in the walls of the aortic arch and carotid arteries. These sensory elements are known as *presoreceptors* (or *baroreceptors*). In mammals, the carotid receptors are located near the junction of external and internal carotid arteries and are known as the carotid sinus pressoreceptors (see Figure 5-40). An increase in arterial pressure, by distending the pressoreceptors, excites the sensory endings and brings about an increase in firing rate of impulses along the nerve fibers. Carotid receptors are innervated by nerve IX, whereas the aortic elements transmit afferent information to the brain along nerve X. Similar receptors have been described in birds.

An increase in pressoreceptor input to the cardiovascular centers of the brain stem has the effect of increasing vagal parasympathetic outflow to the heart, decreasing sympathetic outflow to the heart, and inhibiting sympathetic activity to peripheral arterioles as well as to veins. Such a team effort has the overall effect of returning the arterial pressure toward the norm, as a result of the combination of negative chrono- and inotropism (decreased cardiac output) and reduced peripheral resistance. The decrease in venomotor tone tends to reduce central venous pressure, which secondarily contributes to the decrease in cardiac output. An increase in venous capacity (increased distensibility) as a result of venodilatation promotes a shift in blood volume to the venous reservoir circuit and away from the arterial side. The converse effects are elicited by hypotensive states.

The sympathetic output in response to pressoreceptor stimulation is not uniformly distributed among the organs. Renal, splanchnic, and cutaneous resistances appear to be more affected than other vascular beds. Cerebral vascular resistance is little influenced by sympathetic factors.

The release of catecholamines from the adrenal medulla is partially under sympathetic control. In hypotensive states the sympathetically mediated release of adrenal catecholamines further facilitates cardiac output, peripheral resistance, and venous tone.

Hypotensive states, especially when characterized by arterial anoxemia and elevated P_{CO_2}, evoke further baroreceptor-type sympathetic adjustments and stimulate respiration. Such adjustments are mediated through chemoreceptors such as the carotid and aortic bodies (see Section 5.8.2). Venous return to the heart is augmented under these conditions by the thoracoabdominal pump mechanism.

Restitution of blood volume in hemorrhage is partially accomplished by bulk filtration into the capillaries from interstitial fluid. This physical adjustment to volume-induced hypotension can account for significant blood-volume readjustment in such emergency states.

Responses leading to conservation of volume are prominent in hypovolemic (low-volume) states. In these situations renal vascular resistance is high and blood flow low, so the glomerular filtration rate is reduced. The decreased load of solute presented to the renal tubules for reabsorption is more thoroughly reabsorbed under these conditions. The urine volume and sodium ion excretion falls, indicating both electrolyte and water conservation. Superimposed is the action of antidiuretic hormone, released through the mediation of volume receptors.

The renin-angiotensin system is activated in hypovolemic circumstances with the subsequent release of aldosterone from the adrenal cortex. Sodium reabsorption in the distal tubule is augmented by this hormone, an action promoting further electrolyte and water conservation. In addition, angiotensin further augments the sympathetically mediated rise in peripheral vascular resistance.

Carotid and aortic pressoreceptors are not active below about 60 mm Hg arterial pressure in dogs, and their firing rate reaches a maximum plateau at 160 mm Hg. At pressures below 60 mm Hg, brain blood flow may be severely limited and cerebral hypoxia and hypercarbia ensue. Such a circumstance elicits a profound sympathetic discharge (central ischemic response) of such an extent that glomerular filtration becomes nil, as does urine flow. The central ischemic response may be viewed as an emergency mechanism protecting cerebral blood flow.

The preceding concentration on responses to hypovolemia is useful because a forcing function such as hemorrhage brings to our attention the latitude of the cardiovascular responses, which tend to maintain constancy. A summary scheme of these reactions is presented in Figure 6-40.

6.8.1 Response to Exercise

The transition from rest to exercise is one in which the call for oxygen may increase dramatically. The task of meeting this increased demand for convection of gases falls principally to the heart and ventilatory system. The demand for augmenting blood flow derives primarily from skeletal muscle and secondarily from the heart itself, since an increase in myocardial metabolism must be met by an adequate coronary blood flow. Further demand for blood flow occurs in the skin when heat dissipating mechanisms, dependent on blood flow, are instituted to protect the body from overheating.

Reference to Figure 6-41a will give a general picture of the overall cardiovascular response to various levels of work performance in a human subject. In progressing from rest to heavy work a parallel increase in oxygen consumption occurs. The increment in metabolism is supported by two factors evident in the data. One of these is obviously the increase in cardiac output. However, this increase is not strictly proportional to the oxygen consumption. In this case, the resting blood convection requirement was about 0.55 liter/mmol O_2 consumed. At the highest work load, this figure has fallen to 0.18 liter/mmol O_2. Thus total oxygen consumption increased by around 10 times while cardiac output was elevated only threefold. The explanation for this nonproportionality between oxygen consumption and cardiac output is found

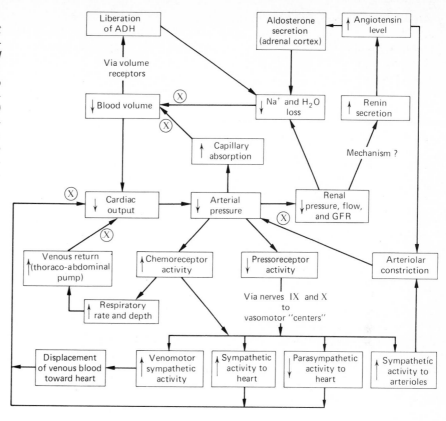

Figure 6-40

Some of the major responses that tend to maintain the arterial pressure, cardiac output, and blood volume following loss of blood. ⊗ indicates an action tending to correct an effect. Circulating catecholamine levels (not shown) are elevated because of stimulation of the adrenal medulla via sympathetic nervous activity. The effect is to further augment cardiac output and arteriolar and venous constriction.

in the marked elevation in arteriovenous oxygen extraction, which rose to over 3 times the resting value. This is an illustration of the utility of maintaining an *extraction reserve* in the blood stores. The increased extraction from the arterial blood occurred in the active skeletal muscles. The greater clearance of oxygen from blood by the active muscles has the effect of minimizing the demand for work by the heart.

The overall peripheral resistance, calculated from the ratio of mean arterial pressure to cardiac output, falls with work load, reflecting the dominant influence of muscular vasodilation. The resistance would be much lower were it not for compensatory sympathetically mediated vasoconstriction in the skin, kidneys, and visceral organs. There was only a mild elevation in mean arterial pressure, a parameter constrained by the baroreceptor reflexes.

It should be noted that the increase in cardiac output is highly rate, rather than stroke, dependent

(rate increased by around 2.6 times resting levels while stroke volume was constrained between 90 and 106 ml). These responses are dependent on sympathetic augmentation of pacemaker activity (chronotropic effect) and myocardial contractility (inotropic effect). The increased contractility of the heart is responsible for delivering a more forceful ejection into the aorta during systole—a factor responsible for the observed rise in systolic pressure.

The augmented cardiac output of the heart must be supported by adequate venous pressure and volume. The thoracoabdominal pump mechanism is more effective during exercise due to the increased depth and frequency of respiration (see Section 6.6.4). An additional "muscle pump" supplied by the contracting skeletal muscles causes compression of the veins of the extremities while the unidirectional venous valves prevent back flow and pooling in the distal segments of the venous tree. Combining with these noncardiac pumps, sympathetically mediated venoconstriction aids by de-

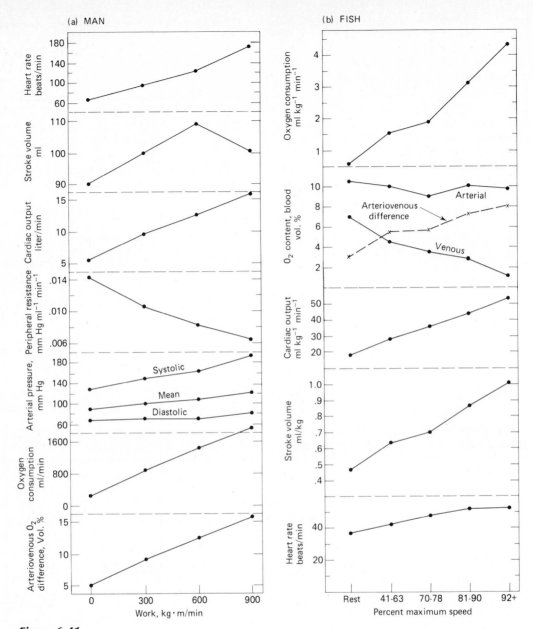

Figure 6-41
Cardiovascular responses to exercise in (a) man and (b) fish. See text for salient points. [*(a) Data from A. Carlsten and G. Grimby*. The Circulatory Response to Muscular Exercise in Man. *Charles C. Thomas, Publishers, Springfield, IL, 1966; (b) data from J. W. Kiceniuk and D. R. Jones.* J. Exp. Biol., **69,** *247 (1977)*.]

creasing venous capacitance (reduces compliance) and steepening the pressure gradient to the heart.

During heavey exercise, especially in a hot environment, the retention of metabolically produced heat induces hypothalamically controlled cholenergic activation of sweat glands and withdrawal of cutaneous sympathetic vasoconstriction. Products released from sweat glands (kinins) also induce cutaneous vasodilation. Massive increases in cutaneous blood flow may limit the heart's capacity to service working muscles. When this occurs, the muscles become hypoxic and anaerobic glycolysis results in liberation of lactate with consequent acidosis and muscle fatigue. In normal individuals the upper limit of exercise is determined by the cardiovascular system. Performance depends on the maximal capacity to increase heart rate and output and reflex control of resistance and capacitance vessels. The organism meets its exercise limit when there is inadequate blood supply to support aerobic metabolism.

In lizards (*Varanus* and *Iguana*) the general pattern of cardiovascular responses to exercise is similar to that of man. Stroke volume appears to be constrained, whereas heart rate and increased oxygen extraction from blood are the dominant contributors to the increased demand for blood gas transport.

Exercising fishes, like other vertebrates, increase both the cardiac output and arteriovenous oxygen differences (Figure 6-41b). A major difference between trout and terrestrial vertebrates is the small dependency on heart rate for augmentation of cardiac output. Instead, fishes appear to rely dominantly on elevations in stroke volume as exercise is intensified. Catecholamines are elevated during exercise in fishes and muscular pumping of veins has been documented. Thus, some combination of Starling's law and inotropic influence of catecholamines may be involved in the response to exercise. Because the relatively undistensible pericardium and its fluid should constrain large increases in diastolic volume, one might doubt a dominant role for the Starling mechanism. We have no information on resting end-systolic volume of the ventricle. Should this be large, one might explain the large increases in stroke volume on the basis of inotropic effects of catecholamines, that is, an increase in force of contraction during systole "bit-ing" into the end-systolic reserve volume, end-diastolic volume remaining rather constant.

We have implied that the limitations on maximum exercise level are largely attributable to the convergence of maximal cardiac output and arteriovenous oxygen extraction by the metabolically intense muscles of locomotion. The maintenance of high arterial oxygen saturation values by lung breathing forms and by fishes at high levels of exercise implies that convection of gases by the respiratory apparatus is not limiting. This generalization may not be safely extended to invertebrates. A study by Rutledge on exercising crayfish indicates that ventilation, rather than circulation, limits active oxygen uptake and scope for activity. The crayfish exhibited a marked decline in the P_{O_2} of post-branchial blood and hemocyanin saturation as exercise intensified.

6.9 ENVIRONMENTAL CORRELATIONS

Alteration of environmental situations imposes upon organisms the necessity of adapting to changes in chemical and physical characteristics of the external world. Here we shall focus attention on the cardiovascular and closely related respiratory adjustments that extend the niches of vertebrates, sometimes over an almost unbelievable range of contrasting environments.

6.9.1 Asphyxic Responses

Representatives of all major vertebrate groups exhibit rather spectacular respiratory and cardiovascular responses when entering or leaving an aquatic environment. Many air breathers have the capacity to remain submerged for long periods of time and are physiologically suited to the diving situation. Some of these forms and their diving times and depths are given in Table 6-6.

A variety of problems face the air breather that submerges below water. Availability of oxygen and elimination of carbon dioxide are obvious problems. But just as great a problem is that of distribution of a limited oxygen supply. Many homeotherms dive into low temperature zones where heat conservation is important. By diving to great depths, pressures of great magnitude are encoun-

TABLE 6-6. Duration and Depth of Diving (Mammals)

Species	Duration (min)	Depth (m)
Platypus	10	
Mink, *Mustela vison*	3	
Harbor seal, *Phoca vitulina*	20	
Walrus, *Odobenus rosmarus*	10	80
Steller's sea lion, *Eumetopias jubata*		146
Gray seal, *Halichoerus grypus*	20	100
Weddell Seal, *Leptonychotes weddelli*	43	600†
Bottle-nosed whale, *Hyperoodon ampullatum*	120	Deep
Sperm whale, *Physeter catodon*	75	900
Blue whale, *Sibbaldus musculus*	49	100
Harbor seal, *Phocaena phocaena*	12	20
Bottle-nosed porpoise, *Tursiops truncatus*	5	
Beaver, *Castor canadensis*	15	
Muskrat, *Ondatra zibethicus*	12	
Most men	1	
Experienced divers	2.5	61*

* Greek sponge diver, Stotti Georghios (1913).
† From Kooyman, G. L., *Science,* **151,** 1553–1554, 1966.
Source: L. Irving. In W. O. Fenn and H. Rahn (eds.). *Handbook of Physiology,* sec. 3, vol. 1, American Physiology Society, Washington, D.C., 1964.

tered (1-atm pressure increase per 10-m depth increase). Coming up can also be a problem, because gases contained within the lung will expand and dissolved blood gases of low solubility (nitrogen) may form bubbles that circulate in the blood (caisson disease, or "bends"). Robert Boyle (1670), who is not especially noted as a comparative physiologist, first observed such effects when he submitted a viper to high pressure and then to low pressure. He reported that when he exposed the snake to very low pressure, a bubble formed within the transparent lidless eye. The danger of bubble formation is much greater for man in a conventional diving suit than for natural divers. In this situation there is a continual acquistion of nitrogen from the air being pumped to the diver. Natural divers can acquire only the nitrogen they take below— that dissolved in body fluids at atmospheric pressure and in the lung air. That natural divers actually face the problem of the "bends" is debatable.

Paul Bert (1870) made one of the earliest observations on cardiovascular adjustments in diving, noting that the heart rate slows when ducks are submerged. Within a short time it was shown that the bradycardia of diving is of vagal origin, because sectioning of the vagi or administration of atropine (which blocks the action of acetylcholine) eliminates the response. Thus far all diving animals studied exhibit bradycardia, although the time of onset and magnitude after submersion is variable (Figure 6-42).

The observation by Scholander that the lactic acid concentration in the arterial blood rises little during diving in seals, but is greatly elevated immediately after emergence, suggests that lactic acid is sequestered in muscle during diving in consequence of restriction of blood flow. Similar lactate patterns have been observed in ducks and alligator when diving, and in grunion (a fish) during excursions to the beach for breeding purposes. Direct visualization of exposed vessels and radiographic studies and measurement of blood flow with flowmeters have verified earlier conclusions that peripheral vasoconstriction is a major component of the response (Figure 6-43). The site of vasoconstriction is unusual and characteristic of both seals and ducks. Instead of arteriolar constriction, it is the larger supply arteries that constrict. The sympathetic adrenergic innervation of these larger vessels is much more dense in the duck than in nondiving birds (turkey) or cats. As pointed out earlier, sympathetic vasoconstriction of the arterioles in animals such as the cat may be overridden by locally produced metabolic dilators, the phenomenon of ascending dilatation. It appears that this is avoided in certain diving animals because the constriction is placed far proximal to the production of local dilator factors. This allows the constriction to hold during long periods of submergence, even though the arterioles may be relaxed. The speed with which arterial constriction occurs and the fact that ganglionic blocking agents or norepinephrine blocking drugs inhibit the response, testifies to the reflex character of this component of diving.

The distribution of vasoconstrictor activity during diving is not dissimilar to that of hemorrhagic responses in nondivers. Muscle blood flow is more profoundly reduced in diving, but the general pattern of renal, splanchnic, and cutaneous vasoconstriction, with adequate maintenance of cerebral

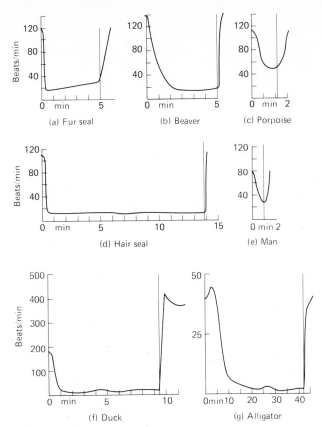

Figure 6-42

*Heart-rate response to diving in various vertebrates. [(a to e) adapted from L. Irving in W. O. Fenn and H. Rahn (eds.), Handbood of Physiology, sec. 3, vol. 1, American Physiological Society, Washington, D.C., 1964; (f) adapted from H. T. Anderson, Acta Physiol. Scand., **58**, 173–185, 1963; (g) adapted from H. T. Anderson. Acta Physiol. Scand., **53**, 23–45 (1961).]*

and coronary blood flow, is common to the two situations.

Even though heart rate falls during diving, the central arterial pressure is maintained because the peripheral resistance is high and the rate of runoff is low. In a California sea lion, trained to immerse its head on command, Elsner and coworkers showed by direct measurement that cardiac output decreased with heart rate, although the stroke volume remained unchanged. These workers noted, in all marine diving mammals examined, the presence of a highly elastic enlargement of the aorta

near the ventricle. This enlargement almost doubles the ascending aorta diameter, and they suggest that rebound energy for maintaining diastolic pressure is derived from this area. This situation is reminiscent of the bulbus function of amphibians and fishes.

Little is known about the afferent sensory components that elicit these complex adjustments. Anderson has shown that, in the duck, the reflex responses to submersion are independent of levels of the central nervous system higher than the medulla. The trigeminal nerve relays sensory information to the medulla and triggers a chain of efferent sympathetic and parasympathetic activity. Wetting areas near the nostrils elicit the diving response, an effect probably mediated by the ophthalmic branch of the trigeminal. These initial responses may be strongly augmented, as the dive proceeds, by the combined effects of elevations in P_{CO_2} and decreased P_{O_2}. Breathing gas mixtures of high P_{CO_2} induces a diving response in ducks, even under pentobarbital anesthesia.

The reduction of cardiac output during diving appears to be a common diving response (present in seals, ducks, and turtles); however, the distribution of the cardiac output in reptiles is considerably different than in mammals or birds. Figure 6-44 shows the striking reduction in the pulmonary stroke flow that occurs with the development of diving bradycardia in the turtle, *Pseudemys scripta*. Prior to diving the stroke output of the heart is approximately 1.6 ml/kg of body weight. Of this output 60% is distributed to the lungs and the remainder to the systemic circuit (left-to-right shunt). During diving bradycardia, cardiac output falls to approximately 5% of the predive level, a situation that persists for more than 1 hr. In this situation pulmonary flow is of the order of 40% of the cardiac output (right-to-left shunt). Pulmonary pressure is well maintained, even at rates of 2 beats/min, indicating that pulmonary vascular resistance increases. The major work of the heart is directed toward systemic perfusion under these circumstances, and it may be that reduced perfusion of lungs depleted of oxygen stores conserves cardiac energy. Alligators also exhibit right-to-left shunting during diving by increasing pulmonary outflow resistance to the extent that right-ventricular pressure exceeds left-aortic pressure.

Diving snakes develop bradycardia when venti-

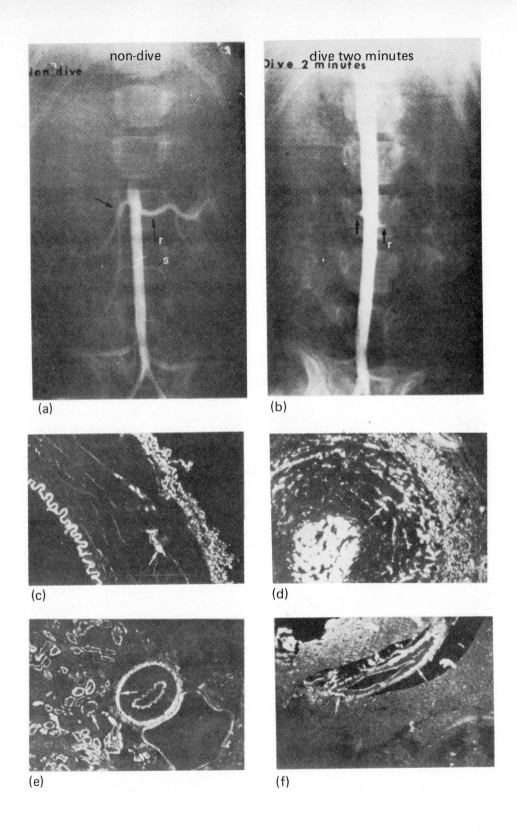

non-dive

dive two minutes

(a)

(b)

(c)

(d)

(e)

(f)

254

Figure 6-43 (see page 254)

*Angiographs of abdominal aorta and its tributaries of a harbor seal (a) prior to and (b) during diving. Arrows indicate renal arteries (r); s indicates paired somatic arteries to back muscles. During diving response, note site of renal arterial constriction near origin of renal arteries and absence of filing of the somatic arteries. Unlike most arteries of nondiving mammals, the supply arteries that undergo constriction show fluorescence characteristic of noradrenaline at sympathetic nerve terminals within the media of the vessels (c, renal artery; d, skeletal muscle artery). Vessels that do not constrict during diving (f, coronary artery) or are deep in the tissue parenchyma (e, small renal parenchymal artery) exhibit sympathetic innervation only at the outer margin (adventitia) of the vessels, just as in the arteries of nondivers. [From F. N. White, M. Ikeda, and R. W. Elsner. Comp. Gen. Pharmac., **4,** 271–276 (1973).]*

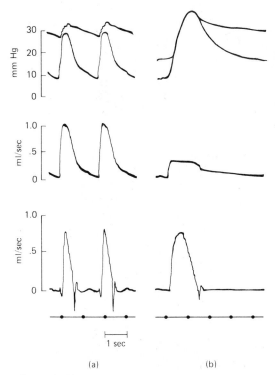

(a)   (b)

Figure 6-44
*Unanesthetized turtle, 3 kg; simultaneous pressures and flows (a) in air and (b) during diving bradycardia. Upper trace: upper line, right aortic pressure; lower line, pulmonary pressure. Middle trace: pulmonary flow. Lower trace: subclavian flow. [From F. N. White and G. Ross. Amer. J. Physiol., **211,** 15–18 (1966).]*

lated with gas of elevated carbon dioxide content, and the postdive reversal of bradycardia in the turtle is blocked if the animals emerge into a carbon dioxide-rich gas phase. Reversal, only seen after taking the first postdive breath, is just as rapid when emerging into pure nitrogen as into room air.

Control of specific gravity by turtles is indicated by the observations that, despite their heavy shells, they maneuver easily through the water, choosing to float near the surface at one moment or sink to the bottom at another. D. C. Jackson has demonstrated that this ability, for *Pseudemys scripta*, depends on the animal's capacity to change reciprocally the volumes of lung air and stored water. The cloacal bursae appear to be the active site for the facultative water storage. The mechanism resembles closely that used in ballasting submarines.

Diving animals might extend their diving time by submerging with a large lung volume from which a greater extraction of oxygen could be derived. However, the lung volumes of diving mammals are not much different than for nondivers except for the large whales, which have about 50% of the lung volume per unit weight of other mammals. The small lung volume of whales, which visit depths of almost 1000 m, may aid in the prevention of caisson disease. Scholander suggests that the lungs of whales may collapse during descent because the thoracic wall does not form a rigid case, the ribs "floating" rather than fused to a sternum. At a depth of 100 m all the lung gas may be in the nonexchange respiratory segments (dead space) and the exchange surfaces completely collapsed. This should effect a decrease in nitrogen diffusion into the blood. On ascent, only a small load of dissolved nitrogen need be disposed of, reducing the danger of bubble formation and its disastrous consequences. Thoracic collapse has been directly observed in the bottlenose porpoise (*Tusiops truncatus*) by utilizing a combination of training and un-

derwater television (Figure 6-45). Some diving forms exhale before or shortly after diving, an act that appears to favor buoyancy control rather than oxygen storage in the lungs. This appears to be the case for Weddell, elephant, and harbor seals, but not for porpoises.

Mammalian divers exhibit a large tidal volume and the lungs are more completely emptied on ex-

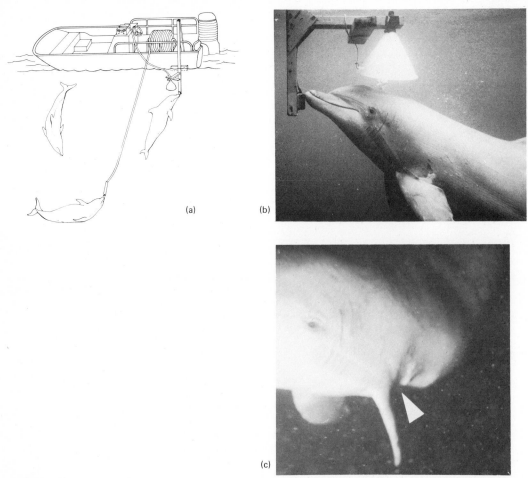

(a)

(b)

(c)

Figure 6-45
(a) The experimental setup for deep-diving experiments in the open ocean. The porpoise dives down when the go signal is turned on. He pushes the plunger on the end of the diving test switch, turning the go signal off, and then returns to exhale into the funnel before surfacing. (b) Porpoise exhales into the funnel just below the surface. (c) The porpoise pushing the plunger of the diving test switch at a depth of 300 m. Thoracic collapse is apparent especially in the area behind the left flipper. The picture was taken with an underwater camera that was triggered when the plunger was pushed. [*From S. H. Ridgway, B. L. Scronce, and J. Kanwisher.* Science, **166**, *1651–1653 (1969).*]

piration. In contrast to man and dog, the ability to increase tidal volume is small, and increased ventilation must be accomplished by increasing rate of respiration. This facet of respiratory dynamics seems to be associated with the capacity to achieve greater lung deflation than is the case for nondivers. It appears, therefore, that oxygen reserves are not accounted for by lung volumes.

There is, however, a tendency for diving animals to have a larger blood volume and in some cases a higher oxygen capacity than nondiving relatives of the same group (Table 6-7). Even within closely related groups, factors that are associated with increased oxygen stores and capacity (elevated blood volume, hemoglobin, and packed red cell volume) may be related to diving habits such as depth and duration of diving. For example, among porpoises it was observed that the total blood oxygen content of the highly active, deep-diving pelagic *Phocoenoides dalli* was about 3 times that for the more coastal *Tursiops truncatus* and about 70% greater than a less active pelagic species, *Lagenorhynchus obliquidens*. Ducks exhibit an unusually large resting-muscle

blood flow with a small $(A - V)$ oxygen difference, suggesting that considerable oxygen reserves are sequestered in the venous system.

Unloading of oxygen from the blood is facilitated in *Alligator* and *Crocodylus acutus* with progressive acidification of the blood. These animals exhibit a more striking Bohr effect than terrestrial lizards such as the gila monster or chuckwalla.

Considerable oxygen is stored in the myoglobin of diving mammals. Unloading of myoglobin is at lower oxygen tensions than for hemoglobin (Section 5.7.2). It has been calculated that nearly 50% of the oxygen reserves of diving seals is in the form of oxymyoglobin. These reserves may be utilized by the muscles, which, it will be recalled, receive little or no perfusion during experimental dives.

Mammalian and avian divers are able to stay submerged up to 4 times longer than their oxygen reserves would last if they continued at their normal prediving rate of oxidative metabolism. Furthermore, upon emerging, seals and ducks do not consume as much oxygen to "repay their debt" as one would predict on the basis of predive oxygen con-

TABLE 6-7. Blood Volume and Oxygen Capacity of Blood of some Diving and Nondiving Species

Animal	Blood volume, % of body weight	O₂ capacity (vol. %)
Pigeon	7.0	21.2
Hen	3.9	11.2
Duck, domestic	10.0	16.9
Guillemot, *Uria troile*	12.3–13.7	26.0
Puffin, *Mormon fratercula*	11.3–12.0	24.0
Penguin, *Pygoscelis papua*	9.0	20.0
Man	6.2–7.0	20.0
Dog	6.2–10.5	21.8
Horse	7.0–10.7	16.7, 14.0
Rabbit	6.5	15.6
Beaver, *Castor canadensis*		17.7
Muskrat, *Ondatra zibethica*	10	25.0
Seal, *Phoca vitulina*	15.9	29.3
Sea lion, *Eumetopias stelleri*		19.8
Porpoise, *Phocaena communis*	15.0	20.5
Blue whale, *Balaenoptera musculus**		14.1
Fin whale, *Balaenoptera plupalis**		14.1
Sperm whale, *Physeter catodon**		29.1

* Blood samples drawn from carcasses up to several hours after death.

Source: H. T. Anderson. "Physiological Adaptations in Diving Vertebrates," *Physiol. Rev.,* **46,** 212–243 (1966).

sumption. Most reports substantiate the contention that metabolism is reduced during diving. Oxidative metabolism is limited by blood flow in most organs as a result of intense vasoconstriction and the fact that anaerobic pathways are predominant in the periphery. Belkin has illustrated the importance of anaerobic processes during submersion in freshwater turtles. Poisoning of anaerobic processes by iodoacetate administration significantly reduced survival time when tested in water or 100% nitrogen environments.

Irving wrote of his experience of apprehensively pushing underwater the head of a recently captured and unfettered large beaver. Its neck relaxed so completely, he reported, that its jaws banged sharply on the bottom of the container. He made the point that facultative relaxation is a general characteristic of divers, in contrast to the frantic motor activity exhibited by rats, cats, or men when forced under water. Divers, when forcibly submerged, are apparently free of the blind terror that dominates nondivers and that hastens asphyxia rather than resulting in coordinated purposive activity.

Sensitivity of the respiratory apparatus to carbon dioxide appears to be lower in divers than in man, cat, rat, or dog. Elevated carbon dioxide does increase ventilation in divers but to a smaller extent than in their nondiving relatives.

As a dive progresses, the P_{CO_2} of the blood may rise to some 2.5 times the resting level and arterial oxygen saturation may fall to about 15%. The pH may drop about 0.4 units. There is thus a tendency toward progressive acidosis, which, according to Anderson, may be divided into three stages: (1) an initial respiratory acidosis, (2) a later combined metabolic (lactic acid) and respiratory acidosis, and (3) an early postdive metabolic acidosis.

The respiratory response of diving vertebrates to lack of oxygen has not been studied extensively. Ducks respond to the inhalation of gas of less than 10% oxygen with large increments in respiratory rate and amplitude. However, ducks submerged for 15 min endure at least the terminal half of such a dive with less than 5% oxygen in the lung gas. This is of the same order of magnitude for the porpoise *(Tursiops truncatus),* trained to dive to a prescribed depth and exhale into a collection device near the surface before air breathing (Figure 6-45a and b). Following a 300-m dive, the expired oxygen level had fallen from a surface value of 13% to around 4%, a value near the P_{50} of the hemoglobin.

The responses characterizing diving are represented in all vertebrate groups. The codfish "dives" when removed from water and exhibits the characteristic bradycardia. Heart rate and cardiac output also decrease in diving frogs, although the peripheral distribution of blood has not been studied. Presumably skin circulation increases as an adjunct to cutaneous respiration. Sea snakes rely heavily on the skin for gas exchange, acquiring a significant amount of oxygen from seawater and eliminating over 90% of their carbon dioxide over this route. Even nondivers, such as man, exhibit the diving response, although to a much smaller degree than the specialists, such as porpoises, seals, and whales. Asphyxia in man and other terrestrial mammals induces peripheral vasoconstriction, but these effects can be overriden by local metabolic factors. It appears that the capacity to perform for extended periods while submerged utilizes mechanisms of long evolutionary standing in vertebrate history. Diving appears to be a special case in which mechanisms of widespread occurrence have been elaborated upon to extremes. These physiologial accomplishments have resulted in the successful invasion by air-breathing animals of aquatic environments and have extended the vertical range of mammals to over 1000 m below the surface of the sea.

It must be realized that our classical concepts regarding the physiological adjustments to diving are based primarily on laboratory studies of constrained animals. Recent studies on unconstrained turtles, marine iguanas, and Weddell seals indicate that most voluntary dives are aerobic in nature, that is, no postdive lactic acid excess is evident in the blood. Fright, or escape dives in the Weddell seal may be extended for a protracted period beyond the routine dives observed, and such prolonged dives are associated with lactic acidosis. However, over 99% of dives recorded under natural circumstances are under about 20 min in duration, a period that can be tolerated by Weddell seals without the appearance of excess lactate. Although bradycardia is evident during these dives, it is far less intense than in forced laboratory dives. The full-blown set of laboratory diving responses, including intense muscle vasoconstriction, may represent an emergency reaction that is life saving in certain natural situations (predator threat, being trapped under ice, and so on). The relatively brief dives and repeated aerobic dives (maximum 10 min) exhibited by freely feeding marine iguanas are protracted to around an hour when the animals are threatened by a SCUBA diver. The recent evidence suggests that most routine natural dives are essentially aerobic in character whereas prolonged emergency dives, while life saving, are associated with intense lactic acidosis and postdive exhaustion.

6.9.2 Response to Temperature

Sudden reductions in environmental temperature elicit physiological changes that center around thermoregulation, including increased metabolism (heat production) and altered insulation (see Section 8.8.2). Patterns of response differ even in human populations. Norwegians resist cooling of the extremities by peripheral vasodilatation, making up heat loss by further increasing metabolism. Australian aborigines, on the other hand, respond by conserving heat through peripheral vasoconstriction; apparently the peripheral tissues are more tolerant of cooling.

Shivering is accompanied by elevations of 2–5 times the resting metabolic rate, and the heart rate increases to about the same degree as in exercise to a similar level of metabolism. Shivering is more intense with the onset of inspiration, an effect that has been related to the rise in arterial pressure induced by alterations in interthoracic pressure (by means of the thoracoabdominal pump). The rise in pressure may exert this influence on shivering through the pressoreceptors, because shivering can be increased by changing the pressures in the isolated carotid sinus. Thus bouts of shivering may involve interactions in both the cardiovascular and thermoregulatory areas of the central nervous system.

Sympathetic activity is a prominent feature of cold response, the cardiovascular components being venoconstriction, elevation of central venous pressure, and peripheral vasoconstriction, especially prominent in the skin. Thus insulation is functionally increased and the distribution of the cardiac output favors maintenance of core temperature.

The danger of frostbite of the extremities is ap-parently minimized in cold-acclimated rats and rabbits because these animals circulate these areas at the expense of added heat loss made good by increased metabolism. Nonacclimated animals are more vulnerable in this respect. The feet of hamsters and birch mice are rosy when in hibernation, and local metabolites may induce this vasodilatation. Upon arousal, vasoconstriction ensues as a result of sympathetic activity.

Major arteries and veins to the extremities are generally parallel and closely adherent to one another. This arrangement favors heat conservation through countercurrent exchange, the warm "core" blood losing heat to the colder venous blood returning from the periphery (see also Section 8.9.3). This principle is utilized to an amazing degree by a number of vertebrates exposed to extremes of cold. An arrangement of veins clustered around a central artery forms the *rete mirabile* (wonderful net), the venous flow being counter to the arterial flow (Figure 6-46). Such arrangements are found in the extremities of whales, seals, and a number of arctic animals in which leg temperatures of 0.5°C are not uncommon. Long-legged wading birds also have rete systems. Examples are found even in the tropics. Animals such as the sloth, armadillo, and anteater are hypersensitive to cold, shivering below 80°F. At night, when the temperature falls, the conservation of heat through countercurrent systems aids in the maintenance of core temperature without the necessity of massive increments in metabolism. The rete is not a feature of all arctic mammals and birds, however, for it is absent in well-insulated forms such as ducks, geese, foxes, and the husky. Because their insulation is so effective, these forms have a problem of heat dissipation when they are active. A rete system would work counter to temperature regulation under such circumstances.

In whales and seals there is parallel to the rete a more usual venous return, and shunts between rete and nonrete circulation can be evoked. Such an arrangement affords a wide range of peripheral temperature regulation.

Hibernating mammals may regulate at body temperatures of 2–5°C (see Section 8.12). In contrast to ectotherms, which rewarm principally on the basis of external factors, hibernators do so by increasing their heat production. This capacity necessitates an adjustable circulation at extremely low

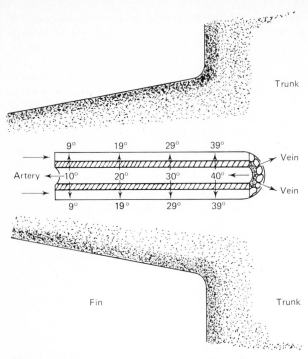

Figure 6-46
Schematic diagram of hypothetical temperature gardients in a concentric countercurrent system. [*From P. F. Scholander and W. F. Scheville. J. Appl. Physiol.,* **8,** *279–282 (1955).*]

Arterial pressure generally falls in hibernation but is variable, even at a constant heart rate, indicating periodic changes in sympathetic vasomotor activity.

Respiration is generally depressed in hibernation, but the patterns are variable in different forms. Hibernating hamsters respire 3–4 times/min, followed by apneic intervals of 2 min or more. At 4.7°C the hedgehog may not respire for nearly 1 hr, at which time breathing may be resumed for about 4 min. Synchrony between heart rate and respiration has been observed in the birch-mouse, in which there was one heart beat per respiration at 7°C. The pattern is similar to that observed in sharks, where peak blood flow to the gills occurred at maximum gill water flow.

Ground squirrels exhibit a fall in hematocrit from 57 to 40% during hibernation. The same trend is reported for bats, in which the spleen has been shown to sequester red blood cells. Upon arousal, splenic contraction occurs with a return of hematocrit to normal levels. Blood viscosity increases with declining temperature, so the trapping of red cells during periods of low oxygen consumption reduces viscosity. As a result, cardiac energy expenditure is less than would be necessary to propel the blood of a normal hematocrit.

Arousal is accompanied by heart-rate and blood-flow changes that precede changes in body temperature. As recovery commences, the front quarters (for example, the brain, forelegs, and liver) are preferentially perfused, the hind end "waking up" later. Blood flow in the arctic ground squirrel is about 16 times greater in the anterior segments than in the hind quarters during arousal.

Brown fat is a thermogenic form of adipose tissue, the metabolism and blood flow of which increases dramatically during arousal. The venous drainage of interscapular brown fat is by way of the azygous vein, and the temperature of the blood entering the heart may be elevated by some 0.6°C over that of vena caval blood in consequence of brown-fat heat production.

These adjustments provide a clear advantage to forms having high metabolic rates when active and provide means of coping with extreme situations in which the food and water supply is not adequate for survival. By reducing metabolism and utilizing endogenous stores for necessary energy production, occupation of climatic zones showing wide variation in temperature and food supply is

temperatures. The heart continues to beat as temperature decreases to near 0°C, in sharp contrast to nonhibernators, which show cardiac arrest at body temperatures between 7 and 15°C. This difference must be due to pacemaker characteristics, because it persists even in isolated hearts. ATP reformation is also much faster at low temperatures for hibernators than for nonhibernators such as the rabbit. The vagal tone is low in hibernation, and the combined influence of sympathetic innervation and adrenal catecholamines drives the heat at its maximum rate for a given temperature during arousal. Arousal apparently depends on hypothalamic driving of metabolism and heart rate.

The distribution of blood flow in hibernation is reminiscent of that of diving mammals and birds, there being a disproportionately large fraction of the cardiac output directed to heart and brain. Cardiac output falls to about one sixty-fifth of the pre-hibernating level in the 13-lined ground squirrel.

achieved. The flexibility of cardiovascular and respiratory adjustments seen in hibernators is an essential component of the overall response to these extreme circumstances.

Further discussion of responses to temperature will be found in Sections 8.9, 8.10, and 8.11.

6.9.3 Adaptations to High Altitude

Most of our knowledge of the effects of acute and chronic exposure to high altitude is derived from studies on man. The major external problem is related to the fact that as one ascends to high altitude, with the drop in barometric pressure, there is a progressive fall in the partial pressure of oxygen (Figure 6-47), which leads to a decline in alveolar P_{O_2} and subsequent problems of hypoxia. At lower altitudes the fall in alveolar P_{O_2} is not so great as the decrease in P_{O_2} of air. Decrements in alveolar P_{O_2} can be offset by increasing ventilation. With higher ascent, however, the fall in alveolar P_{O_2} is greater than that of the atmosphere. The air enter-

Figure 6-47

Effect of altitude on arterial oxygen saturation and alveolar oxygen and carbon dioxide pressures.

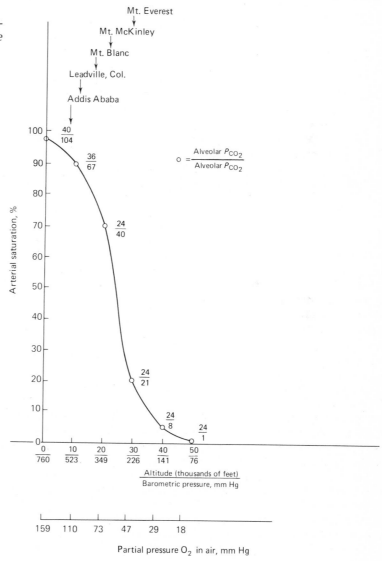

ing the respiratory system is saturated with water vapor before it reaches the exchange surfaces so, at an altitude of 9500 m, water vapor (47 mm Hg) occupies a large portion of the total gas volume. The carbon dioxide has fallen to a plateau of 24 mm Hg at this point, owing to increased ventilation, and nitrogen pressure is of the order of 134 mm Hg in the alveolus. A barometric pressure of 226 mm Hg at this altitude is fractioned in the alveolus as follows: $P_{CO_2} = 24$ mm Hg, $P_{N_2} = 134$ mm Hg, $P_{H_2} = 47$ mm Hg, and $P_{O_2} = 21$ mm Hg. The P_{O_2} of air at 9500 m is 47 mm Hg, so it is apparent that water vapor has diluted both nitrogen and oxygen.

Oxygen utilization continually depletes alveolar oxygen, but hyperventilation offsets the decline up to a limit imposed by maximum ventilatory capacity. Increasing ventilation cannot, however, bring the alveolar P_{O_2} up to that of the air. As alveolar P_{O_2} declines, the resulting decrease in arterial oxygen saturation elicits chemoreceptor-driven respiratory activity. Alveolar ventilation may increase by about 1.5 times normal. The limitation on ventilation is apparently due to the fact that the induced hyperventilation depresses blood P_{CO_2} and elevates pH. Thus the H^+ receptors of the medulla tend to suppress respiration, limiting the level of ventilation. The "defect" in blood pH is then gradually corrected toward normal by renal mechanisms, and, as the blood becomes less alkalotic, the medullary receptors again begin to augment respiratory activity. At this stage of adjustment to altitude, the alveolar ventilation may rise to 7 times normal.

Hypoxia is an effective stimulus for production of red blood cells and hemoglobin. The hormone erythropoietin has been implicated in this response. At 5500 m the red blood cell count may rise from the normal value of 5–7 million/mm³ while hemoglobin levels are elevated from 15 to 22 g%. These changes are associated with hematocrit values as high as 70%. These factors, together with an increase in blood volume of about 25%, may result in a near doubling of circulating hemoglobin. However, man acclimated to high altitude, exhibits a right-shifted hemoglobin dissociation curve (P_{50} low altitude = 26.7 torr; high altitude = 29.8 torr). The reduction in oxygen affinity is associated with higher 2,3-DPG levels. Such a shift may be an aid to unloading at the tissue level; however, it hinders loading of oxygen in the lungs. It

is difficult to see this shift in hemoglobin affinity as adaptive to high altitude. By contrast to man, the high altitude cameloids exhibit left-shifted curves.

Initially, on exposure to high altitude, cardiac output is elevated by 20–50%, but as other adjustments occur, the output returns to nearly normal levels. Cardiac output is increased during exercise at high altitudes but not to the extent that it is at sea level. This limitation is thought to be caused by the combined effects of increased blood viscosity and hypoxic suppression of the pacemaker.

During the early phases of altitude adjustment, blood flow to the skin and kidneys is limited by sympathetic vasoconstriction, while the brain, heart, and skeletal muscle flows are elevated. In time the vascularity of tissues increases, a response that is speculatively said to be due to tissue hypoxia. Whatever the mechanism, diffusion distances are decreased, a factor augmenting gas exchange. It has been shown that the tissues, as such, do not become acclimated to hypoxia—the "call for oxygen" is just as great at altitude as at sea level.

Chemoreceptor driving of respiration reaches its maximum at 5800–6500 m. Further ascent cannot depend on greater pulmonary ventilation. When arterial oxygen saturation falls below 50%, unconsciousness results, and reference to Figure 6-47 indicates that about 7500 m is the critical height.

Mental functions are generally normal up to about 3000 m. At 4000 m hypoxia is severe enough to induce sleepiness, headache, mental fatigue, and often euphoria. Night vision also suffers at high altitude because of hypoxic depression of the rod function. At altitudes above 7500 m, convulsions and finally coma and death may occur as a result of hypoxic suppression of central functions.

The environmental physiologist D. B. Dill remarked about his ascent to 4700 m in South America that his automobile seemed to be more handicapped than the vicuña at that altitude. The primary physiological adaptation of the vicuña and llama appears to be the possession of hemoglobin of high oxygen affinity. The dissociation curve of these animals is to the left of low-altitude animals, thus the oxygen-combining power at the low partial pressures of high altitude is enhanced. The hematocrit is not elevated in these animals as it is in rat or man. However, the hemoglobin content of the red blood cells is greater. At a P_{O_2} of 40 mm Hg, man transports about 14 vol % of oxygen compared to 18 vol % for the vicuña. The hematocrit of these high-altitude cameloids is low by human standards (30%) and the hematrocrit varies very little with altitude. Mammals native to high altitudes apparently meet the challenge of hypoxia by producing hemoglobin of high affinity for oxygen without the necessity of massive stimulation of the bone marrow, which results in a high hematocrit and high blood viscosity. The viscosity factor alone must reduce considerably the work of the heart compared with man at high altitude.

A number of birds have been observed at high altitude. Condors occur at 7500 m and other forms have been seen at 8500 m. Avian adaptations to altitude have not been extensively investigated, but birds can obviously extend vertically into rare atmospheric conditions further than mammals do.

Eleven species of Himalayan frogs are reported to occur above 3700 m, one approaching 5200 m. A lizard, *Leiolopisma ladacense,* was observed at 5500 m. Among snakes, European vipers, garter snakes, and the Pacific rattlesnake all occur between 2800 and 3400 m.

6.9.4 Case of Adaptive "Hypertension"—The Giraffe

Giraffes hold the record for height among living animals, their heads held at lofty levels in excess of 6 m above the ground. If the blood vessels supplying the brain are to have pressure equivalent to other mammals, the aortic pressure must exceed the norm by 2–3 times to overcome the hydrostatic head of pressure represented by the carotid artery fluid column. If this is so, rapid adjustments must also be made when the head is lowered while drinking.

An idea of the vertical distance between heart and brain of a cow and a giraffe is given in Figure 6-48. Heart-to-brain distance in the giraffe is also given in Figure 6-48. Heart-to-brain distance in the giraffe may exceed 160 cm, compared with 65 cm for the cow and 34 cm for man. The hydrostatic pressure of a fluid column 160 cm in height is about 118 mm Hg. If a mean pressure of 90 mm Hg obtains in the cephalic end of the carotid artery, the mean pressure in the aorta must be at least $90 + 118 = 209$ mm Hg. This simple-minded calculation completely ignores losses of energy due to tube dimensions but allows us to predict that the mean arterial pressure at heart level must exceed 208 mm Hg for a vertical giraffe with a heart-to-brain distance of 160 cm.

Actual measurements in the field, in addition to emphasizing the extent to which curiosity drives comparative physiologists, reveal that the mean pressure at brain level is about 90 mm Hg in a quiet, standing giraffe. This value was obtained by surgically implanting in the carotid artery a pressure transducer, the output of which was telemetered to an FM receiver (Figure 6-49). Knowledge of the distance between transducer and brain allowed correction of pressure to the brain level.

Mean pressure in the aorta of a 4.5 m high giraffe has been measured at 220 mm Hg and high carotid

Figure 6-48
Silhouettes of giraffe and cow drawn to scale. Dotted lines indicate heart position. Code: Bl, Hl, and Gl refer to brain, heart, and ground levels, respectively. [From J. L. Patterson, et al. Ann. N.Y. Acad. Sci., **127,** *393–413 (1965).]*

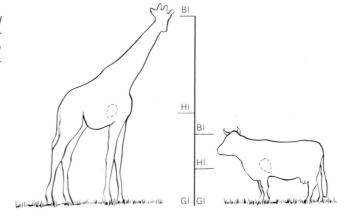

pressure was 90 mm Hg. Left-ventricular systolic pressure varied between 260 and 286 mm Hg and end-diastolic pressure between 10 and 18 mm Hg. Right-ventricular and pulmonary-artery pressures were not remarkable. The cardiac output (Fick method) is similar to that in domestic cattle of equivalent weight. Thus, the high aortic pressures result from forceful emptying of the left ventricle against a high vascular resistance.

Studies of the anesthetized, prone giraffe reveal

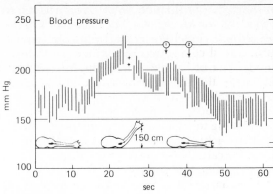

Figure 6-50
*Behavior of blood pressure on elevating the head of an anesthetized giraffe for 150 cm. Miniature manometer in root of aorta. [From R. H. Goetz, et al. Circ. Res., **8,** 1049–1058 (1960).]*

that postural changes are associated with adjustments in aortic pressure that ensure adequate brain perfusion (Figure 6-50). Unanesthetized animals, when changing from a prone to a vertical position, exhibit a transient cardiac slowing, followed by an increase in heart rate when fully upright. The carotid sinus is said to be absent and the reflexogenic basis for these adjustments is obscure.

Correlated with high aortic pressure is left-ventricular hypertrophy. This is characteristic of experimental hypertension in laboratory animals and also of human hypertension. The major arteries of the giraffe are extremely thick, and elastic con-

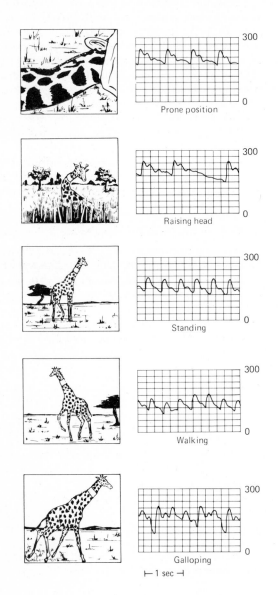

Figure 6-49
*Phasic wave forms of giraffe carotid artery blood pressure during spontaneous activity. Level at which pressure was sampled is indicated by operative site shown in top photograph; blood pressure when lying prone was 280/180 mm Hg. When the giraffe raised its head a transient bradycardia occurred. After the animal had been released to rejoin its herd and was slowly walking about, the blood pressure was as low as 150/90 mm Hg. Recapture required a vigorous chase, during which the blood pressure reached 220/150 mm Hg and the heart rate 170 beats/min. Cyclic drop in blood pressure during galloping was synchronous with the animal's front-hoof beats. [From R. L. Van Citters, W. S. Kemper, and D. L. Franklin. Science, **152,** 384–386 (1966).]*

nective tissue is abundant. The vessels are reported to show greater muscularity, and the lumina, relative to total vessel diameter, are small.

Successful competition with the myriad of grazing animals of the African plains is achieved by the giraffe through a combination of size, speed, and the habit of grazing from the tops of acacia, from which point an almost hawk's-eye view of the surrounding terrain is obtained. This vertical stratification is achieved through association of the long neck with cardiovascular adjustments, assuring adequate perfusion of the brain.

As is characteristic of most physiological problems, many new questions arise from these studies of the giraffe. It is not clear just where in the precapillary vessels the major resistance components lie. A high capillary pressure might be tolerable if plasma oncotic pressure were great enough to offset the hydrostatic pressure. Measurements of the oncotic pressure of giraffe blood reveals, however, that it is comparable to other mammals. It is likely, then, that a relatively higher resistance is to be found in arterioles or in the major supply arteries. The same problem exists in the kidney, where high filtration pressures should promote the formation of a prodigious quantity of filtrate. If glomerular filtration rate is large, the reabsorptive work of tubules would necessitate a high energy expenditure. A more likely solution is that the vascular architecture is such that hydrostatic pressure is considerably reduced before reaching capillary levels.

When running, the impact of the feet hitting the ground is associated with a fall in carotid arterial pressure (Figure 6-49). This raises the question of the effect of such impact on venous pressure and the subsequent influence on heart filling.

The "scientific safari" is becoming more rewarding to comparative physiologists as techniques advance. Modern methodology should allow the solution of many remaining problems in the field under conditions closely approaching those of the natural state.

6.9.5 Swim Bladder and Buoyancy Control

Although the swim bladder has been linked by comparative anatomists with the origin of the vertebrate lungs, in modern teleosts it is more closely associated with buoyancy control than with an accessory respiratory role. In a large fraction of the known teleosts, the swim bladder is absent. Elasmobranchs also lack swim bladders. Teleosts that have a high specific gravity as the result of a heavy skeletal structure usually have well-developed swim bladders. The bladders of freshwater forms tend to be larger than in marine species. By adjusting the gas volume of the swim bladder, a fish can remain suspended at a particular depth with the least muscle effort. With changes in vertical position, gas may be added or removed from the bladder to establish a new equilibrium.

Among the physostomous fishes, which include such forms as the salmon, the swim bladder is attached to the esophagus by a pneumatic duct. Most of these forms must rise to the surface and gulp air initially to charge the bladder. On the other hand, the physoclists, whose bladder gas is derived from the blood, have no connecting duct with the esophagus or can selectively close the duct. This type of bladder is found in many deep-sea fishes, and it is the gas-secreting bladder that is of greatest physiological interest.

A gas bladder, charged with a given volume of gas, will be compressed at greater depths if more gas is not pumped into it. Under the great pressure of the depths of the sea, buoyancy control is achieved by secretion of gas against tremendous pressure gradients; the P_{O_2} of swim-bladder gas may reach values in excess of 100 atm and is likely to exceed 200 atm in forms that live in very deep water. The composition of the gas is variable. Often there is a preponderance of oxygen, but nitrogen pressures of 10–15 atm have been measured in deep-sea fishes in which nitrogen constituted 2–15% of the gas. In the whitefish, *Coregonous,* pure nitrogen is found, deposited against a pressure of 8 atm. Even argon may be deposited against a pressure gradient, but this gas is a minor constituent. When carbon monoxide is added to the water, the toadfish *(Opsanus)* concentrates carbon monoxide in a manner similar to oxygen.

Several pertinent questions come to mind when considering the physiology of the swim bladder: (1) Given that gas pressure in the bladder may exceed many times the partial pressure of blood gases (and because the bladder is perfused with blood), what prevents the rapid depletion of gas from the bladder and its transportation to blood? (2) How is gas reabsorbed from the bladder when fishes ascend from one depth to another? (3) By what mechanism is the gas deposited against a high pressure gradient?

Structurally the swim bladder consists of three layers supplied with nerves and blood vessels. The tunica externa, consisting of dense connective tissue and often containing collagen needles or crystals of guanine, is functionally a diffusion barrier. In coelacanths (*Latimeria*) the tunica externa is rigid, owing to ossification of the layer. The submucosa is composed of loose connective tissue, blood vessels, and nerves that penetrate to the mucosa lining the lumen of the bladder. There is often a layer of smooth muscle underlying the mucosa.

Two functional portions of the swim bladder have been delineated, one secretory, the other storage and reabsorptive. The secretory portion consists of a gas gland and its associated vascular supply, the rete mirabile. Complete separation of the storage from the secretory portion of the bladder may be achieved in some forms, as is *Ctenolabrus* (a wrasse), by contraction of the smooth muscle that forms a constrictive diaphragm between the two areas. The diaphragm is mobile and during the reabsorptive phase of activity, the muscle ring constricts anteriorly near the secretory epithelium of the gas gland. Blood flow to the two areas is under reflex control. Closure of the precapillary vessels to the gas gland with concomitant opening at those of the storage and reabsorptive portion of the bladder occurs during deflation (Figure 6-51). Thus, a reflex-controlled double circulation is present, one portion associated with gas secretion, the other with storage and reabsorption. The gas gland is usually supplied via the coeliacomesenteric artery, and the reabsorptive area receives branches from the dorsal aorta. Venous drainage from the gas gland is via the hepatic portal, that of the reabsorptive portion is via the posterior cardinals.

The maintenance of constant volume is by means of a delicate balance of influx and efflux of gas, an operation that is under autonomic nervous control of the vasculature (Figure 6-51). Two types of reflexogenic behavior have been described—inflationary and deflationary. Inflation may be initiated by experimental withdrawal of gas, increasing depth, replacing saltwater with freshwater, or

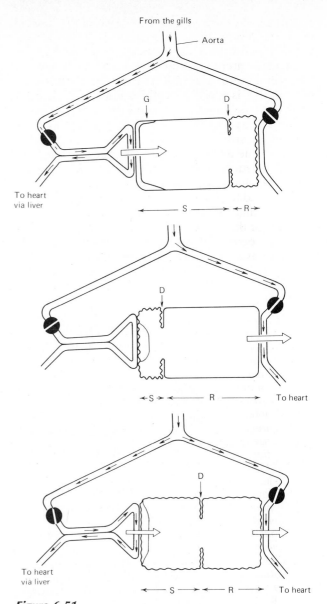

Figure 6-51
Vasomotor responses of swim bladder. **Top:** *blood circulates exclusively through secretory mucosa (S) during gas secretion.* **Middle:** *during gas resorption only resorbent mucosa (R) is supplied with blood.* **Bottom:** *at normal steady state, blood flows through both parts of swim-bladder mucosa. Blood flow is controlled by the precapillary sphincters marked out in the figure and to some extent by mechanical effects of muscularis mucosae, the movements of which are indicated (D) diagrammatically.* [*From R. Fange. Physiol. Rev.,* **46**, *299–322 (1966).*]

weighting the fish. Deflation is initiated by the opposite maneuvers.

Vagotomy blocks the capacity to secrete gas. A secretory center in the diencephalon apparently integrates secretory activity, but the precise location of the area is still uncertain. Cholinergic activity of the vagus (nerve X) is probably the initiating factor in gas secretion, but the relative roles of cholinergic and adrenergic factors are still speculative, as is the location of the receptors that initiate reflex secretion. Suggestions of likely sites for receptors include the organs of balance, proprioreceptors, and the swim-bladder wall itself. It is clear, however, that reabsorption occurs by selective circulation of the caudal portion of the bladder, and that gas secretion is held in abeyance by means of vasoconstriction of the secretory area.

The rete circulation is a natural engineering marvel. The structure consists of a bundle of afferent and efferent capillaries closely packed together in such a manner that maximum surface exposure of afferent and efferent vessles is achieved. The vertex of the rete ends in the gas gland (Figure 6-52). Scholander has characterized the rete of the long-nosed eel—the structure is 1 cm² in cross section and about 1 cm in length. The average capillary diameter is 13 μm with a circumference of 52 μm, giving a surface area of 0.0052 cm²/capillary. All the surface area is available for diffusion because it is closely packed in a checkerboard arrangement. Scholander calculates that the thickness of the walls between two adjacent capillaries is 1.5 μm. There are about 500,000 capillaries/cm². Utilizing a diffusion coefficient of 150 mm³ μm⁻¹ cm⁻² min⁻¹ atm⁻¹,

Figure 6-52

(a) Schematic drawing of capillary rete and gland in relation to the swimbladder wall, which is represented with exaggerated thickness. To the left the rete ends in the gland (e.g. in Antimora, Coryphaenoides); to the right blood vessels radiate from the gland over the surface of the swim bladder (Sebastes, Synaphobranchus). (b) Sections through rete in deep-sea fishes. A: Synaphobranchus, cross section (150×). Each polygon (square) is a single capillary. B: Same as 1, (50×). C: Coryphaenoides, cross section (150×). D: Sebastes, cross section (150×). E: Coryphaenoides, longitudinal section (150×). Many red cells in capillaries. F: Coryphaenoides. Glandular epithelium from "button," (112×). Several capillaries are seen buried among the large glandular cells. (Staining: A, B: Gomori's chrome alum–hematoxylin and phloxine, Bouin; C: periodic acid–Schiff, Bouin; D: periodic acid–Schiff, 80% alcohol; E: hematoxylin and phloxine, Bouin: F: Gomori's chrome alumhematoxylin and eosine, Bouin.) Sections and photographs courtesy of the Department of Anatomy, Harvard Medical School. [From P. F. Scholander. Biol. Bull., **107**, 260–277 (1954).]

(a)
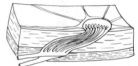
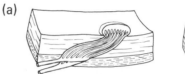

(b)
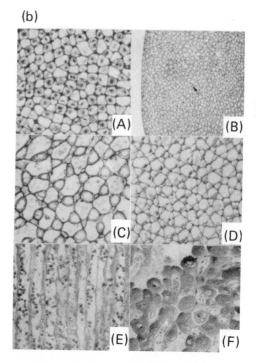

he calculated the diffusion across the length of a capillary to be 0.52 mm³ min⁻¹ atm⁻¹ capillary⁻¹. Since there are 250,000 pairs of capillaries in this rete, the total diffusion capacity is approximately 130 ml min⁻¹ atm⁻¹.

The maintenance of a high partial pressure of gas in the bladder, which is in contact with the capillaries of the distal end of the rete, presents a physical problem because capillary blood should contain oxygen at a pressure no greater than in the surrounding water. Such capillary blood in close contact with the bladder gas would favor rapid washout and consequent deflation. The rete arrangement is the apparent solution to this problem, and, by means of countercurrent exchange, a high partial pressure is maintained near the vertex (Figure 6-53). At extreme depths, the gas tension must drop about 3000-fold along the 1-cm length of rete. The combination of countercurrent exchange with vasoconstriction of the reabsorbent and storage area of the gland keeps gas leakage to a minimum.

The mechanism of gas secretion if still not thoroughly understood, but there is sufficient experimental evidence to formulate a fairly strong theory of secretion. During inflation, an increase in glycolysis occurs in the gas-gland epithelium and there is a large production of lactic acid and carbon dioxide. It is argued that the increased acidity, via the Bohr effect, would favor unloading of the hemoglobin. Many deep-sea fishes, but by no means all, exhibit a peculiarity in the behavior of their hemoglobin. In the presence of an elevated P_{CO_2}, hemoglobin dissociates appreciably at elevated P_{O_2}. This behavior is known as the **Root effect** (this effect also occurs in a number of shallow-water fish species). In addition to the possible role of the Bohr and Root effects, the addition to the blood of any agent reducing the solubility of oxygen would favor its liberation. Such an effect could be produced by liberation of salts, heat, lactate, or ammonium bicarbonate. This idea is referred to as **salting-out**, and lactate is a strong candidate, because during secretion the venous effluent from the rete shows an elevated lactate content.

Kuhn, who demonstrated the operation of a countercurrent multiplication system in the loop of Henle of the mammalian kidney (see Section 7.8.3), has extended the same principle to the swim bladder. The necessary assumption is that the permeability to water is limited between afferent and efferent members of the rete but that nitrogen and oxygen are freely permeable. The circumstance should lead to progressive accumulation of dissolved gases at the vertex as shown in Figure 6-54. By placing a salting-out device at the vertex, gas can be liberated into the bladder at very high

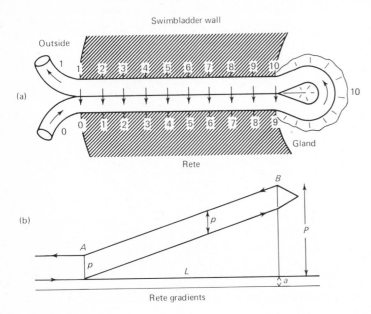

(a)

Swimbladder wall

Outside

Gland

Rete

(b)

Rete gradients

Figure 6-53
Schematic presentation of the countercurrent exchange principle in the swim bladder. (a) Afferent and efferent capillaries. The numbers illustrate the linear gradient set up from the inside of the outside of the structure. The arrows signify the constant transrete diffusion gradient. (b) Graphic representation of the gradients. Code: L, length of the rete; p, transrete gradient; P, gas tension in the swim bladder; a, gas tension of inflowing blood. [From P. F. Scholander, Biol. Bull., **107,** *260–277 (1954).]*

6 CIRCULATION

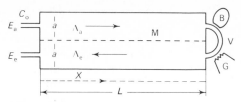

Figure 6-54
Hairpin countercurrent system of the rete is schematically depicted. Channels Λ_a and Λ_e correspond to the afferent and efferent limbs of the retial capillaries. The gas gland (G), is a dosage device through which either salt and/or, for example, lactic acid can be added to the blood passing at the vertex (V). The membrane (M) is permeable for the gas, N_2 or O_2, and impermeable for the solvent (water) and the substances added through G. E_a and E_e are the places where the blood enters and leaves the system. The gases concentrated by countercurrent multiplication may be transferred to the bladder (B) in contact with V as soon as the particular partial pressure exceeds the one inside B. [From W. Kuhn, A. Ramel, H. J. Kuhn, and E. Marti, Experientia, **19**, 497–552 (1963).]

pressures. This combination of countercurrent multiplication and salting-out appear to provide a plausible approach to gas secretion in the swim bladder.

Asphyxia leads to oxygen reabsorption from the swim bladder, but these oxygen stores last only a limited time. In oxygen-poor water, the complication of diffusion of reabsorbed oxygen across the gills and into the water further limits the utility of swim-bladder stores in hypoxic situations. Recharging the physostome swim bladder with air is, however, of utility to swamp dwellers, in which the hemoglobin is relatively insensitive to carbon dioxide and the affinity for oxygen at low partial pressures is great.

Another interesting case of oxygen secretion has been reported for fishes. A rete mirabile is found in the choroid layer of the eyes of a number of teleosts. The similarity of the choroid rete and that of the swim bladder suggested to J. B. Wittenberg and B. A. Wittenberg that the functions might be similar and that the rete of the eye might serve to create high oxygen tensions behind the retina. Such a system would provide a high pressure of oxygen, supportive of the metabolic demand of the relatively avascular retina. Measurements of the oxygen pressure with an oxygen microelectrode inserted into the vitreous chamber revealed, for sea robin, fluke, puffer, jack, and menhaden, oxygen pressures ranging from 250 to 800 mm Hg and arterial P_{O_2} of about 160 mm Hg. Species lacking the rete, such as elasmobranchs and eel, exhibit low tensions, less than 30 mm Hg. There was a correlation of rete size and vitreous oxygen tension with habit, the fast-swimming predators showing a more elaborate rete than the bottom dwellers, which depend more on chemoreceptors for location of food.

ADDITIONAL READING

JOURNALS
Journals that frequently publish research papers on the comparative physiology of circulation: *Journal of Comparative Physiology; Comparative Biochemistry and Physiology; Journal of Experimental Biology; American Journal of Physiology; Acta Physiologica Scandinavica; Circulation Research.*

REVIEWS
See *Annual Review of Physiology* and *Physiological Reviews* for periodic up-to-date overviews of various aspects of circulatory physiology.

MONOGRAPHS, SYMPOSIA AND COMPENDIA
Graubard, M. *Circulation and Respiration—The Evolution of an Idea*, Harcourt, Brace and World, New York, 1964.
Guyton, A. C., and C. E. Jones (eds.). *Cardiovascular Physiology, MTP International Review of Science, Physiology Series One*, vol. 1, University Park Press, Baltimore, 1974.
Hamilton, W. F., and P. Dow (eds.). *Handbook of Physiology*, sec 2, vols. 1–3 *(Circulation)*, American Physiological Society, Washington, D.C., 1962.
Robb, J. S. *Comparative Basic Cardiology*, Grune and Stratton, New York, 1965.
Symposium: "Cardiorespiratory adaptations in the transition from water breathing to air breathing," *Fed. Proc.*, **29**(3), 1118–1153 (1970).

BIOPHYSICS OF BLOOD VESSELS
Bergel, D. H. (ed.). *Cardiovascular Fluid Dynamics*, Academic Press, London, 1972.
Caro, C. G., T. J. Pedley, and W. A. Seed. "Mechanics of the circulatoin." In A. C. Guyton and C. E. Jones (eds.). *Cardiovascular Physiology, MTP International Review of Science, Physiology Series One*, vol. 1, University Park Press, Baltimore, 1974.

Fahraeus, R., and T. Lindqvist. "The viscosity of the blood in narrow capillary tubes," *Am. J. Physiol.*, **96**, 562 (1931).

Fung, Y. C., N. Perrone, and M. Anliker (eds.). *Biomechanics: Its Foundatons and Objectives*, Prentice-Hall, Englewood Cliffs, NJ, 1972.

Fung, Y. C. "Mechanical properties of blood vessels." In P. C. Johnson (ed.). *Peripheral Circulation*, John Wiley and Sons, New York, 1978.

Taylor, M. G. "Hemodynamics," *Ann. Rev. Physiol.*, **35**, 87–116 (1973).

MICROCIRCULATION

Intaglietta, M., and P. C. Johnson. "Principles of capillary exchange." In P. C. Johnson (ed.). *Peripheral Circulation*, John Wiley and Sons, New York, 1978.

Krogh, A. *The Anatomy and Physiology of Capillaries*, Yale University Press, New Haven, 1922.

Zweifach, B. W. "Microcirculation," *Ann. Rev Physiol.*, **35**, 117–150 (1973).

SYSTEMIC CIRCULATION (Various Organs)

Johnson, P. C. (ed.). *Peripheral Circulation*, John Wiley and Sons, New York, 1978.

VENOUS SYSTEM

Shepherd, J. T., and P. M. Vanhoutte. *Veins and Their Control*, Saunders, London, 1978.

CARDIAC DYNAMICS

Sagawa, K. "The ventricular pressure-volume diagram revisited," *Circ. Res.*, **43**, 677 (1978).

INVERTEBRATE CIRCULATION

Johansen, K., and A. W. Martin. "Circulation in the cephalopod, *Octopus dolfleini.*," *Comp. Biochem. Physiol.*, **5**, 161–176 (1962).

Martin, A. W., and K. Johansen. "Adaptations of the circulation in invertebrate animals," *Handbook of Physiology*, sec. 2, vol. 3 (*Circulation*), American Physiological Society, Washington, D.C., 1965.

———. "Circulation in invertebrates," *Ann. Rev. Physiol.*, **36**, 171–186.

FISHES

Johansen, K. "Comparative physiology: gas exchange and circulation in fishes," *Ann. Rev. Physiol.*, **33**, 569–612 (1971).

Jones, D. R., B. L. Langille, D. J. Randall, and G. Shelton. "Blood flow in dorsal and ventral aortas of the cod, *Gadus morhua*," *Am. J. Physiol.*, **226**, 90–95 (1974).

Mott, J. C. "The cardiovascular system." In M. E. Brown (ed.). *The Physiology of Fishes*, vol. 1, pp. 81–108, Academic Press, New York, 1957.

Randall, D. J. "Functional morphology of the heart in fishes," *Am. Zool.*, **8**, 179–189 (1968).

———. "The circulatory system." In H. S. Hoar and D. J. Randall (eds.). *Fish Physiology*, vol. 4, Academic Press, New York, 1970.

Satchell, G. H. *Circulation in Fishes*, Cambridge University Press, London, 1971.

AMPHIBIANS

Brady, A. J. "Physiology of the amphibian heart." In J. A. Moore (ed.). *Physiology of the Amphibia*, pp. 211–250, Academic Press, New York, 1964.

Johansen, K. "Cardiovascular dynamics in the amphibian *Amphiuma tridactylum*," *Acta Physiol. Scand.*, **60**, suppl. 217 (1964).

Johansen, K., and D. Hanson. "Functional anatomy of the heart of lungfishes and amphibians." *Am. Zool.*, **8**, 191–210 (1968).

Jones, D. R., and G. Shelton. "Factors affecting diastolic blood pressures in the systemic and pulmocutaneous arches of anuran amphibia," *J. Exp. Biol.*, **57**, 789–803 (1972).

Morris, R. W. "Function of the anuran conus arteriosus," *J. Exp. Biol.*, **61**, 503–520 (1970).

REPTILES

Bartholomew, G. A., V. A. Tucker, and A. K. Lee. "Oxygen consumption, thermal conductance, and heart rate in the Australian skink *Tiliqua scincoides*," *Copeia*, **No. 2**, 169–173 (1965).

Heath, J. E. "Venous shunts in the cephalic sinuses of horned lizards," *Physiol. Zool.*, **39**, 30–35 (1966).

Kirby, S., and G. Burnstock. "Pharmacological studies of the cardiovascular system in the anesthetized sleepy lizard (*Tiliqua rugosa*) and toad (*Bufo marinus*)," *Comp. Biochem. Physiol.*, **28**, 321–331 (1969).

Millard, R. W., and K. Johansen. "Ventricular outflow dynamics in the lizard, *Varanus niloticus*: responses to hypoxia, hypercarbia and diving," *J. Exp. Biol.*, **60**, 871–880 (1974).

Tucker, V. A. "Oxygen transport by the circulatory system of the green iguana (*Iguana iguana*) at different body temperatures," *J. Exp. Biol.*, **44**, 77–92 (1966).

White, F. N. "Circulation in the reptilian heart (*Caiman sclerops*), *Anat. Rec.*, **125**, 417–423 (1956).

———. "Functional anatomy of the heart of reptiles," *Am. Zool.*, **8**, 211–219 (1968).

———. "Circulation." In C. Gans, and W. Dawson (eds.). *Biology of Reptilia*, Academic Press, New York, 1978.

BIRDS

Berger, M., J. S. Hart, and O. Z. Roy. "Respiration, oxygen consumption and heart rate in some birds during rest and flight," *Z. vergl. Physiol.*, **66**, 201 (1970).

Jones, D. R., and K. Johansen. "The blood vascular system of birds." In *Avian Biology*, vol. 2, pp. 157–285, Academic Press, New York, 1972.

Simmons, J. R. "The blood-vascular system." In A. J. Marshall (ed.). *Biology and Comparative Physiology of Birds,* vol. 1, pp. 345–362, Academic Press, New york, 1960.

Sturkie, P. D. *Avian Physiology,* 2nd ed., Cornell University Press, Ithaca, NY, 1965.

MAMMALS

Berne, R. M., and M. N. Levy. *Cardiovascular Physiology,* The C. V. Mosby Co., St. Louis, MO, 1972.

Folkow, B., and E. Neil. *Circulation,* Oxford University Press, New York, 1971.

Guyton, A. C., and C. E. Jones (eds.). *Cardiovascular Physiology, MTP International Review of Science, Physiology Series One,* vol. 1, University Park Press, Baltimore, 1974.

Rushmer, R. F. *Cardiovascular Dynamics,* W. B. Saunders Co., Philadelphia, 1970.

PHYSIOLOGY OF DIVING

Anderson, H. T. "Physiological adaptations in diving vertebrates," *Physiol. Rev.,* **46**(2), 212–243 (1966).

Bennett, P. B., and D. H. Elliott. *The Physiology and Medicine of Diving and Compressed Air Work,* The Williams and Wilkins Co., Baltimore, 1975.

deBurgh Daly, M., and J. E. Angell-James. "The 'diving response' and its possible clinical implications, Internat. Med., **1,** 12, 1979.

Elsner, R. D., D. L. Franklin, R. L. Van Citters, and D. W. Kenny. "Cardiovascular defense against asphyxia," *Science,* **153**(3739), 941–949 (1966).

Kooyman, G. L., M. A. Castellini, and R. W. Davis. "Physiology of diving in marine mammals," *Ann. Rev. Physiol.,* **43,** 343, 1981.

Rahn, H., and T. Tokoyama (eds.). *Physiology of Breath-Hold Diving and the Ama of Japan,* **Publ. 1341,** National Academy of Science-National Research Council, Washington, D.C., 1965.

Scholander, P. F. "Physiological adaptation to diving in animals and man," *Harvey Lect.,* **57,** 93 (1962).

Slonim, N. B. (ed.). *Environmental Physiology,* pp. 399–436, The C. V. Mosby Co., St. Louis, MO, 1974.

EXERCISE

Gleeson, T. T., G. S. Mitchell, and A. F. Bennett. "Cardiovascular responses to graded activity in the lizards *Varanus* and *Iguana.*" *Am. J. Physiol.,* **239,** R174, 1980.

Jones, D. R., and D. J. Randall. "The respiratory and circulatory systems during exercise." In W. S. Hoar and D. J. Randall (eds.). *Fish Physiology,* vol. 7, Academic Press, New York, 1978.

Nadel, E. R. "Circulatory and thermal regulations during exercise," *Fed. Proc.,* **39,** 1491, 1980.

Rutledge, P. S. "Circulation, oxygen transport, and activity in the crayfish, *Am. J. Physiol.,* **240,** R99, 1981.

HIGH ALTITUDE

Dill, D. B. (ed.). *Handbook of Physiology,* sec. 4, American Physiological Society, Washington, D.C., 1964.

Slonim, N. B. (ed.). *Environmental Physiology,* pp. 350–375, The C. V. Mosby Co., St. Louis, MO, 1974.

SWIMBLADDER

Baines, G. W. "Blood pH effects in eight fishes from the teleostean family Scorpaenidae," *Comp. Biochem. Physiol.,* **51A,** 833–843 (1975).

Berg, T., and J. B. Steen. "The mechanism of oxygen concentraton in the swimbladder of the eel," *J. Physiol. (Lond.),* **195,** 631–638 (1968).

Fange, R. "Physiology of the swimbladder," *Physiol. Rev.,* **46**(2), 299–322 (1966).

Kanwisher, J., and A. Ebeling. "Composition of the swimbladder gas in bathypelagic fishes," *Deep Sea Res.,* **12,** 211–217 (1957).

Kuhn, W., A. Ramel, H. J. Kuhn, and E. Marti. "The filling mechanism of the swimbladder: generation of high gas pressures through hairpin countercurrent multiplication," *Experientia,* **19,** 497–511 (1963).

Root, R. W., E. C. Black, and L. Irving. "Effect of carbon dioxide on the oxygen-combining power of whole and hemolyzed marine fish blood," *Anat. Rec.,* **72,** suppl. 46 (1938).

Malcolm S. Gordon

7

WATER AND SOLUTE METABOLISM

7.1 INTRODUCTION

Another of the groups of regulatory mechanisms essential to the survival of animals controls the water content of their bodies and the concentrations of nongaseous substances in their body fluids. Here again we are dealing with a large and complex field. Different animal groups use different organs to carry out various of the component processes. The detailed mechanisms by means of which these processes are accomplished also vary and frequently are used in differing combinations. Finally, the range of substances regulated is very large, involving compounds such as hormones and vitamins (often present in nanomolar concentrations) as well as the principal osmotically significant solutes such as sodium chloride.

The size of the field forces us to restrict our coverage. The topics we shall discuss are those related to the pathways and mechanisms of water and solute exchanges between animals and their external environments. The solutes of concern will be those comprising 10% or more of the total osmotic concentration of the body fluids in some form. We shall also say a little about the problems of water and solute exchanges between animal body fluids and the cells of their tissues. The Additional Reading list includes sources covering most of the topics not considered here.

7.2 HISTORY

Scientific awareness of the fact that body-fluid compositions of organisms differ from the composition of the waters of the external world apparently developed during the latter eighteenth and early nineteenth centuries in conjunction with the development of methods for determining the physical and chemical properties of solutions. Extensive work in this area seems to have begun only in the latter nineteenth century, initially in France. Paul Bert in Paris was a leader in early studies of the body fluids of animals other than mammals.

Much of the stimulus for research in this field has come from medically oriented studies, especially those directed at understanding how kidneys operate. Even today the largest number of workers and the greatest amount of effort is directed toward understanding of mammalian kidney function in health and disease.

More purely zoological knowledge has, however, been a continuing spin-off from the medical work. Homer Smith in New York worked out many of the basic aspects of water and solute metabolism in fishes during the course of a long and distinguished career in mammalian kidney physiology. August Krogh in Copenhagen contributed extensively to many areas in the field while engaged in studies of much more direct medical import. More recently the development of reliable micropuncture procedures for obtaining tiny fluid samples from single kidney tubules, together with precise and elegant microchemical procedures for measuring quantities of substances dissolved in these samples, has revolutionized our understanding of the functioning of the excretory organs of many animals, not only the mammals. These developments have been achieved as the result of the efforts of many workers, but much of the stimulus derived from studies on frog kidneys by A. N. Richards and colleagues in the 1920s. Richards and most of his successors have been staff members in various medical schools.

Advances in biophysical and biochemical methods and procedures have permitted most major advances in the area of water and solute metabolism. Undoubtedly the most revolutionary set of methodological advances derived from the introduction of isotopes, both stable and radioactive. The earliest applications appear to have been made in Copenhagen in the latter 1930s by George von Hevesy, August Krogh, and various students and collaborators. Since World War II the ready availability of radioisotopes has permitted their use in all phases of the field. Most of what we know about kinetics (for example, fluxes of materials across membranes and rates of uptake or excretion) is based upon radioisotope experiments.

7.3 BASIC PRINCIPLES

7.3.1 Osmotic and Solute Requirements of Living Tissues

Aside from the insects, the vast majority of kinds of animals live in fairly direct association with the oceans. Most insects and a smaller, but still large, number of kinds of other animals live in more or less direct association with the freshwaters of the world. The remainder of animal types live in or depend upon waters of a wide range of fairly unusual compositions. Table 7-1 summarizes data on the compositions of some of the major and also some of the more unusual aquatic habitats with which animals are associated. It is apparent that animals are able to cope with a very broad range of water types.

The ways in which they cope are diverse. However, a uniform general feature is the fact that the range of internal solute concentrations compatible with life is much narrower than the external range. This is particularly true for the commoner inorganic cations: sodium, potassium, calcium, and magnesium.

It is not at all clear why animals should expend the effort they do to control the composition of their extracellular and intracellular fluids. The reasons are presumably related in complicated ways to evolutionary history. Modern advances in knowledge appear to have done away with one of the most appealing theories—that the body-fluid compositions of most animals (which are fairly uniform, especially in terms of relative concentrations of major components) reflect the composition of the primordial oceans in which their ancestral forms first evolved. Put another way, this theory postulates that living systems first evolved their dependencies and sensitivities to solute concentrations and their ratios in an ocean considerably different in composition from today's. The evolution of regulatory systems has been such as to work toward maintenance of this earlier composition in the relevant parts of the bodies of organisms.

Geophysicists and geochemists appear now to be fairly sure that the average composition of the oceans has been very nearly the same as it is today for more than two billion (2×10^9) years. Fossils

TABLE 7-1. Osmotic and Major Solute Concentrations of Some Bodies of Water

Body of Water	Average Osmotic Concentration (mOsm/liter)	Na+ (mmol/liter)	K+ (mmol/liter)	Ca2+ (mmol/liter)	Mg2+ (mmol/liter)	Cl- (mmol/liter)	SO4^2- (mmol/liter)
"Standard" ocean (chlorinity 19°/oo)	1000	470	10	10	54	548	38
"Average" river water	~1	~0.08	~0.01	~0.30	~0.09	~0.05	~0.08
Little Manitou Lake, Canada	~2000	780	28	14	500	660	540
Great Salt Lake, Utah	~6000	3000	90	9	230	3100	150

Source: H. U. Sverdrup, M. W. Johnson, and R. H. Fleming. *The Oceans*, Prentice-Hall, Inc., Englewood Cliffs, N.J., 1942; G. E. Hutchinson. *A Treatise on Limnology*, John Wiley & Sons, New York, 1957.

of recognizably multicellular animals are known from no longer ago than about 600 million (6 × 10⁸) years, so the theory is probably not valid. Further, as this chapter will show, there is a great deal of diversity in the composition of animal body fluids. There is, in fact, much more diversity than the workers who originally made these postulations were aware of. We simply cannot say that the diversity exists only in forms of relatively recent evolutionary origin, whereas uniformity is present in all older forms. The fossil record is not complete enough for that. There is also no reason to assume that Precambrian seas lacked microhabitats such as exist today or that all evolutionary events took place only in the "average" ocean. Life is polyphyletic. Accordingly, it seems more prudent for the moment simply to accept the fact that many processes in organisms are sensitive to changes in amounts or ratios of solutes in their surroundings and to await further information as to the possible historical basis for the sensitivities. In the meantime, studies of the mechanisms involved in these sensitivities form an active and interesting research field.

7.3.2 Quantities and Definitions

There are many ways of expressing the concentrations of solutes in aqueous solutions. We shall concern ourselves only with the major units used frequently in the physiological literature.

The osmotic pressure of a solution is one of its biologically most significant properties. Section 7.4.2 discusses the nature of osmosis and its physical basis. Osmotic pressure is one of the four colligative properties (properties depending on the *number* of solute particles present, not the *kinds* of particles) of solutions. The other three colligative properties are freezing-point depression, vapor-pressure lowering, and boiling-point elevation. Direct measurements of osmotic pressure are difficult, especially on small samples containing unstable substances such as are usually present in biological fluids. All four colligative properties are linearly related to each other, so it is possible to determine the osmotic pressure indirectly by measuring one of the other properties. The most easily measured property is freezing-point depression.

Pure water freezes, by definition, at 0.00°C. A 1-molal aqueous solution (1 mol/kg of water) of

an ideal nondissociating solute freezes at −1.86°C. Thus we say that water has a molal freezing-point depression of 1.86°C. There is no need in most animal physiological studies to convert measurements of freezing-point depression into actual osmotic pressures. Indeed, it is usually more convenient to convert the temperature figures into a concentration equivalent (osmotic concentration or total concentration). The commonest conversion is to define a freezing-point depression of 1.86°C as being equal to an osmotic concentration of 1 Osmolal (1000 mOsmolal). For solutions of concentrations below 1 Osmolal, for most biologically important solutes, an error of no more than a few percent results if molar units are used rather than molal. Thus the units to be used throughout this chapter will be Osmolar (Osm/liter) and milli-Osmolar (mOsm/liter). The quantity 1000 mOsm/liter = 1.86°C, the freezing-point depression of any solution, whatever its composition.

Freezing-point depression measurements have been and continue to be the most widely used method of estimating osmotic concentrations. However, methods perfected in recent years permit rapid and convenient measurement of vapor-pressure lowering of small samples of biological fluids.

Data on total concentrations of the external media of animals are frequently expressed in units other than those related to osmotic pressure. There is little procedural uniformity in the literature in this regard and it is often difficult to convert from the figures given to the osmotic concentrations of interest. A common unit used for freshwater is the total solid content (mg/liter). The most usual unit for seawater and various dilutions of seawater is salinity. The relative proportions of the various solutes in seawater are extremely constant throughout the oceans of the world. Thus it is possible to determine the concentration of one substance and calculate from this result the total concentration and the concentration of most other components. Because nonionized organic substances are usually present in only small amounts, it is also possible to determine salinity from electrical conductivity. Salinities are usually expressed as parts per thousand (ppt; ‰). Many physiological papers use "percent seawater," defining 100% seawater as 35‰ salinity (near the average salinity for the oceans of the world), with osmotic concentrations close to 1000 mOsm/liter.

It is not possible to make a clear distinction between seawater and freshwater. As indicated in Table 7-1, many inland waters can be very concentrated, with compositions often quite different from the sea. Even the more dilute inland waters that we think of as "fresh" are extremely variable in their compositions and total concentrations. As will be seen later, the nature of the solutes present often crucially determines the responses of organisms to the total concentrations they encounter.

Inshore oceanic waters, especially in bays and estuaries, are very variable in concentration. Variations depend on such factors as the balance of local precipitation and evaporation, the flow rates of streams and rivers, and the amount of mixing produced by tides and winds. Ecologists have developed elaborate classifications of brackish-water habitats based upon fine distinctions in environmental salinities. These distinctions often appear to reflect matters of real significance to organisms in these habitats, because distributions and abundances are frequently closely correlated with salinities. Here, however, we shall not concern ourselves with terminology and will discuss osmotic concentrations only in terms of mOsm/liter, salinity, or percent seawater.

Whatever their chemical compositions may be,

two solutions having the same osmotic concentration are said to be *isosmotic*. If two solutions differ in osmotic concentrations, the more concentrated is said to be *hyperosmotic* to the more dilute, and the latter is *hypoosmotic* to the former. The terms *isotonic, hypertonic,* and *hypotonic* are often used as synonyms for isosmotic, and so on. This is not strictly correct, because tonicity is not defined physically, as is osmotic pressure, but is defined biologically, in terms of the responses of living cells to immersion in a solution. A given solution is isotonic to a particular cell if the cell neither swells nor shrinks when immersed. Hypertonic solutions cause shrinkage, hypotonic solutions swelling. Only the physically based terms will be used in this chapter.

Insofar as possible we shall discuss concentrations of specific solutes in terms of millimoles per liter (mmol/liter). For most of the ions or compounds of concern, 1 mmol/liter is osmotically very nearly equivalent to 1 mOsm/liter in the physiological concentration ranges.

Discussions of water metabolism in terrestrial animals must take into account the vapor-pressure lowering of the body fluids. At a given temperature, the greater the osmotic concentration of a solution, the lower the vapor pressure. The differences in vapor pressure of water existing between the interiors of terrestrial animals' respiratory tracts and their environments (vapor-pressure deficits) are among the important determinants of rates of evaporative water loss in these forms. The breathing rate also influences the rate of evaporation (by replacing saturated air inside with unsaturated air from outside), so evaporative water losses are often expressed in units related to oxygen consumption, such as milliliters of H_2O evaporated per cubic centimeter of O_2 consumed.

Aquatic animals can be separated into categories having quite different physiological properties on the basis of their ability to survive in environments of differing osmotic concentrations. Most naturally occurring osmotic fluctuations are due to variations in the inorganic salt content of the medium. Thus most aquatic and amphibious animals can tolerate only relatively narrow ranges of environmental osmotic concentrations. These ranges may involve low absolute concentrations (as in freshwater organisms) or fairly high absolute concentrations (as in marine organisms). Whatever the absolute con-

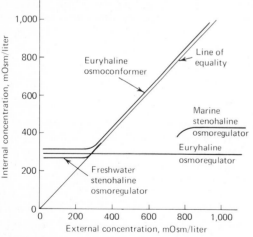

Figure 7-1
Some of the commonest patterns of regulation of plasma osmotic concentration found among aquatic vertebrates and the terminology associated with these patterns.

centration range, animals with limited tolerance for variations in the environment are called *stenohaline.* The relatively few forms that can tolerate large variations in environmental concentration are called *euryhaline.* There is no sharp dividing line between the two groups. Indeed, the degree of euryhalinity shown by an individual animal may vary considerably with age, environmental temperature, details of environmental composition, and other factors.

Whatever the degree of an animal's stenohalinity or euryhalinity, it can be further categorized on the basis of the pattern of changes that occur in its internal, body-fluid osmotic concentration in response to external variations. If internal concentration varies more or less in parallel with external concentration, the animal is called an *osmoconformer.* If internal concentration remains fairly or absolutely constant in the face of external changes, the animal is called an *osmoregulator.* Once again there is no sharp dividing line between the two categories. Euryhaline forms frequently are good osmoregulators over some particular range of environmental concentrations and are equally good osmoconformers over some other range. Figure 7-1 illustrates some of the basic patterns of relations between internal and external concentrations found in animals.

Another phenomenon is usually closely associated with the degree of development of an organism's osmoregulatory capacities. This is its ability to control the water content of its body. Osmoconformity frequently results from a combination of a lack of ability to control body-water content and also a lack of ability to control body-solute content. Conversely, osmoregulation often is a manifestation of a well-developed ability to carry on both processes. For the moment, however, regardless of any correlations that may exist between these processes, the important fact is that some animals change in water content (the usual way of putting it is to say that they change in body volume) in response to changes in environmental osmotic concentration, whereas others do not. Those that change in volume are called *volume conformers;* those that do not are called *volume regulators.*

At this point a question arises: Do animals really try to regulate their volumes and the osmotic concentrations of their body fluids or is the degree of regulation of these quantities simply an automatic result of regulation of the concentrations of

the major solutes present in the body fluids? This question is virtually impossible to answer in general. The detailed answer varies with the species being considered. However, the question points out the existence of still another basis for categorizing animals physiologically.

As will be seen later on, given patterns of osmoregulation, or the lack of it, can be associated with quite different patterns of changes in the relative amounts of important body-fluid solutes. Variations in blood osmotic concentration in one form may be due primarily to changes in sodium chloride concentration, with little or no change in other components. Another species, however, may show equivalent osmotic changes, but these are due to changes in concentration of an organic solute such as urea, with little or no change in sodium chloride or other components. Or the process can be more subtle, as it is in many marine invertebrates. Many such animals have body-fluid osmotic concentrations identical with that of the external seawater. Their internal concentrations are usually 95% + produced by the same inorganic ions as are present in seawater, but the relative amounts of the various ions are different. The control of concentrations of specific solutes, most often inorganic ions, is called *ionic regulation.* Those animals that have powerful mechanisms for doing this are *ionic regulators;* those with weak mechanisms are *ionic conformers.* There is no multicellular organism possessing a distinct body fluid that is known to be completely lacking in *any* ionic regulatory abilities, including those for trace elements.

The external environment for terrestrial animals is most often air. In the present context, the most important variation in the composition of air is variation in the vapor pressure of water. The relationships existing between the body-fluid compositions of terrestrial animals and this external water-vapor pressure are complicated by a wide range of influences, as will be made clear in later discussions. As a result there is no terminology for physiological subgroups of terrestrial animals comparable to that just described for the aquatic forms.

Finally, having said so much about the properties of internal and external environments, we must also describe a few properties of the membranes that keep these environments apart. Living membranes are highly complex and organized entities, varying widely in structure at all levels of the scale, and varying equally widely in physicochemical properties and biological activities. Detailed quantitative discussions of many aspects of membrane properties may be found in several of the books cited in the Additional Reading list.

It will be enough for our purpose to point out that virtually all biological membranes are permeable to some degree to both water and a variety of solutes. There is no biological membrane that is ideally semipermeable (that permits the passage through itself of only solvent molecules, no solute molecules). The ways in which water and solutes move across membranes will be discussed briefly in Sections 7.4.2 and 7.4.3.

The mathematical relationships used to describe and analyze events in membranes are almost as diverse as the membrane processes themselves. However, in many situations one can describe the main features of what is going on by an equation having the general form of the Fick diffusion equation

$$\frac{ds}{dt} = - DA \frac{dc}{dx}$$

where $ds/dt =$ instantaneous rate of movement of substance across the area of membrane being considered, in moles per unit time

$D =$ diffusion coefficient (for simple cases in which thermal diffusion is the major process taking place) or analogous coefficient; this coefficient is a measure of the permeability of the membrane for the subject substance, ordinarily expressed as moles per unit area of membrane per unit concentration or analogous gradient per unit time

$A =$ area of membrane

$dc/dx =$ concentration gradient for substances across membrane (for simple thermal-diffusion situations) or an analogous expression for driving force producing movement.

7.3.3 Methods

Progress in understanding in many areas of the field of water and solute metabolism has hinged upon development of reliable micro and ultramicro techniques for obtaining uncontaminated samples of various animal body fluids and for carrying out precise analyses of microgram or even nanogram quantities of their components. The major stimuli for the development of these procedures have been the increasing interest in understanding mechanisms of osmotic and ionic regulation in various small animals and the need for detailed knowledge of events taking place in such tiny places as the lumina of single kidney tubules. The techniques that have been developed are numerous, diverse, and ingenious. New developments occur at a rapid rate. Many of the newest developments are summarized in such journals as *Analytical Chemistry, Analytical Biochemistry,* and the *Journal of Scientific Instruments.* We shall note here only the fact that successful workers in the field of water and solute metabolism must not only be good physiologists but also good microanalytical biochemists and, increasingly, biophysicists.

Despite the extensive development of the newer procedures, an assortment of older, more traditional, experimental approaches continues to be useful. The use of clearances for studying many aspects of kidney physiology has, if anything, expanded as a result of its combination with newer procedures for micropuncture, stop-flow analysis, and so on.

The clearance approach considers a kidney, or any similarly functioning excretory organ, to be a black box having a fluid input (blood) and a fluid output (urine plus a smaller amount of blood). Three basic processes go on inside the box, the end product of these processes being the urine. The three processes are filtration of part of the incoming blood plasma (in the glomeruli in kidneys), tubular reabsorption of a considerable amount of the filtered water and solutes, and tubular secretion of some waste materials and other solutes.

Consider now a single solute, such as the amino acid glycine. There are several basic quantities one must know to be able to state in detail what the overall performance of a kidney is with respect to glycine. These quantities are

1. The rate of supply of glycine to the kidney. This means knowing the rate of blood flow to the glomeruli and tubules of the organ (the renal

plasma flow or RPF) and the plasma concentration of glycine (assuming it is all transported in solution in the plasma). Let P be the plasma concentration of the solute of interest.

2. The rate of removal of glycine from the circulation by the kidney. This means knowing the rate of urine flow from the organ (V is the urine volume produced during the observation period) and the urine concentration of glycine (U is the urine concentration of the solute).

Only one additional quantity is needed to permit a quite precise statement of the net result of all three of the internal processes going on inside the kidney. This quantity is the rate of filtration of plasma by the kidney glomeruli (glomerular filtration rate, or GFR).

The renal clearance for any substance is defined as the ratio UV/P. This quantity represents the volume of plasma from which the organ would have had to remove 100% of the solute in order to excrete the total amount of solute excreted during the experimental period. For many vertebrate kidneys it is necessary to measure the clearances of only two substances, in addition to that of the solute under study, to specify all the quantities needed.

The two additional substances are

1. A compound that is both filtered by the glomeruli and also virtually quantatively secreted by the tubules. Such a compound will be essentially 100% removed from the entire blood supply to the glomeruli and tubules of a kidney in one pass through the kidney. Its clearance, therefore, is an estimate of the total blood flow to the organ that is significant from the excretory standpoint. A widely used compound for this purpose is p-aminohippuric acid (PAH). (Note that RPF measured as PAH clearance is not necessarily truly an estimate of total renal blood supply. Many kidneys, especially those of the lower vertebrates, have both arterial and venous blood supplies in which only part of the incoming blood actually comes into close contact with glomeruli or tubules.)
2. A compound that is *only* filtered by the glomeruli and neither secreted nor reabsorbed by the tubules. The clearance of such a compound directly measures GFR. The most widely used substance for this purpose is the polyfructoside inulin.

Once these characteristics are known, it is possible to calculate the following important quantities among others:

1. The total supply of solute to the kidney during the experimental period $= P$ (RPF).
2. The fraction of excretorily significant blood flow to the kidney that was filtered (filtration fraction, or FF) $=$ GFR/RPF.
3. The amount of filtered water that was reabsorbed by the tubules $=$ GFR $- V$.
4. The amount of solute that was either secreted or reabsorbed by the tubules $= UV - P$(GFR). Positive values for this difference indicate secretion, negative values reabsorption.

The U/P ratio of a solute is a good measure of the concentrating or diluting ability of an animal's kidney for that substance. The ratio of a solute's clearance to the inulin clearance (GFR) is a measure of the degree to which active tubular processes play a role in the handling of the substance.

There are many complications and pitfalls that need to be avoided in applications of clearance methods to animals other than mammals. The research literature is the best source of information about these.

7.3.4 Theory

The diversity and complexity of mechanisms used by animals in their water and solute metabolism can be more readily understood if placed in the perspective of a few basic ideas borrowed from physical science. One of the most useful of these is the idea of the steady state.

Most organisms can be viewed physically as discrete systems more or less isolated from, but carrying on a number of exchanges with, the outside world. If the outside world is reasonably uniform and constant in time, most organisms not in the midst of major internally directed upheavals (such as molting or pregnancy) will establish a fairly constant set of internal conditions. This means that the amounts of water and various solutes in their bodies will be constant in time, resulting in constancy in body-fluid and tissue composition. The organism will be in a steady state. [It is important to remember that a steady state may be, but is not necessarily, an equilibrium. A great deal of what organisms do physiologically is directed at maintaining steady states (homeostasis) that are very far from thermodynamic equilibria.]

Given a steady state, one can apply what is called the principle of continuity. The amount of any nonmodifiable substance (such as sodium ion) or quality (such as heat) taken in by an organism or one of its organ systems in a steady state during a given period of time must exactly equal the amount of that substance or quality lost during the same period. If the balance is not exact, the organism or organ system cannot be in a steady state and must either be accumulating or depleting its supply of whatever is being exchanged.

This simple formulation has a wide range of implications. Steady-state conditions are obviously the simplest ones to use in beginning a study of any aspect of water and solute metabolism. They are easily defined, readily quantified, and usually reproducible. Their use may enable an experimenter to calculate important quantities that he cannot measure directly. Apparent imbalances between input and output are often the first clues a researcher may find that indicate an organism is using some unusual mechanism or pathway. By studying organisms and organ systems in steady states under appropriate combinations of environmental factors such as salinity and temperature, it is possible to determine the nature of the longer term responses to environmental changes. With care, detailed and exact balance sheets for water and solutes can be written that will immediately permit saying under which environmental conditions drinking is or is not important, under which circumstances specialized extrarenal pathways for salt loss may be needed, or when changes in kidney output result from changes in glomerular filtration.

Once a reasonably detailed series of steady-state pictures have been obtained for the particular function of interest, systematic studies can be undertaken of the effects of perturbations. The kinetics of transient responses to all kinds of environmental fluctuations can be used to fill in the dynamic aspects of physiological adaptations. Steady-state studies establish basic response patterns and can supply information about detailed mechanisms. Perturbation studies establish the kinetic properties

of systems, usually tell more about underlying mechanisms, and provide the basis for tying in laboratory work with free-living animals in nature.

7.3.5 Research Design

Several aspects of research design come up repeatedly in studies of water and solute metabolism. These are fundamentally the same as those underlying most studies of the effects of any physical environmental variable on the physiology of an organism.

First, any ecologically oriented physiological study must begin with extensive knowledge of what the subject organism does while on its own, in nature. This means the gathering of field observations on such items as the range of salinities the animal normally encounters (for aquatic forms) or how often it drinks and the composition of the water it drinks (for terrestrial forms).

Second, the limits of its tolerances and, if possible, voluntary choices of environmental variations must be determined. This may mean, for an aquatic species, determining the minimum and maximum salinities compatible with life under environmentally realistic specified conditions of temperature, oxygen tension, and so on. Salinity preferences would also be determined. For terrestrial forms it is often necessary to know, for example, how long the animal can survive while limited to different specified amounts of water of various compositions under environmentally realistic combinations of humidity, temperature, and so on. The animal's own preferences must also be known.

Third, as indicated in Section 7.3.4, the duration of the animal's exposure to a particular set of conditions is a critical variable. Short-term or acute experiments provide data for perturbation analyses. Long-term or chronic experiments provide data for steady-state analyses. There are usually striking differences between the results of the two types of experiments. Figure 7-2 illustrates these differences for some experiments on the osmotic regulatory abilities of a fish.

Many physiologists have spent considerable time debating the specific words to be used to describe the differences between short-term and long-term responses of animals. A variety of arguments have been advanced over the years in favor of calling the long-term situations acclimations, acclimatiza-

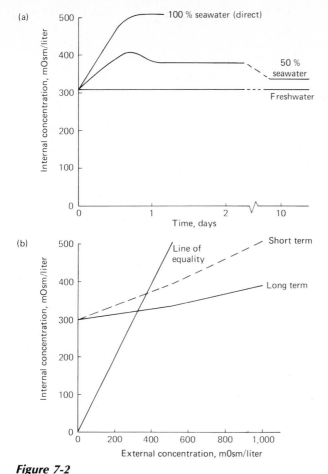

Figure 7-2
(a) Time course of changes in mean plasma osmotic concentration in the brown trout (Salmo trutta) *subjected to sudden changes in environmental salinity at 20°C in winter. These fish can survive transfer to 100% seawater, but only after 7–10 days adaptation to 50% seawater. Direct transfer to 100% seawater from freshwater results in death after 1–2 days. (b) Patterns of regulation of mean plasma osmotic concentration in brown trout 24 hr after direct transfers from freshwater to various seawater dilutions (short-term line) and 8–10 days after such transfers (long-term line). Transfers to salinities above 50% seawater (approximately 500 mOsm/liter) done in two stages (8–10 days in 50% seawater, then 8–10 days in higher salinity). Experiments at 20°C in winter. [Data from M. S. Gordon. J. Exp. Biol.,* **36**, *227–252 (1959).]*

tions, adjustments, or adaptations. In this chapter we shall use the general term *adaptations* for all regulatory processes, trusting that the context in each case will make the nature of the subject clear.

The data obtained from field work, short- and long-term tolerance experiments, and preference experiments establish the basic nature and the outer limits of the physiological capacities of an organism and the types of environmental challenges the organism must contend with. With this as background, it is possible to ask rational questions about how it survives, what the mechanisms are that produce such things as the patterns of regulation of body-fluid concentrations that will be found, how these mechanisms are integrated and controlled, and so on. The application of common sense, insight, skill, care, diligence, and a measure of artistic judgement to these tasks will ultimately result in a reasonable approximation to an understanding of how an animal gets along in its world.

7.4 BASIC STRUCTURES AND PROCESSES

7.4.1 Sites of Osmotic and Solute Regulation

The phenomena of osmotic and solute regulation result from the operation and interaction of a range of processes that are distributed among a number of body organs. The details of which organs do what vary among animal groups. There is also considerable quantitative variation in the relative importance of given organs in different species within a given group. The specifics of these situations will be discussed later.

Among aquatic animals, the organs most often involved in water exchange with the environment are the gills, the gut, and the kidneys. In many forms, notably freshwater groups such as the amphibians, the skin also is a major pathway for water movement.

Internal redistributions of body water also occur in response to environmental changes. Thus a shift in environmental osmotic concentration may produce not only water movements across the gills and skin of an aquatic beast, but also changes in the volumes of body fluid enclosed within tissue cells, in the extracellular spaces between tissue cells, in the blood, hemolymph, or other circulating fluids, and in body spaces such as the peritoneal and pericardial cavities.

Among terrestrial animals, the organs most often involved in water exchanges with the environment are the lungs, gut, and kidneys. The skin sometimes also plays an important role, but this is very variable. The cloacas of reptiles and birds sometimes are also significant in this connection.

Changes in body hydration in terrestrial animals also produce internal redistributions of water similar to those taking place in aquatic forms.

Exchanges of solutes between aquatic animals and their environments occur by means of the same organs that are used in water exchanges plus several others—at least in particular groups. Crustaceans, for example, lose considerable amounts of calcium and magnesium salts, as well as organic matter, every time they molt off their old integument. Sharks and their relatives use special organs called *rectal glands* as an important route for eliminating excess monovalent ions they obtain from their food or elsewhere. The skeletons of most animals, aquatic as well as terrestrial, serve as important storage organs for divalent ions.

Similar situations exist in terrestrial forms. Most reptiles and birds faced with severe problems of water shortage, such as marine and desert forms, compensate for limitations on efficiency of kidney salt excretion mechanisms through the use of highly efficient monovalent ion excretion processes in one or another of the glands *(salt glands)* located in the orbits of their eyes. Many insects resist poisoning by toxic substances they encounter by sequestering these substances in a variety of complex chemical ways and then storing them in specialized organs within their bodies.

7.4.2 Mechanisms of Water Movement

No situations have been studied in adequate detail in which anyone has been able to demonstrate unequivocally the existence of an active molecular-level mechanism for producing movements of water. As far as we can tell at present there is no such thing as biological active water transport. All water movements taking place in organisms appear to result from either molecular diffusion down thermodynamic activity gradients (for most transmembrane movements) or from the operation of various kinds of mechanical pumps (for most bulk movements through macroscopic channels, such as the circulation of blood).

Chapter 6 described the operations of the major mechanical pump possessed by most animals—the heart. Other bulk water movements, such as expulsion of urine from urinary bladders or peristaltic movements along intestines, result from the contractions of striated or smooth muscle cells in other organs.

Water movements by molecular diffusion are restricted in animals almost entirely to movements across membranes. One of the most important categories of transmembrane water movements is that of osmotic movements.

A detailed discussion of osmosis requires much more space than we have available. Several such discussions, written at different levels of mathematical sophistication, are included in the books cited in the Additional Reading list. It seems worthwhile, however, to outline here one of the simplest physical models of osmosis, that based on kinetic theory. Although this model is a highly simplified approximation to reality, it can serve as a basic framework thta will make many complex phenomena more intelligible. This model does not account for all of the phenomena of osmosis. Specifically, it does not account for the fact that water flows across membranes produced by hydrostatic pressure gradients are the same as those produced by equivalent osmotic pressure gradients. Figure 7-3 should assist in visualizing what is being discussed.

Consider a membrane that separates two volumes of fluid from each other. One fluid is a pure solvent (in this case water), the other a solution of an ideal nondissociating noninteracting nonelectrolyte. The molecules of both solvent and solute are perfectly elastic round balls, and the solute molecules are larger than the solvent molecules. The membrane is very thin and ideally semipermeable. That is, it has pores that are large enough to permit free passage of solvent molecules but no passage of solute molecules.

The entire system is maintained at a constant temperature, which is high enough so that all the molecules in both fluids are moving about at fairly high speeds. The molecules move freely in all three dimensions and suffer repeated elastic collisions

Figure 7-3

Possible molecular mechanism of osmosis based on idealized and simplified kinetic theory approach. Arrows indicate direction of net flow of solvent.

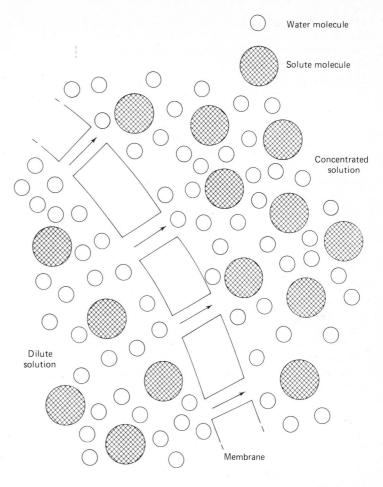

with others of their own kind, with molecules of the other kind present (in the solution), and with the membrane. If the point of collision with the membrane happens to be the location of a pore, a solvent molecule having a component of its velocity in the transmembrane direction will pass right through.

Per unit area and unit time, the membrane suffers a given average number of collisions from solvent molecules on the pure solvent side. A proportion of these molecules passes through the pores. A similar situation exists on the solution side, but the number of solvent molecule collisions per unit time and area is lower than on the pure solvent side because some solvent molecules are deflected from hitting the membrane by colliding with solute molecules. The proportion of solvent molecules

hitting the membrane that pass through the pores is the same as on the pure solvent side but the total number is less. The result is a net diffusion of solvent from the pure solvent (more generally, the more dilute) side to the solution (more concentrated) side. This diffusion of solvent is osmosis.

The real world is obviously much more complex. Molecules are not elastic spheres, and many important solutes are electrically charged. There are solute-solvent and solute-solute interactions of many kinds. There is a spectrum of energies for particles in solution. Membranes have pores of varying sizes, often bearing complex configurations of electrical charges. The model must be greatly elaborated to approach even remotely quantitative agreement with reality.

In recent years there has been a renewal of interest in an alternative approach to providing a possible mechanism for osmosis. This approach originated in the late nineteenth century and was widely discussed by physical chemists in the early twentieth century. Contrary to models like the one just described, which emphasize the activities of solute molecules, this other approach emphasizes solvent properties. In particular it postulates that, at least in water solutions, osmotic phenomena can be accounted for on the basis of changes in the internal hydrostatic pressure of the solvent. Water in solutions is considered to be in a continuous state of tension, thus under negative hydrostatic pressures. The details of the arguments on both sides of this question are too complex for discussion here. Interested students are referred to the Additional Reading list for more information.

Membrane properties are a particularly crucial set of determinants of the rates of solvent movements. It is possible to specify operationally the most important of these properties without needing to investigate the structural and functional details of any particular membrane. We have just considered the extreme situation of a totally impermeant solute. The opposite extreme is a freely permeant solute to which the membrane reacts as if it were solvent (again, water in our case). In the latter situation, transmembrane concentration gradients of such solutes will not result in any shifts of volume across the membrane, but only in a diffusion of solute which is matched in volume by a reciprocal solvent (water) movement. In intermediate cases the osmotic water flow is a fraction of the theoretical ideal (for a totally impermeant solute). This fraction is characteristic for each membrane-solute combination and is called the reflection coefficient (σ). For completely impermeant solute molecules, $\sigma = 1$; for freely permeant solutes, $\sigma = 0$; for solute molecules to which the membrane offers some resistance to diffusion, $0 < \sigma < 1$.

Bulk osmotic movements are not the only kind of water movements that occur across living membranes. It has become clear in recent years that many of the mechanisms producing transmembrane movements of solutes also produce water movements. These water movements often take place in directions opposite to those of the gross osmotic gradients that exist, but are especially apparent in situations in which there is no osmotic

gradient (isosmotic transport). Much remains to be learned about the nature of these solute-related processes, but the main phenomenon in several important cases appears to be what is called *local osmosis.* In local osmosis the active solute-transport processes in the membrane create small regions of higher solute concentration on one side of the membrane than on the other. Water then crosses the membrane in response to the local osmotic gradient, regardless of the macroscopic situation. The routes followed by water in moving across membranes by local osmosis may be different from those followed in the usual bulk osmosis.

7.4.3 Mechanisms of Solute Movement

Detailed explanations of the mechanisms whereby the variety of solutes that move across biological membranes do so are, if anything, more complex than explanations of osmosis. Such discussions, in any case, are more appropriately a part of cellular physiology. Here we shall simply list and briefly define six of the major mechanisms that occur widely.

1. **Passive diffusion.** Classical molecular diffusion down electrochemical activity gradients. Its motive force is the thermal energy of the molecules diffusing.
2. **Active transport.** Movement of substances against their electrochemical activity gradients, apparently most often via reversible combination with some type of intramembrane carrier molecule. The energy required is obtained from metabolism.
3. **Facilitated diffusion.** Movement down electrochemical gradients but at higher rates than for classical diffusion. Sometimes movement is mediated by carrier molecules, sometimes by rapid removal of substance from the downstream end. The second situation is really classical diffusion but with electrochemical gradients made locally steeper than bulk gradients. The energy required is derived from metabolism in both cases.
4. **Exchange diffusion.** Membranes are sometimes virtually impermeable to passage of a solute by classical diffusion but measurably permeable via carrier-mediated transport. Carrier-mediated exchanges not resulting in net movements of solute (unidirectional fluxes are equal) are called

exchange diffusion and require the presence of solute on both sides of the membrane. The energy needed is derived from metabolism.

5. **Pinocytosis.** Movement up or down electrochemical gradients mediated by formation of intramembrane vacuoles on one side of a cell, enclosing some of the outside medium. The vacuoles then migrate across the cell and discharge their contents at the other surface. Unless coupled with processes that modify the composition of the vacuolar contents during its movement, pinocytosis is a relatively nonspecific process. The energy required is derived from metabolism.

6. **Solvent drag.** Movements of solute in the same direction as movements of solvent at rates faster than predicted by solute electrochemical gradients and permeability coefficients. Considered to be the result of transfer of momentum to solute molecules from solvent molecules flowing through pores or channels. The energy involved derives from the process producing solvent flow.

It is impossible to predict which of these mechanisms, or which combination of mechanisms, may be operating in any unstudied membrane. Passive diffusion and some type of active transport are virtually ubiquitous, but their relative significance varies widely. All six processes appear to be involved in the osmotic and ionic regulatory mechanisms used by animals.

One last important physical phenomenon remains to be mentioned. This is the fact that Gibbs-Donnan equilibria also make important contributions to the distributions of solutes that occur on both sides of most living membranes. Most animal membranes separate fluids containing different concentrations of high molecular weight polyelectrolytes (usually proteins) of molecular sizes too large to pass through membrane pores. As long as the membranes are permeable to the smaller sized electrolytes present (such as inorganic ions) transmembrane differences in nondiffusible electrolyte concentrations automatically produce changes in electrochemical gradients that result in important modifications in the distributions of the diffusible ions (see Section 9.4.1). The resulting equilibrium distributions are termed *Gibbs-Donnan equilibria.* The forces leading to establishment of these equilibria are among the most important pro-

ducing passive diffusional movements of ions across living membranes.

Differences in the concentrations of large nondiffusible molecules on the two sides of living membranes also have osmotic effects. Differences in such *colloid osmotic pressures* (also called *oncotic pressures*) are among the most important driving forces for water movements between the various fluid compartments in the bodies of animals (see Sections 6.4.4 and 7.5.2).

7.5 REGULATION OF BODY-FLUID COMPOSITION IN AQUATIC VERTEBRATES

Table 7-2 summarizes data on blood-plasma osmotic and solute concentrations found in representatives of each of the major phylogenetic and physiological categories of aquatic vertebrates. The diversity of patterns of regulation that exists is apparent. Even greater diversity is found among the invertebrates.

7.5.1 Patterns of Regulation

We will now discuss the physiological mechanisms used by each of the vertebrate groups listed in Table 7-2 to produce their characteristic regulatory patterns.

7.5.2 Steady-state Mechanisms in Freshwater Vertebrates

The plasma osmotic concentrations of freshwater vertebrates are uniformly much higher than the concentrations of their external media. The extent to which this circumstance produces problems of osmotic flooding with water and continuing losses of needed solutes varies from one form to another. Some of the major factors producing these variations are differences in absolute and relative amounts of permeable membrane surfaces exposed to the medium, differences in absolute permeability to water and solutes of exposed membranes, differences in degree of development of active solute-uptake mechanisms in exposed membranes, and differences in efficiency of excretory organs (primarily kidneys) in preventing loss of solutes. All

TABLE 7-2. Steady-State Patterns of Osmotic and Solute Regulation in Plasma of Representative Aquatic Vertebrates

Group	Mean Osmotic Concentration, Δ (mOsm/liter)	Na+ mmol/liter	Na+ %Δ	Cl- mmol/liter	Cl- %Δ	Organic Compounds (Comprising 10+ % of Δ) Compound	mmol/liter	%Δ
Freshwater forms								
Lamprey (Lampetra fluviatilis)	245	~110	45	95	39	—	—	—
Primitive bony fish (Erpetoichthys calabaris)	199	91	45	74	37	—	—	—
Advanced bony fish (Perca fluviatilis)	294	154	52	120	39	—	—	—
Cartilaginous fish (Carcharhinus leucas)*	~650	245	38	220	35	Urea	170	26
Amphibian (Rana esculenta)	250	105	42	75	30	—	—	—
Reptile (Alligator mississippiensis)	280	140	50	110	39	—	—	—
Marine forms								
Hagfish (Eptatretus stouti)	950	520	~50	500	~48	—	—	—
Lamprey (Petromyzon marinus)	400	190	48	175	44	—	—	—
"Ordinary" bony fish (Lophius americanus)	350	200	~55	160	~45	—	—	—
Antarctic bony fish (Trematomus bernacchii)	635	285	45	250	39	Glycoprotein supercooling stabilizer	—	Up to ~20
Cartilaginous fish (Squalus acanthias)	1020	290	28	240	24	Urea; trimethylamine oxide	350 60	35 6
Coelacanth (Latimeria chalumnae)	920	197	22	187	20	Urea; trimethylamine oxide	377 122	41 14
Reptile (Chelonia mydas)	~330	160	48	~140	42	—	—	—
Bird (Larus glaucescens)	~320	150	47	~130	40	—	—	—
Mammal (Physeter catodon)	370	170	46	120	32	—	—	—

Source: R. Morris. J. Exp. Biol., **35**, 649–665 (1958); P. Lutz. Copeia. **No. 1**, 119–123 (1975);T. Thorson, C. Cowan, and D. Watson. Physiol. Zool., **46**, 29–42 (1973); E. J. Conway. Physiol. Rev., **37**, 84–132 (1957); R. A. Coulson and T. Hernandez. Biochemistry of the Alligator, Louisiana State University Press, Baton Rouge, La., 1964; W. McFarland and F. Munz. Comp. Biochem. Physiol., **14**, 383–398 (1965); J. Mathers and F. Beamish. Comp. Biochem. Physiol. **49A**, 677–688 (1974); R. P. Forster and F. Berglund. J. Gen. Physiol., **39**, 349–359 (1956); G. H. Dobbs, III and A. L. DeVries. Mar. Biol., **29**, 59–70 (1975); J. Burger and W. Hess. Science, **131**, 670–671 (1960); R. Griffith, B. Umminger, B. Grant, P. Pang, and G. Pickford. J. Exp. Zool., **187**, 87–102 (1974); W. N. Holmes. Arch. Anat. Microsc. Morphol. Exp., **54**, 491–514 (1965); M. Sudzuki. Tohoku J. Exp. Med. **5**, 419–427 (1924).

* See also Table 7-7.

four factors, but especially the latter three, are under some degree of control by the organism and can be varied to permit adjustments to changing conditions. Figure 7-4 illustrates the major routes for water and solute exchanges in various groups.

The nonmigratory, stenohaline species of freshwater lampreys and bony fishes appear to be physiologically similar to each other. There are quantitative differences between the groups, but considerable qualitative similarity. In both groups the general impermeability of the skin (which may derive in large part from its mucous coating) is more than offset by the much larger area of highly permeable surface forming the lining of the pharyngeal cavity and especially the gills (gill surface areas are very difficult to measure, but in most aquatic animals probably are 2–10 times general body-surface area). It seems to be impossible for aquatic animals to produce respiratory surface membranes that are impermeable to water and, to a lesser extent, most small solute molecules. These permeable surfaces permit continuing osmotic entry of water into the body fluids of these forms and continuing outward passive diffusion of needed solutes such as sodium chloride.

Volume regulation in the steady state appears to be excellent in the animals in these two groups, which means that the water uptake must be completely compensated for in some way. This problem is handled primarily by the kidneys, which continually produce large volumes of quite dilute urine. Renal glomerular filtration rates are high and there is relatively little water reabsorption. Most of these animals have no, or only very small, urinary bladders, so the urine is voided almost as fast as it is produced. As one would expect from these circumstances, these forms generally avoid swallowing much of their medium. They do, however, swallow some.

The solute problem involves primarily sodium and chloride ions, since these solutes have the largest concentration gradients to the outside. Diffusional losses, however, are not the whole problem—in fact, they usually seem to constitute 50% or less. Additional major salt losses result from the operation of the kidneys in bailing these animals out. The kidney tubules are quite efficient, but not perfect, water deionizers. The urine they produce contains only low concentrations of sodium and chloride ions, but these concentrations

are high enough so that, in combination with the high urine-flow rates, substantial quantities of salt are lost per unit time.

Plasma osmotic and ionic regulation in the steady state is generally excellent in these animals. This means that the continuing salt losses are usually quite accurately compensated for. The food they eat probably plays a variable, but often small, part in this. Very little quantitative information is available, but it seems probable that herbivorous and plankton filtering forms obtain quite a lot of their needed salt from their food. Life for the predatory, carnivorous types is probably often a matter of long periods between meals. Indeed, some of the freshwater lampreys do not feed at all as adults. Even the herbivores and plankton feeders seasonally will go for long periods without eating, or eating very little. This is especially true in the temperate zones during winter periods of low water temperatures and short day lengths.

In the absence of food, compensation for salt losses is accomplished by means of active uptake mechanisms operating in the gills and perhaps also the pharyngeal membranes. These active pumps are capable of transporting ions, probably both sodium and chloride ions, into the blood streams of these animals from external concentrations as low as 0.02 mmol. The minimum concentrations from which net uptake is possible varies and is probably capable of considerable adjustment by a given animal, as well as considerable adaptation. Thus the minimum concentration of sodium chloride from which a goldfish or rainbow trout can achieve net salt uptake decreases by more than a factor of 10 if the fish is partially depleted of salt by having been kept for some time in running distilled or deionized water. The chances are good that similar differences exist between different populations of given species that live in different streams or lakes having different salt concentrations (this is specifically true for several kinds of amphibians).

Table 7-3 summarizes some of the available information on aspects of water and salt balance in representative freshwater vertebrates having significant areas of permeable integument. The elasmobranchs are omitted, as are the apodans (caecilians). Only incomplete information is available for these groups.

Freshwater elasmobranchs are a lot more common and widely distributed than most people think.

Stingrays and sawfish are found far upstream in most tropical rivers, and several kinds of sharks also occur in various tropical lakes and streams. Our lack of physiological knowledge of these animals is only one of the consequences of the relatively small amount of interest there has been in many aspects of tropical biology.

The greatest part of what we know about water and solute metabolism in freshwater elasmobranchs derives from an expedition to Malaya made by Prof. Homer Smith and his wife in 1928. The work on this expedition was done under primitive conditions and, it now appears, on physiologically abnormal animals. Although there has been an in-

crease of interest in the freshwater sharks of Lake Nicaragua (see Table 7-2), so far, at least, this work has involved other aspects of water and solute metabolism than those studied earlier by the Smiths. Thus, for now, we must rely on the Smiths' results in many important aspects.

The pattern of plasma composition listed for the Lake Nicaragua shark in Table 7-2 occurs also in the Malayan elasmobranchs. The high concentration of urea, added to the sodium chloride present, results in a much higher osmotic concentration than is found in the other groups of freshwater fishes. This large osmotic gradient presumably means equally large rates of water influx, hence

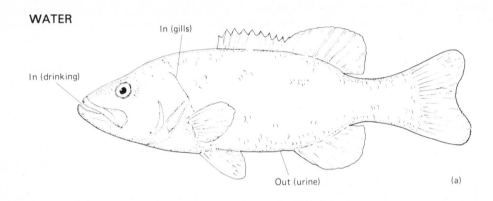

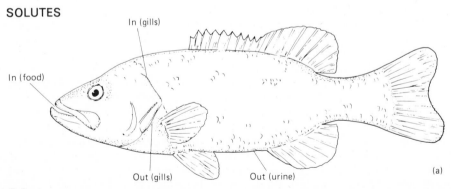

Figure 7-4
Major water and solute exchange pathways in freshwater aquatic vertebrates. Upper series illustrates water movements, lower series solute movements. Groups of organisms are (a) lampreys, bony fishes, and chondrichthyans (Amazon stringrays), (b) amphibians, and (c) reptiles. No data are available for fully aquatic freshwater mammals.

WATER

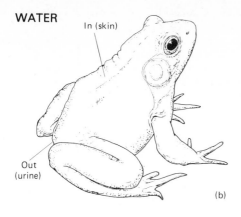

In (skin)

Out
(urine)

(b)

In (drinking)

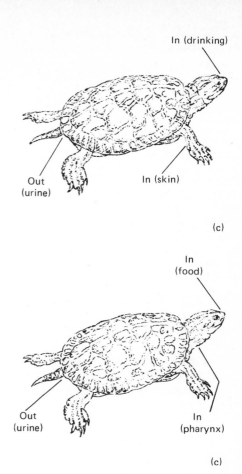

Out
(urine)

In (skin)

(c)

SOLUTES

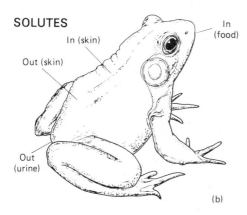

In (skin)

Out (skin)

In
(food)

Out
(urine)

(b)

In
(food)

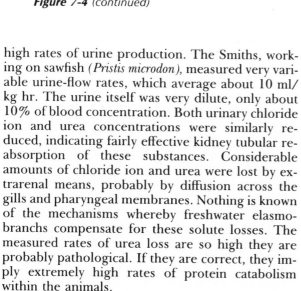

Out
(urine)

In
(pharynx)

(c)

Figure 7-4 (continued)

high rates of urine production. The Smiths, working on sawfish *(Pristis microdon)*, measured very variable urine-flow rates, which average about 10 ml/kg hr. The urine itself was very dilute, only about 10% of blood concentration. Both urinary chloride ion and urea concentrations were similarly reduced, indicating fairly effective kidney tubular reabsorption of these substances. Considerable amounts of chloride ion and urea were lost by extrarenal means, probably by diffusion across the gills and pharyngeal membranes. Nothing is known of the mechanisms whereby freshwater elasmobranchs compensate for these solute losses. The measured rates of urea loss are so high they are probably pathological. If they are correct, they imply extremely high rates of protein catabolism within the animals.

An interesting discovery relating to freshwater elasmobranchs was made in the upper Amazon River. Several genera of stingrays of the family Potamotrygonidae have long been known to inhabit both the Amazon and Orinoco Rivers in northern South America. They are not uncommon at distances as great as 4000–4500 km from the sea. Measurements of osmolality and the concentrations of the major inorganic and organic constituents of the body fluids of stingrays representing the genus *Potamotrygon* (captured far above the mouth of the Amazon River) showed that these elasmobranchs are physiologically virtually identical with freshwater teleost fishes. In particular, they lacked any significant blood concentrations of either urea or trimethylamine oxide (Table 7-7). This discovery demonstrates that freshwater elasmobranchs are

Quantity	Lamprey (Lampetra fluviatilis)	Rainbow Trout (Salmo gairdneri)	Frog (Rana esculenta)	Aquatic Larval Salamanders (Ambystoma spp.)	Alligator (Alligator mississippiensis)
Urine flow (ml/kg hr)	15	4	10	12	0.8
Renal glomerular filtration rate (ml/kg hr)	—	7	~25	25	8.3
Urine Δ (mOsm/liter)	—	30	~80	~15	220
Urine [Cl⁻] (mmol/liter)	0.7	10	1	4	3
Urinary Cl⁻ loss (μmol/kg hr)	10	40	10	50	2.5
Extrarenal Cl⁻ loss (μmol/kg hr)	800	<10	16	—	—
Minimum external [Cl⁻] for net uptake (mmol/liter)	0.02	0.02	0.01	—	—

* All figures are average values.

Source: B.-J. Wikgren. *Acta Zool. Fennica,* **71,** 1–102 (1953); P. O. Fromm. *Comp. Biochem. Physiol.,* **10,** 121–128 (1963); A. Krogh. *Skand. Arch. Physiol.,* **76,** 60–74 (1937); J. Aceves, D. Erlij, and G. Whittembury. *Comp. Biochem. Physiol.,* **33,** 39–42 (1970); R. H. Alvarado and S. R. Johnson. *Comp. Biochem. Physiol.,* **16,** 531–546 (1965); R. A. Coulson and T. Hernandez. *Biochemistry of the Alligator,* Louisiana State University Press, Baton Rouge, La., 1964.

physiologically diverse, and brings into question long held presumptions concerning the requirement, by elasmobranch tissues, for significant urea concentrations in order to function normally. The Amazon stingrays are also relatively stenohaline (maximum tolerable salinity near 21‰).

The only freshwater amphibians that have been studied thoroughly are a few species of frogs and toads and a smaller number of salamanders. Most amphibians are really terrestrial and will be discussed in Section 7.7. The primarily aquatic forms appear to be uniformly similar to freshwater bony fishes, with only two really important modifications. First, urine produced by the kidneys flows into a well-developed bladder where it is stored for some time. The bladder walls are capable of important supplementation of kidney reabsorptive capacities for both water and solutes. Thus, if the circumstances require it (as in a dry period requiring torpor in a mud burrow), aquatic amphibians can cease voiding urine and conserve both water and solutes for metabolic uses, all by means of increasing bladder reabsorption of these substances.

Second, in adult amphibians (frog tadpoles and salamander larvae are like fishes) the entire general integument is permeable to water. Thus most osmotic uptake of water occurs by way of the skin. The skin in these animals is also the site of many of the other processes that take place in the gills of fishes. These include passive diffusional losses of solutes and active uptake of inorganic ions from the environment. The active uptake mechanisms appear usually to be sodium pumps, but a few cases of chloride pumps have been described.

The truly aquatic freshwater reptiles, such as turtles, some snakes, and alligators and their relatives, are poorly known physiologically. Their blood compositions are similar to those of the bony fishes and amphibians (see Table 7-2). Their integuments are almost completely impermeable to water. In addition, many freshwater reptiles drink water.

The few forms studied have relatively low GFRs and produce urine at fairly low rates (see Table 7-3). The urine is either dilute or nearly isosmotic with the blood. The greater part of the urinary osmotic concentration derives from ammonium and bicarbonate ions. Sodium chloride losses via the kidneys are relatively small. Compensation for sodium chloride losses probably occurs most often by way of the food, but the details are not known. Soft-shelled turtles (*Trionyx spinifer*) have been shown to have highly active sodium-uptake mechanisms operative in their pharyngeal membranes and cloacas. The minimum concentration for net uptake of sodium ion by this turtle at 22°C is very low, only 0.002 mmol/liter. Other species of turtles may also use the cloacal bursae (accessory bladders) for sodium ion uptake.

We shall defer until Section 7.8.2 discussion of freshwater birds as they are most similar to terrestrial forms. There is nothing to be said about water and solute metabolism in the truly aquatic freshwater mammals, such as Amazon River dolphins and Lake Baikal seals. What they are doing is unknown, although presumably similar to other mammals. Work on such semiaquatic forms as beavers, muskrats, and nutrias generally fits normal terrestrial mammalian patterns and mechanisms (see Section 7.8.3).

Everything said so far has related to adult freshwater organisms. The eggs, embryos, and larvae of these animals are also faced with problems of osmotic and solute regulation and water and solute turnover. The problems that arise and the mechanisms used for solving them are at least as diverse as those described for the adult forms, but are less well studied and understood.

The sites of development of eggs and embryos can be used as a basis for broadly categorizing the severity of their physiological difficulties. The externally laid, unshelled, and otherwise thinly coated eggs of lampreys, most bony fishes, and most amphibians are probably the most severely stressed. The eggs of many of the elasmobranchs and aquatic reptiles are probably better insulated from the environment by their shells or capsules, although they are also deposited externally. Finally, minimal stresses are imposed on the internally fertilized and developed eggs, embryos, and partly grown young of the live-bearing varieties of fishes, elasmobranchs, reptiles and, of course, all mammals. Some of the more unusually specialized amphibians, such as the Surinam "toad" (*Pipa pipa*), which carries its eggs about in small pockets in its back, are probably in between these categories.

Detailed discussion of water and solute metabolism during development is impossible for most of the groups just listed because of the almost complete lack of information. Much of the little that is known about events in externally laid fish eggs is in the Russian literature and relatively inaccessible to Western workers. Two groups have been studied in great detail. These are the common laboratory species of American and European frogs and salamanders and the common laboratory mammals. For both of these latter groups the literature is

complex and voluminous. It is matched in complexity only by the nature of the problems remaining unsolved. Introductions to important parts of this literature may be found in several of the books cited in the Additional Reading list.

Three additional important topics should be mentioned: (1) the effect of variations in environmental factors, such as temperature and season, on the phenomena we have described; (2) a number of special cases in which freshwater organisms have evolved unusual physiological adaptations to permit survival in unusual circumstances; and (3) the patterns of internal water and solute distributions inside freshwater animals.

Seasonal variations in the physiological capacities of organisms are always difficult to understand, because they result from complex interactions of several simultaneously varying influences. Freshwater animals, just like most other animals, respond in many ways to changes in environmental temperature and length of day, as well as to fluctuations in their own internal states, such as reproductive cycles (which are themselves influenced by temperature and day length). They also are sensitive to a number of changes that are peculiar to their special habitat. These special types of stimuli include the annual patterns of variations in water flow shown by many streams and fluctuations in such chemical variables as water pH and oxygen content. Three examples will illustrate the nature of some of the effects of temperature, one of the best studied variables, on important aspects of water and solute metabolism.

The freshwater stages of a European species of lamprey (*Lampetra fluviatilis*) are subjected during the winter to long periods of water temperatures near 0°C. These winter temperatures are associated with marked decreases in integumentary water permeability and, as would follow from this fact, lower rates of urine production. In addition, rates of active uptake of ions, specifically chloride ion, are greatly inhibited. There appears, however, to be a quantitative imbalance between a large inhibition of uptake and smaller decreases in rates of total chloride loss. Present data make it difficult to understand how these animals can survive over long winters while maintaining reasonably normal blood concentrations—which they apparently do.

The patterns of blood osmotic and ionic regulation in temperate zone freshwater teleost fishes respond to exposures to near freezing, winterlike water temperatures in at least four different ways (Table 7-4). The most uniform feature of all types of responses is a decrease in the total concentration of electrolytes, as compared with levels normal for higher temperatures. The osmotic concentration, however, may or may not follow the electrolyte change. Four different patterns relating these two quantities are distinguishable. These patterns are called (1) no compensation; (2) partial compensation; (3) perfect compensation; and (4) overcompensation.

The brown bullhead catfish (*Ictalurus nebulosus*) shows a response according to pattern (1): serum osmolarity and serum electrolytes both decrease by the same amount. There is no compensation for the electrolyte change with respect to osmolarity. The freshwater adapted killifish (*Fundulus heteroclitus*) responds according to pattern (2): serum osmolarity decreases, but to a lesser extent than the electrolytes; partial compensation with respect to osmolarity results from an increase in the concentrations of organic compounds (especially glucose) in the serum. The carp (*Cyprinus carpio*) during the autumn responds according to pattern (3): serum osmolarity remains constant, although serum electrolytes decrease significantly; the osmotic compen-

sation probably derives from an equivalent increase in the concentrations of still unidentified dissolved organic compounds (possibly not glucose). In winter carp respond similarly to goldfish. The goldfish (*Carassius auratus*) responds according to pattern (4): serum osmolarity increases somewhat, although serum electrolytes decrease; the osmotic overcompensation probably derives from disproportionately large increases in the concentrations of still unidentified dissolved organic compounds (not glucose).

The functional bases for these changes (for example, possible changes in permeabilities to water or electrolytes, possible changes in rates of ion transport) are not yet known.

Temperate-zone frogs subjected to lowered temperatures rapidly increase in weight. Most individuals then stabilize at weights up to 20% higher than their normal (near 20°C) levels. These large-scale weight increases have been shown to be primarily due to decreases in rates of urine production by the kidneys. The Q_{10} (see Section 8.5) for the process is 2–3 in the range 1–24°C. The fall in rate of urine production results from depression of all aspects of kidney water handling, but GFR declines more rapidly than tubular water reabsorption. Water permeability of the skin is relatively insensitive

TABLE 7-4. Responses of Blood Osmotic and Ionic Concentrations of Temperate Zone Freshwater Teleost Fishes to Prolonged Exposures to Near Freezing Water Temperatures*

Serum Concentrations	Brown Bullhead Catfish (Ictalurus nebulosus)		Freshwater Adapted Killifish (Fundulus heteroclitus)		Carp (Cyprinus carpio)		Goldfish (Carassius auratus)	
	10°C	0.5°C	11°C	0.1°C	17°C	4°C	10°C	0.5°C
Osmotic (mOsm/liter)	279	201	372	316	274	277	299	341
Na⁺ (mmol/liter)†	162	101	162	114	130	128	137	130
Cl⁻ (mmol/liter)†	111	82	130	76	125	116	116	104
K⁺ (mmol/liter)†	6	3	4	8	3	3	4	3
Glucose or other organic solute (mmol/liter)	2	4	3	59	10	27	40	99

* All figures are averages for varying numbers of measurements.

† Sources: B. L. Umminger. *Physiol. Zoology,* **44,** 20–27 (1971); B. L. Umminger. *Copeia,* **No. 4,** 686–691 (1971).

to temperature, so the decline in excretory rate results in hyperhydration of the body and the observed weight increase. Blood chloride levels in these hyperhydrated frogs are lowered in fairly close proportion to increases in body water. The relationship persists as long as the weight increase does. This implies a relative balance in the temperature-induced decreases in ion-uptake and -loss rates. Salt balance, however, sometimes is not perfect. The usual imbalance is a slow net uptake. This results in similarly slow continuing swelling caused by water uptake.

Many freshwater organisms are fairly regularly subjected to severe nonseasonal environmental stresses that have major implications for their water and solute metabolism, as well as their overall survival. The most severe stress is undoubtedly the total drying up of the body of water in which they have been living. The majority of species that can survive such dry periods do so in one of two main ways: (1) by means of desiccation-resistant fertilized eggs that remain in the dried bottom deposits until stimulated to hatch upon the return of the water; and (2) by burrowing into the sediment, forming small chambers for themselves, and then going into states of torpor until the water returns. Examples of the first method are the "annual fishes" (mostly members of the family Cyprinodontidae) found in tropical areas. Examples of the second are most aquatic amphibians and reptiles and the well-known African lungfish (*Protopterus aethiopicus*).

The lungfish is capable of surviving periods of complete dryness lasting more than 1 year, although many specimens will die after 8 months or so. As Homer Smith demonstrated in 1930, the fish does this by a combination of structural and physiological adjustments. A major structural feature is the lining of the cocoon in which the fish remains. This is apparently a layer of dried mucous secreted by the animal at the start of its stay, which serves to decrease evaporative water losses. The major physiological adjustments include a reduction in metabolic rate (as compared with nontorpid fish) and a virtual cessation of urine production. Metabolism during torpidity is largely based on proteins. The resulting urea cannot be excreted, hence is accumulated in the body fluids raising their osmotic concentration to quite high levels. The process of accumulation continues until the water returns to the lungfish's lake or stream. The fish then becomes active, breaks out of its cocoon, and rapidly eliminates the accumulated urea by a combination of diffusion across the gills and excretion in the now copious urine. During its periods in water the lungfish appears to regulate its body-fluid composition by the same mechanisms as are used by other freshwater fishes.

Table 7-5 summarizes data on the volumes of the major body-fluid compartments in a series of freshwater vertebrates. Relatively few species in each group have been studied to date, so it is often not possible to be sure whether the figures given are representative of the groups or only of the particular species. Assuming they are representative, it appears that both lampreys and elasmobranchs have significantly larger blood and extracellular fluid volumes than do teleost fishes, the elasmobranchs being intermediate between the other two groups. Reptiles are also intermediate in terms of blood volume, but similar to the teleosts in extracellular fluid volume. Anuran amphibians are very variable, their blood and especially lymph volumes being very sensitive to changes in body hydration. There are no similarly complete data available for freshwater mammals. The functional significance of these differences between groups is not readily apparent.

Few data are available that permit detailed comparisons of the osmotic and solute concentrations of the fluids found in the larger extravascular body-fluid compartments. The common temperate-zone laboratory frogs are a major exception to this statement, at least with respect to comparisons between blood plasma, tissue (especially muscle) extracellular fluid, and tissue (again primarily muscle) intracellular fluid compositions. Table 7-6 summarizes some data on the ionic composition of frog plasma and muscle cell contents. The basic features probably are the same in all vertebrates. The main points are as follows: Evidence (not included in the table) indicates that intracellular and extracellular osmotic concentrations are always equal. With few exceptions, intracellular fluids have high potassium and low sodium and chloride concentrations. Tissue extracellular fluid is quite similar to plasma in composition, but differs in ways that can be accounted for by the existence of Gibbs-Donnan equilibria across capillary membranes. Gibbs-Donnan equilibria are apparently also among the major

TABLE 7-5. Volumes of Major Body-Fluid Compartments in Some Freshwater Vertebrates*

Species	Total Body Water†	Blood†	Extracellular Fluid†
Great Lakes lamprey (*Petromyzon marinus*)	75.6	8.5	23.9
Sturgeon (*Acipenser fulvescens*)	72.7	3.7	20.1
Bowfin (*Amia calva*)	74.5	3.4	18.9
Carp (*Cyprinus carpio*)	71.4	3.0	15.5
Lake Nicaragua shark (*Carcharhinus leucas*)	72.1	6.8	19.7
American alligator (*Alligator mississippiensis*)	72.9	5.1	15.1
Snapping turtle (*Chelydra serpentina*)	72.9	4.7	14.9

* All volumes expressed as percent body mass; all are averages. Some of the variability shown is probably due to variations in body mass between the animals studied.
† Methods used for measurements: Total body water: drying at 105°C; blood volume: Evans blue (T-1824) dye dilution; extracellular fluid volume: sucrose dilution.
Source: T. B. Thorson. *Science,* **130,** 99–100 (1958); T. B. Thorson. *Biol. Bull.,* **120,** 238–254 (1961); T. B. Thorson. *Science,* **138,** 688–690 (1962); T. B. Thorson. *Copeia,* **No. 3,** 592–601 (1968).

determinants of the intracellular ionic composition. More detailed discussions of mechanisms of cellular osmotic and solute regulation may be found in textbooks of cellular physiology.

The smaller extravascular fluid compartments are also physiologically important, each for different reasons. Table 7-7 presents a set of data for three of these smaller compartments, plus the

TABLE 7-6. Ionic Composition of Plasma and Intracellular Water in a Frog (*Rana temporaria*)

Ion	Plasma (mmol/liter)	Intracellular Water (mmol/liter)
Sodium	104	3.6
Potassium	2.2	124
Calcium	2.0	4.9
Magnesium	1.2	14
Chloride	74	1.5
Bicarbonate	25	12

Source: E. J. Conway. *Physiol. Rev.,* **37,** 84–132 (1957).

blood serum, for the Amazon River freshwater stingrays mentioned earlier. It is apparent that, although the osmotic concentrations of these other fluid compartments are near that of the blood, a significant amount of ionic regulation occurs in each one. There is a substantial literature dealing with the compositions and mechanisms of formation of the fluids occupying several of the medically most important of these smaller compartments, such as the intraocular fluids, the cerebrospinal fluid, and the pericardial fluid.

7.5.3 Steady-state Mechanisms in Marine Vertebrates

The number and variety of patterns of regulation of plasma osmotic and solute concentrations used by marine vertebrates matches that just described for the freshwater forms. Table 7-2 summarizes data on representatives of the major groups. The mechanisms used to produce these patterns are even more diverse than the patterns themselves. Table 7-8 presents data on some of these aspects.

TABLE 7-7. Osmotic and Solute Concentrations in Fluids from Several Body-fluid Compartments of Amazon River Freshwater Stingrays (Potamotrygon spp.)*

Quantity	Blood Serum	Perivisceral Fluid	Cranial Fluid	Pericardial Fluid
Osmotic concentration (mOsm/liter)	308	293	289	324
Sodium (meq/liter)	150	103	147	114
Potassium (meq/liter)	6	6	5	3
Calcium (meq/liter)	7	3	4	<1
Magnesium (meq/liter)	4	3	3	<1
Chloride (meq/liter)	149	136	147	180
Urea-N (mg/100 ml)	3	2	2	2

* Figures are average values.
Source: T. B. Thorson, C. M. Cowan, and D. E. Watson. *Science,* **158,** 375–377 (1967).

Marine reptiles, birds, and mammals are omitted from the table and will be discussed separately. Figure 7-5 illustrates the major routes for water and solute exchanges in several groups.

The hagfishes (Myxini) are a purely marine group of primitive fishes that appear to be most closely related to some of the oldest fossil fishes known, the Silurian cyclostome pteraspids. Their patterns of osmotic and ionic regulation are similar to those found in most marine invertebrates. Their blood

TABLE 7-8. Important Features of Steady-State Water and Salt Balance in Some Marine Vertebrates*

Quantity	Hagfish (Eptatretus stouti)	Sculpin (Myoxocephalus octodecimspinosus)	Anglerfish (Lophius americanus)	Dogfish (Squalus acanthias)
Drinking rate (ml/kg hr)	Occurs, variable	4	Occurs	May occur
Urine flow (ml/kg hr)	0.3	0.9	~0.5	1.3
Renal GFR (ml/kg hr)	0.3	1.9	Almost aglomerular	3.5
Urine Δ (mOsm/liter)	~1000	330	310	800
Urine Cl⁻ (mmol/liter)	550	20	10	240
Urinary Cl⁻ excretion (μmol/kg hr)	160	18	<5	290
Extrarenal Cl⁻ excretion (μmol/kg hr)	—	1800	—	Variable

* All figures are average values.
Source: F. Munz and W. McFarland. *Comp. Biochem. Physiol.,* **13,** 381–400 (1964); H. W. Smith. *Am. J. Physiol.,* **93,** 480–505 (1930); R. P. Forster. *J. Cell. Comp. Physiol.,* **42,** 487–510 (1953); R. P. Forster and F. Berglund. *J. Gen. Physiol.,* **39,** 349–359 (1956); J. W. Burger, in P. W. Gilbert, R. F. Mathewson, and D. P. Rall (eds.), *Sharks, Skates, and Rays,* pp. 177–185, Johns Hopkins Press, Baltimore, Md., 1967.

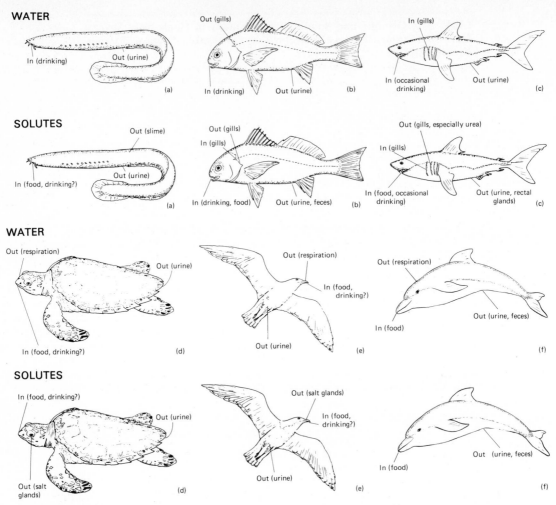

Figure 7-5

As in Figure 7-4, but for marine aquatic vertebrates. Groups of organisms are (a) hagfish, (b) bony fishes, (c) elasmobranchs, (d) reptiles, (e) birds, and (f) mammals. No adequate data are available for other groups such as coelacanths.

is usually isosmotic with, or slightly hyperosmotic to, seawater. Sodium chloride is the major component. A fair degree of ionic regulation occurs, so the concentrations of the other ions present are not the same as in seawater.

Hagfishes produce urine relatively slowly, as would be expected from the small osmotic gradients that occur. In aquaria some hagfishes drink seawater, but it is uncertain how much, if any, salt is absorbed across the gut lining. The role of the gills in ionic regulation has not been established.

The general body slime may, however, be an important pathway for elimination of some ions. The kidneys, at least, are one of the important pathways for ionic regulation. Urine composition differs significantly from that of the plasma for ions other than sodium and chloride.

Little is known about water and solute metabolism in lampreys in the sea. There are no truly marine lampreys, as all forms that have sea-going stages spawn in freshwater. The migratory stages are discussed in Section 7.5.4.

The vast majority of the more than 30,000 species of marine bony fishes appear to use a single basic set of regulatory mechanisms in their water and solute metabolism. All these forms maintain blood osmotic concentrations near 400 mOsm/liter, or slightly above one third seawater concentration. About 90–95% of this blood concentration is due to inorganic salts, about 85% to sodium chloride. As in freshwater fishes, the general body surface is almost impermeable to both water and nongaseous solutes, whereas the gills and pharyngeal membranes are fairly permeable to these substances. The strong outward-directed osmotic gradient across these permeable surfaces results in continuing osmotic dehydration. The concentration gradients for most solutes are inward, so there is some solute invasion by diffusion.

The osmotic water losses appear to be the major water losses. Rates of urine production are low. These low rates are associated with low or no GFRs. Many marine fishes do not use many or any of the glomeruli they have in their kidneys (they are functionally aglomerular). Some groups (pipefishes, toadfishes) are totally structurally aglomerular. Whatever the mode of its formation may be, the urine of marine fishes is isosmotic or only slightly hypoosmotic to the blood. Divalent ions are important contributors to the urinary osmotic concentration; excepting chloride, monovalent ions are present only in low concentrations. Some water reabsorption occurs in the urinary bladders of some marine species having such bladders.

It is probable that the food of marine fishes can compensate only rarely for their continuing water losses. Most marine algae and invertebrates have body fluids isosmotic with seawater, so the many fishes that feed on them cannot obtain help from this source in any case. The sole remaining alternative is drinking seawater, which appears to be a universal phenomenon. The rate of drinking necessarily must equal the sum of the rates of osmotic dehydration, urine formation, and cloacal excretion of the fraction of what was drunk that was not absorbed in the gut. The most reliable drinking rates reported in the literature are in the range of 2–20 ml/kg hr.

Drinking of seawater by itself cannot solve the problem. A complex series of events must follow in the alimentary canal. The details of these events are not at all clear. However, the net result is absorption of the greater part of the ingested water, relatively slight absorption of the ingested divalent ions (these are passed out of the cloaca as a partially crystalline paste composed mostly of calcium and magnesium carbonates), and almost complete absorption of the ingested monovalent ions.

The divalent ions that are absorbed appear to be excreted by the kidneys. The monovalent ions are eliminated by active transport mechanisms in the gills. These ion-transport mechanisms probably are similar in nature to those occurring in the gills of freshwater fishes, but they operate in the opposite direction. Specialized cells called chloride cells appear to be the primary sites for monovalent ion transport.

Marine elasmobranchs are in approximately the same osmotic situation as the hagfishes. The osmotic concentration of their plasma is usually 10–50 mOsm/liter higher than that of seawater. They thus have little need to drink their medium, because they are always taking in small amounts of water by means of osmosis across their gills and pharyngeal membranes. Rates of urine production are fairly low, in keeping with the low water-intake rates. GFRs, however, are relatively high. The urine is generally hypoosmotic to the blood.

The elevated plasma osmotic concentrations of elasmobranchs derive from solute compositions that are fundamentally different from those found in hagfishes. The inorganic ionic components of their plasma are present in amounts quite similar to those found in the plasma of most bony fishes. The additional osmotic concentration is due to the presence of very large quantities of urea and lesser amounts of trimethylamine oxide (TMO). These substances are present in more than twice the concentrations found in the plasmas of most freshwater elasmobranchs (Amazon River stingrays have almost no urea). The urea apparently is produced metabolically from ingested proteins. The TMO is accumulated from food, especially invertebrates, already containing the compound. Chimaeras (Holocephali) differ from other elasmobranchs in having virtually no TMO in their blood.

The high plasma urea concentration creates a strong outward diffusion gradient for this compound across the gills. The urea permeability of the gills is supposedly lowered, as compared with urea permeabilities of bony-fish gills, but there is limited evidence that this is really so. Even if the permeability is relatively low, these fishes suffer continuing diffusional urea losses to the environment across their gills. Their kidneys, on the other hand, are fairly effective in reabsorbing glomerularly filtered urea. As a result, urinary urea losses are relatively small.

The low plasma ionic concentrations in elasmobranchs imply some diffusional uptake of salts from the medium. In addition, many of these fishes feed almost exclusively on salty invertebrates. From time to time all forms probably ingest some seawater while swallowing their food. The kidneys appear to take care of divalent ions that enter the blood, but urinary monovalent ion concentrations are almost as low as those normally encountered in bony fishes. Thus there must be an extrarenal pathway for elimination of monovalent ions.

The gills of elasmobranchs apparently have the outward-directed active transport mechanisms found in marine bony fishes, but they usually function at much lower rates. They also have specialized glands opening near their cloacas, termed *rectal glands,* that produce a secretion isosmotic with their plasma but containing very little urea or TMO. The secretion is, in fact, nearly a pure solution of sodium chloride. The rates of secretion by the rectal glands appear adequate to enable the elasmobranchs to eliminate promptly any reasonable monovalent ion load they may incur.

The coelacanth (*Latimeria chalumnae*) is partly similar to the elasmobranchs in its osmoregulatory pattern. It has comparable plasma concentrations of urea and TMO. However, its total osmotic concentration is below that of normal seawater. Bladder urine obtained from a single living specimen was isosmotic with the blood. The fish was nearly dead at the time urine was sampled, so many details of urine composition could have been changed by stress effects. However, the general pattern of mono- and divalent ion concentrations was similar to that found in elasmobranch urine. The most important, probably real, difference was that urine urea concentration was the same as blood urea level, instead of being much lower than blood level. This indicates that, unlike elasmobranch kidneys, the kidney of the coelacanth cannot reabsorb filtered urea. Anatomical and biochemical data indicate that coelacanths, like elasmobranchs, have rectal glands that are used for excretion of monovalent ions.

Marine reptiles (for example, sea turtles, marine iguanas, and sea snakes), marine birds (petrels, albatrosses, many gulls and terns, and a few salt-marsh-dwelling passerines), and marine mammals (dugongs, seals and sea lions, and whales and porpoises) all maintain plasma osmotic and ionic concentrations that are quite similar to those found in most bony fishes. The integuments of most of these animals are quite impermeable to water and solutes. (Sea snake skins have low, but significant permeabilities to water.) Because they are air breathers they lack the permeable gill "windows" that produce the osmotic stresses in the fishes. They are, however, faced with significant continuing water losses, both from the necessity of producing sufficient urine to eliminate metabolic wastes and, especially for the high-metabolic-rate birds and mammals, from respiratory evaporative water losses.

The details of how the many different kinds of animals in the three groups compensate for these water losses is, in most cases, poorly known. It is not at all certain whether or not marine reptiles and nonpasserine birds regularly drink seawater. It seems, in most cases, that they do not, but in some groups, notably the albatrosses, they apparently do. A salt-marsh-dwelling subspecies of the savannah sparrow (*Passerculus sandvicensis*) from the western United States appears to do this regularly. The marine mammals seem not to drink at all.

Either situation, drinking or nondrinking, has associated problems. The nondrinkers must find enough food to supply their water needs. Nothing is known of the extent to which these animals can satisfy their requirements with water derived from metabolic breakdown of foodstuffs. Those forms that feed on algae and invertebrates must eliminate the salt loads associated with their food. The drinkers must also eliminate the salt they swallow in their water.

Several types of solutions to these problems have been evolved. The kidneys of all the marine reptiles and marine nonpasserine birds apparently cannot produce urines that are hyperosmotic to their plasmas. Their kidneys seem unable to handle very large loads of monovalent ions. As a result, all these animals make use of an extrarenal pathway for elimination of monovalent ions. This pathway is most often a modified circumorbital gland (Figure 7-6). The salt glands of sea turtles are lachrymal glands;

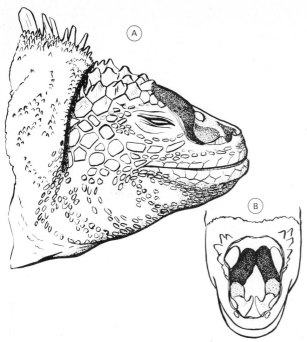

Figure 7-6

The location and size of the nasal salt glands in the Galapagos marine iguana (Amblyrhynchus cristatus). A: Side view showing nasal gland in black lying above the orbit and extending under the nasal passage. B: Top view showing bilateral nasal glands; stippled areas are the portions of the glands under the nasal passages. [*From W. A. Dunson. Am. J. Physiol.,* **216,** *995–1002 (1969).*]

marine iguanas and marine birds use the nasal glands. Different groups of marine snakes use various modified salivary glands: the posterior sublingual gland (beneath the tongue) in the true sea snakes (Hydrophiidae) and some other families

with marine representatives; the premaxillary gland in a marine species in yet another family. These salt glands produce secretions that are similar in chemical composition to, but are frequently much more concentrated than, the secretions of the rectal glands of marine elasmobranchs.

Table 7-9 summarizes data illustrating the relative roles of the kidneys plus gut (cloacal excretion) and salt glands in eliminating acute salt loads from a marine bird. Note especially the fact that the nasal glands were able to eliminate amounts of salt almost equal to those eliminated via the cloaca at a cost of only one half the amount of water.

Table 7-10 lists the usual sodium concentrations found in salt-gland secretions from a variety of species. These ranges appear to be fairly species specific. The rates of secretion from the glands of individual birds can vary adaptively with the average levels of salt-loading the birds encounter. The salt glands of birds of a given species that grow up in marine environments average larger than those in birds of the same species that grow up in freshwater areas.

The salt-marsh subspecies of savannah sparrow mentioned earlier and also a few other passerine species from similarly difficult habitats eliminate their salt loads via the kidneys. These birds lack salt glands. Urine osmotic and sodium chloride concentrations in birds drinking high-concentration sodium chloride solutions are well within the range of salt gland secretion concentrations listed in Table 7-10 (up to 960 meq/liter of chloride in the urine of salt-marsh savannah sparrows). The kidneys of these birds appear to be structurally somewhat different from the kidneys of the usual species, which cannot produce such concentrated urine. Many of their nephrons have long, hairpin loops as parts of their tubules. These loops may

TABLE 7-9. Distribution of Electrolyte Elimination between Cloacal and Nasal Secretion in a Salt-Loaded Double-Crested Cormorant

Quantity	Cl⁻ (meq)	Na⁺ (meq)	K⁺ (meq)	H₂O (g)
Total ingested	54	54	4	50
Cloacal elimination (8 hr)	28	26	3	109
Nasal elimination (8 hr)	26	24	0.3	51
Total elimination	54	50	3.3	160

Source: K. Schmidt-Nielsen, et al. *Am. J. Physiol.,* **193,** 101–107 (1958).

TABLE 7-10. Usual Sodium Concentrations in Salt-Gland Secretions from Various Marine Species

Species	Na+ (meq/liter)
Green turtle (Chelonia mydas)	600–700
Yellow-bellied sea snake (Pelamis platurus)	400–800
Marine iguana (Amblyrhynchus cristatus)	1000–2100
Double-crested cormorant (Phalacrocorax auritus)	500–600
Herring gull (Larus argentatus)	600–800
Blackfooted albatross (Diomedea nigripes)	800–900
Leach's petrel (Oceanodroma leucorhoa)	900–1100

Source: W. N. Holmes. *Arch. Anat. Microsc. Morphol. Exp.*, **54**, 491–514 (1965); K. Schmidt-Nielsen. *Circulation*, **21**, 955–967 (1960); W. A. Dunson. *Am. J. Physiol.*, **216**, 995–1002 (1969).

TABLE 7-11. Aspects of Kidney Function in Two Hawaiian Porpoises*

	Tursiops gilli (160-kg Body Mass)	Steno bredinensis (55-kg Body Mass)
Urine Δ (mOsm/liter)	1600–1800	1700
Urine flow (ml/kg hr)	0.4–0.6	1.4
GFR (ml/kg hr)†	105–110	140
RPF (ml/kg hr)†	195–230	390
Urea clearance (ml/kg hr)	20–30	60

* Individual animals, studied 18–28 hr after feeding.
† GFR: inulin clearance. RPF: *p*-aminohippuric acid clearance.

Source: R. L. Malvin and M. Rayner. *Am. J. Physiol*, **214**, 187–191 (1968).

function as countercurrent multipliers for urine concentration, similar to the loops of Henle in mammalian kidneys (Section 7.8.3).

The kidneys presumably are the major routes for elimination of salt loads from marine mammals, but this is by no means certain in all cases. The urine of these animals is usually strongly hyperosmotic to the blood and, quite often, to seawater. However, most species are carnivores that produce large amounts of waste urea. The largest part of urinary osmotic concentration often derives from this urea. Urinary sodium chloride concentrations higher than seawater levels have been recorded for a few species, but the difficulties of experimentation with animals such as whales have made our understanding of overall salt-balance regimes in most forms very fragmentary. There is a distinct possibility that extrarenal pathways for salt excretion, perhaps the salivary glands, are important for many of these species.

Table 7-11 summarizes some data on aspects of kidney function in two individual Hawaiian porpoises belonging to two different species. The qualitative similarity between the two sets of values is striking. Urine flow rates are low and urine concentrations fairly high. Filtration fractions (GFR/RPF) are relatively high, and renal water reabsorption is very efficient. Urea clearances are relatively high.

This work on porpoises also included time studies of changes in renal function following feeding. These studies produced the unusual (for ordinary mammals) result that, although urine flow rates increased greatly shortly after a meal (up to 25 times minimum rates), there was little change in urine total concentration. Urine total concentrations at highest urine flow rates were reduced by only about 20% below levels at minimum flow rates. Since the major urinary solute during high-flow periods after eating was sodium chloride, this phenomenon indicates a highly developed renal capacity to eliminate ingested salt loads rapidly. In postabsorptive states urea is the major urinary solute, very little salt then being excreted.

The high fat content of the food of many marine mammals probably helps to alleviate their water problems somewhat through metabolic water production. Lactating females, however, must compensate for extra water losses in the milk they feed their young. Nursing females of some of the large whales, such as the gray whale (*Eschrichtius glaucus*), supposedly do not feed during many of the months each year they devote to their extended migrations. Nothing is definitely known about how they satisfy their water requirements during these periods.

Lactating female northern elephant seals (*Mirounga angustirostris*), residents of the north Pacific Ocean, neither eat nor drink during the entire 28-day period they nurse their pups while they are ashore at their breeding locations. The pups gain 100–200 kg body mass during the same period, obtaining all of their water and nutrients from their mothers' milk. The weight composition of this milk changes as the nursing period progresses. During the first 21 days, fat content rises from initial levels near 15% to near 55%; water content falls reciprocally from about 75% initially to about 35%; protein content is stable throughout at 5–12%; reducing sugars are almost absent. All values are stable during the last 7 days of nursing.

These changes in milk composition apparently are closely adjusted to the water and energy needs of both the mother seals and the pups. The only water available to the mother for milk production is derived metabolically from her fat reserves. The decline in water content of the milk greatly reduces the rates at which the females must oxidize these reserves, thus extending the lengths of time they can remain ashore without feeding, and ensuring that they can take care of their pups. The compositional changes cannot go too far, however, since the pups obtain all of their water from the combination of free water in the milk plus the metabolic water they can produce from the fats in the milk. The minimum water content of the milk is thus a measure of the amounts of free water needed by the pups in addition to that which they can produce metabolically.

As they grow, the pups improve their water conservation abilities. They do this in three major ways: reduction in mass specific metabolic rates (partly by improving their thermal insulation by developing blubber layers under their skins); development of breath-hold patterns of breathing similar to those they will use later on during dives; and reduction of rates of urine production in association with reduced rates of protein breakdown (thus reducing water needs for waste nitrogen excretion). These adjustments are essential as the pups themselves undergo extended periods of complete starvation without drinking beginning as soon as they are weaned. This period of pup fasting (the "weaner" stage) lasts 32–52 days. One of its most important characteristics is reduction in rates of water turnover by the pups. Water turnover rates become the lowest known in any mammal (biological halftimes 54 days, compared with only 13–18 days in other species of mammals similarly starved and deprived of drinking water). At the end of these fasts the young go to sea and begin feeding.

The water and solute metabolic problems of the eggs and embryos of marine vertebrates are again at least as diverse as those described earlier for freshwater forms. Once again there is a range of egg types and sites of deposition and maintenance. There is a far greater variety of life histories and environmental stresses. For example, the unshelled, free-floating pelagic eggs of many tropical marine fishes commonly occur in the uppermost few millimeters beneath the sea surface. This *neuston zone,* as it is called, is a far more variable habitat than the waters only a few centimeters deeper down. The immediate surface layer may be heated to quite high temperatures on calm, sunny days, and it is frequently drastically diluted, or even converted to pure freshwater, for varying periods following heavy rain showers. Thus these eggs, and larvae living in the same layer, must not only contend with the osmotic stress imposed by a concentrated environment but must also cope with oftentimes rapid and drastic environmental fluctuations. Several of the books cited in the Additional Reading list discuss a few of the physiological mechanisms used by marine vertebrate eggs and embryos. This is another poorly known field.

Probably the only marine vertebrates that live in an environment that is essentially constant over the entire year are the fishes that inhabit the deep oceans beneath the seasonal thermoclines in the tropics. All the others are subjected to environmental variations of many kinds on time scales varying from minutes to seasons. A few resident species of fishes inhabiting some of the most enclosed coastal bays bordering the Gulf of Mexico in Texas are subjected almost every summer to water temperatures near 40°C and salinities as high as 90‰. Antarctic emperor penguins go without food or water for many weeks while they incubate their eggs during the long winter night. Spring freshets in major temperate-zone rivers often dramatically lower the salinity of the water in which many coastal species live, sometimes for extended periods. The list of factors that can vary is long, and the list of responses made by different marine organisms is still longer. We shall describe only one group of particularly unusual responses.

The seasonal appearance of surface ice on Arctic ocean waters necessarily means that the temperature of the seawater has reached the freezing point. In most areas this temperature is in the vicinity of −1.8°C. The plasma osmotic concentrations of shallow-water Arctic fishes in summer, when there is no sea ice, are normal for marine forms, equivalent to freezing temperatures near −0.8°C. These fishes remain active and survive the winter with no difficulty. They do not freeze, even if they come in contact with some of the surface ice.

A similar set of circumstances exists in the Antarctic, near and under the permanent ice shelves. The Antarctic situation has an additional dimension, however, in the form of ice on the sea bottom (anchor ice; Figure 7-7). Here again, fishes with body fluids strongly hypoosmotic to seawater do not freeze even when in direct contact with ice.

A large amount of complex physical-chemical and biochemical work over the past 20 years has

provided an explanation for this seeming paradox. As shown in Figure 7-8, Arctic shore fishes in winter have plasma salt concentrations somewhat higher than salt concentrations found in the same species in summer. The increase in salt content, however, is much less than the seasonal increase in plasma freezing-point depression. Antarctic fishes living all year round under and near the ice shelves have similar blood compositions to the Arctic shore fishes in winter (Table 7-2). The differences in all these species between the contributions to the plasma freezing-point depressions due to inorganic salts and the measured freezing-point depressions are due to the presence of organic compounds that act either to stabilize a supercooled condition of the plasma or to inhibit body fluid freezing even if it starts. (*Supercooling* is defined as the state of a solution that remains unfrozen at temperatures lower than its thermodynamic equilibrium freezing temperature.) These compounds are unusual in that their antifreeze effects are disproportionately large in comparison to the numbers of molecules involved.

The modes of action of these organic compounds are all similar and remain matters of research and discussion (see Additional Reading list). The chemical natures of the molecules have been established for a number of cases. Most Antarctic species use a mixture of eight glycoproteins, the carbohydrate moieties of which are composed of N-acetylgalactosamine and galactose, the protein moieties of polypeptides containing only alanine and threonine. They have molecular weights of 2,600–33,000. A few Arctic species use similar glycoproteins, but averaging lower in molecular weights (about 10,000) and having somewhat different amino acid compositions of the protein moieties (including arginine and proline). A few Antarctic species and apparently a majority of Arctic species use various polypeptides and small proteins, ranging in molecular weights from about 3000 to 8000.

The presence of these compounds in the body fluids of these fishes has many implications for their water and solute metabolism. One aspect relates to how the kidneys of these fishes avoid excreting large quantities of these compounds. This is a possibility since it is well established that renal glomeruli are permeable to many molecules having molecular weights in the same ranges. The probable answer to this question lies in the fact that the Ant-

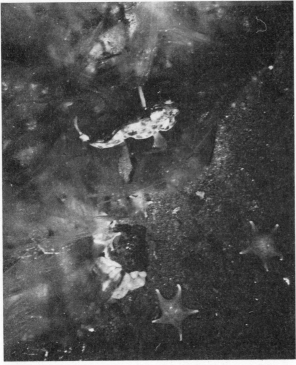

Figure 7-7
Benthic Antarctic fish (Trematomus bernacchii) containing glycoprotein antifreeze substance in its blood resting on mass of anchor ice in 20 m of water in McMurdo Sound, Antarctica. [Photo by Paul Dayton. From A. L. DeVries and D. E. Wohlschlag. Science, **163**, 1073–1075 (7 March 1969). Copyright 1969 by the American Association for the Advancement of Science.]

Figure 7-8

*Plasma chloride concentration (expressed as NaCl) versus plasma freezing point depression in Arctic fishes and invertebrates adapted to various environmental temperatures at different seasons. Diagonal line indicates NaCl solution concentration needed to produce observed freezing point depressions. The vertical distance between this line and the observed NaCl concentrations indicates the magnitudes of the effects of organic supercooling stabilizer substances added to the plasma by the species of shore fishes studied. More recent work indicates that the freezing point depression values used in this graph for plasma from shore fishes may be somewhat higher than the actual values (see text for explanation). [From P. F. Scholander, et al. J. Cell. Comp. Physiol., **49**, 5–24 (1957).]*

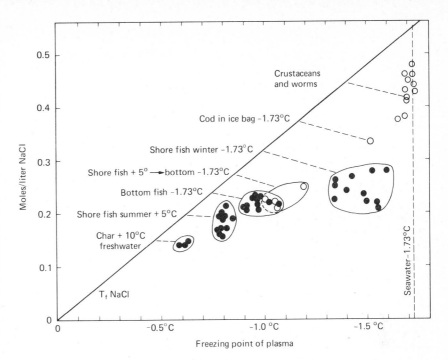

arctic species are all aglomerular, and the Arctic species are probably functionally aglomerular.

To close this section we must say a little about internal distributions of water and solutes in marine vertebrates. Table 7-12 summarizes data on body-fluid-compartment volumes in several diverse species. Plasma osmotic and ionic compositions in most marine vertebrates are similar to those described earlier for most freshwater forms, so it is not surprising that the data in Table 7-6 also can serve as a reasonably accurate description of intracellular fluid composition in the marine forms. There are, however, two major exceptions to this statement—the hagfishes and the elasmobranchs.

As we have seen, both hagfishes and elasmobranchs have plasma osmotic concentrations that are very near that of seawater. In the hagfishes this is due almost entirely to inorganic salts, especially sodium chloride. In the elasmobranchs salts comprise only a bit over one half of the total solutes. The remainder is about 6 : 1 urea : TMO.

Intracellular fluid osmotic concentrations in the tissues of both groups appear to be the same as the plasma concentrations. The solute compositions, however, are quite different. In both groups,

TABLE 7-12. Volumes of Major Body-Fluid Compartments in Some Marine Vertebrates*

Species	Total Body Water	Blood	Extra-cellular Fluid
Nassau grouper (Epinephelus striatus)	71.7	2.6	14.5
Green moray (Gymnothorax funebris)	63.7	2.2	15.8
Great barracuda (Sphyraena barracuda)	70.6	2.8	15.9
Lemon shark (Negaprion brevirostris)	71.1	7.0	21.2
Spiny dogfish (Squalus acanthias)	71.7	6.8	21.2
Green turtle (Chelonia mydas)	64.9	6.5	19.1
Loggerhead turtle (Caretta caretta)	64.0	6.7	19.3

* All volumes expressed as percent body mass; all are averages. For methods see footnotes, Table 7-5.

Source: T. B. Thorson. *Biol. Bull.,* **120**, 238–254 (1961); T. B. Thorson. *Science,* **138**, 688–690 (1962); T. B. Thorson. *Copeia,* **No. 3,** 592–601 (1968).

the intracellular inorganic ions comprise only about 50% of the total. The remainder, in the hagfishes, is composed of several low molecular weight organic compounds, primarily TMO and amino acids. In the elasmobranchs it seems that cell membranes are passively permeable to the urea in the plasma, so the greater part of the remainder is urea.

7.5.4 Mechanisms in Transitions between Osmotically Different Media

Aside from temperature changes, variations in environmental salinity are probably the most widespread and frequent stresses encountered by aquatic animals. Most animals are relatively stenohaline and can tolerate salinity changes of only small amplitude or short duration. Small amplitude for most freshwater species means increases in concentration to levels little higher than isosmotic with their blood (almost one third seawater). For marine species the lower limit is much more variable. It depends strongly on temperature and the divalent ion concentration of the freshwater diluting the native seawater. Many otherwise stenohaline marine fishes (which would die in seawater concentrations below about 50%) regularly go into tropical freshwaters draining from limestone regions. They survive and grow well in these freshwaters. The mechanism is not clear but probably is tied to effects on water permeability of the gills of the fairly high calcium concentrations in these waters.

We do not know why stenohaline animals die when they reach their salinity limits. There probably are many different reasons. Three of the major proposals are (1) breakdown in regulatory mechanisms, permitting fatal changes in internal concentrations; (2) interference with respiratory function of gills or lungs, resulting in fatal asphyxia; and (3) cardiac failure, perhaps as a result of some combination of (1) and (2).

Euryhaline species are found in every vertebrate group. They achieve their euryhalinity in a number of ways. The lampreys, bony fishes, elasmobranchs, amphibians, and reptiles necessarily use primarily physiological means although behavioral salinity selections are frequently used to make environmental changes more gradual. Euryhaline aquatic birds and mammals are usually much more mobile and probably frequently minimize the stresses they must contend with by moving back and forth between habitats of different salinities. Many birds, however, especially ducks, herons, many waders, gulls, and terns, are able to survive either in freshwater or truly marine areas.

Euryhaline lampreys are uniformly anadromous (they spawn in freshwater and carry on much of their physical and sexual growth and maturation in the sea). Marine-stage lampreys are hard to find and have never been studied physiologically. Sexually maturing adults moving through estuaries on their way to their freshwater spawning grounds are easily captured.

The European *Lampetra fluviatilis* on its upstream migration starts off being like a marine bony fish but becomes progressively less tolerant of elevated salinities as it goes along. The integument apparently becomes increasingly permeable to water and the animals lose their ability to compensate for water losses and to eliminate salt loads when in hyperosmotic media. They progressively become stenohaline freshwater fish. Virtually all these animals die after spawning. Lampreys migrating upstream are thus not really euryhaline for long. The young stages that migrate downstream gradually become euryhaline and tolerant of elevated salinities. For reasons not presently understood, but possibly related to the presence of a loop structure in the kidney tubules, those adult *Lampetra* early in their upstream migrations that can still be acclimated to 100% seawater produce, while in 100% seawater, urine that is hyperosmotic to their bood (U/P ratio near 1.5).

The ontogeny of osmoregulatory capacities, and the adult patterns of osmoregulation, appear to be quite similar in North American sea lampreys (*Petromyzon marinus*) to those occurring in anadromous teleosts such as salmonid fishes.

Most euryhaline bony fishes are excellent osmoregulators. Figure 7-9 illustrates the pattern of osmoregulation found in a typical example. Many of these fishes, but not all, can tolerate direct transfers in either direction between freshwater and 100% seawater. The steady-state mechanisms they use for osmotic and solute regulation in hypoosmotic media are the same as have been discussed for freshwater bony fishes. The mechanisms used in hyperosmotic media are the same as those described previously for marine bony fishes.

Table 7-13 summarizes data on the osmotic and ionic concentrations in the blood of a euryhaline

Figure 7-9

Patterns of plasma osmotic and NaCl regulation in the euryhaline mudskipper fish (Periophthalmus sobrinus) *from Madagascar. Upper line joins means for plasma osmotic concentrations in groups of fish fully adapted to salinities ranging from 100% seawater to freshwater. Arrows along abscissa indicate experimental salinities. Vertical lines through points ±2 standard errors of means. Diagonal line, line of equality between internal and external concentrations. Vertical bars, mean plasma NaCl concentrations in same salinities.* [*From M. S. Gordon, et al.* Hvalradets Skrifter Norske Videnskaps-Akad. Oslo, **40**, *85–93 (1965).*]

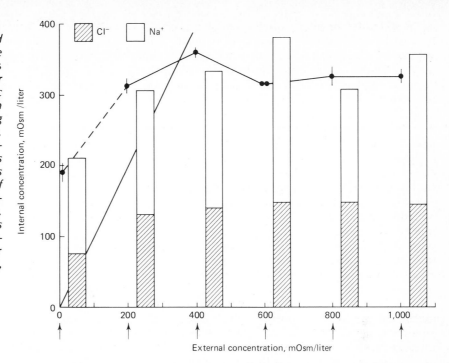

teleost, the common killifish of the coastal areas of the northeastern United States *(Fundulus heteroclitus)*, when fully adapted to freshwater and 100% seawater. The excellent, though not quite perfect, osmotic and ionic regulatory control demonstrated by this fish is readily apparent.

How euryhaline teleosts make transitions between the regimes of hyperosmotic and hypoosmotic regulation is only partially understood. Adjustments in kidney function occur, but in different patterns in different species. In rainbow trout *(Salmo gairdneri)* in full seawater both GFRs and rates of urine flow are reduced to 1–8% of their freshwater levels. In eels *(Anguilla* spp.) both GFRs and urine flows are greatly reduced within a few hours after a direct transfer from freshwater to full seawater, but long-term seawater adaptation produces GFRs comparable to those in freshwater. At the same time urine flow rates are reduced still further. Tubular water reabsorption in the kidneys of these eels must increase greatly in seawater adaptation. The primarily marine but euryhaline aglomerular toadfish *(Opsanus tau)* necessarily uses still another pattern of renal adjustments. Urine flows in toadfishes adapted to freshwater are only 4–5 times larger than flows in fishes in full seawater, whereas urine concentrations remain almost unchanged.

Adjustments also occur in rates of ion uptake

TABLE 7-13. Osmotic and Solute Concentrations in the Blood of a Euryhaline Teleost, the Killifish *Fundulus heteroclitus,* Fully Adapted to Freshwater and 100% Seawater.*

Concentration	Freshwater	Seawater
Osmotic	335	365
Na^+	170	185
K^+	3.0	4.8
Ca^{2+}	2.0	2.3
Mg^{2+}	1.7	2.1
Cl^-	125	145
PO_4^{3-}	5.1	5.3
HCO_3^-	11.8	13.3

* All figures are averages. Fish maintained at 20°C. Osmotic concentrations in mOsm/liter, ionic concentrations in mmol/liter.

Source: G. E. Pickford, F. B. Grant, and B. L. Umminger. *Trans. Connecticut Acad. Arts Sci.,* **43**, 25–70 (1969).

---

or excretion by the gills, as well as in rates of drinking of the medium and absorption of water and solutes by the intestine and, in some species, the urinary bladder. The magnitudes of these adjustments are apparently moderated by decreasing water permeability of the gills and other water-permeable areas in the face of increasing osmotic gradients. Figure 7-10 illustrates aspects of these phenomena in rainbow trout. In these fishes, dehydrating osmotic gradients increase from near zero (in one third seawater) to 540 mOsm/liter (in full seawater). However, between the same two salinities, rates of drinking of the medium and of intestinal water absorption increase by only 3.2 times and 3.8 times, respectively. It is not presently possible to make similar calculations on the hypoosmotic environment side because urine flow rates in trout in one third seawater have not yet been measured.

It is also possible to estimate relative water permeabilities of freshwater- and full-seawater-adapted trout. As indicated in Table 7-3, urine flow rates in trout in freshwater are near 100 ml/kg day. This rate is associated with an osmotic gradient of 300 mOsm/liter. Intestinal water absorption in trout in full seawater is also near 100 ml/kg day, but this is in the presence of an osmotic gradient of 540 mOsm/liter. Thus, the indications are that integumentary water permeabilities in seawater-adapted fish must be only about one half those occurring in freshwater-adapted animals.

The only known case of a bony fish being able to produce urine hyperosmotic to its blood (even though it is still hypoosmotic to the outside medium) is the euryhaline cyprinodont, *Fundulus kansae*. This urinary hyperosmoticity is only a transient stage in the adaptation of this form to suddenly increased salinity.

Nothing is known about how euryhaline elasmobranchs carry out their physiological adjustments to different media. Most freshwater elasmobranchs are really primarily marine but euryhaline species living in freshwater.

There are a very few species of euryhaline amphibians. Only one of these, the crab-eating frog (*Rana cancrivora*) of southeast Asia, lives in a truly marine environment—coastal mangrove swamps. As adults, all euryhaline amphibians are good

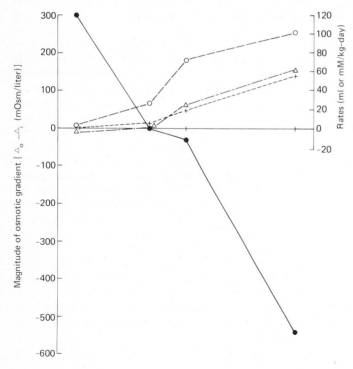

Figure 7-10
Magnitude and direction of the osmotic gradient between blood and external environment (●—●, left hand scale) and (all on right hand scale) rates of intestinal water absorption (O— — —O), intestinal sodium absorption (+· · ·+), and estimated extrarenal sodium excretion (△— — —△) in rainbow trout (Salmo gairdneri) *fully adapted (in steady states) to various salinities. All figures are average values.* [*Data from Z. H. Shehadeh and M. S. Gordon,* Comp. Biochem. Physiol., **30,** *397–418 (1969).*]

osmoregulators between freshwater and about 30% seawater. Above this range they are osmoconformers, but they always maintain a small hyperosmoticity to their medium. Figure 7-11 illustrates this pattern. As a result of this hyperosmoticity they are always in much the same water-and-solute-balance situation as marine elasmobranchs. Their kidneys seem able to handle their normal salt-loading problems, but there is some chance that they have an outwardly directed chloride ion pump in their skins. Adjustments in rates of urine production involve changes in both GFRs and tubular water reabsorption.

The elevated blood concentrations in these animals derive from different sources in different species. *R. cancrivora* adds some sodium chloride to its blood (about 40% of increases above freshwater levels) but mostly urea (the remaining 60% of increases). The green toad (*Bufo viridis*) of Europe, the Middle East, and western Asia increases its blood concentration almost entirely (85% of increases) by sodium chloride. Parallel adjustments in intracellular osmotic concentrations occur in the

tissues of each species. These are produced largely by inorganic ions and urea in *R. cancrivora*, principally by inorganic ions and free amino acids in *B. viridis*.

Salinity adaptation in tadpoles of euryhaline amphibians has been studied only in *R. cancrivora*. These tadpoles are just as euryhaline as the adults but accomplish it in a very different way. The tadpoles are physiologically like euryhaline bony fishes rather than elasmobranchs. This is illustrated in Figure 7-11.

Several species of small (less than 2 g body mass) terrestrial salamanders in California (genus *Batrachoseps*) have also been found to be fairly euryhaline—as euryhaline as some populations of the green toad just described (some populations of these salamanders live adjacent to ocean beaches and may be even more euryhaline). Their patterns of osmotic and ionic regulation are similar to those of *R. cancrivora*. Plasma osmotic concentrations are always higher than environmental concentrations, up to the maximum indefinitely tolerable salinity (which is somewhat above 50% seawater). Sodium chloride accounts for about 55% of plasma osmotic concentration increases above freshwater levels, and plasma urea makes up almost all of the rest. Urinary urea concentrations are well below plasma levels in animals adapted for long periods to higher salinities, implying significant urea reabsorption either in the kidney or urinary bladder, or both. It seems probable that the unusually great salinity tolerances of these salamanders are not primarily adaptations for euryhalinity, but instead are reflections of unusually great abilities to cope with dehydrating conditions in the subsoil habitats where they generally live.

Most euryhaline reptiles and birds appear to be physiologically essentially like their purely marine relatives, but with the ability to switch off or greatly reduce the rate of functioning of their salt glands. As a result they do not deplete themselves of salt while in dilute environments. The diamondback terrapin (*Malaclemys terrapin*) of the salt marshes of coastal southeastern North America adds to this generally distributed mechanism the ability to accumulate substantial amounts of urea in its blood when it is in media more concentrated than 50% seawater (100 mmol urea in animals adapted to 100% seawater).

An interesting recent discovery in Australia is

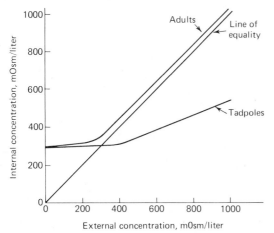

Figure 7-11
*Mean patterns of plasma osmotic regulation in tadpoles and adults of the euryhaline crab-eating frog (Rana cancrivora) of Thailand. The metamorphosis from tadpole to adult involves a metamorphosis in osmoregulatory mechanisms as well as the usual morphological change. See text for discussion. [Data from M. S. Gordon, et al. J. Exp. Biol., **38**, 659–678 (1961); and M. S. Gordon and V. A. Tucker, J. Exp. Biol., **42**, 437–445 (1965).]*

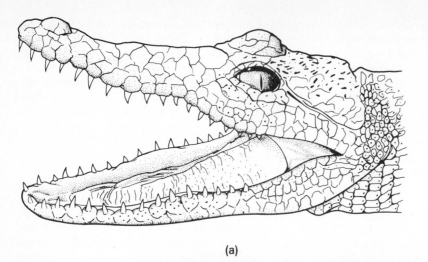

(a)

Figure 7-12
Salt glands in the tongue of the estuarine crocodile (Crocodylus porosus). *(a) Drawing of the head of a crocodile. Dots on the tongue indicate openings of lingual salt glands. (b) Photograph of surface of tongue of a salt loaded crocodile showing droplets of secretion forming at openings of lingual salt glands.* [Both drawing and photograph courtesy of L. E. Taplin and G. C. Grigg, University of Sydney, Australia.]

(b)

that the euryhaline estuarine crocodile *(Crocodylus porosus)*, which occurs widely around the shores of the tropical Indian and western Pacific Oceans, has multiple salt glands in its tongue (lingual glands; Figure 7-12). This crocodile is a perfect osmoregulator over the entire range of environmental salinities from freshwater to 100% seawater. Other crocodilians, including the American crocodile and various freshwater forms, also have lingual glands, but it is not yet known if they function as salt glands.

There are very few euryhaline aquatic mammals (manatees are one example). Their mechanisms of salinity adaptation are completely unknown.

7.6 GENERAL COMMENTS ON AQUATIC VERTEBRATES

The preceding discussion of aquatic vertebrates may have seemed interminable and full of detail, but it did not nearly exhaust the range of important subjects that could legitimately be included. Space limitations permit only the briefest mention of some of these. We shall emphasize areas of future challenge.

Oceanic waters are the major concentrated aquatic environment on earth but they are not the only one (see Table 7-1). The higher concentrations and unusual chemical compositions of many inland saline waters present a multitude of severe physiological challenges. In most cases these challenges are successfully met by an assortment of animals belonging to many different groups. Water and solute metabolism in most of these, even the largest, commonest, and most conspicuous forms, remain completely unstudied.

Similarly, large fields remain to be investigated in the area of the interactions of environmental variables and their combined influences on water and solute metabolism. This is especially true for interactions between salinity and temperature. There are also single variables, most notably hydrostatic pressure, the effects of which are almost unknown. Environmentally oriented knowledge is minimal, in almost every regard, concerning the mechanisms used by animals for handling most minor solutes (including such essential items as calcium, magnesium, and sulfate).

The surface has also only been skimmed lightly in most aspects of the metabolic effects of adjustments in water and solute metabolism. The total energetic requirements for osmotic and ionic regulation appear, in most forms, to be only a few percent of the whole animal metabolic rate. This makes it difficult to answer such a question as: What are the changes in the energy requirements of the excretory organs of animals in response to changes in the amounts of solute these organs excrete? We have also only begun to understand the biochemical bases and mechanisms for the major shifts in end products of nitrogen metabolism (shifts in relative and absolute amounts of ammonia, urea, and uric acid excreted) that occur during growth and development in many species.

Finally, structural and physiological information about osmotic and solute metabolism in the lower vertebrate groups has long served as the basis for arguments about the nature of the aquatic environment in which the earliest vertebrates first evolved. The most careful assessments of this information,

combined with detailed paleontological information, make a marine origin seem very likely. There has been a great deal of subsequent moving back and forth to and from freshwater.

7.7 REGULATION OF BODY-FLUID COMPOSITION IN AMPHIBIOUS VERTEBRATES

The true amphibians are by no means the only amphibious vertebrates. Many kinds of fishes [for example, mudskippers, many gobies, and tree-climbing "perch" (Anabas) of southeast Asia], reptiles (Galapagos marine iguanas and most "water" snakes), birds (rails, gallinules, and water ouzels), and mammals (otters, beaver, and muskrats) can also legitimately be placed in this category. Water and solute metabolic processes in all these animals must be able to cope to a significant degree with the problems of both the aquatic and terrestrial environments.

We have already described adaptations to aquatic environments in each of these groups. Relatively little is known about water and solute metabolism in amphibious fishes while they are out of water (we do not consider the African lungfish to be amphibious). The few forms that have been studied partially (mostly Asian mudskippers) appear to be physiologically similar to anuran amphibians. They stop voiding urine, shift waste nitrogen production away from ammonia and toward more urea, and make internal adjustments of water distributions in ways leading to protection of the hydration state of the central nervous system.

Amphibious reptiles, birds and mammals have not been studied extensively, but what is known about these forms fits well with out understanding of terrestrial representatives of their groups. Our knowledge of the true amphibians is limited to a relatively few species of anurans and salamanders. It is certain, however, that most amphibians must contend with one fundamental problem—their possession of an integument that is almost no barrier to evaporative water loss. Measurements carried out on some of the common temperate-zone laboratory frogs show that they evaporate water from their body surfaces at the same rates as from free water surfaces of the same areas at the same temperatures. Most amphibians apparently have no way of controlling these evaporative losses.

This fundamental situation has been the major reason most amphibians live in freshwater or in areas in which freshwater is available either in bulk or in small quantities in various microhabitats. Behavioral mechanisms for obtaining and conserving water are highly evolved in many forms. The primarily terrestrial species, especially those species living in arid or even desert areas, have also evolved physiological adaptations that are similar in some important respects to those described earlier in euryhaline forms.

Amphibians apparently cannot take up water as vapor, even from saturated atmospheres. Most also seem rarely to drink water. Evaporative water losses are compensated for by osmotic uptake across the skin. So-called pelvic patches, areas of skin on the lower abdomen and inner surfaces of the thighs, are often the primary sites of water uptake through the skin in anurans. The amount of dehydration that can be tolerated is very variable, and it is not necessarily the most terrestrial species that can stand the most. It is also not necessarily the most terrestrial species that can rehydrate most quickly following a given amount of dehydration.

The only water-loss route over which most amphibians have much control is that of their kidneys. The few anuran species that have been studied all show great decreases in rates of urine production when they are out of water. As illustrated in Figure 7-13, the decreases in urine volume result both from decreased GFRs and increased tubular water reabsorption. Voiding of urine stops completely if water supply restrictions persist. Urine concentrations in anurans out of water increase considerably above levels in freshwater. The urine approaches isosmoticity with the plasma, but remains low in sodium chloride. Considerable additional modification of urine volume and composition may take place in the bladder, especially in more terrestrial forms. Many desert species use their bladder, and probably also their lymph spaces, as accessory tanks for water storage during extended dry periods.

Dehydration in species not using accessory tanks, or in individuals of species using them which have had them emptied, does not occur uniformly in all organs. Body-fluid volumes decrease faster than intracellular volumes in all tissues. Water content of the nervous system is maintained at the most constant level of all.

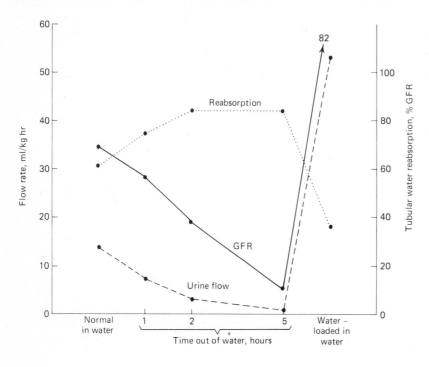

Figure 7-13
Changes in three important aspects of kidney function (urine flow rate, GFR, and tubular water reabsorption) in green frogs (Rana clamitans) *normally hydrated in water, out of water for up to 5 hr, and water-loaded in water. Data points are average values for six frogs each. Experiments at 20–25°C in summer. [Data from B. Schmidt-Nielsen and R. P. Forster.* J. Cell. Comp. Physiol., **44,** *233–246 (1954).]*

Body-fluid concentrations increase both as a result of water loss and as a result of the addition of new solute in the form of metabolic wastes. Urea is the major waste produced by most species when deprived of water. The maximum blood concentration of urea that may be attained depends on metabolic rate, degree of dehydration, and duration of the period of water restriction. Desert species such as the North American spadefoot toad *(Scaphiopus couchi)* spend 6 months or more of each year burrowed deep in sand. Plasma urea levels in these animals may reach concentrations almost double those found in the plasma of *Rana cancrivora* adapted to 100% seawater. New Mexican desert populations of the tiger salamander *(Ambystoma tigrinum)* perform similarly.

Field studies of American southwestern desert populations of spadefoot toads *(Scaphiopus hammondi* and *S. couchi)* and tiger salamanders indicate that, at least in some desert areas, soil water content is high and soil moisture tensions (tensions required to extract liquid water from a soil) are normally low enough for most, sometimes all, of the time when these animals are in their burrows to permit them to function in a near normal manner with respect to both body-water content and solute concentrations. Some populations of toads apparently can extract enough water osmotically from the soil, absorbing it across their skins, to avoid significant dehydration or increases in body-fluid concentrations. They do not appear to use bladder urine as an accessory supply of water in these circumstances. The ability to generate and tolerate very high body-fluid concentrations is, of course, highly adaptive during prolonged dry periods. Under these latter conditions, increased body-fluid osmotic concentrations permit continuing uptake of soil water even in the face of greatly increased soil moisture tensions.

Most, if not all, burrowing species of amphibians (both anurans and urodeles) from arid regions modify the structure and properties of the walls of their burrows so as to reduce substantially their water losses. They often do this by means of mucous-like inner linings.

Careful readers will have noticed that many of the statements just made about amphibian water and solute metabolism have been qualified as applying only to "most" species. The qualifications are made necessary by the discovery, since the late 1960s in subtropical and tropical Africa and South America, of a number of species of arboreal anurans that show evaporative water loss rates only 5–10% as high as those of most anuran species. These forms are capable of tolerating desiccation and starvation to much greater extents than most species (at least one, the southern African *Chiromantis xerampelina,* survives and recovers from losses of 60–65% of initial body weight over periods of 4–5 months). They greatly reduce urinary water losses by producing large fractions (as high as 97%) of their waste nitrogen in the form of uric acid salts (urates), especially during periods of prolonged water deprivation. They also use urate excretion as a means of excreting inorganic ions (especially potassium, sodium, and ammonium ions) almost without water loss. The combination of primarily insect diets, reduced cutaneous evaporative water loss rates, and urate nitrogen and cation excretion makes it at least a physiological possibility for the most arid-adapted of these species (such as *Chiromantis petersi* and the south American *Phyllomedusa sauvagei)* to survive almost indefinitely in the absence of access to free water. The most arid-adapted of these frogs are in these respects physiologically virtually identical with many desert species of lizards.

The basis for the low cutaneous evaporative water loss rates of these animals differs: In *Phyllomedusa* a lipid-based secretion is produced by specialized glands found in the skin. This secretion is water soluble, and must be replaced after the frogs have encountered free water. Some species aid the replacement process by moving their limbs over their body surfaces so as to spread the secretion. In *Chiromantis* dense layers of chromatophore cells in the skin seem to be involved.

Measurements of activities of the purine pathway enzymes, which are responsible for uric acid production primarily in the livers and kidneys, show that these frogs are comparable in these respects to desert lizards.

Most amphibians greatly reduce their salt losses while out of water by reducing their rates of urine production. There is some additional salt loss in the secretions of various glands. The food is probably the principal source of new salt, although the fossorial forms may also use soil water as a source.

Most amphibians, including most of the purely terrestrial species, must have some volume of fairly persistent liquid water available to reproduce. This is not an absolute necessity, however. The barking frogs (Eleutherodactylidae), for example, deposit small numbers of large, encapsulated eggs in crevices in dead wood or under stones. The embryos in these eggs never develop into free-swimming tadpoles, but directly become little frogs, which then hatch out. Many of the South American leptodactylid frogs (the group that includes the urate-excreting frogs just described) lay their eggs out of water in foam nests. The eggs hatch and the tadpoles often remain in the foam for many weeks, until local water conditions become favorable. A number of primarily terrestrial tropical forest species of frogs carry around their tadpoles, which have adhesive organs, attached to their backs. Fossorial salamanders also often lay their eggs in the soil or in crevices in dead wood, where they develop directly into little salamanders. Little is known about mechanisms of water and solute metabolism in these eggs, embryos, and larvae. Several kinds of the partly water-independent frog tadpoles accumulate urea in their body fluids. All other amphibian larvae that have been studied produce ammonia as the primary end product of their nitrogen metabolism.

Metamorphosis from the tadpole to the adult, at least in the familiar laboratory species of anurans, is always associated with a change in the pattern of nitrogen excretion. Tadpoles, as just mentioned, are usually ammoniotelic; the adults are generally primarily ureotelic. The development of urea production and excretion in adult semiterrestrial frogs results from a sudden, dramatic synthesis of the urea-cycle enzymes in the liver during the latter stages of metamorphosis. This appearance is illustrated in Figure 7-14. A comparable phenomenon probably occurs with respect to purine pathway enzymes in the urate excreting frogs.

Chapter 8 describes in some detail many features of the close relationships existing in many animal groups between the body temperatures of animals and their water relations. Amphibians on land are among the animals having the closest coupling of these two aspects of physiological regulation. The development of modern computer modelling techniques has led to several recent efforts to describe comprehensively the water and energy relationships of terrestrial anurans. The Additional Read-

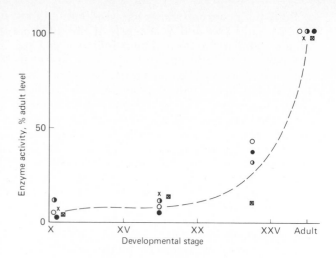

O Carbamylphosphate synthetase

◑ Ornithine transcarbamylase

● Arginine synthetase system

× Argininosuccinate cleavage enzyme

⊠ Arginase

Figure 7-14
Changes in activity of five important enzymes or enzyme systems involved in urea synthesis during tadpole development and metamorphosis in bullfrogs (Rana catesbeiana). *The metamorphic transition from ammoniotelism to ureotelism appears to be correlated with synthesis of new urea cycle enzyme molecules.* [Data from G. W. Brown, Jr., et al. J. Biol. Chem., **234**, 1775–1780 (1959).]

ing list for this chapter includes papers summarizing this interesting work.

7.8 REGULATION OF BODY-FLUID COMPOSITION IN TERRESTRIAL VERTEBRATES

7.8.1 Patterns of Regulation

Table 7-14 summarizes data on blood plasma osmotic and solute concentrations found in representative normally hydrated terrestrial reptiles, birds, and mammals. The uniformity in pattern is appar-

TABLE 7-14. Steady-State Patterns of Osmotic and Solute Regulation in Plasma of Representative Terrestrial Vertebrates*

Species	Osmotic Concentration mOsm/liter	Na$^+$ mmol/liter	%Δ	Cl$^-$ mmol/liter	%Δ
Desert tortoise (*Gopherus agassizii*)	290	120	41	~100	35
Mourning dove (*Zenaidura macroura*)	370	180	49	140	38
Man (*Homo sapiens*)	310	150	48	110	35
Camel (*Camelus dromedarius*)	340	160	47	110	32

* All figures are averages.

Source: W. H. Dantzler and B. Schmidt-Nielsen. *Am. J. Physiol.,* **210,** 198–210 (1966); M. Smyth and G. A. Bartholomew. *Auk,* **83,** 597–602 (1966); K. Schmidt-Nielsen. *Desert Animals, Physiological Problems of Heat and Water,* Oxford University Press, Fair Lawn, NJ, 1964.

ent. These animals all closely resemble the "ordinary" varieties of lampreys, bony fishes, and amphibians. There are no cases known in any of the terrestrial groups in which significant amounts of organic solutes are accumulated in the body fluids. This is true even following severe dehydration, with accompanying elevation in plasma concentrations. The terrestrial amphibians were discussed in Section 7.7.

7.8.2 Mechanisms and Adjustments

Terrestrial animals have three main ways of gaining water; drinking, as part of their food, and by metabolic production from their food. There are four main routes for water loss: urine, feces, evaporation from the integument, and evaporation from the respiratory tract. Over significant periods of time the amounts taken in must equal the amounts lost if the animal is to maintain its state of hydration. The basic circumstance with which almost all terrestrial animals must contend is the need to conserve water in a generally waterless environment. Most of the adaptions we shall discuss are directed toward achieving this end by means of many kinds of variations on the seven basic themes. Figure 7-15 diagrams the pathways used by the major groups.

Solute balance in terrestrial animals involves, on the intake side, drinking and food. Urine, feces, and sweating are the paths for losses. Once again, deviations from balance over significant periods of time lead to changes in solute content. Inorganic salts, especially sodium chloride, and metabolic wastes, especially urea, are usually quantitatively most important. Figure 7-15 also diagrams the major pathways for solute balance.

The relationships existing between water metabolism and other physiological functions, notably respiration and temperature regulation, are extremely close. These relationships are discussed in Chapters 5 and 8.

The mechanism of body-temperature regulation is one of the major factors in the much larger water requirements of birds and mammals as compared with reptiles. Many reptiles maintain quite constant body temperatures, comparable with those of many birds and mammals. They regulate these temperatures for long periods, primarily by behavioral means. The ways in which they rid themselves of excess calories, when this is necessary, generally do not involve the evaporation of much water (see Section 8.7.2). Birds and mammals, on the other hand, even those species that frequently use behavioral thermoregulatory procedures, are always producing heat from their own metabolism. In most

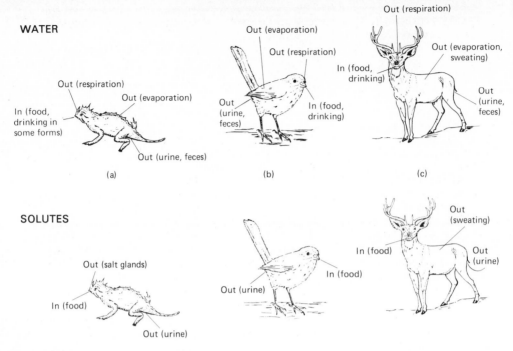

WATER

SOLUTES

Figure 7-15
As in Figure 7-4, but for terrestrial vertebrates. Groups of organisms are (a) reptiles, (b) birds, and (c) mammals. The bird represented is a passerine; nonpasserines differ in some important features.

situations the evaporation of water is one of their main mechanisms for losing whatever part of this heat they do not need. Reptiles additionally seem to have inherent physical limitations on their maximal rates of water loss that keep these below avian and mammalian levels.

Even though reptiles lose relatively little water as a result of thermoregulatory efforts, they sustain considerable losses from their respiratory tracts. These have generally been considered to be the major evaporative losses, but reptilian integuments are by no means impermeable to water vapor. Table 7-15 summarizes data on respiratory and cutaneous water losses in a series of four species of partially and fully terrestrial reptiles, arranged in order of increasing aridity of habitat. Snakes show a similar pattern. An interesting minor point is that mutant scaleless snakes of two species have been found to have the same rates of cutaneous water loss as normal, fully scaled snakes of the same species.

As would be expected, rates of loss by means of each route are much smaller in animals from drier environments. Some desert lizards reduce respiratory water losses by means of nasal countercurrent heat exchangers similar to those found in many small desert birds and mammals (see Figure 7-18). However, all respiratory-loss rates are surprisingly high compared to those found in mammals. Respiratory evaporation in man is about 0.85 mg/ml O_2, and even small nonsweating rodents (for which respiratory evaporation is the major pathway for heat loss) show rates as low as 0.5 mg/ml O_2. It is not apparent why reptiles should lose so much water in this way.

The percentage of total water loss that occurs by way of the skin also declines as the habitat becomes drier. However, even in the desert chuckwalla (*Sauromalus obesus*) at 40°C (a body temperature not too far from the preferred temperature for this form) almost one half of the total loss is cutaneous. The decline in percentage contribution

TABLE 7-15. Respiratory and Cutaneous Water Loss in Various Reptiles in Dry Air*

Species	Respiratory Loss (mg/ml O_2)	Cutaneous Loss (mg/cm² day)	Cutaneous Loss (% of total)
Reptiles at 23°C			
Caiman sclerops	4.9	33	87
Terrapene carolina	4.2	5.3	76
Iguana iguana	0.9	4.8	72
Sauromalus obesus	0.5	1.3	66
Reptiles at 40°C:			
Sauromalus obesus	1.4	3.4	44

* All figures are averages.
Source: P. J. Bentley and K. Schmidt-Nielsen. *Science,* **151,** 1547–1549 (1966).

of cutaneous losses from chuckwallas at 40°C as compared with those at 23°C probably derives from two sources. First, at higher temperatures respiratory evaporation increases because of higher metabolic rate and greater water-vapor capacity of warmer air. Second, although cutaneous evaporation also increases, it does so to a smaller extent, being proportional only to increases in water-vapor pressure.

A surprising point to realize is that total evaporative water losses from the mostly aquatic caiman occur at rates equal to one third to one half loss rates from amphibians under similar conditions. Almost 90% of the caiman's losses are cutaneous.

It is unlikely that reptiles can exert very much direct physiological control over their respiratory and cutaneous rates of water loss. This means that the only water-loss routes upon which they can have significant physiological influence are the gut and kidneys. The cloaca may further modify the water content of the urine and feces by reabsorption. The colloid osmotic pressure of the plasma proteins appears to play a major role in cloacal water reabsorption, at least in some desert lizards.

Little of a quantitative nature is known about the performance of the reptilian gut with respect to water content of the feces. Reptile feces generally appear to be very dry. Reptilian kidneys appear to be very efficient at producing small volumes of urine, even though they cannot produce urine that is hyperosmotic to the blood (urine released from the cloaca may be hyperosmotic to the blood in some species). Many species produce hypoosmotic urine. Urine flow rates in normally hydrated arid habitat lizards are as low as 0.2 ml/kg hr. One of the major bases for these low flow rates is the fact that most reptiles excrete a large part (the proportion can vary with temperature and state of hydration) of their nitrogenous waste as uric acid and urate salts rather than ammonia or urea (Table 7-16). Uric acid is a purine of very low solubility that precipitates out of solution fairly easily. Reptilian urine is often a semisolid paste containing large numbers of small crystals of uric acid and urates. Because of uric acid's low solubility and easy precipitation, it is not necessary for reptilian urine to contain much water to keep nitrogenous wastes in solution at low enough concentrations not to cause osmotic problems (see the discussion that follows of the composition of bird urine). The urates also permit excretion of significant amounts of inorganic cations (especially potassium ions) with little water.

Much of the urate in reptile urine is in the form of minute spheres, similar to those found in bird urine (see page 318).

The uric acid in reptile blood is concentrated in the urine by addition of tubular secretions to the amounts originally filtered in the glomeruli. In *Iguana iguana* about 6% of urinary uric acid is filtered; 94% is added by tubular secretion.

Differences in the maximum urine-concentrating and water-reabsorption capacities of the kidneys of different species of reptiles (lizards) can be correlated with differences in ultrastructure of the cells of the renal tubular walls. Larger numbers of mito-

TABLE 7-16. Principal Nitrogenous Constituents of Dried Urinary Deposits of Some Egyptian Desert Reptiles*

Species	Nitrogenous Constituents†			
	Ammonia	Urea	Uric Acid	Allantoin
Lizards				
Tarentola annularis	1	1	65	0
Mabuya quinquetaeniatus	2	3	43	<1
Agama stellio	2	0	64	3
Uromastyx aegyptia	<1	<1	56	1
Snakes				
Natrix tessellata	5	3	55	0
Psammophis sibilians	2	2	61	0

* Total nitrogen content of deposits ranged from 36 to 48% of dry weight.
† All figures are averages, expressed as percent of nitrogen.
Source: G. Haggag and R. Hassan. *Z. vergl. Physiol.,* **57,** 423–427 (1968).

chondria and greater complexity of folding of basal cell membranes are found in the tubular cells of species having greater capacities for tubular water and salt reabsorption.

Changes in temperature have major effects on many features of kidney functions. In the Australian lizard *Tiliqua rugosa,* rates of urine production increase about 30-fold in uniformly water-loaded animals between 14 and 37°C. Increased GFR is responsible for most of the increase. Tubular sodium reabsorption also increases in this species with increase in temperature.

The limitations on reptilian urinary osmotic concentrations and the small urine volumes produced imply that the kidneys are not very effective in excreting excesses of other solutes, especially salts (but note preceding comment concerning urates). Most terrestrial reptiles appear not to encounter serious salt-loading problems very often. However, requirements for water conservation are sometimes severe enough, especially in desert species, that a number of forms have evolved nasal salt glands that function in the same way as previously described for marine reptiles and birds. Insectivorous and vegetarian species that need to eliminate large amounts of potassium use their salt glands for this purpose. Vegetarian species also excrete excess bicarbonate by this route.

Reptilian water and solute losses are most often compensated for by means of their food. Reptiles living in areas where free water is available will occasionally drink, but most species are carnivorous and live on food that is watery enough (either in actual water content or as a metabolic source) that they do not need to drink very often. There is no evidence for osmotic or other water uptake through the skin.

Although physical drinking is probably rare, some desert species of reptiles swallow water that reaches their mouths by some very unusual routes. An Australian lizard, *Moloch horridus,* has a network of very fine open channels in the outer layers of its skin. If the belly of one of these lizards is placed in water, the water moves through these channels by capillarity and eventually reaches the mouth. When the mouth is reached, the lizard moves its jaws and swallows water. Desert tortoises will drink surface water after sufficiently heavy rains.

Additional information on water and solute metabolism in reptiles may be found in many of the items cited in the Additional Reading list.

Birds are, in many important respects, feathered reptiles maintaining high body temperatures based on high metabolic rates. Allowing for the quantitative differences that result from the high temperatures and metabolic rates, water and salt metabolism in birds resembles that just described for reptiles. The following discussion relates primarily to wild birds. It does not consider the ecologically hard-to-interpret data available on domestic chickens.

Evaporation, primarily from the respiratory tract but also from the skin, is the major route for water loss in birds. Cutaneous losses have generally been considered to be slight because birds are not known to have sweat glands and are usually covered by a fairly tight layer of feathers. However, cutaneous losses can be substantial, in at least a few species [for example, Chinese painted quail (*Excalfactoria chinensis*) and domestic pigeons (*Columba livia*)]. Data on a few other species indicate that this situation is not universal. There probably is a relationship between importance of cutaneous water loss and habitat. The zebra finch (*Poephila guttata*), from arid parts of Australia, is able almost to eliminate cutaneous water losses under conditions of water deprivation. Cutaneous losses from well hydrated birds are 45–74% of total evaporative losses. The mechanism of this adjustment is unknown.

Birds have respiratory water loss rates as high as those listed in Table 7-15 for reptiles. Cardinals (*Richmondena cardinalis*) lose 0.9 mg/ml O_2, and poorwills (*Phalaenoptilus nuttallii*) lose 3.7 mg/ml O_2. The high metabolic rates of birds mean that absolute rates of water loss are much higher than from reptiles. Figure 7-16 illustrates the magnitude of this problem. Some desert-dwelling species significantly reduce their evaporative water losses when deprived of drinking water. The mechanism of this reduction is unknown.

Evaporative water losses from birds are frequently even more severe than indicated in Figure 7-16. Flying requires considerable energy expenditure, hence increased metabolic rates. Another aggravated situation results from the use of evaporation as the major means of controlling body temperatures in warm environments. Increased metabolic rates combine with increased water capacity of warmer air (resulting from somewhat elevated body temperatures) to increase the stress greatly.

For birds, as for reptiles, the gut and the kidneys are the two major physiologically controllable routes for water losses. The cloaca can play a significant role in water reabsorption from the feces and urine especially in somewhat dehydrated birds (Figure 7-17).

Details of the role of the bird gut in water and solute metabolism are unknown. Bird kidneys, with the notable exception of a few species that must cope with continuing salt loads but lack salt glands (see Section 7.5.3), are physiologically similar to reptile kidneys. They produce small volumes of urine, which can, however, reach about 2.5 times

Figure 7-16
Relation between rate of evaporative water loss (EWL) and body mass in birds. Measurements on resting birds at ambient temperatures within or below their respective zones of thermal neutrality. Solid line is the regression line fitted to these data; equation: log EWL (g/day) = 0.61 log mass (kg) + log 24.2. Note that this equation means that small birds lose proportionately more water per unit time than large birds. [Data courtesy of R. C. Lasiewski.]

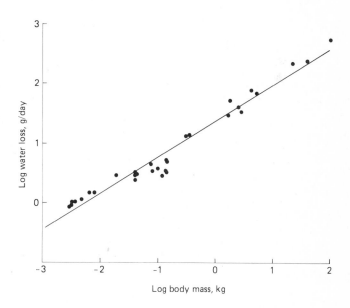

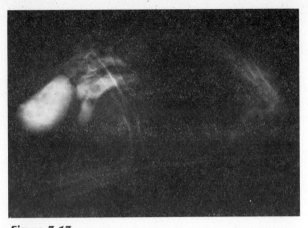

Figure 7-17
*Excretion urograph of a roadrunner (*Geococcyx californianus; *bird's head to right) demonstrating refluxing of postureteral urine from the cloaca into the rectum and intestinal caecae, probably permitting additional, extrarenal water reabsorption from the cloacal contents. Large light area on left the cloaca, smaller light areas to right of this the rectum and caecae. Bird given an intravenous injection of a solution of sodium diatrizoate (an x-ray opaque substance excreted by the kidneys), then photographed 30 min later. [From R. D. Ohmart, L. Z. McFarland, and J. P. Morgan,* Comp. Biochem. Physiol., **35,** *487–490 (1970).]*

the osmotic concentration of the blood. The urine has classically (since about 1811) been considered to be a semisolid paste of uric acid and urate crystals. However, urine from a wide variety of birds from many geographic regions consists largely of minute spherical bodies of fairly uniform size (in single species) varying (between species) from 2 to 10 μm in diameter. These spheres consist primarily of uric acid and urate salts (especially sodium and potassium) plus significant amounts of bound or adsorbed sodium and potassium ions.

The liquid portions of bird urine often contain larger amounts of either dissolved or colloidally dispersed urate than would be expected from the physical properties of the compounds. These other urate fractions also bind or adsorb substantial amounts of inorganic ions. Urinary mucoids (glycoproteins and mucopolysaccharides) apparently help to keep these urate fractions stable in the liquid phase.

The binding or adsorbtion of inorganic ions to

urates that are either precipitated out of solution or maintained in suspension (but not in solution) greatly reduces the amount of osmotic work that bird kidneys must do in order to excrete electrolytes and waste nitrogen. It also reduces the amount of water that must be used in these processes.

The spherical shape of precipitated urate particles is thought to assist in the mechanical process of moving the urine along the ureters.

Little is known about such aspects of kidney function as urine flow rates and GFRs in wild birds. GFRs in normally hydrated mourning doves (*Zenaidura macroura*) are about half those of similar size mammals.

The basis for the ability of bird kidneys to produce urine concentrations significantly hyperosmotic to plasma concentrations derives from the presence in bird kidneys of two types of excretory units (*nephroi*). These are (1) nephroi similar in structure to those found in reptilian kidneys, which lack a loop of Henle and can produce urine that is at most isosmotic with the plasma; and (2) nephroi similar in structure to those found in mammalian kidneys, including a loop of Henle (see Section 7.8.3). Interspecific differences in maximal urine concentrations produced under dehydrating conditions probably relate to differences in relative numbers of reptilian and mammalian type nephroi in the kidneys of the different species (determination of the nature of this relationship is complicated by cloacal water resorption; ureteral urine is hard to collect from birds). Variations in urine concentrations within single species appear to be due to changes in the numbers of reptilian type nephroi operating under different conditions. The structure and function of bird kidneys are both complex; both aspects differ considerably from mammalian kidneys. For more information consult the review on bird osmoregulation in the book edited by G. M. O. Maloiy included in the Additional Reading list.

Birds compensate for their continuing water losses by drinking or by using the water content of their food. The rate of drinking by birds given free access to water relates to body size in a pattern very similar to that shown in Figure 7-16 for evaporative water losses. Careful calculations of the amounts of metabolic water birds can produce from the oxidation of their foodstuffs indicate that this source is almost never adequate for maintenance

318

by itself. The sum of the amount of preformed water included in the food, plus metabolic water formed from the food, is sufficient in a number of xeric adapted species to enable them to go completely without drinking.

With a few exceptions, desert and salt-marsh species of birds are unable to use drinking water that is more saline than the equivalent of about 50% seawater. Thus the occurrence of species such as these in either waterless areas or areas in which all available water is saline usually depends on their use of a method to circumvent the conditions. The main ways of doing this are: flying to suitable water supplies elsewhere (this may mean 50–100 km for very strong and swift fliers such as sand grouse) and feeding on succulent vegetation, meat, or insects. Supplying water to the young is frequently a major problem. There are a number of spectacular instances of elaborate behavioral mechanisms for doing this. Adjustment of breeding periods to coincide with periods of rainfall is the commonest solution.

The only serious solute-loading problems most terrestrial birds are likely to encounter involve inorganic ions. The concentrating abilities of the kidneys seem adequate to handle these problems in all but the few species living in the most severe, driest, saline environments. Some of the most desert-adapted of the nonpasserines, including the ostrich, have functional salt glands operating in much the same fashion as the glands in marine birds. As far as is known at present, there is no terrestrial passerine bird possessing a functional salt gland or other extrarenal pathway for salt excretion. Only the barest beginning has been made on studies of renal-salt-concentrating capacities in the most sedentary, seed-eating species of desert passerines living in areas in which only saline water is available for drinking—if any water is available. In at least one such species, the black-throated sparrow of the American southwest (*Amphispiza bilineata*), renal osmotic and sodium chloride concentrations have been measured that are equivalent to those recorded from salt-marsh savannah sparrows drinking seawater. Thus it seems as if those terrestrial bird species that have evolved direct physiological adaptations to desert life, as opposed to behavioral and nutritional adaptations, have used the same basic adaptive routes for water conservation and elimination of monovalent ion loads

as have marine birds. It will be interesting to see if future work on divalent ion metabolism in both groups demonstrates similar parallels.

Water and solute metabolism in terrestrial mammals differs both qualitatively and quantitatively from the situations just described in birds. Important quantitative differences are: size for size, many mammals have lower rates of respiratory evaporative water loss than birds (the relationship has much the same shape as that for birds); the kidneys of mammals in general are capable of producing higher urinary osmotic and inorganic ion concentrations than the kidneys of birds; and urea is the principal end product of nitrogen metabolism. The difference in urine-concentrating abilities probably relates to a difference in kidney structure (see Section 7.8.3). Other important qualitative differences are (1) many species of mammals (not all) have sweat glands, which are important pathways for cutaneous losses of both water and salts; (2) no terrestrial mammal is known to possess an extrarenal pathway for salt excretion; and (3) mammals lack cloacas or other organs capable of further modifying urinary volume and composition after the urine has left the kidney proper.

These structural and functional differences between the two groups produce a number of adaptively significant corollaries. Size for size, many terrestrial mammals need to drink less water than birds. This is true despite the fact that urinary urea elimination requires somewhat larger volumes of water loss, partially offsetting the water economies resulting from lowered respiratory evaporation and the ability to produce more concentrated urine. The latter two capacities, combined in a variety of ways with other physiological (the ability to become torpid for extended periods, for example) and behavioral adaptations (nocturnal and fossorial habits, for example), permit a high proportion of herbivorous, as well as carnivorous and insectivorous, mammals to get along without ever drinking. Preformed water in the food and metabolically produced water together are adequate to satisfy the water requirements of mammals as different as the gorillas of the high mountain forests of East Africa and a variety of rodents from every major desert on earth.

Table 7-17 summarizes data on maximal urine concentrations in various mammals. The species listed represent a variety of patterns of water intake, as well as a wide range of urinary concentrating ability. The first three species require drinking water, the last four do not. Man and laboratory white rats are relatively intolerant of dehydration and must drink fairly often to maintain hydration. Water loss equal to about 10% of body weight is enough to incapacitate a man; water loss of 12% of body weight results in inability to swallow and impossibility of recovery without assistance. Laboratory rats and most other "ordinary" mammals can tolerate dehydration up to 12–14% of body weight loss. Camels, on the other hand, are amazingly tolerant of dehydration and can go for very long periods without drinking. Resting camels exposed to the full heat of the summertime Sahara desert can tolerate water losses equivalent to more than 25% of their body weight. In cool environments they probably can dehydrate even further. Camels in summer can go for almost 1 week without drinking. Under cooler winter conditions they can go for almost 3 weeks.

The four nondrinking rodents listed also are highly diverse. The antelope ground squirrel and the sand rat both obtain the water they need from succulent vegetation. Such vegetation is usually fairly abundant in their habitats, so it is reasonable to enquire why they possess such efficient kidneys—in the sand rat the most efficient known. The probable reason in the ground squirrel is its pattern of diurnal, aboveground activity, even in midsummer. It lives in the Mojave Desert of California, so this pattern results in increased evaporative water losses associated with body-temperature regulation, hence a need for greater urinary-water conservation. The reason in the sand rat is its use as food of juicy, salt-loving plants having high salt contents in their sap. The plants contain 80–90% water, so the rats must eat huge quantities to obtain sufficient nutrients. Large amounts of salty sap inevitably are taken in, so the rats must continually eliminate large salt loads. They are inhabitants of the Sahara and must contend with significant evaporative water lossses. Thus it is necessary for them to accomplish their task of salt elimination with the use of as little urinary water as possible.

The gerbil and the hopping mouse differ from

TABLE 7-17. Maximal Urine Concentrations in Various Terrestrial Mammals*

Species	Urine Δ (Osm/liter)	Urine Electrolyte† (meq/liter)	Urine/Plasma Osmotic ratio
Man (Homo sapiens)	1.43	460	4.2
White rat (Rattus rattus)	2.9	760	8.9
Camel (Camelus dromedarius)	2.8	1070	8
Antelope ground squirrel (Citellus leucurus)	3.9	1200	9.5
Gerbil (Gerbillus gerbillus)	5.5(est.)	1600	14(est.)
Sand rat (Psammomys obesus)	6.34	1920	17
Australian hopping mouse (Notomys alexis)	9.37	<2000(?)	22

* Figures are the highest available, but do not necessarily represent the concentration limits for the species.
† Electrolyte concentration is an approximation referring, in some cases, to Na concentration, in others to NaCl equivalence from electrical conductivity.
Source: K. Schmidt-Nielsen. *Desert Animals, Physiological Problems of Heat and Water,* table XXVI, Oxford University Press, Fair Lawn, NJ, 1964; R. E. MacMillen and A. K. Lee. *Comp. Biochem. Physiol.,* **28,** 493–514 (1969).

the other two species in that they are capable of surviving on air-dried food, including high-protein seeds. These animals, natives of Old World and Australian deserts, appear to be physiologically similar to the kangaroo rats (*Dipodomys*) of the American deserts. Dietary salt loading in these forms is not so severe as it is for the sand rat, but their environments and their food are very dry and they must conserve water by producing small volumes of highly concentrated urine.

Kangaroo rats and other small mammals have an unusual mechanism for reducing respiratory evaporative water losses. The dimensions of the nasal passages are sufficiently small that the air passing through rapidly comes into thermal and water vapor tension equilibrium with the walls of the passages. The passage walls are, of course, moist. These two factors interact to conserve water as follows.

Inhaled dry air rapidly warms to body temperature (kangaroo rats are fossorial and inactive by day, surface living and active only during the cooler desert nights). Because the air is dry to begin with, considerable water evaporates into it from the passage walls, saturating the air with water vapor and significantly cooling the walls. The warm, saturated air then proceeds to the lungs. With exhalation, the warm, saturated air comes into contact with the cool walls of the nasal passages, cooling it and causing a considerable fraction of the contained water vapor to condense out. This condensation is rapidly absorbed osmotically by the nasal membranes and rejoins the general pool of body water. The still-saturated, but now much cooler air, then leaves the animal, but, because of its lower temperature, with much less water than if it were still at the kangaroo rat's normal body temperature. The arrangement might be termed a nasal tidal countercurrent water conserver. It can only work to a significant degree in species having nasal passages of very small diameter.

Figure 7-18 indicates the magnitude of the water conservation achieved by this mechanism under various conditions of ambient temperature at 25% relative humidity. Note that the mechanism functions in small birds just as it does in small mammals.

The significance of cutaneous water losses in the overall water balance of mammals is usually difficult to assess. Two forms of cutaneous losses occur; diffusion water loss (often called insensible perspi-ration), which is evaporation from the skin surface of the (usually small) amount of water that has reached the surface by diffusion through the cornified layers of the epidermis; and sweating, which is the bulk secretion of liquid from the sweat glands. The chemical composition of sweat varies somewhat with environmental, nutritional, and metabolic conditions, and especially with the rate of secretion. However, it always contains significant amounts of chloride, sodium, and potassium, and frequently contains significant amounts of urea and lactic acid. Thus, especially in conditions requiring high rates of sweat production, the solute losses associated with its release are second in importance only to the water losses.

Cutaneous diffusion water losses vary considerably among species. Rodents, for example, all lack sweat glands, so measured rates of evaporative water losses are due to some combination of respiratory and diffusional losses. Laboratory white rats have nose temperatures as low as those of kangaroo rats, hence lose water by respiratory evaporation at the same rate as kangaroo rats. Respiratory losses account for 95% + of total losses in kangaroo rats but only about 50% in laboratory rats. Rates of diffusion water loss probably cannot be controlled to any significant extent, except by behavioral selection of humid environments.

Sweating rates are, however, very variable and subject to considerable physiological (primarily nervous) control. Our most detailed knowledge relates to man. The major determinant of sweating rate is body heat load, with highest rates occurring at ambient temperatures and humidites close to toleration maxima. Men doing hard physical labor in saturated atmospheres at 35.5°C sweat at rates up to 4.2 liters/hr. The ability to sweat increases with prolonged adaptation to high temperatures.

Sweating, therefore, can be a major factor in water balance in mammals having sweat glands and exposed to high temperatures. This point can be dramatized by noting that the total volume of water in the blood of a man is about 4 liters. Sweating can also be very important in solute balance. During periods of rapid sweating it is possible for men to perspire body sodium chloride at rates of 10–30 g/day, or 6–18% of total body sodium chloride. These salt losses must be compensated for at the same time as the water losses are made up or serious dilution of body fluids can result, producing

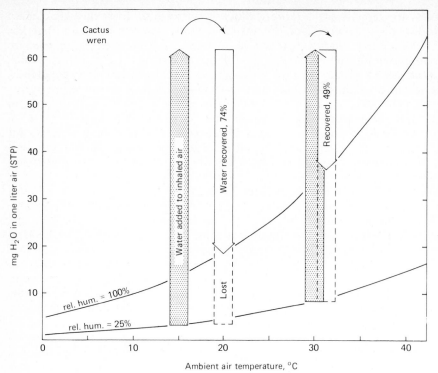

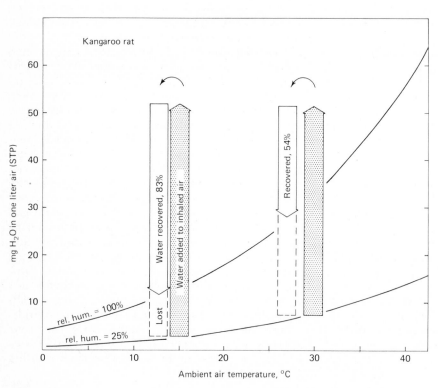

Figure 7-18
Magnitude of water savings achieved in small birds (cactus wren, top) and mammals (kangaroo rat, bottom) through the operation of the nasal tidal countercurrent water conservation mechanism when breathing air of 25% relative humidity at temperatures of 15° and 30°C. See text for discussion. Vertical arrows centered at temperatures of nasal air streams during inhalation (upward pointing arrow) and exhalation (downward pointing arrow). [From K. Schmidt-Nielsen, F. R. Hainsworth, and D. E. Murrish. Resp. Physiol., **9,** *263–276 (1970).]*

cramps and other symptoms of heat prostration.

Sweat glands, at least in man, seem unable to reduce significantly salt concentrations in sweat under conditions of high secretion rates. The kidneys, however, make maximal efforts in this direction under these circumstances, producing urine that is virtually sodium chloride free.

To conclude this discussion of terrestrial mammals it seems appropriate to describe the interactions between the environment and water and solute metabolism as they occur in several species that successfully contend with the direct effects of a number of severe environmental stresses. The species involved are the marine fish-eating bat (*Pizonyx vivesi*) of the desert islands in the Gulf of California, Mexico, and several species of east African plains game, antelopes all belonging to the family Bovidae.

The fish-eating bat contends with virtually all of the water and solute stresses to which mammals may be subjected. The terrestrial habitat it occupies is, for much of each year, hot desert. Its wings contribute to an exceptionally large surface area for evaporative water loss. Its flying requires a major metabolic effort, with associated increased respiratory water loss. It drinks seawater in the laboratory, and probably in nature as well. Despite its name, its food is often largely composed of marine crustacea, virtually all with body fluids isosmotic with seawater. Apparently it successfully overcomes all of these obstacles because it is abundant in its habitat.

Like most bats, *Pizonyx* is nocturnal. It spends its days in crevices between the rocks of talus slopes. These crevices are effective insulation against very high daytime surface temperatures. In summer, early afternoon temperatures in bat-containing parts of crevices usually are well below 30°C. Behavioral adaptations thus permit *Pizonyx* to avoid the thermal extremes of its habitat.

Other important behavioral observations are (1) *Pizonyx* can become torpid (see Section 8.12) if subjected to conditions of low temperature or, probably, food shortage; and (2) there is no indication that these bats normally do become torpid under natural conditions during the warm parts of the year.

Laboratory studies of the major aspects of water and solute metabolism have been carried out on *Pizonyx* under conditions approximating those in the bats' natural environment. From these measurements it is possible to construct a water balance sheet for an average 25-g bat:

Input: Water in crustacean food, plus metabolic water from food: 12.0 ml/day.

Outgo: Evaporation (active, but not flying bats): 3.7 ml/day.
Fecal water loss: 1.3 ml/day.
Urinary excretion of waste nitrogen (urea): 6.4 ml/day.

Nothing is known about how much time *Pizonyx* devotes to flying each night. Evaporation rates associated with flight are so high that the bats go into negative water balance after only 0.5-hr flying time. If, as seems probable from studies of other bats, they actually spend about 2 hr per night flying, flight associated evaporation equals 2.1 ml/day.

On this basis, outgo exceeds dietary input by approximately 1.5 ml/day. This deficit can only be made up by drinking seawater. Nothing is known about how much seawater these bats actually drink. However, the salt concentrating abilities of their kidneys are such that they can just excrete the amounts of salt that would be ingested in the volume of seawater cited, plus the salts eaten in the food, in the volume of water listed above as needed to eliminate dietary waste nitrogen.

It thus seems that the fish-eating bat is able to survive and remain fully active in its habitat by operating within narrow limits. *Pizonyx* must fly in order to eat, but even moderate amounts of flight force it into negative water balance. It then must drink seawater, but the salt loading so incurred, added to the salt in the food, quickly saturates the excretory capacity of the kidneys. If food supplies become scarce enough to force the bat to fly much more than 2 hr per night, or if air temperatures fall low enough to force extra metabolic expenditures (and extra evaporation) for thermoregulation, it has only one alternative: to conserve water and energy by becoming torpid.

The antelopes of the plains and savannas of east Africa also inhabit a harsh environment, though usually not such an extreme one as that of the fish-eating bat. In most areas, however, there are prolonged hot, dry seasons that greatly reduce supplies of drinking water at the same time as they inflict heavy heat loads. Some of the plains game undertake extensive migrations to avoid the worst of these conditions, but others stick it out and remain. Several of the nonmigratory forms also occur in actual, year-round deserts. The reality that faces all forms, wherever they occur, is that they are so large that there are no holes or crevices in which they can hide. They cannot, therefore, evade the full stress of midafternoon heat either by finding cool air and substrate in a burrow or by becoming nocturnally active.

A series of studies has demonstrated that each of the major species of plains antelopes has evolved its own particular strategy for coping with its water problems. We will discuss here only two of these strategies, those utilized by two of the largest of the nonmigratory forms, the eland (*Taurotragus oryx*) and the oryx (*Oryx beisa*). The elands studied averaged about 200 kg in weight, the oryx about 100 kg. Sections 8.9.4 and 8.10.3 describe aspects of the thermoregulatory physiology of these animals.

Both the eland and the oryx can survive indefinitely without drinking, under natural conditions, even at air temperatures near 40°C. In this respect they significantly surpass the dromedary camel (see earlier mention in this section). Laboratory studies in a temperature controlled artificial desert showed that they achieve this independence of liquid water in two different ways. All of the following statements are based on comparisons between fully hydrated animals provided with an excess of water and animals dehydrated to 85% of their initial weight, then given just enough food and water to maintain them at their reduced weights. Figure 7-19 may assist in following the discussion.

The water requirements of both species are low. Under laboratory conditions similar to those of their natural habitat (cool nights, hot days) total water intakes of partially dehydrated animals averaged 5.4 liters/100 kg day for the eland, 3.0 liters/100 kg day for the oryx (the equivalent requirement for the fish-eating bat is 56 liters/100 kg day). Water losses were partitioned: Eland: 60% evaporation, 20% feces, and 20% urine. Oryx: 61% evaporation, 16% feces, and 23% urine. These patterns of water loss are not particularly unusual for ungulates.

The basis for these low water requirements is largely, in both species, the ability to store heat

in the body during the day, resulting in a substantial rise in body temperature. In this they are similar to camels. However, the oryx can become much more hyperthermic than the eland, thus significantly reducing its evaporative water losses below those of the eland. This difference in thermal tolerances is reflected in the behavior of the two species in nature. The eland searches for shade when it becomes hot, while the oryx remains out in the full sun.

The oryx further reduces its daytime evaporative water losses, in comparison with the eland, by being able to suppress sweating completely even when very hot. Both species pant at high temperatures. Both species also reduce their metabolic rates at high temperatures, the oryx to a significantly greater degree than the eland. These metabolic-rate reductions conserve water by reducing the rates of internal production of calories, hence the amounts of evaporative cooling needed to keep body temperatures within tolerance limits.

Possibilities for night-time water conservation are also utilized. Nonsweating cutaneous evaporation is reduced by 30% in the eland, by 60% in the oryx. Respiratory evaporative water losses are reduced in both species in three ways. First, total metabolic rates are again reduced, by about 5% in the eland, by over 30% in the oryx. Second, respiratory rates are proportional to, respiratory efficiencies inversely proportional to, body temperatures. This results in cool animals breathing more slowly than warm ones, and also in cool animals using a greater fraction of inspired oxygen than warm ones. The net result is additional water conservation. Third, the lowered night-time body temperatures automatically mean that saturation levels for water vapor in exhaled air are lower.

Possible differences in urinary system or intestinal tract function between normally hydrated and partially dehydrated animals have not yet been studied.

One final question remains: Since both species do not normally drink, where do they get their water? Field studies indicate that it is in their food, largely as preformed water. Eland eat large quantities of acacia tree leaves. Even during severe droughts these leaves contain an average of 58% water. On the basis of the daily metabolic needs of an eland, calculations indicate that the amount of acacia leaves needed to provide the calories

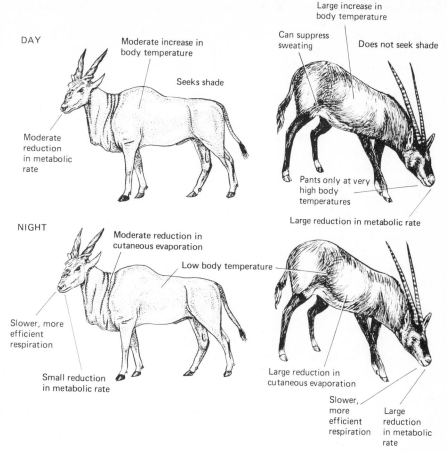

Figure 7-19
Adaptations for water conservation in two nondrinking east African antelopes, (left) the eland (Taurotragus oryx) and (right) the oryx (Oryx beisa). Upper figures for daytime, lower figures for night. See text for discussion. [*From C. R. Taylor,* Sci. Am., **220,** *88–95 (1969).*]

DAY

Moderate increase in body temperature

Seeks shade

Moderate reduction in metabolic rate

Large increase in body temperature

Can suppress sweating

Does not seek shade

Pants only at very high body temperatures

Large reduction in metabolic rate

NIGHT

Moderate reduction in cutaneous evaporation

Low body temperature

Slower, more efficient respiration

Small reduction in metabolic rate

Large reduction in cutaneous evaporation

Slower, more efficient respiration

Large reduction in metabolic rate

would also provide the water, even under drought conditions. Oryx feed mostly on grasses and shrubs, some of which can be fairly succulent. A primary group of food plants, however, are shrubs of the genus *Disperma.* The leaves of these plants fluctuate diurnally in water content, from less than 1% in the afternoon to over 40% in the early morning, before sunrise. Calculations like those made for the eland indicate that the oryx can obtain all of its water needs if it eats food containing an average of 30% water. Thus, by eating *Disperma* only late at night, and more succulent plants at other times, the oryx also can do without drinking.

7.8.3 Mammalian Kidneys

The great importance of the kidneys in water and solute metabolism in vertebrates has been pointed out many times in this chapter. However, aside from the discussion of clearance techniques in Section 7.3.3, we have said very little about the nature of kidney structure and function in the various groups. An adequate discussion of this field requires a book in itself, but we can at least illustrate some of the basic facts and phenomena occurring in the best known of all kidneys—mammalian kidneys.

The **nephron** is the functional unit of all vertebrate kidneys. Mammalian nephroi differ from those of most other vertebrates in their possession of a long, thin middle segment bent in a hairpin turn, called the **loop of Henle.** As will be seen, these loops are the heart of the mechanism permitting mammalian kidneys to produce urine that is strongly hyperosmotic to the plasma.

The special arrangement of the thousands of

nephroi in mammalian kidneys, also of the component parts of the nephroi, is basic to their function. The kidney as a whole is a concentrically layered structure, centered on the renal papilla. The renal papilla supports the openings of all the collecting ducts that drain urine from the nephroi. The papilla is the innermost part of the medulla, or core, of the kidney. The greater part of the mass of the medulla contains loops of Henle, parts of collecting ducts, blood vessels, and supporting tissues. The layer overlying the medulla is the cortex, which contains the glomeruli, proximal and distal convoluted tubules, blood vessels, some endocrine organ tissue, nerves, and supporting tissues. The arrangement is diagrammed in Figure 7-20.

This organizational pattern is by no means universal in vertebrate kidneys. Kidneys in other groups are often more or less layered, but the layering is not so sharply defined. The general organ shape is also different in other groups, ranging from the long, thin, linearly arrayed kidneys of fishes to the complex, highly lobular kidneys of birds. The individual nephroi in most other kidneys wander about seemingly at random, lacking noticeable spatial regularity, but this is not true of bird kidneys. Bird kidneys possess a tubular arrangement for at least a fraction of their nephroi that is equivalent to mammalian loops of Henle.

Our understanding of mammalian kidney function is based upon information derived classically from vast numbers of clearance-type studies on laboratory dogs, cats, rats, and rabbits. Recent years have added to these data crucial measurements obtained by micropuncture technique sampling from different parts of individual tubules. A wider range of species has now been investigated, extending from semiaquatic forms such as the beaver to the most highly desert-adapted rodents, such as kangaroo rats and gerbils. Many other new techniques have been developed and virtually all the classical problems in mammalian kidney physiology now seem well on their way to solution.

The conversion of the plasma filtrate produced by the glomeruli into the urine discharged from the renal papilla obviously involves movements of many substances in both directions across the walls of the tubules. The quantitatively most important of these movements, however, involve primarily three substances—water, sodium chloride, and urea. Under most naturally occurring circum-

stances, amounts of water and sodium chloride are excreted from mammalian kidneys that are somewhere in the range from less than 1% to no more than about 15% of the amounts of these substances filtered. These percentages vary according to the needs of the animal—whether it needs to conserve water but excrete salt, conserve both, and so on. There is a large degree of independence in the handling of the three substances. The mechanism of urine production is probably best understood by reference to Figure 7-21. There are quantitative differences between species. Urea excretion varies very widely, in response to many influences.

The basic features of the mechanism are as follows. The glomerular filtrate that enters Bowman's capsule and the contents of the proximal tubule are isosmotic with the plasma of the general blood circulation. About 75% of the original filtrate volume is reabsorbed by the time the tubular fluid enters the loop of Henle. This water reabsorption occurs osmotically, as a result of dilution of the filtrate due to active uptake of sodium by the cells of the tubule walls.

There is a strong radial osmotic concentration gradient into the depths of the kidney. This gradient, in the tissue fluid surrounding the radially arranged parts of the nephric tubules and the peritubular blood vessels (*vasa recta*), is what produces the final concentration of the urine that leaves the ends of the collecting ducts. The mechanisms producing this radial gradient are complex and not yet fully worked out. The following description is a simplified version of the most widely accepted current theory, which is termed the "passive urinary concentration mechanism." It involves the principles of both countercurrent exchange and countercurrent multiplication.

The varying properties of the cells lining the different parts of the loop of Henle, the collecting ducts, and the vasa recta are crucial to the mechanism. The walls of the descending limbs of the loops are permeable to diffusion of water. The ascending limbs of the loops and the first sections of the distal tubules are impermeable to water. The last parts of the distal tubules and the collecting ducts are variably permeable to water (under the influence of vasopressin, the antidiuretic hormone, abbreviated ADH). All parts of the vasa recta are permeable to water. Sodium chloride can diffuse passively across the walls of the thin sections of

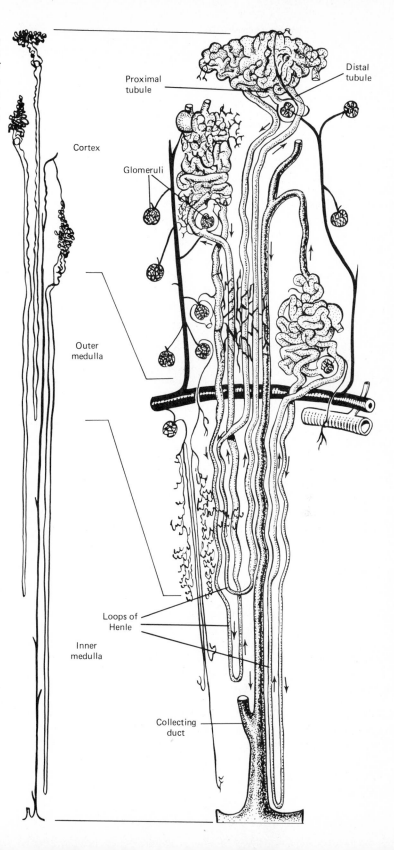

Figure 7-20
*Simplified diagram of major features of the structure of a mammalian kidney (based primarily upon dog and rat). Proportions of main diagram changed for clarity. Three nephrons are shown, two cortical (with loops of Henle of different lengths), one juxtamedullary. Directions of tubular flow indicated by arrows. More realistic proportions of the nephrons indicated by the diagram on the left. [Based upon R. Beeuwkes III and J. Bonventre. Am. J. Physiol., **229,** 695–713 (1975).]*

Cortex

Proximal
tubule

Distal
tubule

Glomeruli

Outer
medulla

Loops of
Henle

Inner
medulla

Collecting
duct

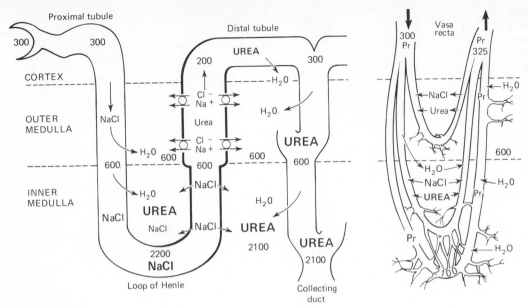

Figure 7-21
*Simplified diagram representing major current theory explaining mechanism of production of hyperosmotic urine by mammalian kidneys. One nephron having a long loop of Henle is diagrammed with its adjacent vasa recta. Numbers represent hypothetical osmolality values. Sizes of lettering indicate relative concentrations of solutes with respect to their concentrations in the outer medulla, but not necessarily in relation to other solutes. Arrows indicate net movements of indicated substances. Based upon R. L. Jamison and R. H. Maffly. New Eng. J. Med., **295**, 1059–1067 (1976).*

the ascending limbs of the loops and of the vasa recta, but not elsewhere. Sodium chloride is actively transported outward across the walls of the thick sections of the ascending limbs of the loops (chloride ion is transported, sodium diffuses with it). Urea can diffuse passively across the walls of the thin sections of the ascending limbs of the loops and of the vasa recta, but not elsewhere. Urea is actively transported outward across the walls of the lower, inner medullary sections of the collecting ducts.

The overall results of these differences in properties of tubular and blood vessel walls are as follows: Urea absorbed from the inner medullary segments of the collecting ducts is trapped and concentrated in the interstitium of the inner medulla by a combination of countercurrent multiplication (based on recycling, with new urea added in each cycle, through the ascending limbs of the loops, the distal tubules, and the collecting ducts) and countercurrent exchange (based on recycling, without addi-

tion of new urea, between the descending and ascending limbs of the vasa recta). The high osmotic concentration of the medullary interstitial fluid thus produced results in water removal from the fluid in the descending limbs of the loops, reducing tubule fluid volumes and concentrating the sodium chloride remaining behind. Much of this sodium chloride passively diffuses outward, into the medullary interstitium, in the thin sections of the ascending limbs. More of the sodium chloride is actively removed by the transport mechanisms in the walls of the thick sections of the ascending limbs. The tubule fluid then delivered to the distal tubules is greatly reduced in volume from that which originally entered the loops, and is distinctly hypoosmotic to the blood.

Passive water absorption next occurs across the walls of the distal tubules and of the collecting ducts. This leads to further volume reductions and the concentration of sodium chloride, urea, and many other solutes in the remaining tubular fluid—

328

which has become the urine. The urine leaving the collecting ducts has a variably small volume and high concentration. The variations are determined by the balances set by the animal in terms of water, sodium chloride, and urea reabsorption.

The vasa recta are apparently the route by which the water, reabsorbed from the tubules, is returned to the general circulation. Blood flow through the vasa recta is usually fairly rapid, preventing complete diffusional equilibration between the blood inside the vessels and the water and solutes in the interstitial fluids outside. Omitting complex details, the net result of this nonequilibrium situation is that the osmotic pressure of the plasma proteins (the oncotic pressure; represented in Figure 7-21 by the letters Pr) results in net water absorption by the blood from the interstitium.

The final volume and concentration of the urine can be finely adjusted to coincide with the water and solute balance needs of the animal by a complex interaction among at least three major factors. These are (1) the steepness of the radial osmotic concentration gradient, specifically the maximum concentration existing in the urinary papilla; (2) the water permeability of the collecting duct walls;

and (3) the rate of active urea removal from the tubular fluid as it flows down the collecting ducts.

To illustrate: A dehydrated and salt-loaded mammal would need to produce as small a volume of as concentrated urine as it could. It would, accordingly, build up the steepness and height of its radial osmotic concentration gradient, maximize the water permeability of the collecting-duct walls, and reduce collecting-duct absorption of urea.

Several additional items should be mentioned before proceeding:

1. There is a strong correlation between gross kidney morphology, specifically the relative lengths of the renal papillae, and maximal urine concentrations elaborated by different species (Table 7-17). Figure 7-22 shows the great papillary hypertrophy associated with the high urinary concentrating ability of the Australian hopping mouse, *Notomys alexis*. Large, long papillae like these are associated with the presence of many nephrons possessing long loops of Henle.
2. As indicated in Figure 7-20, there are two broad categories of nephrons in each mammalian kidney, cortical nephrons and juxtamedullary neph-

Figure 7-22
Sagittal section of a kidney from the Australian hopping mouse, Notomys alexis. *Note the large, long renal papilla.* [*From R. E. MacMillen and A. K. Lee,* Comp. Biochem. Physiol., **28,** 493–514 (1969).]

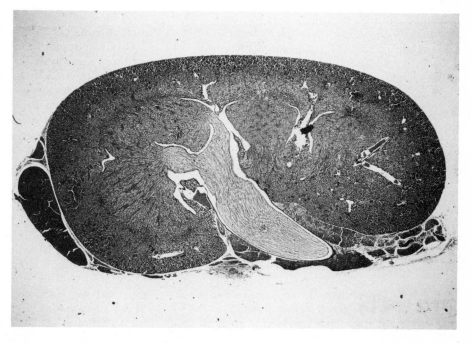

rons. Evidence is increasing that these two categories may differ significantly in function as well as in position and morphology. At least in laboratory rats, cortical nephrons appear primarily to regulate salt excretion. Control of GFRs in these nephrons appears to be largely via the renin-angiotensin system (see item 4 following). Juxtamedullary nephrons in laboratory rats appear primarily to regulate water excretion. ADH may play a role in controlling GFRs in these units. There is a substantial degree of independence in the functioning of the two groups.

3. Countercurrent systems occur in other parts of the kidney in addition to the loops of Henle and vasa recta. Micropuncture studies on the kidneys of laboratory rats indicate that, at least in the most superficial of the cortical nephrons, the flow of fluid in the first portions of the proximal tubules is countercurrent to the blood flow in the peritubular capillaries. The physiological significance of this arrangement is not clear, but it may be related to control of proximal tubule water reabsorption.

4. An important chemical system that plays a large role in controlling many aspects of kidney function is associated directly with the kidneys. This is the renin-angiotensin system (see Section 6.7.2). The system occurs in all vertebrate groups, though its function is not understood in all groups. In mammals the system has several functions. The most important of these is, in the present context, the control of blood flow in the kidney, especially to the glomeruli.

7.9 GENERAL COMMENTS ON TERRESTRIAL VERTEBRATES

Remarks closely analogous to those made in Section 7.6 with reference to aquatic vertebrates could also be made here. We have omitted probably as many significant topics as we have discussed. Our understanding in many areas still only scratches the surface. We have much to learn and will be surprised by many of the things we find. We urge those of you who are interested to investigate for yourselves, beginning with the reviews and books cited in the Additional Reading list.

Three areas that have received a great deal of attention in recent times, and continue to be important, are (1) estimations of total rates of water turnover in animals, including estimates of water and energy budgets for wild, free-living animals in nature, by use of isotopic methods involving doubly labelled water (water labelled in both the hydrogen and oxygen atoms); (2) extensive studies on the water relations of the eggs of both reptiles and birds; and (3) detailed computer based modelling studies of the water and energy budgets of a range of different animals.

7.10 CONTROL MECHANISMS

The preceding discussions have included many examples of situations in which animals adapt themselves to variations in environmental conditions by simultaneous, coordinated adjustments in the modes and rates of operation of several routes for water and solute exchange. The simultaneity and reinforcing nature of these adjustments result from the coordination and control activities of the nervous and endocrine systems. Extended discussions of many aspects of the functioning of both these systems will be found in Chapters 9, 11, and 12. Here we shall only briefly illustrate some of the ways these systems work to control water and solute metabolism.

Direct neural influences have been demonstrated in many important areas. The control of rates of sweat production by mammalian sweat glands was mentioned previously. Changes in blood pressure in mammals, which can be derived from changes in blood volume due to hydration or dehydration, cause reflex changes in renal vascular resistance that are mediated by the renal sympathetic nerves (see also Section 12.4.2). A lowering of blood pressure at the carotid baroreceptors (implying a loss of fluid) causes increased renal vascular resistance (which lowers GFR and rate of urine production). Sectioning the spinal cords of green toads (*Bufo viridis*) leads to significant decreases in salinity tolerance in this euryhaline species.

There are many areas in which it is not possible to differentiate the operations of the nervous system and the endocrine system (see Chapter 12). Many neurons appear to act as endocrine cells (neurosecretory cells), especially in such organs as the neurohypophysis of all vertebrates and the uro-physis of fishes. There are also many circumstances in which neurally transmitted sensory information leads to the release of hormones by other cells. Neuroendocrine relationships of many kinds are particularly prominent in the control mechanisms for water and solute metabolism in the vertebrates.

A wide range of hormones has been shown to have significant effects on water and/or solute (es-pecially inorganic ion) exchanges in various verte-brates under various conditions. These hormones derive from virtually every major endocrine organ and represent virtually every major type of mole-cule having hormonal activity. One or another of the various kinds of neurohypophyseal octapeptide hormones and the adrenocortical steroid hormones seems, however, to be significantly involved in al-most all cases. The major possible exceptions are the neurohypophyseal hormones of fishes, for which no one has yet conclusively demonstrated such a role.

The hormones that control water and solute me-tabolism have more than one effect. In virtually all cases they significantly affect both water perme-ability of membranes and rates of active monova-lent ion (especially sodium) transport by the same membranes. In general, however, the increased wa-ter-permeability effects are quantitatively most im-portant for the neurohypophyseal hormones, and the active transport stimulatory effects are most important for the adrenocortical hormones.

One illustration of the functioning of a neurohy-pophyseal hormone, vasopressin (ADH), has al-ready been given (Section 7.8.3; see also Section 12.4.2). Another clear example, one illustrating how a single hormone can produce coordinated, mutually reinforcing activities in several different organs at once, is the action of arginine vasotocin on water balance in anuran amphibians.

Anurans injected with arginine vasotocin and al-lowed to remain in water rapidly gain in weight. This effect was first observed by Brunn in 1921, using crude extracts of pituitary glands. It is thus called the *Brunn effect*. Detailed studies over many years have shown that this weight gain results from increased water content of the animals. The water is accumulated as a result of hormone-induced in-creases of water permeability in three membranes; the skin (resulting in increased rates of osmotic uptake from the dilute environment), the distal convoluted tubules of the kidneys (resulting in in-creased osmotic reabsorption from the dilute urine), and the urinary bladder (resulting in addi-tional osmotic reabsorption from the urine). The net result of these changes is increased water up-take and decreased urinary water loss, resulting in the observed body-weight increase.

It is not possible to present such a clear-cut illus-tration of the action of the adrenocortical hor-mones. Aldosterone has stimulatory effects on so-dium transport by amphibian skins and urinary bladders. There seem to be no special target organs for this or the other adrenocortical hormones. In-deed, it seems possible that movements of water and inorganic ions through all the living mem-branes in the bodies of vertebrates may be regu-lated by these hormones.

The Additional Reading lists for this chapter and for Chapter 12 include several articles and books that provide excellent introductions to the complex field of comparative endocrinology.

ADDITIONAL READING

GENERAL

Andreoli, T. E., J. F. Hoffman, and D. D. Fanestil (eds.). *Membrane Physiology,* Plenum Publishing Corp., New York, 1980.

Assenmacher, I., and D. S. Farner (eds.). *Environmental Endocrinology,* Springer-Verlag, New York, New York, 1978.

Crowe, J. H., and J. S. Clegg (eds.). *Dry Biological Systems,* Academic Press, New York, 1978.

Dacke, C. G. *Calcium Regulation in Sub-mammalian Verte-brates,* Academic Press, New York, 1979.

Drost-Hansen, W., and J. S. Clegg (eds.). *Cell-Associated Water,* Academic Press, New York, 1979.

Fitzsimons, J. T. *The Physiology of Thirst and Sodium Appetite,* Cambridge University Press, New York, 1979.

Giebisch, G., D. C. Tosteson, and H. H. Ussing (eds.). *Membrane Transport in Biology,* vols. 1–4, Springer-Ver-lag, New York, New York, 1978.

Gilles, R. (ed.). *Mechanisms of Osmoregulation in Animals: Maintenance of Cell Volume,* Wiley-Interscience, Somer-set, NJ, 1979.

Gupta, B. L., R. B. Moreton, J. L. Oschman, and B. J. Wall (eds.). *Transport of Ions and Water in Animals,* Aca-demic Press, New York, 1977.

Hammel, H. T., and P. F. Scholander. *Osmosis and Tensile Solvent,* Springer-Verlag, New York, New York, 1976.

Hargens, A. L. (ed.). *Tissue Fluid Pressure and Composition,* Williams and Wilkins. Baltimore, 1980.

Hoffman, J. F., D. C. Tosteson, Y. A. Ovchinnikov, and

R. Latorre (eds.). *Membrane Transport Processes*, vols. 1–2, Raven Press, New York, 1977.

Hollwich, F. *The Influence of Ocular Light Perception on Metabolism in Man and in Animal*, Springer-Verlag, New York, New York, 1978.

Jørgensen, C. B., and E. Skadhauge (eds.). *Osmotic and Volume Regulation*, Academic Press, New York, 1978.

Jungreis, A. M., T. K. Hodges, A. Kleinzeller, and S. G. Schultz (eds.). *Water Relations in Membrane Transport in Plants and Animals*, Academic Press, New York, 1977.

LaPointe, J. (ed.). "Neurohypophysial hormone action," *Am. Zool.*, **17**, 275–851 (1977).

Loewenstein, W. R. (ed.). "Epithelia as hormone and drug receptors," *J. Membrane Biol.*, **40**, 1–354 (1978).

Maloiy, G. M. O. (ed.). *Comparative Physiology of Osmoregulation in Animals*, vols. 1–2, Academic Press, New York, 1979.

Rechcigl, M. (ed.). *Nitrogen, Electrolytes, Water and Energy Metabolism*, Karger, Basel, Switzerland, 1979.

Schafer, J. A. (ed.). "Water transport in epithelia," *Fed. Proc.*, **38**, 119 (1979).

Schmidt-Nielsen, K., L. Bolis, and S. H. P. Maddrell (eds.). *Comparative Physiology: Water, Ions and Fluid Mechanics*, Cambridge University Press, New York, 1978.

Schultz, S. G. *Basic Principles of Membrane Transport*, Cambridge University Press, New York, 1980.

Wessing, A. (ed.). "Excretion," *Prog. Zool.*, **23**, 1–362 (1975).

AQUATIC ANIMALS

Hardisty, M. W. *Biology of the Cyclostomes*, Halstead Press, J. Wiley and Sons, New York, 1979.

Lahlou, B. (ed.). *Epithelial Transport in the Lower Vertebrates*, Cambridge University Press, New York, 1980.

Oglesby, L. C. "Volume regulation in aquatic invertebrates," *J. Exp. Zool.*, **215**, 289–301 (1980).

Pang, P. K. T., R. W. Griffith, and J. W. Atz. "Osmoregulation in elasmobranchs," *Am. Zool.*, **17**, 365–377 (1977).

Robertson, J. D. "Chemical composition of the body fluids and muscle of the hagfish *Myxine glutinosa* and the rabbit-fish *Chimaera monstrosa*," *J. Zool. London*, **178**, 261–277 (1976).

Schmidt-Nielsen, B. M. (ed.). "Biology of the chloride cell: Jean Maetz Memorial Symposium," *Am. J. Physiol.: Regulatory, Integrative and Comparative Physiology*, **238**, R139–R276 (1980).

AMPHIBIOUS AND TERRESTRIAL ANIMALS

Andersson, B. "Regulation of water intake," *Physiol. Rev.*, **58**, 582–603 (1978).

Andreoli, T. E., J. J. Grantham, and F. C. Rector, Jr. (eds.). *Disturbances in Body Fluid Osmolality*, American Physiological Society, Bethesda, 1977.

Arlian, L. G., and M. M. Veselica. "Water balance in insects and mites," *Comp. Biochem. Physiol.*, **64A**, 191–200 (1979).

Beeuwkes, R., III. "The vascular organization of the kidney," *Ann. Rev. Physiol.*, **42**, 531–542 (1980).

_____ and B. M. Brenner. "Kidney," in P. C. Johnson (ed.), *Peripheral Circulation*, J. Wiley and Sons, New York, 1978.

Carey, C. (ed.). "Physiology of the avian egg," *Amer. Zool.*, **20**, 325–484 (1980).

Edney, E. B. *Water Balance in Land Arthropods*, Springer-Verlag, New York, New York, 1977.

Jamison, R. L., and R. H. Maffly. "The urinary concentrating mechanism," *New Engl. J. Medicine*, **295**, 1059–1067 (1976).

Kaissling, B., and W. Kriz. "Structural analysis of the rabbit kidney," *Adv. Anat. Embryol. Cell Biol.*, **56**, 1–123 (1979).

Katz, A. I. and M. D. Lindheimer. "Actions of hormones on the kidney," *Ann. Rev. Physiol.*, **39**, 97–134 (1977).

Lameire, N. H., M. D. Lifschitz, and J. H. Stein. "Heterogeneity of nephron function," *Ann. Rev. Physiol.*, **39**, 159–184 (1977).

Mantel, L. H. "Terrestrial invertebrates other than insects." In G. M. O. Maloiy (ed.), *Comparative Physiology of Osmoregulation in Animals*, vol. 1, Academic Press, New York, 1979.

Nagy, K. A., and W. R. Costa, "Water flux in animals: analysis of potential errors in the tritiated water method," *Am. J. Physiol.*, **238**, R454–R465 (1980).

Packard, G. C., C. R. Tracy, and J. J. Roth. "The Physiological ecology of reptilian eggs and embryos, and the evolution of viviparity within the class reptilia," *Biol. Rev.*, **52**, 71–105 (1977).

Prakash, I., and P. K. Ghosh (eds.). *Rodents in Desert Environments*, Dr. W. Junk b.v., The Hague, Netherlands, 1975.

Schmidt-Nielsen, B. M. (ed.). "Comparative physiology of the urinary tract," *Fed. Proc.*, **36**, 2479–2503 (1977).

Shoemaker, V. H., and K. A. Nagy. "Osmoregulation in amphibians and reptiles," *Ann. Rev. Physiol.*, **39**, 449–471 (1977).

Stephenson, J. L. "Countercurrent transport in the kidney," *Ann. Rev. Biophys. Bioeng.*, **7**, 315–339 (1978).

Thurau, K. (ed.). *Kidney and Urinary Tract Physiology II*, University Park Press, Baltimore, 1976.

Tracy, C. R. "A model of the dynamic exchanges of water and energy between a terrestrial amphibian and its environment," *Ecol. Monogr.*, **46**, 293–326 (1976).

Welch, W. R., and C. R. Tracy. "Respiratory water loss: a predictive model," *J. Theor. Biol.*, **65**, 253–265 (1977).

Wright, F. S. (ed.). "Feedback control of glomerular filtration rate," *Fed. Proc.*, **40**, 77–115 (1981).

_____ and J. P. Briggs. "Feedback control of glomerular blood flow, pressure, and filtration rate," *Physiol. Rev.*, **59**, 958–1006 (1979).

George A. Bartholomew

BODY TEMPERATURE AND ENERGY METABOLISM

8.1 INTRODUCTION

Most individual biochemical reactions are extremely sensitive to temperature. A convenient approximate quantification of this fact is given by van't Hoff's generalization, which states that the rates of most biochemical reactions approximately double with each 10°C increase in temperature. Because temperature affects chemical reaction rates so profoundly and because most organismic responses involve chemical transactions, it follows that temperature is a major parameter in virtually all biological activities. Furthermore, temperature is an easy parameter to measure and its relationships are somewhat simpler than those of many other factors in physiology.

Temperature versus Heat. Before considering the interplay between body temperature and energy metabolism we must have a clear perception of the difference between temperature and heat. *Heat* is the form of energy that is transferred into or out of a system as a consequence of temperature differ-

ences. It is a function of the *total* kinetic energy of all the molecules in a system, and all systems with temperatures above absolute zero contain heat. *Temperature* is a function of the *mean* kinetic energy of the molecules in a system and it determines the direction of heat flow. When two systems with different temperatures are placed in contact, energy in the form of heat flows from the one with the higher temperature to the one with the lower temperature. However, the quantity of heat that will flow between them cannot be determined from the difference in their temperatures alone. Their specific heat capacities must also be known. The *heat capacity* of a system is the ratio of the change in heat dH, to the product of its mass m, times the temperature difference dt between it and its surroundings; heat capacity $= dH/m\,dt$. Although specific heat capacity varies slightly with temperature, for practical purposes the specific heat of water, which is set equal to 1.0, can be taken to be 1 cal g^{-1} °C^{-1}. The specific heat of protoplasm is usually assumed to be approximately 0.82.

Despite the central role of heat in the biological

333

world, animals are not equipped with sensors for detecting heat but with sensors for the detection of temperature. The heat content of a substance must be inferred by integrating the perception of its temperature with additional sensory information about its mass and rate of temperature change. The relation between temperature and heat is further confused by the words "hot" and "cold" and by the physiological factors associated with their perception and description. Temperature is a continuum, but the human thermometer, or temperature sense, is conditioned to measure it as a dichotomous entity. That is, we recognize things as being hot or cold by relating them either to the temperature of the surfaces of the exposed parts of our bodies or to the rate of temperature change from radiation, convection, and conduction between our bodies and the environment. There is clearly a high level of subjectivity involved in the perception of the degrees of hotness or coldness. We can, however, minimize the subjectivity and ambiguity by avoiding the words hot and cold and speak instead of specific temperatures.

8.2 RANGE OF ENVIRONMENTAL TEMPERATURES

The range of temperatures in the universe extends from near absolute zero ($-273°C$ or 0 Kelvin) upward over many orders of magnitude; for example, the surface temperatures of our sun are about 6000 K and the core temperatures of some of the hotter stars are many millions Kelvin.

As far as is known, chemical systems of sufficient complexity, versatility, and stability to be characterized as living are all based on carbon. The range of temperatures over which the carbon compounds characteristic of most living systems are stable, but still sufficiently active to function as parts of an active, living system, extends over only about 50°C. The lower limit of this range is a degree or so below the freezing point of pure water. The upper limit is near the temperature at which proteins begin to denature—usually between 45 and 50°C, although a few algae and bacteria live at temperatures above 70°C. Thus organisms can exist over only a minute sector of the total range of environmental temperatures that occur in the cosmos.

It is a conspicuous feature of the temperature range of biological activity that it lies very close to the lower limit of possible environmental temperatures. Life is a low-temperature phenomenon. Other things being equal, below-normal temperatures are less damaging to the biochemical integrity of organisms than above-normal temperatures. Unless crystallization occurs—unless the water in tissues freezes—the chemical substances in living systems are relatively unaffected by extremely low temperatures. However, the cells of many organisms in their normally active state are only a few degrees centigrade removed from the permanently damaging effects of elevated temperatures. For example, the typical body temperatures of active song birds lie between 40 and 42°C. This is only about 3–5°C below the level at which protein denaturation begins to occur.

All animals must either find external environments or maintain internal environments that do not transgress the narrow range of temperatures in which their chemical systems can maintain integrity and the capacity for response. Temperatures within this range are commonplace on earth, so the physical environment restricts the range of temperatures with which organisms have to contend. Thus, from the point of view of temperature regime, as well as that of water availability, our planet is particularly suitable for the support of life. This matching of the physical environment to the conditions of temperature appropriate for life, as we know it, is most obvious in the aquatic environment. Temperatures in natural bodies of water that are not heated by vulcanism extend from about -2 to $+40°C$. This is a range in which the chemical systems of most organisms maintain their viability. Consequently, the thermal problem facing most aquatic organisms is not the avoidance of biochemical damage from extremes of temperature but the maintenance of effective organismic integrity by regulating the balance between the rates of various chemical activities.

For animals that live in a terrestrial environment, thermal problems are more acute. Air has a much lower specific heat than water and it is also much more transparent to solar radiation. The radiant energy to which terrestrial organisms are exposed has a substantial thermal effect. Because of incoming solar radiation, environmental temperatures can be lethally high, on the one hand, and because of radiational loss to space, lethally low, on the other. Air temperatures of 50°C and soil and substratum temperatures of 70°C occur commonly in the horse-latitude deserts. Both these temperatures exceed the limits within which most living systems can remain active. Outside the lowland tropics and subtropics, temperatures below freezing are commonplace. Air and surface temperatures of −65°C have been recorded in a number of different places and air temperatures of −50°C are routine at high latitudes. Therefore, the transition from aquatic to successful terrestrial life represents a quantum step in terms of physiological adjustments to temperature, and the accommodations of terrestrial animals to very low and very high temperatures present adaptations which, from the biological point of view, differ qualitatively and quantitatively from the adjustment to temperature that aquatic organisms face.

8.3 THERMOREGULATORY TAXONOMY

Like most aspects of biology, the study of body temperature is plagued with a distressing abundance of descriptive terms, some of which are exasperatingly ambiguous and carry connotations that are not rigorously acceptable in a physiological context. However, many of the terms, despite their inadequacy, are in common use and must be examined briefly. The two most familiar are *warm-blooded* and *cold-blooded.* These terms are subjective, inaccurate, unscientific, and enjoy almost universal usage. It happens that the body temperatures of human beings and other mammals average very high as temperatures of biological systems go and, even though the hands and other parts of the surface of the human body are cooler than the core or deeper parts, we find that most animals feel cold to the touch. Therefore, we designate them cold-blooded. The most familiar exceptions to this are birds and mammals, which often, but not invariably, feel warm to the touch and are hence designated warm-blooded. This simple and direct approach to the body temperatures of animals is quite unsatisfactory; the information content of the sensory perception on which this subjective judgement is based is small. It merely reveals that the temperature of the surface of the organism is lower or higher than that of the surface of the observer at the time the determination is made. It would facilitate communication if the terms cold-blooded and warm-blooded were expunged from the scientific literature.

Terms that have somewhat greater precision and also more information content are *poikilothermic* and *homeothermic* (homoiothermic). A poikilothermic animal has a relatively variable body temperature; a homeothermic one has a relatively constant body temperature. These terms are useful but they became generally established prior to the development of a biologically adequate appreciation of the complexity of animal body temperature, and they communicate no information about mechanism.

Body temperature is the result of two processes, heat gain and heat loss, and neither of these is simple. As long as any animal is alive it continuously produces heat. For the most part this heat production has no perceptible effect on the animal's temperature because it is lost to the surroundings as rapidly as it is produced. Most animals have such high rates of thermal conductance and such low rates of heat production that the external environment acts as an infinitely capacious heat sink. Consequently the body temperatures of animals are independent of the heat produced by their oxidative metabolism and are determined exclusively by heat acquired from the environment. Therefore, they are described as *ectothermic.* That is to say, the heat that determines their temperature is obtained from the environment rather than from their own oxidative metabolism. In contrast to ectotherms, which include the overwhelming majority of animal species, there are a few atypical groups that produce sufficient heat by their own oxidative metabolism and have a sufficiently low thermal conductance that the heat which contributes to their body temperature is largely derived from their own oxidative activity. Such animals are called *endothermic.* The only continuously endothermic animals are found among birds and mammals. However, by no means all birds and mammals are continuously endothermic nor is endothermy confined to the classes Aves and Mammalia. Large reptiles are from time to time at least partially endothermic, as are some large, fast-swimming fishes. Some small terrestrial animals, for example, certain insects including bumble bees, beetles, heavy-bodied moths, and dragonflies, are conspicuously endothermic during their periods of activity and maintain body temperatures substantially above those that would normally be imposed by the environment. Even some plants such as *Philodendron* and the skunk cabbage may temporarily elevate their temperatures 10°C or more above environmental temperatures by endogenous heat produced during flowering.

Because of the existence among both vertebrates and invertebrates of species, which during activity, are capable of endothermic heat production, but which are not characterized by a sustained constant level of body temperature, an additional term is useful. This term is **heterothermy.** Heterothermic animals are facultative endotherms. They are capable of physiological temperature regulation but they do not regulate precisely at all times. By convention the term heterothermy is usually confined to the description of birds and mammals that are not continuously homeothermic. However, it can validly be applied to certain insects that are ectothermic while at rest, but by means of endothermy, maintain high and regulated body temperatures during activity.

By appropriate combination of the terms poikilothermic, homeothermic, endothermic, ectothermic, and heterothermic, one can obtain adequate descriptive designations for the patterns of body temperature of most animals.

It is important to recognize that aside from certain marine situations, for example, the abyssal depths below the photic zone, the physical environment is thermally complex. Many animals behaviorally exploit this thermal complexity and often demonstrate impressive themoregulatory capacities that may go unnoted in strictly physiological treatments of the topic of the control of body temperature.

The physiology of animals living under natural conditions frequently violates the conventions that have gradually developed from laboratory investigations. Therefore, in the discussion that follows we shall attempt to approach the topic of body temperature not from the standpoint of the laboratory physiologist but from the standpoint of a person attempting to understand the physiological functions of intact animals living under natural conditions.

8.4 ENERGY EXCHANGES BETWEEN ORGANISM AND ENVIRONMENT

The exchange of energy between organism and environment depends not only on nutritional and metabolic activities but on an exceedingly complex and variable array of physical transactions. Four processes need to be considered:

Conduction. The movement of heat from regions of high temperature to regions of low temperature by interaction of adjacent molecules and without mass motion of the medium through which the energy transfer takes place.

Convection. The movement of heat through a fluid (either liquid or gas) by mass transport in currents.

Thermal Radiation. Energy transfer by means of electromagnetic radiation; radiation travels at the speed of light and needs no medium of propagation. Radiative heat transfer depends on the fourth power of the absolute temperature (K^4).

Change of State. Energy transfer by means of evaporation or condensation; in biological systems the process primarily involves the change of water from a liquid to a gas or vice versa.

In the terrestrial habitat, because of its physical complexity and resultant thermal and radiational complexity, animals live in an environment in which there can be no simple way of characterizing the energy flux. In the depths of the ocean and in subterranean habitats the thermal environment of an organism may be described with reasonable completeness merely by specifying the ambient temperature. However, no single measurement can characterize the thermal environment of a terrestrial animal. This statement is particularly applicable to air temperature, which is the measurement that is routinely used to characterize thermal conditions in the terrestrial environment despite the fact that thermal flux is affected by many components. The exchange between the environment and the animals that live in it is too complicated and changes too rapidly to allow more than the most general of descriptions. For any system, either living or nonliving, a simple equation for energy balance can be written.

energy in + energy generated = energy out
+ energy stored

The major components in the energy balance of organisms are obvious—radiation, conduction, convection, evaporation, metabolism, and change in temperature. They can be introduced into a rearranged form of the preceding equation

$$S + R + L\dot{E} + G + C + M + s = 0$$

where S is solar radiation; R thermal radiation from and to substratum, atmosphere, and objects in the environment; L the latent heat of evaporation of water; $\dot{E}$ the rate of evaporation; G conduction to or from the ground; C convectional exchange with the air; M the heat of metabolism; and s, a term for indicating gains and losses in energy, is related to changes in body temperature. Each of the components in the equation varies continually, and for some of the components the range of variation is enormous. Under natural conditions partial control over the magnitude of each of the components can be exerted by individual animals through behavior, habitat selection, daily and seasonal cycles of activity, or postural adjustments. In addition, through evolutionary adaptations the various taxa of animals have evolved special structures or special physiological mechanisms for augmenting or diminishing each of the terms in the equation. For example, the effective magnitude of solar radiation (S) will depend on the transparency and reflectivity of the animal's tissues; that of thermal radiation (R) will depend on the emissivity of the animal's surface; and that of evaporative heat exchange will depend on insensible perspiration, respiratory water loss, and sweating, the latter two being under physiological control.

The control of conductive and convective heat exchanges will be dealt with in detail in Sections 8.9 and 8.10, but some of the relevant aspects of thermal and solar radiation can most conveniently be considered in the present context. Any physical object with a temperature above absolute zero emits thermal radiation. All functioning animals have surface temperatures between about 271 and 320 K and emit radiation in the electromagnetic spectrum with peaks in the infrared at a wavelength of about 10 μm. All objects on earth, living or nonliving, are exchanging thermal radiation with their surroundings. The net direction of the exchange depends on surface temperatures, with energy passing from the object with the higher surface temperature to that with the lower surface temperature.

As a result, animals have an extremely complex radiative exchange with their surroundings, and by appropriate behavioral and physiological activities they can control the rate, and often the direction, of flow of radiative energy. Under many conditions this control of rate and direction is of critical importance in thermoregulation. A qualitative picture of the complicated pattern of thermal exchanges experienced by a mammal under moderately warm conditions can be obtained by examining Figure 8-1. It should be clear that the details will vary with the species, habitat, time of day, latitude, cloud cover, humidity, and season, to name a few of the variables.

Most of the heat that all animals continuously produce is lost passively to the environment through radiation, conduction, or convection. However, when ambient temperatures equal body temperature these routes for dissipating endogenous heat are unavailable. When ambient temperatures exceed body temperature, the direction of heat movement is reversed and heat flows from the environment into the animal. Under such circumstances, a terrestrial animal can avoid unfavorable elevation of body temperature by the evaporation of water, but an aquatic animal can do nothing.

The conversion of water from a liquid to a gas requires energy. Whenever water is evaporated from any surface of an animal the temperature of that surface falls. Consequently the respiratory and integumentary systems of terrestrial animals are always sites of some heat loss, and under appropriate circumstances they can be converted to sites of major thermoregulatory significance, by local augmentation of the rate of evaporation.

The heat of vaporization of water varies with temperature, but for most physiological purposes one can assume that the evaporation of 1 g of water requires approximately 580 cal. If an animal can economically mobilize water for evaporative cooling, it can tolerate a positive heat load; however, except for the larger mammals, only a few species are able to depend on this type of thermoregulation except for short periods (see Section 8.10).

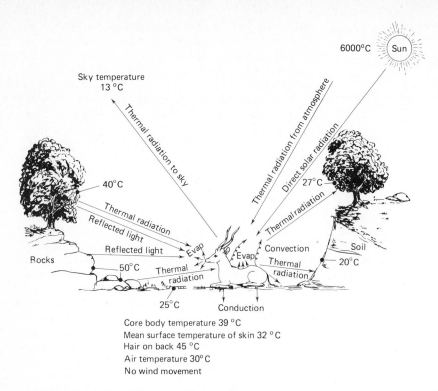

Figure 8-1
Diagrammatic representation of the energy exchanges between an animal and its environment under moderately warm conditions.

Sky temperature 13 °C

Thermal radiation to sky

6000°C Sun

Thermal radiation from atmosphere

Direct solar radiation

40°C

Thermal radiation

Reflected light

Reflected light

27°C

Thermal radiation

Rocks

Evap

Evap

Convection

Soil

50°C

Thermal radiation

Thermal radiation

20°C

25°C

Conduction

Core body temperature 39 °C
Mean surface temperature of skin 32 °C
Hair on back 45 °C
Air temperature 30° C
No wind movement

8.5 BODY TEMPERATURE, ENERGY METABOLISM, AND ENVIRONMENTAL TEMPERATURE

Although the special physiological adaptations by which animals meet the challenge of extremely hot or extremely cold climates invite immediate attention, these cannot be intelligently considered until they are placed in an appropriate perspective by examining the general responses of body temperature and energy metabolism to a moderate range of environmental temperatures. We shall, therefore, examine the responses of these two physiological parameters to a range of environmental temperatures such as is commonly faced by many homeothermic and poikilothermic animals. A wide variety of animals are active in environmental temperatures between 0 and 35°C. It is, therefore profitable to examine the response of oxygen consumption and body temperature of a hypothetical homeotherm and a hypothetical poikilotherm to this range of temperatures. The responses of such

theoretical animals are shown in diagrammatic form in Figure 8-2 and 8-3. Over the restricted temperature range being considered, the hypothetical poikilotherm has a body temperature that is indistinguishable from that of the environment, whereas the hypothetical homeotherm holds its body temperature at some closely regulated level (between 35 and 42°C depending on the species), which is independent of environmental temperature.

The poikilotherm's temperature passively follows environmental temperature, so it expends virtually no energy on thermoregulation. The homeotherm's temperature is independent of that of the environment, and to maintain this independence it must continuously expend energy. Consequently, if oxygen consumptions are plotted against environmental temperature (Figure 8-3), the responses of the poikilotherm and the homeotherm follow different patterns. The poikilotherm's rate of oxygen consumption behaves like that of an *in vitro* biochemical system, and doubles for each 10°C increase in temperature, in general conformance with

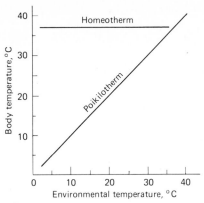

Figure 8-2
Relation of body temperature to environmental temperature in homeotherm and poikilotherm.

van't Hoff's rule. That is, oxygen consumption of the poikilotherm follows temperature, but its rate of change with increasing temperature resembles an exponential curve. In contrast, the oxygen consumption of the homeotherm decreases linearly with increasing environmental temperature to some critical point and then is independent of envi-

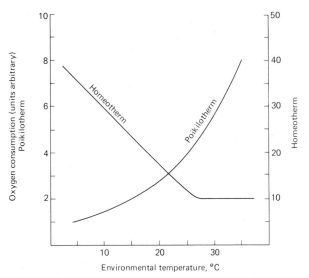

Figure 8-3
Relation of oxygen consumption to environmental temperature in homeotherm and poikilotherm.

ronmental temperature over a considerable range. That is, the lower the temperature, the more work the homeotherm does to maintain a uniform body temperature; but at some specific temperature, or limited range of temperatures, the amount of energy required for thermal regulation reaches a minimum level. This level of minimum work is called the **basal metabolic rate.**

Before examining the physiological aspects of the contrasting patterns of energy metabolism shown by ectotherms and endotherms, it is appropriate to put the two patterns into ecological perspective. First, at all temperatures the energy metabolism of the homeotherm is roughly an order of magnitude greater than that of the poikilotherm (see Section 3.10). The metabolic cost of living is high for homeotherms because approximately 80–90% of their oxidative energy is utilized for the maintenance of thermal homeostasis. Thus although homeotherms are more independent of the environmental temperature than are poikilotherms, they pay a very high price for their independence. Second, at all ambient temperatures the metabolic rate of the homeotherm is part of a closely regulated feedback system (see Section 8.8.3). In contrast, the rate of the energy metabolism of poikilotherms is not part of such a system, and those aspects of its rate that are temperature-dependent are externally controlled. Third, it can be inferred that, if a poikilotherm is to exert control over its body temperature, it must do so primarily by behavioral means rather than by means of endogenous heat production. It must seek out situations in which environmental temperatures appropriate to its activities exist or in which sources of thermal energy are available in the environment from which, through behavioral means, it can acquire exogenous heat, or lose heat to the environment to modulate body temperature.

Heat exchange in aquatic animals is largely by means of conduction and convection. Therefore they offer a convenient material with which to begin an examination of body temperature.

8.6 AQUATIC ECTOTHERMS

The rate of metabolic heat production by gill-breathing aquatic ectotherms is low when compared with their rate of heat exchange with the surrounding water. Therefore (except for a few special cases examined in Section 8.9.7), they cannot continuously maintain a significant difference between body temperature and environmental temperature. Because their body temperature varies directly with that of the environment, they are ideally poikilothermic. The physiological factors involved in poikilothermy are readily isolated in the relatively simple thermal environment that exists in natural bodies of water. Because of the high specific heat of water and the readiness with which it transports heat by the mixing caused by wind, gravity, density currents, and convection, large bodies of water, particularly the open ocean, comprise one of the most stable and uniform thermal environments in the biosphere.

8.6.1 Fishes

For simplicity let us consider a fish swimming in a leisurely manner in the open ocean several meters below the surface at a level where virtually all the infrared radiation in the solar spectrum has been absorbed. Such a fish is in an environment characterized by a very high specific heat and a high level of thermal uniformity. Although the fish is producing heat by the muscle activity associated with its swimming, there is no distinguishable difference between its core temperature and the temperature of the surrounding water. The tissues of the fish are completely permeated by its circulatory system. No part of its muscular system is more than a fraction of a millimeter from a blood vessel. Any heat being produced in the muscle heats the blood flowing through it. The heat is carried rapidly away from its point of production and eventually reaches the gills. The arterioles, capillaries, and venules in the gills are in intimate contact with the water, and the entire gill apparatus is adapted for the exchange of dissolved gases between blood and environment (see Chapter 5). All the physical factors that facilitate the exchange of dissolved gases between fish and environment—large surface areas, and thin-walled blood vessels—also facilitate the conduction of heat between fish and environment. As a result, the metabolic heat being produced by the fish is continually lost to the environment and

its body temperature remains indistinguishable from that of its surroundings.

In the absence of special conditions, such as sustained intense activity, very large size, or countercurrent vascular mechanisms (see Sections 8.9.3 and 8.9.7), which prevent heat from passing rapidly from muscle to gills, all fishes, and, for that matter, all aquatic animals other than mammals and birds, have body temperatures that do not vary significantly from that of the water in which they are floating. For an animal continuously to maintain an internal temperature significantly different from that of the environment, it must be able to exert control of the temperature of the respiratory surface. Because of the high specific heat of water, such a situation is substantially impossible for gill-breathing forms. An additional factor that contributes importantly to the poikilothermy of gill-breathing aquatic animals is the low oxygen content of water as compared with air. The amount of oxygen dissolved in natural waters rarely exceeds 1 part of oxygen per 100 parts of water. This is about one twentieth of the ratio that exists in air. Consequently, very large amounts of water must continuously bathe the respiratory surface. This, of course, is an ideal physical circumstance for heat exchange between organism and environment.

Even this simple qualitative diagnosis of the physical circumstances in which aquatic poikilotherms exist allows one to make a series of predictions that are amenable to testing by observation and measurement.

1. The matching between body temperature and water temperature will be closer in small species than in large ones.
2. Under conditions of sustained intense activity, large forms will have transient increases in body temperature.
3. Fishes will be readily victimized by any rapid drastic changes in environmental temperature.
4. Aquatic poikilotherms will have sensitive temperature receptors that will allow them to recognize and select water masses with temperatures appropriate for their physiological functions (Figure 8-4).
5. Fishes that live in areas where drastic temperature fluctuations occur either erratically or seasonally, for example, in shallow bodies of fresh-

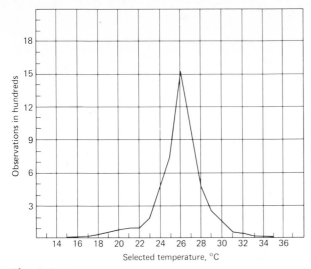

Figure 8-4
Temperature selection by 39 individuals of a shore fish, the California opal-eye (Girella nigricans). [From K. S. Norris, Ecol. Monogr., **63,** 23–62 (1963).]

water or in the surface layers of the sea in temperate latitudes, will be relatively eurythermic—relatively independent of temperature in their physiological adjustments.

6. Fishes that live under conditions where water temperatures do not vary, for example, in the tropics or at great depths at any latitude, will be relatively stenothermic—extremely sensitive to changes in environmental temperature.

7. Fishes that normally experience gradual seasonal changes in temperature will have evolved biochemical adjustments that allow them to maintain normal functions despite changes in body temperature.

8. The distribution of surface fishes in the sea will be in large measure determined by temperature.

9. In areas where the temperature of the open sea varies seasonally there will be seasonal migrations in the fish fauna.

8.6.2 Q_{10} and Temperature Compensation

The relationships shown in Figure 8-2 indicate that the body temperature of a poikilotherm in a steady state is passively determined by that of its sur-

roundings. However, one cannot infer that poikilotherms are metabolically completely the captives of their body temperatures, because temperature is not the only factor that determines chemical rate. Different enzymes have different temperature sensitivities, and different biochemical systems have different temperature constants. One can, therefore predict a priori that different poikilotherms will have different sensitivities to temperatures and that they will differ with respect to the temperatures at which their physiological activities function most effectively. One can further predict that a given poikilotherm should, as a part of its metabolic repertory, have alternate patterns of biochemical sensitivity to temperature; by changing one set of rate-determining reactions for another, it should be able to compensate for at least some of the changes in environmental temperature to which it is exposed at different times of year. Such biochemical temperature compensation should be best developed in aquatic poikilotherms, for the range of temperatures to which they are exposed is restricted by the nature of the aquatic environment. Moreover, the high specific heat of water results in relatively slow rates of change in environmental temperature. In contrast, terrestrial poikilotherms are exposed to rapid and drastic changes of temperature and might be expected to show a more limited dependence on the relatively slow acquisition of biochemical compensation for temperature differences. Furthermore, the thermal flux of the terrestrial environment offers alternative modes for adjustment to seasonal and daily changes that are not available in the aquatic environment (see Sections 8-4 and 8-7).

The effects of temperature on the rates of physiological processes are often described in terms of a temperature coefficient, Q_{10}. A process that shows the classical van't Hoff doubling (Section 8-1) when temperature increases by 10°C is said to have a Q_{10} of 2.0. If the rate changes by a factor of 3, a process is said to have a Q_{10} of 3.0. No change in rate with temperature (that is, temperature independence) is a Q_{10} of 1.0.

The situation is complicated by the fact that Q_{10} varies with temperature. The rate of a metabolic process usually approaches zero as temperature nears the freezing point of water. Consequently, at temperatures near 0°C the Q_{10} is usually high.

Conversely, at temperatures in the upper range of tolerance of an organism, Q_{10} is usually low. The value of the Q_{10} of a process is readily computed if its rates, k_1 and k_2, are measured at any pair of temperatures, t_1 and t_2:

$$Q_{10} = \left(\frac{k_1}{k_2}\right)^{10/(t_1 - t_2)}$$

In some poikilotherms, particularly intertidal invertebrates that are subject to rapid, recurrent, and drastic temperature changes of as much as 15 or 20°C by the rise and fall of the tide, temperature compensation is essentially instantaneous. For example, standard oxygen consumption of a variety of forms including sea anemones, polychaete worms, snails, mussels, and barnacles shows a Q_{10} of approximately 1.0 over the range of environmental temperatures they normally encounter. Even some terrestrial species show a plateau in their standard metabolism in the midrange of temperatures in which they are usually active (Figures 8-15 and 8-17).

A more familiar and longer studied type of temperature compensation does not occur instantaneously but develops slowly during a period of days following a change in temperature regime. Two related terms have been used to describe this situation: **acclimation** and **acclimatization.** The former refers to compensatory changes that are induced under laboratory conditions. The latter refers to compensatory changes that develop under natural conditions. The two terms are often used interchangeably and the distinction between them may not be worth maintaining.

For half a century students of muscle physiology have distinguished "winter frogs" from "summer frogs" in their experimental protocols because the temperature sensitivity of frog muscle varies between summer and winter. The existence of this seasonal variation in the response of muscle tissue indicates that biochemical temperature compensation occurs at the cellular level. Its existence at the organismic level has been abundantly demonstrated in a number of taxa of poikilotherms during the last two decades. Seasonal temperature compensation is conveniently demonstrated (but not explained) by contrasting the total energy metabolism of an animal during summer and winter (Figure 8-5). From the two curves several things are apparent.

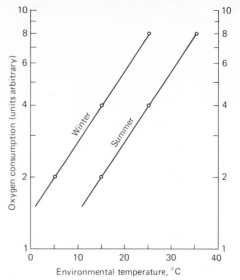

Figure 8-5
Temperature compensation in the oxygen consumption of a hypothetical fish. In winter the metabolic rate at 5°C equals the metabolic rate in summer at 15°C.

1. During both winter and summer, energy metabolism is sensitive to environmental temperature and Q_{10} remains the same.
2. Throughout the range of temperatures tested, the oxygen consumption of the fish at any given temperature is greater in winter than in summer.
3. The vertical distance between these curves is a measure of the degree of temperature compensation shown by the standard metabolism of the fish.
4. In the diagram, metabolic rate in winter at 5°C is the same as it is in summer at 15°C; therefore, this hypothetical animal has compensated by its metabolism for a difference in environmental temperature of 10°C. Thus it can be inferred that this species has a degree of metabolic control that should allow it to be as active as 5°C during the winter as it would be at 15°C during the summer, and so partially escape the constraints imposed by environmental temperature.

It should be noted that temperature compensation is usually only partial. Animals can acclimate both to high temperatures and to low temperatures.

The transition from the cold-adapted to warm-adapted animal is not immediate but usually requires a period of days to go to completion. (Limited temperature acclimation, however, can sometimes be shown after only a few hours exposure to changed temperatures.) Such a gradual adjustment to the declining temperatures of fall and early winter gives natural populations a limited degree of freedom from strict control of their levels of activity by environmental temperature. The reversal of the process during the spring should help reduce the metabolic requirements for a high level of food intake. Temperature acclimation can play an important role distributionally as well as seasonally. The various populations of a species that has broad latitudinal or altitudinal distribution, or occurs in habitats characterized by different temperature regimes, can show marked differences in their responses to temperature that have clear adaptive value. Some of these differences are not genetic but are due to acclimation.

In the laboratory, acclimation can be induced merely by changing the level of the temperature at which experimental animals are maintained. Its extent can be determined by measuring the rate of some physiological function (oxygen consumption, nerve conduction velocity, or enzymatic activity) at both the original temperature and the acclimation temperature. The extent of acclimation can also be measured by examining some physiological function at a temperature intermediate between two temperatures to which experimental groups have been acclimated. In this case the cold-acclimated animals often show a higher rate than the warm-acclimated ones.

In some cases acclimation results in a vertical translation of the rate curve as shown in Figure 8-5 without a change in Q_{10}. In other cases it results in a change in Q_{10}, and the slope of the rate curve increases or decreases. A vertical translation can be interpreted as a quantitative change in the amount of enzyme activity, whereas a change in Q_{10} (that is, a change in slope) can be interpreted as a qualitative or quantitative alteration of enzymes or cofactors. However, acclimation is easier to demonstrate at the organismic level and to interpret ecologically than it is to explain biochemically at the cellular level. Nevertheless, a substantial body of knowledge concerning temperature compensation and acclimation has been developed in the recent past. The biochemical details of these mechanisms are beyond the scope of the present treatment. Interested readers should refer to Hochachka and Somero [1973] and Hazel and Prosser [1974] for comprehensive reviews.

The phenomenon of temperature acclimation has been studied most intensively in fishes. It occurs not only in intact fishes but to varying degrees in the isolated tissues of individuals that have been acclimated. It also affects various nervous functions of intact fishes. For example, in goldfishes (*Carassius*) acclimated to 30°C, spinal reflexes are blocked below 10°, but when the acclimation temperature is 15°, the blocking temperature is reduced to 1°C. Temperature acclimation has been demonstrated in complex functions of the central nervous system. By using light as the conditioned stimulus, goldfishes can learn to interrupt their opercular movements. If, after conditioning, the temperature of the water is gradually decreased, a temperature can be found below which the response disappears. In individuals conditioned at 30°C the response disappears near 20°, but when individuals are conditioned at 15° the response remains until the temperature is reduced to 10°.

8.6.3 Extremes of Body Temperature of Active Aquatic Ectotherms

The patterns of activity of individual aquatic ectotherms are largely controlled by the environment, but various groups of aquatic ectotherms have evolved adaptations that allow them to occupy and exploit the full range of temperatures existing in natural bodies of water (excluding some hot springs). Some species of fishes are as active at 5°C as others are at 30°C. Fishes in shallow tropical waters live at temperatures near 40°C, whereas those in dense, highly saline, polar waters live at temperatures between −1.0 and −2.0°C, which is below the freezing point of the body fluids of most vertebrates. Some of these cold-water fishes apparently live in a permanently supercooled state; others prevent freezing by addition of antifreeze to their body fluids (see Chapter 7). Because of the low temperatures the amount of oxygen dissolved in the water is often very great, and some antarctic fishes of the family Chaenichthyidae can satisfy the oxygen requirements of their tissues even though they possess no respiratory pigments whatsoever.

Although some fishes live in hot springs, they usually do so by exploiting local variations in temperature and remaining in the cooler parts of the springs. There are no convincing data on fishes remaining permanently in water more than 1 or 2 degrees above 40°C. Some water beetles and chironomid larvae, however, live in water as hot as 50°C, and there are algae that can live and reproduce under controlled conditions at 70°C.

8.6.4 Behavioral Control of Body Temperature

Although fishes occupy the full range of naturally occurring aquatic temperatures, in a given species at a given season the individuals of a population occur in a narrow range of water temperatures. In fact, despite their ectothermy, fishes maintain a remarkable uniformity of body temperature by behavioral means (Figure 8-4). Moreover, it has been repeatedly demonstrated that fishes of many different taxa select characteristic temperatures when placed in a temperature gradient under controlled conditions in the laboratory. Thus, there is strong circumstantial evidence that fishes, like other vertebrates, have receptors capable of precise temperature discrimination; these receptors transduce temperature to behavioral responses that result in a positive regulation of body temperature. However, the existence of such a receptor system has been demonstrated in only a few species. For example, cooling, by perfusion with water at 1°C, of tiny thermodes (thin-walled stainless steel tubes) implanted in the telencephalon, or forebrain, of Arctic sculpins (*Myxocephalus*) that have been adapted to water at 5°C causes them to remain in water at 20° despite a marked elevation in body temperature. Perfusion of the thermodes with water at 25° causes the fishes to leave the 20° water and return to 5° water. From these data one can conclude that responses that result in temperature control can be elicited by changes in receptors in the forebrain of bony fishes. Similar patterns exist in other major groups of vertebrates, including sharks, reptiles, and mammals.

8.7 TERRESTRIAL ECTOTHERMS

Aside from mammals, birds, and a few insects, all terrestrial animals are ectothermic. In view of the multiplicity of kinds of terrestrial ectothermic animals, the variety of habitats they occupy, and the range of thermal environments to which they have accommodated themselves through evolutionary change, it is not surprising that many patterns of adaptation to temperature exist. Some appreciation of the nature of these patterns can be obtained by considering amphibians and reptiles. During periods of activity each reptile and amphibian has a characteristic and restricted range of body temperatures, its *temperature preferendum,* that it maintains by choice of habitat and by selective exploitation of the spatial and temporal variety in the conditions of temperature and radiation in its general environment. Consequently, the macroclimate of the habitat within which a species occurs may be quite misleading with regard to its adaptations to, and physical responses to, temperature.

8.7.1 Amphibians

Present-day members of the class Amphibia are morphologically different from the Paleozoic forms, which were transitional between fishes and reptiles. Nevertheless, they demonstrate clearly the high level of success with which animals are sometimes able to exploit a harsh and demanding environment, despite physiological adjustments that appear at first glance to be modest and ineffective.

Water moves freely in both directions through the skin of amphibians. In a gaseous environment they lose water rapidly through evaporation. Consequently it has been generally assumed that while on land amphibians, at least from the standpoint of body temperature, function like wet-bulb thermometers. However, like most simple physical analogies in physiology, although heuristically useful, this idea must be applied cautiously. At least one anuran, the bullfrog (*Rana catesbeiana*), may be able to control its rate of evaporative cooling. The secretory activity of the mucus glands in its skin, the product of which is highly aqueous, is under nervous control, and the rate of secretion varies directly with deep body temperature and with the temperature of the hypothalamic area of the brain.

In general, amphibians are less tolerant of high body temperatures than are reptiles, birds, or mammals. Their sensitivity to elevated temperatures and their high rate of cutaneous water loss would seem to make them poorly adapted for terrestrial life. Nevertheless, in many parts of the world, particularly in the humid tropics and subtropics, they are by far the most abundant of terrestrial vertebrates. Their success, despite apparent physiological handicaps, is related to behavioral adjustments, to the capacity for thermal acclimatization, and to their ability to remain inactive and forego eating during periods when environmental conditions are unfavorable.

Behavioral Thermoregulation. The behavioral control of body temperature that amphibians exercise is generally less precise than that shown by reptiles (see Section 8.7.2), but it is nonetheless impressive in view of the fact that they appear to be exclusively ectothermic. The low metabolic rates of amphibians allow them to go for long intervals without eating and to become active only during limited periods when conditions of temperature and water availability are appropriate. For example, such burrowing forms as *Heleioporus* and *Cyclorana*, which live in the arid parts of Australia, may remain dormant underground for 2 years or more during periods of drought.

Those few salamanders (order Caudata) that are permanently aquatic exert temperature control in the same manner as freshwater fish. Their body temperatures are always approximately in thermal equilibrium with the water around them, so their temperature control consists primarily of habitat selection. They merely seek out local situations in which water temperatures are appropriate. The thermal situation for semiterrestrial salamanders, and for completely terrestrial species such as members of the family Plethodontidae, as well as for most frogs and toads (order Anura) is more complicated. The plethodonts go through their entire cycle in the complex terrestrial thermal environment. Most other amphibians are terrestrial as adults but as larvae live in the more thermally stable aquatic environment.

The red-bellied newt *(Taricha rivularis)* has been tested under controlled conditions for temperature preferences. It has the capacity for behavioral control of body temperature throughout its life cycle. Within a week of hatching the aquatic larvae select water temperatures between 20 and 26°C. If terrestrial adult *Taricha* are well hydrated, temperature appears to be more important than soil moisture

in their initial habitat selection. Aquatic adults avoid water temperatures above 18–20°C but make little effort to avoid temperatures near 0°C. Typical ranges of body temperature of active salamanders of the family Plethodontidae are shown in Figure 8-6. The low levels of body temperature are noteworthy; temperatures below 20°C are the rule rather than the exception. This contrasts strongly with the situation in reptiles.

The body temperatures of frogs and toads show a broad overlap with those of salamanders, but generally reach higher levels, and those of several families approach the lower limits maintained by mammals (Figure 8-7). At night or on cloudy days the body temperatures of anurans are approximately the same as their environment (substratum, air, or water). However, a number of species are heliothermic. By basking in the sun they can acquire heat from solar radiation and may raise their temperatures 10°C or more above that of the air or water immediately adjacent to them.

Among anurans, behavioral temperature regulation has been most thoroughly studied in the bullfrog. During normal daytime activity in southern California, body temperature in this species ranges between 26 and 33°C with a mean of 30°C. Body

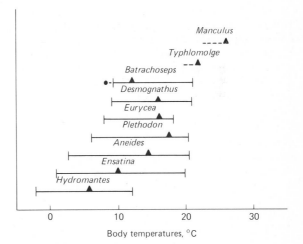

Figure 8-6
Body temperatures in nine genera of terrestrial salamanders of the family Plethodontidae. The horizontal lines indicate the ranges. The triangles indicate the means. [From B. H. Brattstrom. Ecology, **44**, *238–255 (1963).]*

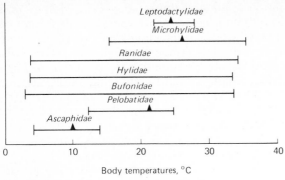

Figure 8-7
*Body temperatures in seven families of frogs and toads. The horizontal lines indicate the ranges. The triangles indicate the means. [From B. H. Brattstrom. Ecology, **44**, 238–255 (1963).]*

temperature is controlled by changes in location and modifications of posture. Solar and thermal radiation are used as heat sources. Pond water is used as a heat source on overcast days and as a heat sink on clear sunny days. Under natural conditions bullfrogs do not act as wet bulb thermometers; their body temperatures correlate more closely with dry bulb than with wet bulb air temperatures. The extent of thermoregulatory control displayed by bullfrogs is demonstrated by the narrow range of variation shown in continuous records of body temperature obtained by radio telemetry from free-ranging individuals (Figure 8-8).

Maintenance of elevated body temperature is incompatible with the high rates of cutaneous water

loss of amphibians. Frogs, which maintain elevated body temperatures by basking either make frequent trips to water, or while basking keep the lower parts of the body in the water and thus are able to remain in a positive water balance. Terrestrial amphibians as well as aquatic amphibians depend to a large extent on dermal respiration. An obligatory correlate of this, even in highly terrestrial species, is extensive cutaneous water loss. However, the degree to which amphibians depend on evaporative cooling under natural conditions during the terrestrial phase of their activities has never been systematically documented. The potential importance of evaporation, at least as an emergency measure during heat stress, is shown by the fact that the frog *Rana pipiens* can maintain a body temperature of 35°C for over 3 hr while kept at 50°C in dry air.

Thermal Acclimation. Like fishes and reptiles, amphibians have the capacity to compensate in part for the effects of temperature on metabolic rate. Also like fishes and reptiles they show another pattern of thermal acclimation related to the tolerance of high body temperatures. A conventionalized measurement called the ***critical thermal maximum*** offers a convenient index to this capacity. It is defined as the temperature at which an animal's locomotor capacity is so reduced that it loses the ability to escape from the thermal conditions that will lead to its death. The critical thermal maximum varies from species to species. It is also dependent on the prior thermal history of the individual being tested. Hence it is a measure of thermal acclimation. If a salamander or frog has been acclimated

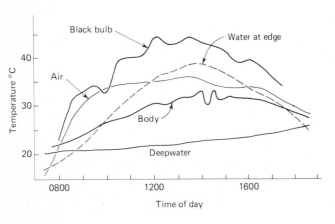

Figure 8-8
*The relation of telemetered body temperature of a free-ranging, 205-g, female bullfrog to the temperatures in its habitat, a small pond and its environs. [Data from H. B. Lillywhite, Copeia, **No. 1**, 158–168 (1970).]*

to 5°C, its critical thermal maximum is lower than if it has been acclimated to a higher temperature. The speed of acclimation increases with increasing temperature, and some acclimation is usually demonstrable after only a few hours (Figure 8-9). The rapidity and extent of the acclimation under laboratory conditions strongly suggest that it should be of selective advantage even under conditions of rapidly changing environmental temperatures, such as are common at midlatitudes during spring.

Rapid acclimatization as the result of exposure to the natural daily temperature cycle has been demonstrated in the salamanders *Ambystoma maculatum* and *Notophthalmus viridescens*. As a result of the slightly elevated temperatures experienced on warm sunny days individuals of *Notophthalmus* show more resistance to heat incapacitation in the afternoon than during the morning or at any time of day in cloudy weather (Figure 8-10). This response

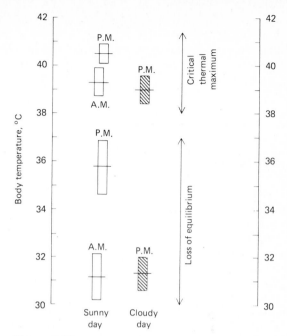

Figure 8-10
Natural rapid acclimatization of thermal incapacitation in the salamander Notophthalmus viridescens *as a result of exposure to the natural temperature cycle in its habitat on sunny days. The cross bars indicate the mean. The boxes enclose twice the standard errors. [Data from F. H. Pough. Comp. Biochem. Physiol., 47A, 71–78 (1974).]*

suggests the possibility that acclimatization in amphibians may be rapid enough to have survival value during the drastic diurnal temperature cycles that occur at high altitudes and in some deserts.

8.7.2 Reptiles

In Sections 8.6 and 8.7.1 we examined the patterns of adaptation to temperature shown by poikilothermic vertebrates that are wholly or partially buffered by aquatic habitats. These organisms modulate the effects of environmental temperature by habitat selection, by restriction of periods of activity, by limited behavioral exploitation of short-term environmental variations of thermal flux, by instantaneous, or at least very rapid, biochemical temperature compensation, by acclimation, and in some cases by evaporative cooling.

Reptiles, aside from secondarily aquatic forms

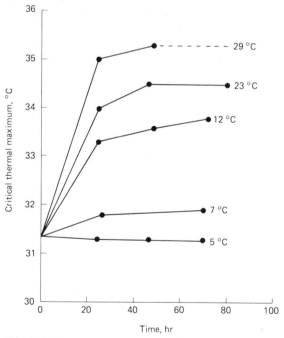

Figure 8-9
Acclimation of the critical thermal maximum in the frog (Rana pipiens). *The animals were held at 5°C for several weeks and then acclimated to temperatures of 7, 12, 23, and 29°. The CTMs were then determined at the intervals shown. [From B. H. Brattstrom and P. Lawrence, Physiol. Zool., 35, 148–156 (1962).]*

such as turtles and crocodilians, face the thermal challenges of complete terrestriality. They show the same general adaptive patterns found in fishes and amphibians, and also a series of physiological capacities that anticipate the thermoregulatory adaptations of birds and mammals.

Behavioral Thermoregulation. If one puts a lizard in a cage in the laboratory under constant conditions, in a short time its body temperature is indistinguishable from that of its surroundings. Thus it is clearly ectothermic and poikilothermic. However, with such treatment the animal in question is not a lizard in any biologically useful sense of the term, because it has been arbitrarily isolated from the environmental conditions appropriate to its essential functions. Nor is it even an adequate physiological preparation, for its integrity has been inadvertently destroyed without the observer realizing it. Such a lizard is really only a biological artifact, for a reason that is simple but, until recently, has not been obvious. The procedure of placing an organism, or preparation, in a constant temperature follows a long-established and usually essential physiological technique, but in this case it is poor biology and leads to erroneous conclusions. The reasons for this are related to the thermal complexity of the terrestrial environment.

As discussed in Section 8.4 and shown schematically in Figure 8-1, there is no single temperature that describes the thermal environment of a terrestrial animal. The diversity of specific heats of the materials in the environment, plus the complications imposed by the heat gain and heat loss from solar and thermal radiation, conduction, convection, and wind movements, and changes of state between solids, liquids, and gas, create a complex state of thermal flux. From this thermal flux, an ectothermic animal during its period of daily activity can, by relatively straightforward postural and behavioral adjustments, maintain any of a series of levels of body temperature. Each of these levels will differ from that of the air, or the substratum, or probably from any other single measurement of temperature in the animal's environment.

The disparity between body temperature and the temperatures in the environment is apt to be greatest when direct solar radiation is intense but air and substratum temperatures are low, as is the case in deserts or high mountains shortly after sunrise.

A dramatic example is offered by a common genus of small South American lizards, *Liolaemus,* some species of which live on the altiplano of Peru at altitudes of 4600 m or more. At this altitude, even within a few degrees of the equator, nighttime temperatures regularly fall several degrees below freezing. During the wet season, mornings are usually clear but the ground is often snow covered. By noon the sun has melted the snow but clouds have begun to gather. In midafternoon, snow or hail starts to fall and remains on the ground until the following morning.

Liolaemus spends the night in a burrow where it is protected from freezing. Shortly after sunrise, when its body temperature is 5°C or less, it crawls slowly to the surface, climbs onto a tuft of dry grass, and orients itself with its back to the sun. It is rapidly heated by the intense radiation of the equatorial sun even though the air temperature may be at, or below, 0°C. The rate of heating can be as rapid as 1°C/min, and the lizards frequently attain temperatures that are as much as 30°C above that of the air. Indeed, the activity temperature of this high-altitude species is essentially the same as those of many lizards that live in the intensely hot deserts of southwestern North America, namely, near, or slightly below, 35°C.

Liolaemus is merely a striking example of a common adaptive pattern in reptiles. Except for a few forms (notably dwellers in tropical rain forests), most lizards and many snakes maintain core body temperatures at a characteristic level by behaviorally exploiting the thermal variety of their environment. The mechanisms of behavioral thermoregulation employed by lizards involve a variety of motor patterns that offer attractive analogies with the temperature-control systems used in engineering.

For example, on a sunny day when the air temperature is lower than its temperature preferendum, a lizard will move into the sun and gain heat. As soon as body temperature rises above the temperature preferendum, the lizard moves back into the shade and loses heat. By shuttling back and forth between sun and shade it can maintain its body temperature within a limited range. It is as if the lizard were employing an *on-off control* system consisting of one heater and two thermostats, each with a different set point. The thermostat with the low set point turns on the heater. The one with the high set point turns it off. When the lizard is in the shade it loses heat. The decrease in body temperature activates the thermostat with the low set point, which turns on the heater (the animal moves from the shade into the sun). The solar radiation impinging on the animal causes it to gain heat. Body temperature then rises until it reaches a level that activates the thermostat with the high set point and the heater is turned off (the animal moves from the sun into the shade). The precision of regulation in such a system will clearly be related to the interval between the set points of the two thermostats. If the set points are far apart, the animal will shuttle infrequently and its body temperature will oscillate widely. If the set points are close together, the animal will shuttle frequently and the oscillations in its body temperature will be small.

More precise control of body temperature can be obtained with a system of *proportional control* of temperature, in which the output of the heater bears a continuous positive relation to the deviation of body temperature from the set point. Such a system will damp out oscillations. Many lizards employ a pattern of behavioral thermoregulation that has the properties of a simple proportional control system. On a clear day with air temperature lower than its preferred temperature, a lizard can control its rate of heat gain from solar radiation by variations in orientation with reference to the sun and by changes in the shape of its body. Many lizards can flatten their bodies and increase their surface by spreading the ribs. Horned lizards (*Phrynosoma* spp.) of the American deserts are a familiar example. With the ribs appressed and the long axis of the body parallel to the rays of the sun, the area exposed to direct solar radiation (and heating) is minimal. With the ribs spread and the flattened dorsal surface oriented at right angles to the rays of the sun, the area exposed to solar heating is maximal. In *Phrynosoma* the surface area that is exposed to direct sunlight is inversely related to body temperature. If body temperature falls below the set point the ribs are spread. If body temperature rises above the set point they are appressed. Thus changes in body area allow the lizard to employ a proportional control system.

Under natural conditions lizards employ an elaborate pattern of thermoregulatory behavior, many aspects of which are so subtle as to be generally unappreciated. Some indication of the variety of activities involved in the thermoregulatory behavior of a single genus is shown in Figure 8-11.

As in fish (Section 8.6.4), the presence of behavioral thermoregulation in reptiles strongly implies the existence of physiological temperature sensors. It is, therefore, of particular interest that behavioral thermoregulatory responses of the sort involved in an on-off control system have been invoked by local heating and cooling of thermodes implanted in the hypothalamus of the blue-tongue lizard (*Tiliqua scincoides*), a large Australian skink. This is the same part of the brain that activates physiological thermoregulatory responses in endotherms when its temperature is changed.

Body temperatures during activity are usually the same in all the lizards of a given species, even though the populations sampled may live in very different climates. Conversely, sympatric species of different genera may maintain mean body temperatures that differ by several degrees centigrade. The fact that body temperatures during activity are controlled by the animals themselves and not just imposed by the environment implies that physiological adjustments to specific temperatures must exist in reptiles and that these adjustments differ from group to group. This has been demonstrated to be the case at a variety of levels—enzyme, tissue, organ system, and intact organism.

Physiological Functions and Preferred Body Temperature. The characteristic levels of body temperature maintained by behavioral thermoregulation in different lizards have sufficient stability and sufficient antiquity to have associated with them an elaborate suite of physiological adjustments to the various temperature preferenda. This is clearly shown by the desert iguana, *Dipsosaurus dorsalis*, a common lizard of the deserts of southwestern United States and northern Mexico, which is active during the intense heat of midday. This species is remarkable for its adjustments to high temperature and strikingly illustrates the inadequacy of the phrase cold-blooded. Much of the time its body temperature equals or exceeds that characteristic of birds and mammals. The mean body temperature of active individuals in nature is slightly above 42°C. Body temperatures at which this species is normally active sometimes exceed 45°C, a level that is lethal to many lizards, many birds, and nearly all mammals. The tolerance to high temperatures

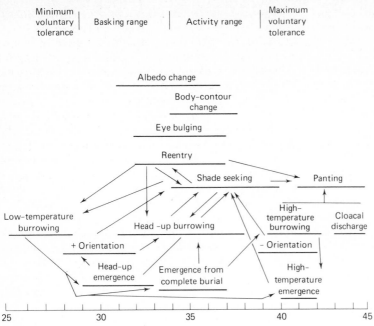

Minimum voluntary tolerance | Basking range | Activity range | Maximum voluntary tolerance

Albedo change

Body-contour change

Eye bulging

Reentry

Shade seeking → Panting

Low-temperature burrowing

Head-up burrowing

High-temperature burrowing | Cloacal discharge

+ Orientation

- Orientation

Head-up emergence

Emergence from complete burial

High-temperature emergence

Body temperature, °C
25   30   35   40   45

Figure 8-11
Relations between different patterns of behavioral thermoregulation and body temperature in the horned lizard (Phrynosoma coronatum). [*From J. E. Heath,* Univ. Calif. (Berkeley and Los Angeles) Publ. Zool., *64, 97–136 (1965).*]

of both the intact *Dipsosaurus*, and its tissues *in vitro*, is the most extreme reported for any terrestrial vertebrate. For example, (1) there is no heat suppression of oxygen consumption at body temperatures as high as 45°C, (2) body temperatures as high as 47°C are tolerated, and (3) spontaneous beating of excised heart auricles continues *in vitro* at temperatures between 46 and 47°C. Obviously these adaptations to high temperature fit the species nicely to its desert habitat, but they are not a necessary part of the adaptation of lizards to desert life. Other species of reptiles occupy the same habitat as *Dipsosaurus*, even though they are much less heat tolerant. They avoid high temperatures by burrowing and by the crepuscular and nocturnal timing of their surface activity. Thus the striking physiological adaptations of *Dipsosaurus* to high temperatures are not a condition for existence in hot deserts but a capacity that allows it to exploit the desert environment during the daytime hours of intense heat when few other local reptiles are active.

It is apparent that no matter what aspect of reptilian physiology is being considered, the role of the preferred level of body temperature cannot be overlooked with impunity. Thyroid function is a case in point. It has long been known that thyroxine is an important determinant of the rate of oxygen consumption in birds and mammals. The thyroid gland is well developed in reptiles, but investigations of the metabolic effects of thyroxine have yielded confusing and often negative results, apparently because investigators have often assumed that their experiments could be carried out at room temperature as would be the case if the experimental animal were a bird or mammal. However, it turns out that in reptiles the effects of thyroxine depend on the body temperatures of the animals to which it is administered. This is shown nicely by experiments on a skink, *Eumeces fasciatus*, which during activity maintains its body temperature between 28 and 35°C. In this species when the animals are kept at 30°C the administration of thyroxine increases oxygen consumption and thyroidectomy reduces it, but these effects are abolished when the animals are kept at 20°. Positive results are also obtained when the animals are kept at 30° for 10 hr/day and at 20° for the remaining

350

hours—a temperature regime which approximates that normally experienced by these animals during the summer.

The importance of behaviorally maintained levels of body temperature to a variety of physiological functions in lizards is well documented. At temperatures between 20 and 40°C the relation of heart rate in resting lizards to body temperature is rather uniform. The Q_{10} typically remains quite constant between 2.0 and 2.5. However, during intense activity the values for this temperature coefficient are much more variable; in most of the species for which data are available, the difference between resting and maximal levels of heart rate is greatest at, or slightly above, the thermal preferendum. Similar relationships have been demonstrated for skeletal muscle. Conspicuous interspecific differences exist in the temperature dependence of the contractile performance of skeletal muscle in lizards, and these differences correlate nicely with temperature preferenda (Figure 8-12).

The temperature preferenda of lizards are also expressed at the subcellular level. For example,

the temperature dependence of ATPase activity of myosin extracted from skeletal muscle of different species correlates with their preferred body temperatures (Figure 8-13). The temperature peaks of enzyme activity rank in the same order as the preferred body temperatures of the species from which they were obtained, and the levels of enzymatic activity at low temperatures rank in inverse order.

The thermal dependence of other reptilian physiological functions also correlates with the preferred temperatures of the species involved. Examples are (1) resistance of testicular tissue to thermal damage, (2) sensitivity of the inner ear to sound, (3) photoperiodically induced testicular recrudescence, (4) rate of digestion, (5) rates of development of antibodies, and (6) capacity of the kidney for sodium retention. A recent review with bibliographic citations is available in Dawson [1975].

Despite the correlations discussed, it should be remembered that, in reptiles as in other poikilotherms, biochemical temperature compensation and acclimation offer the possibility of stabilizing reaction rates over a range of body temperatures

Figure 8-12
The relation between the highest temperature at which twitch tension of skeletal muscle is maximal and the preferred temperatures of lizards as determined in a thermal gradient. If the correlation were perfect, all the points would fall on the diagonal line. The unshaded circles are North American species. The shaded circles are Australian species. Family Gekkonidae: Phyllurus *(1),* Diplodactylus *(2),* Gehyra *(3). Family Skincidae:* Egernia *(4),* Eumeces *(5). Family Agamidae:* Physignathus *(6),* Amphibolurus *(7, 8). Family Iguanidae:* Dipsosaurus *(9),* Uma *(10),* Sceloporus *(11). Family Anguidae:* Gerrhonotus *(12). [Modified from P. Licht, W. R. Dawson, and V. H. Shoemaker. Z. vergl. Physiol.,* **65,** *1–14 (1969).]*

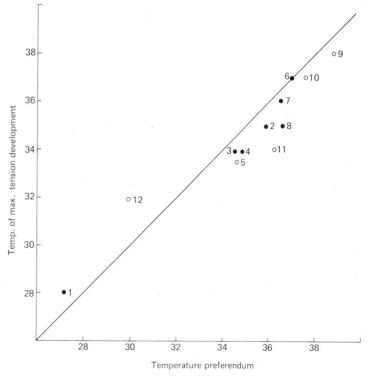

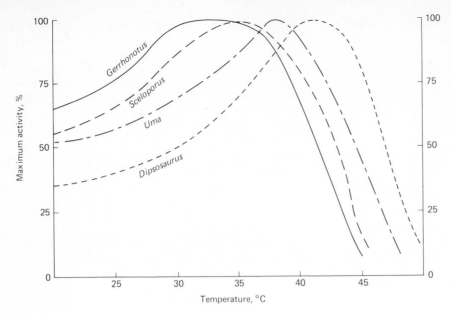

Figure 8-13
Influence of temperature on the activity of ATPase from the skeletal muscles of lizards with different activity temperatures. Under laboratory conditions their temperature preferenda are as follows: Dipsosaurus dorsalis, 38.8°C; Uma notata, 37.5°C; Sceloporus undulatus, 36.3°C; Gerrhonotus multicarinatus, 30.0°C. [*Modified from P. Licht.* Comp. Biochem. Physiol., **12**, 331–340 (1964).]

and that there is no *a priori* necessity that the rate of a given physiological function peak at or near the temperature preferendum of the species being studied. For example, maximum aerobic metabolic scope tends to coincide with preferred body temperature in agamid and iguanid lizards (Figure 3-10), but it does not do so in varanid lizards nor in the turtles *Pseudemys scripta* and *Terrapene ornata*.

Energetics and Body Temperature. The relation of energy metabolism to body temperature in reptiles was briefly discussed with regard to metabolic scope (Section 3.13). The typical response of oxygen consumption during rest and during activity at moderate temperatures is similar to that shown by the Galapagos marine iguana, *Amblyrhynchus cristatus* (Figure 8-14). From about 20° to about 40°C, standard metabolism usually shows a fairly uniform thermal dependence with a Q_{10} of more than 2 but less than 3. When an animal is stimulated to maximum activity over the same range of temperatures, oxygen consumption shows a strong thermal dependence below the preferred body temperature often with a Q_{10} of 3 or even 4, and it plateaus or even declines at higher body temperatures. This pattern of aerobic scope increasing with temperature and peaking at or near the preferred body temperature has usually been interpreted as the

central adaptive aspect of behavioral thermoregulation. The preferred body temperature has been considered to be the temperature at which a species can most readily mobilize energy for the capture of prey, escape from predators, dig nest holes, challenge territorial intruders, or engage in energy demanding activities of any sort.

Although the aerobic metabolism of reptiles always shows temperature dependence, the magnitude of the dependence varies from species to species. At temperatures below 15°C the values for Q_{10} of most lizards that have high temperature preferenda (*Dipsosaurus, Crotyphytus, Uma*) range from 3.5 to more than 6, whereas the values for species normally active at lower body temperatures (*Xantusia, Gerrhonotus, Coleonyx*) are near 3 or even less. The converse occurs at temperatures above 30–35°C.

Instantaneous temperatures compensation is not as important in reptiles as in some fishes and invertebrates, but there are a number of cases where, over restricted ranges of temperature, an inverse relationship to temperature develops; for example, the heart rate of the marine iguana at body temperatures above 35°C (Figure 8-17). Probably the clearest case of reptilian temperature compensation is shown by a Canadian population of the viviparous red-sided garter snake, *Thamnophis*

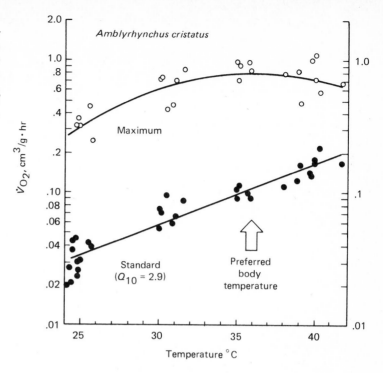

Figure 8-14

The relation of oxygen consumption to body temperature in Galapagos marine iguanas with a mean weight of 489 g. Maximum rates were elicited by electrical stimulation. The regressions are fitted by the method of least squares. [Data from A. F. Bennett, W. R. Dawson, and G. A. Bartholomew. J. Comp. Physiol., **100**, 317–329 (1975).]

sirtalis, that occurs near the northern limit of the range of the species where summers are short and highly variable in temperature. These animals are normally active at body temperatures between 18 and 30°C, but seek shelter and become dormant when body temperatures fall below 17°C. This pattern of behavior shows a remarkably close correlation with both standard energy metabolism and *in vitro* oxygen consumption of liver tissue (Figure 8-15). The Q_{10} values of oxygen consumption are 2 or less in the activity range, but 8 or more in the lower part of the dormancy range. At the temperatures where the animals normally seek shelter and in the upper part of the dormancy range, oxygen consumption actually increases as temperature decreases. This reversal of temperature dependence presumably allows the animal to find adequate shelter and to maintain adequate levels of essential physiological functions such as digestion and gestation until temperatures suitable for normal activity again become available. The extremely high values of Q_{10} at temperatures below 10°C should facilitate energy conservation during protracted cold spells in summer and during the pro-

longed hibernation which extends from fall through the winter to spring.

Physiological Temperature Control. One of the responses that lizards show in relation to body temperature illustrates with particular clarity a pattern of physiological activity that is often important to animals under natural conditions, and, from both ecological and physiological points of view, is as interesting as the more familiar steady-state responses associated with the thermal homeostasis of birds and mammals. Many animals under natural conditions neither achieve nor maintain a steady state with regard to even important physiological parameters. However, even though their physiological control is limited, it allows them to modify rates of change so as to extend periods of activity under unfavorable conditions, or to prolong survival under conditions that cannot be permanently tolerated. Crocodilians, lizards belonging to at least five families (Iguanidae, Gekkonidae, Varanidae, Teidae, and Agamidae), and some snakes as well as some turtles and the tuatera *Sphenodon* can control the local distribution of heat and the rate of change

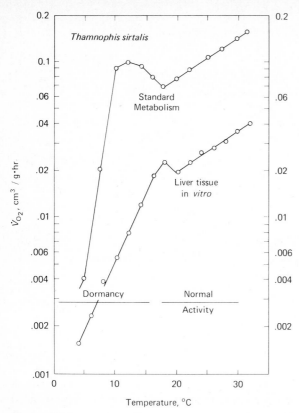

Figure 8-15
Whole animal and tissue oxygen consumption in relation to temperature in the red-sided garter snake. [Date from M. Aleksiuk. Thermal Physiol., **1***, 153–156 (1976).]*

of body temperature by cardiovascular adjustments, evaporative cooling, and, to a lesser extent, by variations in the rate of energy metabolism. The net effect of this pattern of physiological control is to accelerate heating and retard cooling. The type of thermoregulatory control shown by these reptiles probably never yields a steady state, but it damps the oscillations in body temperature and increases the fraction of the day during which they can keep the body temperature near the preferred level.

As is often true in comparative physiology, an extreme instance helps one to understand a general pattern. The Galapagos marine iguana *(Amblyrhynchus cristatus)*, which occupies a thermal environment that for a lizard is unique, is a case in point. It spends most of its time on barren rocky shores

exposed to high ambient temperatures and the intense radiation of the equatorial sun. However, it feeds on marine algae in the relatively cool waters of the Peruvian current, which bathes the Galapagos Islands. During the day, while ashore, it maintains body temperature close to 37°C, primarily by behavioral means. When it enters the water to feed, it enters an environment that is 10–15°C below its temperature preferendum. Because of the temperature sensitivity of most physiological processes it would be advantageous to the marine iguana to minimize its rate of cooling while in the water and thus extend the time during which body temperature remained near the preferred level. Conversely, it should be advantageous to heat up as rapidly as possible after leaving the water to facilitate digestion of the algae it consumes during its dives. It is of interest, therefore, that the cooling rates of *Amblyrhynchus* are about one half the heating rates (Figure 8-16). This is the greatest capacity to control the rate of change of body temperature that has so far been demonstrated among reptiles.

The major contribution to the difference between heating and cooling rates comes from cardiovascular adjustments. Heart rate at a given temperature is generally more rapid during heating than during cooling (Figure 8-17). Other things being equal, the rate of heat exchange between lizard and environment will depend on the volume of blood flowing per unit time between core and periphery. The difference in heart rates during heating and cooling suggests that circulation augments heat exchange during heating and diminishes it during cooling. The effect is further increased by peripheral vasodilatation during heating and peripheral vasoconstriction during cooling. Studies using the ^{133}Xe clearance technique (radioactive xenon in saline is injected subcutaneously and its rate of washout is measured with a scintillation detector) have shown that cutaneous blood flow increases during heating and decreases during cooling in at least three iguanid lizards, *Amblyrhynchus, Dipsosaurus,* and *Iguana,* and the teid *Tupinambis.*

A further example of cardiovascular contributions to heat exchange is suggested by Figure 8-17. Not only is heart rate at a given temperature higher during heating than during cooling, but heart rate diminishes with increasing body temperature above 35°C. This diminution can be inter-

Figure 8-16

Heating and cooling rates of the Galapagos marine iguana (Amblyrhynchus cristatus) *in water and air.* ΔT *is the difference between body and ambient temperatures* (T$_a$). *During heating* T$_a$ = 40°C; *during cooling* T$_a$ = 20°C. [*From G. A. Bartholomew and R. C. Lasiewski.* Comp. Biochem. Physiol., **16**, *573–582 (1965).*]

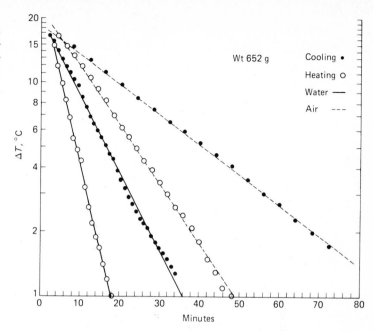

preted as a mechanism for reducing the rate of temperature increase as the lizard approaches its preferred body temperature. Similar decreases in heart rate at temperatures near or above the temperature preferenda have been reported in other iguanid lizards.

Evidence for physiological control of body temperature is also available for turtles. Although the role of temperature in the biology of turtles is poorly understood, at least two freshwater turtles, the snapping turtle *(Chelydra sepentina)* and the cooter *(Pseudemys floridana)* have some physiological control of body temperature. Both species when submerged in water heat about 25% faster than they cool and, in both, cutaneous blood flow increases during heating and decreases during cooling. Turtles are the oldest surviving group of reptiles and, from the morphological point of view, living forms are very similar to types which lived 200,000,000 years ago in the early Triassic.

The tuatara *(Sphenodon punctatum)*, a relict form surviving only in New Zealand and the sole living representative of the reptilian order Rhynchocephalia, is very similar morphologically to fossil forms from the Triassic. *Sphenodon* has the capacity to heat rapidly and cool slowly. Since both the tuatera and some turtles can exert limited physiological control over the rate of change in their body temperature, it is possible that at least a rudimentary capacity for physiological thermoregulation evolved very early in the history of terrestrial vertebrates.

The partial physiological control of body temperature in reptiles is not confined to cardiovascular mechanisms. It is also apparent in endogenous heat production and in evaporative cooling.

A certain amount of evaporation of water necessarily accompanies aerial respiration. Amphibians because of their dermal respiration inevitably evaporate water from their skin. Whether or not the heat loss associated with this evaporation of water is ever more than a continuous passive process in amphibians has been shown only for bullfrogs, but reptiles do use evaporative cooling for thermoregulatory purposes.

In most reptiles, breathing rates increase with increasing body temperature. At very high body temperatures the mouth is held open and the breathing movements become deep and labored. Reptiles never respond to heat stress with the abrupt acceleration of breathing that is commonly observed in panting mammals. However, in the genus *Varanus*, which includes the largest living lizards, when body temperature exceeds 38°C power-

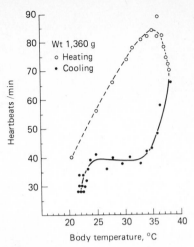

Figure 8-17
Relation of heart rate to body temperature in the Galapagos marine iguana during heating and cooling in water. [From G. A. Bartholomew and R. C. Lasiewski. Comp. Biochem. Physiol., **16,** *573–582 (1965).]*

ful gular pumping motions of large amplitude develop. These pumping motions involve the floor of the mouth and the entire neck region and appear to be the functional equivalent of panting. Little quantitative information about the physiological effectiveness of panting as a mechanism for evaporative heat loss is available for reptiles, but at least one species, *Dipsosaurus dorsalis,* can lose an amount of heat by panting at least equal to that produced by its own metabolism, even at very high body temperatures. It is clear that in their control of evaporative cooling, reptiles show some capacity for physiological thermoregulation.

The standard metabolic rate of reptiles is much lower than the basal rate of mammals of the same (Figure 3-6) size even if the measurements are made at the same body temperature (37°C). The endogenous heat production of reptiles at rest has no significant effect on body temperature. However, under conditions of intense activity, metabolism is greatly increased and endogeneous heat production does affect body temperature, particularly in large forms. For purposes of emphasis let us consider an extreme case. The standard metabo-

lism of a medium-sized (700-g) monitor lizard (*Varanus*) at a body temperature of 37°C is 0.13 cm³ O_2 $g^{-1}hr^{-1}$ (Table 3-5). This is about one-sixth the calculated basal rate of a mammal of the same size at the same body temperature. However, the maximum sustained metabolic rate of the *Varanus* exceeds 1.0 cm³ O_2 $g^{-1}hr^{-1}$ which surpasses the basal rate of the mammal by 30%. Thus *Varanus* bridges the metabolic gap between reptiles and mammals. It is not surprising, therefore, that active monitor lizards can, by endogenous heat production alone maintain body temperatures 2°C or more above ambient for long periods of time. These large and active lizards show indications of incipient endothermy. However, this is not true of iguanid and agamid lizards of similar size.

Although no reptiles are permanently endothermic, at least one snake, the Indian python (*Python molurus*), maintains an elevated body temperature by means of endogenous heat production for long periods of time while incubating its eggs. The female python coils about its eggs and by means of spasmodic contractions of its muscles generates enough heat to maintain body temperature as much as 7°C above ambient temperature. The effect of environmental temperature on the oxygen consumption of a nonincubating python is like that on any ectotherm, but incubating pythons show a pattern similar to that of endotherms. Above 33°C metabolic rates of incubating and nonincubating pythons are the same, but at environmental temperatures below 33°C the oxygen consumption of the former increases whereas that of the latter decreases (Figure 8-18). The functional significance of this pattern of endothermy is obviously that it contributes to the maintenance of a high and uniform thermal environment for the eggs encircled by the coils of the snake's body.

From the preceding discussion it is clear that reptiles have evolved a flexible, effective, and energetically economical pattern of thermal control that exploits with impressive virtuosity the thermal variability of the physical environment and allows them to maintain a closely regulated level of body temperature during their periods of activity. Even though all reptiles are primarily ectothermic, they also possess the essential physiological components on which the physiological thermoregulation of endotherms depends. They have cardiovascular

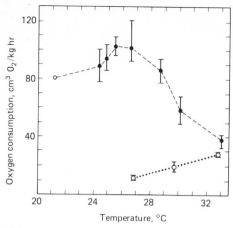

Figure 8-18

Oxygen consumption of a 14-kg Indian python (Python molurus) *as a function of environmental temperature. Upper curve, while brooding eggs; lower curve, during non-brooding period. Vertical lines show range. Circles indicate means.* [*From V. H. Hutchison, et al. Science,* **151,** *694–695 (1966).*]

control over rates of heat gain and heat loss from the environment. They can produce effective amounts of heat by endogenous means. They employ evaporative cooling. Their metabolic response to the thyroid hormone is similar to that of endotherms (at least at high temperatures), and they have hypothalamic temperature sensors that are involved in behavioral temperature regulation. The complete separation of systemic and pulmonary circulations is missing (except in crocodilians), and, of course, no reptiles have an exoskeletal covering of feathers or hair. It should be emphasized that the possession of such an insulating layer would probably be maladaptive for an ectotherm because it diminishes energy exchanges with the environment. The reptiles of the Paleozoic and early Mesozoic clearly set the stage for endothermy and two of the groups (birds and mammals) derived from them have developed it to a high level of refinement. Modern reptiles have evolved a less energetically expensive strategy of energetics than their contemporary mammalian and avian relatives. They should not be thought of as unsuccessful or primitive endotherms but as highly successful and

precisely adapted exploiters of an alternative life style—ectothermy.

8.8 BODY TEMPERATURE AND ENERGY METABOLISM IN ENDOTHERMS

As discussed in the preceding sections of this chapter ectotherms show a complicated array of responses that allow them to find, maintain, and, to a limited extent, assemble by their own efforts, temperature regimes appropriate to their functions and modes of life. However, because they depend almost exclusively on exogenous sources of heat, these animals are subject to major environmentally imposed constraints that limit the timing and spatial patterns of their activity. Birds and mammals (and a few representatives of other groups) have escaped these environmental constraints by evolving a thermoregulatory system in which the central element is a high rate of endogenous heat production and in which rates of heat production and heat exchange are largely under physiological control. Temperature regulation has been more intensively studied in birds and mammals than in other endotherms and we shall, therefore, base our initial examination of endothermy on these two taxa. Birds and mammals are descended from different groups of reptiles, and these groups were already separate in Paleozoic times. Therefore, the vertebrate pattern of endothermic temperature regulation must have evolved independently in the class Aves and the class Mammalia. Nevertheless, the mechanisms of thermoregulation found in birds and mammals are so similar that they can be considered together.

8.8.1 Energy Metabolism and Ambient Temperature

The relationships summarized diagrammatically in Figure 8-19 offer a general theoretical base from which the physiology and ecology of temperature regulation in birds and mammals can be examined.

If one places a bird or mammal in a respirometer and leaves it there until it calms down and reaches a steady level of energy metabolism, the relationship of its oxygen consumption to the external tem-

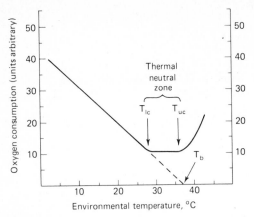

Figure 8-19
Relation of oxygen consumption to environmental temperature in a hypothetical mammal.
T_{lc} = *lower critical temperature;* T_{uc} = *upper critical temperature;* T_b = *core body temperature.*

peratures to which it is exposed will in most cases be described by a curve qualitatively similar to that shown in Figure 8-19. The precise shape of the curve and the slope of its arms will, of course, depend on the type of animal tested, the general pattern of its adaptations, and its mode of life. As indicated in Figure 8-3 and shown in more detail in Figure 8-19, the response of the energy metabolism of homeotherms to ambient temperature is not simple. Over some range of temperature (27–35°C in the case of the hypothetical animal being considered) oxygen consumption is minimal and virtually independent of temperature. This area of minimum oxygen consumption is referred to as the **thermal neutral zone.** At its lower end the thermal neutral zone is bounded by the **lower critical temperature** and at its upper end by the **upper critical temperature.**

As usual we are faced with some terminological confusion. Although we shall use the phrase **critical temperature** only with reference to thermal neutrality, it is sometimes used to designate the lowest environmental temperature at which a homeotherm can maintain its body temperature at the normal level. In the older literature the zone of thermal neutrality is often referred to as the zone of physical thermoregulation and the region below thermal neutrality as the zone of chemical ther-

moregulation. These terms are unsatisfactory and will not be used.

Below the lower critical temperature, oxygen consumption increases linearly as temperature decreases. If the slope of this line is extrapolated to zero oxygen consumption, it intersects the horizontal axis at approximately the body temperature of the animal being measured, in this case at 37°C. Above the upper critical temperature, oxygen consumption increases rapidly as ambient temperature rises, but the rate of increase is usually not linear. The relationships between oxygen consumption and ambient temperature obviously result from the interaction of many factors. Figure 8-19 describes only the metabolic cost of maintaining body temperature in different ambient temperatures. However, when one understands the physiological factors that combine to produce the relationships described, a graph such as that shown can yield a large amount of information about the process of thermoregulation. In addition, by comparing such graphs based on data from a variety of species, one can obtain some understanding of the nature of physiological adaptation to climatic conditions.

Certain commonsense deductions can be made from Figure 8-19 without further information. For example, it is clear that in the thermal neutral zone the animal is expending a minimum of energy on thermoregulation. It is equally obvious that below the lower critical temperature and above the upper critical temperature an animal must do additional metabolic work beyond the resting level to maintain body temperature constant. Further, because metabolism does not fall to its minimum value when ambient temperature and body temperature are equal, one can conclude that at all temperatures a homeotherm produces a large amount of heat that is continuously being lost to the environment. To go beyond these elementary deductions one must employ some of the simpler ideas borrowed from classical thermodynamics and heat engineering.

8.8.2 Newtonian Cooling, Heat Transfer, and Endothermy

There is an enormous literature on heat transfer and thermoregulation in endotherms, including man, but for purposes of an introductory discussion we can develop our ideas with regard to some simplified models and thus avoid many of the confusing complications that have intrigued investigators in this active field. In biology it is, of course, invariably difficult to find a simple model that is theoretically satisfactory. In the case of energy metabolism and body temperature, this is certainly true. For example, homeotherms maintain body temperature constant and vary the rate of heat loss so that the familiar physical models used to demonstrate Newtonian cooling are only indirectly applicable. Although it does considerable theoretical violence to conventional physical procedures, a convenient way of developing an understanding of the fundamental relations involved in thermoregulation in animals is to start with Newtonian cooling, and then modify it until it is applicable to biological systems.

The change of heat in a body *(dH)* per unit-time *(dt)* depends on its surface area *(A)* and difference between its surface temperature *(T_s)* and the ambient temperature *(T_a)*. This can be stated,

$$dH/dt = hA(T_s - T_a) \qquad (8\text{-}1)$$

where h, the coefficient of heat transfer, is a complicated function that includes convection, conduction, and radiation. Equation (8-1) is a linear approximation of overall heat transfer between an object and its environment and is applicable only over a limited range of conditions. However, it is a reasonable description of heat exchange in the circumstances that exist under laboratory conditions when temperatures are moderate (0–40°C) and the radiative temperature of the environment (T_r) is the same as the air temperature.

However, this simplified equation is inadequate for convenient application to endothermic thermoregulation because birds and mammals can change their surface area by postural changes and can alter the temperature of the body surface by vasomotor activity and changes in the arrangement of hair and feathers. Moreover, local variations in the quality and thickness of plumage and pelage cause heat loss to vary regionally over the body. Consequently, neither the surface area or the surface temperature of an endotherm are constant. In equation (8-1), h is obviously a general property and if the body in question is inanimate h is a constant. If the body in question is a living animal, however, h is by no means constant. In fact, it is a kind of an umbrella beneath which are crowded

many of the parameters that affect the rate of heat exchange between organism and environment—rate and pattern of blood circulation, posture, piloerection, rate of evaporation of water, and regional differences in T_b to name a few.

To fit endothermic regulation to the "Newtonian" cooling described by equation (8-1), the equation has not been made more realistic, but instead has been simplified even further

$$dH/dt = C(T_b - T_a)$$

where, instead of surface temperature, deep body temperature (T_b) is used and C, called **thermal conductance,** is used instead of h. Thermal conductance specifies the net rate of heat transfer of an organism per degree centigrade difference between body temperature and ambient temperature. Clearly C will vary with physiological state. If one considers C in the perspective of the annual cycle through which an animal goes rather than in terms of minute-to-minute adjustments, it is equally apparent that thermal conductance will also vary with the seasonal changes in the thickness of subcutaneous fat and seasonal changes in the density and insulative properties of the plumage or pelage.

Now let us apply the ideas of Newtonian cooling to the energy metabolism versus temperature curve shown in Figure 8-19. It is obvious that for an animal to maintain a constant body temperature, the rates at which it is acquiring heat and losing heat must be identical. It is also obvious that at all ambient temperatures lower than body temperature, an animal loses heat to the environment. To maintain a constant body temperature under such circumstances, an endotherm must produce heat by its own metabolism at a rate exactly equal to the rate that heat is being lost to the environment.

In relating endothermic temperature regulation to Newtonian cooling, investigators have (1) assumed that body temperature remains constant and is uniform throughout the body, (2) measured minimum energy metabolism at various temperatures below the thermal neutral zone, and (3) plotted energy metabolism against ambient temperature. Newtonian cooling has been said to apply if the data for oxygen consumption fit a straight line that, at zero heat production, extrapolates to an environmental temperature equal to body temperature. As the critical criteria are the linearity and the extrapolation, obviously the concept has relevance only below thermal neutrality and in cases of strict homeothermy. However, this application of Newtonian cooling has a broad range of usefulness, because it allows us to isolate the factor of thermal conductance and to relate it to a series of physiological parameters from which we can build a generalized picture of the processes of physiological thermoregulation. As long as a bird or a mammal remains homeothermic and environmental (ambient) temperatures are lower than body temperature, its energy metabolism must be calorically equal to its heat loss (see Chapter 3). From these relationships, a series of simple equations can be written:

$$\text{heat loss} = \text{heat gain} = \text{heat production}$$
$$= \text{metabolic rate } (\dot{E}_m) \qquad (8\text{-}2)$$

$$\text{heat loss} = C(T_b - T_a) \qquad (8\text{-}3)$$

$$\dot{E}_m = C(T_b - T_a) \qquad (8\text{-}4)$$

$$C = \frac{\dot{E}_m}{T_b - T_a} \qquad (8\text{-}5)$$

As a result of the relationship summarized in equations (8-1) to (8-5), we can obtain considerable additional information from Figure 8-19 by relating energy metabolism to thermal conductance and measuring one in terms of the other. If we consider thermal neutrality in terms of thermal conductance, we find that our understanding of this zone is enhanced and that we are able to offer new definitions for upper and lower critical temperatures. The thermal neutral zone becomes the range of ambient temperatures over which a homeotherm can vary its thermal conductance in an energetically inexpensive manner and on a short time scale. Indeed, over the range of thermal neutrality the metabolic expense of increasing or decreasing thermal conductance is so slight as to be indistinguishable against the background of resting metabolism. That is, thermal conductance can be varied by changes in the supply of blood to superficial areas, by fluffing or compressing the feathers or fur, or by simple postural adjustments. These changes, although negligible from the standpoint of energy metabolism, are sufficient from the standpoint of heat flow to allow the animal to regulate its body temperature.

The lower critical temperature can be defined with regard to thermal conductance as the temperature below which metabolically inexpensive variations in thermal conductance are no longer sufficient to keep heat production and heat loss in balance. Below its lower critical temperature the animal can no longer decrease its thermal conductance sufficiently to compensate completely for the enhanced rate of heat loss as the difference between body and ambient temperature increases. Therefore, it must increase the rate of heat production. Otherwise, its body temperature will decrease. Except for temperatures a few degrees below the lower critical temperature, most homeotherms show an approximately linear increase in metabolism with decreasing temperature. We can therefore infer that at the lower limit of the zone of thermal neutrality, conductance has been reduced virtually to its minimal level. Consequently, below the lower critical temperature, the thermal conductance C becomes a constant. From this it follows that heat production, HP, will increase linearly with increases in difference between body and ambient temperatures, if strict homeothermy is maintained. Thus

$$HP = C(T_b - T_a)$$

Because HP is measured in terms of oxygen consumption, C can be expressed directly as cubic centimeters of oxygen per gram per hour per degree centigrade and, if the caloric equivalent of the oxygen being consumed is known, C can be expressed as calories, or joules, lost per degree centigrade difference between T_b and T_a. However, both of these usages do some violence to established terminology, because in physical systems heat transfer is expressed per unit surface area rather than per unit mass. In the interests of clarity it is preferable therefore to use the phrase, mass-specific conductance, when referring to C.

Since both body temperature and energy metabolism remain constant with increasing temperature throughout the zone of thermal neutrality and body temperature increases only slightly, if at all, one can assume that over this range thermal conductance increases with increasing environmental temperature. At the upper critical temperature, thermal conductance should, therefore, have reached its maximum. It is obviously necessary that the upper critical temperature be slightly below the operational body temperature; otherwise heat could not be lost to the environment. It is equally obvious that when ambient temperature exceeds body temperature the animal can no longer lose heat by passively transferring it to the environment through conduction, convection, or radiation, but can do so only by means of the evaporation of water. Further, from the fact that the rate of energy metabolism increases sharply at the upper critical temperature, it is clear that readily measurable amounts of work are required to mobilize the water used in evaporative cooling.

As discussed in Section 8.10, the amount of heat that an endotherm loses through evaporation of water increases directly with increasing ambient temperature. Since the evaporation of water plays an important part in the heat budget of animals (see Section 8.4) it contributes importantly to thermal conductance. For this reason it is helpful under some circumstances to distinguish between *total conductance* and *dry conductance.* Dry conductance is calculated by subtracting from total conductance the heat dissipated by evaporation.

The preceding discussion of conductance and "Newtonian" cooling offers only a crude approximation of the thermoregulatory complexities actually dealt with by animals in the natural world. It completely ignores two of the important elements in the heat balance equation (Section 8.4), namely level of solar (and thermal) radiation and air movement. Both the conditions of radiation and the conditions of convection must be quantified before one can establish the thermal equivalence between laboratory and outdoor environments. This gap can be bridged by means of a temperature index, the *standard operative temperature* (T_{es}), which in effect uses the animal itself as a thermometer and the net heat flow between animal and environment as the "temperature indicator." Discussion of this procedure is beyond the scope of this book. Interested readers should see Bakken [1980] for an analysis of the method and Chappell and Bartholomew [1981] for an example of its application.

8.8.3 Thermoregulatory Control Systems

The process of thermoregulation in birds and mammals invites the construction of model control systems based on feedback theory. Such models, although helpful in relating the various thermoregulatory processes, cannot themselves yield an adequate basis for physiological understanding. Sound biological data are needed for the critical step of relating model to structure. Fortunately, there are now available analytical data on peripheral and central receptors and on the integrative mechanisms of the central nervous system (see Chapters 10 and 11) that allow feedback theory and physiological information to combine and supply at least a qualitatively satisfying model.

Even a simple model for endothermic thermoregulation must have a number of elements if it is to approximate physiological activity. Each of these minimal elements consists of a complex suite of interrelated physiological systems. Even if we treat each of the elements as a black box, the situation is not simple. The black boxes in our minimum model must include (1) control of rate of heat production; (2) control of rate of heat loss; (3) either one or two thermostats in the central nervous system. If two, one has a high-temperature set point and one has a low-temperature set point; (4) a series of peripherally located thermostats, some with high set points and others with low set points; (5) machinery for coordinating the signals from the various thermostats; (6) machinery for allowing the central thermostats to override the peripheral ones; (7) channels of communication that allow the thermostats to activate, in a coordinated manner, the mechanisms of heat production and the mechanisms controlling heat loss; and (8) appropriate integrative machinery to allow all the above systems, which are at the automatic level, to be reinforced by behavioral adjustments, at least some of which are anticipatory.

For simplicity, consider a system with one central thermostat and assume that the environmental temperature is near the midpoint of thermoneutrality. If heat loss runs ahead of heat gain, body temperature falls. This triggers the thermostat (actually a center in the hypothalamus), that activates the machinery for minimizing heat loss. If heat loss continues to exceed heat production and heat loss has already been reduced to a minimum, machinery for augmenting heat production is progressively activated until body temperature is returned to normal. If heat gain runs ahead of heat loss, body temperature rises. This triggers the thermostat that

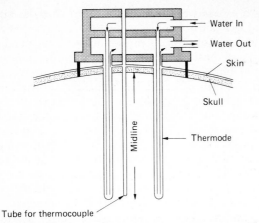

Figure 8-20
Diagram of thermode used to heat and cool the preoptic nuclei in the hypothalamus. [*Modified from H. C. Heller and H. T. Hammel.* Comp. Biochem. Physiol., **41A**, *349–359 (1972).*]

In very small endotherms, present evidence suggests that peripheral temperature sensors have relatively little effect on rates of heat production whereas even slight changes of the temperature of the thermostat in the central nervous system evoke major changes in metabolic rate. Thus, the peripheral sensors can be ignored and a simple thermoregulatory model with only a central thermostat can be experimentally tested. In mammals the central regulator of body temperature is located in or near the preoptic nuclei of the hypothalamus (see Chapter 11). The temperature in this region of the brain can be experimentally manipulated by surgically implanting tiny thermodes (Figure 8-20) made of stainless steel tubing through which warmed or cooled water can be pumped. When hypothalamic temperature, T_{hy}, is increased above the set point, T_{set}, heat loss responses are evoked. When it is reduced below the set point, heat production and heat retention responses are evoked. These responses are summarized in the equation

$$R - R_0 = \alpha(T_{hy} - T_{set})$$

where R_0 is the thermoregulatory response when $T_{hy} = T_{set}$, R is the response when T_{hy} differs from

activates machinery for increasing heat loss. If heat gain continues to exceed heat loss when thermal conductance is maximal, devices for evaporative cooling (either sweating or enhanced respiratory water loss) are mobilized and body temperature is returned to normal.

Figure 8-21
The response of energy metabolism in a kangaroo rat to artificially controlled hypothalamic temperatures. See text for details. [*Modified from S. F. Glotzbach and H. C. Heller.* Am. J. Physiol., **228**, *1880–1886 (1975).*]

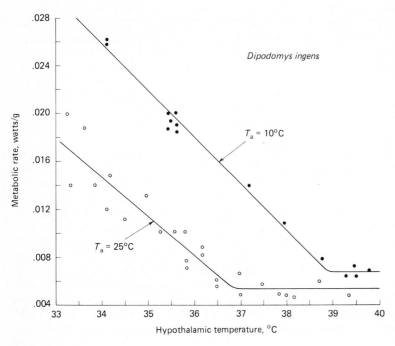

T_{set}, and α is a proportionality constant; α is positive for heat loss responses and negative for responses involving heat production and retention.

The general validity of this model has been established in a number of species and is shown with particular clarity in a kangaroo rat, *Dipodomys ingens*, a nonhibernating rodent with a normal deep body temperature near 36°C. In this species oxygen consumption increases rapidly and linearly with decrease in hypothalamic temperature below the threshold temperature (T_{set}) which is somewhat higher at low ambient temperatures than at high ambient temperatures (Figure 8-21). In this species the sensitivity of the hypothalamus to changes in temperature disappears almost completely when the animals are asleep at an ambient temperature of 30° (about 6° below normal body temperature).

Most birds that have been studied also have a central thermoregulatory center in the spinal cord as well as one in the hypothalamus, with the spinal center being the more important with regard to control of deep body temperature. Further complications will not be pursued, but the complexity of the interactions are readily appreciated if one considers the potential interactions among even a minimal list of control elements such as presented above. When some of the complicating factors considered in the following sections are fitted into the thermoregulatory process, the situation becomes staggeringly complex.

8.9 ADAPTATIONS OF ENDOTHERMS TO LOW TEMPERATURES

Organisms are functionally indivisible. The conventional compartments of morphology, physiology, behavior, genetics, and all the myriads of subdivisions do not exist in animals but only in the minds of investigators, students, and university administrators. Each of the subdivisions is only an arbitrary aspect of the organism, defined and maintained for the convenience of human beings. It is the organism as a whole that deals with the physical environment, so in examining a topic such as "adaptations to low temperature," there is no justification for distinguishing between physiology, morphology, and behavior. As students of organisms we must deal with all three, for they form a single operational system.

From what you have read here, and from your information from other sources, the main ways in which it is possible for a homeotherm to adapt to low temperatures should be obvious.

1. The most direct method is to decrease conductance by increasing the effectiveness of insulative mechanisms. The diminution in thermal conductance will result in a significant downward extension of the lower critical temperature. This, in turn, means that the metabolic cost of living in a cold environment will be minimized or, in some cases, virtually eliminated. As discussed later, such a decrease in conductance can involve seasonal morphological changes that are under endocrine control, short-term changes in cardiovascular patterns, and of course, changes in posture and other more complex patterns of behavior.

2. A second kind of adjustment involves the augmentation of capacity to increase heat production. This offsets heat loss at very low temperatures after further reductions in conductance are no longer possible. In the final analysis such augmentation in heat production must involve the control mechanisms of intermediary metabolism and may be on a permanent or a seasonal basis.

3. Many animals adjust to low temperatures in an essentially negative manner, for example, by avoidance. Such avoidance responses are exceedingly varied and often involve highly complex patterns of behavior. They may involve long-distance movements such as the migration of birds, bats, marine mammals, and some of the large ungulates, with all the attendant complications of timing, orientation, navigation, and habitat exploitation. They may involve construction of shelters, the net effect of which is to decrease an animal's effective thermal conductance, or they may involve changes in patterns of habitat utilization that minimize heat loss.

4. Finally, many mammals, and a number of birds, minimize the entire problem by seasonal and/or daily relaxation of thermal homeostasis so that they allow body temperature to drop 20–35°C below the levels characteristic of normal activity. The details of this heterothermy are incompletely understood, but it is best thought of as a special adjustment of energy metabolism

rather than as inadequate homeothermy (see Section 8.12).

8.9.1 Conductance and the Lower Critical Temperature

The problem of expanding the zone of thermal neutrality by reducing the lower critical temperature is essentially one of decreasing thermal conductance [C in equations (8-3), (8-4), and (8-5)], or, in more familiar operational terms, increasing the effectiveness of the insulative shell so as to decrease the rate of heat flow from body core to environment.

As discussed in Section 3.7.2, in virtually all groups of organisms, mass-specific energy metabolism decreases as body size increases. In homeotherms there are no major regular trends in the level of core body temperature with size; the slight reduction of body temperature in large birds is not significant in this context. Therefore, other things being equal (which they rarely are in biology), it is obvious that large birds and mammals must have lower mass-specific conductances than small ones. Therefore, it should be less energetically expensive per kilogram for a large homeotherm to live in a low temperature than it would be for a small one to do so. This relationship is the rational basis of *Bergmann's rule*, the earliest and easily the most durable of the various "bioclimatic laws" that were promulgated during the nineteenth century. Bergmann's rule, which is philosophically related to the surface law (see Section 3.7.3), states that, in any closely related group of birds or mammals, individuals from populations that live in areas characterized by low temperatures will be larger than those which live in areas of higher temperature.

This generalization has attracted argument from the time it was promulgated a century and a quarter ago. Although it has been repeatedly demonstrated that mean body size in different populations within species and genera often shows a negative correlation with environmental temperature, the assignment of causality for this situation to thermoregulatory relations has remained almost exclusively inferential (and therefore controversial). As more data have become available on the heat exchanges of closely related endotherms from widely different climates, it is clear that Bergmann's rule can be physiologically validated at least in some groups

of small body size that do not depend extensively on evaporative cooling for thermoregulation at high temperatures. Wood rats of the genus *Neotoma* offer one of the best documented examples. In these animals body size is inversely related to mean environmental temperature, and lethal ambient temperature and specific conductance show a positive correlation with mean environmental temperature. The relationship described by Bergmann's rule, however, is by no means all pervasive. For it to be clearly expressed within a taxon, other patterns of climatic adaptation primarily involving physiology and behavior must be of minor importance; this is rarely the case. In any event, it is clear that not all birds and mammals that live under cold climatic conditions are large. Therefore, there must be ways of circumventing the relationship between size and thermal conductance. In fact, the circumvention of this relationship is a major evolutionary and ecological theme that runs through adaptation to cold environments.

An obvious way to decrease thermal conductance is to increase the quantity and insulative quality of fur and feathers. That homeotherms in cold climates have thick, fluffy pelage and plumage, particularly during winter, is too familiar to require extended discussion, but some of the corollaries of this familiar situation merit examination. The insulative effectiveness of pelage and plumage increases with thickness, but there is obviously a practical limit to how thick the fur or feathers of a small homeotherm can be. It is not physically possible for a small endotherm to carry as thick an insulation as a large one (consider a mouse and a bear, a chickadee and a goose).

Seasonal changes in insulative effectiveness of the fur of arctic mammals range from 12 to 52%, the magnitude of change being greatest in the larger species. We can therefore assume that primary dependence on the insulative effectiveness of the exoskeleton is possible only for relatively large homeotherms—probably foxes and ptarmigans are near the minima for mammals and birds, respectively.

There is also a size above which insulation becomes of reduced importance. The total heat production of very large animals is so great in relation to their surface/volume ratio that prevention of heat loss ceases to be a major problem. Consequently an insulative cover of hair, fur, or wool becomes of limited significance with regard to the control of heat loss. Whales and porpoises are completely naked and so represent an extreme case (see Section 8.9.7). A more homely example is offered by pigs. Bare-skinned swine reared in Alaska have a lower critical temperature of approximately 0°C. They minimize their thermal conductance by using the superficial parts of their body as an insulative shell. They allow the temperature of the skin and subcutaneous tissue to fall to near freezing and their own cold shell acts as a barrier to heat loss (see the next section). They prevent overcooling of the shell by local and phasic relaxation of vasoconstriction, which temporarily warms the surface and prevents damage by low temperature.

8.9.2 Regional Heterothermy

It is mechanically impossible to insulate all parts of the body of even a large homeotherm with equal effectiveness. The legs, tails, ears, eyes, and muzzle of mammals, and the legs, wings, beak, and eyes of birds obviously cannot have insulation as effective as that carried on the thorax (Figure 8-22). If poorly insulated peripheral structures are not to be major sources of heat loss, only one adjustment is possible. They must be allowed to fall to a temperature much lower than that of the core, for the rate of heat loss depends on the difference between body surface and environment.

Such regional hypothermia is inescapable from the point of view of thermodynamics, but its achievement in a biological system presents many difficult problems. For example, the tissues involved must be able to function at temperatures near freezing, and they must still remain under thermoregulatory control so freezing can be avoided. Clearly, peripheral heterothermy involves a whole suite of physiological adjustments the delineation of which could involve many lifetimes of experimentation at the level of tissues and cells. Some of the exciting facets of the phenomenon of regional heterothermy that have received at least preliminary investigation are enumerated:

1. In herring gulls *(Larus argentatus)* acclimated to environmental temperatures between −1.0 and 6.0°C, core body temperature was 38–41°C, that of the tibia (feathered) was 33–41°C, and that of the naked tarsometatarsus was 6–13°C. The

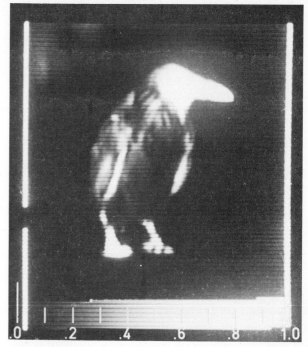

Figure 8-22
Black and white copy of an infrared radiogram showing the variations in surface temperature of a raven (Corvus corax) in an ambient temperature of 18°C. The lighter the shade, the higher the temperature. Each division along the horizontal axis represents a difference of 2°C. [Courtesy of J. Veghte.]

axons of the tibial nerve extend from the spinal cord to the feet. Thus individual nerve cells were exposed simultaneously to temperatures ranging from 6 to 41°C. *In vitro* measurements of cold resistance of the tibial nerve show that conduction in the feathered tibial sector is blocked at 11.7°C, whereas conduction in the tarsometatarsal sector continues down to temperatures as low as 2.8°C. Thus different parts of the same nerve cells have different temperature sensitivities. Presumably this is a general pattern in aquatic birds that occur at high latitudes; otherwise effective use of the feet would be impossible in the ice water in which they swim.

2. Gulls in winter are unaffected by walking on ice that may be as cold as −30°C, but if acclimated to warm laboratory conditions and then allowed to walk on ice their feet freeze. Thus

there must be marked seasonal adjustment in the blood circulation and fluid constituents of the tissues of the feet.

3. Fat from the soft tissues of the distal parts of the legs of boreal mammals such as the Alaskan red fox *(Vulpes vulpes)* and the caribou *(Rangifer tarandus)*, has a melting point that is more than 30°C lower than that of fat from core areas. The fat in the footpads and hooves melts at 0°C or less. The softness of the fat in the extremities, even when severely chilled, apparently allows the feet to remain supple and flexible despite low operating temperatures. The decrease in melting points correlates with shift in the fatty acid composition distally along the legs. A similar diminution in melting point has been found in the legs of some species that live in temperate and tropical areas, so the phenomenon should probably be thought of as a general mammalian adaptation to regional heterothermy rather than as a specific biochemical adjustment to extremely cold climates.

4. The dark coloration of the paws, nose, tips of the ears, and end of the tail found in many mammals does not develop if the animals are reared in a warm environment where the peripheral tissues are not subject to low temperatures.

5. *In vitro* cultures of cells from the skin of harbor seals *(Phoca vitulina)* and Steller sea lions *(Eumetopias jubata)* survive storage at 4°C for as long as 6 months, which is as much as 10 times longer than the length of survival of cell cultures from human skin and lung under the same conditions. It is possible that this tolerance of prolonged hypothermia by the cells of the marine mammals is adaptive since these animals live for months at a time in water near freezing. It is of interest that the seal skin cells do not undergo mitoses at culture temperatures less than 17°C. This poses some interesting probems for the repair of superficial wounds in these animals.

8.9.3 Countercurrent Heat Exchange

The problem of allowing peripheral cooling, particularly of the appendages, and at the same time maintaining an adequate blood supply without excessive heat loss has been met by the development of a pattern of geometric arrangement of blood vessels that allows countercurrent heat exchange.

Countercurrent exchangers have evolved many times and exist in a variety of configurations that serve a variety of functions. For example, they occur in the gills and swim bladders of fishes, in the kidneys of most vertebrates, and in the appendages of a variety of homeotherms including man (see Chapter 5). The underlying physical principle of a countercurrent system is shown with particular clarity in regard to heat exchange in the thermoregulation of birds and mammals. Countercurrent heat exchangers are commonly used by engineers. A simple and familiar case is the use of exhaust gases from a boiler to preheat incoming air prior to its introduction into the combustion chamber, thus saving heat that otherwise would be lost.

The superficial parts of homeotherms that live in cold climates or are adapted to aquatic life are often 30°C or more lower than core temperature. Nevertheless these hypothermic areas and structures remain functional and must be kept adequately supplied with blood. Obviously if the blood supplying these areas started out at 37°C and was then chilled to 5°C or less in the peripheral areas and subsequently returned to the core, the rate of heat loss would be excessive and thermal homeostasis would be lost. This awkward situation does not develop because of the geometry of the vascular system supplying these hypothermic regions.

A schematic countercurrent exchanger is shown diagrammatically in Figure 5-8. In the context of thermoregulation its essential feature is that its geometry allows the directions of heat flow and blood flow to be uncoupled. The heat in the arterial blood is shunted into the venous blood and carried back into the core instead of being transported to the surface of the shell. Artery and vein are in contact with each other. Arterial blood at a high temperature flows from the core toward the periphery, whereas venous blood at a low temperature flows in the opposite direction from the periphery toward the core. Heat moves by conduction from regions of high temperature to regions of low temperature. The artery and vein, which lie side by side, contain blood of different temperatures. Consequently, heat moves laterally from the warm arterial blood to the cool venous blood and is shunted back into the core without ever reaching the periphery.

The effectiveness of a vascular countercurrent

heat exchanger can be enhanced by its anatomic configuration. For example, in the flippers and flukes of porpoises the major arteries are completely surrounded by venous channels, allowing a maximum surface for heat exchange at right angles to the direction of blood flow. However, there are alternative routes over which venous blood can return. The flippers are equipped with separate superficial veins that are physically remote from the arteries. The existence of this alternative route for venous return allows the countercurrent system to be bypassed when it is necessary for the animal to lose metabolic heat to the environment. By appropriate vasomotor adjustments the volume of blood returned through each of the venous systems can be regulated and precise differential control of rate of heat loss can be attained.

In addition, many homeotherms have complex *retia mirabilia* in the proximal parts of their appendages (see Chapter 5). These retia are complex, tangled, and intertwined bundles of arteries and veins, which, whatever else they do, necessarily function as countercurrent heat exchangers that sequester metabolic heat in the trunk and prevent its loss to the periphery.

Countercurrent heat exchangers are not confined to endotherms that live in the water or in the Arctic. Many tropical and warm-temperature terrestrial mammals have well-developed retia in their appendages. Instead of a clearly defined femoral or brachial artery, in some mammals the vascular supply to the limbs consists of a complex network of small arteries and veins grouped together in a bundle. Bundles of this sort are found in monotremes, several edentates (sloth, armadillo, and anteater), and in a primitive primate, the loris. If the foot of a sloth is chilled in ice, the heat exchange between afferent and efferent blood flow is so efficient that, even after more than 1 hr, no measurable cooling occurs above the elbow. In such a case the longitudinal temperature gradient along the arteries is as much as 1.5°C/cm. Humans have no such elaborate retia in their limbs, but many of the arteries in our arms have pairs of veins closely appressed to them in an arrangement suitable for countercurrent heat exchange. Under conditions of moderate cold stress, temperature gradients of 0.3°C/cm have been observed along the radial artery in man.

8.9.4 Increased Heat Production

Although probably present to a greater or lesser extent at all times in every endotherm and conspicuously present in those species exposed to cold stress, regional heterothermy and insulation are insufficient to account for the adjustment of most endotherms to low temperature. Even though some mammals, such as the Arctic fox *(Alopex lagopus)*, which has a lower critical temperature of −40°C or less, show a negligible metabolic response even to very low temperatures, most endotherms can maintain a high core body temperature in low ambient temperatures only by greatly increasing metabolic heat production.

As an index to the degree of adaptation to low temperatures, it is possible to use (1) the temperature of the lower limit of the thermal neutral zone (T_{lc}), and (2) the rate of increase of energy metabolism with decreasing temperature below the thermal neutral zone. The lower the lower critical temperature and the flatter the slope of the increase of metabolism with decreasing temperature, the better the adaptation to low temperatures. The slope of the curve relating heat production to ambient temperatures below the thermal neutral zone often extrapolates to a temperature equal to the core body temperature, so items (1) and (2) are usually correlated—the lower the T_{lc}, the flatter the slope of the heat-production curve. However, let us momentarily ignore climatic adaptations and examine the distribution of these two parameters among endotherms in general.

Because small animals have relatively greater surface areas than than do large ones, mass-specific thermal conductance is inversely related to body size. This relationship can be described by the allometric equation $Y = aX^b$ (see Section 3.7). The exponent for both birds and mammals is near −0.5, but the proportionality constant varies from taxon to taxon. For example, from equations (8-6) through (8-8) it is clear that relative to their size bats (Order Chiroptera) have higher conductances and heteromyid rodents (kangaroo rats and pocket mice) have lower conductances than mammals in general.

All mammals

$$C = 0.76M^{-0.426} \tag{8-6}$$

Order Chiroptera

$$C = 1.54M^{-0.57} \qquad (8\text{-}7)$$

Family Heteromyidae (Order Rodentia)

$$C = 0.62M^{-0.44} \qquad (8\text{-}8)$$

where C is in cm^3O_2 $g^{-1}hr^{-1}$ $°C^{-1}$ and M is grams body mass.

The general relation of body size to the lower critical temperature, T_{lc} in small to medium sized mammals is approximated by the empirical equation $T_b - T_{lc} = 4M^{0.23}$. Because T_b is essentially independent of mass (M) in mammals, as mass decreases T_{lc} approaches T_b. For a mouse with a mass of 20 g, the calculated value of T_{lc} is 29°C—which is far above the average temperature encountered by a small nocturnal mammal even in the tropics, let alone in the Arctic during winter.

Once a general pattern has been established to give a base for reference, the deviations of individual species become of particular interest to the comparative physiologist because they are often indicative of special adaptations or special physiological capacities.

In this regard, allometric relationships, such as those shown for mass-specific thermal conductance, are particularly useful. Size based analysis identifies functional attributes that are primarily the outcome of body mass and distinguishes them from those that are related to the ecological circumstances to which an animal is adapted. It is convenient to think of the former as "first order evolutionary responses," related primarily to physiological mechanism, and the latter as "second order evolutionary responses," which represent adjustment of a physiological mechanism to specific ecological situations or modes of life. Most mammals have conductances that lie close to the regression line that scales this attribute to body mass; their conductances represent first order responses. A few mammals such as flying foxes and the naked mole rat (Figure 8-23) have conductances that diverge widely from the values predicted on the basis of body mass; their conductances represent second order responses—that is, special adaptations to their modes of life.

If physiological adaptation to climate were simple, ideally birds and mammals should show a graded series that correlates with the physical conditions they encounter under natural conditions with regard to these two parameters. Such a situation is not always apparent, but it can be demonstrated if appropriate examples are selected. For clarity and simplicity we shall consider only a few clear-cut examples (Figure 8-23). Both the Arctic fox *(Alopex lagopus)* and the Alaskan red fox are active on the surface in the intensely cold Arctic winter. They are extremely well insulated and have a marked tolerance of peripheral hypothermia. Both can maintain core temperature at 37°C with no increase in endogenous heat production even in the face of extremely low environmental temperatures. The thermal neutral zone of the red fox extends to −10°C and its metabolic rate increases only slowly with decreasing environmental temperatures below its lower critical temperature. The thermal neutral zone of the Arctic fox extends well below −40°C. It is probably unstressed by any temperatures it ever encounters. At the opposite extreme is the flying fox *(Dobsonia minor)* a nocturnal, nectar-feeding, tropical species of the Megachiroptera that lives in the lowlands of New Guinea where it never contends with extremes of temperature, either high or low. It has limited control of its conductance and its thermal neutral zone extends only from about 33 to 35°C. Both above and below this restricted range oxygen consumption increases rapidly with changes in environmental temperature.

The naked mole rat *(Heterocephalus glaber)* is a special case. This aberrant, colonial, burrowing rodent is completely fossorial, completely naked, and maintains its body temperature at an unusually low level near 32°C. It is confined to the drier parts of equatorial Africa. Its BMR is less than half that of most placentals of its size and it has very limited capacity for homeothermy. It is not surprising, therefore, that its lower critical temperature is near 31°C and that its oxygen consumption rises precipitously at all temperatures below this point.

The eland *(Taurotragus oryx)* is the largest of the African antelopes (weight 100–500 kg) and occurs widely in the savannahs of tropical East Africa. Despite its very large size, its lower critical temperature is between 20 and 25°C and it has an extremely high conductance. Indeed its lower critical temperature and the slope of its oxygen consumption below the thermal neutral zone are remarkably sim-

Figure 8-23

Response of metabolic rate to environmental temperature in birds (broken lines) and mammals (solid lines) adapted to different climates. To facilitate comparisons metabolic rates are plotted as percent of basal. [Sources of data: Arctic fox, P. F. Scholander, et al. Biol. Bull., 99, 237–258 (1950); red fox, L. Irving, et al. Physiol. Zool., 28, 173–185 (1955); ptarmigan, R. E. Johnson, Comp. Biochem. Physiol., 24, 1003–1014 (1968); eland, C. R. Taylor and C. P. Lyman. Physiol. Zool., 40, 280–295 (1967); crowned sparrow, J. R. King. Comp. Biochem. Physiol., 12, 13–24 (1964); frogmouth, R. C. Lasiewski, et al. Condor, 72, 332–338 (1970); Dobsonia, G. A. Bartholomew, et al. Z. Vergl. Physiol., 70, 196–209 (1970); mole rat, B. K. McNab. Ecology, 47, 712–733 (1966).]

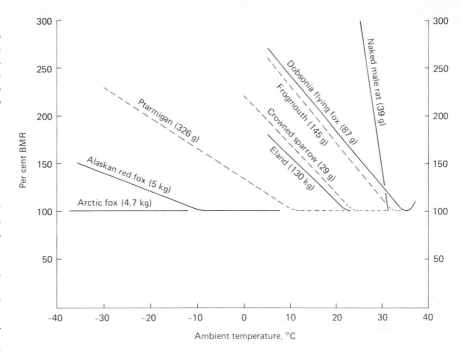

ilar to those of the white-crowned sparrow which weighs only about 1/4000 as much. The oryx (*Oryx beisa*), another large antelope of tropical Africa, is very similar to the eland with regard to lower critical temperature and conductance. The primary thermoregulatory adjustments of both these antelopes are to high environmental temperatures and seasonal aridity (see Section 8.10.3).

The metabolic response of the birds shown in Figure 8-23 parallels that of the mammals discussed. The white-tailed ptarmigan (*Lagopus leucurus*) occurs in the Arctic and in alpine areas of western North America where it is active throughout the winter. These birds are very well insulated and their feathers cover even their legs and feet. Their thermal neutral zone extends to 10°C and even at −30°C they need only increase oxygen consumption to about twice BMR to remain homeothermic. They avoid the most extreme winter conditions by burrowing beneath the snow. The little papuan frogmouth (*Podargus ocellatus*) lives in New Guinea and the tropical parts of northern Australia and never contends with low temperatures. Like other

members of the order Caprimulgiformes (see Section 3.8) its basal metabolism is much lower than predicted on the basis of body size. This low BMR (40% less than most birds of its size) reduces endogenous heat production and minimizes the amount of heat that must be dissipated during thermoregulation in high ambient temperatures. The white-crowned sparrow (*Zonotrichia leucophrys*) can be considered typical of small birds of cool temperature regions. The population shown in the figure breeds in Alaska and winters in the state of Washington. Its metabolic responses to temperature are intermediate between the tropical and boreal species shown.

All of the data on which Figure 8-23 is based, and most of the data on conductance in the literature, are difficult to place in an adequate ecological context because the information relates to the performance of fasting animals at rest under controlled conditions. For satisfactory interpretation, it is obviously necessary to know the magnitude of conductance under specified conditions of activity as well as specified environmental conditions. Such data are exceedingly scarce. However, in at least one mammal, a chipmunk *(Eutamias merriami)*, it has been found that running causes conductance to increase to more than twice the resting level (Figure 8-24).

Patterns of metabolic adaptation to temperature similar to those in mammals have been abundantly demonstrated in birds. Because of their aerial habits, birds cannot avoid stressfully low temperatures as readily as can small mammals. Many birds live much of their lives in environmental temperatures that are significantly below their thermal neutral zones. The evening grosbeak *(Hesperiphona vespertina)*, for example, a common winter bird in the northern United States and southern Canada, has a lower critical temperature of approximately 16°C, which is far higher than the environmental temperatures in which it is normally active during the winter months. This bird, and presumably most other species, adjusts to short-term decreases in

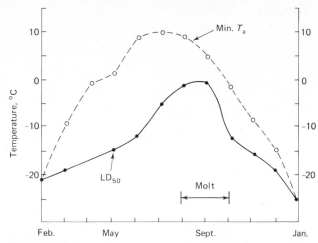

Figure 8-25
Limit of temperature tolerance (LD_{50}) of the house sparrow, and minimum air temperatures during a year in central Illinois, U.S.A. [*Data from L. B. Barnett.* Comp. Biochem. Physiol., *33, 559–578 (1970).*]

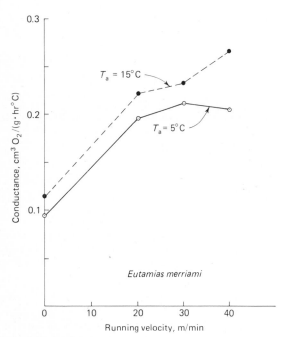

Figure 8-24
The effects of locomotor activity on mass-specific conductance in Merriam's chipmunk. [*Data from B. A. Wunder,* Comp. Biochem. Physiol. *33, 821–836 (1970).*]

environmental temperature by increasing its rate of metabolic heat production; but most birds, like other vertebrates, accommodate to the long-term seasonal changes in temperature by energetically less extravagant means. The most obvious of these is an increase in the effectiveness of insulation (that is, a decrease in conductance) prior to the onset of winter. In a population of house sparrows *(Passer domesticus)*, for example, weight of body feathers increased 70% as a result of the fall molt, and the lower limit of temperature tolerance (defined in this case as LD_{50}, the minimum temperature at which 50% of the birds could survive) was −25°C in January but rose gradually to 0°C by late summer (Figure 8-25).

The seasonal change in the capacity of American goldfinches *(Spinus tristis)* to tolerate low ambient temperatures is even more remarkable (Figure 8-26a). Despite their small size (12–17 g), during the winter goldfinches freshly captured in Michigan remained homeothermic for 8 hr while perching quietly in temperatures of −70°C. Their aerobic metabolism held steady at about 300 cm³ O_2/hr— about 6 times BMR—until their fuel reserves were depleted. In the summer, however, the capacities of freshly captured birds for elevating aerobic me-

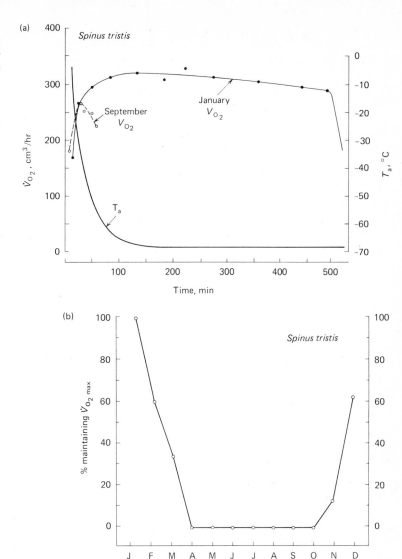

Figure 8-26
(a) The response of the aerobic metabolism of individual American goldfinches to very low ambient temperatures on the day after capture. (b) Seasonal changes in the maximum aerobic capacity of American goldfinches freshly captured in Ann Arbor, Michigan. [Data, courtesy of W. R. Dawson and Cynthia Carey.]

tabolism were much reduced. In less than an hour they became hypothermic at ambient temperatures 50°C higher than tolerated in winter by birds from the same population. The seasonal changes in cold tolerance are not closely associated with seasonal changes in plumage. The times of molt do not correspond with the times when the birds experience the change in aerobic metabolic capacity (Figure 8-26b).

8.9.5 Exercise, Shivering, and Nonshivering Thermogenesis

The most conspicuous mechanism by which endotherms increase heat production is the augmentation of muscular activity, either by locomotion or by shivering. The familiarity of all human beings with both of these activities makes a prolonged general discussion unnecessary, but several aspects of this topic will be considered. It is a familiar experience that a warmly dressed person can readily become overheated, even in severely cold weather, by sustained vigorous exercise. Indeed, sweating inside of one's clothes can be a serious hazard in cold weather. Yet the data presently available indicate that, in contrast to the situation in large mammals such as man, in small mammals sustained vigorous exercise does not necessarily make available supplementary heat that is of thermoregulatory significance. Several factors contribute to this paradox. Exercise eliminates shivering so that the quantities of heat produced by shivering and by exercise are not additive. The movements used in locomotion probably disturb the insulative integrity of the pelage sufficiently that the heat produced by the muscular activity is not retained. The increased peripheral circulation that accompanies vigorous exercise also increases the rate of heat loss.

Shivering need not be visible for it to increase the rate of heat production. Muscle tremors too slight to be visible to the human eye are clearly revealed by a great increase in electrical activity that can readily be picked up by either surface or intramuscular electrodes and displayed as an electromyogram on an oscilloscope.

Shivering can be evoked either by a slight decrease in core temperature or by a decrease in skin temperature. Thus its control can be either central or peripheral or both. The conspicuousness and importance of shivering varies from animal to animal. Among mammals, the most obvious shiverers are probably the fruit bats or flying foxes of the Old World (order Chiroptera; suborder Megachiroptera). Some flying foxes of the genus *Pteropus* shiver strongly and visibly at all temperatures below their thermal neutral zone. They shiver in cool environments even though their body temperature may be kept several tenths of a degree centigrade above the normal resting level by so doing.

Nonshivering Thermogenesis. It has been demonstrated in several kinds of mammals that heat production can be augmented without increased muscular activity or muscle tonus. All living tissues, of course, produce some heat, and the higher the metabolic rate the more the heat produced. It is, therefore, of interest that in birds the status of nonshivering thermogenesis in thermoregulatory adjustments at low ambient temperatures remains uncertain. It may exist in hatchlings, but neither its site nor its mechanisms have been clearly identified. Birds, of course, shiver strongly and conspicuously, and shivering is the predominant means of thermogenesis. As in mammals, the onset of shivering in birds may be either centrally or periphally controlled, or both.

Nonshivering thermogenesis plays an important role in the acclimation of mammals to low temperatures and in arousal from hibernation and daily torpor (see Section 8.12). In laboratory rats acclimated to 30°C, exercise can substitute for shivering as a heat source at ambient temperatures above 10°C, but at temperatures below 10°C the amount of heat produced by sustained, forced exercise is insufficient to replace heat loss and hypothermia results. However, when acclimated to 6°C, the level of nonshivering thermogenesis is elevated sufficiently that forced exercise can replace shivering as a heat source down to temperatures as low as −20°C. It is clear that the capacity for nonshivering thermogenesis acquired during acclimation to cold allows rats to be active at much lower temperatures than would be the case if they had to depend on shivering alone, for exercise eliminates shivering but does not affect nonshivering thermogenesis. Exercise and nonshivering thermogenesis are additive; exercise and shivering are not. Nonshivering thermogenesis also significantly increases metabolic scope for activity. In cold-acclimated rats with a high level of nonshivering thermogenesis, the metabolic scope below 30°C is almost twice that of warm-acclimated rats at the same temperature.

It is reasonable to assume that the increase in nonshivering thermogenesis associated with cold adaptation in mammals involves shifts in the pathways and relationships of intermediary metabolism. Although the situation is still incompletely understood, it seems probable that cold-acclimated rats have increased capacity for the formation and utilization of glucose. It also seems clear that some nonshivering calorigenesis is evoked in part through hormonal stimulation of lipid catabolism. For example, the calorigenic response to injection of L-norepinephrine in cold-acclimated rats involves a marked increase in oxygen consumption and a rise in body temperature. Both these responses can be related to an increased capacity of the liver to form and mobilize nonesterified fatty acids. The thyroid hormone, thyroxin, plays an important role in the long-term development of mammalian nonshivering thermogenesis.

One of the important sites of nonshivering thermogenesis is brown fat. This highly vascular, multilocular, brown adipose tissue occurs in the young of most species of mammals and is particularly well developed in mammals that hibernate. It usually increases in mass during chronic exposure to low environmental temperatures. The *in vitro* oxygen consumption of brown fat is high relative to that of other mammalian tissues. The deposits of brown fat are localized around the neck, thorax, and major blood vessels in such a way that heat generated in them is transported to the heart and brain, the temperatures of which are, of course, particularly critical. The heat production of brown fat is shown with particular clarity in relation to daily torpor and hibernation (see Section 8.1.2).

8.9.6 Facilitation of Heat Loss

Adequate thermoregulation for a large endotherm living in a cold climate is not just a matter of augmenting heat production, maintaining a broad zone of thermal neutrality, and achieving a low lower critical temperature. An effective means of rapidly increasing heat loss is also of critical importance. Otherwise under conditions of high metabolic heat production, such as sustained vigorous muscular activity, heat could accumulate in the animal to such an extent that death, or damage from overheating, could occur.

That such overheating is not just a theoretical possibility is spectacularly demonstrated by adult male northern fur seals, *Callorhinus ursinus*. Bull fur seals, like all mammals that live in the water, face a chronic problem of heat loss to the environment. In this species, the adult males rarely venture south of the Aleutian Islands and are pelagic for two thirds of the year. The difference between their core temperature and water temperature is 30°C or more at all times. This profound difference is offset by a heavy waterproof insulation of fur and a thick subcutaneous layer of fat. This insulation, which is needed for the maintenance of a high core temperature when the bulls are in the water, presents an acute physiological problem when they engage in their intense and prolonged territory-maintaining activities while on shore during the breeding season, even though their breeding grounds in the Pribilof Islands are characterized by almost continuously cloudy skies and air temperatures that rarely rise above 12°C. While they are in the water, the large quantities of heat generated by sustained rapid swimming are readily lost to the environment through the large naked, and richly vascular flippers. However, on land, because of the low specific heat of air, the effectiveness of the flippers as heat exchangers is greatly reduced. Indeed, the activity of the bulls falls off markedly whenever the air temperature rises above 12°C. When the animals are forced to walk overland for any distance, as during sealing operations, when they are herded inland from the rookeries to areas used for sorting and killing, death from heat prostration is commonplace, even though air temperatures remain below 10°C. In fact, so common is death from heat prostration that the Aleut sealers have a special name, "roadskins," for animals that die of heat prostration while being driven overland.

The situation facing large Arctic homeotherms is analogous to that of a man bundled in warm clothing and forced to exercise. Obviously animals living in areas of intense cold must be able to vary their total conductance rapidly and accurately. They must be equipped with thermal windows that can be opened or shut depending on circumstances. The thinly insulated extremities supply such thermal windows. When the peripheral structures are near 0°C they form an effective part of the insulative shell. If they are warmed by being flooded with blood they become devices for heat dissipation. The skin beneath the fur of well-insulated Arctic mammals is probably never more than 10°C below that of the core, whereas the temperature of the appendages may be 35°C or more below that of the core. Consequently when the appendages are flooded with blood during vasodilation they serve effectively for the rapid dissipation of heat.

The requirements for simultaneous possession of good insulation and effective thermal windows faced by a large mammal may be even more extreme at high altitudes than in the Arctic. During the day at high altitudes solar radiation is often intense and air temperatures may be high, but at night high winds and rapid heat loss by radiation to space can produce intensely chilling conditions. The guanaco (*Lama guanacoe*) occurs at altitudes as high as 4500 m in Peru. The back of this wild

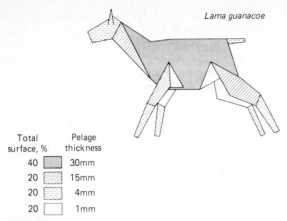

Figure 8-27
Schematic representation of distribution of wool of differing thickness in the guanaco, a South American member of the family Camelidae. The white areas are thermal windows of almost bare skin. [Modified from P. Morrison. J. Mammal., **47,** *18–23 (1966).]*

Total surface, %	Pelage thickness
40	30mm
20	15mm
20	4mm
20	1mm

Lama guanacoe

groups of large fast-swimming fishes—bony fishes of the family Scombridae (mackerel, bonito, skipjack, tuna) and sharks of the family Lamnidae (mako, great white, and porbeagle)—have independently evolved effective systems of regional endothermy in the axial musculature. These systems present remarkably similar solutions to the physically difficult problems of heat loss posed by aquatic endothermy. In both groups the distribution of isotherms in the body is similar to that in Figure 8-28, with the highest temperature, which may be 10° or even 15°C above ambient water temperature, located centrally in the heaviest parts of the body. In both, heat loss at the respiratory surface is minimized by countercurrent heat exchanges in the circulatory system that sequester heat in the deeper parts of the axial muscles and prevent it from reaching the gills where it would be rapidly lost to the water.

relative of the llama, is covered by a thick, tangled mat of wool. The head, neck, and outer sides of the legs carry a thinner covering of wool, and the inner surfaces of the upper legs and the adjacent flanks are almost bare and act as thermal windows that cover almost 20% of the surface of the body (Figure 8-27). By appropriate postural adjustments these thermal windows may be completely closed, or partially to completely open. Even in still air, postural adjustments can cause a fivefold change in thermal conductance. This insulative flexibility underscores the variability of heat transfer across the surface of endotherms and documents one of the inadequacies of the "surface law" discussed in Section 3.7.3.

8.9.7 Aquatic Endothermy—An Extreme Case

Fishes. In Section 8.6 it was pointed out that the high specific heat and low oxygen content of water make it impossible for gill-breathing animals to exert effective control over the temperature of their respiratory surfaces. This in turn makes it difficult for them to maintain a significant difference between their core temperature and that of the water around them for more than short intervals. Nevertheless, representatives of two quite unrelated

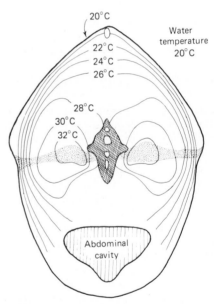

Figure 8-28
Distribution of temperature in the muscle of a freshly captured 70-kg big-eye tuna (Thunnis obesus). Water temperature 20°C. Isotherms are plotted at 2°C intervals. Stippling indicates the distribution of red muscle. [From F. G. Carey and J. M. Teal. Proc. Nat. Acad. Sci. U.S., **56,** *1464–1469 (1966).]*

8 BODY TEMPERATURE AND ENERGY METABOLISM

For simplicity we shall concentrate our description on scombrids, but at least in principle it is applicable to lamnid sharks also. Most scombrids do not use their branchial muscles to pump water over their gills. Instead, they depend on *ram ventilation.* They swim with mouth open and use the forward motion imparted by their locomotor muscles to force water through the mouth and across the gills for respiratory exchange. They must swim continuously in order to respire. Therefore, large amounts of heat are continuously being generated particularly by the red muscle that is used for sustained cruising and which, *in vitro,* consumes oxygen about six times faster than does white muscle (see Chapter 4).

The vascular heat exchangers that retain the heat continuously being produced in the muscles are of two sorts, lateral and central. The lateral heat exchangers consists of segmentally arranged pairs of arteries and veins with associated retia mirabilia (see Chapter 6) that function as countercurrent heat exchangers. Heat generated in the muscles enters the venous blood, and, because the venules are in intimate contact with the arterioles in the rete, the heat is immediately transferred to the cooler arterial blood and carried back into the muscles.

The central vascular heat exchanger consisting of the dorsal aorta, the posterior cardinal vein, and a large rete is located in the haemal arch of the spinal column. The dorsal aorta is enclosed within the posterior cardinal vein and the associated rete so it is surrounded by vessels carrying venous blood warmed by the continuous activity of the red muscles. Heat in the warm venous blood flowing from the muscles is transferred to the cool arterial blood flowing toward the muscles. Some scombrids like the tuna, *Katsuwonus pelamis,* have four sets of lateral exchangers plus a central exchanger, whereas others like the black skipjacks, *Euthynnus lineatus,* have only a central heat exchanger.

In lamnid sharks, the viscera as well as the muscles can be kept warm. A countercurrent heat exchanger consisting of a network of small arteries is located in the hepatic sinus. Venous blood from the liver bathes these arteries and heat is transferred from the venous blood to the arterial blood and is carried back to the viscera. In addition, there is a large venous channel that bypasses this arterial network and is equipped with a sphincter, which

allows it to be shut off. Thus, venous blood can either flow among the small arteries, which lie within the hepatic sinus, or be shunted past them, providing a mechanism for controlling visceral temperature.

The selective advantage of endothermy in these fishes is related, at least in part, to the increased power generation and swimming speeds that it allows to be supported aerobically. Each 10°C rise in temperature increases the aerobic power output available from vertebrate muscle about threefold. However, the capacity for endogenous heat production may offer other benefits as well. In the sea temperatures decrease rapidly with depth. Large, strong-swimming, predatory fishes commonly pass through steep temperature gradients when they ascend or descend. For example, free-swimming swordfishes *(Xiphias gladiator),* to which temperature telemeters have been attached, experience a change of 20°C during a period of 2–3 hr. Unlike tunas, swordfishes do not have elevated muscle temperatures. When fishes are chilled the first indications of malfunction usually appear in the central nervous system, particularly in the brain. Such temperature-induced problems are minimized in swordfishes because they can maintain brain temperature as much as 8°C above that of the surrounding water (and also that of their own axial muscles) by means of local nonshivering thermogenesis in a highly specialized organ that lies immediately beneath the brain. This organ, apparently derived from the extrinsic eye muscles, is functionally analogous to the brown adipose tissue of mammals. It is equipped with vascular heat exchangers that prevent the heat it produces from being lost into the general venous circulation. Blood going to the brain passes through this organ, picks up heat and transfers it to the brain. Additional heat passes from the organ into the brain by conduction.

Mammals and Birds. The evolution and maintenance of effective regional endothermy by scombrids, swordfishes, and some sharks, despite dependence on aquatic respiration, is one of the adaptive peaks of thermoregulatory virtuosity; but the precision of thermal homeostasis achieved by aquatic birds and mammals, particularly boreal species, is much greater. They hold core temperatures close to the high and uniform levels characteristic of their terrestrial relatives even when this requires the maintenance of a temperature differential of 35–40°C between themselves and the surrounding water.

Water conducts heat 20–26 times as rapidly as air. However, a submerged animal does not lose heat 20–26 times faster in water than in air because heat exchange between animal and environment occurs only at the surface between the two. The limiting factor thus becomes the rate of heat flow in the animal from core to surface, and this, in endotherms and some ectotherms, is amenable to physiological control. The conductance of small mammals, such as mice and shrews, is 5–10 times greater in water than in air; in seals the ratio is about 2:1. Nevertheless, water that is near 0°C—a seasonal commonplace in many regions and a permanent state in Arctic, Antarctic, and at great depths in the sea—presents problems of impossible difficulty for most birds and mammals. For example, a human being submerged in ice water can remain in thermal homeostasis for only a few minutes, and incapacitating hypothermia develops in 15 or 20 min. Nevertheless, some seals spend most of their lives, and some porpoises spend their entire lives, in ice water—and they are of the same order of size as humans. Even more extreme cases are presented by the muskrat (*Ondatra zibethicus*), which, despite its small size, is active during winter in ice water, and particularly by the water shrew (*Sorex palustris*), which weighs only 10–15 g, but even in winter forages underwater in temperatures at or near freezing. These two small semiaquatic species manage this difficult feat by entrapping air in their fur. Consequently, except for the naked areas on feet, snout, and tail, their skin is never wetted and they are completely surrounded by a layer of air that insulates them from the water. The same is true of virtually all aquatic birds, otters (*Lutra*), sea otters (*Enhydra*), and fur seals (*Callorhinus* and *Arctocephalus*). However, many marine mammals, such as walruses (*Odobenus*), seals of the family Phocidae, polar bears, and all cetaceans, are either naked or lack waterproof hair and must depend on other insulating mechanisms for heat retention—either the cooled superficial tissues of their bodies or a layer of still water trapped in the pelage next to the skin.

The pelage of both polar bears and seals sequesters a layer of water next to the skin. In polar bears this layer of still water may be as much as 10 cm thick and contributes significantly to the animal's total insulation. The thermal conductance of still water is about the same as muscle and 3 times that of blubber. In seals the trapped layer of water is only a couple of millimeters thick and is of negligible importance in comparison with their thick layer of subcutaneous blubber. The entire shell of hair seals (phocids), walruses, and cetaceans functions in a manner analogous to the thinly covered or naked appendages of terrestrial homeotherms. For example, the skin temperature of harbor seals (*Phoca vitulina*) varies directly with water temperature down to 0°C (Figure 8-29), so only a negligible

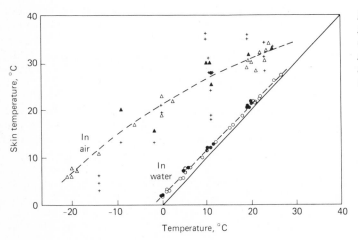

Figure 8-29
Skin temperature of the harbor seal as a function of air temperature and water temperature. Solid symbols are winter measurements. Hollow symbols are summer measurements. [*From J. S. Hart and L. Irving.* Can. J. Zool., **37**, *447–457 (1959).*]

8 BODY TEMPERATURE AND ENERGY METABOLISM

difference exists between the temperatures of environment and body surface. Thus the seal's own tissues form an insulative shell. The effectiveness of this shell of chilled tissue as an insulator is shown by the fact that the harp seal, *Phoca groenlandica*, has a thermal neutral zone in water that extends to 0°C. The extent of peripheral cooling, and consequently the steepness of the temperature gradient from core to surface, is under physiological control. Thus, during exercise the harp seal, unlike the fur seal (discussed in Section 8.9.6), does not have to depend, exclusively on its appendages for loss of excess heat but can also use its main body surface for this purpose by employing either local or general vasodilatation.

The superficial parts of these aquatic endotherms show the most remarkable insensitivity to temperature of any vertebrate tissues. They function adequately and are apparently adequately nourished at temperatures near 0°C. Yet within a few minutes they can be warmed from near 0°C to 35°C by vasodilatation and still function normally. It is safe to say that no aquatic ectotherm has tissues the function of which shows similar temperature independence.

8.10 ADJUSTMENTS OF ENDOTHERMS TO HIGH TEMPERATURES

From the standpoint of thermoregulation, the most difficult of the major terrestrial habitats are the Arctic, Antarctic, high mountaintops, and low-latitude deserts. The thermoregulatory challenge presented to endotherms by the sustained aridity and intense heat of low-latitude deserts is in many respects more acute than that presented by boreal and mountain regions, because the control of heat loss to a cold environment is physiologically easier than the control of heat gain from a hot environment. Normally active birds and mammals have core temperatures between 35°C and 42°C, which is distinctly higher than the air temperatures that prevail in most terrestrial environments. Consequently, mammalian and avian temperature regulation is primarily geared to ecological circumstances in which heat is continually being lost to the environment. Two of the fundamental aspects of thermoregulation are control of the rate of heat loss

to the environment and augmentation of endogenous heat production. These are coupled in such a manner as to prevent any significant change in the amount of heat stored in the interior of a homeotherm's body, because only in this way can a relatively uniform core temperature be maintained. Birds and mammals living in extremely cold environments face only an accentuated form of the general problems of temperature regulation confronted by homeotherms in most situations. Maintenance of a high and uniform body temperature requires merely the quantitative enhancement of the various physiological mechanisms available to all endotherms for controlling heat loss and increasing endogenous heat production. Given adequate resources of food, adaptively effective responses to extreme cold appear to have been readily evolved by both birds and mammals. There are many species in a number of orders of birds and mammals that can maintain adequate levels of core body temperature, however cold the environment.

However, the problems of temperature regulation for mammals and birds living in the heat and drought of low-latitude deserts appear to have been more difficult to solve physiologically. When ambient temperatures exceed body temperature, heat moves from the environment into the animal, not vice versa, and mechanisms for limiting the rate of heat loss become inappropriate for the maintenance of a constant body temperature. Moreover, endotherms produce large amounts of heat, so the problem is not simply one of preventing environmental heat from entering the body; endogenous heat must also be transferred against the thermal gradient from the body to the hot surroundings. Because of the direction of the temperature gradient when environmental temperatures approach, equal, or exceed body temperature, adequate heat loss can be achieved only by evaporation of water. However, deserts are by definition areas where water is in short supply, and the problem of physiological temperature regulation under conditions prevailing in them thereby becomes singularly intractable. Nevertheless, despite the almost overwhelming thermoregulatory demands of severe low-latitude deserts, a number of endotherms have evolved limited solutions to the physiological problems they present. In fact, under appropriate circumstances very large populations of herbivorous mammals and granivorous birds exist in desert regions and form the base for a flourishing fauna that includes many predatory mammals, birds, and reptiles.

8.10.1 Role of Behavior

Before considering the thermoregulatory adjustments of vertebrates to high temperatures, it is important to emphasize that, from the standpoint of animal adaptation, regions of intense heat, such as the low-latitude deserts, are very different from what they appear to be to the casual observer. Most major deserts are far enough away from the equator to have marked seasonal changes in thermal regime; consequently few desert animals have to face stressfully high temperatures every day of the year. Even during the season of highest temperatures there is a marked daily temperature cycle that makes the heat stress periodic rather than continuous. All the physical factors that favor high environmental temperatures during the day also favor rapid heat loss after the sun sets, so that relatively undemanding temperatures exist in most deserts at night. Thus the thermal stress of the desert is ameliorated on both a daily and a seasonal basis. As a result, no vertebrates need to cope continuously with the extreme physical conditions associated with low-latitude deserts.

Vertebrates as a group are well endowed with a complex central nervous system. A functional corollary of this is the capacity for varied behavior in which choice is a conspicuous component. Consequently, they are able to select with remarkable precision the particular aspects of the total environment with which they will contend. They seek out and utilize those special and restricted facets of the total physical environment within which their anatomic and physical attributes can function adequately for survival and reproduction. Consequently, a terrestrial vertebrate can assemble its own environment about itself through behavioral means—an environment fitted with exquisite precision to its physiological capacities.

In view of the diversity of vertebrates that live in climatic regimes characterized by high temperatures, it is to be expected that the environmental challenge posed by severe externally imposed heat loads will have been met by a variety of adaptive patterns. When dealing with complex systems such as those involved in integrative aspects of biological activity, all classifications are inadequate. However, analysis is greatly facilitated by at least informal classification. In the present discussion we shall arbitrarily divide the responses of desert endotherms to thermoregulatory problems into three categories: (1) relaxation of the limits within which homeostatic control is maintained, (2) avoidance of the problem of thermoregulation by behavioral adjustments, and (3) special developments and modifications of structural and functional attributes that are already of general occurrence within the taxa.

The patterns of thermoregulatory adaptation to high environmental temperatures not only vary between endotherms and ectotherms but also between birds and mammals, so it is appropriate to consider the two groups of terrestrial endotherms separately.

Although each species occupies only a restricted sector of the total array of possible environments, a large animal such as a camel or an ostrich lives in a climate that in terms of temperature, humidity, radiation flux, and air movements, is a reasonable approximation of that dealt with by the meteorologist. However, for most animals this is not the case. It is important to bear in mind that the mass of most terrestrial vertebrates is less than 250 g and that of the majority is less than 50 g. For these small creatures standardized meteorological measurements have only indirect relevance. Their universe is one of holes and crevices, tunnels and nests, one where distances are measured in meters rather than in kilometers, one where the differences between winter and summer, night and day, or shadow and sun may represent the difference between life and death.

For most desert endotherms, the fundamental parameters of thermal regulation are not the control of heat loss and gain, but the matching of the physiological control of these two functions to the exigencies of the physical environment by precise orientation in space and the precise patterning of activity in time. Not only can they select from the conspicuous physical variability of deserts both temporally and spatially, but they can greatly increase the availability of favorable physical circumstances by their own efforts—for example, by burrowing beneath the surface of the substratum or by building nests. Obviously any ecologically meaningful understanding of thermoregulatory physiology of an animal living in a difficult environment must involve precise and detailed information about its behavior.

8.10.2 Nocturnal Mammals

The heat gained from the environment by an animal will be roughly proportional to its body surface. Small animals have a greater surface, relative to their mass, than do large ones, so it is obvious that small mammals will gain relatively more heat from the environment than will large ones. As previously indicated, the only way an animal can lose heat against a thermal gradient is through the evaporation of water. One can use body surface as a basis for calculating the amount of water a 50-g kangaroo rat (*Dipodomys*) would have to lose in evaporative cooling to remain in thermal balance while fully exposed to the sun during the daytime in a desert region of the southwestern United States. The figure comes to the startling total of 13% of the animal's body mass per hour. When one realizes that many desert rodents are much smaller than kangaroo rats, the improbability of primary diurnality for small mammals in a hot environment becomes obvious. Nevertheless, all but the most barren of the world's deserts characteristically support substantial populations of small mammals, and most of these are rodents.

If number of species, variety of habitats occupied, and total population of individuals are used as criteria, the order Rodentia is by far the most successful order of mammals. Most rodents are small, nocturnal, and fossorial. In those parts of the world with enough rainfall to support continuous plant cover, nocturnality and burrowing are probably not necessary conditions for allowing rodents to thermoregulate adequately under the conditions of radiation, air and substratum temperatures, and water availability that exist in their environments. Indeed, is it probable that nocturnality and burrowing have their primary ecological significance not in thermoregulation but in minimizing mortality from predation. However, the transfer of the adaptive suite involving the physiology, anatomy, and behavior associated with burrowing and nocturnality to the desert is a beautiful example of the way in which an already established complex can preadapt the members of a mammalian taxon to a demanding environment.

The primary adaptations of rodents for a nocturnal and fossorial life have allowed them to occupy desert regions while remaining in circumstances in which they need cope only with those aspects of the total physical environment for which they are suited physiologically. Paradoxically, an abundant rodent fauna lives in the desert by the simple device of avoiding desert conditions. Their adaptive pattern of nocturnality-plus-burrowing allows them to evade the dilemma of too much heat and insufficient water. They spend the hot daytime hours underground in the relatively equable environment of their burrows and nest chambers and are active on the surface of the ground only after sunset, when temperatures have fallen. Thus they can uncouple their thermoregulation from evaporative cooling. The major physiological parameters of this mode of adjustment to life in the desert are the minimization of water loss incidental to excretion, defecation, and respiration, and the behavioral avoidance of the need to use evaporative cooling for thermoregulatory purposes (see Chapter 7). If water loss is kept at minimal levels and temperature stress is avoided, many small rodents appear capable of obtaining adequate supplies of water, either preformed in their food or produced by the oxidation of hydrogen incidental to their own metabolic processes.

An interesting result of this situation is that for many nocturnal desert rodents, particularly those that occupy areas with relatively severe winters such as the Great Basin of North America, the primary thermoregulatory problem is not coping with high temperatures during the summer but adjusting to very low nighttime temperatures during the winter.

8.10.3 Diurnal Mammals

Very large mammals cannot escape high temperatures by going underground. Although the systems of thermoregulation employed by large mammals exposed to high temperatures vary from species to species, it is safe to say that none of them have patterns of thermoregulation that differ qualitatively from those found in mammals living in more temperate environments. Their adaptations to high temperatures consist of a series of interrelated alterations in the balance between the various physiological processes usually involved in mammalian thermoregulation. The effectiveness of the rebalanced variations of the basic mammalian thermoregulatory system is assured by behavioral adjustments that fit the various patterns to the physical demands of the situation and to the ecological tolerances of the species.

Adaptive Correlates of Large Size to Heat Stress. Although size makes it impossible for large mammals living in hot climates to avoid heat stress, large size has a number of functional correlates that help to offset this constraint. The same factors that tend to minimize the mass-specific heat loss of large mammals in low temperatures tend to minimize mass specific heat gain from the environment under conditions of high temperature. Moreover, mass-specific endogenous heat production is less in large mammals than in small ones. Therefore, the amount of metabolic heat that must be lost to the environment against the thermal gradient per unit mass is inversely related to body size. In addition, when a mammal is sufficiently large, the physiological possibility of heat storage rather than temperature regulation becomes available. Because a very large animal has a high thermal inertia, it can store large quantities of heat with only a slight rise in body temperature. (Effective heat storage is, of course, impractical except for very short periods of time for animals of small size because of their limited capacity for heat storage and their rapid production of heat.)

Perhaps even more important than either of the above is the increased mobility that goes with large size. Large mammals can travel many kilometers to reach surface water. As long as they can find water holes they can visit even once every several days, they are not restricted either to the oxidative water they produce by their own metabolism or to the preformed water they obtain in their food. They can, therefore, afford to employ evaporative cooling to buffer, at least in part, the heat stress under which they frequently operate. However, those large mammals, such as human beings and horses, that closely regulate body temperature and in hot environments must routinely depend on extensive evaporative cooling, cannot operate for more than a few hours or at most a day or so away from free water in hot weather. From this it follows that most successful large desert mammals do not place primary dependence on evaporative cooling in their thermoregulatory processes. Instead they employ evaporative cooling as a subsidiary thermoregulatory process or as a mechanism for meeting crises.

Although few large mammals can escape the heat by going underground, their mobility allows them to reach areas where the externally imposed heat load is minimal. This they do by the obvious stratagem of seeking shade, from vegetation, cliffs, large boulders, or caves. They have an additional and significant resource at their disposal in that their large size allows them to carry with them an effective radiation shield in the form of their pelage.

Pelage and Thermoregulation. Like mammals adjusted to cold climates, those adjusted to hot climates use their hair and fur to minimize the rate of heat flow—in this case, heat flow from the environment into the animal. Hair is nonliving and is undamaged by temperatures that would be injurious or lethal to living tissues. It is therefore unharmed by the heat associated with exposure to direct solar radiation. On the dorsal surface of large ungulates, where the hair is heavy and thick, startlingly high surface temperatures have been reported—as high as 70°C in the dromedary camel and 85°C in the Merino sheep. The core body temperature of a large ungulate approximates 40°C. Therefore, when such an animal is exposed to the sun, an extremely steep temperature gradient exists between the surface of the pelage and the skin, and a shallower gradient of opposite direction will exist between core and skin if sweating occurs.

Maintenance of these two opposing temperature gradients is an essential feature of the thermoregulatory process in a hot environment. The insulation supplied by the pelage, of course, not only retards heat movement from environment to skin but prevents large amounts of the heat associated with incident solar radiation from reaching the mammal's skin. This heat is returned to the environment from the surface of the hair by reradiation and convection. The reduction in surface temperature of the pelage by air movement is very significant and, under conditions of forced convection, as in a light breeze, large quantities of heat are removed from the surface of the pelage. This decreases the steepness of the thermal gradient across the pelage and consequently reduces the heat load on the animal.

The steepness of the gradient from surface of pelage to skin is also affected by the quality of the hair. Typically, at least during the seasons of highest temperatures, many large mammals living in arid regions have sleek, glossy, and light-colored pelages that reflect many of the wavelengths of sunlight and thus reduce solar heating. This is dramatically demonstrated by populations of Grant's gazelle *(Gazella granti)* that live in the intensely hot and dry northern parts of Kenya and Uganda. The light-brown pelage of these animals is so glossy that under some lighting conditions they appear silver in color.

A mammal's pelage is not uniformly distributed. In mammals, such as the guanaco (Figure 8-27), living in regions of intense solar radiation or high air temperatures there are usually sparsely haired, sometimes completely naked areas, in the axilla, in the groin, on the scrotum, and on the mammary glands. Moreover, many members of the order Artiodactyla, regardless of the climate in which they live, have horns, and these are often richly vascularized. In appropriate circumstances any of these naked or incompletely shielded areas can function as thermal windows through which, under conditions of local vasodilatation, some of the heat gained from solar radiation or produced endogenously can be lost by convection and conduction. It is also possible that these thermal windows can also serve as sites of heat loss by reradiation to the shade of the animal's own body or to cool spots within the environment (see Figure 8-1).

The horns of members of the family Bovidae have many functions—defense, intraspecific aggressive behavior, courtship display, species recognition. In addition they have a role in thermoregulation, a topic that has been examined in the domestic goat *(Capra hircus)*. The bony core of the goat's horn is supplied with a rich plexus of blood vessels that vasodilate in response to heat stress and exercise and vasoconstrict in response to cold. In the winter, when a goat's body carries its maximum insulation of wool, the blood vessels of the horns can assist those of the legs in dissipating heat produced during short periods of intense activity. In warm weather the horns can contribute to a loss of endogenous heat through convection. At an ambient temperature of 30°C, a goat with horns of average size can lose about 3% of its basal heat production through them. When T_a is less than T_b the horns also appear to assist in a heat exchange mechanism that selectively cools the brain.

It is of interest that in bovids superficial structures such as the scrotum and udders contain temperature sensors that can initiate general evaporative cooling in response to local heating before the increased heat load imposed by the environment causes a change in the core temperature itself. Thus their superficially located sensors allow them to anticipate developing heat loads and initiate appropriate thermoregulatory activities prior to significant increases in core or brain temperatures.

The effectiveness of the heat shield supplied by integumentary derivatives is obviously reinforced by behavioral means. For example, appropriate orientation towards the sun can minimize the quantity of incident radiation. An animal may stand quietly in one place so that the shade of its body reduces the amount of back-radiation by preventing solar heating of the substratum. Or, an animal may lie or crouch on the ground, thus minimizing the area of its surface exposed to solar radiation and back-radiation. This has the additional advantage of allowing conductive heat loss from the animal's ventral surface to the substratum, which remains relatively cool in the shade of the animal's body.

As mentioned in Sections 3.8, 8.9.6, and 8.9.7, aquatic mammals have a high BMR and are adapted for heat conservation and may experience difficulty in losing heat to the air even at moderate ambient temperatures. It is not surprising, therefore, that most seals and sea lions come ashore to breed either in the Arctic and Antarctic, or on fog-shrouded islands in cool temperate waters. The northern elephant seal *(Mirounga angustirostris)*, one of the largest of the world's seals, is an exception. Its principal rookeries are on the desert islands off the west coast of Mexico where air temperatures are often high and solar radiation can be intense. One of the devices it employs to avoid overheating is sand-throwing. With its flexible front flippers the seal scoops up sand and throws it onto its back, often covering itself with a continuous layer of sand several inches thick. This behavior not only supplies the seal with a shield against solar radiation but it excavates a depression into the cool, moist, deeper layers of sand. The animal lies in the depression, shielded from the sun by a layer of sand

Figure 8-30
A young northern elephant seal responding to heat stress by throwing sand onto its back. [*Photo by F. N. White and D. K. Odell*, J. Mammal., **52,** 758–774 (1972).]

on its back and losing heat by conduction from its chest and belly to the cool moist sand beneath it (Figure 8-30).

Heat Storage and Evaporative Cooling. No matter how effective and elegant from the engineering point of view the thermal shields and thermal windows associated with the structure and pattern of the pelage may be, homeotherms faced with a positive heat load from the environment must either store heat and accept the resultant rise in body temperature or employ water for evaporative cooling. If water is in short supply, extensive use of evaporative cooling can be avoided if an animal relaxes the precision of its homeostatic mechanisms and allows heat to accumulate in its body. A large mammal may have sufficient thermal inertia so that it can safely store much of the excess heat acquired during the day and then lose it by passive heat transfer at night. Although such a process involves tolerances of hyperthermia, it requires no metabolic work. Its only cost is associated with a small

van't Hoff effect caused by an increase of a few degrees centigrade in T_b. Such heat storage with associated hyperthermia during the hours of high heat load is a characteristic feature of the thermal adjustment of the dromedary camel (Figure 8-31).

When the oryx and the eland (two large African antelopes that, as previously mentioned in Section 8.9.4, have surprisingly high lower critical temperatures) are subjected to heat stress, they show patterns of hyperthermia similar to that of the dromedary camel. The oryx, which occurs in some extremely severe desert regions, is particularly noteworthy in this regard. When dehydrated and subjected to heat stress it allows body temperature to rise to remarkably high levels, and it can tolerate a deep body temperature of 45°C for several hours (brain temperature remains lower, as discussed later). Except under extraordinary environmental conditions this extreme hyperthermia keeps body temperature above air temperature. Consequently, the animal can passively lose by convection its endogenous heat and the heat acquired from solar

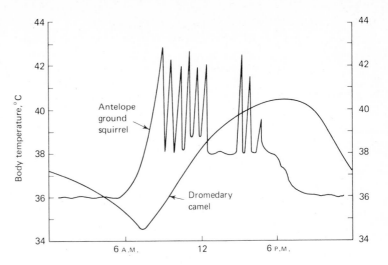

Figure 8-31
Diagrammatic representation of the daily patterns of body temperatures in a large and a small mammal subjected to heat stress under desert conditions. [*From G. A. Bartholomew. Symposia of the Society for Experimental Biology, No. 18, pp. 7–29, Academic Press, Inc., New York, 1964; original data from K. Schmidt-Nielsen, et al. Am. J. Physiol.,* **188,** *103–112 (1957), and J. W. Hudson, Univ. Calif. (Berkeley and Los Angeles) Publ. Zool.,* **64,** *1–56 (1962).*]

and thermal radiation, thus minimizing dependence on evaporative cooling. The water saving is obviously of critical importance in regions of aridity.

There is, of course, a limit to the extent of hyperthermia that mammals can tolerate. Under conditions of severe heat load, body temperature would exceed the tolerable level, which is usually 43°C or less except in extraordinary cases such as the oryx. Damaging hyperthermia under such circumstances can be avoided only by evaporative cooling. The flow of sweat onto a surface of an animal receiving large amounts of heat from the environment is a very inefficient use of evaporative cooling. A large part of the water thus dissipated would only get rid of environmental heat that had not yet entered the animal's body. If the evaporation can take place underneath the thermal shield made available by the pelage, the physical situation is such that metabolic heat can be lost from the surface of the body by evaporation of sweat while the pelage affords protection from the heavy heat load associated with direct and indirect radiation as well as high air temperatures. This arrangement is the one found in camels, horses, and some bovids.

All terrestrial vertebrates continuously lose water incidental to respiration, and when exposed to an increase in heat load they usually elevate respiratory evaporative cooling by increases in breathing rate. Among mammals such thermal panting is most important in species that are large enough to depend routinely on evaporative cooling but do not have an abundance of sweat glands. As a mechanism for evaporative cooling thermal panting is energetically more expensive than sweating because of the work that must be done by the muscles of the respiratory pump to move large volumes of air across the moist surfaces of the respiratory tract. This, of course, adds to the heat load which must be dissipated. Moreover, increases in tidal volume that accompany panting result in hyperventilation of the lungs which can cause alkalosis (see Chapter 5).

Panting is particularly conspicuous in members of the order Carnivora. It has been studied most extensively in domestic dogs which like wild canids generate large amounts of heat during activity but have few sweat glands. In domestic dogs at least two mechanisms have been identified that minimize the energy cost of panting. A dog tends to pant at a constant frequency that closely matches the resonant characteristics of its respiratory system and thoracic region. Thus, the respiratory pump is driven at its most economical rate and the energy spent on moving air over the evaporative heat exchanger is minimized. In addition, when a dog pants, it usually inhales through the nose and exhales through the mouth. This unidirectional flow of air maximizes evaporation. If a dog breathed both in and out through its mouth the warm saturated exhaled air would pass over surfaces that has been cooled by evaporation during the prior inhalation. As a result heat that had previously been lost by evaporation during inhalation would be regained by condensation during exhalation and the total heat loss would be reduced.

During bursts of sprinting when muscular effort is at its peak, cursorial mammals produce heat more rapidly than they can lose it by evaporation. At low ambient temperatures this situation presents few problems because the internally generated heat can be lost by radiation and convection. In a hot environment, however, where the mammal is already under an externally imposed heat load, the internal heat load generated by intense muscular effort presents a significant challenge. Consider the cheetah, *Acinonyx jubatus,* probably the fastest of runners, with a top speed of about 100 km/hr and enormously rapid acceleration. Cheetahs capture their prey by a high speed sprint. If they do not capture their prey within 45–60 sec they abruptly terminate the chase. At rest cheetahs are excellent evaporative coolers. By thermal panting they can keep body temperature at 40°C even when air temperature is 50°C. A sprinting cheetah produces heat at more than 50 times its resting rate, and during a sprint only a portion of this heat can be lost by evaporation. Therefore, the heat has to be temporarily stored which, of course, causes a sharp rise in body temperature. Captive cheetahs refuse to continue to run after deep body temperature reaches 40.5°C. Thus, the distance over which a cheetah will pursue its prey appears to be limited by the time required for its body temperature to rise to the highest level voluntarily tolerated. If a cheetah does not overtake its prey within this brief interval it must stop and cool off until its body temperature returns to near its normal level of about 37°C.

Mammals, like antelope (Family Bovidae) that sweat as well as pant can increase evaporative cooling while running, and thus partly avoid the necessity for storage of the heat generated by intense muscular activity. Consequently, they can sustain high-speed locomotion in a warm environment without overheating for longer periods than forms which depend exclusively on thermal panting for enhancement of evaporative cooling. Moreover, as described elsewhere, bovids have special adaptations for keeping brain temperature below core body temperature which helps them to tolerate extreme hyperthermia and thus makes it possible for them temporarily to store heat produced by intense muscular activity.

The patterns of adaptive strategies for evaporative cooling in mammals that have been favored by natural selection are illustrated with particular clarity by macropod marsupials (kangaroos and wallabies), which employ three different mechanisms for enhancement of evaporative heat loss. When faced with an externally imposed heat load from solar radiation or high air temperatures they pant and also salivate copiously and spread the saliva over the front legs and belly. While they are hopping rapidly they do not pant or spread saliva, but they sweat heavily and their respiratory evaporative heat loss is elevated incidental to the increased pulmonary ventilation required by the high levels of oxygen demand. As soon as locomotion ends, they stop sweating and immediately begin to pant rapidly and to salivate and spread saliva over themselves.

From the preceding discussion it is clear that sweating provides a supplementary surface for evaporative cooling that facilitates heat loss during sustained high levels of locomotor activity and facilitates intense muscular activity with a minimal amount of overheating even in the presence of an external heat load.

The scarcity of small day-active desert mammals may be considered an indication of the difficulty of the problems they face. The rapidity of their metabolic turnover, their relative lack of mobility and consequent inability to seek surface water, and their small size and resultant limited capacity for heat storage have forced them to a precarious behaviorally maintained balance between physiological capacity and environmental stress.

Among the day-active desert mammals that weigh 100 g or less, only one, the antelope ground squirrel, *Ammospermophilus leucurus,* has been systematically examined from the standpoint of temperature regulation and water economy. It is of widespread occurrence in western North America. It neither hibernates not estivates and must, therefore, cope with the problems of desert life at all seasons. When ambient temperatures exceed the acceptable levels of body temperature, the heat of metabolism cannot be lost except by the evaporation of water, and in arid regions water is too scarce to squander, particularly for small mammals. One solution to this problem, and the one evolved by *A. leucurus,* is the acceptance of hyperthermia so that body temperature rises until it exceeds ambient temperature, and metabolic heat can be lost passively by conduction, radiation, and convection without dependence on evaporative cooling. Captive *A. leucurus* show a labile body temperature that increases regularly with ambient temperature (Figure 8-32). Because it tolerates body temperatures

as high as 43°C, *A. leucurus* can lose heat passively to the environment even when ambient temperatures are as high as 42°C. To maintain body temperature at or near the resting level of 38°C rather than develop hyperthermia (assuming dry air and an ambient temperature of 42°C), the antelope ground squirrel would have to evaporate a quantity of water equal to 13% of its body weight per hour, obviously an impossibility for an animal that may never have access to drinking water.

Antelope ground squirrels exploit their tolerance of hyperthermia in another and intriguing way. Although they may sometimes be seen sitting quietly in the shade or perching in open, sparsely leafed, creosote bushes *(Larrea tridentata),* they characteristically dash about at high speed, often traveling a hundred meters or more from their home burrows. The furious activity, the hot soil and sand (the temperature of which may exceed 70°C), and the intense solar radiation soon cause them to heat up. Periodically throughout the day, therefore, the hyperthermic animal must unload its accumulated heat. This it does by going down a burrow, flattening itself against the relatively cool floor and losing heat to the substratum by conduction, and to the walls by radiation, until its body temperature approaches its usual resting level. Thereupon the squirrel returns to the surface and forages until its hyperthermia again reaches the limits of tolerance, and heat is again unloaded underground. The antelope ground squirrel has a regulatory pattern in which behavior and physiology allow its temperature to oscillate over a range of 4 or 5°C with a periodicity measured in minutes rather than hours (Figure 8-31). The critical element of this performance is the restriction of body temperature to a series of oscillations rather than to a fixed level. The lower limit of these oscillations is set by the physiology of the mammalian temperature-regulatory complex; the upper limit is not regulated by physiology but by behavior.

Local Differences in Body Temperatures. Regional heterothermy is more pronounced in mammals exposed to very low temperatures than in those subjected to heat stress. In the latter situation it occurs primarily in association with evaporative cooling. The tissues of endotherms vary in their sensitivity to high temperature, the brain and the testes being particularly prone to heat damage. In most verte-

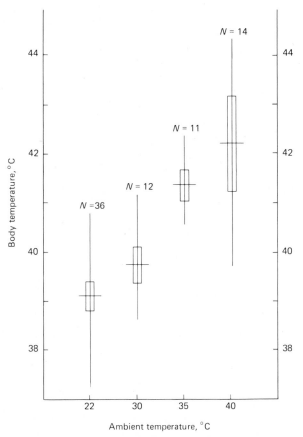

Figure 8-32
Effect of environmental temperature on body temperature in the antelope ground squirrel. The vertical line indicates the range. The horizontal line indicates the mean. The rectangle encloses the interval ±2 standard errors of the mean. [*From G. A. Bartholomew.* Symposia of the Society for Experimental Biology, *No. 18, pp. 7–29, Academic Press, Inc., New York, 1964; original data from J. W. Hudson.* Univ. Calif. (Berkeley and Los Angeles) Publ. Zool., **64,** *1–56 (1962).*]

brates, spermatogenesis is inhibited by high temperatures, and cell division in the seminiferous tubules is adversely affected by temperatures in the range typical of core temperatures in birds and mammals. Although the subject has not been investigated on a broadly comparative basis, in one bird, the house sparrow, mitotic activity in the testes is greatest during the night when body temperature is at the lowest level of the daily cycle. In most mammals the testes are located, either permanently or seasonally, outside the abdominal cavity in the scrotum where they remain several degrees below core temperature. In at least some species, flying foxes of the order Megachiroptera for example, the testes are usually abdominal but descend into the scrotum during heat stress. In others, such as many rodents, the testes are scrotal only during the breeding season.

In some members of the family Bovidae, such as the oryx whose tolerance of elevated T_b was previously discussed, special cardiovascular mechanisms allow the brain to remain cooler than the rest of the body during periods of hyperthermia. In bovids most of the blood supplied to the brain comes from the carotid arteries. Just beneath the brain is a large space, the **cavernous sinus,** in which the carotids divide into a complex network of small arteries that subsequently coalesce and continue on into the brain. The moist surface of the mucosa of the nasal passages is a major site of evaporative cooling in bovids. The venous system draining blood from the cool nasal region passes through the cavernous sinus where its vessels mingle with the subdivided carotid arteries to form a complex that functions as a heat exchanger. The net effect of this heat exchange is to reduce the temperature of the blood flowing to the brain. In bovids exposed to heat stress, the temperature of the brain may be as much as 3°C lower than that of the blood leaving the heart. When air temperature is below body temperature, the horns of bovids may also contribute to cooling the blood in the arterial supply to the brain because the veins that drain the rich vascular supply of the horns also pass through the cavernous sinus.

8.10.4 Birds

From the standpoint of adaptation to high temperatures, birds are particularly interesting, not only because of their small size and high rate of heat production, but because most species are neither nocturnal nor fossorial, and so must meet the challenges of the hot physical environment directly.

The body temperatures of most birds when alert and active often exceed 40°C, and during flight they may be even further elevated. Deep body temperatures of starlings, *Sturnus vulgaris,* reach about 44°C during flight in an air tunnel at environmental temperatures ranging from 0 to 28°C. The high levels of their body temperatures preadapt birds to live in areas of intense heat. There are few environmental situations in which temperatures are higher than the body temperatures of birds. Conse-

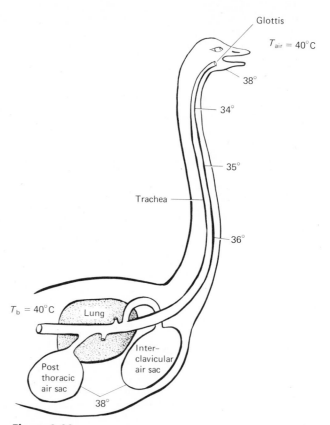

Figure 8-33
Typical distribution of temperatures in the respiratory system and buccal cavity of a panting ostrich when deep body temperature and air temperature both equal 40°C. Only part of the air-sac system is shown. [Data from K. Schmidt-Nielsen, et al. Condor, **71,** *341–352 (1970).]*

quently, they can almost always lose heat passively by radiation, conduction, or convection.

When, despite their high body temperatures, they cannot passively lose heat to their surroundings, birds, like other vertebrates, resort to evaporative cooling. Birds do not sweat, and insensible water loss through the skin is minimized by a thick insulative covering of feathers. For evaporative cooling, most species pant and thereby lose water from the buccal area and from the respiratory system (Figure 8-33). Panting, which in birds may be defined as an increase in the rate and usually a decrease in the amplitude of breathing, is not, however, an unmixed blessing. It requires work and thereby adds significantly to the total heat load that must be dissipated to the environment. When environmental conditions make passive heat loss difficult birds almost invariably develop hyperthermia (Figure 8-34). This hyperthermia is usually self-limiting, because the rise in body temperature reestablishes the temperature difference between body and environment. This difference again makes passive heat loss possible and thermal homeostasis is again established but at a new and higher level of body temperature. This type of hyperthermia is very widespread in birds faced with heat stress. The functional significance of avian hyperthermia contrasts with that shown by the dromedary camel (Section 8.10.3). Avian hyperthermia facilitates heat loss, whereas that of the camel allows heat storage.

Under conditions of very low humidity a number of kinds of birds can actually maintain body temperature below ambient temperature by vigorous, sustained panting. Panting is energetically more expensive than sweating. No birds sweat, but a number have evolved a method of evaporative cooling that requires less energy than the breathing movements of panting. This method is *gular flutter.* The gular area, the floor of the mouth, and the anterior part of the esophagus are thin and highly vascular. In some species, conspicuously in the order Pelecaniformes, it is enlarged into a pouch. The gular area has little mass and can be fluttered, either by contractions of muscles in the tissue itself or by movements of the hyoid apparatus. By holding the beak agape and fluttering the gular area, birds can evaporate water with little energy expenditure and without involving the respiratory system (Figure 8-35). Gular flutter is employed by members of several orders—goatsuckers, owls, herons, pelicans, boobies, cormorants, doves, many gallinaceous species, and colies.

Both panting and gular flutter show two patterns of response: a gradual increase in rate with increasing heat load and a fixed rate that is independent of heat load. In most birds, the breathing rate increases gradually with the heat load (as shown in

Figure 8-34
Schematic representation of the daily cycle of body temperature in the brown towhee (Pipilo fuscus), *showing the marked hyperthermia characteristic of most birds when exposed to high ambient temperatures.* [From G. A. Bartholomew, Symposia of the Society for Experimental Biology, No. 18, *pp. 7–29, Academic Press, Inc., New York, 1964; original data from W. R. Dawson.* Univ. Calif. (Berkeley and Los Angeles) Publ. Zool., **59,** 81–124 (1954).]

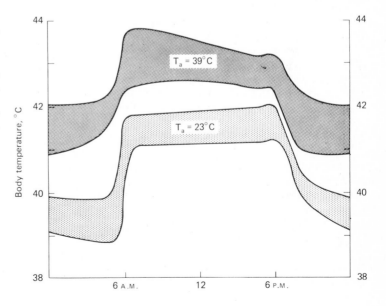

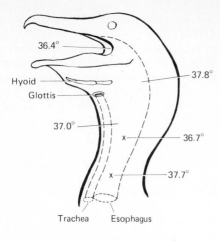

Figure 8-35

Temperatures on the evaporative surfaces in the buccal cavity and esophagus of the double-crested cormorant (Phalacrocorax auritus) *during gular flutter. [Modified from R. C. Lasiewski and G. K. Snyder. Auk, **86,** 529–540 (1969).]*

Figure 8-36), but the rock dove, *Columba livia,* and the ostrich, *Struthio camelus,* have two apparently nonoverlapping breathing rates, one of which is relatively slow and is used at low and moderate temperatures and another which is fast and is used for panting. Presumably in these species the panting rates are fixed by the natural resonant frequency of the thoracic cage, as is the case in dogs. Thus, while panting, the birds drive their ventilating apparatus at the most economical rate possible—its resonant frequency. A similar situation exists with regard to gular flutter. Some birds, such as quail, increase flutter rate with increasing heat load. Others, such as cormorants, pelicans, poorwills, and nighthawks, flutter at a fixed frequency presumably determined by the resonant characteristics of the gular apparatus (Figure 8-37).

A convenient evaluation of the energetic cost of evaporative cooling can be obtained by simultaneously measuring water loss and oxygen consumption. One can then calculate the ratio of heat lost by evaporation (EHL) to heat produced (HP) by oxygen consumption (EHL/HP), by assuming that the oxidation of oxygen yields 4.8 cal/cm³ and

the evaporation of water requires 0.58 cal/mg. The ratio, EHL/HP, is equal to 1 when an animal dissipates by evaporation an amount of heat equal to that produced by its metabolism. For an animal to keep its body temperature below ambient temperature obviously requires that the ratio exceed unity. Passerine birds depend on panting for enhancement of evaporative cooling, and can achieve parity between EHL and HP only when breathing air of low humidity. Partly because of their exceedingly high mass-specific metabolism, ratios significantly above unity are rarely reached in this group. The highest EHL/HP ratio so far recorded from a species that depends on panting is 1.8, a value attained by a frogmouth *(Podargus ocellatus)* maintained in an air temperature of 46.8°C and a water vapor pressure of 10–18 mm Hg. The high ratio is due in part to the fact that this species has an extremely large mouth and pharynx; however, it is closely linked to the fact that (like all other members of the order Caprimulgiformes) this species, which lives in New Guinea and tropical Australia,

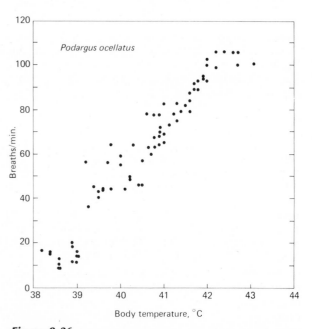

Figure 8-36

*The relation of breathing rate to body temperature in the little papuan frogmouth, showing the increase in panting rate with increase in heat load. [Data from R. C. Lasiewski, et al. Condor. **72,** 332–338 (1970).]*

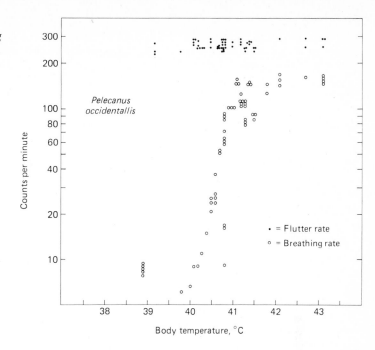

Figure 8-37
The relation of rates of gular flutter and breathing to deep body temperature in the brown pelican. [Data from G. A. Bartholomew, et al. Condor, **70**, 31–34 (1968).]

has an extremely low metabolic rate. Its basal rate is approximately 60% of the value predicted on the basis of its mass (150 g). Another member of this order, the poorwill (*Phalaenoptilus nutallii*), which depends on gular flutter for evaporative cooling, has a BMR less than half of that predicted on the basis of its size (50 g). The poorwill can achieve an EHL/HP ratio of 3.5 in air with a water vapor pressure of 10–15 mm Hg and can, therefore, readily dissipate by evaporation not only its endogenous heat but also the heat it acquires from the environment by radiation, conduction, and convection. In an ambient temperature of 45°C and a relative humidity of 20% it can hold its deep body temperature stable at 42°C for many hours.

Some large birds use a system that is wasteful of water but is more economical energetically than either panting or gular flutter. Under conditions of heat stress several species of storks, some cormorants, and also the turkey vulture (*Cathartes aura*) urinate on their legs. The evaporation of the urine cools the blood flowing in the vessels just beneath the skin of the unfeathered tarsometatarus and feet. It also leaves a whiteish caking of urates all over their legs.

8.11 ONTOGENETIC ASPECTS OF ENDOTHERMIC TEMPERATURE REGULATION

The thermoregulatory ability of newly hatched birds and newly born mammals varies markedly from species to species. In some, the young are well insulated by pelage or plumage and are effective homeotherms at the time of birth or hatching. In others, they are naked, blind, helpless, and poikilothermic.

Birds that are in an advanced state of development at hatching are called *precocial.* Those that are in an early stage of development are called *altricial.* From the standpoint of thermoregulation, the altricial state is of particular interest. In altricial species the development of thermoregulatory capacity and the transition from poikilothermy to homeothermy is readily investigated. Because birds are particularly instructive in this regard, we shall confine our attention to them.

Altricial species may be large or small and occur in many different orders of birds. Although the young of altricial species have limited thermoregulatory capacity, their temperatures are usually restricted to narrow limits by the brooding behavior of the parents. The adults compensate behaviorally for the thermoregulatory inadequacies of the chicks. Thus, for altricial chicks, the thermoregulatory unit is not a single animal but a system, involving adult plus young.

In the masked booby (*Sula dactylatra*), a large pantropical sea bird of the order Pelecaniformes, the naked newly hatched chicks are almost completely ectothermic. Nevertheless their body temperatures are maintained near 38°C by the brooding behavior of the adults. If the hatchlings are removed from the nest and exposed to the full sun, they will die of overheating in 15 or 20 min. If they are placed in the shade, body temperature soon falls to ambient levels. Their capacity for homeothermy increases with increasing mass (Figure 8-38), and during the period of developing homeothermy the adults gradually reduce the attentiveness of their brooding.

The effectiveness of the interplay between the behavior and the physiological capacities of adults and chicks in maintaining high and relatively uniform body temperatures in the latter is shown with particular clarity if one compares the responses of different kinds of birds to the same environmental stress. For example, on Angel de la Guarda, a desert island in the Gulf of California, severe heat stress exists almost daily during the nesting season. Yet aquatic birds of three species, western gulls (*Larus occidentalis*), great blue herons (*Ardea herodias*), and brown pelicans (*Pelecanus occidentalis*), the young of which have profoundly different thermoregulatory capacities, nest side by side in fully exposed situations on the shores of this barren island. The gulls are precocial. They are fully covered with down at hatching, and within a few minutes of leaving the egg can move about and seek shelter from the blazing sun; their parents are inattentive. The pelican chicks are extremely altricial. They are naked, helpless, and, like the booby described previously, have almost no capacity for thermoregulation. During the hours of intense heat their survival depends on their being shielded from the sun by the adult birds, which brood them closely and are

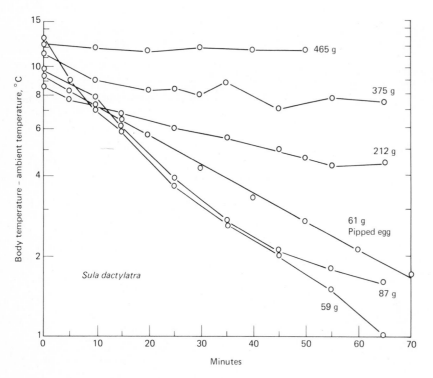

Figure 8-38
*Cooling rates in relation to size in chicks of the masked booby. The chicks were in the open shade, sheltered from the breeze, and exposed to ambient temperatures from 27 to 29°C. [From G. A. Bartholomew. Condor, **68**, 523–535 (1966).]*

extremely attentive. The young herons are altricial but less so than the pelicans; their parents are of intermediate attentiveness. Thus each of these species meets the heat stress of the same demanding physical environment by virtue of the adult compensating behaviorally for the differing physiological capacities of the chicks.

Virtually all members of the order Passeriformes are altricial. In these song birds, as in other groups, altriciality is not to be equated with primitiveness. Rather, it represents a highly adaptive system that allows rapid development at a minimal cost in energy. In the vesper sparrow *(Pooecetes gramineus)* the highly altricial young hatch after only 12 days of incubation. Their mass at hatching is 2 g. Nine days after hatching they attain 18 g, are fully feath-

Figure 8-39

Relation of body temperature and oxygen consumption to air temperature in nestling vesper sparrows of various ages. Note that the younger nestlings follow the ectothermic pattern, whereas the older nestlings follow the endothermic pattern. [From W. R. Dawson and F. C. Evans. Condor, **62,** 329–340 (1960).]

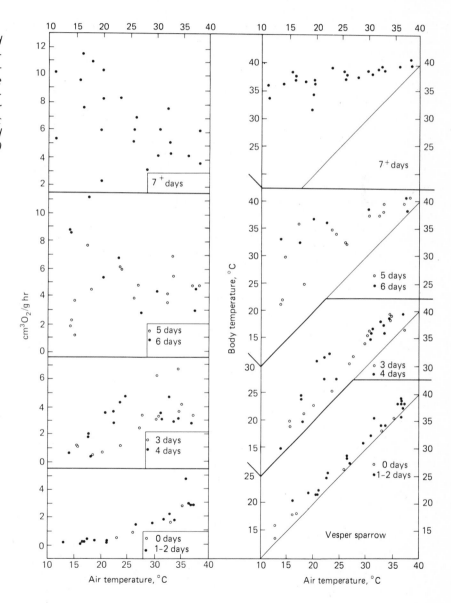

ered, and leave the nest. During the first 4 days after hatching, while they are almost completely poikilothermic, their growth rate is approximately 40%/day. A rapid and dramatic transition from poikilothermy to homeothermy, and from ectothermy to endothermy, is essentially complete within 1 week (Figure 8-39). The acquisition of homeothermy in altricial species involves a complex array of developmental processes involving neural and endocrine integrative systems, the capacity for a high level of endogenous heat production, and the capacity for control of thermal conductance. In addition, during the process of becoming homeothermic, continuous support, in terms of food and heat, must be supplied by the parents. The evolution of such a complicated system of posthatching development strongly suggests that the altricial state must have selective value.

The altricial state is clearly significant from the standpoint of energetics. The immaturity of the hatchling and the short incubation period allow the egg to be small and to have a minimum of food storage. Both are energetically advantageous. The altricial system also minimizes the maintenance costs for the young birds during the nesting stage. The parents brood the young at a trifling energetic cost to themselves because they are continually losing heat. Some of this waste heat can be used to contribute to the maintenance of a high and uniform body temperature in the young during their ectothermic stages. The altricial young do not have to employ endogenous heat for temperature regulation, so a larger percentage of the energy from their food can be used for growth than would be the case if they were precocial.

From the discussion in Section 3.7.2 and the curves in Figure 3-7 it is apparent that birds and mammals weighing less than 10–15 g have extravagantly high rates of mass-specific oxygen consumption. If the values of oxygen consumption of the infants of these species lay along the same curve as those of the adults, their metabolic rates would be impossibly high. This situation, of course, does not occur because the infants of birds and mammals weighing less than 20 g are, apparently without exception, ectothermic at time of hatching or birth. Gradually, as their size increases they develop the physiological capacities and insulating layers of fur and feathers necessary for effective endothermy.

8.12 HETEROTHERMY AND ADAPTIVE HYPOTHERMIA IN BIRDS AND MAMMALS

The availability of free energy that animals can use for doing physiological work varies in both time and space. The problems associated with a patchy distribution of free energy are further complicated by seasonal and daily variations in environmental temperatures, and fluctuations in the availability of water. Clearly, natural selection will favor the evolution of temporal patterns of energy utilization that allow animals to survive periods when energy resources are in short supply, when water is scarce, or environmental temperatures are either very low or very high.

Endothermy is metabolically expensive. The continuous maintenance of high and uniform body temperatures by endogenous heat production requires sustained high levels of energy utilization and food consumption. Consequently, an obligatorily high rate of energy metabolism exacerbates the problems inherent in the temporal and spatial variations in availability of free energy in the environment. Because the rate of specific energy metabolism is inversely related to size, small birds and mammals must face especially acute challenges.

Whereas a small ectotherm in a cool environment may survive for weeks or months without eating, under the same conditions a small mammal such as a shrew will starve to death in a few hours. Very small birds or mammals often must consume food equal to their own body weight every day to maintain homeothermy. A small shrew produces feces equal to one half its body mass daily. In view of the high energy cost of endothermy and the high energy cost of activity, it is not surprising that a few small birds and many small mammals periodically become inactive, abandon homeothermy, and for intervals, sometimes measured in hours, sometimes in days, allow their body temperatures to fall to levels equal to or only slightly above those of their surroundings. These intervals of adaptive hypothermia are not the result of imperfect or poorly developed homeothermy but are precisely monitored, clearly adaptive states of hypothermia from which the organism can arouse by endogenous heat production. They represent a highly refined state of endothermy in which rates of heat production and levels of body temperature are controlled on a much more flexible basis than in those birds and mammals whose thermostats are set at a relatively fixed level.

The most familiar of these patterns of temporal heterothermy is the hibernation, or winter dormancy, of mammals in cool temperate and boreal regions. Strictly speaking, the word "hibernation" refers to winter dormancy and does not specify any particular group of organisms or any one type of physiological response. Unfortunately in the literature of vertebrate physiology it has been used to designate the state of dormancy and low body temperatures that some endotherms enter during winter in cold temperate and boreal regions. This usage is doubly unfortunate. It forces a special physiological connotation on a descriptive natural history term and it makes the implicit assumption that the adaptive hypothermia of endotherms is fundamentally a response to low temperature, which at best is only partly true. Moreover, it has led many physiologists to treat adaptive hypothermia as if it were only a special case of thermoregulation, whereas, as indicated, it is both biologically more realistic and physiologiclly more useful to relate adaptive hypothermia to the broader problem of animal energetics.

Adaptive hypothermia is not a single entity nor is it primitive. It encompasses a variety of responses, many of which have been evolved independently in different taxa in response to ecological and physiological conditions that exert widely diverse conditions for the operation of patterns of natural selection favoring a common end product—saving energy by the temporary abandonment of homeothermy. Although no sharp boundaries can be drawn between them, several overlapping patterns of adaptive hypothermia may be distinguished.

1. **Regulated Shallow Hypothermia.** Body temperature reduced below the normal resting (normothermic or euthermic) level and regulated at some lower limit that varies with the taxon and ambient temperature but is usually within about 10°C of normothermia. Animals may be lethargic but are usually capable of coordinated movements and responses.

2. **Torpor or Profound Hypothermia.** (1) Body temperature falls to within 1°C or less of ambient temperature; (2) oxygen consumption is greatly reduced, to as little as 5% of the BMR; (3) prolonged periods of suspended respiration (apnea) develop—breathing rates may be reduced to one breath or less per minute; (4) a condition of torpor or dormancy occurs that is much more profound than deep sleep; (5) heart rate is markedly reduced; (6) the ability to arouse is retained and, by mobilizing the mechanisms of endogenous heat production and the mechanisms of heat conservation, the high body-temperature levels characteristic of normally active endotherms can be reestablished.

(a) **Daily torpor.** Cycles of entry into, and arousal from, hypothermia are on daily basis, with the period of torpor or dormancy occurring during the time of inactivity and sleep.

(b) **Seasonal torpor.** Cycles of torpor last several days and are seasonally restricted.
(1) **Hibernation.** Dormancy (torpor) is associated with cold period of year.
(2) **Estivation.** Dormancy (torpor) is associated with warm, or warm and dry, periods of year.

The differences between these patterns are more quantitative than qualitative. At present it seems wisest to treat them as nodal points along a continuum that extends from a decline of a few degrees in body temperature without a marked diminution in alertness or coordination to prolonged periods of profound dormancy accompanied by drops of as much as 35°C in body temperature. In all cases the animals normally retain the capacity to arouse, and by production of endogenous heat are able to return to normal levels of body temperature and activity.

The distinctions between the various patterns are sometimes blurred by fluctuations in the body temperature and activity state of essentially normothermic animals—the relation of deep sleep to the lethargy associated with marked hypothermy is clear but its study is still in an early stage. Distinctions may be further confused by variations in the ambient temperature to which the animals are exposed during episodes of hypothermia and by the fact that animals may pass imperceptibly from one pattern to another—daily torpor during the autumn may merge into hibernation during the winter.

Daily Torpor. Small birds and mammals have very high mass-specific metabolic rates and can store only limited amounts of energy in the form of fat. As long as they maintain high body temperatures, they are never more than a few hours from death by starvation, particularly at low ambient temperatures. This situation is particularly acute for small bats and for hummingbirds, not only because of their small size but because of the large amounts of energy required for flight. Moreover, both groups have restrictive feeding adaptations that limit their periods of feeding (daytime in hummingbirds, nighttime in bats). Both have evolved the same solution—daily periods of torpidity. Both maintain normally high body temperatures when active but undergo profound drops in temperature and oxygen consumption during their periods of inactivity. While active they profit from the independence of environmental conditions allowed by homeothermy; while inactive they profit from the economy of energy expenditure made possible by poikilothermy. In this way they are able to exploit a variety of habitats that would not otherwise be available to them.

Although the metabolism of a 3.2 g Costa hummingbird *(Calypte costae)* during sustained flight (hovering) is approximately 42 cm^3 O$_2$ g^{-1}hr^{-1}, in a torpid hypothermic state at 20°C its oxygen consumption is only about 1/100 as great (0.39 cm^3 O$_2$ g^{-1}hr^{-1}), which is approximately the same as the standard metabolism of a 3-g lizard at the same temperature. Data on oxygen consumption of very small bats during flight are not available, but their metabolic rate during torpor is only about 1/20 that during activity under conditions of captivity.

The importance of daily torpor for a small endotherm is very striking even under laboratory conditions at temperatures near thermal neutrality. Under natural conditions at temperatures below thermal neutrality, the energy saving for unconfined animals will be even greater than that shown diagrammatically in Figure 8-40 for animals inside respirometer chambers.

The cycle of entry into and arousal from daily torpor has been most thoroughly studied in the California pocket mouse, *Perognathus californicus,* a heteromyid rodent weighing about 20 g. In this species, entry into torpor can occur at any temperature below the lower critical temperature (32.5°C) by a cessation of thermoregulation and a reduction of metabolism to the basal level. With its thermostat shut off and its thermal conductance maximal, heat loss exceeds heat production and the mouse's body temperature falls. Decreasing body temperature is accompanied by decreasing heat production until equilibrium is reached at or slightly above environmental temperature (Figure 8-41). Under these circumstances the mouse cools very much as would an ectotherm. During torpor at temperatures above 15°C, this species does not thermoregulate; body temperature fluctuates directly with ambient temperature although remaining slightly above it. During arousal from torpor, thermal conductance is made minimal, the animal shivers violently, and oxygen consumption becomes maximal. Heat production exceeds heat loss and body temperature rises at a rate of more than 0.5°C/min.

Perognathus californicus cannot mobilize enough heat production to arouse from torpor if its body temperature falls below 15°C, nor does it regularly become torpid in captivity when food is present in excess. However, at any season of the year, if the daily food ration is reduced slightly below the level of *ad libitum* consumption, it undergoes a regular daily cycle of torpor. The duration of the daily cycle of torpor is inversely related to the daily food ration. The mouse adjusts its period of torpor in such a way as to compensate for the degree of food deficiency—the smaller the ration of food, the longer the daily period of torpor. Thus this species has a feedback mechanism that couples the metabolic cycle and the associated thermoregulatory system to the availability of food in such a way that a balance is struck between energy supply and energy expenditure.

A similar cycle has been demonstrated in one of the desert-dwelling white-footed mice *(Peromyscus eremicus),* which, unless it is maintained in an environment of high humidity, requires water either to drink or in its food. It is of interest that some individuals of this species can be induced to enter torpor even in the presence of excess food merely by placing them in a negative water balance.

Figure 8-40

Schematic representation of the daily pattern of metabolism of two small endotherms (humming-bird and bat) which experience torpor, and one (shrew) which does not. [*Data from O. P. Pearson.* Ecology, **28,** 127–145 (1947); *P. R. Morrison, et al.* Physiol. Zool., **32,** 256–270 (1959); *R. C. Lasiewski.* Physiol. Zool., **36,** 122–140 (1963).]

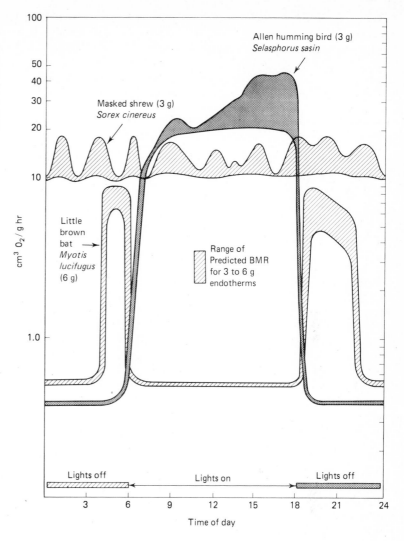

The data available on *Perognathus californicus* allow a precise estimate of the energy saving made possible by a cycle of torpor. If one of these mice were to enter torpor at 15°C and then immediately arouse, the process would take 2.9 hours. The cost of maintaining normal body temperature for this length of time would be 11.9 cm³ O₂/g. The oxygen consumption during entry into and immediate arousal from torpor is 6.5 cm³ O₂/g. The latter figure is only 55% of the former. Thus from the standpoint of energetics, even the shortest possible period of torpor is profitable for this species and presumably for other small endotherms as well.

If the mouse were to remain torpid for 10 hr at 15°C the energy cost of the entire torpor cycle would be only 10% of that required for maintenance of normal body temperature for the same period.

In view of the energy saving made possible to small endotherms even by brief episodes of torpor, it is not surprising that the capacity to undergo daily periods of adaptive hypothermia is found in many mammals (including monotremes, marsupials, and placentals) and in at least six groups of birds (the swifts, goatsuckers, hummingbirds, sunbirds, manakins, and colies).

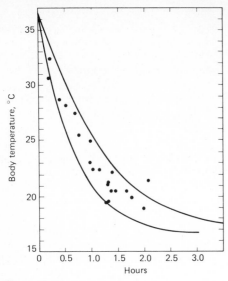

Figure 8-41
Body temperature of the California pocket mouse during entry into torpor (circles). The upper curve shows the theoretical course of body temperature, assuming minimum heat production and minimum heat loss. The lower curve shows the theoretical course of body temperature, assuming minimum heat production and maximum heat loss. [From V. A. Tucker. J. Cell. Comp. Physiol., 65, 405–414 (1965).]

The phenomenon of daily torpor is much more widespread than was formerly appreciated. Not only does it occur commonly among humming-birds, bats, and a variety of rodents, but in some cases it intergrades with seasonal dormancy, and in others it has a strong seasonal component. The mastiff bat, *Eumops perotis*, shows daily torpor during the winter but not during the summer. The birch mouse, *Sicista betulina*, and the little pocket mouse, *Perognathus longimembris*, undergo daily cycles of torpidity during summer and early fall but hibernate during the winter. Many bats have daily cycles of torpor during the summer and prolonged ones during the winter. Some species, such as the pigmy phalanger *(Cercaertus nanus)*, the ground squirrel *(Spermophilus mohavensis)*, and the kangaroo mouse *(Microdipodops pallidus)* show a continuous intergradation between daily and seasonal torpor.

Kangaroo mice are heteromyid rodents weighing 10–14 g. They occur only in aeolian sand dunes in the deserts of the Great Basin of western North America. Their food is mainly seeds which they glean from the sand and store in their burrows. In their habitat, temperatures fluctuate widely both daily and seasonally. However, because of their nocturnal and fossorial habits they need never face environmental temperatures above 30 or below 0°C. In most deserts seed production is limited to a brief part of the year and sometimes may not occur for several consecutive years. Consequently, small desert mammals must be able to carry out all essential functions on a limited and fixed energy supply. This situation makes strict homeothermy and sustained high levels of activity self-defeating. Kangaroo mice meet this situation by storing food in their burrows, and by a pattern of facultative hypothermia that is intermediate between daily torpor and the prolonged dormancy of estivation and hibernation. In this species the duration and frequency of torpor is dependent on both the availability of food and the ambient temperature. The animals integrate these two variables in such a way that they can accumulate food stores and maintain body weight under many different combinations of food availability and temperature. When food is available in excess and temperatures are high kangaroo mice do not become torpid. When food is in short supply the amount of time they spend in torpor is inversely related to temperature. (Their periods of torpor may last for only a few hours or for several consecutive days). The more time they spend in torpor the less food they need. During torpidity metabolic rate decreases directly with temperature, so the energy savings made available by prolonged torpidity are greatest at low temperatures (Figure 8-42).

Hibernation. In areas where the low temperatures of winter represent a regular and acute adaptive challenge, many medium- or small-sized mammals undergo prolonged periods of hypothermia and dormancy. In contrast to the situation in those mammals that commonly experience daily torpor, entry into prolonged periods of seasonal dormancy often requires a period of preparation and adjustment. Many small rodents also store large quantities of food in their burrows prior to entry into hibernation. In nearly all species that enter long periods of dormancy, enormous amounts of body

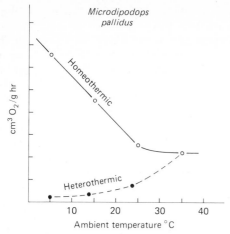

Figure 8-42
*The effects of environmental temperature on the oxygen consumption of kangaroo mice when maintaining normally high body temperatures and when in torpor. The points are mean values. [Data from J. H. Brown and G. A. Bartholomew. Ecology, **50,** 705–709 (1969).]*

fat are deposited prior to entry into torpor, the animals experience progressively increasing periods of lethargy, and they undergo a series of increasingly deep and prolonged periods of hypothermia followed by return to normal body temperature.

In at least some hibernators, the rate of entry into torpor is not determined, as it is in *Perognathus,* by the relation between minimum heat production and maximum heat loss. Instead the rate of decline in body temperature is regulated by linking it with controlled rates of heat production caused by bouts of shivering.

There is an enormous literature on the physiology of mammalian hibernation, and an almost embarrassing wealth of descriptive information on cardiovascular activity, spontaneous and induced nervous activity, biochemistry, and endocrinology is available. Two of the few general conclusions that can be drawn from this plethora of data are (1) that the patterns of regulation of adaptive hypothermia are both complex and variable, and (2) that it is of polyphyletic origin.

In at least some species the level of body temperature is not passively determined as it appears

to be in torpid bats, some hummingbirds, and pocket mice, but represents a loosely regulated, or at least monitored, state.

In most of the larger hibernating rodents, if body temperature falls dangerously low either the machinery for arousal is activated and they return to normothermia or they begin to thermoregulate and maintain body temperature somewhat above the danger point. In the golden-mantled ground squirrel, *Spermophilus lateralis,* a central nervous regulator involved in the control of body temperature during the profound hypothermia of hibernation is located in the preoptic nuclei and hypothalamus, and it appears to consist of the same population of cells that control body temperature during normothermia (see Section 8.8.3). During hibernation the deep body temperature of *S. lateralis* passively follows ambient temperature when the latter is slowly decreased. Body temperature continues to decline with decreases in ambient temperature until at some critical point near 0°C it reaches the set point (T_{set}) of the hypothalamic regulator; whereupon, the hypothalamic regulator is activated. This causes an increase in metabolic heat production that in turn causes an elevation in body temperature (Figure 8-43) despite the continued fall in ambient tem-

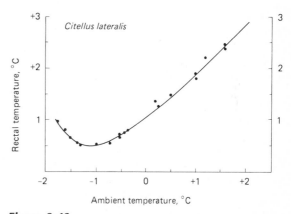

Figure 8-43
*The course of rectal temperature (T_r) in a hibernating golden-mantled ground squirrel subjected to a slowly decreasing ambient temperature. T_r passively followed T_a until the set point of the hypothalamic regulator (about 0.5°C) was reached, whereupon thermogenic activity was initiated and T_r increased. [Modified from H. C. Heller and G. W. Colliver. Am. J. Physiol., **227,** 583–589 (1974).]*

perature. Thus, in this species (and presumably others) even the profound hypothermia of torpor represents a regulated physiological state.

Although many hibernating mammals remain relatively inactive and stay inside the burrow for months at a time, apparently all of them arouse from dormancy at least every couple of weeks (usually oftener), remain at normally high body temperatures for a few hours or days, eliminate their accumulated metabolic wastes, and then return to dormancy.

During arousal from hibernation, both shivering and nonshivering thermogenesis are employed and heat production in the depots of highly vascularized deposits of brown fat is quantitatively important, particularly in the early stages. Not only are the mechanisms for heat production maximally activated, the heat they produce is usually sequestered in the anterior parts of the body, where it warms the heart and the central nervous system. This anterior concentration of heat due to the localization of depots of brown fat and to cardiovascular shunting is particularly apparent during the early part of arousal. At this time the hindquarters may be as much as 15°C cooler than heart and head regions (Figure 8-44).

Estivation. Although most students of mammalian hibernation have approached seasonal adaptive hypothermia as a pattern of response primarily related to low temperatures, this is too restricted a frame of reference to allow the phenomenon to be placed in adequate biological perspective. Seasonal dormancy can also be an adjustment to periods of drought or heat or to any other prolonged and recurrent pattern of seasonal stress. This situation is shown with unusual clarity by ground squirrels of the genus *Citellus (sensu lato)*, which are of circumpolar distribution and have representatives in Arctic, temperate, and tropical areas. The patterns of adaptive hypothermia shown by this widespread and successful group are various: (1) they hibernate where winters are cold; (2) they estivate in periods of prolonged seasonal drought; (3) they both estivate and hibernate with the two merging, in northern arid regions where precipitation is seasonal and restricted to winter and spring, but not in areas of regular summer rainfall; and (4) some species in warm temperate and tropical areas neither hibernate nor estivate. So viewed, adaptive hypothermia offers a primary physiological key to understanding the abundance and success of these animals in a variety of habitats and climates ranging from tundra to subtropical deserts and covering more that 50° of latitude.

Estivation has attracted much less investigation than hibernation. This is partially because most physiologists work in regions where cold rather than heat and drought represents the major seasonal climatic stress, and partially because the behavior of many small endotherms shows surprisingly little change during moderate hypothermia. They just seem somewhat lethargic. Estivation takes place at relatively high temperatures and, in

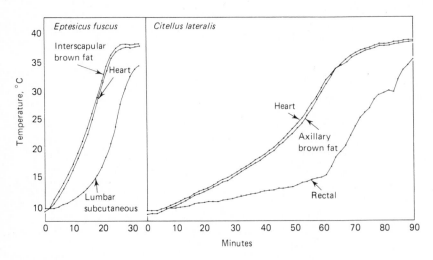

Figure 8-44
*Patterns of body temperature during arousal in the big brown bat and the golden-mantled ground squirrel. Environmental temperature, 6–8°C. [From J. S. Hayward, et al. Ann. N.Y. Acad. Sci., **131**, 441–446 (1965).]*

the absence of measurements of body temperature, it often goes unnoticed. For example, the pigmy possum *(Cercaertus nanus)*, a marsupial estivator, eats and moves in its customary manner at body temperatures as low as 32°C, and even at 28°C behaves in a manner that appears to be essentially normal. The little pocket mouse *Perognathus longimembris* will eat while its body temperature is less than 25°C and can vocalize loudly when it is only 10°C (Figure 8-45).

Entry into torpor at high environmental temperatures should be facilitated if the lower limit of the thermal neutral zone is above the level of prevailing ambient temperatures. In such a situation, with heat production maintained at the basal level and environmental temperature less than the lower critical temperature (T_{lc} in Figure 8-19), heat loss exceeds heat production. The animal becomes hypothermic and enters into torpor. It is of interest that most estivators have high thermal neutral zones

and many have basal metabolic rates slightly below the levels predicted on the basis of body mass.

Estivation is known to occur in a number of endotherms. It is more common in rodents than in other mammals, but it has also been observed in insectivores and marsupials. Representatives of at least two groups of aerial-feeding insectivorous birds, the swifts and the goatsuckers, experience episodes of hypothermia and torpidity during periods of inclement weather that are similar to the dormancy of estivating mammals.

Body Mass and Heterothermy. All heterothermic birds and mammals are small. No species with a mass of more than 50 g regularly experience daily cycles of torpor. Marmots (rodents of the family Sciuridae), which are the largest mammals known to undergo true hibernation, range in mass from 3.0 to 7.5 kg. Some very large mammals like bears den up during the winter, become inactive and live off accumulated deposits of body fat, but they do

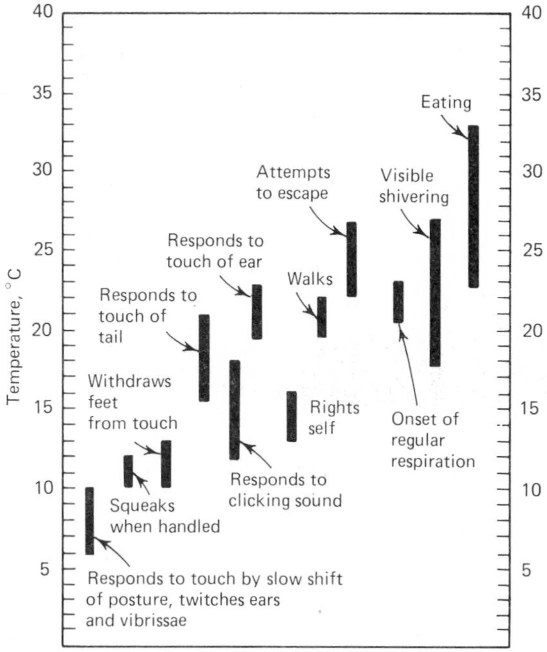

Figure 8-45
Range of minimal body temperatures for various patterns of behavior by the little pocket mouse during arousal from hibernation. [*From G. A. Bartholomew and T. J. Cade.* J. Mammal., *38, 60–72 (1957).*]

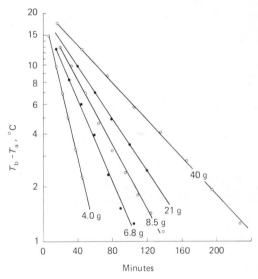

Figure 8-46
Rates of cooling during entry into torpor of four humming birds ranging in mass from 4 to 21 g, and a 40-g poorwill. On a semilogarithmic plot the exponential curves become straight lines with slopes that are inversely related to body mass. Ambient temperatures, 20–23°C. [*Modified from R. C. Lasiewski and R. J. Lasiewski.* Auk, *84, 34–48 (1967).*]

not become torpid and their body temperatures do not fall more than a few degrees.

During entry into torpor the course of body temperature of both birds and mammals approximates a "Newtonian" cooling curve, which is to say that the log of the difference between T_b and T_a is linearly related to time, and at any given ambient temperature the rate of decline of body temperature is inversely related to body mass (Figure 8-46). During arousal from torpor the rate of warm-up of heterothermic birds and mammals, of course, increases directly with ambient temperature, but at any given ambient temperature the rate of warm-up, like the rate of cooling, is inversely related to size (Figure 8-47).

The rapid cooling of small heterotherms is related to their high surface to volume ratios and associated high thermal conductance. Their rapid warm-up during arousal from torpor is related to their high rates of mass-specific energy metabolism. It has been calculated that, for a mammal with a mass of 200 kg torpid at 10°C, to warm up to its normothermic T_b would require the endogenous generation of an amount of heat equal to its normal

24-hr energy budget. In contrast, a typical hummingbird requires only a little more than 1% of its normal daily energy budget to warm itself from 10° to 40°C. Moreover, a small heterotherm can warm up to its normal operating body temperature in about 15 min, whereas a medium size heterotherm requires several hours (Figure 8-47).

The dependence of rates of cooling and warming on size makes cycles of daily torpor energetically advantageous for small birds and mammals, but it presents a major obstacle to the evolution of either daily or seasonal torpor in large birds and mammals. At least a few birds have developed partial solutions to this constraint by combined limited hypothermia during the night with a heliothermy similar to that employed by many ectotherms. The best documented case is that of the roadrunner (*Geococcyx californianus*), a 300-g terrestrial cuckoo of southwestern North America. In the dark at ambient temperatures of 6–10°C roadrunners allow T_b to fall about 4°C, but after sunrise they warm up to their normothermic level of 38°C at no metabolic cost by basking. A basking roadrunner faces away from the sun, holds its back at right angles

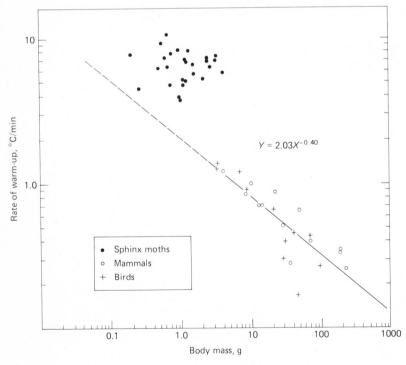

Figure 8-47
The relation of rate of warm-up to body mass in heterothermic birds, mammals and moths. In all cases both ambient temperatures and initial body temperatures were between 20 and 25°C. The regression line is fitted to the points for birds and mammals only. [*Modified from B. Heinrich and G. A. Bartholomew. J. Exp. Biol.,* **55,** *223–239 (1971).*]

to the incident radiation, droops its wings, elevates the feathers on its neck and back and exposes to the sunlight a specialized heat-absorbing area of heavily pigmented black skin. Normothermic birds can also substitute solar energy for endogenously produced heat. In the dark, the lower critical temperature of roadrunners is about 27°C, but when given access to radiation of the same intensity as the early morning sunlight in their natural habitat, their thermal neutral zone is greatly expanded and the lower critical temperature is extended downward to 9°C. The energy savings to the normothermic roadrunner from absorbing this radiant energy is about 550 cal/hr (0.64 W), which is about 40% of its basal metabolic rate.

8.13 INSECT ENDOTHERMY

Endothermic temperature regulation requires that rates of heat production and rates of heat loss be precisely balanced. Other things being equal (a rare circumstance in physiology) the rate of heat loss from an object is proportional to its surface area. All objects of similar shape, no matter how different their size, have surface areas that are proportional to the two-thirds power of their volumes, so the smaller an animal the greater its surface in comparison to its volume. In small mammals and birds the rate of mass-specific heat loss is inversely proportional to body mass, and one can assume that this inverse relationship is largely explained by the physical circumstances summarized above. Rates of specific heat production, however, depend primarily on biochemical rather than on physical factors. At the cellular level all birds and mammals are fundamentally alike, and there are clearly upper limits to their capacities for heat production.

If the maximum rate of specific heat production is fixed while the rate of specific heat loss continues to incease as size decreases, there must be some minimum size below which birds and mammals cannot produce heat rapidly enough to balance heat loss. At or below this critical size endothermic homeothermy should be difficult if not impossible. There are no birds or mammals that, as adults, have a mass of less than 2–3 g. Moreover, most birds and mammals that weigh less than 5–10 g are heterothermic and maintain normothermic temperatures only during activity or when free energy is abundantly available.

In view of the 2- and 3-g cut-off in the size of vertebrate endotherms, it is remarkable that many insects of a variety of types (moths, bees, flies, beetles, and dragonflies), most of which have a mass of less than 2 g and some with a mass of less than 100 mg, maintain body temperatures of 35–40°C during periods of activity. All endothermic insects are heterotherms and, with a few exceptions, such as honey bee workers during cold weather, incubating bumble bees, stridulating katydids, and large beetles during sustained terrestrial activity, maintain elevated body temperatures only during the intense activity associated with flight or the preparation for flight. At other times, their T_b is not significantly different from T_a. (Many diurnal insects, of course, achieve high body temperatures by means of behavioral thermoregulation resembling that of the ectothermic vertebrates previously considered.)

The phenomenon of insect endothermy has been most extensively studied in moths, beetles, and bees and we shall confine our discussion to these groups. A more broadly based review is available in Heinrich [1981]. It has long been known that a colony of honey bees can produce enough metabolic heat to raise the temperatures in their hive. Heat production by insects was first demonstrated in 1831, and the correlation between elevated T_b and the activity of the flight muscles was established in 1837. Nevertheless, detailed analyses of insect thermogenesis and the mechanisms of thermoregulation have become available only during the last decade.

Insect flight muscles are metabolically the most active tissues known. Like the muscles of vertebrates they have a mechanical efficiency of 20% or less. Thus, at least four fifths of the energy metabolism of the flight muscles appears as heat whereas only about one fifth appears as mechanical work. As a result, large amounts of heat are continuously produced during flight. Some of this heat, which is an obligatory by-product of sustained flapping flight, is retained in the thorax and is the source of the high body temperatures observed in some large, strong-flying insects. Nevertheless, the elevated temperature characteristic of the insect flight motor is not just a consequence of its activity, it is also a prerequisite for flight. The flight muscles cannot contract with sufficient speed, power, and frequency to keep the insect airborne unless they are at a high temperature. The muscle temperatures necessary for flight vary from species to species, and depend on body weight, wing area, and wing-beat frequency. An insect with a small body and large wings may be able to fly with slow wing beats and remain airborne with thoracic temperatures as low as 5–10°C. An insect with a heavy body and small wings, like a bumble bee, a sphinx moth, or a large beetle must elevate body temperature to about 35°C before it can generate enough lift to sustain itself in free flight. Both bumble bees and sphinx moths are covered by a dense layer of long scales that is the functional analog of fur and feathers. This insulating cover greatly reduces the rate of heat loss from the thorax, and large bumblebees and sphinx moths maintain thorax temperatures of 35–40°C even when flyng at air temperatures of 3–5°C.

Endothermic insects must undergo a preflight warm-up to bring the flight motor up to operating temperature. The functional necessity for a preflight warm-up is particularly obvious for moths, most of which are nocturnal. The rate of increase in body temperature during preflight warm-up in moths (and also other endothermic insects) is more rapid than that achieved by even the smallest of heterothermic birds and mammals during arousal from torpor. When rate of warm-up is scaled on body mass, the values for moths lie substantially above the regression line based on the values for birds and mammals (Figure 8-47).

Moths are "synchronous fliers." That is to say, there is a fixed relationship between the nerve impulses to the flight muscles and the contraction of the muscles. During warm-up the upstroke and downstroke wing muscles contract almost simultaneously and so work against each other. The result resembles the shivering of a bird or a mammal. The wings vibrate with small amplitude and negligible aerodynamic effectiveness, and virtually all of the power they generate appears as heat. This heat rapidly raises the temperature of the thorax. As soon as thoracic temperature is high enough, the upstroke and downstroke muscles being to contract alternately. Wing beats of normal amplitude are generated, and the insect takes off and flies away.

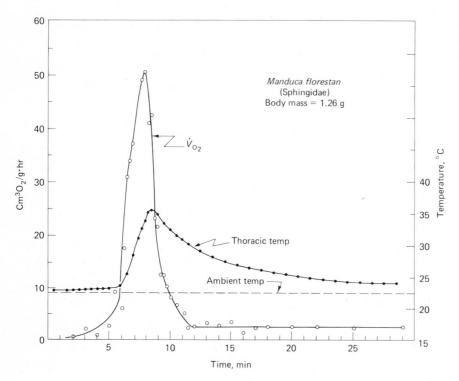

Figure 8-48
*Oxygen consumption and thoracic temperature during preflight warm-up, 15 sec of hovering flight, and postflight cooling in a sphinx moth. [Modified from G. A. Bartholomew, et al. J. Exp. Biol., **8, 90**, 17–32 (1981).]*

It maintains high thoracic temperatures as long as it continues to fly, but as soon as it alights, its temperature starts to fall (Figure 8-48).

During flight the thoracic temperature shows a substantial dependence on wing loading (weight/wing area) in both bumblebees and moths. In any given family of moths the heavier the wing loading the higher is the thoracic temperature (Figure 8-49). A similar dependence of body temperature on wing loading has been elegantly demonstrated in bumblebees by allowing them to imbibe varying amounts of honey syrup, thus changing their wing loading. The bees were then flown for several minutes; their thoracic temperatures were measured, and then their abdomens were weighed. As shown in Figure 8-50, the bees with the heaviest abdomens, and hence with the highest wing loading, had the highest thoracic temperatures.

The heat which raises the thoracic temperature is a by-product of flight muscle activity. Therefore, in insects neither the maintenance of body temperatures above air temperature during flight nor even the maintenance of a uniformly elevated temperature during continuous flight in a constant ambient temperature demonstrates temperature regulation. However, the maintenance of an elevated

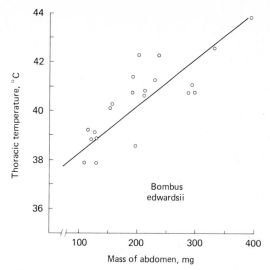

Figure 8-50
*Thoracic temperatures of queen bumblebees as a function of abdominal mass, which closely reflects the amount of honey carried in the honey stomach. Wing loading increases directly with abdominal mass. Ambient temperature, 16–22° C. [Modified from B. Heinrich, J. Comp. Physiol., **96,** 155–166 (1975).]*

and uniform level of thoracic temperature during flight over a wide range of ambient temperatures does show that body temperature is being regulated. Such regulation is well documented by both laboratory and field measurements, particularly in sphinx moths and bumblebees.

Two primary physiological mechanisms of temperature regulation are available to endothermic animals; (a) varying the rate of metabolic heat production and (b) varying the rate of heat loss to the environment. Both mechanisms are employed by heterothermic insects.

Regulation of Heat Production. The simplest way for an insect to vary its heat production is to fly intermittently, as do bees when they are foraging. By varying the ratio of time spent in flight to the time spent perching on flowers, a bumblebee can control its rate of heat production. In addition, while perched it can produce heat specifically for thermoregulation by means of periodic bouts of shivering of the thoracic muscles. It is possible that

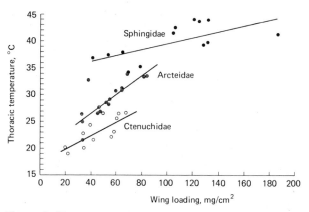

Figure 8-49
*The linear regressions of thoracic temperature during flight on wing loading in three families of moths. Each point represents the mean value for one species. Animals captured as they flew into a light trap in a rain forest in Costa Rica. Ambient temperature, 15–17° C. For conversion to SI units, see Table 3-2. [From G. A. Bartholomew and B. Heinrich. J. Exp. Biol., **58,** 123–135 (1973).]*

some insects can modify their rates of heat production during sustained flight by varying the amount of time spent gliding or by varying wing beat frequency, but documentation is not yet available.

Regulation of Heat Loss. Sphinx moths, like hummingbirds, feed while hovering and during sustained flight they control body temperature primarily by means of variation in rates of heat loss. During preflight warm-up most of the heat produced by the wing muscles of moths is sequestered in the thorax and the temperature of the abdomen remains near ambient. The abdomen is more lightly insulated than the thorax and during flight it serves as a heat exchanger by which the rates of heat loss to the environment can be regulated.

The hemolymph (blood) is the vehicle for transferring heat from the thorax where it is produced to the abdomen where it is lost to the environment. Insects have an open circulatory system and the hemolymph, which freely bathes the internal organs, is pumped from posterior to anterior by a large dorsally located, tubular heart that extends from the abdomen into the thorax. Insects respire by means of a tracheal system. Consequently, blood is not involved in the transport of oxygen or carbon dioxide. Since the hemolymph is not involved in gas transport, the rate of circulation can be varied over a wide range to control the rate of heat transfer from the warm thorax to the cool abdomen.

During flight at low air temperatures, heart beats are weak and irregular and the flow of hemolymph from abdomen to thorax is restricted. During flight at high air temperatures, however, the pulsations of the heart are strong and rapid. It pumps cool blood from the abdomen to the thorax while warm blood flows from the thorax to the abdomen (Figure 8-51) where the heat is lost to the environment through the thinly insulated body wall. By appropriate variations of the rate of flow of hemolymph between thorax and abdomen, sphinx moths can maintain high and uniform body temperatures over a wide range of ambient temperatures even though rates of heat production remain essentially constant.

Queen bumblebees incubate their brood by means of metabolic heat. The shivering of their thoracic muscles produces heat that is transferred by circulating hemolymph to the abdomen. The ventral surface of the abdomen is pressed against

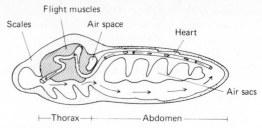

Figure 8-51
Schematic longitudinal section through the body of a sphinx moth. Arrows indicate the direction of haemolymph flow [Modified from B. Heinrich. Science, **168**, 580–582 (1970).]

the brood cluster and heats it by conduction. Under laboratory conditions when continuously supplied with food in the form of sugar syrup a queen bumblebee can incubate uninterruptedly for many days, maintaining brood temperature within a few degrees of 30°C even though ambient temperatures may fall as low as 3°C.

ADDITIONAL READING

The literature on body temperature is enormous and growing exponentially. There is no way one person can keep up with the new publications and at the same time acquire a mastery of the already existing works. However, the list below can open the door to the literature. It includes a number of recent specialized reviews each of which have extensive bibliographies. For example, there are over 1000 references, mostly on mammals, in Bligh [1973], over 500 references on birds in Calder and King [1974], and almost 300 on temperature compensation of poikilotherms in Hazel and Prosser [1974], over 130 on preferred body temperatures of reptiles in Dawson [1975], and over 500 on insect temperatures in Heinrich [1981]. The list also includes references which have appeared since the most recent reviews and references dealing with topics that received attention in the text but are not cited in the figure legends.

Bakken, G. S. "The use of standard operative temperature in the study of animal energetics," *Physiol. Zool.,* **53**, 108–119 (1980).
Bartholomew, G. A. "Aspects of timing and periodicity of heterothermy." In F. E. South, et al. (eds.). *Hibernation and Hypothermia, Perspectives and Challenges,* pp. 663–680, Elsevier Publishing Co., New York, 1972.

Bartholomew, G. A. "Physiological thermoregulation." In C. Gans and F. H. Pough (eds.). *Biology of the Reptilia,* vol. 12, *Physiological Ecology,* Academic Press, New York, 1981.

Bligh, J. "Temperature regulation in mammals and other vertebrates," *North Holland Research Monographs, Frontiers of Biology,* vol. 30, American Elsevier Publishing Co., New York, 1973.

Bradley, S. R., and D. D. Deavers. "A reexamination of the relationship between thermal conductance and body weight in mammals," *Comp. Biochem. Physiol.,* **65A,** 465–476 (1980).

Brown, G. W., Jr. (ed.). *Desert Biology,* Academic Press, New York, 1968.

Brown, J. H., and A. K. Lee. "Bergmann's rule and climatic adaptation in woodrats (Neotoma)," *Evolution,* **23,** 329–338 (1969).

Calder, W. A., "Temperature relations and underwater endurance of the smallest homeothermic diver, the water shrew," *Comp. Biochem. Physiol.,* **30,** 1075–1082 (1969).

Calder, W. A., and J. R. King. "Thermal and Caloric Relations of Birds." In D. S. Farner and J. R. King (eds.). *Avian Biology,* vol. IV, pp. 259–413, Academic Press. New York and London, 1974.

Carey, F. G., J. M. Teal, and J. W. Kanwisher. "The visceral temperatures of mackeral sharks," *Physiol. Zool.,* **54,** 334–344 (1981).

Chappell, M. A., and G. A. Bartholomew. "Standard operative temperatures and thermal energetics of the antelope ground squirrel *Ammospermophilus leucurus,*" *Physiol. Zool.,* **54,** 81–93 (1981).

Dawson, W. R. "On the physiological significance of the preferred body temperatures of reptiles." In D. M. Gates and R. B. Schmerl (eds.). *Perspectives of Biophysical Ecology, Ecological Studies,* vol. 12, pp. 443–473, Springer-Verlag, New York, 1975.

Dawson, W. R., and J. W. Hudson. "Birds." In G. C. Whittow (ed.). *Comparative Physiology of Thermoregulation,* vol. 1, pp. 223–310, Academic Press, New York, 1970.

Dawson, T. J., D. Robertshaw, and C. R. Taylor. "Sweating in the kangaroo: a cooling mechanism during exercise, but not in the heat," *Am. J. Physiol.,* **227,** 494–498 (1974).

Dizon, A. E., and R. W. Brill. "Thermoregulation in tunas," *Am. Zool.,* **19,** 249–265 (1979).

Feltz, E. T., and F. H. Fay. "Thermal requirements *in vitro* of epidermal cells from seals," *Cryobiology,* **2,** 261–264 (1966).

Fry, F. E. J. "Temperature compensation," *Ann. Rev. Physiol.,* **20,** 207–224 (1958).

Gates, D. M. *Energy Exchange in the Biosphere,* Harper & Row, Inc., New York, 1962.

Graham, J. B. "Heat exchange in the yellowfin tuna, *Thunnus albacares,* and skipjack tuna, *Katsuwanus pelamis,*

and the adaptive significance of elevated body temperatures in scombrid fishes," *Fish. Bull.,* **78,** 219–229 (1975).

Hammel, H. T., S. B. Strømme, and K. Myhre. "Forebrain temperature activates behavioral thermoregulatory response in Arctic sculpins," *Science,* **165,** 83–85, (1969).

Hardy, R. N. *Temperature in Animal Life,* University Park Press, Baltimore, 1978.

Hart, J. S., "Insulative and metabolic adaptations to cold in vertebrates," *Symposia of the Society for Experimental Biology,* no. 18, pp. 31–48, Academic Press, New York, 1964.

Hazel, J. R., and C. L. Prosser. "Molecular mechanisms of temperature compensation in poikilotherms," *Physiol. Revs.,* **54,** 620–677 (1974).

Heinrich, B. "Thermoregulation in endothermic insects," *Science,* **185,** 747–756 (1974).

Heinrich, B. *Insect thermoregulation,* John Wiley and Sons, New York, 1981.

Heller, H. C., and G. W. Colliver. "CNS regulation of body temperature during hibernation," *Am. J. Physiol.,* **227,** 583–589 (1974).

Herreid, C. F., and B. Kessel. "Thermal conductance in birds and mammals," *Comp. Biochem. Physiol.,* **21,** 405–414 (1967).

Hochachka, P. W., and G. N. Somero. *Strategies of Biochemical Adaptation,* W. B. Saunders Co., Philadelphia, 1973.

Hudson, J. W., W. R. Dawson, and R. W. Hill. "Growth and development of temperature regulation in nestling cattle egrets," *Comp. Biochem. Physiol.,* **49A,** 717–741 (1974).

Irving, L. *Arctic Life of Birds and Mammals,* Springer-Verlag, New York, 1972.

Kahl, M. P. "Thermoregulation in the wood storks, with special reference to the role of the legs," *Physiol. Zool.,* **36,** 141–151 (1963).

Kendeigh, S. C. "Energy responses of birds to their thermal environments," *Wilson Bull.,* **81,** 441–449 (1969).

Licht, P., and A. G. Brown. "Behavioral thermoregulation and its role in the ecology of the red-bellied newt, *Toricha rivularis,*" *Ecology,* **48,** 598–611 (1967).

Lyman, C. P., and R. C. O'Brien. "A comparison of temperature regulation in hibernating rodents," *Am. J. Physiol.,* **227,** 218–223 (1974).

Maher, M. J. "The role of the thyroid gland in the oxygen consumption of lizards," *Gen. Comp. Endocrinol.,* **5,** 320–325 (1965).

May, M. L. "Insect thermoregulation," *Ann. Rev. Entomol.,* **24,** 313–349 (1979).

McNab, B. K. "The economics of temperature regulation in neotropical bats," *Comp. Biochem. Physiol.,* **31,** 227–268 (1969).

Norris, K. S. "Color adaptation in desert reptiles and its thermal relationships." In W. W. Milstead (ed.).

Lizard Ecology a Symposium, University of Missouri Press, Columbia, MO, 1965.

Ohmart, R. D., and R. C. Lasiewski. "Roadrunners: energy conservation by hypothermia and absorption of sunlight," *Science,* **172,** 67–69 (1971).

Pough, F. H. "The advantages of ectothermy for tetrapods," *Am. Nat.,* **115,** 92–112 (1980).

Richards, S. A. "The biology and comparative physiology of thermal panting," *Biol. Rev.* **45,** 223–264 (1970).

Schmidt-Nielsen, K. *Desert Animals: Physiological Problems of Heat and Water,* Oxford University Press, Fair Lawn, NJ, 1964.

Scholander, P. F. "Counter current exchange: a principle in biology," *Hvalradets Skrifter Norske Videnskaps-Akad., Oslo,* **44,** 1–24 (1958).

Taylor, R. C., and C. P. Lyman, "A comparative study of the environmental physiology of an East African antelope, the eland, and the Hereford steer," *Physiol. Zool.,* **40,** 280–295 (1967).

——, and V. J. Rountree. "Temperature regulation and heat balance in running cheetahs: a strategy for sprinters?" *Am. J. Physiol.,* **224,** 848–851, (1973).

Walsberg, G. E., G. S. Campbell, and J. R. King. "Animal coat color and radiative heat gain: a reevaluation," *J. Comp. Physiol.,* **126,** 211–222 (1978).

Wang, L., and J. W. Hudson, (eds.). *Strategies in Cold: Natural Torpidity and Thermogenesis,* Academic Press, New York, 1979.

White, F. N., and J. L. Kinney. "Avian incubation," *Science,* **186,** 107–115 (1974).

Whittow, G. C. (ed.). *Comparative Physiology of Thermoregulation,* Academic Press, New York and London, 1970, 1973.

Wilson, K. F., and Lee, A. K. "Changes in oxygen consumption and heart-rate with activity in the tuatara, *Sphenodon punctatum,*" *Comp. Biochem. Physiol.,* **33,** 311–322 (1970).

Alan D. Grinnell

INFORMATION PROCESSING IN SENSORY AND NERVOUS SYSTEMS

9.1 INTRODUCTION: THE NEED FOR INFORMATION

A preeminent factor in natural selection is the capacity for obtaining and coordinating information, for the race is often to those of the species that can learn the most about their environment and act most effectively on this knowledge. The types of information are many and diverse. We are well aware of many of them: the visual information telling us where objects are, what is moving, what might represent danger or food or shelter; the auditory information permitting communication and the detection of sounds produced anywhere in the environment (again perhaps representing danger or food); the tactile information telling us of objects we are in contact with and their shape and texture; the olfactory and gustatory information helping us identify and choose foods and environments and

sometimes determining our moods. All these sources of information must be assessed simultaneously, because all may occur at the same time, and all may be meaningful. Moreover, the information from different sources may conflict, necessitating a decision favoring one source.

When all this is taken into account, it is only the beginning, for most of what we have mentioned deals only with the external environment. Just as necessary is information about the state of the organism. Some of this is conscious, such as the presence of hunger, thirst, or pain, all of which have powerful influences on behavior. Even more vital information is being gathered constantly and used unconsciously: the position of the head and limbs, the rate and direction of movement of the body, the state of contraction of muscles, the surface and core temperatures of the body, blood pressure, and the levels of oxygen and carbon dioxide in the blood. The list could be made much longer, but the point is obvious. In the course of evolution, animals have developed sensory organs capable of collecting all this information. At the same time has come the necessary development of neural processes to conduct the coded information to analysis centers, and central nervous ganglia to decode and integrate the incoming information and distribute instructions to muscles, glands, and other effector systems.

Competition for survival has provided the selective pressure to develop these sophisticated structures as organisms increased in structural and behavioral complexity and invaded new niches requiring new types of information. Initial sensitivity was probably chemical and photochemical, soon accompanied by mechanical irritability. As organisms grew more complicated and their environments more varied and variable, they evolved organelles, cells, or tissues whose exclusive function was that of responding with great sensitivity not only to the presence or absence of certain stimuli but to small changes in the magnitude of these stimuli: to the amplitude and changes in amplitude, the duration, the rate of onset and decay, the timing of the stimulus with respect to earlier stimuli, and even more complex stimulus parameters (color, saturation, presence of several wavelengths, frequencies, concurrent odors, and so on). Olfactory and gustatory organs took the place of simpler chemoreceptors. Light-sensitive organs developed means of discriminating patterns as well as intensities. Pressure and movement-sensitive organs evolved to monitor both external and internal environments as it became necessary to know rates of movement, body position, orientation to the ground, and whether or not anything was moving or vibrating nearby. To detect vibration at a distance, sensitivity to sound developed.

The evolution of a stable "internal milieu" permitted the invasion of harsher and harsher environments, but only at the expense of constant surveillance. Moreover, it became imperative that information from every source be integrated to provide the fastest, most appropriate action. Hence the specialized sense organs assumed even greater prominence, and the central analyzing centers became larger and more complex to process more rapidly and accurately the available information.

Commonplace as these concepts are, the products of this development are anything but commonplace, as the following chapters will show. For almost all sensory modalities zoological examples exist of sensitivities approaching or matching those of the most sensitive instruments designed by man, with versatility and dynamic range far beyond our present level of engineering sophistication. It goes without saying that the natural processes of perception, analysis, and integration of sensory input in most cases quite defy our understanding, let alone imitation.

Our increasing appreciation of the nature and importance of evolutionary processes has led to several convictions: that sensory and nervous capabilities are likely to be much greater than they first appear before we learn what specific questions they are designed to answer, that information obtainable by a receptor is useful to the organism (that neural development keeps pace with receptor evolution), and that no consistently observed morphological or physiological features of nervous or sensory systems are unessential.

This does not mean that all receptors for a given sensory modality are equally useful. The animal's selected environment greatly affects which senses will be emphasized and which neglected. If light is absent, as in caves and subterranean environments, vision is often reduced or absent. Similarly, restrictions in size and neural complexity, or successful adaptation to a restricted niche, tend to perpetuate the form of relatively primitive receptor systems.

Conversely, an animal's environment is defined and limited by its senses. What is perceived and experienced is the information provided by the senses. This information can be profoundly different in different animals. Highly evolved as the human senses are, a man walking his dog is quite unaware of the wealth of odors that intrigue the dog or of the high-frequency sounds the dog might be hearing. The dog, on the other hand, sees no color and is much less sensitive to low-frequency sounds. In general, any animal's senses ignore most of the ambient energy in the environment and only selectively react to certain features of stimuli they do sense. Color is meaningless to a man who is color blind. It is entertaining to imagine how different the world would seem if we could "see" into the infrared or directly sense radio waves; how different the world must be for a fish that lives with approximately neutral bouyancy at such great depths that sunlight cannot penetrate its environment, or for the larva of a fly, feeding industriously inside the larva of a moth. In most animals, simplicity of receptors and nervous system forces simplicity of behavior. In others, behavior may be made incalculably complicated by the interaction of several sensory systems of high acuity and the presence of subtle stimuli to which we are either insensitive or unconscious.

The nervous system is involved in all levels of an organism's behavior from the unconscious regulation of intracellular activity to coordination of movement of the whole organism and to the elusive phenomena of learning and memory. Our principal interest, in this text, is the way in which sensory organs and analyzing centers have, in the course of evolution, become adapted to perform their specialized functions. Before discussing the nervous system at this level, however, it is necessary to consider the fundamental properties of all sensory and nervous elements and the principles involved in receiving, transmitting, and integrating information. A thorough knowledge of these fundamentals is essential to an understanding of the diversity of sensory function and the limits within which a system can adapt and change its performance.

9.2 STRUCTURAL ELEMENTS

Information is normally carried from one part of an organism to another in one of two ways. The most primitive of these is the diffusion or circulation of an active chemical having a strong effect on specific other parts of the system. This is the principal means of communication between different components of single cells (and unicellular organisms) and, in the form of hormones, is of enormous importance in even the largest multicellular animals. A much faster and more specific means of information transfer is necessary, however, in animals having millions of cells specialized into tissues with very different functions. Nervous systems have evolved to fill this need.

9.2.1 Nerve Cells

The functional unit of the nervous system is the *neuron* (Figure 9-1). In vertebrates, the vast majority of these neurons are located within the spinal cord and brain, collectively known as the *central nervous system* (CNS). (The only exceptions are the retina of the eye and the cells of the autonomic nervous system, a collection of neurons that act relatively independently of the brain in unconscious regulation of visceral activity.) Neurons exist in a bewildering variety of shapes, but all share common features. There is always a cell body *(soma)* containing the nucleus and most of the cell's metabolic and synthetic machinery. Usually, two or more processes leave the soma. These are functionally specialized either to receive inputs from other cells or to provide synaptic output to other cells. In almost all cases, if one of these processes is severed from the cell body, it degenerates, but the cell body is capable of regenerating a new process in its stead.

The classical neuron is multipolar, with several dendrites at one end, an axon on the other. The *dendrites* are relatively short, thick, branching processes of irregular shape that receive most of the input to the cell. The *axon* is a thinner, much longer process, normally without synaptic input and with few branches (collaterals) until near its point of termination. The axon runs for long distances to act on other cells. Axons may reach several meters length and contain most of the cell's axoplasm. Other neurons are unipolar or bipolar; that is, the cell body has one process leaving it that branches and extends for considerable distances in more than one direction, or two axonlike branches that leave from different points of the cell soma. These branches are not morphologically distinguishable, but normally the terminals of one receive input while the other acts upon cells. Needless to say, there are many intermediates and exceptions to this overly neat categorization.

The vertebrate peripheral nervous system consists of bundles of axons, or nerve fibers, emanating from each segment of the spinal cord and from the base of the brain. A large fraction of them are the afferent processes of bipolar sensory neurons whose cell bodies are located near the spinal cord, in the dorsal root ganglia. These carry information from the receptor sites to the CNS. The rest of the fibers in the peripheral nerves are the axonal processes of multipolar motoneurons. These carry efferent commands to muscles, glands, and other effector organs. Between the sensory input and the efferent outflow is the CNS, sorting, comparing,

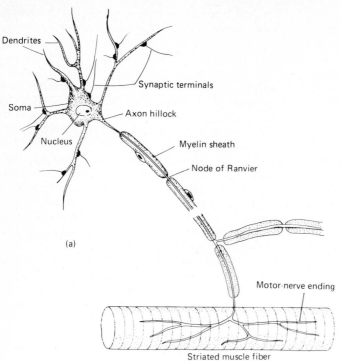

Dendrites

Synaptic terminals

Soma

Axon hillock

Nucleus

Myelin sheath

Node of Ranvier

(a)

Motor-nerve ending

Striated muscle fiber

Figure 9-1
Structure of neurons. (a) Diagrammatic figure of a neuron (vertebrate motoneuron) showing a small fraction of the afferent input to the cell, and one termination on a muscle fiber. An indefinite length of axon has been omitted. (b) Spectrum of types of vertebrate neurons, arranged according to function and level in the nervous system. Invertebrate neurons have similar shapes, but there is a higher proportion of monopolar neurons (type 3), and the nervous organization is different. [*From D. Bodian.* Cold Spring Harbor Symp. Quant. Biol., **17**, 1–17 (1952).]

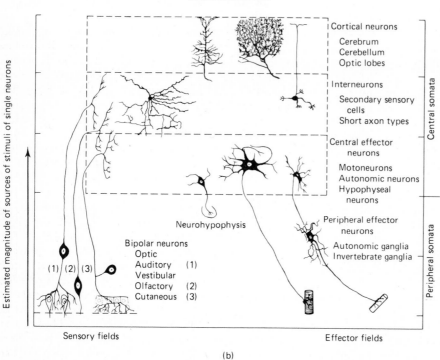

Estimated magnitude of sources of stimuli of single neurons

Cortical neurons

Cerebrum
Cerebellum
Optic lobes

Interneurons

Secondary sensory
cells
Short axon types

Central effector
neurons

Motoneurons
Autonomic neurons
Hypophyseal
neurons

Central somata

Neurohypophysis

Peripheral effector
neurons

Autonomic ganglia
Invertebrate ganglia

Peripheral somata

Bipolar neurons
Optic
Auditory   (1)
Vestibular
Olfactory   (2)
Cutaneous   (3)

(1) (2) (3)

Sensory fields

Effector fields

(b)

and integrating the sensory messages and formulating the complex reactions that constitute behavior.

An important part of the sensory input to the CNS is that providing continuous feedback information about the effects of efferent commands, for example, muscle movement. As muscles contract, sensory receptors (*proprioceptors*) like miniature strain gages maintain a flow of information about amount and rate of change of stretch and tension. This information allows constant readjustment of muscle activity, even after movement has started (see Section 10.3.5). An important part of the efferent outflow, by the same token, is devoted to maintaining receptors in the desired state. Muscle stretch receptors are mounted on specialized muscle fibers whose tonic contraction can be adjusted to maintain the receptors at their state of greatest sensitivity or help initiate and maintain a given degree of muscle contraction. This servocontrol mechanism is just one of countless many that exist throughout the nervous system, governing excitability, channels and speed of analysis, and efferent outflow.

Invertebrate nervous systems clearly operate on the same principles with neuronal units and a central analyzing and controlling system. Invertebrate sensory cells are usually located peripherally, however, and there is a high proportion of unipolar cells in which the cell body is far removed from the site of interaction between cells.

The cell membranes of axon, soma, and dendrite appear much the same, even at the resolution of the electron microscope, but they differ in many properties, just as the enclosed cytoplasm differs. Both axons and dendrites contain mitochondria and characteristic neurotubules (microtubules) of unknown function; axons also contain smaller neurofilaments (see Figure 9-2). These structures may be involved in the movement of substances away from the soma, which has been shown to occur at the rate of a few millimeters per day (see Section 11.6.3). There is also movement of substances against this current toward the soma, so it must not be concluded that the sole function of nerves is to conduct impulses, or that the sole contact between nerves and other tissues is electrical, or even associated with electrical activity. As we shall see in Chapter 11, important trophic influences operate between nerve and end organ independent

of electrical activity, and the same is probably true of transneuronal interaction.

9.2.2 Neuroglia

In a few invertebrates, all of the axons are naked, with no surrounding sheath. This is unusual, however. In most invertebrates and all vertebrates, a large proportion of the axons are wrapped in one or more layers of satellite (*glial*) cell tissue. The most extreme example is the thick myelin sheath found surrounding a large fraction of vertebrate peripheral axons (see Figure 9-2). In both vertebrates and invertebrates, the cell bodies in the CNS are normally also surrounded by large numbers of glial cells, most of which are not involved primarily in forming insulating sheaths. Their function, however, is poorly understood. They are often huge cells, especially in invertebrates, with large membrane potentials and low (electrical) resistance contacts from one cell to another. They are unquestionably important as structural supports for neurons, and there is good evidence for their involvement during degeneration and regeneration of nervous elements. Moreover, there are now well-documented cases in which neurons appear to use glial cells as guides or markers as they move from one area of the nervous system to another during development.

Although glial cells do not form a barrier to free diffusion of substances through the extracellular space, it is possible that a primary function is nutrient and waste-product exchange because they are, in general, rich in mitochondria, glycogen and lipid granules, and endoplasmic reticulum. Because the glial cells are highly and selectively permeable to K^+, it is possible that one of their roles might be to remove extracellular K^+ from areas of high neuronal activity and redistribute it in less active areas. Often neuroglia are seen surrounding synaptic terminals, perhaps insulating one synapse from others nearby. In a few cases, glia, in close association with large numbers of active neurons, exhibit slowly rising and long-lasting depolarization. They appear not to produce action potentials, and there is little or no evidence for their involvement (except for the sheath cells) in the functional electrical activity of the nervous system. However, such potentials almost certainly affect extracellular recordings

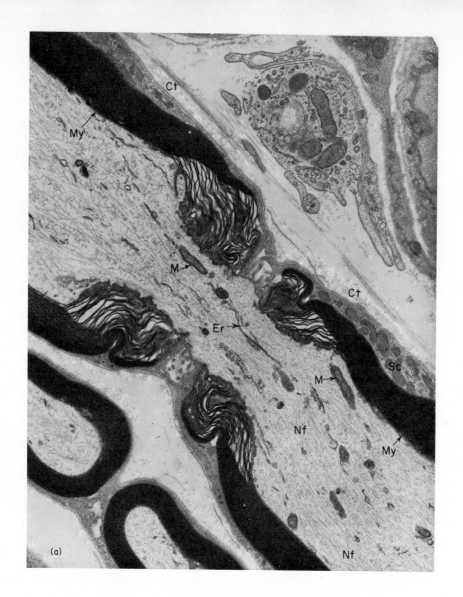

(a)

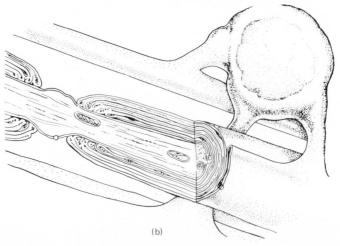

(b)

Figure 9-2

Axon sheaths. (a) An electron micrograph of mouse sciatic nerve showing a node of Ranvier where the myelin sheath is interrupted and excitation takes place during saltatory conduction (see Section 9-6-3). Note the fine structure within the axon. Code: Nf, neurofilaments; Er, endoplasmic reticulum; M, mitochondria; My, myelin sheath; Ct, connective tissue; Sc, Schwann cell cytoplasm. [From K. R. Porter and M. A. Bonneville. An Introduction to Fine Structure of Cells and Tissues, *2nd ed., Lea & Febiger, Philadelphia, 1964.]*

(b) Diagrammatic representation of a myelin sheath and node in the CNS, with the sheath cut away to show the way it is wrapped around the axon. In the CNS a single oligodendroglial cell can form myelin sheaths around several axons. In the peripheral nervous system, Schwann cells form single internodal sheath segments around a single axon. [From M. B. Bunge, R. P. Bunge, and H. Ris. J. Biophys. Biochem. Cytol., **10,** *67–94 (1961).]*

(c) Cross section through adult rat sciatic nerve showing a myelinated axon (top) and a Schwann cell (with prominent nucleus) providing a loose wrapping for several unmyelinated axons (Ax_1, Ax_2, etc.). [From A. Peters, S. L. Palay, and H. Webster. The Fine Structure of the Nervous System, *Harper & Row, New York, 1970.]*

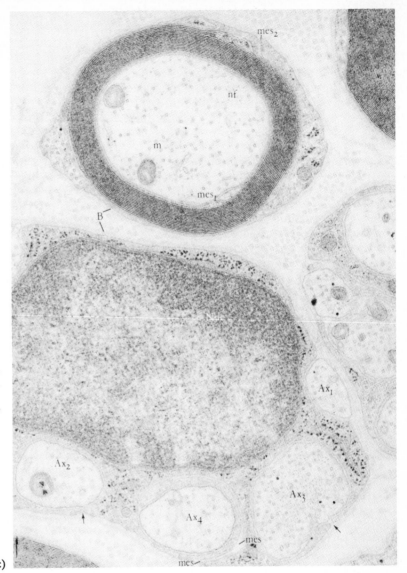

(c)

designed to interpret the electrical correlates of nerve activity.

9.2.3 Sensory Receptors

A stimulus goes no farther than the receptor. Light is not itself carried by way of the optic nerve to the brain and, even if shone directly on the optic centers of the brain, would have no effect. It is the job of the receptors to extract all the information they can and pass this on to central analyzers. All sensory information passes to the CNS by way of the primary sensory neurons. Most commonly, processes of these neurons are the sensory elements themselves. In the simplest case, the nerve terminals of certain touch- and temperature-sensitive fibers end freely in epithelial tissue, with no consistent morphological organization of the surrounding cells (Figures 10-4 and 10-26). In at least one tissue (tadpole tail), they have been shown to move slowly from one point to another. Terminals of different cells often overlap, but apparently do not make functional connections with each other. Each of these terminal fields may apparently be selectively sensitive either to temperature or to mechanical stimulation, or in some cases to both, but often no corresponding structural differences can be demonstrated.

Primary sensory fibers usually terminate in a more characteristic and complicated way, in association with specialized surrounding tissue (see Figure 10-4). These specialized surrounding structures often magnify a stimulus and affect its dynamics, and they usually impose specificity on the system. Most surface touch receptors, for example, are associated with hairs or specialized cells in the dermis; olfactory-nerve terminals end in precise association with nasal epithelial cells; and the most sensitive temperature receptors are located on thin membranes having extremely low heat capacities. A large number of sensory terminals are enclosed in connective tissue capsules, ranging in complexity from simple joint receptors to Pacinian corpuscles (Section 10.3.4). Although it has been customary to correlate specificity of response with type of capsule in such endings, recent studies have revealed such a variety of surrounding capsules that there is no longer widespread agreement on specific structure-function relationships.

Most complicated of all, of course, are the eye and ear, in which the role of transducing the stimulus has been taken over by highly specialized cells (secondary sensory receptors) in contact with the nerve endings. The morphological basis for specificity of response is quite obvious in these receptors, in contrast to simpler ones. Often, however, there can be astonishing degrees of similarity of function in very different-looking receptors. Yet each receptor unit, be it a free nerve ending or a retinal cone, is evolved to be particularly sensitive to only one form of energy (the "adequate stimulus") and will respond, if at all, only to much more intense stimuli of other types. To its appropriate energy form, each receptor is much more sensitive than are other types of receptors or nerve cells, and the receptors of vision, hearing, and smell approach the theoretical limits.

One of the most remarkable findings to emerge from electron microscopic studies of the fine structure of receptors is the widespread use of modified cilia in reception. Among invertebrates and vertebrates, cilia have been adopted quite independently in many different receptors, both as the receptor terminals of primary sensory nerves and as the receivers of mechanical energy in secondary sensory cells. For example, in insects, where cilia do not otherwise occur, the sensory nerves in mechanosensory hairs terminate in a modified cilium that attaches to the hair, and sound-sensitive terminals in the tympanic organ are similar modified cilia (Figure 10-15). In vertebrate hair cells, each cell has protruding from it and coupled with some mechanical transducer a group of cilialike hairs and one typical *kinocilium* (Figure 10-12). Moreover, the terminal hairs on most chemosensory nerve endings appear to be modified cilia, and the vertebrate visual receptor cells (rods and cones) are derived from cilia (Figure 10-32). The advantages of using modified cilia for reception are not clear, but they must be considerable.

9.3 ELECTRICAL EVENTS AND INFORMATION PROCESSING

As an introduction to the vocabulary of electrophysiology and to some of the basic principles of sensory function, it might prove useful to describe information transfer in one of the best-known mechanoreceptors—the crustacean stretch receptor, used to monitor the position of different parts of the exoskeleton. This organ consists of two tiny modified muscle bundles lying parallel to the skeletal muscle joining two segments of carapace. One of these bundles is slightly more tightly stretched than the other. Attached to the midportion of each are the dendrites of a sensory neuron (Figure 9-3). If electrodes are placed on each of the axons of these sensory cells when the muscle bundles are at resting length, it is found that the cell on the more tightly stretched muscle is spontaneously producing a regular train of identical spikes; the other is silent. If the two pieces of carapace are slightly separated, stretching both muscles, the first cell sharply increases its rate of firing, then adapts to maintain a new rate higher than the initial, while the second cell produces only a brief burst of spikes

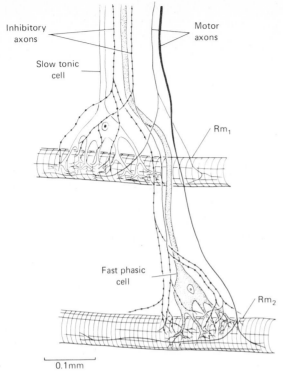

Figure 9-3
Crustacean stretch-receptor organ, with slow and fast receptor neurons on their specialized muscles. Receptor muscle 1 (Rm₁) is stretched more tightly than Rm₂. Both sensory neurons receive inhibitory innervation from central neurons, and both muscles receive motor in-nervation that can adjust the amount of their tonic contraction. [*From T. H. Bullock and G. A. Horridge.* Structure and Function in the Nervous Systems of Inver-tebrates, *vol. 1, W. H. Freeman and Company, San Francisco, 1965.*]

Labels in figure: Inhibitory axons; Motor axons; Slow tonic cell; Rm₁; Fast phasic cell; Rm₂; 0.1mm

(Figure 9-4). These extracellularly recorded spikes are a reflection of all-or-none potentials within the axons, conducting coded information from the re-ceptor neurons to the CNS. The nature of the code is obvious. One cell (called the *slow receptor*), is typical of *phasic-tonic receptors*, in which a certain level of sustained activity is increased or decreased to reflect changes in stimulus intensity and dura-tion. The relationship is such that the number of spikes increases approximately in proportion to the logarithm of the stimulus intensity and, as in almost all receptors, there is a transient phasic response

(either "on" or "off") of greater magnitude that depends on the rate of change of the stimulus. This is somewhat more prominent in most receptors than it is in the crustacean slow cell. The second, *fast receptor* cell is typical of *phasic receptors,* in which a normally silent cell responds sharply to a change in stimulus, but "adapts out" readily, pro-viding information only about the dynamic changes in a stimulus.

The way in which a graded stimulus can be trans-lated into a series of all-or-none action potentials does not become apparent until one can record inside the sensory cell, near the dendrites them-selves. This was made possible by the development in 1951, by Nastuk and Hodgkin, of electrodes suit-able for intracellular recording from small cells. A tiny glass micropipette is employed with a tip diameter of less than 1 μm and filled with 3 M KCl or a similarly conductive fluid. As such elec-trodes penetrate the crustacean stretch receptor neurons, they reveal that in both the potential in-side is approximately 70 mV negative with respect to the outside. This is the "resting" or membrane potential. In the fast cell, the inside potential re-mains essentially constant. In the slow cell, how-ever, the resting potential undergoes changes such as are shown in Figure 9-4b, decreasing gradually toward a threshold (about 15 mV depolarized from the most negative potential) at which point an ac-tion potential arises suddenly. At the peak of the spike, the potential briefly reverses polarity, the inside becoming 30–40 mV positive with respect to the outside. After each action potential, the membrane potential falls back toward its most neg-ative value, then begins to depolarize again. When a stretch is applied, the depolarization to threshold is more rapid, but the spikes still look the same. Because there is a greater flow of depolarizing cur-rent, however, there can be more spikes per unit time. As the looser muscle bundle begins to stretch, the membrane potential inside the fast cell also falls. If the stretch is large enough, threshold is reached and one or more spikes are initiated. If the stretch is less, the depolarization reaches a pla-teau below threshold. The rate and extent of depo-larization toward threshold are found to vary with the logarithm of stimulus intensity in both cells, although the fast one, because of its mechanical coupling, requires a much larger stimulus before it reaches threshold.

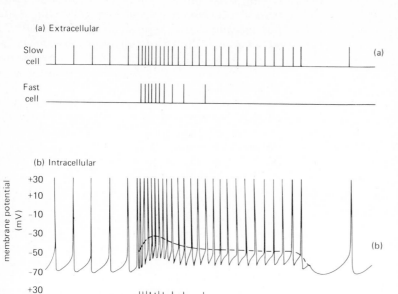

(a) Extracellular

Slow cell

Fast cell

(a)

(b) Intracellular

Slow cell–membrane potential (mV)

+30
+10
−10
−30
−50
−70

(b)

Fast cell–membrane potential (mV)

+30
+10
−10
−30
−50
−70

(c)

Stimulus

0.1 sec

Figure 9-4

Firing patterns of crustacean stretch-receptor neurons. (a) Extracellular records from axons of both fibers. Note the regular spontaneous firing of the slow (slowly adapting) receptor. When a stretch is applied, the slow receptor responds with a sustained increase in firing rate, the fast receptor with a brief burst of spikes that quickly adapts out. (b) Intracellular potential changes recorded at a point (near the axon hillock, for example) where both generator potential and spike can be recorded. The slow cell undergoes slow spontaneous depolarization until threshold for spike initiation is reached. After each spike, the membrane potential falls toward its most negative value, then begins depolarization again. When a stimulus (stretch) is applied, both slow and fast neurons are depolarized for as long as the stimulus remains. In the slow cell this results in sustained firing at an accelerated rate, whereas in the fast receptor, despite the continued presence of a generator potential, the burst of spikes ends very quickly. This adaptation is the result of accommodation (see Section 9-9-2). Following each spike there is a period of refractoriness during which excitability is entirely eliminated and then, for a brief time, much reduced. In both receptors, the rate of firing is linearly related to the rate or amount of depolarization up to a point of saturation, at which the refractory period following each spike makes faster firing impossible (see Section 9-6-1). Note the constant amplitude reached by each spike. The dotted line represents the generator potential one might record if the action potentials were blocked with tetrodotoxin (see Section 9-5).

The graded depolarization is called a ***generator potential,*** because it somehow causes the action potentials. It is found, moreover, that the frequency of action potentials produced is directly proportional to the rate of depolarization (or the amount of depolarization above threshold). The true extent of the generator potential and its time course in the absence of all-or-none spikes can be seen by applying certain drugs (procaine, tetrodotoxin) that selectively block action potentials. It is found in such preparations that the generator potential is maintained at almost the same amplitude as long as a stimulus is applied. In the slow cell, the generator potential continues to produce spikes at a rate proportional to its amplitude. In the fast cell, however, the firing rate falls rapidly, despite the continued presence of a generator potential. This is the result of *accommodation* at the locus where spikes are initiated, raising the threshold for firing (see Figures 9-20 and 9-24 and Section 9.9.2).

Thus in this typical mechanoreceptor, the information (size and time course) in a stimulus is twice translated—first into a graded intracellular depolarization, the size and duration of which reflect the stimulus intensity and duration. This graded potential presumably results from the distortion and stretching of dendritic membranes. The second translation is from generator potential into all-or-none action potentials in direct proportion to the amount of depolarization. The two receptors differ in behavior, and presumably in function, because of differences in stretch of the muscle bundles to which they are attached and accommodation in the spike-generating locus of the fast cell.

It is obvious that some information is lost in the second translation. Uncertainty is greatest near threshold, for either a long weak stimulus or a brief intense one might produce a single spike. Even at a high intensity there may be a range of durations that would result in, say, four action potentials but not five. The higher the possible spike frequency and the greater the change in frequency with change in stimulus intensity, the greater the accuracy of information transfer. Redundancy of input, the parallel functioning of many different receptor units (and even different receptor organs), helps accomplish the same increase in certainty.

Like other sensory receptors, the crustacean stretch receptor is subject to central control. Its excitability is governed by the activity of an inhibitory nerve ending on the dendrites of both cells. This nerve releases a chemical, γ-aminobutyric acid (GABA), that, by increasing cell-membrane conductance to chloride ion, prevents receptor-cell depolarization to threshold.

The electrical events associated with sensory and nervous activity are the usual, and sometimes the only, measure of response, and they probably represent the principal means of information coding, so we shall discuss these events in greater detail.

9.4 MEMBRANE POTENTIAL

9.4.1 The Origin of the Resting Membrane Potential

All living cells are enclosed by a semipermeable membrane and contain large quantities of cytoplasmic protein molecules too large to penetrate the membrane. These protein molecules typically bear a net negative charge. Smaller ions that can diffuse across the membrane, such as K^+ and Cl^-, tend to be distributed in a way that compensates for this: K^+ predominantly inside, Cl^- outside. The distribution approximates a *Gibbs-Donnan equilibrium*, which, however, is not a sufficient description of the system in most cells, because there are permeant ions that do not assume the expected distribution (see following paragraphs). Table 9-1 shows the observed distribution of the principal ions for the two cells from which we have derived most

TABLE 9-1. Distribution of Ions in Squid Axons[1] and Frog Muscle Fibers[2] and Their Surrounding Fluids (in mM)

Ion Species	Seawater	Squid Blood	Squid Axon	Frog Blood	Frog Muscle
K^+	10	20	400	2.5	140
Na^+	460	440	50	120	9.2
Cl^-	540	560	40–150	120	3–4
Ca^{2+}	10	10	0.4		
Mg^{2+}	53	54	10		
Anions$^-$	—	—	350 (isethionate 250) (aspartate 75)		

Sources: (1) A. L. Hodgkin. *The Conduction of the Nerve Impulse,* Charles C. Thomas, Publisher, Springfield, Ill., 1964; (2) B. Katz. *Nerve, Muscle, and Synapse,* McGraw-Hill Book Company, New York, 1966.

of our knowledge: the squid giant axon and frog skeletal-muscle fibers (see also Table 7-6).

Ions to which the membrane is permeable tend to diffuse down their concentration gradients, for example K^+ from inside to outside, Cl^- from outside to inside. However, because of the opposing need to keep equal numbers of positive and negative charges on either side of the membrane, movement of diffusible ions down their concentration gradients continues only until the imbalance of charge on either side of the membrane builds up a potential sufficient to counteract their concentration gradients. This potential exists as an electromotive force across the membrane resistance, which charges the membrane capacitance.

This is essentially the hypothesis proposed by the German physiologist Bernstein, who was building on the work of Ostwald and Nernst when he suggested, in 1902, that selective permeability to one ion (K^+) and impermeability to others could explain the membrane potential. Potassium would flow out, down its concentration gradient, leaving an excess of negative charge inside to build up a negative membrane potential that would eventually halt any further net efflux of K^+. It was soon shown that, as Bernstein predicted, the resting potential varied linearly with the logarithm of external K^+ concentration (written $[K^+]$), being essentially eliminated if the $[K^+]$ outside was made as high as that inside. The membrane potential could even be reversed if the external $[K^+]$ exceeded the internal concentration. Subsequent use of radioactively labeled ions confirmed that movement of K^+ could explain these effects, and the membrane potential was shown to vary directly with the absolute temperature. Thus to a first approximation, the membrane potential was shown to be determined by the balance between two forces:

1. The tendency of K^+ ions to diffuse out of the cell, down their concentration gradient, a force described by

$$RT \ln \frac{[K^+_{inside}]}{[K^+_{outside}]}$$

where R is the gas constant and T the absolute temperature.

2. The tendency for K^+ ions to be pulled back by the buildup of negative charge, an electrical force described by the expression zFE, where z is the valence of the ion in question ($+1$ for K^+), F the Faraday number, and E the membrane potential.

At equilibrium, these form the Nernst equation, as specifically stated for K^+:

$$zFE_K = -RT \ln \frac{[K^+_{in}]}{[K^+_{out}]}$$

or

$$E_K = -\frac{RT}{zF} \ln \frac{[K^+_{in}]}{[K^+_{out}]}$$

At 20°C, using base-10 logarithms, the Nernst equation reduces to $E_K = -58 \log [K^+_{in}]/[K^+_{out}]$, where E_K is the equilibrium potential for K^+ in millivolts. This indicates that, for every tenfold concentration change, the K^+ equilibrium potential changes by 58 mV. As can be seen from Table 9-1, $[K^+_{in}]/[K^+_{out}]$ is approximately 20 in the squid axon, 55 in frog muscle. This leads to values of E_K of -75 and -90 mV.

It is apparent from Table 9-1 that Na^+ and the divalent cations behave very differently than K^+. They are located predominantly outside, despite concentration gradients tending to drive them inside and a net negative potential tending to pull them in. The membrane is, in fact, essentially impermeable to these ions. If the membrane were selectively permeable to Na^+ instead of K^+, the expected membrane potential would be $E_{Na} = 58 \log 9 = +55$ mV. Na^+ is 125 mV out of equilibrium.

E_K should be identical to the membrane potential, if this is the only ion involved. In fact, in most cells the membrane potential is somewhat lower (less negative) than E_K; this discrepancy is explainable by Cl^- permeability and a slight leakiness of the membrane to Na^+, which moves into the cell, allowing more K^+ to diffuse out. To keep the inside from filling up with Na^+, metabolic work must be done to pump it out. This energy-consuming process has been termed the sodium pump. Although the permeability to Na^+ (or other ions out of electrochemical equilibrium, such as Ca^{2+}) is normally so low that metabolic poisoning of a cell does not result in rapid depolarization of the membrane, still the metabolic pumps are necessary over long periods of time. The "resting" membrane potential is at best a stable potential, not an equilibrium condition.

The membrane permeability to Cl^- ions is often nearly as high as that to K^+. Because Cl^- has the opposite concentration gradient from K^+ and the opposite charge, its equilibrium potential lies close to the same value as E_K. One of the major importances of this high Cl^- permeability is that it, like the K^+ permeability, acts as a membrane-potential stabilizer. Whenever the potential changes slightly, K^+ moves out and Cl^- moves in (or vice versa) to counteract the change, restoring equilibrium. In fact, a more rigorously correct expression for the membrane potential E_M must include all major ions with their permeabilities and concentration ratios. This is expressed in the Goldman-Hodgkin-Katz equation:

$$E_M = -\frac{RT}{zF} \ln \frac{P_K[K_{in}^+] + P_{Na}[Na_{in}^+] + P_{Cl}[Cl_{out}^+] + \cdots}{P_K[K_{out}^+] + P_{Na}[Na_{out}^+] + P_{Cl}[Cl_{in}^+] + \cdots}$$

where P represents membrane permeability to a given ion.

In resting squid axons, if we define $P_K \equiv 1$, then $P_{Na} \doteq 0.013$ and $P_{Cl} \doteq 0$. Clearly the resting membrane potential could be anywhere between E_K and E_{Na}, depending on the relative permeabilities to the various ions. In frog skeletal muscle, $P_K \equiv 1$, $P_{Na} \doteq 0.03$, $P_{Cl} \doteq 2$, and the membrane potential closely approximates both E_K and E_{Cl} (See Figure 9-5a).

A familiar electrical model of an excitable cell's membrane is shown in Figure 9-5b, where C_M denotes the membrane capacity, the batteries E_K, E_{Na}, and E_{Cl} represent the equilibrium potentials of the major ions, and the variable resistances represent the changeable membrane permeabilities to various ions. Because this is an electrical model, for which values are normally obtained by electrical measurements, ionic conductances (g_K, etc.) are used instead of permeabilities. For our purposes, membrane permeability and conductance can be taken as essentially equivalent; but they are not identical. They can differ significantly as a function of the charge environment of the membrane and

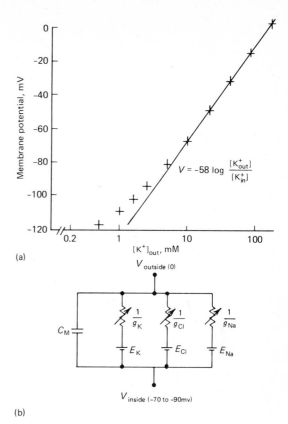

(a)

(b)

Figure 9-5

*(a) Plot of the effect of external potassium ion concentration on the resting membrane potential of frog skeletal muscle fibers. These measurements were made in the absence of Cl^-, to which the membrane is also permeable (SO_4^{2+} being used instead). The points correspond over much of this range to a line (shown) with a slope of 58 mV per tenfold change in [K^+] (a "Nernst slope"). At low [K^+], however, there is a marked deviation. This is caused by the Na^+ in the solution. Although the membrane is only 1–3% as permeable to Na^+ as to K^+, the Na^+ begins to become important at very low [K^+]. [After A. Hodgkin and P. Horowicz. J. Physiol., **148**, 127–160 (1959).] (b) Equivalent circuit diagram of nerve or muscle cell membrane.*

the specific mechanism of permeation. Moreover, conductance depends not only on permeability but on the availability of the appropriate ions to reveal the permeability.

In a few cases the resting potential is established

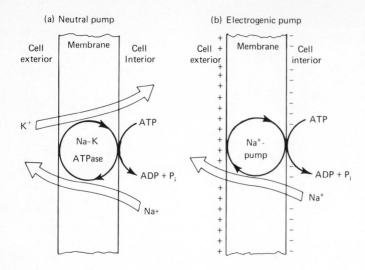

(a) Neutral pump

Cell exterior | Membrane | Cell Interior

K^+

Na–K ATPase

ATP

ADP + P_i

Na^+

(b) Electrogenic pump

Cell exterior | Membrane | Cell interior

Na^+-pump

ATP

ADP + P_i

Na^+

Figure 9-6

Cation pumps may be (a) neutral or near neutral, in which energy from splitting ATP is used by an intrinsic membrane protein to move Na^+ outside the cell and K^+ inside, both against their concentration gradients; or (b) electrogenic, where the "pump" moves only one ion (Na^+ in the example shown) across the membrane, or an unequal number of ions in opposite directions.

in large part by active ionic pumps, such as the Na^+ pump mentioned previously. Active ion transport may be "electrogenic" or neutral (Figure 9-6). In an electrogenic pump, one ion species is actively transported across a membrane either with ions of opposite charge following passively or in exchange for a smaller number of ions of the same charge. A neutral pump is one in which equal numbers of ions of positive and negative charge are transported simultaneously, or equal numbers of ions of the same charge exchanged, generating no net potential. A neutral pump could appear to be electrogenic if in its pumping it significantly depleted some ion species (for example, K^+) on one side of the membrane, changing the concentration gradient and equilibrium potential for that ion. Although there is no evidence that it is of general importance, in at least one case electrogenic ion transport is known to contribute as much as one third of a cell's resting potential. Na^+ pumps are by no means the only ion transport systems of importance. Cl^- pumps are known to exist in some membranes, and most cells must possess a strong Ca^{2+} pump to compensate for an electrochemical imbalance of 200 mV or more.

9.4.2 Biophysical Basis for Selectivity of Permeability

The selective permeability properties of the membrane are critical to the establishment of transmembrane potentials. Much attention has been focused on cell membranes, with electron microscopic, x-ray diffraction, and sophisticated chemical analysis, but the morphological basis for selective permeability is still not well understood. There are always a number of connective tissue layers (basement membrane, or *ectolemma*) on the surface of the fiber, surrounding the actual plasma membrane *(plasmalemma),* which apparently forms the main diffusion and potential barrier. The plasmalemma is less than 100 Å thick, and consists basically of a double layer of lipid associated with protein. Where it was once envisioned that the membrane is a relatively static structure, with layers of protein fixed to the polar groups of the lipid bilayer, it is now clear that the membrane is highly dynamic (Figure 9-7). Both lipid and protein components show high turnover rates. The protein appears to be organized into functional aggregates, differing on the two sides of the membrane, incorporating specific receptors and their associated enzymes, such as adenyl cyclase and Na^+–K^+ ATPase. These aggregates in most cases appear to be able to move with considerable freedom from one point to another on the cell surface. Where components are fixed in position, such as in the subsynaptic membrane of the neuromuscular junction, there is commonly a thickening of the membrane. The proteinaceous components of the outside of the membrane, including glycoproteins with long, highly charged carbohydrate chains, sometimes

Figure 9-7

Schematic representation of the plasma membrane (modified fluid mosaic model) showing three-dimensional and cross-sectional views. The large globular bodies represent integral proteins, assembled into functional aggregates of component proteins. These can be localized to one side of the phospholipid bilayer or extend through it. Many have long glycoprotein chains (GP) with special reactive groups that are important to cell-cell interactions, immune recognition, etc., extending into the extracellular environment. The exposed surfaces of each protein aggregate would bear most of the ionic residues, whereas the nonpolar residues would largely be imbedded in the matrix. Some of the membrane proteins are free to diffuse laterally in the membrane, others are held in place by extracellular glycosaminoglycans (GAG), and/or anchored to deeper structures such as microfilaments (MF) and microtubules (MT). [*From G. L. Nicolson.* Current Topics in Developmental Biology, **13,** *305–338 (1979).*]

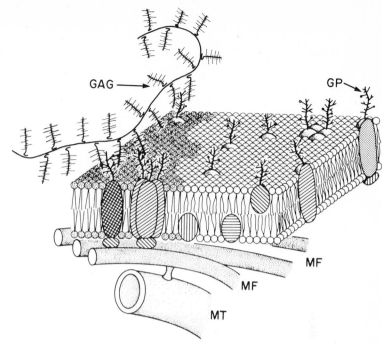

extend far out from the surface and probably are important both in enzymatic interaction with the environment and in specific interactions with hormones, antigens, and other cell surfaces. Some complex transmembrane proteins even extend through the lipid to the other side of the membrane. The lipid layer is sufficiently constant to form a good insulator, however, and most or all of the potential difference between inside and outside exists as a gradient across it. By maintaining a 70–90 mV capacitive charge, the plasmalemma has to withstand a field across it of approximately 10^5 V/cm. Its resistance to ionic flow is 100 million times greater than that of the fluids within or outside the membrane, and its permeability, even to K^+, is so low that fewer than 1 in 10^8 colliding ions penetrate it.

It was once widely assumed that the differences in permeability to different ions, for example the monovalent cations, were governed simply by the diameter of membrane pores and the effective diameter of the ion. For example, the apparent hydrated radius of K^+ is slightly less than that of Na^+ (2.4 Å versus 3.3 Å), so that a resting membrane might have pores of about 3 Å radius. The fact that the same membrane, during an action potential, becomes much more permeable to Na^+ than to K^+ led to the suggestion that permeation in the active state was in the nonhydrated form, where the Na^+ radius (0.95 Å) is significantly less than that of K^+ (1.33 Å). This explanation became untenable with the demonstration that other monovalent alkaline cations, Li^+, Rb^+, and Cs^+, did not fit predictions.

Major advances have been made in our understanding of the basis for selectivity of permeability to different ions, and the mechanisms whereby ions permeate biological membranes. It is clear now that most, if not all, major ion permeabilities associated with neuronal signal processing do involve aqueous channels (see Figure 9-9). The permeability through these channels is so great, when they are open, that it is not feasible that molecular carriers could be responsible. However, the determination of what will go through a channel depends only in part on the dimensions of the channel and its opening. In addition, an important feature of such channels—and the property that gives each its ionic selectivity—is the presence of one or more charged groups lining the channel. Given the basic ability for an ion to fit into the channel, and through its narrowest constriction, the main factor determining permeability appears to be the relative strength of coulombic interactions between the ion and water (hydration energy) on the one hand, and between the ion and charged membrane sites on the other. If the field strength of the membrane sites or carriers is very weak, ion interactions with the dipole and quadrupole charges of water will predominate, and the ions most strongly interacting with opposite charges will experience greatest difficulty in binding with membrane sites. Among the monovalent cations, this results in the permeability sequence $Cs > Rb > K > Na > Li$, since this is the reverse order of ionic radius; and the smaller the shell of electrons around the positively charged nucleus, the stronger is the interaction with external negative charges. If the membrane charges are so strong that they overwhelm ionic interaction with water, then the permeability sequence is $Li > Na > K > Rb > Cs$, with the smallest ions of a given charge now binding most strongly with membrane sites. With membrane sites of intermediate field strength there are intermediate sequences, but only 9 compared with the 118 that could occur by random combination. The 11 observed sequences are those predicted to result from the fact that the coulombic force of attraction between an ion and water falls as the third or fourth power of ionic radius, whereas the force of attraction between the ion and a membrane negative charge falls only as the square of the radius. Thus the larger ions are more strongly influenced by changes in membrane field strength than are the smaller ions.

In biological membranes the most common charged groups are usually carboxyl and phosphoric acid groups. The field strengths of these groups are strongly affected by inductive influences of other molecular groups in their immediate environment. Changes in selectivity could result from changes in these inductive influences or from exposure of new sites of different field strength. Carboxyl and phosphoric acid groups are not the only potential membrane sites, however. Recent work has shown that there are large macrocyclic molecules (some of them antibiotics, for example valinomycin) that can strongly interact with cations and, if present in a membrane or added to it, can increase permeability to specific ions by as much as

10^8 times. These molecules characteristically are uncharged but surround a ring of oxygens that form a strongly electronegative site with which cations interact readily. Such molecules apparently stack together to form channels for permeation in the membrane. Current hypotheses of ion permeation through channels envision a series of sites of interaction between ions, associated water molecules, and membrane charges in such a way that both charge interaction and channel size are important.

9.5 GENERATOR AND RECEPTOR POTENTIALS

A membrane potential is found in all cells, the inevitable result of separating intracellular contents from a different extracellular environment with a semipermeable membrane. This potential may well be involved in the function of most cells—in transport of nutrients and secretory products or in the movement of cell processes, for example. Only in nerve, muscle, and sensory cells, however, is the membrane potential known to be necessary to the principal function of the cell. In these cells the membrane potential constitutes a source of stored energy, and the effect of a stimulus is merely to trigger the response—to cause the controlled release of a larger amount of stored energy.

The trigger mechanism in most sensory receptors is a permeability change at the site of stimulation. Over a wide range of stimulus intensities, the more intense the stimulus the greater the change in permeability and the greater the flow of ions down their concentration gradients. The nature of the permeability change is poorly understood, but in most cases there is an increase in permeability to Na^+, perhaps to all ions, or to all cations, as is the case in the neuromuscular junction. K^+ and Cl^-, as we have seen, are nearly in equilibrium at the resting membrane potential, but Na^+ is approximately 125 mV out of equilibrium. Consequently it rushes into the cell, down its concentration gradient, and the cell is proportionately depolarized. If all Na^+ is removed from the surrounding medium, depolarization is much reduced and there are no action potentials in the axon. It is not clear what ions, in the absence of Na^+, are responsible for the remaining receptor potential, but Ca^{2+} is known in some cases to be the primary carrier of current, and it can strongly influence permeability to other ions, including Na^+ and K^+. Often Ca^{2+} is necessary even in the presence of Na^+ for obtaining normal Na^+ potentials. Usually the change in permeability lasts as long as the stimulus persists, quickly returning to the resting value at its termination. Thus the stimulus-induced depolarization has no "threshold," is graded in amplitude, and accurately reflects the stimulus duration (as we saw to be the case in the crustacean stretch receptor dendrites).

The terms *generator potential* and *receptor potential* are often confused, which is not surprising, since they denote sometimes identical, sometimes different, phenomena. The initial graded potential response to a sensory stimulus is always a receptor potential. If the stimulus acts directly on the terminals of an excitable axon, this potential is also a generator potential because, when it reaches a certain magnitude by passive conduction (electrotonic spread) into the adjacent excitable axon, it can cause the initiation of all-or-none action potentials. The specialized secondary sensory cells (for example visual receptors, the hair cells of the statoacoustical system, and taste buds) apparently do not themselves produce action potentials. Their receptor potentials are not generator potentials but instead cause, or are associated with, transmitter release that produces postsynaptic potentials (generator potentials) in postsynaptic sensory-nerve terminals.

Some caution is necessary in interpreting the importance of receptor potentials in secondary sensory-receptor cells, for the stimulus may cause transmitter release by some mechanism only indirectly related to membrane potential. Visual receptors, for example, behave in ways quite unlike the receptor terminals of a mechanosensory axon. Vertebrate photoreceptors are hyperpolarized by light, with an associated *increase* in membrane resistance (see Section 10.6.3), whereas the response of most invertebrate photoreceptors is an increase in conductance to certain ions.

The receptor and generator potentials, being the medium of coding of sensory stimuli, are of immense importance. Mechanisms of production of these potentials will be discussed in detail in the context of the various sensory receptors.

9.6 ACTION POTENTIALS

The most familiar activity of nerve cells is the nerve impulse, or *action potential.* If the organism were only a few hundred microns in size, so that receptor potentials could directly influence the next higher integrative or effector level, then the action potential probably never would be missed. Evolution has carried animals far beyond this stage, however. Analyzing centers have been removed several centimeters, even meters, from receptors, and just as far again from effector organs. Generator potentials cannot work over these distances. The nerve fiber is cablelike, with a low-resistance core enclosed by a membrane of higher resistance that may even be conspicuously insulated in layers of lipid-rich myelin. Nevertheless, axoplasm does offer some resistance to ionic flow, and the membrane is leaky. Any potential change tends to be countered by transmembrane movements of K^+ and Cl^-, with the result that a typical receptor or generator potential disappears altogether within 1 mm or so of its site of origin. Its reduction with distance depends on the ratio $\sqrt{R_m/R_i}$ (the *length constant,* λ), where R_m is the membrane resistance per unit length and R_i is the internal resistance per unit length. If the fiber is not in a volume conductor, the resistance to current flow outside the axon may become important, and the expression becomes $\sqrt{R_m/(R_i + R_e)}$ where R_e is the resistance of the external medium per unit length. R_i, as we shall see, is largely a function of fiber diameter, larger fibers having lower internal resistance. To carry the essential information centimeters or meters quickly and with no loss, the axon (but not the sensory terminals or the subsynaptic membrane) has evolved the property of *electrical excitability,* the capability of producing an all-or-none action potential. Its sole known purpose is to translate graded nonpropagating generator potentials into an accurate representation that can be transmitted without loss of information, then reconverted at the other end into a graded influence on the postsynaptic cell.

9.6.1 Mechanism of Production

Action potentials have long puzzled neurophysiologists. Muscle fibers were the first subjects of study because of their large size. Bernstein, having hypothesized that selective permeability of the membrane to different ions could explain the resting potential, suggested that a temporary breakdown of this selectivity, which allowed all ions to redistribute themselves, even to the point of eliminating the transmembrane potential, could explain the action potential. The largest action potential would thus be equal in amplitude to the previous resting potential, in cases when the membrane potential fell to zero.

This hypothesis was generally accepted until 1939. Long before that, however, evidence was accumulating that urged a different explanation. For example, as early as 1892, the measurable action potential was found (in frog muscle) to be larger than the largest measurable resting potentials. In 1914 Adrian demonstrated that the action potential is an all-or-none phenomenon, always having the same amplitude in a given cell under the same conditions, independent of stimulus intensity. In 1902, in the same journal issue in which Bernstein proposed his membrane-potential hypothesis, Overton reported that when Na^+ was removed from the external medium, the muscle membrane potential remained but the action potential was abolished. He proposed that Na^+ must be entering the fiber either during the action potential or during contraction in exchange for K^+ that had to leave the fiber to maintain electrical equilibrium within. He was quick to point out the major difficulty of this hypothesis, however: With repeated activity, soon the fiber would fill up with Na^+, and the mechanism should fail unless some independent way was found for Na^+ to get out again. This remained a serious difficulty until the metabolic processes involved in the sodium pump were discovered. In support of his findings came several reports that, in both nerve and muscle, there is a net exchange of Na^+ for intracellular K^+ during activity, and that the resting K^+ is somehow regained during rest.

The rediscovery of the giant axons of squids, by J. Z. Young in 1936, and the development of micropipette electrodes that could be inserted even into smaller nerve and muscle cells without causing apparent damage, permitted direct testing of the hypotheses of Bernstein and Overton. In 1939 Curtis and Cole, passing current across the squid axon membrane, showed that the electrical conductance (inverse of resistance) rose sharply during the action potential. This was consistent with both hypotheses, and clearly established that during activity the membrane became more permeable to ions. Also in 1939 Cole and Curtis, and in England, Hodgkin and Huxley, measured electrical activity across the membrane. Both groups found that during the action potential the membrane potential did not merely fall to zero, it actually reversed, with the inside becoming briefly positive with respect to the outside. The positivity approached that expected from the calculated E_{Na}. Bernstein's hypothesis could no longer be accepted in detail.

The presently accepted explanation of the action potential comes principally from the work of Hodgkin, Katz, and Huxley. It is enormously complicated in detail, but in concept marvelously simple, depending on the following membrane properties, now well established by the experiments of Hodgkin and Huxley:

1. Not only does the differential permeability of the membrane to Na^+ and K^+ govern the electrical activity of the membrane, as we have seen, but the permeabilities are governed by the electric field across the membrane. In particular, the permeability to Na^+, which is almost negligible normally, increases (temporarily) in proportion to the reduction in membrane potential (see Figure 9-16b). The circular interdependence of permeability and potential is just what is needed to produce an explosive all-or-none action potential. As the membrane potential is slightly reduced, Na^+ permeability increases, Na^+ flows into the fiber, which further reduces the membrane potential, Na^+ permeability increases again, and so forth.

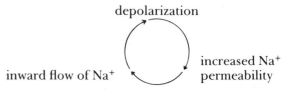

depolarization

inward flow of Na^+ increased Na^+ permeability

If the inward flow is slow enough, it can be counteracted by outward flow of the still more mobile K^+ ions. But if the membrane becomes sufficiently depolarized for the Na^+ current to equal and begin to exceed the maximum rate of K^+ outflow (which usually happens at about 10–20 mV depolarization), the compensating K^+ efflux can no longer keep up, *threshold* is reached, and an explosive action potential occurs. Cyclical positive feedback causes the process to accelerate until the Na$^+$ permeability is 30 times greater than K$^+$ permeability and the membrane potential has approached the Na$^+$ equilibrium potential.

2. For some unknown reason, Na$^+$ permeability shuts off near the peak of the action potential (*sodium inactivation*) (Figure 9-8a). Even if the membrane potential is held (by the experimeter) in a depolarized state, Na$^+$ permeability falls to its resting level. This permits a return to normal resting membrane potential as internal K$^+$ moves outward. Inactivation, which may be viewed as an uncoupling of g_{Na} from membrane potential, is maximal in the squid axon at about the time of maximal increased K$^+$ conductance (see below), then gradually returns to normal. Since extracellular Ca^{2+} has a strong influence on the coupling of g_{Na} to membrane potential, one possible mechanism of Na$^+$ inactivation might depend on the local buildup of Ca^{2+} within the membrane during a spike.

3. The change in membrane potential due to Na$^+$ influx also causes an increase in K$^+$ permeability (*delayed rectification*). This occurs only after some delay, however, so that K$^+$ permeability does not reach its peak until after Na$^+$ permeability has substantially declined. There is evidence that in some cells this delayed increase in K$^+$ permeability is caused by Ca^{2+} ions that have entered during the earlier portion of the action potential. The increase in K$^+$ permeability speeds the return of the membrane potential toward E_K, a case of negative feedback. A graphic representation of these changes is shown in Figure 9-8. Note that the recovery phase of the action potential may actually fall to a more negative membrane potential than in the resting condition, because of maintained increased K$^+$ permeability, driving the potential toward E_K. The temporary inactivation of Na$^+$ permeability explains the observed *absolute refractory period*—a time after an action potential when it is impossible to initiate another. This is followed by a *relative refractory period* when an unusually large current is necessary to increase Na$^+$ permeability sufficiently to overcome the high K$^+$ permeability and remaining slight Na$^+$ inactivation to reach threshold. We have seen the practical results of these refractory periods in intensity coding in the crustacean stretch receptors (Figure 9-4).

Of major importance in confirming these concepts was the development of the *voltage clamp technique,* in which a large electrode inserted into the cell (for example, a squid axon) is used to pass enough current to maintain the membrane potential at any desired value. A separate recording electrode in the cell senses any deviation from the desired (electronically set) membrane potential and feeds back commands to the current amplifier to increase or decrease the amount of current passed so that it remains sufficient to maintain the membrane potential at a constant level. The direction and magnitude of the current passed can then be used as a criterion of changes in membrane conductance with time. These changes can be assigned to particular ions by appropriately changing the external ionic environment.

The great value of the technique is that it uncouples the normal link between conductance change and membrane potential; changes of conductance can be studied without the interference of capacitative currents associated with membrane potential changes. Without clamping, the measured membrane current consists of both a resistive current, dependent on the membrane potential and membrane conductance, and a capacitative current, dependent on the membrane capacitance and the rate of change of membrane potential. By eliminating changes in potential, the capacitative current is eliminated and conductance can be studied directly. Moreover, spatial current spread within a cell is eliminated by clamping the whole cell, or a large segment of it, at one potential. Normally, during an action potential, different parts of the cell are at different potentials and accurate interpretation of ionic currents is difficult.

The voltage clamp technique reveals that a suprathreshold depolarization causes an immediate brief inward current, followed by a gradual and sustained outward current. The inward current is normally carried by Na$^+$, reversing direction at the Na$^+$ equilibrium potential, although Li$^+$ can substitute. The delayed outward current is carried predominantly by K$^+$. It has now been shown that tetrodotoxin (TTX), a naturally occurring poison produced in puffer fish, blocks the initial inward current with remarkable selectivity at low concen-

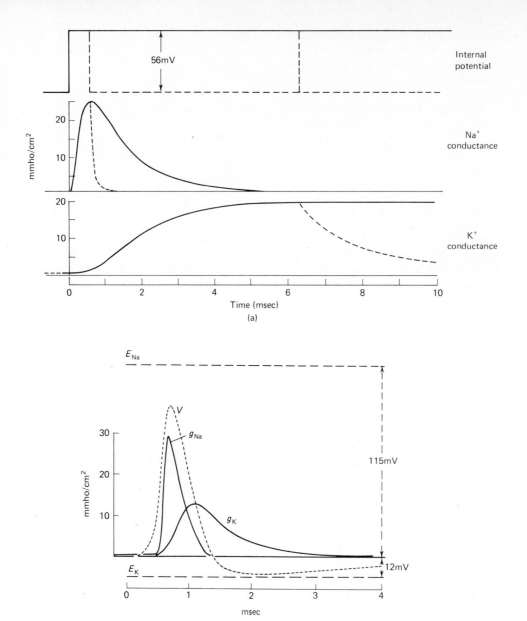

Figure 9-8
*Potential-permeability relations in the squid axon. (a) Changes in sodium and potassium conductance resulting from a step depolarization of 56 mV, applied at time 0, temperature 8.5° C. When the depolarization is maintained, Na⁺ conductance reaches a maximum and then declines (Na⁺ inactivation). K⁺ conductance rises after some delay and remains high. If the depolarization is terminated, the conductances change as shown. (b) Calculated changes of Na⁺ and K⁺ conductances (g_{Na} and g_K) during an action potential (V) in a squid axon. Note the delay in onset of increased g_K and the negative after-potential resulting from the increased g_K when g_{Na} has returned to normal. There is an absolute refractory period during the time of Na⁺ inactivation, a relative refractory period until the K⁺ conductance has returned to normal. [(a) from A. L. Hodgkin. Proc. Roy. Soc. (London), **B148,** 1–37 (1958); (b) from A. L. Hodgkin and A. F. Huxley. Cold Spring Harbor Symp. Quant. Biol., **17,** 43–50 (1952).]*

tration, with virtually no effect on the delayed outward current; that is, it blocks the increase in Na^+ conductance with depolarization. Complementing this useful chemical is tetraethylammonium chloride (TEA), which has been shown to be almost equally effective in specifically blocking the delayed K^+ conductance increase.

Another technique, **axon perfusion,** has also contributed importantly to acceptance of the postulated membrane properties responsible for action-potential production. Control of the medium inside as well as outside the membrane is accomplished simply by squeezing out the axoplasm and replacing it with the desired solution. It is evident from these studies that the electrogenic properties of the axon are truly membrane properties and do not depend on the complicated fine structure of axoplasm. A growing number of experiments employing voltage clamp or internal perfusion techniques on different cells have shown that the principles of membrane excitability found in squid axons

appear to apply, with minor variability, to virtually all excitable cells tested.

The voltage-gated channels through which Na^+ flows during an action potential, and the separate channels for K^+ permeation, are the focus of much study. The chemical nature of the channel is unknown in either case, but the density of channels of any given type may be as high as $10,000/\mu m^2$. The Na^+ channel appears to be an intrinsic membrane protein of about 250,000–300,000 daltons molecular weight, with a pore of 0.4–0.6 nm diameter lined by a number of charged groups. Near the membrane surfaces, the channel contains other charged groups placed so that, with changes in transmembrane potential, they change conformation to open or close the channel. Each channel, once opened, tends to remain fully open for a very short time, then close completely during the "inactivated" state (Figure 9-9).

Return to the original membrane potential after a spike does not mean a return to the original con-

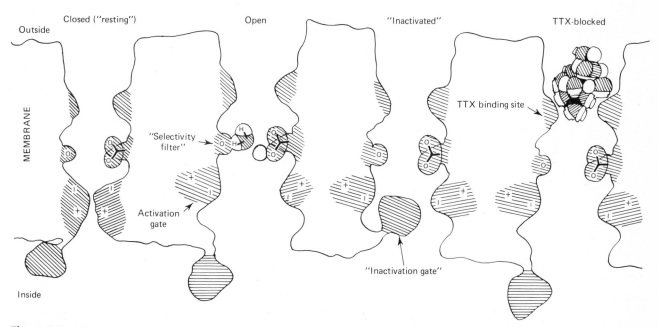

Figure 9-9
Cartoon model of the Na^+ channel of excitable membranes, incorporating current concepts of how conformational changes triggered by potential changes are thought to open the channel, selectively pass Na^+ in association with one water molecule, and become inactivated. Tetrodotoxin blocks from the outside by selective binding to a separate site in the channel. [Courtesy of Dr. S. Krasne.]

dition, of course. Some Na$^+$ has moved into the fiber, some K$^+$ out of it. The amounts are not large. The fact that membrane potential has briefly reversed does not mean that ionic concentration gradients have been reversed, or even changed greatly, only that the membrane capacity has been discharged and momentarily reversed. This requires the transfer of only a small number of charges. One can calculate the number of Na$^+$ ions necessary to cause the 110 mV change observed during the action potential. Using the relationship $q = CV$ [charge (q) = capacitance $(C) \times$ voltage (V)] it turns out that if only Na$^+$ ions were involved the required number would be about 1.5×10^{-12} mole/cm^2 of membrane, corresponding to an average transfer of one ion at points 200 Å apart on the membrane. Tracer experiments measuring the uptake of labeled Na$^+$ or the loss of labeled K$^+$ show that the actual exchange is about 2–3 times this great, an inefficiency reflecting the fact that the increased Na$^+$ and K$^+$ permeabilities overlap. Because the plasmalemma of all nerve cells is apparently much the same, having a capacitance of about 1 μF/cm^2 (microfared per square centimeter) and maintaining much the same membrane- and action-potential amplitudes, approximately the same number of ions must be transferred across a given area of unmyelinated membrane during an action potential in any cell. It follows that the smaller the fiber volume, the more important the exchange will be. In the squid axon, the net gain of Na$^+$ per impulse is about 1/300,000 of the internal [Na$^+$]. Even in normal mammalian axons, the net influx from 10,000 impulses would little more than double the already low [Na$^+$]. On the other hand, there exist tiny axons, approximately 0.1 μm in diameter, that can have their internal Na$^+$ doubled with as few as ten action potentials. In these the need for outward sodium pumping is most important and may be a limiting factor in their rate or duration of activity. Even in larger fibers, the amount of Na$^+$ removed by the Na$^+$ pump in 1 sec is about the same as enters during one impulse. Fibers conducting impulses at rates of 100 per sec or more clearly must rest occasionally, or else the rate of pumping must increase, if excitability is to be maintained.

The "Na$^+$ pump" is actually a large protein molecule (about 275,000 daltons, with dimensions of approximately 6 $\times$ 8 nm) that can use the energy stored in one ATP molecule to move 3 Na$^+$ from inside the cell to outside, in exchange for 2 K$^+$ (see Figure 9-6). There may be anywhere from 100 to 1000 of these molecules per square micrometer of membrane, and each, working at its maximum rate, can remove about 200 Na$^+$ per second. The sodium pump can be reversibly blocked by certain metabolic inhibitors (for example, strophanthidin and cyanide) and, when blocked, can be temporarily revived by the injection of ATP or organic phosphate into the fiber. The sodium pump of plasma membranes does not act by expelling Na$^+$ alone. Instead, it simultaneously extrudes Na$^+$ and takes in K$^+$. Thus the rate of pumping of Na$^+$ is greatly reduced if K$^+$ is removed from the external medium, and during pumping there is a simultaneous change of the two concentrations in the opposite direction. For this reason it is perhaps more accurately called a *cation pump*.

The job of maintaining ionic gradients in the face of leaky membranes and electrical signaling is an enormous one. The brain represents only about 2% of the total body weight in a human being, but it accounts for approximately 20% of the total resting oxygen consumption. This rate is fairly constant, day and night. Moreover, whereas other tissues are able to operate for short times in the absence of oxygen, the brain depends entirely on oxidative metabolism. Consciousness is lost within a matter of seconds and permanent brain damage results if the flow of oxygenated blood to the brain is interrupted, or if blood glucose falls below a certain level. It is no wonder there are many circulatory system adaptations for ensuring an adequate blood supply, regardless of what demands are being made elsewhere in the body.

9.6.2 The Ca^{2+} Spike and Other Variations on the Theme

Similar ionic events appear to be involved in electrical activity of muscle and nerve throughout most of the animal kingdom. Not all systems are identical, however. Some cells (for example certain elasomobranch muscles) have membrane potentials determined by anion, rather than cation, concentration differences. In some nerves there is probably no increase in K$^+$ permeability to speed recovery after an action potential; whereas in frog skeletal muscle the high Na$^+$-conductance phase of the action potential is usually entirely over be-fore the K$^+$ conductance increases. Certain plant cells *(Chara, Nitella)* have conducted action poten-tials that depend on the regenerative *exit* of intracel-lular Cl$^-$, rather than Na$^+$ entry.

Of special interest are the many instances in which the active inward current in excitable cells is carried by Ca^{2+} rather than Na$^+$. The muscle fibers of barnacles and other crustaceans can pro-duce Ca^{2+} action potentials, and similar potentials have been found in *Paramecium*. In vertebrate heart muscle, the action potential begins as a Na$^+$ spike, but then is prolonged by a large increase in Ca^{2+} (or Ca^{2+}-dependent) conductance. A comparable combination of Na$^+$ and Ca^{2+} spikes is seen in *Am-phioxus* skeletal muscle, and certain molluscan gan-glion cells *(Aplysia, Helix)* have been found to have Ca^{2+} spikes, or both Na$^+$ and Ca^{2+} spikes. As will be described in detail later, the terminals of verte-brate motor nerves (and the presynaptic terminal of the squid giant synapse) may show all-or-none Ca^{2+} spikes in the absence of Na$^+$. This is probably a common feature of transmitter-releasing termi-nals. It may also prove to be the case that excitation-contraction coupling in muscle is achieved through a Ca^{2+} spike in the sarcoplasmic reticulum. Indeed, one is tempted to speculate that Na$^+$ spikes have been employed for the conduction of coded infor-mation, Ca^{2+} channels when the electrical event must accomplish some more complicated function, such as secretion or movement.

Mention should be made also of the large spec-trum of tissues that show only graded depolariza-tion, no action potentials. This includes most inver-tebrate muscles, vertebrate smooth muscle, some vertebrate slow skeletal-muscle fibers, a large frac-tion of the neurons in the vertebrate retina (see Section 10.6.3), and perhaps other small nerve cells in the central ganglia of vertebrates and inverte-brates. The ionic events in these tissues have been little studied, but probably correspond to normal graded generator potentials, causing a localized response in proportion to the magnitude of depo-larization (see Section 9.8).

9.6.3 Conduction of the Action Potential

The all-or-none nature of the action potential is of immense importance, for it means that despite the leakiness of the membrane and despite the de-crease in amplitude of a potential spreading only electrotonically down an axon, the action potential is constantly being amplified back to maximum size. This function is directly comparable to the amplifi-cation at frequent intervals of a signal crossing the Atlantic Cable. The current generated by an action potential spreads to adjacent membrane areas, de-polarizing them to threshold; at this point they be-come active, generate an explosive action potential, and depolarize the membrane farther on (Figure 9-10). Hodgkin showed in 1937 that the process of conduction is electrical and that it depends criti-cally on the **length constant**, λ, which was defined on page 423 as equalling $\sqrt{R_m/(R_i + R_e)}$, but which is also the distance along the fiber at which an imposed potential change has fallen by a given amount: to $1/e \times$ the imposed value, or 37%. These two definitions are reconciled by the formula for exponential decay of a steady-state potential with distance along a cable: $E_x = E_0 e^{-x/\lambda}$, where E_x is the potential measured at a distance x from the point of origin of the potential E_0. When $x = \lambda$, $E_x = E_0 e^{-1} = E_0(1/e) = 0.37 E_0$. The higher the resis-tance of the axoplasm and external medium, the greater is the tendency for current to leak out of the axon across the membrane [$\sqrt{R_m/(R_i + R_e)}$ is smaller, and consequently the distance, x, at which the potential has fallen to 37% of its original value, is similarly reduced]. Because the surface area of a cylinder increases only linearly with an increase in diameter (decreasing R_m), whereas the volume increases as the square of diameter (decreasing R_i), the length constant (and conduction velocity) in-crease in proportion to the square root of the diam-eter (see Figure 9-11).

The potential also shows a progressively slower rise time with distance because of the capacitance of the added membrane. The **time constant**, τ, is that time at which the potential has risen to $1-1/e$ (63%) of the eventual value. It is determined by the product of the capacitance, C, of the membrane, and the resistance of the entire circuit, R. The equa-tion relating the potential at time $t(E_t)$ to the poten-tial at final value (E_∞), is $E_t = E_\infty (1 - e^{-t/RC})$.

The action potential was developed to transmit information over long distances without loss. It is evident that the faster this information is carried the better, at least for critical responses. Two meth-ods have been devised for increasing the speed

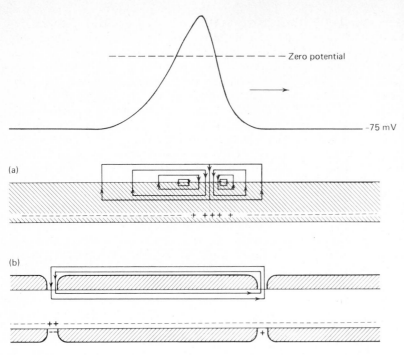

Figure 9-10
Local circuits during conduction of an action potential down (a) an unmyelinated axon or a muscle fiber, and (b) a myelinated fiber. The action potential associated with (a) is also shown. Note that the current density is greater in front of the peak than during the decay of the action potential. Refractoriness ensures that an action potential continues to travel in only one direction and does not bounce back upon reaching the end of a fiber. An action potential initiated in the middle of a fiber, on the other hand, will be conducted in both directions. Conduction in the normal direction is orthodromic. Conduction in the opposite direction, for example, from neuromuscular junction to motoneuron soma, is antidromic. The length of the fiber depolarized at any given instant depends on the length constant of the membrane.

of conduction: decreasing the internal resistance of the axon by increasing fiber diameter, and increasing the transmembrane resistance by insulating the membrane with satellite cells. In all animals there is a gradation of fiber size. Those involved in alarm systems or fastest responses are normally largest. The best examples of this are the "giant" fibers of certain invertebrates, especially annelids, arthropods, and mollusks.

Vertebrates have evolved the most sophisticated and successful method of increasing conduction velocity. This is the development of the myelin sheath, a lipid-rich wrapping of high electrical resistance. Current, being unable to pass through it, spreads far down the fiber until it reaches a localized area where the myelin sheath is absent (the node of Ranvier) and where it can cross the membrane (Figure 9-2). The current density is very high at the node, so the threshold is quickly reached. The result is an extremely fast form of conduction, termed *saltatory conduction,* in which the current generated by the activity at one node spreads electrotonically with little loss to the next node (as much as 1–2 mm away), where activity is in turn produced. As in unmyelinated axons, the greater

the fiber diameter, the lower the internal resistance and the higher the conduction velocity. In myelinated fibers the velocity is found to be directly related to diameter, because internodal distance is proportional to the diameter (see Figure 9-11).

9.6.4 Ontogeny of Excitability

The property of selectively changing membrane permeability in response to membrane polarization is one that is observed even in the eggs of certain organisms. In other cases, cells through the first few divisions show only passive membrane characteristics. Even where excitability exists in the earliest stages, these properties change with differentiation and maturation. Very few cases have been studied, but it now appears that the initial condition is often one in which there are two or more inward voltage-dependent ionic channels, only one of which survives into the adult. The egg cells of a certain tunicate, for example, show both Na^+ and Ca^{2+} channels, whereas only the Ca^{2+} is present in the muscle cells of the adult. The eggs of a starfish that has been studied show two different Ca^{2+} channels, one of which is like the Ca^{2+} channel

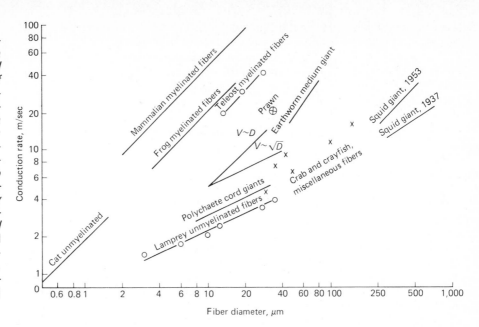

Figure 9-11
Conduction velocity (V) *versus axon diameter* (D). *Data collected from myelinated and unmyelinated fibers of many vertebrates and invertebrates. Unmyelinated fibers generally conform to a square-root relationship* (V ~ √D), *whereas myelinated fibers more closely approximate a linear relationship* (V ~ D). *Note the great increase in conduction velocity resulting from myelination itself.* [*From T. H. Bullock and G. A. Horridge.* Structure and Function in the Nervous Systems of Invertebrates, *vol. 1, W. H. Freeman and Company, San Francisco, 1965.*]

in the adult and in other animals, the other is like none previously described. Rat muscle at the myotube stage has both Na^+ and Ca^{2+} channels, but loses the latter during subsequent differentiation. Moreover, the specific voltage dependence, kinetics, inactivation properties, and presence of the K^+ channel associated with delayed rectification often change or appear during later development. The field is still too little explored to allow any generalizations, but on the basis of initial reports an ontogenetic approach to membrane biophysics looks extremely promising.

9.7 SYNAPTIC TRANSMISSION

The functional junction between a nerve cell and another cell is called a *synapse.* Although there are many billions of nerve cells packed together within the brain of a higher animal, they are connected to one another only in highly specified and astonishingly precise ways.

9.7.1 Synapses

These connections are all-important to the organism's function, so it is essential that these be the only functional interactions between cells. The insulation around nerve fibers is thus important not only in speeding action-potential conduction, but also in keeping each fiber separate and independent of its neighbors.

It is important to note that the postsynaptic cell obtains only quantitative information. Qualitative information, about the type of stimulus that produced a given train of potentials, is built into the system by the specificity of the receptor and the specificity of connections between nerve cells. Of these, only the specificity of connections is infallible. Any sensory receptor, or nerve, can be excited in abnormal ways—by electrical stimulation if no other way. The resulting stimulation may seem abnormal, but if the stimulated receptor or nerve is gustatory, the sensation will be taste; if it is auditory, the sensation will be sound. Everyone is familiar with the visual sensation resulting from a severe blow to the head. A most dramatic example of the *specific energy* of sensory nerves is the phenomenon of phantom pain—localized, for example, to a certain part of a leg that has been amputated and is no longer there, or an itch localized to some no-longer-existent patch of skin.

In the simplest case there need be only two neurons involved in the chain from detection of a stimulus to causing a response. This is the so-called *monosynaptic reflex,* such as the familiar knee jerk. In the vast majority of cases many interneurons are involved, input from several sources is mixed, conscious control may be exerted, and the situation becomes complicated beyond any possible rigorous explanation at today's level of understanding. Indeed, one might say that synapses exist to permit integration. We shall now see how this is accomplished.

Two questions immediately arise: How does the action potential, reaching the terminal of a fiber, influence the postsynaptic cell? And how is this influence integrated with other inputs? Ever since the triumph of the view that neurons are discrete cells, physically separated from the cells they interact with (a view championed principally by the great anatomist, Cajal, and subsequently confirmed in all but very special cases by electron microscopy), there has been disagreement on the mechanism of interaction. Otto Loewi had demonstrated in 1921 that impulses in the vagus nerve caused the release of a substance, "vagusstoff," that inhibited the heart's activity and could, by simple transference of perfusate to a second heart, inhibit it as well. This substance was subsequently identified as acetylcholine (ACh), and an antagonist substance was found to be released by the stimulation of sympathetic fibers. Thus it was established that chemical release could affect postsynaptic cells in some cases, albeit rather slowly. Neuromuscular and transneuronal transmission were much faster, however, with delays of only 1–2 msec, and with no sign of diffusion of an active chemical to nearby cells. Consequently, during most of the development of our knowledge of membrane and action potentials, it was felt that excitation jumped across the synapse electrically, that current generated by the action potential in the axon terminal must be sufficient to depolarize the postsynaptic membrane to threshold. The difficulty with this scheme was obvious. The presynaptic terminal in most cases is miniscule compared with the postsynaptic process it acts upon. Even a full action potential in so small an area would not be expected to depolarize the large postsynaptic volume significantly, surely not to threshold. One of the most severe examples of this inequality is the neuromuscular junction. Hence in 1936, when Dale and his colleagues reported that ACh is released by the motor-nerve terminal at the neuromuscular junction, the

importance of this discovery was immediately apparent.

9.7.2 Postsynaptic Events

It was soon found that there is a postsynaptic potential (the *endplate potential,* or EPP) restricted in location to the end plate, where the motor nerve ends on it, and having a totally different time course than the nerve action potential. It begins about 0.5 msec after the arrival of the impulse in the nerve terminal and normally rises quickly to reach threshold for a muscle action potential. It can be reduced below threshold by the muscular paralyzing agent, curare, however, to reveal a long-lasting, slowly declining potential (see Figure 9-15). Other drugs were found to affect it as well, without influencing the nerve impulse or the muscle-action potential (when one was produced). Today it is accepted that most synapses are chemically mediated. The chemical transmitters are different in different cases (Table 9-2), but the fundamental mechanisms seem to be much the same. Most of our knowledge of these mechanisms comes from the careful study of the neuromuscular junction, especially by Bernard Katz and his coworkers.

Electron microscopy shows that a 200–500 Å space separates the two cells in most synapses (Figures 9-12 and 9-13). The presynaptic terminal is typically enlarged and filled with 200–500 Å vesicles, so characteristic of synapses that they are called *synaptic vesicles.* The subsynaptic membrane is usually thick and darkly staining. In the neuromuscular junction, the subsynaptic membrane area is vastly enlarged by the presence of *synaptic folds.* Although little is known about the composition of the subsynaptic membrane, it must contain receptor sites that can interact with the transmitter to cause permeability changes. Other molecules, which perhaps mimic the transmitter somehow, may block or react with the same sites without causing these permeability changes. The molecular basis for receptor specificity, as for the permeability changes themselves, is not known; but it is clear that the subsynaptic membrane is a highly specific chemosensory element, and the permeability change is analogous, perhaps identical, to that causing the receptor potentials of peripheral receptors. One of the blocking agents is *d*-tubocurarine chloride, the active component of the South Ameri-

TABLE 9-2. Chemical Transmitters and Other Synaptically Active Substances

Cholinergic depolarizing agents:

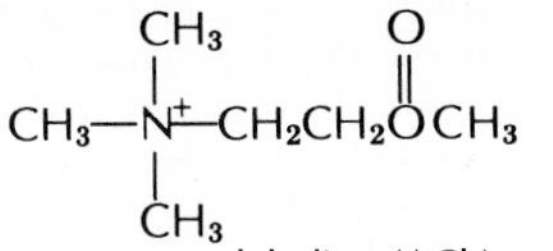

acetylcholine (ACh)

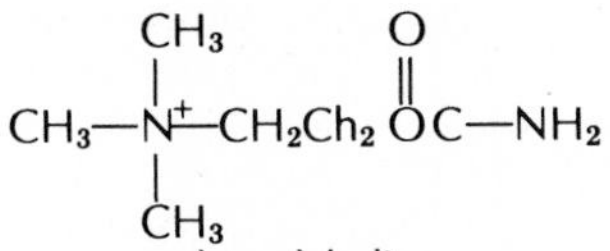

carbamylcholine

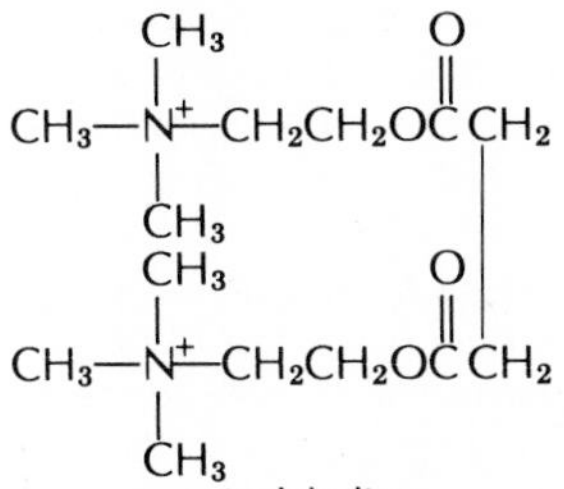

succinylcholine

ACh blocking agents:

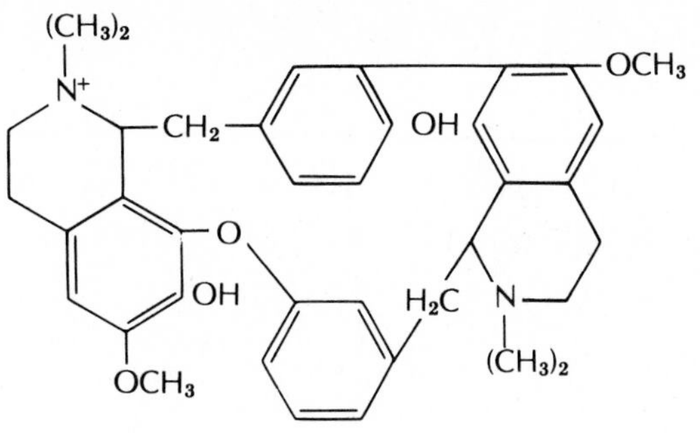

d-tubocurarine (dTC)

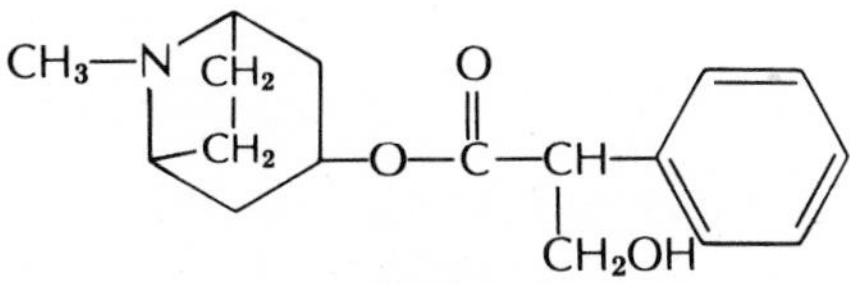

atropine

Adrenergic synaptic agents:

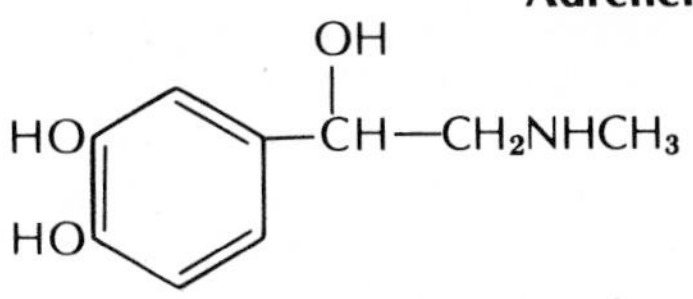

epinephrine

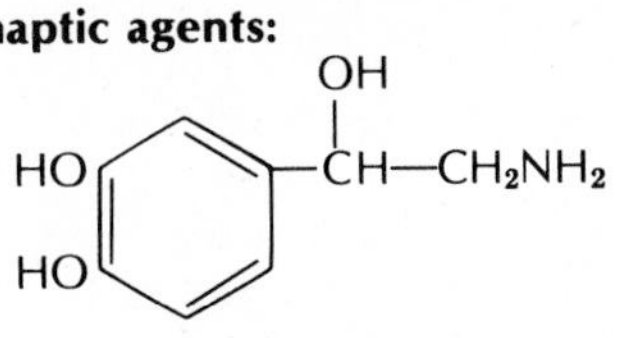

norepinephrine

Amino acids:

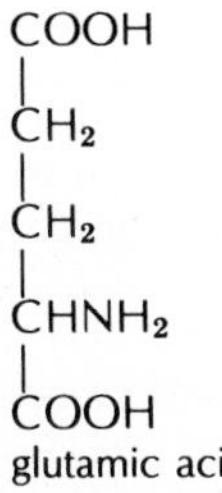

glutamic acid

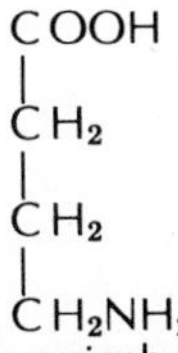

γ-aminobutyric acid (GABA)

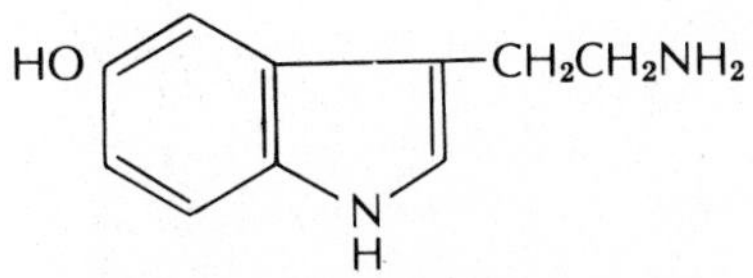

5-hydroxytryptamine (5-HT)

can Indian arrow poison, curare. Another, which has the advantage that it reacts irreversibly with the ACh receptor, is α-bungarotoxin, one component of the toxin of a Formosan cobra. Because it binds irreversibly, this toxin has been used in attempts to extract and purify the receptor. This work is still incomplete, but it appears now that the receptor is a large protein (molecular weight about 320,000) that contains four different subunits, may exist as a monomer or dimer, and consti-

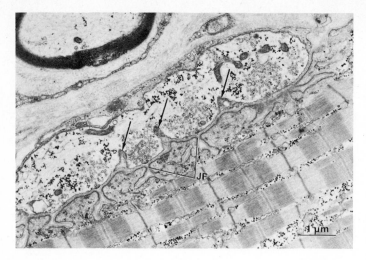

(a)

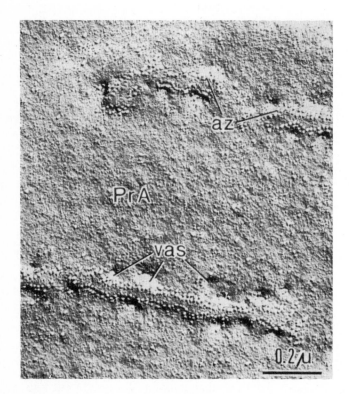

(b)

Figure 9-12

Structure of the neuromuscular junction.

(a) Electron micrograph of a vertebrate neuromuscular junction, showing a nerve terminal containing mitochondria and large clusters of synaptic vesicles. Overlying the terminal is a Schwann cell process, fingers of which also encircle the terminal and invade the space between it and the muscle surface (arrows). Separating the pre- and postsynaptic membranes is a diffuse basement membrane. Below each cluster of vesicles is a junctional fold (JF). Acetylcholinesterase is located throughout the membrane forming these junctional folds, and the ACh receptor is concentrated on either side at the top. Note the slight increase in density of the presynaptic membrane opposite each fold. Release of vesicles apparently occurs on either side of this dense band. Several myofibrils can be seen inside the muscle fiber. [From U. J. McMahan, N. C. Spitzer, and C. K. Poper. Proc. Roy. Soc., **B181,** *424–430 (1972).]*

(b) Longitudinal view of the inside of the plasma membrane of a short length of motor nerve terminal, obtained by freeze-fracture of the terminal. In favorable cases this splits the membrane apart along the internal lipid bilayer. Embedded in the inner membrane are dense bands of material, lined by particles possibly representing Ca^{++}-channels. Two are shown in this figure. These dense bands correspond to presynaptic densities, and each represents a transmitter release (active) zone (az), based on the observation that, when the terminal is quick-frozen in a high release state, depressions are seen that probably result from vesicle fusion with the presynaptic membrane at vesicle attachment sites (VAS). [From K. Peper et al, Cell Tiss. Res. **149,** *437–455, 1974.]*

(c) Tangential section through a frog motor nerve terminal just interior to the terminal membrane, showing three double rows of synaptic vesicles lining successive active zones. Note the alignment of active zones with junctional folds. [From R. Couteaux & M. Pecot-Dechavassine, C. R. Acad. Sci. **D279,** *291–293, 1974.]*

(d) Schematic three-dimensional reconstruction of a frog motor endplate based on thin section and freeze-fracture electron microscopy. The nerve terminal (Nt) is surrounded by a Schwann cell (S) which encircles it with finger-like processes (f). Active zones (az) are associated with clouds of vesicles (sv) inside the terminal and junctional folds (jf) postsynaptically. At the upper rims of the junctional folds are particle aggregates (pa) corresponding to ACh receptor molecules. A basement membrane separates pre- and post-synaptic structures. [From K. Peper et al, Cell Tissue Res. **149,** *437–455, 1974.]*

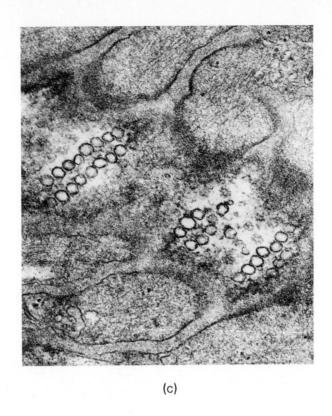

(c)

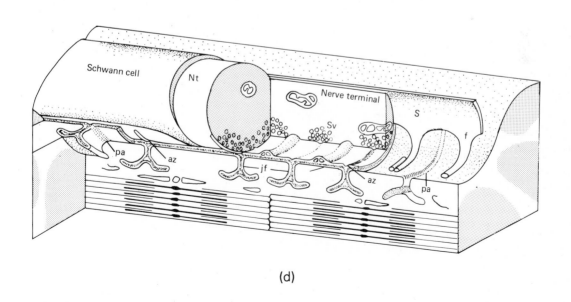

(d)

435

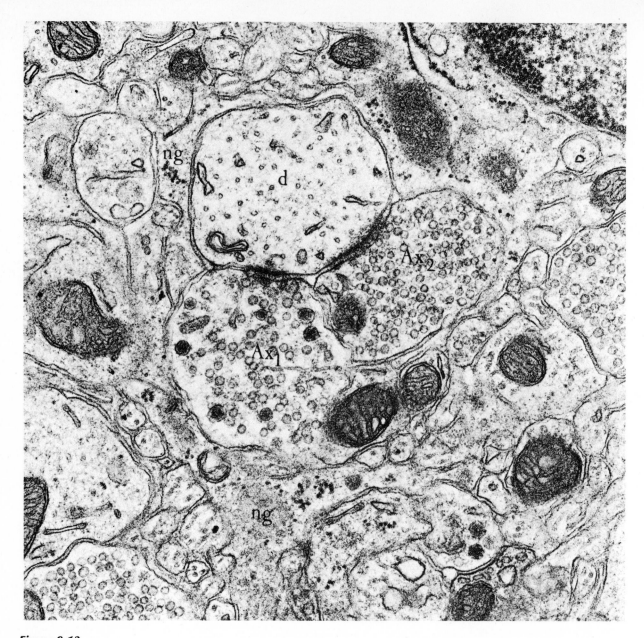

Figure 9-13
Structure of chemical synapses. Electron micrographs of two characteristic types of synaptic structures from the vertebrate CNS. Two axonal boutons (Ax_1 and Ax_2) ending on a dendrite in the rat brainstem. The difference in vesicle size and presence of large dense-core vesicles only in Ax_1 indicate that the two boutons originate from different sources. Both show characteristic pre- and postsynaptic densities and a dense matrix between the pre- and postsynaptic cells. Note that the dendrite and both boutons are mostly surrounded by neuroglial profiles (ng). [From S. L. Palay and V. Chan-Palay, Cold Spring Harbor Symp. Quant. Biol. **40,** *1–16, 1976.*]

tutes most of the protein in the subsynaptic membrane at the neuromuscular junction. Estimates from toxin binding, as well as direct electron microscopy, place the density of receptor molecules at between 10,000 and 20,000 per μm^2.

The experiments of Fatt and Katz, and subsequently A. and N. Takeuchi, have demonstrated that the effect of the transmitter at the neuromuscular junction is a simultaneous increase in permeability to both Na^+ and K^+. Na^+ rushes in, K^+ rushes out, and the membrane is depolarized toward an equilibrium potential near zero, far above threshold. The permeability change is actually considerably greater than that occurring during an action potential. At the synapses between sensory-nerve terminals and motoneurons in the spinal cord, the effect of the excitatory transmitter (of which there are probably several, none of them ACh) is the same. It seems likely that this is the normal mechanism of action of excitatory transmitters. Figure 9-14 shows that the *excitatory postsynaptic potential* (EPSP) equilibrium (or reversal) potential in these cells is approximately -10 mV, essentially the same as in muscle. This would be predicted if the permeabilities to K^+ and Na^+ became approximately equal during this time. If the membrane potential is displaced toward this equilibrium potential by intracellularly passed current, the depolarizing effect of the transmitter is less but the change in permeability is the same. If the membrane potential is made more positive than the equilibrium potential, the effect of the transmitter is hyperpolarization.

The EPP, although much longer than an action potential, is abbreviated by the presence of an enzyme, acetylcholinesterase (AChE), that hydrolyzes ACh, destroying its transmitter properties (Figure 9-15). The hydrolysis also produces choline, which appears to be rapidly taken up by the axon terminal and used for the synthesis of more ACh.

AChE is a large protein molecule synthesized predominantly or entirely by the muscle cell, but only when a nerve has made functional contact with it. At that location, the enzyme is inserted into the basement membrane, a complex noncellular coating of collagen and other molecules that is secreted immediately above the muscle fiber surface. The basement membrane is present in each synaptic fold, interposed between the nerve terminal and the postsynaptic membranes (see Figure 9-12). AChE is, in fact, one of the major morphological

indicators of the end plate because it is easily stained for light microscopy. The receptor molecules are located at the tops of the postsynaptic junctional folds, the cholinesterase over the entire surface of the folds (see Figure 9-12a). ACh is effective only on the outside of the membrane. ACh injected into the muscle fiber at the end plate has no effect.

It is critical to neuromuscular function that the transmitter be destroyed. Diffusion away from the receptor may be fast enough at some synapses to reduce transmitter levels quickly, but not at the relatively huge neuromuscular junction. Indeed, the active components of several potent poisons are anticholinesterases, among them the acylfluorophosphate compounds used in "nerve gas." Whether other chemical synapses employ a similar transmitter-hydrolyzing enzyme to shorten transmitter action has not been established. It is known that sympathetic nerve terminals abbreviate the action of the noradrenalin they release by having a specific reuptake mechanism that returns noradrenalin from the extracellular space to the presynaptic cytoplasm.

The subsynaptic membrane clearly must be different than the rest of the postsynaptic membrane. Not only are there receptor molecules and perhaps enzymes to break down transmitter, but the nonspecific cation-permeability increase has no regenerative link with the membrane potential. The same change in membrane conductance is seen regardless of the membrane potential. It may be that there are mixed patches of normal and receptor membrane, even subsynaptically, but a certain measure of electrical inexcitability is unquestionably an important property of receptor membranes in general.

Not all chemical synapses are excitatory. In fact, inhibition of activity is fully as important to nervous function as is excitation. There are no established morphological criteria for distinguishing excitatory from inhibitory synapses, both of which exist on most nerve cells. Nor are the transmitters known, in most cases. The mechanism of action is apparent, however. The transmitter causes a temporary increase in permeability to certain ions (including K^+ and/or Cl^- but not Na^+), essentially nailing the postsynaptic membrane potential to the resulting equilibrium potential. In most cases the effect is a slight hyperpolarization.

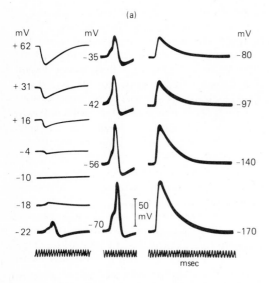

(a)

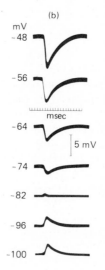

(b)

Figure 9-14
*Equilibrium potentials for transmitter action. The effect of a synaptic transmitter is to change the permeability of the subsynaptic membrane to certain ions, changing the potential at which ions are at equilibrium across the membrane. These synaptic equilibrium potentials can be demonstrated by displacing the postsynaptic membrane potential with intracellularly passed current, and finding the potential at which the effect of transmitter action is reversed (the reversal point). In (a) this was done with a frog sympathetic ganglion cell and the reversal point for the excitatory transmitter (ACh) is seen to be about −10 mV. Note that with severe hyperpolarization or severe depolarization from the resting membrane potential (−70 mV) the EPSP does not elicit a spike. In (b) the reversal point for an inhibitory transmitter IPSP is seen to be approximately −80 mV in records from mammalian motoneurons. Probably the effect of the transmitter is to increase selectively permeability to K^+ and Cl^- ions. [(a) from S. Nishi and K. Koketsu. J. Cell. Comp. Physiol., **55**, 15–30 (1960); (b) from J. S. Coombs, J. C. Eccles, and P. Fatt, J. Physiol. (London), **130**, 326–373 (1955).]*

One of the best known inhibitory synapses in the vertebrate CNS is that between Renshaw cells and motoneurons. Motoneuron axon collaterals excite nearby Renshaw interneurons, causing prolonged firing. (ACh is the transmitter at these synapses.) Renshaw-cell axons in turn feed back onto the motoneurons, inhibiting their further activity for as long as the interneuron is firing. This is an important form of negative feedback, controlling the activity of motoneuron pools. Whether a given transmitter is excitatory or inhibitory is determined by the receptor involved and the ionic contents of the receptor cell. For example, ACh is excitatory to vertebrate skeletal muscle but inhibitory to cardiac muscle. More dramatic examples are certain ganglion cells of the nudibranch, *Aplysia*, activity of which causes EPSPs in one type of postsynaptic cell, inhibitory postsynaptic potentials (IPSPs) in

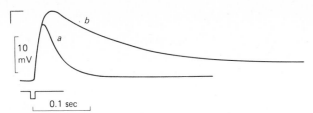

Figure 9-15
*Effect of acetylcholinesterase. Records showing the EPP (a) before and (b) after application of an anticholinesterase (prostigmine) to the bath. When the action of cholinesterase is blocked, the EPP is enlarged and prolonged. Usually the prolonged presence of ACh at the end plate leads to reduced sensitivity to addition of more transmitter (a phenomenon known as desensitization). [From J. del Castillo and B. Katz, J. Physiol. (London), **128**, 157–181 (1955).]*

another, with ACh the transmitter at both synapses.

In other molluscan synapses, electrophoretically applied ACh mimics the normal transmitter in causing a complex postsynaptic potential; this first shows a depolarization due to selective increase in Na^+ conductance, then a longer inhibitory potential resulting from selective increase in conductance to Cl^-. Apparently the two actions are mediated by different receptors. In some cases there is a third action, of much longer time course, effecting inhibition by increasing the membrane conductance to K^+. These inhibitory synaptic actions are not susceptible to known cholinergic blocking agents, and are still poorly understood. Normally, any given cell releases only one transmitter. On the other hand, it is known that in many instances more than one *substance* may be released. For example, ATP and a specific protein are released at the same time as norepinephrine from sympathetic terminals, and a protein of unknown function has been reported to be released at the same time as ACh from motor nerve terminals. Such substances may have a potent postsynaptic effect, but there is no convincing evidence of this at present. Norepinephrine (NE), like ACh, can be either excitatory or inhibitory in different postganglionic sympathetic synapses. A more detailed discussion of what is known about neurotransmitters in the CNS can be found in Section 11.5.

It has been proposed that some long-term changes in membrane potential may result from transmitter-induced activation or inactivation of electrogenic pumps. Reported instances of the phenomenon to date, however, have proved erroneous or ambiguous. Nevertheless, the possibility is real and would provide an important mechanism for effecting long-term changes in the efficacy of neural networks.

9.7.3 Presynaptic Events

Chemical synapses impose a delay (of at least 0.2–0.3 msec) on information transfer. (Synaptic delay is defined as the interval between the peak of the action potential in the presynaptic terminal and the onset of the postsynaptic potential.) Several processes contribute to synaptic delay: the release of transmitter, its diffusion to the subsynaptic membrane, the necessary reaction time with receptor molecules, and the turning on of conductance. Most of this delay is in the first step. In frog sartorius muscle preparations cooled to 2–5°C, Katz and Miledi find that ACh electrophoretically ejected through a micropipette at the end plate causes a response at as short a delay as 0.1–0.2 msec. The diffusion distance is surely no less in this case than for naturally released transmitter, and the receptor reaction should be the same. Yet at these temperatures normal neuromuscular transmission may take 5–10 msec or more. Most of the delay must be in release.

External Ca^{2+} is known to increase the amount of transmitter released per impulse, whereas Mg^{2+} reduces the amount. By using low Ca^{2+} concentrations or by adding Mg^{2+} to the solution, release can be sharply reduced. When this is done, it is found that the small amount of transmitter still released is in the form of packets of approximately uniform size, for the resultant EPP is always a multiple of one minimum size. It is now apparent that evoked release of ACh at the neuromuscular junction is in quantal packets, normally so numerous (several hundred) that the presence or absence of a few more is undetectable. Even when there is no nerve impulse, there are frequent "miniature" EPPs (up to about 2 mV in amplitude) in the muscle fiber, corresponding to the effects of one of these quanta of transmitter. From the demonstration that electrophoretically applied ACh produces responses this large only when tens of thousands of ACh molecules are being ejected and that the potential produced is finely graded with current passed, it can be concluded that each naturally occurring quantum represents the synchronous release of a large number of molecules, at least several thousand, of ACh. Apparently packets of this size are normally poised near the membrane surface, and occasionally, spontaneously, one breaks loose. Figure 9-16a shows records of such miniature EPPs. When the nerve terminal is depolarized (by adding K^+ to the external medium or passing current through the terminal), the frequency (but not the size) of the miniatures increases. There is an approximately tenfold increase in transmitter released for every 15 mV of depolarization (Figure 9-16b and 16c). When an action potential invades a terminal, the sudden depolarization causes the synchronous release of large numbers of packets. Although the terminal membrane appears to be capable of producing a regenerative spike, in the cases studied, release does not depend on the entry of Na^+ or on the absolute current flow across the membrane. It depends instead on the level of depolarization achieved and the amount of Ca^{2+} that enters the terminal (see discussion later).

It is important to realize that not all of the ACh in a nerve terminal is packaged into vesicles or released as quantal events. It is estimated that the ACh released under resting conditions as spontaneous miniature EPPs represents only about 1% of the total spontaneous release. The rest is released, or leaks out, in nonquantal form. However, there is enough of it to cause a significant depolarization of the postsynaptic cell, which can be reversed by curare. Moreover, with sophisticated modern electrophysiological techniques, it is possible to demonstrate the postsynaptic effect of single ACh channels opening and closing. Each channel opens in an all-or-none manner, with a mean open time of 1–1.5 msec, has a current-carrying capacity of about 10^{-11} amps/msec, and causes a depolarization, under normal conditions, of about 0.2 μV. Each miniature EPP is caused by the opening of about 1500 channels.

There are uniformities of structure and physiology that suggest that quantal presynaptic release is accomplished by the same mechanism in most or all chemical synapses. In all cases that have been studied, Ca^{2+} enhances and Mg^{2+} depresses transmitter release. Miniature postsynaptic potentials have been reported in excitatory chemical synapses in both vertebrate and invertebrate nerve and muscle, including the squid stellate synapse (see following paragraph). Spontaneous quantal inhibitory postsynaptic potentials have also been reported in vertebrate motoneurons, and it seems probable that all inhibitory transmitter release is quantal, although the evidence is much less extensive.

It is widely accepted now, although not yet rigorously proved, that the *synaptic vesicles* contain the quantal packets of transmitter. It is still not understood how these vesicles release their contents. As Figure 9-12 shows, there are clouds of vesicles over

Figure 9-16
Characteristics of transmitter release. (a) Intracellular record from within a muscle fiber at the end plate (left) and 2 mm away. Lower traces, at low amplification, show an EPP growing into a muscle action potential at the end plate, only the action potential arising at somewhat greater delay at other positions. At higher gain (upper traces) there are seen to be frequent miniature EPPs, restricted to the end plate region. These are about 1 mV or less in amplitude and represent random release of single packets of synaptic transmitter. Calibrations: 3.6 mV and 46 msec for top traces, 50 mV and 2 msec for bottom traces. (b) The relationship between frequency of miniature EPP release and depolarization of an axon terminal is plotted (solid line) with a comparison curve (dashed line) showing the observed relationship between Na^+ conductance (g_{Na}) and membrane potential in the squid axon. (c) A curve showing the relationship between presynaptic spike amplitude and transmitter release (amplitude of EPSP) in the squid giant synapse. Presynaptic spike amplitude was varied by polarizing the presynaptic membrane. Amplitude was measured from the original resting potential. It was found that spike amplitude was reduced by depolarization and increased by hyperpolarization of the terminal. This effect becomes very important in the phenomenon of presynaptic inhibition (see Section 9-7-5). [(a) from P. Fatt and B. Katz. J. Physiol. (London), 117, 109–128 (1952); (b) from B. Katz. Proc. Roy. Soc. (London), 155, 455–477 (1962); (c) from S. Hagiwara and I. Tasaki. J. Physiol. (London), 143, 114–137 (1958).]

each subsynaptic junctional fold at the vertebrate neuromuscular junction. Each cloud is associated with a presynaptic densification of the membrane, and release apparently takes place at the borders of this denser region. Probably the vesicle membrane somehow fuses with specific sites on the plasma membrane of the terminal, causing a transient increase in terminal membrane area. Subsequently part of the membrane invaginates endocytotically and separates from the membrane to form

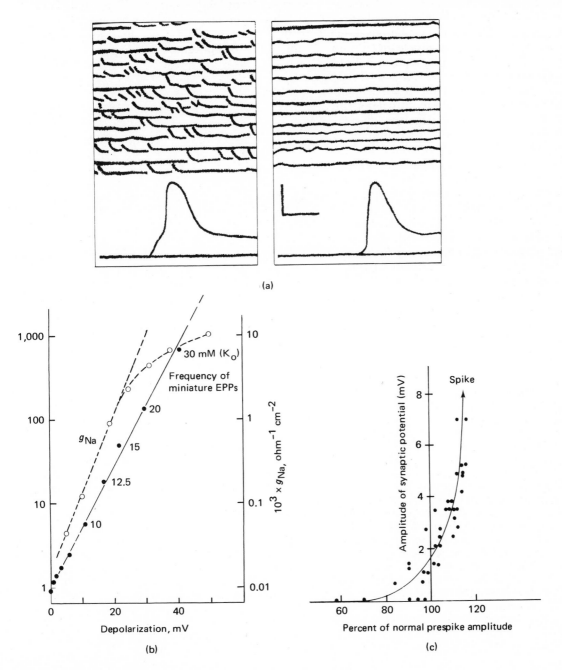

(a)

(b)

(c)

another vesicle. Evidence for this is the observation that large membrane-impermeable molecules such as ferritin or horseradish peroxidase are selectively loaded into "synaptic vesicles" in proportion to the amount of transmitter release at the neuromuscular junction.

Characterization of the vesicle membrane has revealed some features that fuel speculation about the mechanism(s) of release. The vesicle membrane is rich in lysolecithin, for example, a phospholipid that promotes membrane fusion. Once a vesicle is brought sufficiently close to the plasma membrane, fusion might occur spontaneously. How the vesicle is brought into contact with the plasmalemma is another problem. There is a pronounced tendency for vesicles to be associated with the dense matrix material near the immediate presynaptic membrane. Biochemical tests have revealed a protein, stennin, with many of the properties of muscle myosin, that is associated with, or part of, synaptic vesicles. A filamentous protein, neurin, very similar to actin, is associated with the inner surface of the plasma membrane. It has been suggested that these two proteins can interact (neurostennin) to cause an actomyosinlike contraction, pulling the vesicle into the plasmalemma. Also, it has been suggested that the movement of vesicles into the area near the terminal surface where other mechanisms may become operative depends on mi-

crotubules and neurofilaments, since substances (colchicine, cytochalasin) that disrupt these structures block neurosecretion.

The experiments of Katz and Miledi and others on the squid stellate giant synapse, have greatly increased our understanding of the role of Ca^{2+} in transmitter release. In this synapse, microelectrodes can be inserted presynaptically to pass current and record potential (see Figure 9-17a insert).

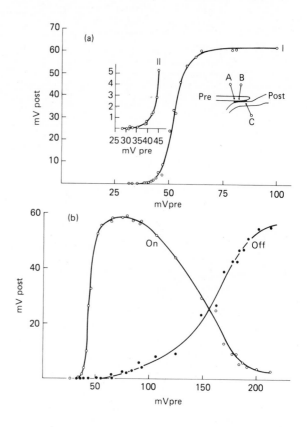

Figure 9-17
(a) Relationship between presynaptic depolarization and postsynaptic potential in squid stellate ganglion synapse. Insert drawing shows electrode arrangement, with electrode A (filled with TEA) passing current, electrode B recording presynaptic potential, and electrode C recording postsynaptic potential. I, relation using 1 msec current pulses; II, start of curve in greater detail. (b) Relationship found using long (19 msec) current pulses and larger presynaptic depolarizations. Note the shift from responses only at the onset of the current pulse to response predominantly at the "off" of the pulse. Initial level of presynaptic membrane potential in this experiment was −69 mv. (c) Relationship between current passed presynaptically and presynaptic depolarization. A, before TEA injection; B and C, after successive doses of TEA, showing the sudden jump from "subthreshold" to "suprathreshold" levels of depolarization. [(a) and (b) from B. Katz and R. Miledi. J. Physiol., **192**, 407–436 (1967); (c) from B. Katz and R. Miledi. J. Physiol. (London), **203**, 459–487 (1969).]

Katz and Miledi used tetrodotoxin (TTX) to block spikes both presynaptically and postsynaptically, and they injected TEA presynaptically to block increased K^+ conductance on depolarization. Under these conditions they were able to produce large sustained changes in the membrane potential of the presynaptic terminal. They found the previously reported steep exponential relationship between presynaptic potential and transmitter release (Figure 9-16c), but discovered that, on further depolarization, the release of transmitter (as judged by the postsynaptic potential) plateaued, then decreased with terminal depolarization greater than 100 mV. (Figure 9-17b). At the same time, however, a postsynaptic potential appeared and grew at the termination of the presynaptic current pulse. Their interpretation is that Ca^{2+} or a positively charged Ca^{2+} complex must enter the terminal for transmitter release and that with depolarization greater than about 150–200 mV, the equilibrium potential for Ca^{2+} is exceeded; therefore, what Ca^{2+} does move across the membrane is moving outward. Only when the membrane potential falls below the equilibrium potential again can Ca^{2+} flow into the terminal to initiate release, even though permeability has been high throughout. Because the permeability change lags slightly behind membrane potential changes, Ca^{2+} would be expected to enter at the end of the depolarization but not at its onset; this explains the "off-response" without an "on." The ratio of known external $[Ca^{2+}]$ to probable intracellular $[Ca^{2+}]$ is consistent with a reversal potential at about the level observed.

The calcium hypothesis is strengthened by evidence that transmitter release is unimpeded in the presence of TTX, which blocks active Na^+ inflow, or in the absence of external Na^+, with only Ca^{2+} present. In fact, in Na^+-free Ringer, with TEA injected presynaptically, a regenerative action potential of variable amplitude and duration is found localized to the terminal (Figure 9-17c). Clearly Ca^{2+} can carry a significant amount of inward positive current in the terminal. The regenerative event seen in the absence of Na^+ suggests comparisons with the Ca^{2+} spike of crustacean muscle, which is also TTX resistent. Closely similar Ca^{2+}-dependent phenomena are seen at the frog neuromuscular junction, where, however, it is not possible to measure the presynaptic potential intracellularly.

How Ca^{2+} causes release, once it gets into the terminal, is not known, although it has been shown that four Ca^{2+} ions somehow interact to effect the release of one packet. There is growing evidence that Ca^{2+} may act via a small intracellular modulator protein, *calmodulin.* Calmodulin is present in virtually every eukaryotic cell, and is one of the most highly conserved proteins known. Where it has been studied, from marine coelenterates to mammals, it consists of 148 amino acid residues, of which only six show any substitution. It has four binding sites for Ca^{2+}, and is very similar in structure to troponin C, the protein associated with actin that interacts with Ca^{2+} to help regulate muscle contraction (see Chapter 4). Calmodulin has been shown to play an important role in regulation of adenyl cyclase and phosphodiesterase activity, in protein phosphorylation, in glycogen metabolism, in microtubule assembly, in Ca^{2+} pumping, and in various forms of secretion. In the latter category is the report that calmodulin is present in nerve terminals and necessary for neurotransmitter release in response to Ca^{2+} influx.

9.7.4 Facilitation

It is characteristically observed in chemical synapses that the second of two apparently identical action potentials invading a synaptic terminal at a few milliseconds interval causes the release of more transmitter than the first. Katz and Miledi have also shown that the calcium hypothesis of presynaptic function can explain this phenomenon of *facilitation.* It is postulated that for a short time after an initial presynaptic spike, in the presence of Ca^{2+}, there is a residue of active Ca^{2+} inside the terminal that can add to the Ca^{2+} entering in a subsequent spike to enhance transmitter release. Other experiments have shown that Ca^{2+} must be present during the first action potential if facilitation is to occur.

The relationship between Ca^{2+} entry and transmitter release is very steep in most preparations. At the frog neuromuscular junction, transmitter release appears to vary with the fourth power of the external $[Ca^{2+}]$; in the squid giant synapse the exponent is approximately 2.7. This implies that, in the neuromuscular junction, for example, four Ca^{2+} ions interact at some site to cause the release of one quantum of transmitter. Thus as the $[Ca^{2+}]$ that entered the terminal or terminal membrane during an action potential is reduced by a Ca^{2+}

pump, it falls exponentially to a level where it will no longer cause transmitter release (for example to 30%, where probability of release would be 0.3^4 or 0.8% of that at the peak of the spike). If a second spike initiates another inward rush of Ca^{2+} at this time, however, the probability of release rises to 1.3^4 or nearly 3 times the release to the earlier spike. Although this seems a very probable explanation for facilitation at most synapses, it is also possible that the Ca^{2+} that enters during an action potential might transiently decrease K^+ permeability or increase the Ca^{2+} influx caused by a second action potential. (Unfortunately for this hypothesis, Ca^{2+} influx appears to increase K^+ conductance and/or decrease Ca^{2+} conductance, where this has been examined.)

9.7.5 Tetanic Depression and Posttetanic Potentiation (PTP)

If a series of stimuli are given presynaptically at a chemical synapse, one characteristically sees a large build-up of transmitter release to each stimulus in the train during the first few stimuli, then a decline until each stimulus causes release of only a fraction of the original number of quanta. This is synaptic *depression.* The cause of this depression is not understood, but it appears to involve both a decrease in the probability of release of any given quantum and a depletion of the number of quanta available to be released.

At some delay after such a train of stimuli, even where there has been considerable depression, the amount of transmitter released by a test stimulus increases to exceed greatly the original amount. This potentiation, called *posttetanic potentiation* (PTP), may last several seconds or even minutes after a very long train. The explanation for this phenomenon is also unknown, although it may involve long-term changes in the levels of Ca^{2+} in the terminal (external Ca^{2+} must be present for PTP to occur) or an increase in presynaptic spike amplitude due to terminal hyperpolarization, resulting from a long-term increase in K^+ permeability or to activity of electrogenic pumps removing the Na^+ that entered during the tetanus. Such hyperpolarization has been observed in several neurons after tetanic stimulation.

9.7.6 Presynaptic Inhibition: The Block of Transmitter Release

There is another form of inhibition, recently recognized, that is fully as effective as the postsynaptic inhibition discussed (Section 9.7.2) and much more specific; it is capable of blocking one input without affecting others. This is *presynaptic inhibition,* in which an "inhibitory" nerve ends on or near the presynaptic terminal of the synapse to be inhibited. Inhibition may be accomplished in different ways, but the underlying mechanism is a reduction in the amplitude of the spike invading the presynaptic terminal. This results in a sharp reduction in the amount of transmitter released (see Figure 9-16). In vertebrates the effect of inhibitory input appears to be a subthreshold depolarization of the presyn-

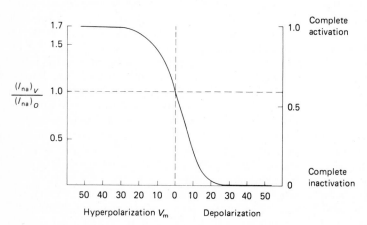

Figure 9-18
*Schematic diagram showing how the sodium current (I_{Na}) and degree of Na^+ activation of the squid axon change as a function of a conditioning hyperpolarization or depolarization. At the normal membrane potential ($V_m = 0$), approximately 60% of the Na^+ channels can be opened by a large step depolarization. With a conditioning hyperpolarization, this value rises to 100%, whereas with only 30 mV conditioning depolarization, the membrane becomes totally inactivated. The left ordinate shows the sodium current, $(I_{Na})_v$, that results from a given voltage step with such conditioning changes in V_m, compared with the $(I_{Na})_o$ for the same step normally. [Adapted from A. L. Hodgkin and A. F. Huxley, J. Physiol., **116,** 497–506 (1952).]*

aptic terminal, which produces slight Na$^+$ inactivation (see Section 9.9.2 and Figure 9-18). An action potential, invading the partly depolarized terminal, is reduced in total amplitude, and transmitter release is much reduced. In the crayfish neuromuscular junction, GABA (see Table 9-2) is the presynaptic inhibitory transmitter; the GABA causes a large increase in conductance of Cl$^-$, which depresses the amplitude of the spike that spreads electronically into the terminal.

It has been discovered, on the other hand, that presynaptic inhibition in some cells (including vertebrate dorsal root ganglion cells and certain neurons of the nudibranch, *Aplysia*) is achieved by a specific, transmitter-mediated decrease in Ca^{2+} channel conductance. Other inputs, in the *Aplysia* system, can cause presynaptic facilitation by increasing the voltage-dependent Ca^{2+} conductance.

9.7.7 Electrical Synapses

As one synapse after another proved to be chemically mediated, it was generally assumed that all synapses were chemical (where once they had all been considered electrical). More recently, however, a substantial number of electrical synapses have been described. Many are cases in which speed of transmission and lack of a synaptic delay are of obvious selective advantage to the animal. These are not amplifying synapses. Presynaptic activity must provide all the current for postsynaptic depolarization, not merely a trigger (chemical transmitter) that causes the release of energy stored in the postsynaptic membrane potential. Electrical synapses, therefore, are uniformly ones in which (1) the presynaptic terminal is of comparable size or larger than the postsynaptic process, and (2) there is morphological evidence for electrical coupling: Either the presynaptic terminal surrounds the postsynaptic process in a calyx (as in the chick ciliary ganglion), or there is close apposition or partial fusion of the cell membranes (Figure 9-19). The latter most commonly are associated with "gap junctions," in which the extracellular space decreases to about 20 Å. Within the gap there is a regular hexagonal array of channels apparently allowing cytoplasmic communication between the two cells. Gap junctions can be formed very quickly in culture (within 60 sec) and can perhaps be "bro-

ken" as easily; hence they may be both a very potent and potentially labile form of cell interaction. (Gap junctions should be distinguished from what are called "tight junctions," in which there is fusion of the membranes of the two cells and no evidence of low resistance connections between them. Such tight junctions are commonly found ringing cells in epithelial layers, forming an effective block to diffusion of substances across the epithelial layer.) Unlike their chemical counterparts, most electrical synapses can transmit in either direction.

Some electrical synapses appear to offer little or no resistance to current flow and interpose no synaptic delay. The septa joining single-cell segments of the annelid and crustacean giant nerves and the intercalated discs of vertebrate heart muscle are examples of this. In other cases, transmission is slowed and may even be unidirectional (or rectifying), as in the synapse between the lateral giant fiber and the motor giant fiber of the crayfish. Synapses with electrical transmission were first described in invertebrates, but there are now many known in vertebrates as well. The medullary neurons controlling firing of the electric organs of some fish (see Section 10.4) are exactly synchronized with each other via electrical coupling; and there apparently are two types of electrical synapses (one of them inhibitory) on the huge Mauthner cells of teleosts. The chick ciliary ganglion synapse is mediated initially by chemical transmitter, then, as the presynaptic process grows to surround more of the postsynaptic cell, by shorter latency electrical transmission as well. Less potent but perhaps no less important electrical interactions between motoneurons and between motoneurons and afferent sensory nerves have also been described in amphibians and mammals. Gap junctions have been seen joining neurons of several brain nuclei, and it is now well documented that there are effective low resistance junctions between several like populations of cells in the vertebrate retina (receptor cells, horizontal cells.) There is even a form of electrical inhibition, described again for the fish Mauthner cell, in which the axon hillock of one cell is surrounded by a spiral process of a collateral axon from the contralateral cell. When the surrounding axon is active, it causes a local increases in positivity around the postsynaptic axon, effectively hyperpolarizing that membrane and inhibiting impulse generation.

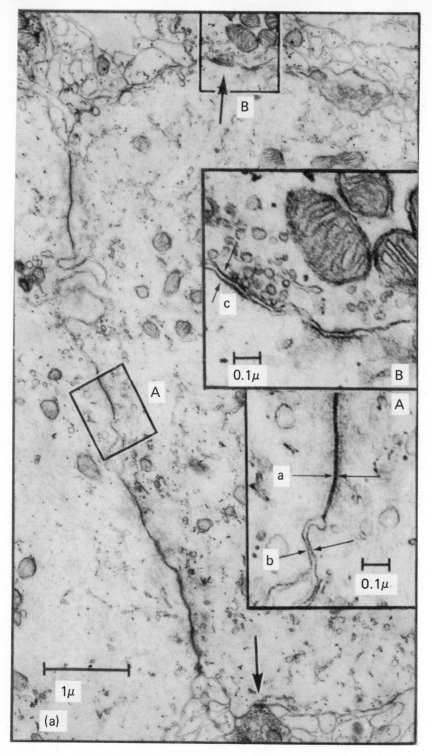

Figure 9-19a

(a) Electrical synapses. An electron micrograph showing gap junctions (inset A) between dendritic processes of the medullary motoneurons controlling the electric organs of mormyrid fishes (see Section 10-4). Note the apparent absence of intercellular space in the presumed electrically transmitting junctions. Normal chemical synapses are also present (inset B). [From M. V. L. Bennett, E. Aljure, Y. Nakajima, and G. O. Pappas. Science, **141**, 262–264 (1963).]

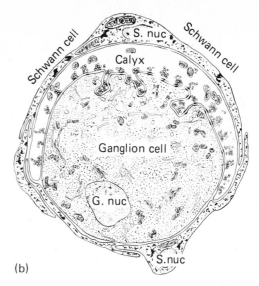

Figure 9-19b

Electrical synapses. Diagrammatic cross section through a chick ciliary ganglion synapse, showing the way in which the presynaptic terminal (calyx) comes to enclose the postsynaptic cell. The axon terminal itself is enclosed in a Schwann-cell sheath. This synapse may also contain tight junctions, but an intercellular space is normally present and chemical transmitter is released. The synaptic area is so large relative to the postsynaptic cell surface, however, that electric coupling is sufficient to cause excitation. [From A. J. de Lorenzo. J. Biophys. Biochem. Cytol., **7**, *31–36 (1960).]*

(b)

9.8 INTEGRATION

The process whereby the postsynaptic cell responds to all of its input—both excitatory and inhibitory, chemically and electrically transmitted—is termed *integration* (see Figure 9-20).

9.8.1 Interaction of Inputs and Importance of Cytoarchitecture

This is the most important function of higher nervous systems. Most cells probably receive hundreds or thousands of inputs, some of them excitatory, other inhibitory, and the relative preponderance of the two determines the postsynaptic cell's behavior. Usually, each individual synapse is weak by itself, that is, it apparently releases very little transmitter, so many inputs must be fired simultaneously to have a significant effect. Obviously, because of the need for many inputs to reach threshold, there may be *spatial summation,* with the effect of simultaneous input from two sources sometimes being much greater than the sum of these influences alone. Similarly, a subthreshold response to one input may summate with a small subsequent input to reach threshold *(temporal summation)*. Analogous phenomena apply to whole populations of cells as well.

In most vertebrate neurons, synaptic input is predominantly on the dendrites and cell body. In mo-toneurons, and perhaps in most neurons, spikes arise first in the axon hillock where, for some unknown reason, the threshold for firing is much lower than in the cell body itself. The cell body is then invaded by the action potential, but it probably does not get far into the dendrites, which are thought to be incapable in most cases of regeneratively conducting an action potential. (In a few cells, on the other hand, there is good evidence that large dendrites can conduct action potentials.) In at least one respect it would be highly disadvantageous for the dendrites to be capable of generating an all-or-none spike. If massive input on one dendrite could reach threshold, it could fire the whole cell, nullifying the influence of other inputs on other dendrites or the cell body.

It is generally felt that the cell body integrates the input from all sources and that the net effect is read out at the axon hillock. Histological studies employing degeneration of inputs from different sources (for example, ipsilateral versus contralateral sensory input) to central neurons reveal that these inputs are often segregated in clear-cut ways to different portions of the complicated dendrite "tree" (Figure 9-21). Such spatial distribution of input is undoubtedly very important to integration, because synaptic input would be expected to be more or less effective depending on its position with respect to dendrite branches and the soma. Most effective would be the inputs on the soma.

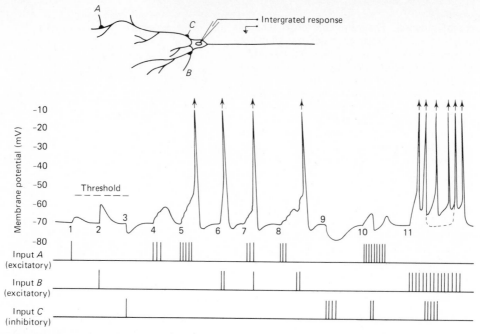

Figure 9-20

Integration of convergent inputs in a single neuron. A hypothetical arrangement showing two excitatory (A and B) and one inhibitory (C) ending on a cell's dendrites with the integrated result in the postsynaptic soma. In stimulus condition (1), (2), and (3) each afferent is fired at an interval long enough to ensure no interaction, the result being simple postsynaptic potentials. The EPSP from (B) is larger and faster rising because it originates near the soma. In (4), (5), and (9) the effects of temporal summation on weak and strong synapses may be seen. In (7), the postsynaptic effects of (A) and (B) are spatially summated; and in (8) there is both temporal and spatial summation. Conditions (10) and (11) show the integration of weak and strong excitatory input with inhibitory input. Were the (B) afferent located near the base of the same dendrite as (C), one might have expected to see a silent period (dashed line) during the firing of input (C). The elements shown would actually constitute only a tiny fraction of the synaptic complement of most cells. Usually there are more dendrites and many more synapses on each dendrite.

It is perhaps no accident that inhibitory endings appear to be located predominantly on the soma, where they would be more effective in "short-circuiting" excitation than if they were located further out on dendrites.

The precise timing of antagonistic inputs can also be of immense importance. For example, there are certain cells in the medullary auditory nuclei of the mammalian brain that receive input from both ears. If one arrives just microseconds before the other, the cell will fire; if they arrive simultaneously or in opposite temporal sequence, the cell will not fire. Such behavior is presumably involved in binaural discrimination of direction.

Sensory inflow of information and the outflow of commands both proceed through hierarchies of neural levels. For example, in the visual system (see Chapter 10) excitation passes from the receptors (rods and cones) to bipolar cells to retinal ganglion cells to lateral geniculate cells to the primary visual cortex, and then through an unknown number of steps beyond. At each level one incoming axon may branch to supply several cells, a phenomenon called **divergence.** Conversely, a given cell nor-

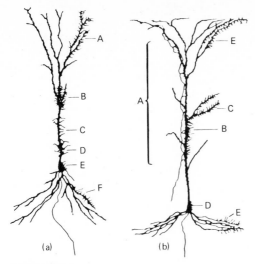

Figure 9-21
Two large pyramidal cells, from the hippocampus (a) and the cerebral cortex (b), showing complex dendritic cytoarchitecture and segregation of input. Each letter corresponds to a region receiving presynaptic input predominantly from a different source within the brain, some excitatory, some inhibitory. Postsynaptic action potentials are generated, in both cases, near the point where the axon (bottom) emerges from the cell body. (After M. E. Scheibel and A. B. Scheibel. Intern. Rev. Neurobiol., **13,** *1–27 (1970).]*

mally receives afferents from many cells in the preceding level, illustrating **convergence.** These inputs may be inhibitory or excitatory.

Even within any given neural level, there is often strong interaction between adjacent cells. This is specially prominent in the sensory pathways where strongly activated elements characteristically inhibit less strongly excited ones, usually through interneurons. This phenomenon, known as **lateral inhibition,** has the effect of exaggerating differences in excitation level of analyzing neurons at successive levels of sensory pathways, thereby enhancing contrast and accuracy of stimulus analysis. Because of such interaction, a lighted square on a dark background appears to be brighter at the edges than at the center. Figure 9-22 shows schematically one way in which such sharpening of contrast can occur. In sensory systems the part of the receptor organ

that provides input to a given cell at any level is called the **receptive field.** This may be excitatory or inhibitory and applies equally well to parts of the retina in vision, of the basilar membrane in hearing, or skin areas in touch. There is a pronounced tendency for adjacent elements in a receptor to affect adjacent populations of cells at higher neural levels, a phenomenon known as **topographic localization.**

Frequently stimuli to different receptors, for example, hearing and vision, may be strongly facilitatory to each other or strongly inhibitory. Or input to one of a bilateral pair of receptor organs may dominate input to the other, as in vision. In a condition known as "squint," the two eyes cannot be brought to focus on a common point, and the dominance of one eye becomes so complete that the other may become functionally blind. The nature of integration of sensory information is probably best understood in the visual system. A more complete discussion of integration will be given in this context in Chapter 10.

One of the basic accomplishments of integration, at least in certain cells, is pattern recognition. These cells, either by innate circuitry or by "learned" associations, come to respond selectively to a certain balance of inputs from different sources—from many different receptors in the same modality or from receptors of different modalities. The specificity of response can be quite fantastic, as we shall see in subsequent descriptions of specific sensory pathways.

9.9 COMMON FEATURES OF INFORMATION PROCESSING IN SENSORY SYSTEMS

9.9.1 Sensitivity, Regulation, and Dynamic Range of Receptor Organs

It is evident that with only a few exceptions, the following generalizations may be made about information processing in sensory pathways:

1. Receptors respond only to highly specific "adequate" stimuli, usually determined by the receptor's surrounding tissue.
2. The energy in the response, even at the receptor

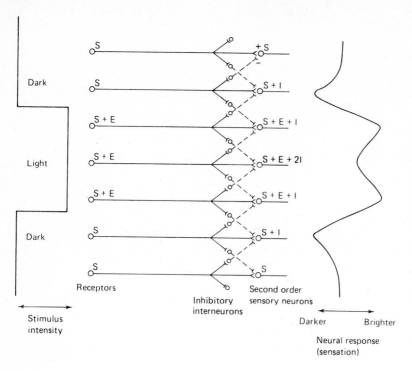

Figure 9-22
Schematic diagram illustrating the principle of lateral inhibition. Three populations of cells are shown: receptors, which are excited by the stimulus (light), second-order sensory neurons, excited by the receptors, and interneurons, excited by the receptors and inhibiting adjacent second-order neurons. Both first- and second-order neurons are assumed to be spontaneously active with second-order neuron activity capable of being increased or decreased. Because of the lateral inhibition, there is exaggeration of differences in activity of second-order cells on either side of the border of the receptive field. S, spontaneous activity; E, excitation: I, inhibition. Obviously the phenomenon depends on the relative strength of excitation and inhibition and on the accuracy of lateral interconnections. Many other forms of lateral inhibition exist that could work as well: presynaptic inhibition, inhibition of second-order cells by activity of other second-order cells, and so on. To be realistic, the scheme should also show much more divergence than is indicated.

level, is often far greater than that of the stimulus, which acts only to trigger the response. The minimum power needed to elicit a threshold visual, auditory, or olfactory response, for example, is approximately 10^{-17} to 10^{-18} W-sec/sec, and the energy in a single sensory nerve impulse is approximately 10^{-11} W-sec.

3. Information is coded first into a graded potential, then into a series of all-or-none spikes that are sent at high speed without loss to distant, higher level cells. Here the spikes are retranslated into a graded release of transmitter, which may prove excitatory or inhibitory to the postsynaptic cells that integrate input from several such sources.

4. The sensitivity of a receptor pathway is usually controllable by efferent outflow from the CNS.

5. The dynamic range of receptor organs is typically enormous, spanning 10^{10} or greater variation in stimulus strength. This range, and accurate discrimination of stimulus parameters within the range, are achieved predominantly in two ways: (a) response of the receptor is usually logarithmically related to stimulus intensity, that is, $R \propto \log S$ (where R is response amplitude and S represents stimulus intensity), so that a tenfold increase in stimulus intensity causes only a twofold increase in generator potential; and (b) different populations of receptors in each receptor organ have thresholds that may differ by as much as four or five orders of magnitude.

9.9.2 Adaptation and Spontaneous Activity

The transduction of stimulus parameters into spike-train response is highly dependent on another important phenomenon, common to varying extent in most receptor pathways. This is the phenomenon of *adaptation*. A large fraction of receptors respond only briefly to a constant stimulus, and even tonic receptors usually have a brief dynamic response before settling into a new sustained firing rate (Figure 9-4). Adaptation is of immense value to the organism. Because of it we do not stay blinded in bright light, we are not constantly aware of the touch of our clothing, and our senses do not become saturated at stimulus intensities slightly above threshold. Nevertheless, small changes in stimulus intensity remain readily detectable.

There are two conspicuous causes of adaptation: (1) a drop in the generator potential; and (2) accommodation at the site of impulse generation. Adaptation of the generator potential is a property of the receptor itself, for example, the rapid bleaching of receptor pigment in the eye, with the resulting temporary drop in sensitivity, or the reduction in the effect of a mechanical stimulus as tissues surrounding the receptor belatedly "give" with the stimulus. Such changes in coupling between receptor and environment appear sufficient to explain adaptation of the receptor potential in most, if not all, cases. Adaptation of the receptor potential is insufficient to explain the fast adaptation of response in most phasic receptors, however. The crustacean stretch receptor organ is a case in point. With a sudden maintained increase in length of the muscle on which the receptors are located, the receptor potentials of the fast and slow neurons are closely similar; both show a sharp onset and rapid decline to a sustained depolarization. The slow cell continues to produce action potentials throughout the stimulus in proportion to the depolarization, whereas action potentials "adapt out" readily in the fast cell.

Nakajima and Onodera have shown that the adaptation of the receptor potential in both cases is explainable by changes in tension with time at the constant length, that is, mechanical readjustment of viscoelastic components in a constant-length stimulus reduces the effective length change and tension at the site of the nerve terminals. When a feedback control was used to maintain constant tension, the adaptation of the generator potential was essentially eliminated. Sample records are shown in Figure 9-23. With such a constant generator potential, the fast cell was still strongly phasic. Its fast and complete adaptation is explained by the phenomenon of *accommodation,* a poorly understood property of spike-generating membranes wherein a maintained depolarization causes a gradual inactivation of Na^+ conductance (and possibly a slight increase in K^+ conductance). As a result, a depolarization must be greater than usual to cause a sufficient increase in Na^+ conductance to

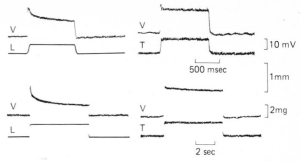

Figure 9-23
Receptor potentials (V) recorded from a slowly adapting crustacean stretch receptor neuron in the presence of tetrodotoxin, when stimulated by constant length (L) and constant tension (T) changes in the muscle on which the neuron was located. The base lines correspond to the resting muscle length and zero tension. [From S. Nakajima and K. Onodera. J. Physiol. (London), **200,** *187–204 (1969).]*

equal K^+ conductance. In effect, the threshold rises. If there is a maintained hyperpolarization, the opposite change occurs and the threshold for spike initiation falls. Varying degrees of accommodation, coupled with the adaptation of receptor potentials, can in this way explain many of the different patterns of response seen to a constant stimulus (Figure 9-24, see also Figures 9-18 and 10-8).

The importance of accommodation becomes even more apparent when it is realized that this phenomenon affects excitability throughout the nervous system. The postsynaptic potentials of CNS neurons are equivalent to receptor potentials, and their size and time course markedly affect the response. It is found in cat motoneurons that accommodation is much faster in the initial segment of the axon than in the soma, so that slowly rising depolarizations initiate impulses in the soma first. Indeed, in most nerve fibers, accommodation is very rapid, and one needs to postulate an unusual absence of accommodation in the important sensory nerves (and CNS neurons) that show sustained firing rates and measure absolute stimulus values, for example, the receptors measuring arterial blood pressure or deep body temperature.

Many sensory and central neurons fire spontaneously at rates up to 200 per sec or more. In some, such as the crustacean slow cell, the firing is due to sustained low-level resting stimulation. In others

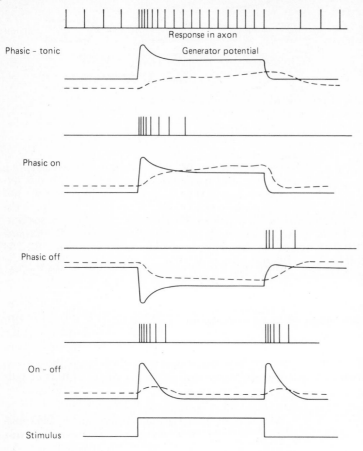

Phasic - tonic — Response in axon / Generator potential

Phasic on

Phasic off

On - off

Stimulus

Figure 9-24
*Typical response patterns, showing the impor-
tance of accommodation. In all cases the stimulus
has a sharp onset and constant amplitude, as
shown. For each response type, the receptor or
generator potential is shown (middle) with the
resulting firing pattern of its axon (top) and a
hypothetical curve (dashed line) depicting fluctuations in threshold (relative to the generator potential) resulting from accommodation at the
spike-generating locus. The firing rate is assumed
to be proportional to the amount by which the
generator potential exceeds threshold. In phasic-
tonic receptors, there is usually some adaptation
of the generator potential, probably resulting
from changes in coupling to surrounding tissues,
but there is apparently little accommodation.
Thus one sees an initial adaptation of the response, followed by a constant firing rate, and
only a very brief silent period following the "off"
of the stimulus. The initial response amplitude
usually depends on the rate of onset of the stimulus (see also Figure 10-8) and, with a slow enough
rise time, results in a purely tonic response, having no initial adaptation. Phasic "on" responses
are characterized by very fast accommodation,
lifting threshold back above even large generator
potentials. Phasic "off" responses result from the
reverse changes—a hyperpolarizing generator
potential, often caused by inhibition from a
nearly excited element, and accommodation that
lowers the threshold. In "on-off" responding
cells, there is usually fast adaptation of the generator potential. An example is the Pacinian corpuscle (see Section 10-3-4).*

there is firing in the absence of any obvious stimulus. This is believed to reflect a slight leakiness
of the receptor membrane, coupled with random
fluctuations either in threshold or leakiness. If the
noise is slight and the leak constant, the spontaneous firing rate is nearly constant, often governed
by the time course of decay of K^+ conductance
after each spike. Such behavior occurs in *pacemaker*
cells, whose activity excites and synchronizes the
activity of populations of other cells, as in the vertebrate heart or the lobster cardiac ganglion. If
the fluctuations in noise are large, the timing of
spikes is seemingly quite random and without relation to other cells in the same population. There

are at least two major advantages to this spontaneous activity, both of which are apparent from Figure
9-25. These advantages are (1) an increased sensitivity to changes in stimulus intensity, and (2) the
capacity to show both positive and negative
changes in intensity. A cell firing in the midrange
of its response curve will respond more sensitively
to small changes than will a cell that is either near
threshold or already firing at the fastest possible
rate. Also some cells, having just fired, will be refractory to a stimulus change but others, being already partially depolarized toward threshold, will
be unusually sensitive. In its midrange of response
a cell can also slow its rate of firing as well as in-

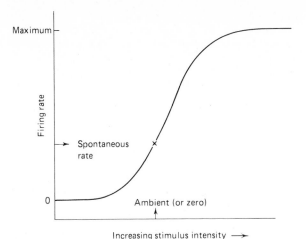

Figure 9-25
Virtues of spontaneous activity. The sigmoid curve relating firing rate to stimulus intensity is typical of input/output curves, whether in spontaneously firing cells or not. It is apparent that there are two regions of stimulus intensity at which the firing rate is insensitive to any change in stimulus: when the stimulus is subthreshold and when the cell is already firing as fast as possible.

In nonspontaneous receptors, the membrane potential is normally well below threshold, a large stimulus is necessary to initiate firing, and relatively large increments in stimulus intensity are necessary to change the response significantly. Such behavior is characteristic of receptors exposed to intense stimuli. In spontaneously firing cells, on the other hand, the membrane potential is always near threshold and the cell is almost maximally sensitive to changes in stimulus intensity. Moreover, if the spontaneous activity is really due to ambient or random stimulation, a reduction in stimulus intensity can be signaled as well as an increase.

vous systems, particularly at periperal sensory levels.

ADDITIONAL READING

Aidley, D. J. *The Physiology of Excitable Cells,* 2nd ed., Cambridge University Press, New York, 1978.

Baker, P. F., A. L. Hodgkin, and T. I. Shaw. "Replacement of the axoplasm of giant nerve fibres with artificial solutions," *J. Physiol. (London),* **164,** 330–354 (1962).

Bekesy, G. von. *Sensory Inhibition,* Princeton University Press, Princeton, NJ, 1967.

The Brain, A Scientific American Book, W. H. Freeman & Co., San Francisco, 1979.

Bullock, T. H., R. Orkand, and A. D. Grinnell. *Introduction to Nervous Systems,* W. H. Freeman & Co., San Francisco, 1977.

Cajal, S. Ramon y. *Histologie du Systeme Nerveux de l'Homme et des Vertebres.* Consejo Superior de Investigaciones Cientificas, Institute Ramon y Cajal, Madrid, 1911.

Cooke, I., and M. Lipkin, Jr. *Cellular Neurophysiology,* Holt, Rinehart & Winston, New York, 1972.

Davis, M. *Function of Biological Membranes,* Chapman & Hall, London, 1973.

Diamond, J. M., and E. M. Wright. "Biological membranes: the physical basis of ion and nonelectrolyte selectivity," *Ann. Rev. of Physiol.,* **31,** 581–646 (1969).

Eyzaguirre, C., and S. W. Kuffler. "Processes of excitation in the dendrites and in the soma of single isolated nerve cells in the lobster and crayfish," *J. Gen. Physiol.,* **39,** 87–119 (1955).

Fatt, P., and B. Katz. "An analysis of the end-plate potential recorded with an intracellular electrode," *J. Physiol. (London),* **115,** 320–370 (1951).

Furshpan, E. J., and D. D. Potter. "Transmission at the giant motor synapses of the crayfish," *J. Physiol.,* **145,** 289–325 (1959).

Hagiwara, S., and L. Byerly. "Ca channels," *Ann. Rev. Neurosci.,* **4,** (1981).

Heuser, J. E., and T. S. Reese. "Evidence of recycling of synaptic vesicle membrane during transmitter release at the frog neuromuscular junction," *J. Cell. Biol.,* **57,** 315–344 (1973).

Hille, B. "Ionic channels in nerve membranes," *Prog. Biophys. Mol. Biol.,* **21,** 1–32 (1970).

———. "Ionic selectivity, saturation and block in sodium channels. A four barrier model," *J. Gen. Physiol.,* **66,** 535–560 (1975).

Hodgkin, A. L. "The Croonian lecture: ionic movements and electrical activity in giant nerve fibers," *Proc. Roy. Soc. (London),* **B148,** 1–37 (1958).

crease it, for example to signal a darkening of the environment instead of a brightening.

Clearly, a single, spontaneously firing sensory input could be very misleading. The central analyzing center would not know whether an unusually short or long interval between spikes represented information or merely an unusually great random fluctuation. When there is redundancy in the system, however, with hundreds or thousands of units carrying much the same information, any synchrony of response will be readily apparent. Redundancy is a prominent feature of all vertebrate ner-

————. *The Conduction of the Nerve Impulse,* Liverpool University Press, Liverpool, 1964.

————, and P. Horowicz. "The influence of potassium and chloride ions on the membrane potential of single muscle fibers," *J. Physiol.,* **148,** 127–160 (1959).

————, and B. Katz. "The effect of sodium ions on the electrical activity of the giant axon of the squid," *J. Physiol. (London),* **108,** 37–77 (1949).

Junge, D. *Nerve and Muscle Excitation,* 2nd ed., Sinauer Press, Sunderland, MA, 1980.

Katz, B. "The Croonian lecture: the transmission of impulse from nerve to muscle, and the subcellular unit of synaptic action," *Proc. Roy. Soc. (London),* **B155,** 455–477 (1962).

————. *Nerve, Muscle and Synapse,* McGraw-Hill Book Company, New York, 1966.

————. *The Release of Neural Transmitter Substances,* The Sherrington Lectures X, Liverpool University Press, Liverpool, 1969.

Kuffler, S. W., and J. G. Nicholls. "The physiology of neuroglial cells," *Ergeb. Physiol.,* **57,** 1–90 (1966).

————. *From Neuron to Brain,* Sinauer Press, Sunderland, MA, 1976.

Martin, A. R. "Quantal nature of synaptic transmission," *Physiol. Rev.,* **46,** 51–66 (1966).

Mellon, D., Jr. *The Physiology of Sense Organs,* W. J. Freeman & Co., San Francisco, 1968.

Mountcastle, V. B. (ed.). *Medical Physiology,* 13th ed., vol. II, C. V. Mosby Co., St. Louis, MO, 1974.

Peters, A., S. L. Palay, and H. deF. Webster. *The Fine Structure of the Nervous System,* Harper & Row, New York, 1970.

Singer, S. J., and G. L. Nicholson. "The fluid mosaic model of the structure of membranes," *Science,* **175,** 720–731 (1972).

The Synapse, Cold Springs Harbor Symposium of Quantitative Biology, vol. 40, Cold Spring Harbor, NY, 1967.

Alan D. Grinnell

10

SENSORY PHYSIOLOGY

10.1 INTRODUCTION

The pressures of natural selection have produced an immense variety of sensory receptors. Organs of each modality have evolved and become specialized independently in many different animal groups. The wealth of variety has stimulated numerous alternative receptor classifications. We prefer distinctions based on the energy form of the stimulus: chemical, mechanical or electromagnetic.

Few generalizations can be made about sensory receptors, even within a given modality. They tend to be concentrated on the body surface, as would be expected and, in bilaterally symmetrical animals, toward the anterior end of the organism (see Section 11.2). Probably all produce receptor potentials in response to an appropriate stimulus, but this has not been demonstrated. The receptor potentials that have been studied differ significantly in their ionic mechanisms. Animals that are highly evolved for a restricted environment often show extreme development of one type of receptor, relative neglect for others. Echolocating bats, for example, have large ears and hypertrophied auditory

nervous systems but poor vision (see Section 10.3.9). Similarly, the sensitivity of a receptor is more likely to be associated with the importance of the organ to an animal than with the animal's phylogenetic position. There are many cases of convergent evolution of receptor organs, where central nervous analyzing centers remain quite different. In other cases, however, very dissimilar sense organs apparently provide essentially identical information. In general, "highly evolved" animals excel more in central nervous development and versatility than in receptor capabilities.

Obviously a great many questions can be asked about a given receptor. How sensitive is it? How specific is its response? How does it encode amplitude, position or location, temporal features, and more complex aspects of a stimulus? What are the patterns of central representation? There are two predominant methods of studying receptor properties: behavioral studies and neurophysiology. Both have advantages and disadvantages. Behavioral tests are restricted by the skills of the trainer, the condition and trainability of the animal, and the naturalness of the surroundings. Answers obtained are always subject to reservations about the ani-

455

mals' level of motivation. Conversely, neurophysiological studies, in which a receptor is usually examined in an anesthetized animal (or even *in vitro*), are beset with sampling problems. Granted, if the animal is capable of detecting some change in stimulus parameter, the nervous system must be carrying the information. However, the probability of recording from the appropriate component, and recognizing the significance of the finding, may be statistically nil. Moreover, one sometimes finds neurophysiological responses to stimulus changes that apparently are undetectable to the awake, trained animal. Clearly, one must rely finally on behavioral answers to questions of receptor capabilities (and they often indicate far greater discrimination or sensitivity than can be demonstrated neurophysiologically). Information about mechanism of function, on the other hand, comes more easily from neurophysiological experiments, where stimuli can be better controlled, more animals used, and animal training is unnecessary. A clear picture of sensory function comes only from a combination of all available techniques of investigation.

10.2 CHEMORECEPTION

Chemoreception is a nearly universal sense. The most conspicuous sensory response of bacteria or primitive protozoans is **chemotaxis,** the use of chemical stimuli to locate food or adequately oxygenated water and to avoid noxious substances. *E. coli* responds to 10–20 different classes of attractants (amino acids, sugars, inorganic ions) and nearly as many classes of repellants (including H^+, fatty acids, alcohols, and others). Probably each class involves a separate receptor in the membrane. Most cells in the bodies of the largest and most complicated vertebrates are coordinated or controlled by chemical substances: levels of carbon dioxide, oxygen, and nutrients in the blood, hormones, or the chemical transmitters released at nerve terminals. (Indeed, chemical communication between cells has not been replaced by the development of a nervous system, merely speeded up and localized.) Neural transmitters and receptors have already been discussed, and the regulatory action of hormones will be considered in Chapter 12. In this section we shall restrict our attention to the chemosensory mechanisms for obtaining information about the external environment.

Vertebrates possess clearly distinguishable senses of taste and smell. Taste receptors are normally associated with feeding structures (or the skin and fins of fish), olfactory receptors with respiratory structures. Although the mechanism of reaction to stimulating substances is probably similar, the distinction between taste and smell is justified because the taste receptors differ morphologically, are generally less sensitive and more restricted in response, and send information to different parts of the brain. Olfaction has been of greater adaptive importance, sampling as it does information coming from greater distances on wind or water currents.

Among invertebrates, the insects, which also possess clearly distinguishable organs of taste and smell, have perhaps the most highly developed use of chemoreception. Moreover, their economy of structure and stereotyped behavior make them particularly valuable as objects of study. For example, blowflies sense the chemical nature of their food via sensory hairs (contact chemoreceptors) on their mouthparts and legs (Figure 10-1). Each hair has two to four chemosensory cells (plus one mechanosensory cell to record hair bending). Normally one of the chemosensory cells is selectively sensitive to water, another to salts, a third to sugars. Although complicated interaction between them exists, in the simplest case activation of the sugar receptors on the feet reflexly causes protrusion of the proboscis. Contact with chemosensory hairs on the end of the proboscis causes opening, and stimulation of receptors inside the proboscis elicits sucking. Only those carbohydrates that evoke impulse activity in the sugar receptors trigger proboscis extension, and feeding continues until the receptors are adapted, the sugar concentration falls below threshold for the sugar receptor, or the system is inhibited by central receptors reporting satiation. Curiously, the sugar receptor of the fly and the sweetness receptor of the human tongue show the same sugar selectivity (fructose > sucrose > glucose). The sensitivity and response patterns of individual cells can be studied and their inhibitory and facilitatory interaction with others potentially understood.

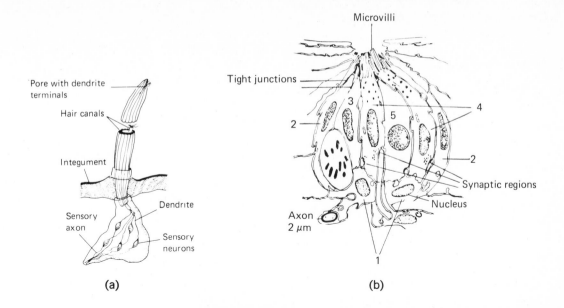

(a)

(b)

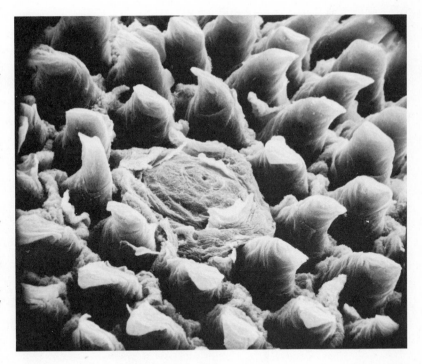

Figure 10-1
Chemosensory taste receptors. (a) Chemosensory hair from the mouthparts of the blowfly, Phormia. *(b) Schematic low power electron micrograph of a rabbit taste bud, showing the continuous development and replacement of cells. A basal cell (1) divides to produce daughter cells (2) that elongate and gradually develop taste microvillae (3), finally maturing into taste bud cells (4). These eventually lose their microvillae and degenerate (5). The mature taste bud cells are joined distally by tight junctions and are innervated by a single sensory nerve fiber. (c) Scanning electron micrograph of the surface of the rat tongue, showing several filiform papillae (nonsensory) and one fungiform papilla, at the center of which is the tiny pore of a taste bud (×550). (Courtesy of P. P. C. Graziadei.)* [*(a) from M. Wolbarsht and F. Hanson.* J. Gen. Physiol., **48,** *675 (1965); (b) from E. T. Burt.* The Senses of Animals, *Wykeham Publications, London, 1974.*]

(c)

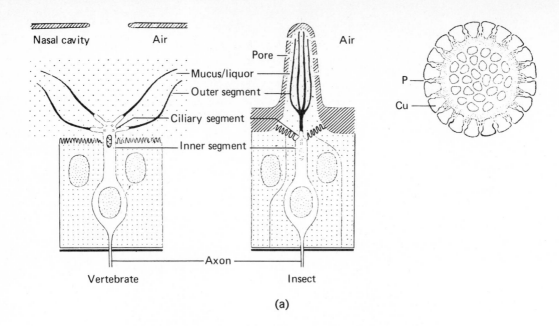

Nasal cavity   Air   Pore   Air

Mucus/liquor

Outer segment

Ciliary segment

Inner segment

P

Cu

Axon

Vertebrate   Insect

(a)

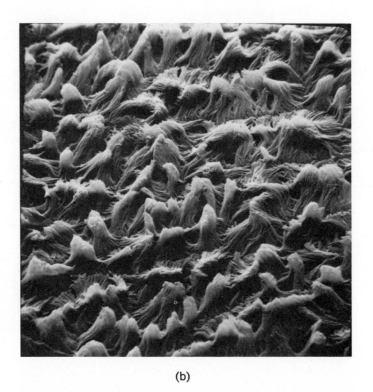

(b)

(c)

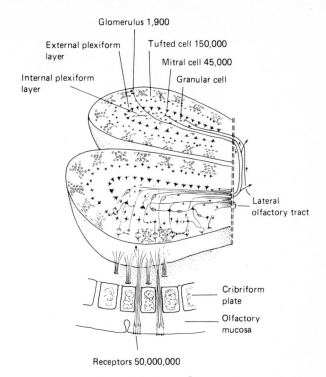

Glomerulus 1,900

External plexiform layer

Tufted cell 150,000

Internal plexiform layer

Mitral cell 45,000

Granular cell

Lateral olfactory tract

Cribriform plate

Olfactory mucosa

Receptors 50,000,000

Figure 10-2
Olfactory receptors. (a) Schematic comparison between vertebrate and insect olfactory receptors, showing possibly analogous structures. To the right is a cross section through an insect sensory hair, showing the dendrites exposed to the external environment via pores (P) in the cuticle (Cu). (b) Respiratory mucosa of Gopher tortoise nasal cavity, magnified approximately 2000×, showing the large olfactory cell cilia grouped in waves reflecting their normal metachronal rhythm of movement. (Courtesy of P. P. C. Graziadei.) (c) Schematic representation of the primary olfactory system of vertebrates, showing primary receptors and cells of the olfactory bulbs. The estimate of number of cell types applies to a single nasal cavity and olfactory bulb of a rabbit. [(a) from R. A. Steinbrecht. "Comparative Morphology of Olfactory Receptors," pp. 4–21. In C. Pfaffman (ed.). Olfaction and Taste III, The Rockefeller University Press, New York, 1969.]

10.2.1 Chemoreceptor Morphology

All invertebrate and most vertebrate chemoreception takes place on the terminals of bipolar nerve cells. As a more recent phylogenetic development, secondary sensory receptor cells have been evolved in vertebrate taste organs, and apparently also in the carotid sinus, where levels of carbon dioxide and oxygen in the blood are monitored. In taste buds, columnar cells differentiate from the adjacent epithelium and congregate at a nerve ending, with hairlike processes extending freely into the external environment (surface of tongue) at one end and synaptic contact with sensory endings of the chorda tympani or glossopharyngeal nerve fibers at the other (Figure 10-1). These modified epithelial cells—not the nerve terminals—are the receptors; but the adaptive value of interposing a secondary sensory cell is not obvious. The sensitivity is about four orders of magnitude less, in general, than that of olfactory-nerve terminals, and the specificity probably no greater. Perhaps the exposure of the taste organs to violent mechanical agitation has ne-

cessitated the interposition of a relatively easily replaceable sensory element. (The average lifetime of a human tastebud cell is about 250 hr.) About 10% of the cells of each taste bud are chemosensory; the rest are supporting cells or basal cells that divide to form the other types. Each bud is innervated by several nerve fibers, each of which also innervates 4–5 other taste buds.

Vertebrate olfactory-nerve terminals and most invertebrate chemoreceptors are somewhat more protected. Most invertebrate endings are associated with hairs or papillae located where they will come in contact with food or other substrate chemicals or, in some cases, with pits exposed to airborne substances. The tips of the sensory nerves are often subdivided into folds or processes only about 0.02 μm in diameter, shielded from the outside world by a thin layer of cuticle that probably has special permeability properties or pores. Vertebrate olfactory-nerve endings are located at the back of the nose, under a thin sheet of mucous, along the sides of the turbinate bones. Humans have 10–20 million endings covering about a 5 cm² area, dogs perhaps

10 times that many. Each ending is intimately associated with supporting epithelial cells and terminates in up to 20 cilialike filaments, 0.1 μm thick and up to 250 μm long (Figure 10-2). The receptor sites are probably located on these filaments, the total area of which is enormous, perhaps exceeding that of the entire body surface. Unlike other sensory nerve cells, the olfactory neurons send axons directly to the forebrain. After only one synapse, with mitral or tufted cells in the olfactory bulb, information is sent directly to the cortex, much of which in submammalian vertebrates is devoted to olfactory analysis. There are also numerous endings of a different nerve, the trigeminal, in the nasal epithelium. These are free nerve endings and have been found to be chemosensory, in part to different substances than the olfactory endings.

Vertebrate olfactory receptors are placed off the main respiratory stream. It is estimated that only about 2% of the inspired air normally passes over the receptor endings. To enhance the sampling size, most animals sniff, causing eddies of air to circulate over the olfactory epithelium, and some animals can actively enlarge the openings to the appropriate parts of the nose.

10.2.2 Taste

Taste and smell are extremely difficult to study, partly because there is no simple way to quantify or even describe the response, partly because human beings are relatively deficient, especially in smell, and cannot appreciate the chemosensory experience of other animals.

In man, there are classically four types of taste sensations: sweet, sour, salt, and bitter. The tip of the tongue is most sensitive to sweet, the side to sour, and the back to bitter substances, with salt sensitivity distributed over most of the tongue. Individual taste buds are indistinguishable morphologically, but usually respond preferentially to one type of stimulus. However, within a given category (for example, sweet) several different effective substances (sucrose, glucose, and lactose) have slightly different tastes. In general, sweet sensation appears to be associated with organic compounds: carbohydrates, alcohols, glycols, amino acids, and saccharine, for example. A pure salt sensation is achieved only with NaCl, although other salts, such as KCl, NH_4Cl, NaBr, NaF, and $CaCl_2$, all taste somewhat salty (often with a bitter component). Sour sensation is clearly associated with $[H^+]$, and most acids, when presented in appropriate concentrations, are indistinguishable from one another. Bitter taste is more complicated, being elicited by compounds of many chemical classes. Quinine is the classic bitter stimulus, but many substances with NO_2 groups or sulfide bonds taste bitter.

Most vertebrates and even many invertebrates appear to taste the same substances we do with comparable thresholds. Honeybees, for example, can be trained to discriminate between the same substances that we can. Thresholds for sucrose are 20 mM in man, 5–6 mM in bees; for HCl the threshold concentration is 1.25 mM in man, 1 mM in bees; for guanidine, it is 1.5 mM in man, 0.8 mM in bees, but 1.25×10^{-6} M in aquatic beetles. Cats appear not to have sweet receptors, and dogs can taste sugars but not saccharine. Chickens apparently are insensitive both to sugars and saccharine, and pigeons respond to saccharine but not to sucrose or quinine. On the other hand, electrophysiological records show that several vertebrates, including frogs, rabbits, cats and monkeys, have receptors most sensitive to distilled water, which man does not.

Some clue to mechanisms of taste may come from study of taste-altering substances. For example, there is a glycoprotein, "miraculin," of about 44,000 molecular weight, derived from an African plant, that somehow modifies taste so that sour substances are perceived as sweet! Bitter, salty, and true sweet responses are unaffected. (Miraculin might be a great commercial success except that the sensation persists for too long a time.) Another plant product, gymnemic acid, somehow suppresses taste sensitivity to sweet substances temporarily.

10.2.3 Olfaction

The sense of smell is much more complicated. Skilled humans are able, with practice, to distinguish thousands or tens of thousands of odors, and it is readily apparent that many other animals can do much better. One of the most potent stimulants for man is butyl mercaptan, which we can smell in concentrations as low as 10^7 molecules/cm^3 of air (averaging out to about one molecule per sensory ending). It is this substance that is mixed with odorless natural gas to provide warning of gas leaks. The best documented case of great sensitivity is the attraction of the male silkworm moth, with its huge antennae, to one of the four isomers (the naturally occurring form) of the female sex attractant bombykol,

$$CH_3(CH_2)_2CH{=}CHCH{=}CH(CH_2)_8CH_2OH,$$

in concentrations of 200 molecules/cm³. Approximately 40 molecules, presumably acting on 40 different receptor cells (1/1000 of the total) during a 30-sec period, are sufficient to elicit the response. Certainly in this case, and probably in humans as well, one molecule is sufficient to excite a receptor ending—the theoretical maximum in sensitivity. Any changes in stereospecificity, chain length or side groups causes a 100–1000-fold or greater reduction in sensitivity.

It is generally believed that, in vertebrates, the transduction takes place on the terminal cilia of olfactory fibers (and on the free nerve endings of trigeminal terminals), but the nature of the transduction process is far from known. Molecules of different sizes and composition may smell identical, but closely related chemical substances, even different isomers of the same molecule (for example, *d*- and *l*-methanol, or *d*- and *l*-carvone) may have quite different odors. Dozens of theories have been proposed to explain olfaction, but most have fallen victim to a surfeit of obvious exceptions. Although no hypothesis is fully acceptable yet, one of the most promising appears to be a stereochemical theory based on the work of Pauling, Moncrieff, and Amoore.

The major premise of this theory is that molecules having similar stereochemical shapes smell the same. Presumably there are specific molecular traps or sites on the receptor membrane, into which the odorant can fit. (The sites need not be preformed; they might form as a conformational change in interaction with the odorant molecule.) The better the fit, the more effective the stimulus.

A promising approach to the identification of such specific receptors is the search for individuals who have specific genetic olfactory deficiencies (anosmias) and the determination of which substances they cannot detect. If other individuals are found with the same deficits, it probably results from the absence of some specific category of receptors, and

an analysis of the molecules affected may provide important insight into the mechanisms of olfactory transduction.

With an understanding of what features of a molecule are important in giving it a certain odor, it should be possible to predict successfully the odor of newly synthesized molecules and to explain differences in odors of isomers, or in sensitivity to them, which may differ by several orders of magnitude. Almost certainly there will be several important factors involved: volatility, solubility in the mucous layer and in lipid, molecular shape, rigidity, presence of polar groups and dipoles, number and placement of hydrogen-bonding sites and of attached water molecules, and levels of vibrational and rotational energy. One can only speculate about the actual source of transduction energy, but one possibility is that molecules having accessible regions that are of the appropriate size and shape, or have the right charge or hydrogen-bonding properties, or otherwise have the right properties to interact with a given receptor site can cause a conformational change in membrane components that can sharply increase or decrease permeability to some ion.

10.2.4 Electrical Correlates of Chemosensation

Graded receptor potentials can be recorded near taste-bud cells, olfactory epithelium, and invertebrate chemosensory hairs or antennae—wherever there are chemical receptors (Figure 10-3). These slow potentials may be either positive or negative, depending on the stimulating substance and its concentration. The rate of firing of the sensory nerve fibers is usually increased when it is negative. It seems highly probable that the positive component is associated with the inhibitory effect of an increase in membrane permeability to K^+ or Cl^-, and the negative component (which is highly dependent on $[Na^+]$ in the medium) probably represents receptor depolarization resulting from an increase in permeability to Na^+. These permeability changes have not been directly established, nor is there always a clearcut causal relation between potential amplitude and nerve firing. In invertebrate receptors the recorded potential presumably is a generator potential but, in vertebrate taste buds, receptor and generator potentials (probably sepa-

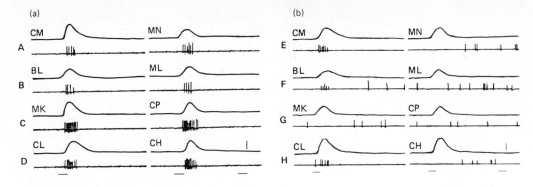

Figure 10-3
Slow-wave electroolfactograms (EOGs) from the olfactory mucosa, and spike discharges from two neurons of the frog olfactory bulb in response to stimulation by brief puffs of air carrying eight different odorant substances: forms A–D in the "camphoraceous" category and E–H in the "minty" category. In all cases the integrated slow wave response shows excitation. The single unit in (a) showed an excitatory response to all eight odors, but much stronger to some than to others. Neuron (b) was excited by three of the four camphoraceous odors, inhibited by all of the minty substances. Thus single neurons can be highly discriminatory in their response patterns. Horizontal bars below the bottom spike records indicate the time and duration (1sec) of odor applications. (From Higashino, S., H. Takeuchi & J. E. Amoore, in Olfaction & Taste III, ed. C. Pfaffman, Rockefeller Univ. Press, 1969).

rated by a chemical step) are not easily distinguishable. There are also questions about how a generator potential could get to a spike-generating membrane, if it arises from a conductance change near the end of a 200 μm- long olfactory cilium.

Chemosensory nerves tend to be of the phasic-tonic type. They respond at 3 msec or longer latency with a burst of spikes which rapidly drops to a much lower level that is sustained throughout the stimulus. In practical experience this means that one quickly loses awareness of a substance, even if it continues to be present in high concentration. Sensitivity to other substances remains high. There is considerable evidence for response by a single receptor to several substances, so it may be possible for one type of receptor site to adapt independent of other receptor sites, leaving virtually the whole population of fibers free to respond to other substances. The size of the population may be important, for the basis of odor discrimination is not known. It is possible that different types of receptor sites (assuming these exist) are distributed unevenly over the olfactory epithelium, and that the spatial pattern of excitation is the clue to odor identification. The most obvious alternatives are receptor terminals sensitive only to one class of

substances (which may exist, but must be rare) or a firing pattern that reflects the nature of the odorant.

10.2.5 Use of Chemoreception in Behavior

We are all familiar with the highly developed use of smell in mammals other than ourselves for detecting and procuring food, for marking out territories, for recognition of each other, for initiation of mating behavior, and for detecting enemies (also, in at least one familiar case, for defense against enemies by exposing them to a highly unpleasant odor). Even in humans, odors are important—not merely to the perfume industry, but in other ways as well, even if reactions to them are unconscious. For example, volatile aliphatic acids produced in late follicular phase by young women is probably responsible for the synchrony of menstrual periods found to occur in dormitory situations. Simpler vertebrates and invertebrates are often far more dependent on the sense of smell. Most birds are considered to be very deficient in the sense, but frogs and toads often locate food, and can apparently locate home ponds from miles away by means of smell. Reptiles have developed a special organ of smell in addition to the usual olfactory area, perhaps because normal breathing does not provide sufficient amounts of environmental air. This is the Jacobson's organ, a pair of small, richly innervated pits lying in the roof of the mouth, into which the forked tongue is inserted after being extended from the mouth to pick up airborne molecules. The sensitivity of fish to chemicals in water is sometimes truly remarkable. Sharks, for example, are famous for their response to trace amounts of blood in the water. Equivalent sensitivity, coupled with strongly imprinted chemosensory memory, is the usual explanation for the ability of salmon to migrate from the sea hundreds of miles upstream to the exact spot in which they spawned.

Insects display highly chemosensory-directed behavior. The reflex control of feeding in flies and the powerful influence of female sex attractant in mate selection by male silkworms have already been described. It is often found that adult insects will lay their eggs on the same species of foodplant on which they fed as larvae, regardless of what they may have ingested as adults or what other equally acceptable food plants are available.

It is in the social insects, however, that the most elaborate use of chemical senses is found. A good example is an ant colony, in which the principal form of communication between individuals is by secreted chemical substances that act as regulatory hormones for the social organism. Such intraspecific chemical stimuli are called *pheromones.* Among fire ants (*Solenopsis saevissima*), odor trails laid by worker ants act as guides and releasers of hunting behavior in sister workers. Secreted chemicals also are used in recognition of other ants of the colony, in providing alarm, in eliciting aid, in stimulating oral grooming and clustering, in initiating and organizing colony migration, and in eliciting disposal of dead ants. The latter behavior is a good example of the strong control such chemical transmitters have over insects. A dead ant, as it decomposes, produces a necrotic substance that causes other ants to recognize the dead ant for what it is and carry it away to a distant dump. This substance has been identified and can be directly applied to a living ant—which is promptly carried, struggling, to the dump. As long as it keeps coming back, it is picked up and rediscarded, until finally the odor wears off.

10.3 MECHANORECEPTION

Mechanical irritability is a general phenomenon of most single-celled organisms. Even in the specialized vertebrate body many nonreceptor cells respond to such stimuli. For example, smooth muscle cells contract in response to stretch, cardiac muscle contracts more strongly when stretched (Starling's law, see Chapter 6), and any nerve can be caused to fire by strong mechanical stimulation. In these cases, however, the effective stimulus must be several orders of magnitude greater than that to which a specialized receptor will respond.

10.3.1 Generalities Concerning Receptor Structure

Like chemoreceptors, mechanoreceptors may be primary or secondary. In all invertebrate and most vertebrate receptors, the stimulus acts upon the sensory-nerve terminal itself (as in the crustacean stretch receptor), although usually these are associated with complicated and important surrounding tissue structures, such as hairs, bulbs, and stretched membranes. Among vertebrates the secondary hair cells have evolved and adapted to perform a wide variety of functions, especially in vibration reception. In both types, a receptor's specificity resides principally in its location and in its mechanical coupling with the environment. Thus a nerve terminal lying against a hair probably responds to membrane distortion in much the way a nerve terminal in a tendon does, but the effective stimuli are very different and need not be of the same order of magnitude.

The mechanism of transduction is poorly understood. The receptor membrane is coupled to structures that move on stimulation, transmitting the energy of the stimulus and causing a stretch or compression of the receptor membrane. This distortion causes an increase in permeability to certain ions that flow down their concentration gradients, depolarizing the cell. In receptor-nerve terminals, this constitutes the generator potential. In secondary sensory cells, membrane distortion is presumed to cause similar changes in permeability, producing a receptor potential and the graded release of an excitatory transmitter that in turn stimulates synaptically coupled sensory-nerve terminals.

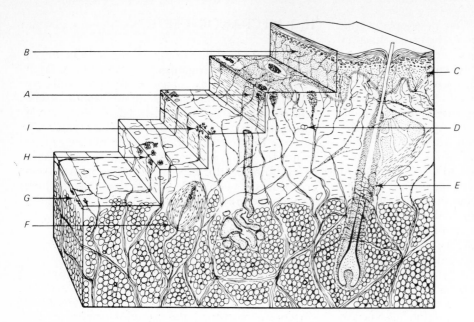

Figure 10-4
Cutaneous sensory receptors. A three-dimensional construction of mammalian skin, showing a variety of receptors, including: A, Meissner's corpuscles (touch); B, net of free nerve endings (perhaps pain); C, Merkel's disks (touch); D, beaded nerve net around blood vessels (probably pain); E, nerve terminals around sheath of hair (movement of hair); F, Pacinian corpuscle (pressure); G, Ruffini endings (warmth); H and I, groups of Krause's end bulbs (cold). [*From G. Weddell,* Brit. Med. Bull., **3**, 167–172 (1945).]

10.3.2 Superficial Touch Receptors

The morphological spectrum of touch receptors is immense. Many are free nerve endings, as described in Chapter 9. Most, however, are in the form of specialized bulbs or corpuscles (for example, Meissner's corpuscle, Figure 10-4, and the Pacinian corpuscle, Figure 10-7), or are closely associated with hairs or vibrissae that by their leverage magnify a stimulus and confer directional sensitivity on the receptor (Figure 10-4). Some 640,000

Figure 10-5
Schematic outlines of body representation in the ventrobasal thalamus of rabbit, cat, and monkey, showing the relative increase in representation of the limbs in cats and monkeys. In all three, the face and tongue are of paramount importance. [From J. Rose and V. B. Mountcastle. In H. W. Magoun and J. Field (eds.). Handbook of Physiology, *sec. 1, vol. 1, American Physiological Society, Washington, D.C., 1959.]*

separate sensitive spots are found on the surface of the human body. Most of these are concentrated into a few areas, especially the tip of the tongue, the lips, and the pads of the fingers.

The preponderance of receptors on the face and the increasing importance of limb extremities in exploring and manipulating the environment is reflected in the relative volume of central analyzing centers devoted to processing information coming from these parts of the body. Figure 10-5 shows an approximate representation of the relative importance of different parts of the body surface of a rabbit, cat, and monkey in the part of the thalamus devoted to tactile sensation (the ventrobasal complex). The equivalent analyzing areas of submammalian vertebrates would, of course, be very different. In fishes, for example, the head, fins and barbels would presumably predominate. The central organization of cutaneous receptor pathways is such that fine discriminations of shape can be made on the basis of touch. Strongly excited endings inhibit less strongly excited neighbors (lateral inhibition) and a central analyzing center "recognizes" the spatial distribution of endings with sufficient accuracy, for example, for the astonishingly rapid discrimination of different letters in the braille alphabet (Figure 10-6).

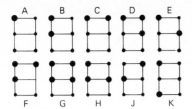

Figure 10-6
Several letters of the Braille alphabet.

10.3.3 Pain

In addition to cutaneous touch-pressure and temperature receptors, there are receptors for pain. The existence of special pain receptors has long been disputed. It is now apparent that they do exist, although it is undeniable that excessive stimulation of any sensory nerves can be painful, as in a "blinding" light or a "painfully loud" sound. It is possible that pain fibers may carry meaningful information about touch or temperature near threshold and that the rate of firing is important in determining whether a sensation is painful or not. The true pain fibers are slow-conducting and appear to end freely with little specialization of the surrounding epithelium. They show little specificity of response to a single type of stimulus. In the spinal cord, nociceptive (pain) information is brought in mainly via sensory C fibers which appear to employ an 11-amino acid peptide, substance P, as their neurotransmitter (see Section 11.5).

The fact that painful stimuli usually are ones that are capable of producing tissue damage has led to the suggestion that the pain fibers might be secondarily stimulated by some substance released by damaged cells. Histamine was once considered a good candidate, since it is released by damaged tissues and can activate small pain fibers. Concentrations as low as 10^{-8} M, injected into the skin, produce a pricking or stinging sensation. However, antihistamines do not counter the sensation. It now appears more likely that injury causes the release of a proteolytic enzyme that can cause the breakdown of plasma globulins to form polypeptides similar to bradykinin; these polypeptides are potent excitors of pain receptor endings. Pain fibers adapt very slowly, as would be expected for such a sense.

It is not desirable to neglect any stimulus that can cause extensive tissue damage.

10.3.4 Pacinian Corpuscle

The sensation of deep touch or vibration is provided mainly by Pacinian corpuscles. These are located in deep layers of the skin, in the connective tissue surrounding muscles, tendons, and joints, and in the mesenteries supporting visceral organs. The structure of these receptors is shown diagrammatically in Figure 10-7. They are relatively huge, with 0.5 mm or more of unmyelinated sensory-nerve terminal surrounded by an onionlike series of 30–50 connective tissue lamellae. The mechanism of touch-pressure transduction has been better studied in these than in any other receptors. Mechanical deformation of the capsule causes distortion of the receptor ending, a generator potential is produced and, if it is great enough, an action potential arises from the first node of Ranvier (still within the capsule). The structure of the onion is such that only the dynamic components of a stimulus are transmitted to the core; sustained mechanical deformation is simply compensated for by a readjustment in corpuscle shape by the sliding of lamellae over one another. Because of this, the capsule is admirably suited for detecting changes in pressure but adapts readily to a sustained stimulus. It is thus capable of again responding to a change in condition, either an increase or a decrease in pressure (Figure 10-7).

Techniques have been devised for removing the surrounding lamellae from the nerve terminal to permit study of the characteristics of the receptor membrane itself. The generator potential is produced only in the localized region of the membrane being stimulated and is proportional to the area stimulated. The terminal is oval in cross section and constructed so that stimuli normally flatten it even more, stretching the membrane and causing a depolarization. If the stimulus is presented in such a way that the terminal becomes more circular in cross section (reducing the membrane area necessary to enclose the volume), a hyperpolarization can be seen; this suggests that permeability to ions producing the generator potential can be increased or decreased from resting level by increasing or decreasing the stretch on the membrane. The gen-

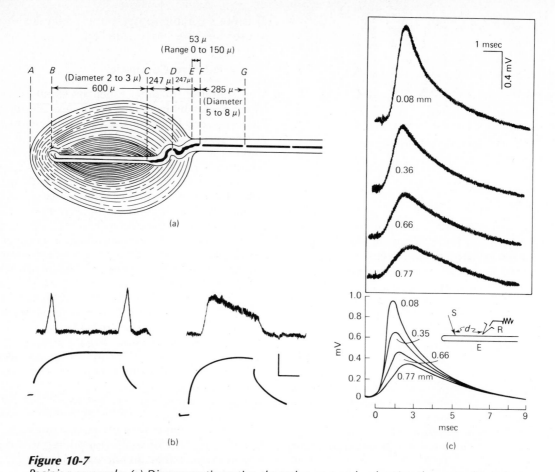

Figure 10-7

*Pacinian corpuscle. (a) Diagrammatic section through a corpuscle, showing the unmyelinated nerve terminal (B to C) enclosed in the onionlike lamellae. D, F, and G, the first three nodes of Ranvier. (b) Because of sliding of the lamellae over each other, a stimulus of long duration evokes rapidly adapting receptor potentials (left) at the "on" and the "off" of the stimulus and short bursts of spikes in response to each. When the lamellae are removed, the receptor potential is maintained. Calibration bars: 10 msec and 50 µV (upper trace). (c) Traces of responses to an equal-intensity stimulus (S) applied along the terminal (E) at different distances from a recording electrode (R). As the stimulus is progressively removed, the amplitude of the potential falls and rise-time increases, indicating that the generator potential is produced only at the point of stimulation. [(a) from T. A. Quilliam and M. Sato. J. Physiol. (London), **129,** 167–176 (1955); (b) W. R. Loewenstein and M. Mendelson. J. Physiol., **177,** 377–397 (1965); (c) W. R. Loewenstein, Ann. N.Y. Acad. Sci., **94,** 510–534 (1961).]*

erator potential decays exponentially with distance from the point of stimulation, indicating that under normal circumstances there is no regenerative event in the terminal itself; however, other experiments have shown that the terminal is capable of being antidromically invaded. Potentials sum temporally and spatially but show "desensitization" when one point is stimulated repetitively at a high rate. Unstimulated portions of a membrane are unaffected.

Interestingly, in the desheathed preparation a sustained stimulus produces a relatively constant generator potential, which, however, still produces a short burst of spikes in the axon. Apparently even without the surrounding capsule, accommodation would make the Pacinian corpuscle a highly phasic receptor. Indeed, the high rate of adaptation of the Pacinian corpuscle and superficial touch receptors of the skin correspond so closely to the rate of accommodation of their respective axons that the limiting factor in adaptation may well simply be axonal accommodation. Fast adaptation is characteristic of most mechanoreceptors, but there are many exceptions, notably the baroreceptors measuring blood pressure in large vessels.

10.3.5 Proprioceptors

It is common experience that, without looking at our hands, we know their location and position. With eyes closed we can bring our arms together behind our backs to touch index fingers. This operation is a mild example of the sophisticated coordination of movement in animals, all of which requires a sense of what position the body is in, the state of contraction of muscles, the angle formed at each joint. This is the so-called *kinesthetic muscle sense,* or *sixth sense*—one that is absolutely essential to adaptive posture and normal movement. The information about state of muscles and position of bones is provided principally by three types of proprioceptors: joint receptors, Golgi tendon organs, and muscle spindle receptors.

Joint receptors are complex nerve terminals located in the connective tissue capsule of the joints. Any one ending is sensitive to movement within an angle of 10–15°, with the population of receptors distributed so that excitatory angles for different endings are in different positions covering the range of joint movement. These are slowly adapting receptors, giving a burst of impulses at the onset of movement that reflects the rate and degree of movement, then settling down to a maintained discharge reflecting the position of the joint (Figure 10-8). Joint receptors are probably also the principal sensors of substrate vibration in vertebrates.

The Golgi tendon organs are likewise high-threshold, slowly adapting receptors. In these, terminal branches of the sensory axon end on the ligaments attaching muscles to bone. Being in series with the muscle, stretch of the tendon due either to muscle contraction or passive stretch of the muscle causes distortion of the receptor membranes and sustained firing of the fibers. Each receptor fiber ends on a tendon formed by the ends of muscle fibers from a few different motor units, but not all. Hence, contraction of some or all of these motor units is a much more effective stimulus than stretch of the whole muscle or contraction of other motor units. It is probable that the selective nature of the information provided by such Golgi endings is useful in motor coordination.

Distinguishing between passive stretch and active contraction and measuring the degree of contraction, are the muscle spindles, many of which are present in each vertebrate muscle (with the experimentally useful exception of certain frog toe muscles that may have only one spindle). Vertebrate muscle spindles are similar in many respects to the crustacean stretch receptors discussed in Chapter 9. In both, there are sensory nerve endings on modified muscle fibers lying parallel to the normal muscle fibers. The modified fibers (intrafusal fibers) are striated and contractile at their ends, but have an expanded, nucleated, bulbous midregion where the sensory terminals end in chains of enlargements invaginating into depressions in the muscle fiber surface (Figure 10-9). An increase in length of the intrafusal fiber causes a sharp increase in firing rate of the sensory axons, a decrease in length causes a drop in firing rate or its cessation altogether.

In the first careful studies of a single receptor's generator potential, Katz showed that the frog spindle generator potential has two components equivalent to the responses of the two sensory neurons of the crustacean stretch receptor. There is an initial large response, an increase in firing on stretch, a decrease on contraction which reflects accurately the rate of change in length of the fiber, followed by adaptation to a nearly constant firing level proportional to the absolute length. In the intact animal this second phase is brief because of reflex readjustments coordinated in the central nervous system (CNS). One of these is the familiar *stretch reflex.* When the spindle stretch receptor is excited, it monsynaptically activates the α motoneurons of adjacent somatic muscle fibers causing contraction that reverses the stretch. Through interneurons, the stretch-receptor input inhibits the motoneurons of antagonistic muscles. In cases of extreme stretch, when added contraction might be damaging, the Golgi tendon organ response excites inter-

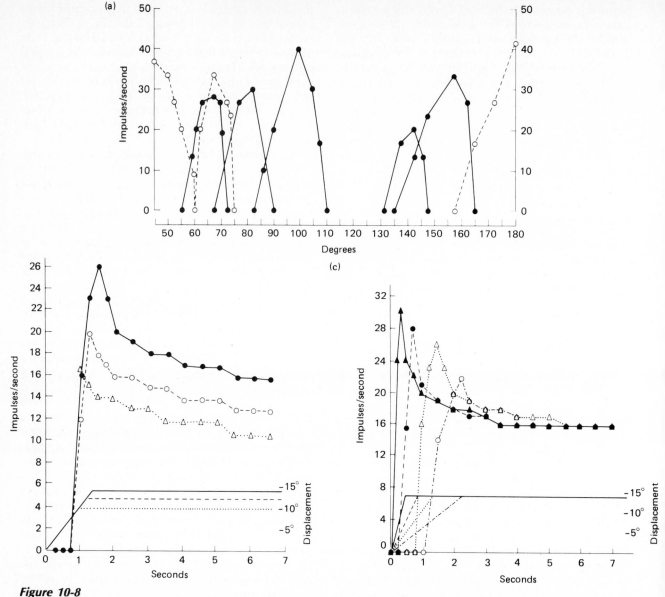

Figure 10-8

*(a) Graph of impulse frequency versus joint angle for eight slowly adapting joint receptors in the knee capsule of a cat. Data are taken from two experiments (solid and dotted curves). A majority of receptors are located so that they fire maximally near points of full flexion or full extension. (b) and (c) Graphs of impulse frequency in a typical joint receptor as a function of joint displacement at a constant rate, and as a function of rate of displacement. Note that the slower the rate of movement, the later and smaller is the peak of activity. [(a) from S. Skoglund. Acta Physiol. Scand. suppl., **124**, 1–101 (1956); (b) and (c) from I. A. Boyd, and T. D. M. Roberts, J. Physiol. (London), **122**, 38–58 (1953).]*

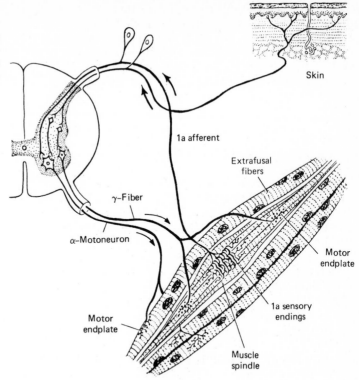

Figure 10-9
Schematic representation of a vertebrate stretch receptor, with stretch-sensitive sensory endings (1a endings) providing monosynaptic input to α motoneurons in the spinal cord, and γ motoneurons innervating the contractile regions of intrafusal fibers. One input to the α motoneuron is shown.

Skin

1a afferent

Extrafusal fibers

γ-Fiber

α-Motoneuron

Motor endplate

1a sensory endings

Motor endplate

Muscle spindle

neurons that inhibit the α motoneurons from contracting.

There is also important feedback control of the sensitivity of the stretch receptors themselves, accomplished by special motor fibers (γ efferents) that govern the degree of tonic contraction of the intrafusal fibers. By changing their firing rate, the γ efferents can increase or decrease the tension on the midregion of the intrafusal fibers, altering the firing rate and sensitivity of the stretch receptors. The γ motoneurons are not directly excited or inhibited by stretch receptor activity, however. They are controlled principally by descending inputs from the brain and by afferents from other areas, for example, from skin (see Figure 10-9).

It has been suggested that most movements, especially postural movements, may be initiated by the γ efferents, that is, that the first change in the system is an increase in γ-fiber firing, which causes increased contraction of the intrafusal muscle fibers, increasing the stretch receptor fiber activity, which in turn reflexly excites α motoneurons to

the same muscle, causing muscle contraction. In several cases, however, it is now clear that the α motoneurons are active before the stretch receptors during contraction. Hence, probably the α and γ motoneurons are excited together to initiate contraction. Nevertheless, the feedback loop described is critical to achieving the desired amount of movement. It permits what may be called "servoassisted," as opposed to "servocontrolled," movement. The α motoneurons cause the muscle as a whole to change length, and the stretch receptors feed back information to the CNS about the extent of the change in length: but it is the γ motoneurons, by independently setting the length of the intrafusal fibers, that determine when the loop reaches the desired state.

The γ motoneurons cause a change in length of intrafusal fibers that is virtually independent of the load on the muscle as a whole. The degree of shortening of the whole muscle, however, is highly dependent on the load. Hence a given level of α motoneuron activity may not be adequate to achieve the "desired" shortening, that is, the amount necessary to bring the stretch receptors, mounted on intrafusal fibers that have changed length because of the change in γ-efferent activity, back to the rest level of firing. If the muscle remains too stretched, there continues to be a barrage of stretch receptor input increasing the level of excitation of α motoneurons to help achieve the additional contraction. Figure 10-10 shows schematically how the efferents and afferents interact.

As we have seen (Section 9.3), invertebrates also possess stretch receptors capable of providing a measure of "muscle sense." Many invertebrates with hard exoskeletons have highly successful joint receptors as well, in the form of hair plates located where one part of the exoskeleton joins another. These plates contain large numbers of protruding cuticular hairs, each innervated by a single, nonadapting sensory fiber that responds to bending of the hairs, whenever one part of the exoskeleton approaches or touches another, effectively monitoring the relative position of body elements (Figure 10-11). Receptors of this sort have been shown to be important in the orientation of insects, where their information is coupled with that of muscle receptors to determine head-to-body and body-to-ground angles. These sources of information in turn are coordinated with visual input to determine such parameters as sun angle (in bees) or the direction in which to strike for prey (in mantids).

10.3.6 Equilibrium, Gravity, Acceleration, and Vibration: Hair-Cell Receptors

The mechanosensory receptors discussed so far provide information principally about the state of the body, the contraction of muscles, bending of joints, and points of contact with the environment. A great deal more information, especially concerning the external environment, is available to mechanoreceptors. Perhaps most important is knowledge of the direction of the ground (the direction of gravity) and the measurement of movement in space, especially angular acceleration. In addition, the presence of mechanical coupling through a surrounding medium (air or water) has permitted the evolution of organs capable of detecting vibrations from distant objects or other individuals in the environment. In its highest development, this sensi-

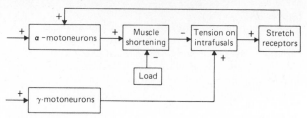

Figure 10-10
γ-loop "servo assistance" of movement. The box diagram shows the interrelationship of variables influencing contraction. Note that the γ motoneurons can set the desired length by independently varying the tension on the intrafusal fibers. Not shown are the Golgi receptors, which provide negative feedback to the α motoneurons when strongly stimulated by muscle shortening.

tivity becomes the phenomenally sophisticated sense of hearing. In the vertebrates, all these functions are performed by extremely versatile and successful secondary sensory cells, called **hair cells.** All hair cells have the same embryological origin and similar structure, characterized by the frog inner ear hair cells shown in Figure 10-12. Protruding from the exposed surface of the cell are 20–50 "hairs," all but one of which are composed of a membrane-surrounded core of highly organized actin filaments that arises from a tapered root inside the cell. These "hairs" are called **stereocilia** and are usually embedded in the cuticular plate, a region of dense material just below the cell membrane. The remaining hair is a true **kinocilium**, having the typical core of nine peripheral pairs of filaments and two central fibers. The plasma membrane of the sensory cell is continuous with the membrane enclosing the cilium, which is always located on one side of the hair bundle. It has been shown that depolarization of the cell results from deflection of the hairs in the direction of the kinocilium, hyperpolarization from deflection in the direction of the stereocilia. Recent electron microscopic studies of vertebrate hair cells reveal that the kinocilium is attached near its distal end to the adjacent stereocilia by a number of fine processes, and that the cell membrane near the base of the kinocilium looks relatively thin. Movement of the stereocilia might cause an up and down plunging movement of the plasma membrane, alternately depressing it and stretching it outward. The membrane could undergo corresponding changes in permeability. Or, slight bending of the stereocilia might cause some change in their internal structure or membrane conformation that directly changes permeability or induces a potential change.

Vertebrate hair cells are extremely small, and good intracellular recordings have been obtained from them only recently. Figure 10-12 a and b shows the experimental arrangement used for studying hair cells in the frog sacculus. The statolith membrane is removed, allowing access to the hair-bearing surface of the cells, where, under direct microscopic observation, a small stylus is used to stimulate the hairs mechanically and a microelectrode is inserted into the cell. Figure 10-12 c and d shows the types of potentials obtained, with amplitudes up to 5–10 mV, saturating with movements greater than $\pm$ 1–2 μm. There is strong rectification, with much greater depolarization resulting from movement in one direction than hyperpolarization from movement in the other. The depolarization is associated with a decrease in membrane resistance that can be shown to encompass most small cations, although K^+ carries most of the current *in vivo*. Interestingly, in this preparation, the kinocilium can be physically dissociated from the stereocilia bundle, and each moved independently (Figure 10-12 e and f). It is found that movement of the kinocilium is ineffective; it is movement of the stereocilia that causes changes in permeability and potential.

At the other end of the hair cell are synapses with both afferent and efferent nerve terminals. There are many variations in this structure, especially in the form of the hair bundle and in the arrangement of synapses with nerve endings, but the general formula is the same. Furukawa and Ishii have demonstrated that transmission of excitation from hair cells to sensory nerve is chemical. However, the afferent synapses on these cells appear to operate along different principles than the chemical synapses described in the last chapter. Maximal release is achieved within 5–10 mV of the resting membrane potential, and the postsynaptic neuron can be driven by changes much smaller than this, sometimes less than 1 mV. Moreover, postsynaptic hyperpolarization is sometimes seen when a hair cell is hyperpolarized. Either transmitter release is mediated by some means not reflected in membrane potential, or very large changes in transmitter release, probably either an increase or a decrease from some tonic level, can be achieved by very slight membrane potential changes around the resting level. Direct evidence that the latter may be possible is the equivalent change in transmitter release from electroreceptors caused by vanishingly small changes in electrical field around these cells. (See Figure 10-24 and Section 10.4 for a description of these modified hair cells, which have been adapted for electroreception.) It is not known what transmitter is released from hair cells. The postsynaptic afferent nerve terminals have been shown to be depolarized by ACh, an action that is blocked by curare. It seems probable that the natural link between receptor and sensory nerve is not ACh, however, because the normal depolarization occurs in response to a natural stimulus even when the effect of ACh is blocked.

10.3.7 Lateral Line Receptors

Elasmobranch and teleost fishes and all larval amphibians have rows of hair cells in the walls of canals extending the length of the animal and over the head, either as open grooves or as tunnels opening only occasionally to the outside. These are the lateral line organs. In general, the rougher the water in a fish's habitat, the more protected are its lateral line receptors, and the more active the fish, the better developed its lateral line system. The hair cells are commonly aggregated into **neuromasts** (Figure 10-13), with their hairs embedded in a common gelatinous cupula. Movement of the cupula, reflecting movement of water in the canal, causes a shearing distortion of the hairs, depolarizing the hair cells and stimulating the lateral line nerve fibers synapsing with them. Electrical recording on branches of this nerve reveal spontaneous firing with superimposed responses to water movements. Flow of water through the canal in one direction has a stimulatory effect on one population of receptors, an inhibitory one on another, and flow in the opposite direction has the reverse effect. The difference in effect is explained by the presence of two populations of hair cells having their kinocilia on opposite sides of the hair bundles. Sensitivity to flow of water suggests a rheotaxic role (detection of velocity relative to water) for the lateral line organ but, where available, vision is probably more important for this information. There are also

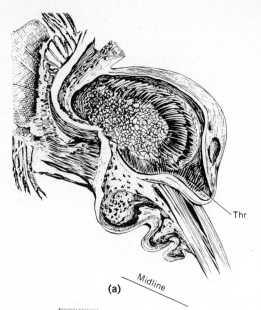

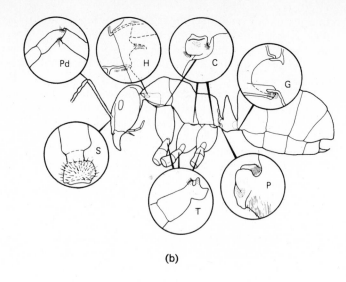

(a)

Midline

(b)

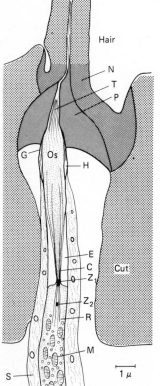

Distal nerve process

(c)

Figure 10-11

Invertebrate hair-plate receptors. (a) Lobster statocyst, seen from a dorsal view, with its connections to the brain. The heavy statolith is supported by four rows of hair cells, the inner three of which are normally in contact with it. In the chamber, but not in contact with the statolith, is a separate plate of finer hairs (Thr) that respond to motion of the fluid around them. (b) Several of the hair plates bordering joints in the exoskeleton of an ant. Deflections of hairs at these joints provide information about body conformation and orientation. (c) A detail from an electron micrograph, showing how each hair in such a hair plate is innervated. The nerve terminal shows unmistakable ciliary structure at point C. [(a) from M. J. Cohen. Proc. Roy. Soc. (London), B152, 30–49 (1960); (b) from H. Markl. Umschau Wiss. Tech., 6, 185–188 (1965); (c) from U. Thurm. Cold Spring Harbor Symp. Quant. Biol., 30, 75–82 (1965).]

Figure 10-12

Vertebrate hair cells. (a) Experimental arrangement for studying bullfrog saccular hair cells (HC), the hair bundle (HB) of which is embedded in a fenestrated otolithic membrane (OM). Supporting cells (SC) separate the hair cells. A stimulating probe (SP) is used to stimulate the specific hair cell from which a microelectrode (ME) is recording responses. (b) A scanning electron micrograph of a single hair cell bundle showing the stereocilia, with a bulbous-tipped kinocilium at one side. (c) Intracellularly-recorded responses from such a hair cell, in response to the three triangular wave-form stimuli shown below. (d) A graphical display of the hair cell responses as a function of displacement, showing the pronounced rectification of the response, and the saturation at displacements greater than about ±1 μm. (e) and (f) scanning electron micrographs of a hair bundle before and after the kinocilium was mechanically separated from the stereocilia. Movement of the kinocilium then evoked no response, while movement of the stereocilia gave a normal response. [(a, c, and d) from J. Hudspeth and D. Corey, Proc. Nat. Acad. Sci., 74, 2407–2411 (1977); (b, e, and f) from A. Hudspeth and R. Jacobs. Proc. Nat. Acad. Sci., 76, 1506–1509 (1979).]

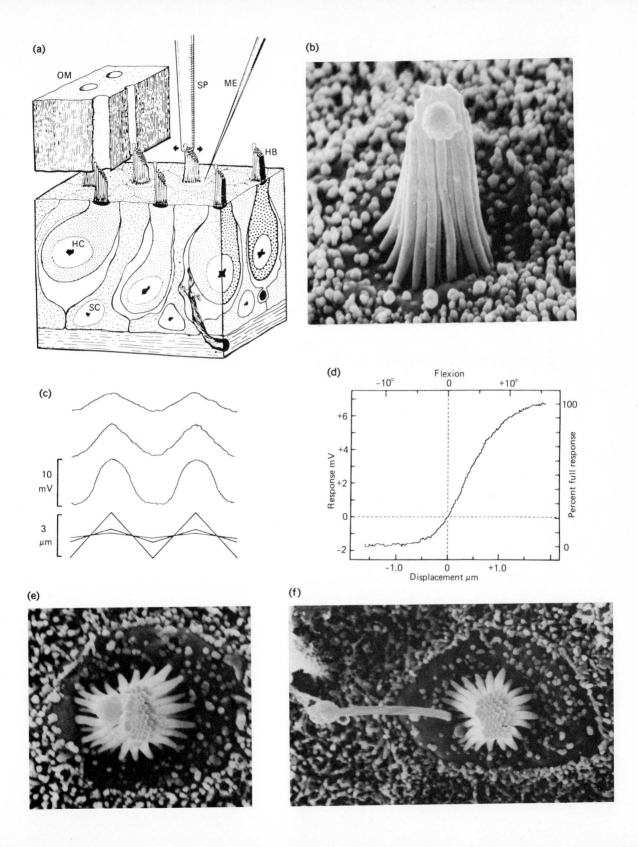

(a)

OM

SP   ME

HB

HC

SC

(b)

(c)

10
mV

3
μm

(d)

Flexion

−10°   0   +10°

+6

+4

+2

0

−2

Response mV

100

0

Percent full response

−1.0   0   +1.0

Displacement μm

(e)

(f)

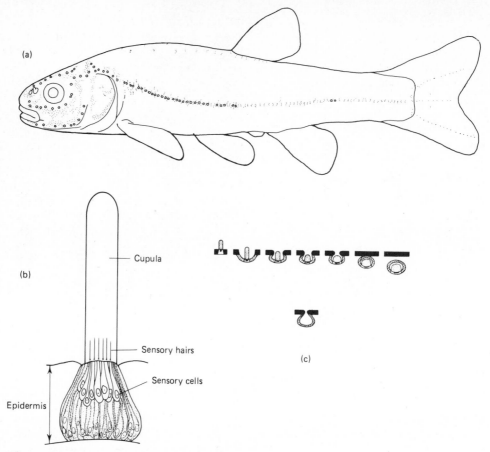

Figure 10-13
*Lateral line system. (a) Side view of the lateral line of a representative fish having both exposed neuromasts (•) and neuromasts inside a lateral line canal (○). (b) a diagrammatic cross section through a free neuromast, showing sensory cells, their hairs, and the common cupula. (c) A series of transverse sections through lateral line canals, showing the progressive development of a closed canal system. In "closed" canals, there are openings to the surface between neuromasts, as seen below. [(a) from S. Dijkgraaf. Z. vergl. Physiol., **20**, 62–214 (1934); (b and c) from S. Dijkgraaf. Experientia, **8**, 205–216 (1952).]*

bursts of responses to vibrational stimuli in the water, up to 100–200 Hz (cycles per second) with two bursts of spikes per cycle. There is behavioral evidence, however, that the major function of this vibration sensitivity is not in hearing but in the detection and localization of nearby moving or stationary objects on the basis of waves reflecting off them. Consistent with this is the high degree of development of the lateral line system in deep-sea fishes and the adaptation of some of these recep-

tors, in certain fishes, for the detection of nearby objects by electrical signals (see Section 10.4).

In elasmobranch fishes the lateral line system has evolved a special type of structure in which the sensory cells (probably specialized hair cells) are deeply sunk into a pit and embedded in a gelatinous substance. These specialized receptors are called **ampullae of Lorenzini** and are distributed over the head. They were first thought to be temperature receptors, because their spontaneous firing is highly sensitive to temperature change, reflecting differences as small as 0.05°C. Another role is suggested by the demonstration that they are highly sensitive to changes in hydrostatic pressure. Still another function, as salinity detectors, was proposed when it was shown that the ampullary nerves respond to changes as small as 100% → 97% seawater. Most intriguing, however, is the fact that these receptors show remarkable sensitivity to weak electrical fields. Behavioral tests show that sharks can detect gradients as small as 0.1 μV/cm, less than the 0.2 μV/cm gradient that would be induced in a fish moving east or west in the earth's magnetic field at 1 m/sec. Of more obvious behavioral significance is the demonstration that these elasmobranchs can detect the electrical fields generated by hidden prey. At present, this electroreceptor function appears to represent the main role of the ampullae of Lorenzini.

10.3.8 Equilibrium Organs

The membranous labyrinth of the vertebrate inner ear develops from an enlarged anterior portion of the lateral line system. This labyrinth system consists of two principal bony chambers, the *sacculus* and the *utriculus,* with three semicircular canals arising from the utriculus and oriented perpendicular to each other (Figure 10-14). Within these bony chambers are a series of interconnecting ducts and sacs filled with a viscous fluid, *endolymph,* having a composition much like that of intracellular fluid. This fluid system is surrounded by a less viscous fluid, *perilymph,* continuous with the cerebrospinal fluid.

At one end of each semicircular canal is an enlarged ampulla, in which there is a crista, consisting of rows of hair cells with their cilia embedded in a single gelatinous cupula extending across the canal. Rotational movement causes the bone and cupula to move, whereas inertia tends to keep the fluid in the canal where it was. Thus the hairs are displaced and the sensory nerves stimulated. All hair cells in an ampullary hair-cell organ have the kinocilium oriented in the same direction; movement in that direction is excitatory. Cells of the two vertical canals have their kinocilia oriented toward the canal; in the horizontal canal the kinocilia

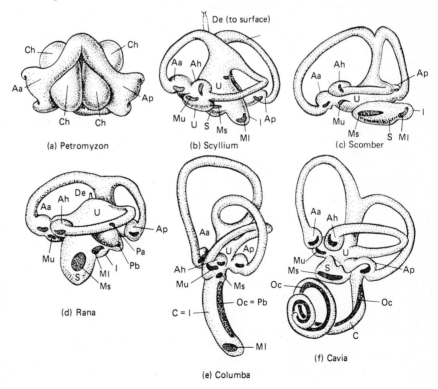

Figure 10-14
Inner ear of vertebrates. The membranous labyrinth of a lamprey (a), shark (b), teleost fish, (c), frog (d), bird (e), and mammal (f). Sensory hair patches are shown as if the membrane were transparent. Code: Aa, Ah, Ap, ampullae of the anterior, horizontal, and posterior semicircular canals; C, cochlear duct; Ch, chambers of the lamprey ear lined with a ciliated epithelium; De, endolymphatic duct; l, lagena; Ml, macula of the lagena; Ms, macula of the sacculus; Mu, macula of the uticulus; Oc, organ of Corti; Pa, amphibian papilla; Pb, basilar membrane (papilla); S, sacculus; U, utriculus. [From A. S. Romer. The Vertebrate Body, 3rd ed., W. B. Saunders and Company, Philadelphia, 1962.]

(a) Petromyzon

(b) Scyllium

(c) Scomber

(d) Rana

(e) Columba

(f) Cavia

are on the side of the utriculus. The three canals are oriented perpendicular to each other, so their relative levels of excitation are an accurate reflection of the direction of rotation as well as the angular acceleration. The arrangement seems so perfect, in fact, that it is somehow dissatisfying to find that lampreys operate with only two semicircular canals and, worse still, that hagfish get along with one!

In the larger chambers below the semicircular canals, there are three more patches *(maculae)* of hair cells, each bearing a mineralized concretion, or *otolith.* The otolith in the utriculus has been shown to be concerned with gravitational stimuli and body position. Thus in vertebrates the receptors for angular acceleration and for gravity are different.

This is generally not the case in invertebrates, although the similarity between vertebrate receptors and the equilibrium organs in many invertebrate phyla is striking. In most cases, the invertebrate organs, called *statocysts,* are derived from cuticular hair plates and hence do not involve secondary sensory cells (Section 10.3.5). The principle of operation is almost identical, however. The hairs line the inside of a sphere, which also encloses a stony concretion or *statolith.* This may be cemented onto a restricted patch of hairs or free to move about within the sac. In the latter case, hairs cover most of the sac surface and any change in the animal's position changes the spot on which the statolith lies, stimulating a different population of hairs, or bending them in a different direction. When the animal is oriented horizontally to the ground, the statoliths lie on the ventral surface of their sacs; when it is desirable to swim upward, the animal turns its body until the statoliths lie on the caudal poles of their sacs, and so forth. When the statolith is attached to certain hairs, a change in orientation relative to gravity causes a change in shearing force of the weight against the hairs. Angular or linear acceleration are measured by the same receptors. (There was considerable reluctance to accept so simple an explanation for control of orientation in animals as complicated, say, as lobsters, but that it is so was experimentally demonstrated with great ingenuity by the Austrian scientist, Kreidl, in 1893. Kreidl studied shrimp, whose statocysts are located at the base of the first pair of antennules and open to the outside through fine slits. At each molt the old statoliths are lost and new ones formed with sand from the environment. Kreidl replaced the sand in his acquaria with iron filings, and when the new statolith was formed, showed that he could orient the animals any way he wished with a strong magnet.) In some invertebrates there are even statocyst organs with three patches or rows of hairs set in planes at right angles to each other, reminiscent of the vertebrate semicircular canals. This has been shown in several decapod crustacea (including crabs), and is especially highly developed in cephalopod molluscs. The cephalopod statocysts, moreover, have secondary hair cells looking almost indistinguishable from vertebrate hair cells, except that all of the hairs are kinocilia.

Resuming our discussion of the vertebrate ear, the sacculus contains two otolith organs, the function of which is not entirely clear. In fish they both appear to be useful in hearing and vibration reception. In amphibia and most reptiles there is developed another hair-cell patch, the rudimentary beginning of the *basilar membrane.* In crocodilian reptiles the lagena, one of the patches, moves to the end of a prominent outgrowth of the sacculus, with the basilar membrane lying along its length. In birds and mammals, this outgrowth gets longer and longer and the basilar membrane becomes associated with surrounding tissues in the *organ of Corti,* a hair-cell organ marvelously adapted for analytical response to high-frequency vibrations. In birds the cochlear duct is long and slightly curved, with the basilar membrane stretching its length and the lagena at its end. In mammals the lagena is missing and the duct, now called the *coclea,* is very much longer and coiled into a spiral of two to four turns (Figures 10-14 and 10-16). With the increase in length of choclear duct has come an increase in the number of receptor cells and in complexity of structure. These changes are clearly correlated with increasing usefulness of the sense of hearing. Probably more than any other sense, hearing is a recent phylogenetic development. We shall trace the stages in this development in the following section.

10.3.9 Hearing

Many animals can hear, but few communicate with sound. Sound coming from any direction can be heard, so it is a good source of warning, even in animals that cannot themselves produce sound. In species where sound is used for communication, production is often restricted to one sex (usually the male), and the principal function of sound is mate attraction and maintenance of territory. Among invertebrates, communication sounds are limited to a few groups of insects, especially the Orthoptera, Hymenoptera, and Cicadidae, and perhaps some crustaceans. The first two groups produce sound by stridulation, rubbing a toothed structure across a ribbed plate. This usually results in a burst of pulses, extending up to 40–50 kHz in frequency. Cicadas, on the other hand, produce sound by vibrating a thin section of cuticle; and some decapod crustaceans make sound by thumping the substrate with their pincers.

Hearing is more widespread and is accomplished most commonly by a modified tracheal structure exposed to the surface as a delicate, taut membrane with the terminal cilia of sensory nerves attached (Figure 10-15). This is called a *tympanic organ* and may be located on the legs (Orthoptera, Arachnida, and Crustacea), thorax or abdomen (Lepidoptera and Hemiptera), or antennae (Diptera). The number of receptor neurons varies from 2 (moths) to 70 or more (locusts), and usually they all respond with greatest sensitivity at the resonant frequency for the tympanic membrane, although the membrane can, depending on its structure and attachments, resonate differently at different points. Although frequency discrimination is probably poor or absent within a given tympanic organ, sensitivity may be very great and intensity discrimination is possible. The principle functions of these organs appear to be the detection of a sound, its recognition, apparently by the number of bursts per second, and its localization (probably done by centrally comparing the levels of excitation in tympanic organs on opposite sides of the body). Second-order neurons show some evidence of frequency discrimination, apparently due to integration of input from other vibration-sensitive organs (other tympanic organs or sensory hairs) as well.

Among vertebrates the most sophisticated use of sound for communication is in birds and mammals. Several teleost fishes rub gill or pectoral structures together to produce drumming, scraping, or grunting sounds, whereas others vibrate their air bladders. These sounds probably subserve simple communicative functions, especially in in-

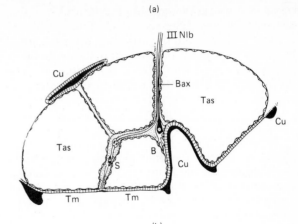

(a)

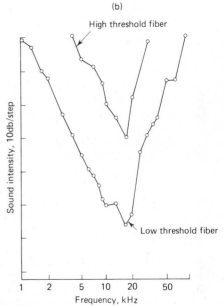

(b)

Figure 10-15
Tympanic organs of insects. (a) Diagrammatic section through the tympanic organ of a noctuid moth. The two sensory cells (S) are attached to a taut tympanic membrane (Tm), exposed at the surface of the moth's thorax. A third and much larger cell (B) also sends its axon (Bax) through the tympanic nerve (III Nlb). The function of this cell is unknown, but it is almost surely sensory. Additional code: Cu, cuticular skeletal elements; Tas, tympanic air sacs. (b) "Response areas" of the two auditory neurons in the ear of a moth. The sound intensity at threshold is plotted as a function of stimulus frequency.[(a) from A. E. Treat, and K. D. Roeder, J. Insect Physiol., 3, 262–270 (1959); (b) A. N. Suga, Jap. J. Physiol., 11, 666–677 (1961).]

traspecific alarm and schooling calls and in reproductive behavior. There is a severe limit to the usefulness of these sounds, however, for most fishes hear very poorly, responding only to low frequencies (up to about 1000 Hz) at high intensity. This is understandable in view of their lack of a specialized cochlea or any means of concentrating acoustical energy on the otoliths. In certain fishes (the Ostariophysi), on the other hand, processes of several vertebral bones have been adapted as a coupling device (the Weberian apparatus) between the sacculus and the air bladder. The air bladder, because of its different refractive properties, very effectively takes the place of a tympanic membrane, picking up vibrations in water that would otherwise pass right through the fish. These fish hear sounds up to several thousand cycles and are several thousand times more sensitive than fish without Weberian ossicles.

Terrestrial vertebrates have found much more complicated uses for sound, paralleling their increased capability to analyze sound. Sounds are produced in a variety of ways but almost all involve the controlled movement of air through respiratory passages, from the simple hiss of reptiles to the resonating vocal sacs of amphibia, the syrinx of birds, and the larynx and vocal cords of mammals, with lips and tongue helping in man. Nevertheless, most amphibians do not make sounds (frogs being notable exceptions); and among the reptiles apparently only the lizards of the family Gekkonidae utilize sounds for intraspecific communication. Even among birds and mammals, nearly all of which can hear well, there are some that are usually mute.

Frogs have, in addition to the lagena and macula of the sacculus, a rudimentary basilar membrane and a single middle-ear ossicle to transduce sound energy mechanically from the large tympanic membrane to the sacculus. Sensitivity is still restricted to 3000–4000 Hz and under, however, and it is doubtful that frequency discrimination is very accurate. In the bullfrog, response is greatest in the 100–200 Hz range, where most of the energy of the male frog's call is.

An elongated basilar membrane separated by perilymphatic ducts is first found in certain reptiles, especially crocodilians. Snakes and most lizards have less well-developed ears and respond only to loud sounds of a few hundred cycles frequency.

Geckos respond above 10,000 Hz, but sensitivity is still much below that of birds and mammals.

The cochleas of birds and mammals are both highly developed, but differ in several fundamental ways. It is most surprising, in fact, to find that both seem about equally capable of auditory analysis. Before considering these differences, it will prove useful to describe in some detail the mechanism of sound analysis in the mammalian auditory system.

Mammalian Ear. As we are all aware, sound has many informative parameters: loudness, pitch (or frequency), timbre (a function of the complexity of the sound), and direction of incidence. We can hear frequencies between about 20 and 20,000 Hz (10 octaves), and, within that range, distinguish between frequencies only 0.3% apart. At our "best" frequencies (those to which we are most sensitive: 500–2000 Hz), the range of audible intensities covers about 12 orders of magnitude [120 db (decibels) on a log scale where 10 db represents a tenfold increase in intensity] between threshold, near the level of background thermal noise, and the level at which painful loudness is reached. Differences as small as 0.1–0.5 db are distinguishable. Most naturally occurring sounds are complex, with overtones and noise transients, all of which are involved in making the sound recognizable and carrying information. The sound of the vowel "a," for example, is easily recognizable whether spoken by a man or a woman, although the dominant frequency components are quite different in the two cases.

As we have seen, the hair cells of the basilar membrane are much like those of the lateral line or equilibrium organs. Their capabilities in audition depend on how the stimulus is brought to them. Elegant studies of cochlear function, principally by von Békésy, have demonstrated how these accurate analyses can be performed.

External Ear. Most mammals, with the notable exception of the cetaceans (whales and porpoises) and pinnipeds (seals), have evolved complex external ear structures (pinnae) that help select sound from certain directions and concentrate it into the external ear canal, through which it passes to the tympanic membrane. Other vertebrates do not have such external ear structures, although the feathers of birds undoubtedly contribute to creating comparable directional sensitivity of hearing, especially in owls.

Middle Ear. The all-important hair cells are in a liquid medium, so airborne vibrations must be mechanically transduced into vibrations in the liquid. This is no trivial feat, for the compressibility and viscosity of the two fluids are very different. To achieve a comparable amplitude of movement in the endolymph as in air, a much amplified force is necessary. The necessary impedance matching is achieved by the middle-ear ossicles that at one end fit against the tympanic membrane, absorbing energy over a large area (50–90 mm^2 in man), and transfer this energy with little change in amplitude of movement to the much smaller area of the oval window of the cochlea (above 3.2 mm^2 in man) (Figure 10-16). The ratio of the two areas (the transformation quotient) is a measure of the amplification of force pushing the oval window against the perilymph on the other side. The middle-ear ossicles are derived from primitive hyomandibular and jaw bones, and only reach full development in mammals. Even birds have only one ossicle (the columella) compared with three in mammals (malleus, incus, and stapes). There appear to be two main advantages to the greater number of ossicles: they fit together in such a way that additional mechanical leverage is achieved to increase the transformation quotient, and they are able to "buckle" at very high sound intensities, reducing the transmission of energies that might be damaging to the more delicate structures beyond. There are also middle-ear muscles that, by contracting, dampen the ossicles and reduce energy transmission, especially at low frequencies. These muscles contract reflexly in response to loud sounds or during speech, yawning, swallowing, chewing, and so on.

Cochlea. Sound energy enters the cochlea at the oval window and is converted to compressional waves in the perilymph. As Figures 10-16 and 10-17 show, the perilymph fills two long tubes (the scala vestibuli and scala tympani), separated from one another by a membrane-enclosed endolymphatic chamber (the scala media) except at the end, where the scala media terminates blindly and the perilymphatic chambers are joined by the helico-

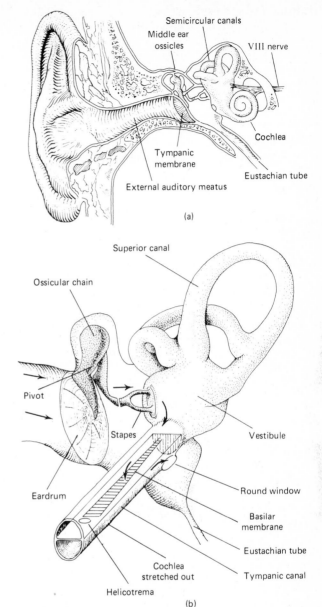

Figure 10-16
Mammalian ear. (a) Realistic drawing of ear structures in man. (b) A simplified construction of the ear, showing the pathways of sound to the inner ear, with the cochlea shown straightened out. [After G. von Békésy. Symposia of the Society for Experimental Biology, No. 16, Academic Press, Inc., *New York, 1962, pp. 267–288.]*

trema. These liquids are essentially incompressible, so any inward movement of the stapes must be compensated for by a movement outward of the round window membrane exposing the lower perilymph chamber to the air of the middle ear. If the inward movement is slow enough, the perilymph in the scala vestibuli can move through the helicotrema into the scala tympani and thence to the round window. Vibrations at a higher frequency must find a shorter route. They short-circuit across the scala media and in so doing cause displacement

of the two partition membranes: Reissner's membrane and the basilar membrane.

It is the basilar membrane that is of foremost interest, for on it are located the sensory hair cells with their supporting cells (Figure 10-17). The basilar membrane stretches between a bony shelf in the center of the cochlea and the lateral surface. Extending the length of the basilar membrane are four rows of hair cells: one row inside and three outside the supporting rods of Corti. The hairs are embedded in another membrane (the tectorial

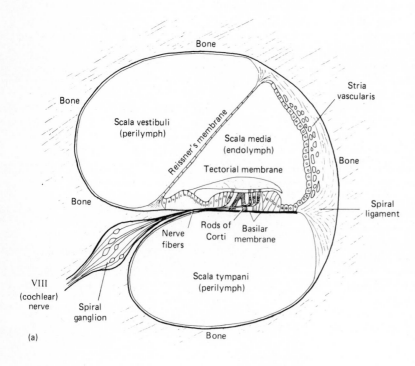

Figure 10-17

Mammalian cochlear structure. (a) Transverse section through the cochlear canal. The organ of Corti, containing the hair cells, their supporting cells, and the tectorial membrane, is located on the basilar membrane stretched between the scala media and the scala tympani. (b) Reconstruction of the fine structure of inner (left) and outer (right) hair cells, from electron micrographs. Code: Ane, afferent nerve endings; C, centriole (apparent basal body of missing kinocilium); Dc, Deiters cells; Ene, efferent nerve endings; H, stereocilia; Ip, internal pillar (or rod) of Corti; M, mitochondria; N, nucleus. [From J. Wersäll, A. Flock, and P. -G. Lundquist. Cold Spring Harbor Symp. Quant. Biol., **30,** *115–132 (1965).]*

membrane), which also extends the length of the cochlea. Together, these structures are known as the *organ of Corti.* As displacement of fluid in the scala vestibuli causes displacement of the scala media toward the scala tympani, the basilar membrane rotates around its point of attachment along its inner margin. Because the tectorial membrane is anchored at a different pivot point, any movement of the cochlear partition causes a shearing force on the embedded hairs. Unlike other hair cells, the basilar-membrane receptors lose their kinocilia during development, retaining only their basal bodies (or centrioles). The stereocilia and basal bodies are still precisely oriented, however; and, from experiments with frog saccular hair cells, we know that it is movement of the stereocilia that is important in generating a response (see Figure 10-12). The inner hair cells have a simple row of stereocilia; the outer hair cells have their stereocilia arranged in a V or W-shape (see Figure 10-18). In all cases the basal body of the missing kinocilium faces outward toward the heavily vascularized stria vascularis at the lateral side of the scala media.

The movement of the cochlear partition is accompanied by an extracellularly recorded "cochlear microphonic," an electrical potential that accurately mirrors the signal. The microphonic seldom exceeds a few hundred microvolts in amplitude (a few millivolts at its largest, within the scala media), but it is thought to be a reflection of the intracellular receptor potential because it reverses polarity at the level of the hair-bearing surface of the hair cells and is lacking when the hair cells are lacking. At high intensities, this microphonic is seen to ride on a direct current "summating potential," probably representing an intracellular direct current bias, which may be polarizing or depolarizing, resulting from nonlinearity in the cochlear microphonic.

Von Békésy showed very elegantly that the microphonic was not merely the transduction of mechanical into electrical energy: the response contains more energy than the stimulus and hence represents modulation of some existing energy source. Moreover, he demonstrated that the voltage change is a response to displacement rather than to the velocity of displacement. The microphonic is therefore viewed as resulting from hair-cell depolarization and repolarization with changes in permeability of the hair-bearing surface membrane as the hairs are deflected toward and away

from the basal body of the missing kinocilium. The driving force for this variable flow of ions is generally accepted to be the cell's membrane potential (assumed to be about −70 mV), supplemented by a large positive potential in the endolymph that bathes the hair-bearing surface of the hair cells. Compared to the surrounding tissue and the perilymph bathing the other surfaces of the hair cells, the endolymph is at +80 mV. Thus across the top hair-cell surface the difference in potential is presumed to be close to 150 mV, and unusually large driving force. The endolymph potential is apparently a secretory potential arising in the stria vascularis and maintained by metabolism. It falls rapidly on exposure to metabolic poisons or during anoxia.

This scheme is not without its difficulties, however, for it has been shown that the composition of the endolymph is very much like that of cytoplasm, rich in K^+ with little Na^+. The perilymph is close to normal extracellular fluid in composition. Thus across the hair-bearing surface of the cell, where the change in resistance is thought to take place, there are nearly symmetrical K^+-rich solutions. Unless the permeability of that membrane is extremely low to K^+ and Cl^- (its electrical resistance extremely high), it will be depolarized, with little or no driving force to take advantage of a given change in permeability. If the resistance is high, however, it is very difficult to explain the very short time-constant of the membrane involved—capable of producing microphonics at frequencies of 100 kHz/sec or higher. The time constant of the membrane involved should be not more than about 10 μsec, requiring phenomenally high permeability if the membrane capacitance is around 1–10 $\mu F/cm^2$, as in most membranes studied. These problems have not yet been resolved. It should be noted, however, that recent experiments suggest that Ca^{2+} may be the most important ion for carrying current in ciliated cells as disparate as *Paramecium* and the lateral line hair cells of *Necturus.* Moreover, the main Ca^{2+} permeability may be in the cilia membrane. The role of Ca^{2+} in the inner ear has been little studied, and it may be quite asymmetrically distributed.

Frequency Resolution. Early histologists found that the basilar membrane was actually composed of 20,000–30,000 tiny cross strands, joined by connective tissue, and that the membrane increased

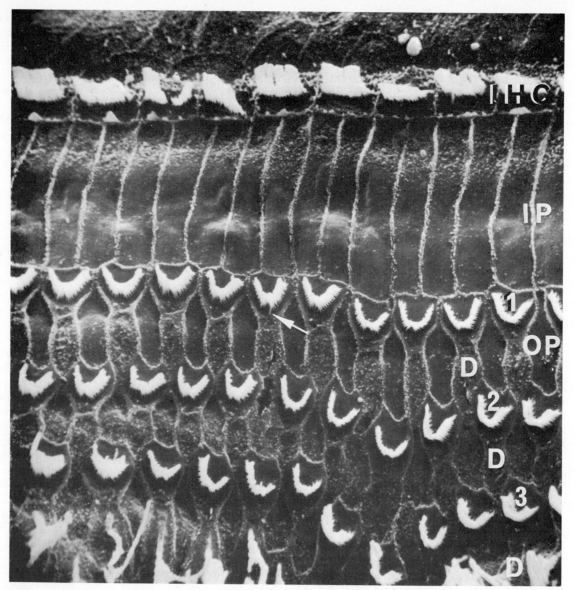

Figure 10-18
Scanning electron micrograph of the surface of the hair cell region of the organ of Corti of a 16-day old kitten. One row of inner hair cells (IHC) is shown at the top, three rows of outer hair cells (1, 2, 3) at the bottom. One of the cells of the first row of outer hair cells still retains its kinocilium (arrow). IP, inner pillar cells; OP outer pillar cells, D, Deiter's cells. Magnification, 2200×. From Lindemann, H., H. Ades, G. Bredberg & H. Engstrom, Acta Otolaryng., **72**, *229–242, 1971.*

in width by about 4 times from the base of the cochlea to the apex, at the helicotrema. On the basis of these findings, Helmholtz in 1868 proposed what became the basis of the resonance theory of hearing: that each cross strand resonates at a different frequency, ranging from highest at the base to lowest at the apex, and that excitation is restricted to that small point. The locus of excitation would then describe the stimulating frequency.

The brilliant research of von Békésy subsequently established that this is only partly the case. Because each cross fiber in the membrane is joined to the adjacent ones, it cannot resonate alone. Instead, compressional waves arising anywhere in the perilymph set up a *traveling wave* in the basilar membrane moving from the base toward the apex and reaching a maximum amplitude of deflection where the elasticity of the membrane is appropriate for that frequency (Figure 10-19). The peak of deflection at any given frequency is quite broad, especially toward the basal end of the cochlea, although it is rapidly damped out when it reaches the more flexible parts of the membrane specialized for lower frequencies. By dropping tiny metal crystals on the basilar membranes of cadavers and experimental animals, and using stroboscopic illumination, von Békésy was able to observe directly the movement at different frequencies and make measurements such as those shown in Figure 10-19. At low frequencies (below about 50 Hz) the whole membrane moves in synchrony to each cycle of a sound, but at progressively higher frequencies the peaks of displacement are localized closer and closer to the narrow basal end.

The elasticity of the membrane is such that the relation between distance along the membrane and most effective frequency is approximately logarithmic. This relationship is shown for a cat in Figure 10-19. As this figure indicates, cats can hear up to 50 kHz. This extension of sensitivity to higher

Figure 10-19
Cochlear mechanics. (a) Pattern of maximal displacement of the basilar membrane caused by a traveling wave initiated by a low-frequency sound. Three different phases of the traveling wave are shown. (b) Envelopes of maximal displacement of the basilar membrane when stimulated at different frequencies, shown for the cat cochlea. The relationship between sound frequency and position of peak deflection is shown below. The extent of deflection is enormously exaggerated relative to membrane length.

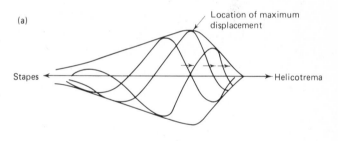

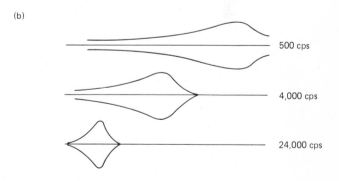

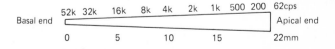

frequencies (and an accompanying loss to low frequencies) is a result of the smaller dimensions of the cat cochlea. In general, the smaller a mammalian species, the higher the frequencies to which it will respond. Certain bats can hear frequencies as high as 150–200 kHz. Elephants, at the other extreme, have their greatest sensitivity at frequencies near the lower limit of our hearing. Cetaceans are exceptional, hearing up to 100–150 kHz, despite their large size. They have large cochleas, but the basilar membrane is extremely narrow and taut, appropriate to the higher frequencies.

The spatial separation of response to different frequencies has led to the general acceptance of the place theory of pitch discrimination; that is, when a given cochlear nerve fiber is excited, its input is recognized by the CNS as signalling a cer-tain frequency, regardless of its firing pattern. Below about 50Hz, where there is no discrete peak of vibration on the basilar membrane, and probably even up to 500–1000 Hz, frequency can be signaled by the synchronous firing of units all along the membrane at the frequency of the sound. The broadness of any given peak of deflection at higher frequency makes it difficult to understand how frequencies as close together as 0.3% can be discriminated. Recent studies have shown, in fact, that the "tuning" of primary auditory neurons is sharper than would be predicted on the basis of basilar membrane deflections alone. There apparently is a mechanical or electrical "second filter" responsible for this sharpening. Moreover, as we shall see in the following paragraphs, neural interactions in more central auditory nuclei contribute impor-

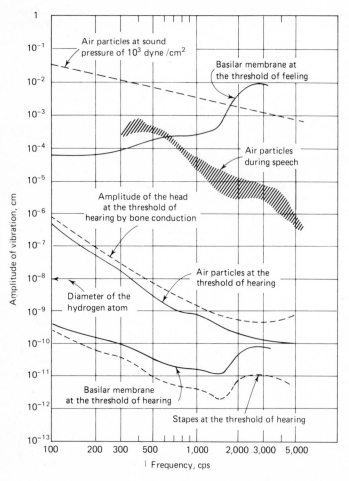

Figure 10-20
Vibration amplitudes of air particles and ear structures at different sound levels, from threshold to pain levels. Note for comparison the diameter of a hydrogen atom. Threshold values are derived by extrapolation from measurements of vibration amplitude at high sound intensities. [*After G. von Békésy.* Symposia of the Society for Experimental Biology No. 16, *Academic Press, New York, 1962, pp. 267–288.*]

tantly to the sharpening of tuning of responses that makes possible accurate frequency resolution.

Absolute Sensitivity. The sensitivity of the mammalian auditory system is truly remarkable, comparable in threshold energy (10^{-17} to 10^{-18} W-sec/sec) to the olfactory sensing of a few tens of odorant molecules or to a visual sensation to 5 or 10 quanta of light. This sensitivity is the more remarkable in the auditory system, however, because the threshold energy must move the tympanic membrane, middle-ear ossicles, and cochlear partition before it reaches the receptors. Von Békésy, using an ingenious direct optical method, measured the displacement of the tympanic membrane, stapes, and basilar membrane when the ear was stimulated with painfully loud sounds or driven mechanically, and calculated the approximate displacement of these structures at threshold, assuming, as seems to be the case, that the response is linearly related to displacement even at the lowest levels of stimulation (Figure 10-20). The maximum amplitude of vibration of the basilar membrane to the loudest sounds is only about 1 μm. At threshold, the calculated amplitude of vibration is about 10^{-11} cm, three to four orders of magnitude less than the diameter of the hydrogen atom! This is a minuscule displacement compared even to the random Brownian movements of molecules, but it should be borne in mind that the calculated value is for the synchronous movement of the whole membrane. Nevertheless, the effectiveness of this displacement is difficult to believe. More recently others, using the much more sensitive Mossbauer or laser reflection techniques, have measured vibration down to moderate sound intensities directly, confirming von Békésy's calculations.

Neural Integration of Auditory Information. The responses of cochlear nerve cells reflect the mechanical transducing properties of the cochlea. Most of the primary neurons appear to fire spontaneously at rates up to 100 or 200 per second and respond with an increased firing rate to any sound that deflects their part of the basilar membrane. One frequency will be most effective, but any lower frequency, eliciting a traveling wave that continues beyond that point, will be moderately effective. As a result, one normally gets response (or tuning) curves for single cells that look like those in Figure

10–21a. Cells tuned to the same frequency may have thresholds differing by 60 db or more and responses ranging from continuous firing during a sound to only a few spikes at the onset of the stimulus. Interestingly, although there are three times as many outer hair cells as inner hair cells, the inner row receives about 95% of the auditory nerve fibers, in a highly restricted innervation pattern. The outer hair cells, which might be expected to be more strongly excited at low stimulus levels, are multiply innervated by a few nerve fibers, each of which runs a considerable distance along the basilar membrane.

The fundamental roles of the two types of hair cells are poorly understood. There is even some question whether any recordings have been made from nerve fibers innervating the outer hair cells, since these fibers tend to be much smaller (as well as less numerous) than those innervating the inner hair cells. One clue to the role of the outer hair cells comes from the use of certain ototoxic drugs (such as the antibiotic streptomycin), which selectively destroy the outer hair cells, leaving the inner ones apparently intact. Under these conditions, single auditory nerve fiber sensitivity falls sharply. Yet recordings are almost certainly from the inner hair cells. It is possible that the main role of the outer hair cells is somehow to modulate response of the inner hair cells. There is convincing evidence for efferent control of the responses of primary neurons by means of the efferent synaptic endings on hair cells or the sensory nerve terminals. These efferent terminals appear cholinergic.

Most of our knowledge of auditory neural integration concerns the processing of auditory information as it passes from the cochlea to the auditory cortex. There are about three times as many cells in the second-order nuclei (the cochlear nuclei) as there are primary neurons, yet each primary fiber is thought to terminate on 75–100 second-order cells. Obviously there is a great deal of overlap. The cochlear nuclei send their output to other centers in the medulla, predominantly on the contralateral side and, with successive enlargements of responding populations, on to the auditory cortex. At each level, the cells responding to different frequencies are spatially organized, just as they are on the basilar membrane, but often with two or three maps in parallel. Figure 10-22 shows a typical map of best frequency for single cells in the line

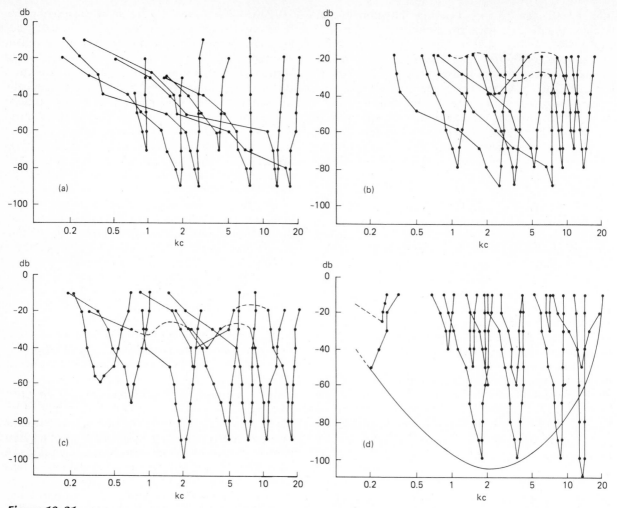

Figure 10–21
Tuning curves (response areas) of auditory neurons at successively higher levels of the auditory nervous system, from the primary sensory neurons of the cochlear nerve (a) to second-order cells of the dorsal cochlear nucleus (b), and the higher-order neurons of the trapezoid body (c) and inferior colliculus (d). All data are from cats. The threshold intensity (in decibels with respect to the arbitrary reference level of 0.0002 dyne/cm²) is shown at each frequency. Dotted portions of curves were not actually measured, and the smooth curve in (d) shows the average human threshold for comparison. [From Y. Katsuki, T. Sumi, H. Uchiyama, and T. Watanabe, J. Neurophysiol., 21, 569–588 (1958).]

of one microelectrode penetration through the anterior and posterior cochlear nuclei of a cat. This spatial arrangement is called *tonotopic localization.*

As information is processed and passed centrally, the response patterns of cells become more specific. There is inhibition by cells responding best to nearby frequencies, to earlier stimuli, or to sounds presented to the other ear. (After the sec-

ond neural level, and perhaps even there, input from the two ears is mixed.) Response tuning becomes sharper and sharper (Figure 10-21b–d). Von Békésy has shown that a comparable focusing of excitation exists in the touch nerves of the skin. With a mechanical model of the stretched-out cochlea that could be laid along one's outstretched arm and driven mechanically at appropriately

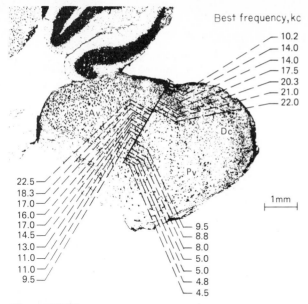

Best frequency, kc
— 10.2
— 14.0
— 14.0
— 17.5
— 20.3
— 21.0
— 22.0

22.5
18.3
17.0
16.0
17.0
14.5
13.0
11.0
11.0
9.5

9.5
8.8
8.0
5.0
5.0
4.8
4.5

Av.

Dc.

Pv.

1mm

Figure 10-22
Tonotopic localization. A sagittal section through the left cochlear nuclei of a cat, showing the track of a microelectrode and, along the track, the frequencies at which individual cells were most sensitive. Along the track there were two maps of frequency, from low to high and then the reverse. Each tonotopic map represents a population receiving input from cells excited along the corresponding portion of the basilar membrane. [From J. E. Rose, pp. 116–136, in G. L. Rasmussen and W. F. Windle (eds.), Neural Mechanisms of the Auditory and Vestibular System Charles C Thomas, Publisher, Springfield, Ill., 1960.]

scaled-down frequencies, he found that the broad peak of deflection was felt as a sharply localized vibration at the point of maximum displacement. With a change in driving frequency, this point moved as predicted.

Because of the importance of inhibition, an increasing proportion of auditory neurons at higher levels respond phasically—only at the beginning of a sound. At these higher levels, cells are found that will respond only to highly specific and complex stimuli, or only to sounds arriving slightly earlier at one ear than at the others. These cells are adapted as filters to pass only selected information, usually about sounds important to an animal.

Sound Localization. Just as important as the hearing and recognition of a sound, often, is the deter-

mination of its direction. This is especially true for animals that detect prey or predators by the sounds they make. There are two basic mechanisms for angular discrimination, both requiring the use of two receivers (both ears). These are (1) the measurement of an arrival-time difference (or phase of each cycle) of the same sound component at the two ears, and (2) the measurement of the relative intensity at the two ears. Arrival-time differences are most useful to large animals having widely separated ears. Trained humans, for example, can detect changes as small as 1–2° away from the midline on the horizontal plane on the basis of arrival-time differences of only 10 μsec. The same angular discrimination in a shrew would require the measurement of time differences with an accuracy of 1 μsec. The necessary measurement probably takes place at third-order auditory neurons in the olivary nuclei of the medulla, which receive input from both ears. Certain of these cells have been shown, in cats, to change response with changes of only 10–20 μsec in differences in arrival time of input from the two sides. Sensitivity to changes of 1 μsec would require an order of magnitude more precision in conveying the timing of the stimuli through two synapses, but even this is not impossible. This degree of accuracy has recently been reported for echolocating bats. Smaller animals, on the other hand, tend to be sensitive to higher frequencies, to which their head and external ears cast strong "sound shadows." For these, the apparent relative intensity of a sound may change considerably with changes of only 1–2° in angle. It is found in man that if a sound arrives later at one ear, the time delay may be counteracted by making the later sound louder (about 18 μsec time difference per 1 db intensity difference). Most mammals can localize accurately only near the midline on the horizontal plane, less well farther to the side, and, without extensive head movement, scarcely at all in the vertical plane.

Hearing in Birds. Behavioral and electrophysiological studies of hearing in birds reveal that they have approximately the same absolute sensitivity, differential frequency and intensity discrimination, and dynamic range as mammals. Sensitivity may be slightly greater at low frequencies, and is restricted to under 12–15 kHz, but in general it is astonishing that audition is so similar considering the differences in cochlear structure of the two classes. Since there are no good explanations for the equivalence of function, the problem will merely be stated by summarizing the apparent differences.

The bird middle ear has one ossicle, rather than three, but the transformation ratio is approximately the same, 25. A small patch of hair cells, the lagena, is present in birds at the end of the cochlear duct. The scala vestibuli is very narrow and compressed, in contrast to the large mammalian chamber. The delicate Reissner's membrane is replaced by a gross, heavily vascularized tegmentum vasculosum, containing most of the blood vessels that, in the mammal, are located in the stria vascularis. A helicotrema is absent, which might affect low-frequency hearing, and the endolymphatic potential (+15 mV) is much smaller. In contrast to the long, coiled, mammalian basilar membrane, the bird equivalent is short (0.5 cm maximum except in owls, where it reaches 1 cm) and only slightly curved. A major difference is in the change in basilar membrane width from base to apex. The change is about 400% in mammals, 25% in birds. There are no supporting rods of Corti in birds, and only about 1200 transverse fibers in the basilar membrane, in contrast to the 20,000–30,000 in mammals. Where in the mammalian cochlea there are four hair cells across the width of the membrane, clearly distinguishable as inner (one) and outer (three), with different sensitivity and innervation, in birds there are 30–50 hair cells extending cross the width of the membrane, with no apparent differences between them.

Echolocation. With the exception of human communication, probably the most remarkable use of sound by animals is in echolocation. This is a method of acoustic orientation evolved by several animals, but most successfully by one large suborder of bats (Microchiroptera) as they evolved to fill their niche, the nighttime air. The story of our scientific understanding of this phenomenon is a fascinating one, beginning in 1793 with the discovery by the great Italian physiologist, Spallanzani, that bats, unlike owls, could orient perfectly in pitch darkness, and even when blinded could return across the city of Pavia to their "home" church tower to lead normal lives. Spallanzani established that hearing was necessary, but it was more than

a century later that Griffin, in 1938–1940, first demonstrated that echoes of emitted sounds were being used for guidance.

The principle is very simple and was invented for sonar and radar before it was discovered in bats. The skill with which bats extract information from echoes, however, is quite fantastic. By analyzing return echoes, they not only detect the presence of all the objects in their environment but determine the distance and direction and much about the dimensions and surface characteristics of each. Many bats are insectivorous and catch mosquito-sized prey at rates of up to 2 per sec on the wing.

The emitted orientation sounds are ultrasonic (above our range of hearing, which explains some of the delay in discovering the phenomenon), usually with components up to 50 or 100 kHz or higher, having wavelengths short enough to reflect strong echoes off even small targets. Moreover, in all known cases, the orientation sounds are of wide bandwidth—probably a necessity for the discrimination of target distance and direction.

The largest and most widespread family of bats (Vespertilionidae) uses loud frequency-modulated (FM) pulses, cries that sweep downward through approximately an octave in the range 100–20 kHz. These cries are emitted at a "cruising" rate of 10–20 per sec, increasing to 100–200 per sec after detection of an obstacle to be avoided or a target to be caught, while the cruising duration of 2–5 msec falls to 0.5–1.0 msec during the buzz (Figure 10-23). (The mechanism of producing independent, structured pulses at rates this high must involve phenomenal properties of the laryngeal muscles, a conclusion supported by the extraordinary development of the sarcoplasmic reticulum in these muscles.)

Other bats emit pulses that are longer and of constant frequency until the end, when there is a brief downward sweep. The constant frequency may be as high as 145 kHz, and can be maintained constant, within one pulse and from one pulse to another, with phenomenal accuracy. In the well-studied European horseshoe bat, *Rhinolophus ferrumequinum,* for example, the constant frequency component of a given bat is maintained within 30–50 Hz of a basic frequency of about 83 kHz. This is regulation to within approximately 0.05%. What makes this particularly remarkable is that the bats can alter the frequency in a controlled way with

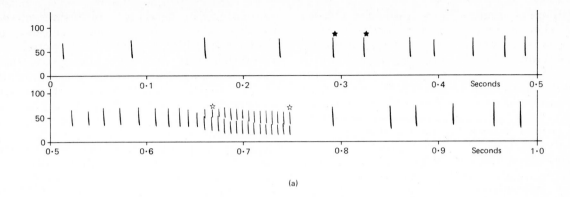

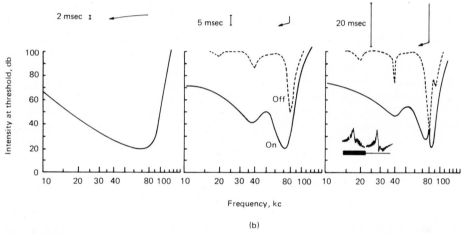

(a)

(b)

Figure 10-23

Bat echolocation sounds and auditory adaptations. (a) Sound spectrograph analysis of the frequency modulated (FM) pulses emitted by the small brown bat, Myotis lucifugus, *during an interception maneuver. Frequency in kilohertz is plotted against time during the continuous 1-sec record. Filled stars indicate typical loud pulses near the time of detection of the target, open stars indicate the terminal part of the "buzz." (b) Schematic "audiograms" characteristic of bats employing purely FM pulses (left), pulses with a constant frequency component shorter than 10 msec ending in an FM sweep (middle), and pulses with a longer constant frequency component (right). Each is representative of many species that have been studied to date. The sensitivity of the animal (judged by behavioral testing or neurophysiological recording) is plotted against frequency. Over each audiogram is a plot of frequency versus time for a characteristic emitted pulse, with time running from top to bottom. Purely "FM-bats" tend to have a smooth audiogram similar to those of other mammals but displaced to higher frequencies and centered at the emitted frequency band. The bats with constant frequency components have much more specialized audiograms, reflecting filter properties of the cochlea and very accurate frequency analysis in the narrow band of emitted frequencies. [(a) from G. Sales and D. Pye.* Ultrasonic Communication by Animals, *Chapman and Hall, London, 1974; (b) from A. D. Grinnell, J.* Acoust. Soc. Am., **54,** *147–156 (1974).]*

comparable accuracy over a range of 3 kHz or more. They do this, during flight or when a target is approaching them, to compensate exactly for the Doppler shift introduced by the relative velocity of the bat and target, ensuring that the return echo is always at the same frequency. The accuracy of control of emission in this system, based on precise frequency analysis, exceeds that of any other biological system reported to date, with the possible exception of the regular discharges of certain electric fish (see Section 10.4). Still other bats emit FM pulses with four or five harmonics. Despite the considerable differences in frequency pattern and intensity of emitted sounds, all species tested exhibit nearly equal skill in avoiding natural obstacles or wires down to 0.1 mm diameter in laboratory tests.

Although blind humans can sometimes use ambient noise or the tapping of a cane to orient with surprising skill, the abilities exhibited by bats require adaptations of the auditory system that most animals clearly do not enjoy. Obvious morphological evidence of this is the enormously exaggerated development of external ear structures and brain auditory nuclei. An obvious physiological adaptation is sensitivity to frequencies as high as 100–150 kHz. Another is the extremely fast recovery, and even short-term (2–20 msec) facilitation, of the auditory nervous system to sounds having the same characteristics as one just heard. Sound travels at 34.4 cm/msec, so echoes at 2–20 msec delay represent objects 1–10 ft distant, in the range of greatest importance to bats. Most animals, including man, appear to be adapted instead to be insensitive to echoes and reverbations at these intervals.

The bats employing constant frequency pulses show other adaptations. Their auditory systems are extraordinarily sharply tuned to the region of the emitted constant frequency and within this range there are even more sharply tuned responses ("off-responses") to the termination of sounds of certain frequencies. These appear to be properties originating in the cochlea, where the basilar membrane and related structures are highly specialized for analysis of the appropriate narrow frequency band, but equivalent emphasis on frequencies just above the emitted high frequency is seen at all neural levels. Thus, a large proportion of the auditory nervous system is sharply tuned to the narrow band of frequencies that the bat, by Doppler-compensa-tion, ensures in the echoes. Small fluctuations in the amplitude and frequency of these echoes, resulting especially from wing-beating of insect prey, provide information for discrimination of targets.

Echolocation reaches its highest development in bats, where it is a successful substitute for vision, but simpler versions are seen to complement vision in other animals. For example, the whirligig beetle, *Dytiscus,* creates surface waves as it swims and detects nearby objects by the reflections of these waves. Lateral line organs of fish very likely function in a comparable way and are particularly well developed in deep-sea fishes (Section 10.3.7).

Two genera of birds (the Asian cave swift, *Collocalia,* and the South American oil bird, *Steatornis*) produce audible tongue clicks for orientation in dimly lit caves, and a similarly produced and useful click has been independently evolved by one of the visually orienting Old World fruit bats, *Rousettus.* The only system that might rival that of microchiropteran bats, however, is the echolocation of certain cetaceans, especially dolphins and porpoises. Using clicks of 1 msec or shorter duration and rich in ultrasonic frequencies, these animals are impressive in their ability to avoid or find small objects and perform difficult discriminations on the basis of subtle features of returning echoes. In all echolocators other than microchiropteran bats, however, vision is the preferred means of orientation, where sufficient light is available.

Owls do not actively echolocate, but they are superb at passive localization of the sources of environmental sounds. Their skulls are asymmetrical, facilitating target localization in the vertical plane; and it has now been shown that neurons in a certain midbrain nucleus respond only when a sound arises from a restricted point in space. The cells of this nucleus are systematically arranged according to azimuth and elevation so that their receptive fields form a physiological map of auditory space.

Countermeasures to Echolocation. The hearing of insects has been mentioned. In Noctuid moths, which are natural prey for bats, the tympanic organ has developed extreme sensitivity to the frequencies emitted by bats. These tympanic organs (Figure 10-15) are marvelously simple, providing a model auditory system for neurophysiological study. Each "ear" contains two auditory neurons, both responding in the same frequency range (10–

150 kHz) but with different thresholds, one being about 20 db more sensitive than the other. It is easy to evoke firing patterns that could explain moths' evasive behavior on hearing ultrasound of increasing intensity. First they fly away from the source of sound, then as it gets louder (the bat nearer) they begin to fly irregularly, and, finally, when the sound is very loud (perhaps saturating all four neurons) and at a high repetition rate, they dive directly at the ground. Arctiid moths have carried their defense further and can themselves produce ultrasonic pulses that deter attacking bats. It seems unlikely that the moth's sounds can be jamming the bats, so they might instead be a warning of bad taste.

As a footnote to the story of adaptation and counteradaptation, mention can be made of a moth ear mite, a tiny parasite living inside and destroying a moth's tympanic organ. These mites for some reason are never found in both ears of a moth, and, if placed in the second ear of an already infested moth, will crawl to the dorsal surface and apparently follow a scent trail to the already attacked organ.

10.4 ELECTRORECEPTORS

Bioelectric potentials have been known to man since antiquity in the form of discharges from electric fishes such as the marine electric rays *(Torpedo)* and the freshwater electric catfish *(Malapterurus)*. These discharges arise from the synchronized transmitter-induced depolarization of large numbers of flat muscle fibers (electroplaques) arranged in series, each with the same surface modified to be one large end plate. This organ, in *Torpedo,* produces brief voltages up to 50 V at several amps current for as much as 6 kW of power output. Such a discharge, being tolerable (for unknown reasons) to the fish producing it, is of obvious offensive and defensive value. Some freshwater fish, however, produce extremely weak electric signals, useless either in offense or defense. These include the New World knife fishes (Family Gymnotidae) and the Old World elephant-nose fishes (Family Mormyridae), as well as the electric eel, *Electrophorus electricus,* which is capable of large discharges as well. The

weak electric signals usually are only a few milliseconds in duration and range in repetition rate from three to several hundred per second. In elephant-nose fishes the rate is varied with activity, sometimes stopping altogether; in the knife fishes it is remarkably constant (varying by no more than 0.01–0.02% for long periods of time), although some can change it over a wide range and maintain a new rate with equally astonishing accuracy.

With behavioral experiments, Lissman and others have established that these signals can be used for orientation. A momentary electric field is produced around the animal with each discharge, and the shape of that field varies with distortions introduced by objects in the field that have a different conductivity than the water around them. Changes in field shape change the magnitude of voltage gradients along the lateral lines. The receptors are located near the lateral line and exhibit phenomenal sensitivity. These fish can detect currents of approximately 3×10^{-15} amp, or a longitudinal field of 0.05 μV/cm. (Thus they could, in principle, detect the electrical field generated by a 1.5 V flashlight battery having one pole in Boston, the other in New York City.) They can easily distinguish between a glass rod and metal rod of the same dimensions at several centimeters distance, and can apparently use this sense to orient successfully in dark, murky waters. Interestingly, equivalent or even greater sensitivity to electrical signals is found in certain marine elasmobranchs, especially sharks. These animals employ their ampullae of Lorenzini for detection of the DC and low frequency electrical fields produced by potential prey. There are many other sources of low frequency electrical fields in the oceans large enough to be detected by elasmobranchs and potentially usable, especially for orientation: the fields produced by mixing of fresh with salt water, by waves and tides, and by the fish themselves as they move through the earth's magnetic field.

The receptor cells are specialized secondary cells, probably derived from lateral line receptors. There appear to be two general classes of receptor organs: (a) the tonic-firing ampullary type, used in reception of low frequency (< 10 Hz) and direct-current electrical fields, and (b) the phasic-responding tuberous receptors, sensitive at higher frequencies and used in active "electrolocation." In both, the receptor cells are located in pits or in a canal.

The ampullary receptors, of which the elasmo-branch ampullae of Lorenzini are examples, have their receptor cells mostly embedded in the walls of the chamber, which is connected through an open channel to the outside. In tuberous receptors, the receptor cells protrude into the lumen of the cavity, which is connected to the outside only via the intercellular clefts between cells in a plug of epithelial tissue at its mouth. Each organ is inner-vated by a single sensory-nerve fiber, and the or-gans tend to be distributed widely over the surface of the animal, with only a slightly greater concen-tration in the lateral line itself. These receptors are most sensitive to higher frequencies, in the range 60–2000 Hz. Their sensitivity, too, can be phenomenal. As Figure 10-24 shows, transmitter release can be sharply increased or decreased by imposed potential changes of less than ±1 mV.

Electrophysiological recordings from the lateral line nerve in several gymnotids confirm the effec-tiveness of these receptors in detecting changes in an electric field generated by their own dis-charge. In fish with low-frequency discharges, the intensity of the field is coded into the number of sensory impulses following each pulse of the elec-tric organ: the greater the voltage gradient, the larger the response. In fish with fast discharge rates, the intensity of the field affects the probability of firing a single time to a given discharge. If a conduc-tor or nonconductor is placed in different positions near the fish, the response rate (or probability) is sharply and oppositely affected by the two objects (Figure 10-25). The change in firing rate is exagger-ated by movement, which causes a large phasic dis-charge or inhibition of discharge.

Recordings from second-order neurons in the medulla show first stages of integration of this in-formation. Many cells respond like presynaptic fi-bers, but others respond phasically to movement. This response can be opposite for movement in different directions, or independent of direction of movement. In either case, the responses in some units are identical for conducting and nonconduct-ing objects, in others opposite. Thus units with tonic responses provide information about the na-ture and position of objects near the fish, whereas phasically responding units provide information about movement, direction of movement, position of the front edge of an approaching object, and size of the object. In fish that exhibit active electro-location, it is apparent that the cerebellum is adapted for analysis of the electroreceptor input. The cerebellum is larger, compared with the rest of the brain, than in any other vertebrates, and often comes to overly the whole brain, much as the cerebrum does in mammals.

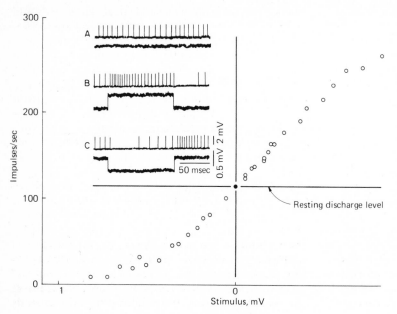

Figure 10-24
Input-output relations at the chemically transmitting synapse between tonic elec-troreceptor (a modified lateral line hair cell) and afferent nerve in the electric fish, Gym-notus carapo. Inset are records showing (A) tonic discharge, (B) response to a weak anodal stimulus, and (C) response to a weak cathodal stimulus. [From M. V. L. Bennett. "Similarities Between Chemically and Elec-trically Mediated Transmission," pp. 73–128. In F. D. Carlson, (ed.). Physiological and Biochemical Aspects of Nervous Inte-gration, Prentice-Hall, Englewood Cliffs, N.J., 1968.]

Figure 10-25.

Electroreceptor properties. (a) Primary receptor neurons of a knife fish fire a burst of impulses in response to each of its own electric discharges, the magnitude of the response changing as a conductor (silver) or nonconductor (paraffin) is placed at different positions near its body. In this way information about its environment is initially coded. (b) At higher neural levels (medullary neurons), integration of input from primary receptors allows much more versatility of response and accurate detection of direction of movement of an object in the environment. This figure shows the asymmetry of response depending on direction of movement. Many receptors, unlike the one shown, respond oppositely to motion of silver and plastic plates. [(a) from S. Hagiwara, and H. Morita. J. Neurophysiol., 26, 551–567 (1963); (b) from P. Enger and T. Szabo. J. Neurophysiol., 28, 800–818 (1965).]

(a)

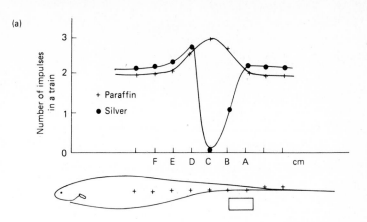

(b)

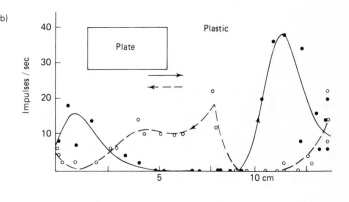

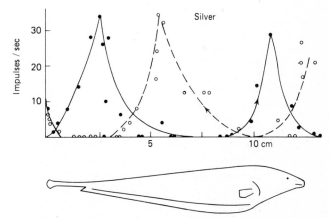

10.5 TEMPERATURE SENSING

The intricate biochemical machinery that constitutes living organisms is highly sensitive to its working temperature and is usually adapted to work best within a temperature range much narrower than that of the environment (see Chapter 8). For this reason, most animals, both poikilotherms and homeotherms, have to regulate their temperatures within certain limits. In poikilotherms, this is done

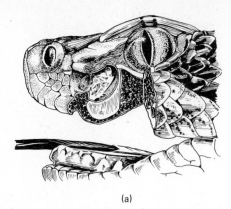

(a)

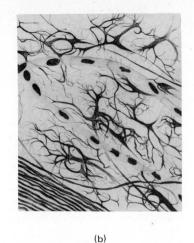

(b)

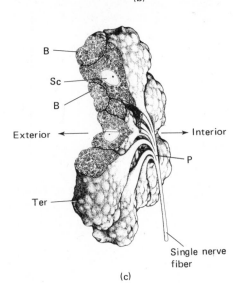

B
Sc
B

Exterior ←   → Interior

P

Ter

Single nerve fiber

(c)

Figure 10-26
Temperature receptor of pit vipers. (a) Side view of the head of a crotalid snake, with part of the upper jaw removed to show the location of the pit organ. The thin (10–20 μ thick) membrane near the base of the pit is densely packed with bulbous trigeminal nerve terminals. (b) Drawing from light micrography of the pit membrane, showing myelin-stained nerve terminals densely packed together. Each palmate terminal corresponds to the unmyelinated portion of a single nerve ending, as shown in (c). (c) A reconstruction of a single nerve terminal in the pit membrane. The nerve fiber loses its myelin sheath in the course of spreading out in a palmate array, and then proliferates into an enormous number of tiny nerve branchlets (B) surrounded by the cytoplasm of Schwann cells (Sc). Thousands of such terminals (Ter) constitute most of the pit membrane, the rest being an external cuticular covering, basement membranes, Schwann cells, and, in some species, pigment granules. The dense packing of nerve terminals may make this a particularly favorable preparation for the study of nerve-terminal properties. [(a) from T. H. Bullock, and F. P. J. Diecke. J. Physiol. (London), 134, 47–87 (1956); (b) from T. H. Bullock, and W. Fox. Quart. J. Microsc. Sci., 98, 219–234 (1957); (c) courtesy of S. Terashima.]

primarily by behavioral adaptations, especially the selection of the right microenvironment; in homeotherms this is done by active metabolic adjustments as well. These behavioral and physiological adaptations all require accurate information about body and ambient temperatures and radiation levels, supplied by temperature receptors.

Relatively little is known about temperature receptors. Specificity of function is particularly difficult to establish, because the same temperature changes that influence temperature-sensing cells have an inevitable effect on all cells. One must look for receptors showing an unusually large effect, or ones that are clearly involved in behavioral response to temperature changes. Those that have actually been demonstrated and studied are few.

Presumed thermoreceptors are found in a wide variety of locations on invertebrates, with the greatest concentration, usually, on the antennae, mouthparts, or legs. In some insects temperature measurement is so good that essentially homeothermic regulation to within 3–4°C is possible (for example, in Sphingidae). Blood-sucking insects and ticks are sensitive to changes as small as 0.5°C in their pursuit of warm-blooded prey.

Vertebrates, too, have temperature receptors distributed widely over their bodies, usually as apparently unspecialized endings of small myelinated and unmyelinated fibers. There is evidence that the lateral line system of fish serves a temperature-receptor function, for severing the lateral line nerves greatly reduces the fishes' normal sensitivity to water temperature changes as small as 0.03–0.05°C.

The most sensitive temperature receptor known occurs in Crotalid snakes (pit vipers). These snakes are often active at night, feeding on small mammals. In the absence of visual cues or adequate hearing, they have evolved remarkable pit organs on each side of the head, about 2 mm wide at the opening (in a 1 m long snake), 5 mm deep, and 5 mm wide at the base, with a 10 μm thick membrane extending across it near the bottom. This membrane is densely innervated with branching, flattened terminals of trigeminal sensory fibers (Figure 10-26). Exposure of this membrane to long infrared radiation (0.5–15 μm wavelength) causes an increase in the resting discharge of the sensory nerves. The electrophysiological threshold is very close to that established behaviorally, to changes of only 0.001–0.003°C. This would correspond to the detection in 0.5 sec of a small rodent 10°C warmer than its environment at 40-cm distance. Because there are two pits, each directionally sensitive, the information they provide can almost certainly be used not only for prey detection, but for localization as well. Certain other snakes (boas, for example) have similar temperature receptors in shallow pits between scales. These appear to be about three times less sensitive, however, and much less directional than the pit-viper receptors.

Among homeotherms, thermal receptors are distributed over the body as free nerve endings, with the greatest concentration being on the tongue. As in other temperature receptors studied, a steady level of activity is increased or decreased with an increase in temperature depending on whether the endings are "hot" or "cold" receptors. Cold fibers are more numerous and occur more superficially. Warmth receptors respond to warming with a transient sharp increase in firing rate followed by gradual adaptation to a new steady discharge rate that

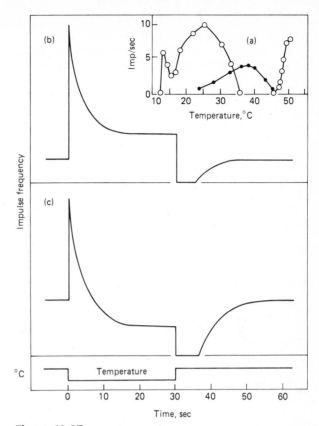

Figure 10-27

Temperature-receptor response patterns. (a) Responses of single "warm" (·) and "cold" (°) fibers of the mammalian tongue, as a function of temperature. (b) and (c) Schematic representations of response amplitude of "cold" fibers to a brief temperature drop when the starting temperature is (b) above and (c) below the temperature of maximum resting discharge. "Warm" receptors would show comparable changes on warming, the opposite changes on cooling. [From Y. Zotterman, Ann. Rev. Physiol., **15,** *357–472 (1953).]*

depends on the final temperature and is maximal at 38–43°C. Threshold changes are about 0.15°C. On cooling, these units may stop firing altogether, and eventually establish a new reduced rate. Cold fibers, on the other hand, respond to cooling with a sharp increase in firing rate, then adapt to a steady rate that is maximal between 20 and 34°C, often silent between about 36 and 45°C, then paradoxically high at temperatures above 45°C (Figure 10-27). Receptors of the ampullae of Lorenzini of elasmobranchs are cold receptors; those of the Crotalid pit organs are warmth receptors. The biophysical mechanism whereby temperature changes have these opposite effects on the two types of receptors is unknown, and to date there have been few attempts even at speculation.

In homeotherms, impulses from the surface temperature receptors must be integrated with information about internal body temperature if accurate temperature regulation is to be maintained. This integration takes place in the hypothalamus, which receives peripheral sensory information and itself contains cells sensitive to changes of less than 0.1°C in blood temperature (see Section 8.8.3). Regulation of body temperature, by sweating or shivering, is initiated first by action of the peripheral receptors, only in more severe conditions by changes in core temperature. Probably the hypothalamic regulatory center is involved only in adjustments involving the whole body (violent shivering, or increases in metabolism due in part to release of noradrenalin and thyroxin into the blood).

10.6 VISION

The surface of the earth is bathed in electromagnetic energy, including x rays, ultraviolet and visible light, heat (infrared), and radio waves. This energy is in the form of quantized packets traveling at 186,000 mi/sec and behaving in many ways like waves, with a wavelength (λ) inversely proportional to the energy in the packet (Figure 10-28). The source of these quanta is changes in the structure of molecules and atoms, mostly occurring on the sun or in other celestial bodies. The shortest wavelength quanta result from the breakdown of nuclear structure and ionization of atoms (up to λ of about 100–200 nm). Changes in electronic orbit around a nucleus produce less energy-rich quanta (150–2500 nm); still weaker quanta result from changes in vibrational (2000 nm to 30 μm) or rotational structure of molecules (30–600 μm). Absorption of one of these quanta by a molecule can produce an equivalent change in structure in a stoichiometric 1 quantum/1 molecule way, and it is this energy that powers life on earth. Only a tiny part of the available energy spectrum is useful, however. Quanta of wavelength shorter than about 300 nm are too energetic. They break bonds in the molecule that absorbs them. Wavelengths longer than about 1000 nm are too weak to produce significant changes in molecular energy. Fortunately, those wavelengths that are energetic enough to cause changes in molecules without being damaging are precisely the wavelengths that reach the earth's surface with least atmospheric interference. Ozone in the outer atmosphere absorbs most of the ultraviolet and shorter wavelengths, whereas wavelengths longer than about 1000 nm are absorbed by water vapor in the atmosphere. Hence plants have adopted pigments (carotenoids) that absorb especially well in the range between 300 and 800 nm for photosynthesis, and animals utilize similar pigments to absorb directly these wavelengths (now called light) and trigger the excitation of sensory nerves. The animal visual pigments are also carotenoids, all obtained from plants, for as far as is known no animal can synthesize carotenoids.

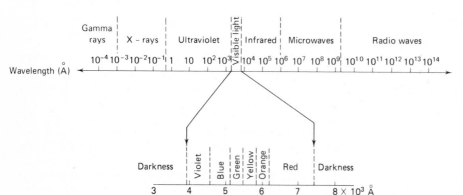

Figure 10-28
The electromagnetic spectrum, with the visible portion expanded. The energy in a quantum of radiation is related to the wavelength by the following expression:

$$E = \frac{2.854 \times 10^8}{\gamma \text{ (in Å)}} g\text{-cal/mol}$$

Thus the visible spectrum lies between the energies of approximately 30 kcal/mol and 100 kcal/mol of quanta.

These pigments are associated with protein molecules and typically occur in dense aggregations on precise arrays of membranes, either in the form of flattened disks or rods. Ontogenetically, these highly oriented membrane structures arise in one of two ways: as microvillae of epithelial cells or as infoldings of the membranes of flagellae or cilia. Although many exceptions exist, most invertebrate photoreceptors are derived from microvillae, vertebrate receptors from modified cilia.

Generalized photosensitivity (the dermal light sense) unrelated to "eyes" exists in many animals, from protozoa to some fishes and amphibians and even birds. Development of some moth pupae is dependent on direct responses to light photoperiod through a transparent patch of cuticle over the head ganglia, and some birds similarly show photoperiod hormonal regulation effected by light reception directly through the skull. Little is known about the photoreceptors in these cases, however. The most primitive specialized visual receptors are probably the simple eye spots on the flagellae of some protozoans, such as *Euglena.* These apparently only detect the presence or absence of light and crudely measure its intensity. Much more information than this is available in the light reflected from the environment, and natural selection has, several times, independently produced eyes of great sophistication to utilize it. In three groups, especially, mollusks, arthropods, and vertebrates, mechanisms have been developed for forming an image of the environment, most commonly by the adoption of lenses of refractile material (Figure 10-29). Methods have evolved for moving the lens or changing its shape to permit focusing at different distances. In some eyes, a reflecting pigment (a tapetum) or air layer (tracheoles in some moths) exists behind the receptor pigment, so that light that gets through the receptor layer is reflected back for another pass. Other pigments have been adapted as masking against certain wavelengths or extremely bright light. The visual pigments themselves are often modified to absorb different wavelengths, permitting color vision. The variety of different optical arrangements is enormous, and sometimes the degree of convergent evolution can be remarkable. The eyes of squids and octopus, for example, are almost identical, optically, to the vertebrate eye (Figure 10-30). The receptor elements are on the surface of the retina, however,

unlike the vertebrate retina, and are of the usual invertebrate microvillar type. We shall consider in detail only the vertebrate eye, with a few comparative observations about arthropod ommatidia.

10.6.1 Vertebrate Eye

Optics. There are a number of striking similarities between the highly evolved vertebrate eye and the technologically evolved camera. In both, a lens focuses an inverted image of its visual field on a thin layer of photosensitive pigment. The amount of light admitted is governed in both by an iris, and the lining of the inside is coated with a black substance to reduce obscuring reflections. Eyelids act as lens covers. In many vertebrates, accommodation (focusing at different distances) is brought about, as in a camera, by moving the lens back and forth. In most fishes, amphibians, and snakes, the eye is normally focused at infinity and the lens is actively moved forward to focus on near objects. Birds, mammals, and reptiles other than snakes accommodate in an entirely different way, by increasing the curvature of the lens during contraction of the ciliary muscle. This method has some disadvantages, for with age the lens becomes less elastic. Where a child can focus at 2–3 in. distance, most adults cannot focus closer than 6–8 in. Actually, the lens is used only for fine adjustments of focus. Most of the focusing is done by the cornea, which has greater curvature than the lens and which, at its interface with air, provides the greatest change in refractive index for incoming light. The refractive index of air is 1, of the cornea 1.38, of the lens 1.45, and of the fluids between cornea, lens, and retina, 1.33. (Birds and some lizards achieve increased accommodation by deforming the cornea as well as the lens.)

It is difficult to build a perfect lens. In both camera and eye, the periphery of the lens tends to focus more sharply than the center, causing spherical aberration. In cameras this difficult is overcome by using several apposed lenses. The eye, too, is remarkably well corrected—in two ways: by reduced curvature of the cornea at its margins, and by an increase in density of the lens at its core. Chromatic aberration, however, is more serious. Shorter wavelengths are refracted more strongly than longer ones, so they focus nearer the lens. Camera lenses are compensated by apposing in the

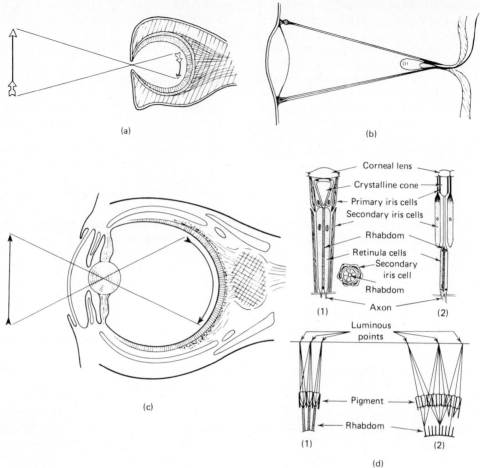

Figure 10-29

Types of eyes, all from invertebrates. (a) Pinhole type of focusing mechanism, found in the mollusk Nautilus. This eye is in focus at all distances, but admits relatively little light. (b) The lens-type eye of the arthropod Copelia. A single receptor element is reportedly moved back and forth, scanning the image formed by the lens. (c) The eye of the squid Sepia. In this eye the image is formed by a cornea and lens, and the shape of the lens can be changed to focus at different distances. This eye is remarkably similar to that of vertebrates, except that the receptor elements are different in structure and are on the front surface of the retina rather than at the back surface. (Compare with Figure 10-30.) (d) The two classical types of arthropod ommatidium: (1) apposition, in which light reaches the sensory retinula cells via only one corneal lens and crystalline cone, and (2) superposition, in which light is focused upon a single set of retinula cells by several ommatidial lenses. Generally appositional eyes are found in diurnal species, superposition eyes in nocturnal species. [(d) from V. B. Wigglesworth, Principles of Insect Physiology, Methuen & Co., Ltd., London, 1965.]

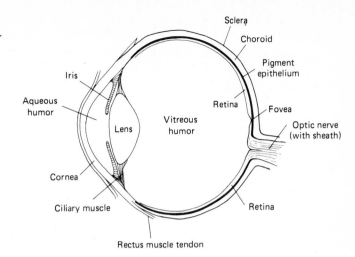

Figure 10-30
Vertebrate eye. Diagrammatic section through the human eye.

lens crown and flint glass, which have widely different refraction and dispersion properties. The eye apparently cannot produce two such different materials, for the biological solution to the problem is simply to eliminate the shortest wavelengths. The lens itself is yellow, filtering out ultraviolet and violet light, and in primates, there is, in addition, a yellow pigment (the macula lutea) lying immediately over the fovea, the tiny patch of photosensitive retina on which the lens focuses for nearly all detailed vision. It is this tiny area, subtending only about 1.7° out of the total 240° of visual field that most of the receptors (cones) for color vision are located.

Vertebrates in special environments possess optically specialized eyes. In fishes, for example, the refractive index of the cornea is so close to that of water that it can contribute little to focusing; therefore it is normally quite flat, improving the hydrodynamic structure of the head. Animals active in near darkness, on the other hand, frequently have corneas that not only are huge but protrude far from the orbit, collecting light from a larger area. Often the lens is almost perfectly spherical, focusing light equally well from any direction on different parts of the retina. Such animals usually lack a fovea. Other adaptations involve protection of the cornea from damage and desiccation. In most mammals this is accomplished by the secretion of lacrymal fluid and frequent blinking. In birds, however, there is another transparent membrane, the nictitating membrane, that covers the cornea during flight. It closes from side to side rather than vertically. Snakes and certain lizards protect the cornea with an unmoving transparent window formed by the fusion of eyelids. This window is shed with the animal's skin when it molts.

Many vertebrates have little or no overlap of visual field for the two eyes. In these, the eyes are normally moved independently, although where there is binocular vision, the eyes can be brought to focus on a common point. Some birds (for example, hawks and swallows) have two foveas: one in the center of the retina, for monocular use, the other near the temporal edge of the retina, for binocular focusing.

Retina. After passing through the focusing apparatus of the eye and the fluid-filled chambers, light falls on the retina. As Figure 10-31 shows, the retina is a highly complicated tissue. It contains not only the receptor cells, filled with visual pigment, but four major populations of nerve cells (horizontal, bipolar, amacrine, and ganglion cells), arranged in distinct layers and synaptically connected to perform intraretinal integration. (A small fifth population of unexplored function has recently been described.) It is, in fact, a bit of brain tissue moved peripherally to perform the initial analyses of sensory input before sending the information via the ganglion cell axons (optic nerve fibers) to the lateral geniculate body in the midbrain and thence on to the cortex (see Figure 10-42). The outer segments of the receptor cells themselves are deriva-

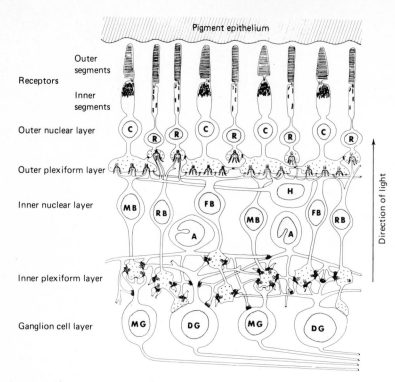

Receptors
Outer segments
Inner segments
Outer nuclear layer
Outer plexiform layer
Inner nuclear layer
Inner plexiform layer
Ganglion cell layer

Pigment epithelium

Direction of light

Figure 10-31

Vertebrate retina. Diagrammatic reconstruction from electron microscopy of the principal structures and pathways of the vertebrate retina (near the fovea of a mammal having color vision). Note that light passes through the whole system before it is absorbed by the receptor outer segments and the pigment epithelium. Where in the past it was assumed that the pathway of excitation involved only receptors, bipolar neurons, and ganglion cells, it is now recognized that the horizontal and amacrine cells of the plexiform layers are not mere supporting structures but are involved in complicated synaptic networks. Especially interesting are the numerous "triad" synaptic contacts, in which three elements converge in a synapse, often with vesicles and receptor sites evident in all three. Such a structure implies direct feedback interaction within a single synapse, and could be an important structural component of lateral inhibition. Code: R, rod; C, cone; MB, midget bipolar; RB, rod bipolar; FB, flat bipolar; H, horizontal cell; A, amacrine cell; MG, midget ganglion; DG, diffuse ganglion cell. [From J. E. Dowling, and B. B. Boycott, Proc. Roy. Soc. (London), B166, 80–111 (1966).]

tives of cilia, which, because of the way the retina develops, are located at the back of the retina, absorbing light only after it has passed through the successive layers of neural tissue in addition to all the optical apparatus. It sometimes seems truly remarkable that our visual resolution is so good.

On the other hand, the "grain" is exceedingly fine. The receptor elements themselves are minute and extremely numerous. Throughout the vertebrates there are two distinguishable types: *rods* and *cones.* Both are the distal processes of sensory cells. Rods are relatively long and narrow, cones conspicuously shorter and thicker. The specific shape of both may enable them to function as waveguides to concentrate light on the visual pigment inside. The human retina contains approximately 125,000,000 rods and 6,000,000 cones, converging in a complicated synaptic network on a smaller number of *bipolar* neurons. The bipolars provide input to approximately 1,000,000 *ganglion cells.* The rods and cones, in effect, represent two separate (but

not independent) photoreceptor systems. The rods are more than two orders of magnitude more sensitive than cones, and are the only source of vision in dim light. Resolution is poor with rods, however, in part because many rods over a large area provide input to a single ganglion cell. In bats (which have no cones) as many as 1000 rods may be connected to one ganglion cell. Cones respond only to brighter light and in most animals are concentrated in the fovea (where there are few if any rods). Often a single cone provides the sole input to a ganglion cell. Hence images are sharply defined. Unlike rods, which all contain the same pigment, different cones contain different pigments, permitting color discrimination (see Figure 10-36).

In submammalian vertebrates, the rods and cones are often contractile in the myoid region between inner and outer segments. When this region "relaxes," the outer segment extends into the pigment epithelium where the receptor pigment is partially masked; when it shortens, the outer segment is pulled back out of the pigment epithelium. In this way the amount of light reaching the outer segment (and hence the sensitivity) can be varied and scatter in bright light much reduced. Another means of reducing scatter and chromatic aberration is the incorporation of red or yellow oil droplets into the inner segments of cones, a feature common to many amphibian and reptilian and some birds' eyes.

10.6.2 Photochemistry

Both rod and cone outer segments are densely packed with flat lamellae, formed at least in part by invagination of the outer limiting membrane (Figure 10-32). Located in these membranes are the photosensitive pigments, constituting as much as 35% of the dry weight of the structure, 90% of the protein. The pigment molecule is a complex one, with several chains that extend all the way across the membrane. The lamellae, or discs, are entirely enclosed within a plasma membrane in rods, but arise as deep invaginations of the plasma membrane in cones. The functional importance of this difference is not yet understood.

The receptor pigment in the rods of all land vertebrates and marine fishes is a red substance called *rhodopsin,* first extracted and carefully studied by Kühne in 1876. Light, if absorbed, causes this pigment to undergo a series of structural changes and eventually to split into two component parts: a colorless protein, opsin, and a yellow carotenoid, retinaldehyde, or *retinal,* (previously called retinene), identified by Wald in 1933–1935 as a vitamin A derivative (Figure 10-33). Retinal can exist in several isomeric forms, depending on the planarity of its double bonds and the twist of the conjugated isoprenoid chain. The most stable conformation, by far, is the all-*trans* form, although several slightly less stable forms, bent at different points in the chain (carbons 9 and 13) would be predicted to exist. Because of the methyl group at carbon 13, however, the 11-*cis* form is "hindered," and, by all the rules of stereochemistry, should not exist. This is the isomer that is found, bound to opsin, in rhodopsin. The only known effect of light in vision is the isomerization of this molecule into the all-*trans* form. Thereafter, even in the dark, the rhodopsin molecule undergoes several changes, depending only on thermal reactions until finally

the retinal and opsin split, a phenomenon known as *bleaching.* Each of the intermediate states has a slightly different absorption curve against wavelength, allowing spectrophotometric study of its formation and disappearance. Figure 10-34 shows, diagramatically, a current view of how the rhodopsin molecule is embedded in (and across) the disc membrane.

Visual excitation, which involves a change in permeability of the receptor membrane, presumably occurs at some point along this chain, almost certainly before the last step which is a relatively slow hydrolysis. The isomerization and subsequent changes expose two sulfhydryl groups and one negatively charged residue, but it is not clear how these could be effective. Perhaps a more likely explanation is that slight conformational changes in the opsin, or even in charge distribution within the opsin, could transform it into an ion permeation channel (ionophore) or into a potent enzyme, capable of triggering further chemical reactions. Such a phenomenon *(allosteric transition)* is thought to be involved in the action of many regulatory enzymes and may explain the fact that hemoglobin becomes a stronger acid when combined with oxygen.

Obviously, to maintain visual sensitivity, the photosensitive pigment must be reconstructed. This occurs partially by the photochemical reisomerization of all-*trans* to 11-*cis*-retinal in the light and, in both light and dark, by the enzymatic reconversion of the all-*trans* to the 11-*cis* form, either directly or by way of the alcohol form, vitamin A (Figure 10-35). The equilibrium established between the rates of bleaching and regeneration of pigment helps determine the level of light adaptation.

As was mentioned above, rhodopsin is found in all land vertebrates and marine fishes. For some reason, freshwater fishes or fishes spawned in freshwater utilize a different pigment, porphyropsin, identical to rhodopsin except for the presence of an extra double bond in the ring and the property of absorbing maximally at a slightly longer wavelength (522 nm). The light and dark reactions of porphyropsin are apparently identical to those of rhodopsin. Brackish-water fishes (for example, *Fundulus*) often have both rhodopsin and porphyropsin, whereas fishes that move from freshwater to saltwater or to land change their pigments. Frog tadpoles and urodele larvae change from having

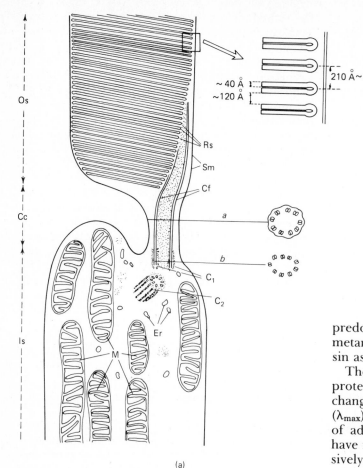

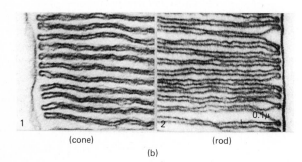

(a)

(b)

(cone)   (rod)

Figure 10-32
*Fine structure of vertebrate photoreceptor cells.
(a) Diagram showing the fine structure of a mammalian rod, including part of the outer segment
(Os), the connecting cilium (Cc), and part of the
inner segment (Is), filled with mitochondria (M).
Note the conspicuous ciliary structure at a and
b. Code: C_1 and C_2, two centrioles; Rs, rod lamellae bearing the visual pigment; Cf, ciliary filaments; Sm, surface membrane; Er, endoplasmic
reticulum. (b) High-magnification electron micrographs of cone (1) and rod (2) lamellae of a rhesus
monkey. Note the separation from the plasma
membrane, the slightly greater density of the
cone disks, and the difference in spacing between
the membranes of the disks. [(a) from E. de
Robertis, J. Gen. Physiol., **43** (Suppl. 2), 1–13
(1960); (b) from J. E. Dowling, Science, **147,** 57–
59 (1965).]*

predominantly or entirely porphyropsin before
metamorphosis to having predominantly rhodopsin as terrestrial adults.

The greatest variability in pigments involves the
protein opsin. Slight differences in structure
change the wavelength of maximum absorption
(λ_{max}) by 20–30 nm. In certain cases this is clearly
of adaptive value. Deep-sea fishes, for example,
have rhodopsin with sensitivity displaced progressively toward the blue with increasing depth, compensating for the decreasing penetration of longer
wavelengths.

It is noteworthy that among the three animal
phyla having the best developed eyes, the mollusks,
arthropods, and vertebrates (each of which has its
own type of eye, apparently evolved quite independently of the others), the visual pigment is invariably retinal, and in every case yet analyzed it is
the 11-*cis* isomer. There is good reason for this.
It is one of the smallest organic molecules having
a conjugated side chain to absorb energy, and the
hindered form is not only more easily isomerized
by light than other isomers but the change in structure with isomerization to all-*trans* is greater than
for any other form.

Role of Cones. Most mammals that have been
tested appear to be color-blind. Primates and a few
other species (including squirrels and perhaps cats)

Figure 10-33
Chemical structure of vitamin A_1 and retinal$_1$ isomers. The 11-cis form is the one that occurs naturally in combination with opsin to form the visual pigments. Retinal$_2$ is identical but for a second double bond in the 3,4 position of the ring.

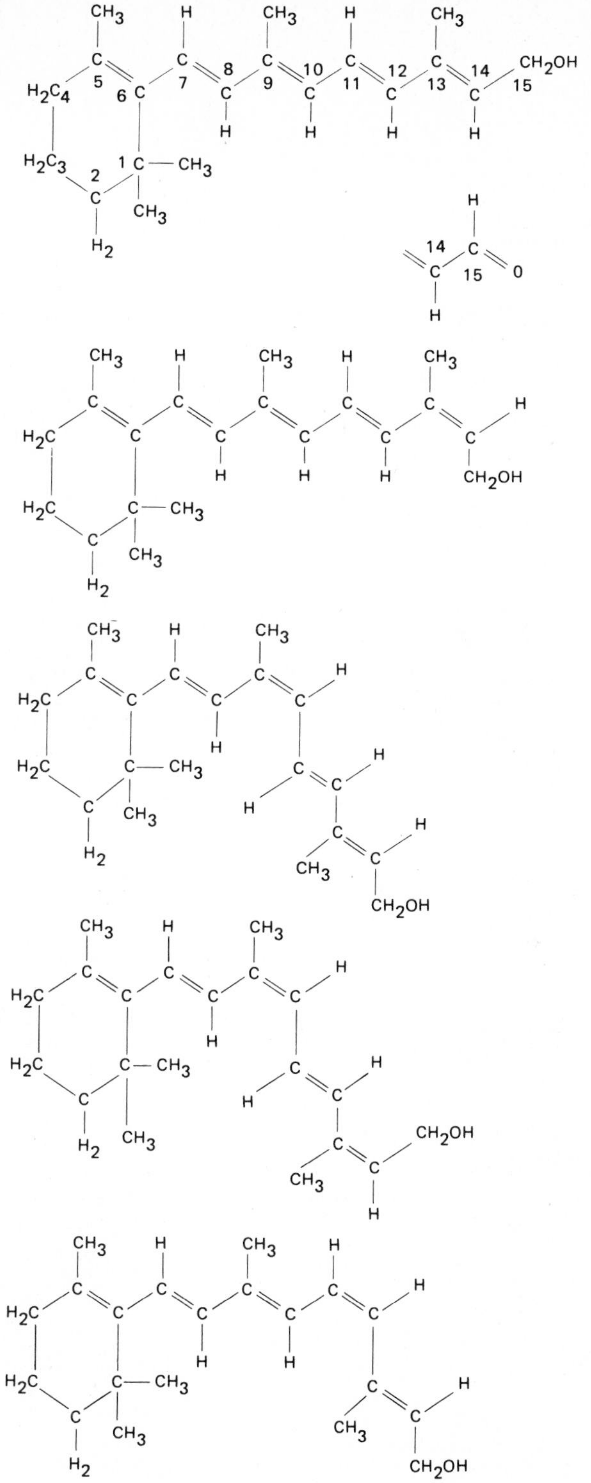

all-*trans* Vitamin A_1

Retinaldehyde$_1$

13 – *cis*

9 – *cis*

9, 13 – dicis

11 – *cis*

503

(a)

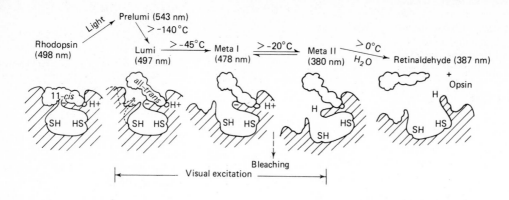

(b)

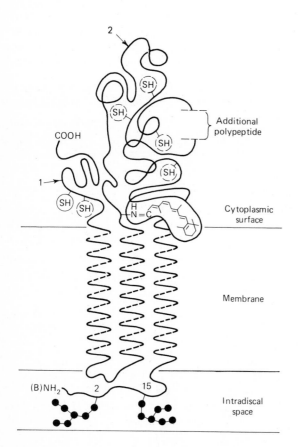

Figure 10-34

*(a) Simplified diagrammatic sequence of the steps in the bleaching of rhodopsin, with a corresponding theoretical scheme to explain the changes in configuration of retinal and opsin and exposure of active groups. The only action of light is the isomerization of 11-cis to all-trans retinal. Thereafter thermal reactions cause conformational changes in the opsin and eventually hydrolysis of retinal from the opsin. There are other steps in the metarhodopsin sequence, with visual excitation apparently occurring at one of these steps. With each change in configuration the wavelength of maximum absorption changes as shown. (b) More accurate current model of the organization of the rhodopsin molecule in the photoreceptor disc membrane. An extensive and complicated polypeptide chain (only partly shown) extends into the cytoplasm. Embedded in this and oriented parallel to the membrane surface is the 11-cis-retinal chromophore. At least three helical segments of the polypeptide chain extend across the membrane, with a number of sugar monomeres (filled circles) extending into the intradiscal space. Isomerization of the retinal causes a succession of conformational changes in the protein, resulting in excitation, possibly by conversion of the molecule into an ionophore, or ion permeation channel. [(a) from G. Wald, and P. K. Brown. Cold Spring Harbor Symp. Quant. Biol., **30**, 345–359 (1965); (b) from W. L. Hubbell and M. D. Bownds, Ann. Rev. Neurosci., **2**, 17–34 (1979).]*

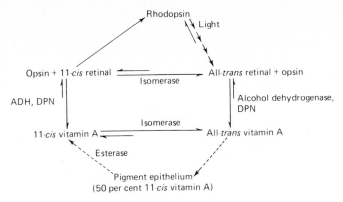

Figure 10-35
Pathways of bleaching and reconstruction of rhodopsin.

can discriminate colors. On the other hand, birds, lizards, turtles, frogs and teleost fishes have all been demonstrated to have color vision. Why most diurnal mammals do not share the same capability is a puzzle of evolution.

Perception of color is apparently accomplished, much as Young, Helmholtz, and Maxwell theorized in the nineteenth century, by comparing the levels of excitation of three different interdispersed populations of cones with different absorption spectra.

The existence of such pigments has been demonstrated in highly exacting experiments measuring the absorption spectra of single rods and cones (Figure 10-36). The technique is to view a portion of retina through a microscope focusing on the layer of outer segments and projecting the image through the microscope onto a screen. The image of a single rod or cone is then allowed to pass into a spectrophotometer, after the surrounding area is masked off. The difference between the absorption so obtained and that of light transmitted through an adjacent part of the retina without a rod or cone provides the correct pigment absorption curve. The chemistry of these color pigments is not known, but 11-*cis*-retinal is needed for their regeneration when bleached, so the opsin apparently confers wavelength specificity in these pigments as in all others studied. There are many types of color blindness, apparently arising from the lack of one, two, or all the pigments.

In dim light, only rods are functional and no colors are distinguishable. Sensitivity to different wavelengths corresponds exactly to the rhodopsin absorption curve (Figure 10-37). In bright light much of the pigment in the rods is bleached and their sensitivity much reduced, and the cones provide visual information. The shift from rod to cone vision is known as the ***Purkinje shift,*** and is characterized by a relative increase in sensitivity to longer

Figure 10-36
Absorption spectra of five individual human cones normalized to an absorbance of 1 at the peak. Relative absorbances are shown in the inset. Each curve was obtained by subtracting the absorbance of a "bleached" cone from the same cone dark-adapted. In these measurements, as in others with monkey and fish cones, the cones fall into three populations, lending support to the trichromatic theory of color vision. [From G. Wald, and P. K. Brown. Cold Spring Harbor Symp. Quant. Biol., **30,** *345–359 (1965).]*

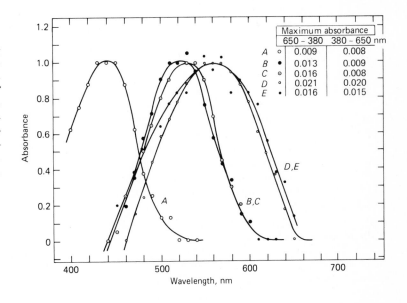

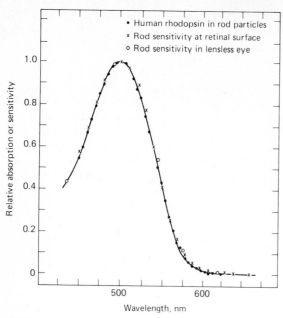

Figure 10-37

Absorption spectrum of human rhodopsin from a suspension of rod outer segments, compared with human visual sensitivity. The latter values can be obtained either by correcting for absorbance properties of the lens (which has yellow-filter characteristics), or by direct measurement in an individual having no lens. There is exact correspondence between the rhodopsin absorbance and behavioral sensitivity in both cases. [From G. Wald, and P. K. Brown, Science, **127,** *222–226 (1958).*]

wavelengths. After exposure to bright light, cones in the dark return to full sensitivity within about 5 min, whereas rods require 30–40 min for dark adaptation. The return of sensitivity parallels exactly the restoration of visual pigment. There is a wide range of light intensities (approximately four orders of magnitude), however, within which there is no net bleaching of pigment (the light is too dim to reduce the amount of pigment present), and yet rapid adaptation takes place. Although some of this adaptation to light may be neural in origin, possibly occurring in the bipolar layer, there is evidence that much of it is a still unexplained property of the receptors themselves.

The fully dark-adapted eye is phenomenally sensitive to light. It is well established that a single quantum is sufficient to excite a single rod, because at threshold luminosity a single rod absorbs an average of approximately one quantum every 38 min. Even at 1000 times threshold, a rod absorbs a quantum less than once every 2 sec on the average. Absolute threshold for a brief exposure appears to require the excitation of six to seven rods. Why rods are so much more sensitive than cones is not known. The photochemical reaction may be much more sensitive, but it is also clear that a large number of rods interact, potentially increasing their effect.

Because of the absence of rods in the fovea, dim-light sensitivity is lower there than at other points on the retina. Hence to detect faint lights, it is necessary to look slightly away from them, letting the light fall on a nonfoveal part of the retina. Many diurnal birds have retinas consisting mainly of cones, allowing extremely accurate image resolution but poor vision in dim light. They have enlarged foveas that may extend into a horizontal band, with cone density reaching 10^6 per mm^2, about 7 times greater than in man and permitting considerably sharper visual acuity. Other birds, such as owls, have essentially all-rod eyes, with correspondingly poor daylight vision.

The existence in chickens and pigeons of retinas that consist primarily of cones permitted the extraction and characterization of the first cone pigment, iodopsin. This pigment was found to consist of retinal (the carotenoid of rhodopsin), plus a special cone opsin. Freshwater fishes having color vision were then found, as predicted, to have a cone pigment consisting of retinal₂ (the porphyropsin group) plus cone opsin (cyanopsin).

10.6.3 Neural Events in Vision

It is the job of the visual nervous system to interpret the temporal and spatial patterns of excitation in the hundreds of thousands of receptors stimulated whenever the image of an object falls on the retina. Forms must be recognized regardless of where or in what orientation on the retina they fall, their absolute size, or brightness. In addition, movement, color, and depth must be discriminated, using the same input. All this information is obtained by the integration of activity at successive neural levels between the receptors and the cortex. Figure 10-42 shows the principal pathways involved.

Retina. Until recently, the only recordings from single retina cells were from ganglion cells; our knowledge of electrical correlates of more peripheral events was based on the summed response of the whole retina, the *electroretinogram* (ERG). This consists of several waves of different polarity attributable with less than satisfactory accuracy to the various cell populations of the neural retina and the pigment epithelium. A major component of it is now thought to represent depolarization of a purely glial population, the Müller's cells. However, the perfection of microelectrode techniques and development of new preparations have made it possible to record intracellularly from the receptors and all of the types of neurons in the retina. In fact, the vertebrate retina is now the most thoroughly studied part of the CNS.

The first major surprise was the observation that the receptors, and apparently also the bipolar and horizontal cells (see Figure 10-38), respond to light with hyperpolarizing responses of up to 10–20 mV. Both rods and cones have low resting potentials in the dark and are repolarized (hyperpolarized) by light in a response that increases with the logarithm of external $[Na^+]$. Further evidence that Na^+ is the principal ion involved is the fact that, with artificially produced hyperpolarization of the cells, the light-induced hyperpolarization is increased; with depolarization, the response to light is reduced to amplitude and eventually reverses to become a depolarizing response at a membrane potential that depends on the external $[Na^+]$. This is just the opposite of what one would expect of a light-increased membrane conductance to Na^+. It would be better explainable as a light-activated Na^+ pump, hyperpolarizing the cell by actively expelling Na^+. Such a pump is known to be present, but not necessary to the response, since ouabain does not eliminate the response, as long as the $[Na^+]$ gradient is maintained. Instead, it is found that there is an increase in membrane resistance with light; that there is a constant "dark current" flowing from distal to proximal side of the receptors, which is reduced in the light.

The postulated explanation for these findings is that there is, in the dark, a constant Na^+ leak into the receptor at one end, and that a metabolic pump expels it at the other (creating a "dark current"). This current is so large that if the metabolic pump is blocked the ionic gradients across the receptor

membrane run down within about 90 sec. When light is absorbed, it somehow causes a decrease in the Na^+ inflow without affecting the pump. The effect is hyperpolarization. Measurements of the change in current during response to light suggests that there are between 1000 and 2000 Na^+ channels in a receptor outer segment, approximately as many channels as there are discs. (There are about 10^8 rhodopsin molecules per receptor.)

The mechanism whereby light triggers this change in permeability is not fully understood, although it seems probable that light somehow causes the release of Ca^{2+} that has been sequestered by the rhodopsin-bearing discs. The Ca^{2+} then diffuses to the plasma membrane and blocks Na^+ permeability. On the other hand a decrease in permeability is not a direct, inevitable result of the isomerization of rhodopsin because the same photochemical event in the receptor cells of invertebrates results in large depolarizing responses, due to increased conductance to Na^+ and perhaps Ca^{2+}. Single retinula cells of the horseshoe crab, *Limulus,* even show quantal humps in the graded receptor potential, each of which may represent the absorption of a single quantum of light.

Microelectrode recordings from single units in the retina of lower vertebrates have added greatly to our understanding of retina integration. Figure 10-38 shows the form of the responses of the several cell types in a very simple retina. A more detailed picture of their anatomical relationships is shown in Figure 10-31. Note that the horizontal cells, like the receptor cells, respond with slow hyperpolarization, and that the first three types of neurons in the pathway all respond with only graded potentials. Of the bipolars, however, about half respond to a small spot of light with hyperpolarization, the others with depolarization.

The receptor cells innervate both bipolar and horizontal cells, most prominently together in the form of specialized compound synapses (Figure 10-39). These take the form of invaginations in the receptor cell surface, with processes of one bipolar cell and two horizontal cells (cone) or several of each (rod) postsynaptically, and a dense presynaptic ribbon, surrounded by vesicles, opposite them in the receptor cell. These presynaptic ribbons are reminiscent of the presynaptic densities seen in hair-cell receptors in the vertebrate inner ear. As in the case of auditory receptors, transmitter re-

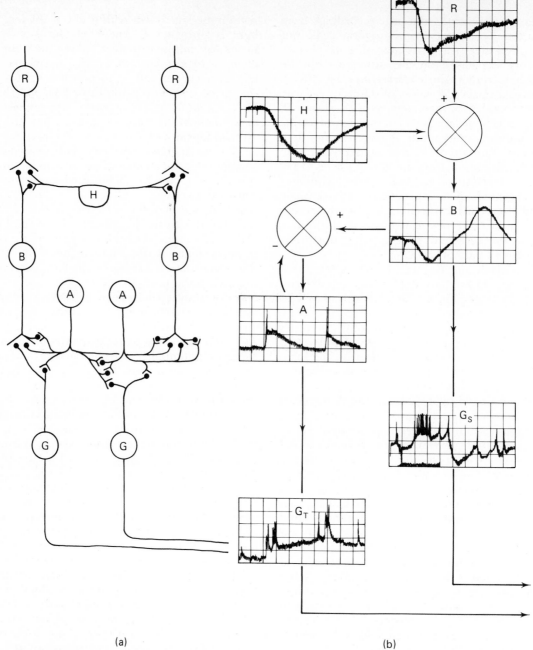

(a)   (b)

Figure 10-38

*(a) Schematic summary of synaptic interactions in the retina of Necturus, revealed by electron micros-
copy (see Figure 10-31). (b) Summary of characteristic electrical responses of Necturus retina cells,
and of their synaptic organization. It is postulated that the input from the receptor cells to the
bipolars is modulated by the activity of horizontal cells, also excited by the receptors. Bipolar cells
may drive ganglion cells directly or by way of amacrine cells, which have feedback influence in
bipolars. Ganglion cells may fire either in the pattern of bipolars, with concentric fields and sustained
responses, or like amacrine cells, with transient responses at the "on" and "off" of a stimulus.
The first evidence of lateral inhibition is seen in the horizontal cells, and surround inhibition is
strongly pronounced in the bipolar and more central neurons. This influence of the surround is
apparently mediated principally by the horizontal cells, whereas dynamic response properties may
be determined largely by the amacrine cells [(a) from J. E. Dowling, and F. S. Werblin. J. Neurophysiol.,
32, 315–338 (1969); (b) from F. S. Werblin, and J. E. Dowling. J. Neurophysiol., **32**, 339–355 (1969).]*

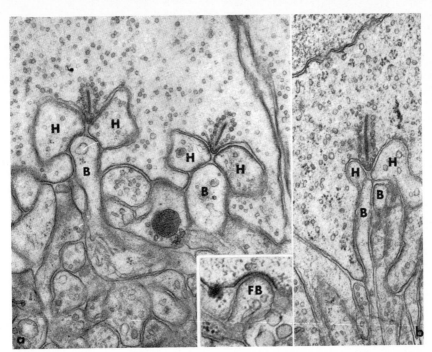

Figure 10-39
*Electron micrographs of characteristic synapses between receptor cell terminals (large presynaptic processes) and horizontal (H) and bipolar (B) cell processes. (a) In some terminals, there are invaginations, each with two horizontal cell processes laterally and one bipolar process centrally. (From monkey fovea.) (b) Rod terminals typically have only one invagination, with more processes, including those from several bipolar cells (cat rod). The inset in (a) shows a superficial contact between a frog cone and bipolar cell. Note the dense presynaptic "ribbons." (See also Figure 9-13C.) [From J. Dowling. Invest. Ophthalmol., **9**, 655–680 (1970).]*

lease around a tonic level is exquisitely sensitive to membrane potential, showing changes in release with potential changes as small as 5 μV. The biophysics of transmitter release from these receptors remains an interesting problem for future research.

Rods and cones innervate separate populations of bipolars. Rod bipolars appear to form one population, and are each innervated by a large number of rods (30,000 in humans). Cone bipolars are more complicated, especially in higher vertebrates. Some are innervated by several (5–7) cones, others receive input from only one cone. The latter synapses are usually of the invaginating type, whereas the multiply innervated cone bipolars usually form simpler contacts with the cones. Horizontal cells, being large cells oriented parallel to the surface of the retina, receive inputs from enormous numbers of receptors. Some contact only rods or cones and appear to be selective, in the latter case, for a given color-specificity type of cone. Others, for example in humans, have a small dendritic field receiving input from only cones (approximately 50 contacts with about 10 cones), but they also have a long process (called an axon, although it does not produce action potentials) that forms an extensive arborization some distance away where it receives input from up to 2500 rods.

The principal interest in the retina is as a fortuitously placed piece of the brain in which the complex cellular interactions involved in information processing can be understood. Such interactions are seen at all levels. Although in *Necturus* it appears that receptor cells are not affected by stimulation of adjacent receptor cells, in most other animals studied both rods and cones are interconnected in such a way that there is summation of response over many receptors. The receptive field for a rod extends over about 0.5 mm², encompassing about 10,000 rods. Each rod is connected by 4–6 gap junctions to each of its immediate neighbors. Each cone appears to be connected to about 100 other cones, in a much smaller receptive field. Moreover, cones specialized for a given color form electrotonic connections only with other cones of the same color specificity. There is no good evidence at present for electrical interconnections between rods and cones.

Cones, but probably not rods, receive inhibitory feedback from horizontal cells. Since the horizontal cells are also electrotonically coupled and are large cells extending laterally over long distances compared with a single receptor, this inhibitory feedback is most effective when a large area of retina surrounding the receptor is illuminated. Thus a small spot results in a large hyperpolarization of cones, whereas a larger spot elicits a slightly reduced response. However, much stronger lateral inhibition is seen at the level of the bipolars, where a small spot produces a strong response and an annulus stimulating the area surrounding the spot produces a response of the opposite polarity. This effect of the surround is seen in both cone and rod pathways.

The relationship between transmitter release and membrane potential in these cells with no spikes is not well understood. It is even possible that release is governed primarily by events not reflected in the membrane potentials, especially in receptors. However, it is now widely believed that transmitter is released in proportion to the level of depolarization, that transmitter is being released constantly from receptors, and that this release is reduced when the cell is hyperpolarized by light. If the transmitter is normally depolarizing to the horizontal cells, a reduction of release will also cause their hyperpolarization. At the next level, it must then be assumed that some bipolars respond to the receptor transmitter (assuming there is just one) with depolarization, the others with hyperpolarization. The bipolars that are depolarized by transmitter from the receptors are apparently hyperpolarized by transmitter released from the horizontal cells, whereas the other bipolars respond in the opposite way.

The final two neuron types in the retina are the amacrine and ganglion cells; of these, the ganglion cells behave like conventional neurons, responding with sharply rising depolarizations that can generate bursts of action potentials. Amacrine cells, like horizontal cells, extend laterally for considerable distances across the retina and have homogeneous rather than concentric receptive fields. They have no true axons, but via their many processes interact with other amacrines, bipolars, ganglion cells, and even horizontal cells. Both amacrine and ganglion cells receive input from bipolars via complex synapses again featuring a dense presynaptic ribbon. The postsynaptic elements may be a ganglion cell and an amacrine cell, or two or more amacrine cells. The amacrine cell processes then synapse

with each other, either in serial synapses (unidirectional from one to another) or reciprocal synapses (in both directions). Amacrine cells are particularly sensitive to movement or to change in light conditions, typically showing a large phasic depolarization at the "on" and "off" of a light, with small spikes riding on top of the depolarizations. It seems likely that the amacrine cells are involved in analysis of specialized stimulus features, which is especially important in the retinas of lower vertebrates, where much of the visual information processing takes place in the retina. In higher vertebrates, there is a higher proportion of direct bipolar to ganglion cell synapses.

Ganglion cells have been extensively studied, especially in frogs and cats. Each ganglion cell is spontaneously active and has a roughly circular receptive field of up to 4–8° (a circle of about 1–2 mm diameter on the retina of the cat, representing an immense number of receptors and bipolar cells). Many of the ganglion cells respond much like bipolar cells, with receptive fields that are spatially subdivided into distinct excitatory and inhibitory components arranged either with an excitatory center, surrounded by an inhibitory area (on-center) or the other way around (off-center). These are subdivided into two categories, both morphologically and in response characteristics. X cells are small, tonically active, with slower conducting axons and sustained responses. Y cells are larger, with larger receptive fields and more phasic responses. As Figure 10-40 shows, even these simple response patterns permit considerable specificity to response.

The response seems to be essentially a summation of excitatory and inhibitory influences, with the center having a disproportionate effect. Normally after strong excitation there is a reduced rate of spontaneous firing, whereas after inhibition there is a rebound increase in spontaneous firing. It is particularly noteworthy that a small light in the right spot is much more effective than a brighter diffuse light. Thus within the retina there are both sophisticated lateral inhibition mechanisms to enhance contrast and specify response, and neural mechanisms for adjusting visual sensitivity independent of photopigment concentrations.

There is another population of ganglion cells, called W cells, which show more complicated response specificities, probably indicative of inputs predominantly from amacrine cells. Among these

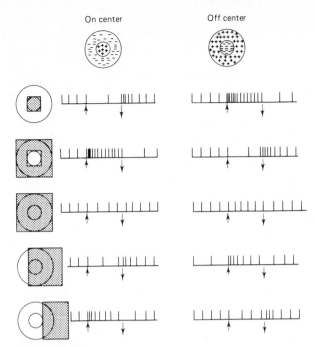

On center   **Off center**

Figure 10-40

Receptive fields and response patterns of retina ganglion cells. Each covers a large area of retina and is organized into a concentric pattern with the center either excitatory or inhibitory and a surrounding area having the opposite effect. Shown in the figure are the different patterns of response observed in on-center and off-center units when different stimuli are given, as indicated on the left. The shaded area represents darkness, the unshaded area light. The stimulus is represented as going on at the upward arrow (↑), off at (↓). Note that there is relatively little response either to illumination or darkness covering the whole field. A moving shadow would elicit a more complicated response, but one that could be accurately predicted from the responses shown.

are color-opponent cells (with the center excited or inhibited by one wavelength, the surround influenced in the opposite way by a different wavelength, and little or no effect of white light on the cell), direction-sensitive cells (also with concentric fields but excited only by movement of a signal in one direction), local edge detectors (which are excited only by a dark-light border in the receptive field), and uniformity detectors (which fire continuously in the dark, but are inhibited by almost any stimulus in the center of the receptive field).

Vertebrates without a highly developed visual cortex tend to perform elaborate visual analyses within the retina or in the first central visual nucleus, the optic tectum. This has been particularly well demonstrated in the frog. Recording from optic-nerve fibers as they enter the midbrain nucleus, investigators find several distinct and spatially separated populations of fibers responding preferentially to highly specific stimuli. For example, only convexly curved edges moving in a certain direction across the receptive field might be effective; or a black spot (decoy fly) moving against the background of a photographed natural environment, while moving the background with the fly has no excitatory effect. Curiously, rabbit ganglion cells show much more complicated receptive fields than do those of cats or monkeys, particular in specificity of response to the direction of movement of objects. The physiological and adaptive explanations of this difference are not clear.

The retina bases for color vision have been beautifully demonstrated in studies by Tomita and his coworkers and by Kaneko. Intracellular recordings from single fish cones reveal the existence of three populations having maximal sensitivity in the red (80%), green, and blue, each responding only with hyperpolarizing potentials of a few millivolts to light of the appropriate wavelength. Single horizontal cells receive inputs from one or more receptor types in mutually antagonistic fashion to produce either hyperpolarizing or depolarizing responses (Figure 10-41). Bipolar cells are the first to show concentric color-specific receptive fields. Each has a center of about 200 μm that is either depolarized or hyperpolarized by one wavelength (red) and a large surround (1–1.5 mm) in which a different wavelength causes the opposite effect. Amacrine cells again show simple spatial summation of effect rather than center-surround organization and are either of the transient-responding variety or show steady depolarizing responses to one wavelength, hyperpolarizing to another (usually red). The size of the center of the ganglion cell receptive fields corresponds approximately to the size of the bipolar receptive fields; the ganglion cell surround approximates the size of the amacrine receptive fields.

Lateral Geniculate Body. In submammalian vertebrates, apparently without exception, the optic-

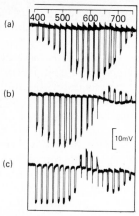

Figure 10-41
Wavelength dependence of retina potentials in the carp. Responses recorded at different points in the retina, probably all in the layer of horizontal cells surrounding the bipolar neurons. (a) Luminosity type, responding with the same polarity at all wavelengths. (b) Biphasic chromaticity type, reversing polarity once as the wavelength is altered across the spectrum. (c) Triphasic chromaticity type, with two reversals across the spectrum. [*From T. Tomita.* Cold Spring Harbor Symp. Quant. Biol., **30**, *559–566 (1965).*]

(P) cells, which provide the output to the visual cortex, and interneuronal (I) cells, which are axonless cells that appear to provide lateral inhibitory input to the P cells. Each geniculate neuron receives afferents from several ganglion cells in a single retina. One of these is usually much more effective than the others, causing firing each time it is active. A single optic-nerve fiber usually innervates many geniculate cells. The response patterns are essentially the same as those of retina ganglion cells, with one important difference. The importance of the surrounding area is much greater, so that diffuse light has less effect, either inhibitory or excitatory.

Visual Cortex. In mammals, most of the analysis of visual input is done in the cortex. As might be anticipated, even the primary sensory cortex is tremendously complicated. On the other hand, research in several modalities has revealed the cortex to be a beautifully and precisely ordered part of the nervous system. Research on the primary visual cortex, especially by Hubel and Wiesel, has made this the best understood part of the cortex, if not of the brain altogether.

The primary visual areas, like most of the rest of the cortex, show layering in the cells of the grey matter forming the surface of the brain. Incoming fibers from the lateral geniculate body end exclusively in layer IV (near the middle) mostly on small stellate cells. These in turn interact mostly with other neurons (pyriform and pyramidal cells) above and below in vertical columns. Single cells in these columns show response properties that appear to represent three different levels of complexity, each the result of convergence and interaction between inputs from the next simpler level. Hubel and Wiesel classify the main neuronal types as simple, complex, and hypercomplex. *Simple cells* are probably the first receiving geniculate input and show behavior easily derived from the concentric receptive fields of geniculate neurons. The important change is that the border between excitatory and inhibitory area is no longer circular, but linear. Figure 10-43 shows characteristic receptive fields. The orientation of the border is shown the same in every case, but other cells having receptive fields in all orientations are found.

Complex units are intermixed with simple ones, probably receiving afferent input from them. Like

nerve fibers all cross to innervate contralateral visual nuclei. In mammals having at least partial binocular vision, an increasing proportion stay on the ipsilateral side. In primates approximately one half of the fibers decussate in the optic chiasm with the components from the same half of each retina going to the same side of the brain to synapse in the lateral geniculate nucleus (see Figure 10-42). This nucleus is arranged in three (cat) to six (primate) layers, receiving input alternately from the contralateral and ipsilateral half-retinas. The half-retina in each case is topographically represented with the layers in register. There are two neuron types in the lateral geniculate nucleus: the principal

Figure 10-42

Primary visual pathways. From the retina (see Figure 10-31), optic nerves pass to one or the other of the lateral geniculate nuclei, with a few fibers branching off to the superior colliculi where they mediate subcortical reflexes. In primates, approximately half of the fibers decussate to the opposite side of the brain, with the same part of both retinas projecting to the same side. In lower vertebrates there is complete decussation. The lateral geniculate is organized into a number of layers with alternate layers receiving input exclusively from the ipsilateral (I) or contralateral (C) eyes. Lateral geniculate output then projects to the primary visual cortex in the occipital lobe of the cortex. The visual cortex itself is organized into at least three areas, with most of the input to areas II and III coming from I.

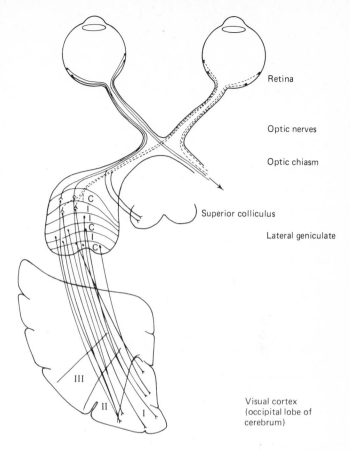

Retina

Optic nerves

Optic chiasm

Superior colliculus

Lateral geniculate

Visual cortex (occipital lobe of cerebrum)

simple ones, they are sensitive to light only on certain parts of the retina, and prefer patterns with linear properties. It is not possible to localize excitatory and inhibitory areas, however, and although the orientation of an edge is critical, its location in the receptive field is not. These cells respond preferentially to an edge or rectangle at a given angle to the vertical, but do not seem to care what fraction of the receptive field is light or dark. Cells having *hypercomplex* receptive fields appear to be excited by converging input from several complex cells and respond specifically to oriented stimuli that end within the receptive field or show two or more different edges, each at a particular orientation. Figure 10-43 illustrates the simple and complex response characteristics. None of these cell types responds to diffuse illumination.

It thus begins to be possible to envision the recognizable patterning of activity of different sets of simple, complex, and hypercomplex cells when a pattern of any shape is presented anywhere on the retina. This is only the beginning, however, for cells of the primary visual cortex (approximately sixth-order neurons) project to a number of other sites, to secondary visual cortex, to association areas, and to motor cortex. Increasingly complex stimulus parameters are coded, including color and movement, in ways not yet worked out.

In addition, in the cortex for the first time, cells receive input from corresponding points on both retinas. Efferents from the two lateral geniculate nuclei do not end on the same neurons; they continue to be segregated, each ending in layer IV of one small part of the cortex. On either side of this (about 0.5 mm either way) are inputs from the other lateral geniculate, with visual input from the same or immediately adjacent parts of the visual field. Each input tends to dominate within the verti-

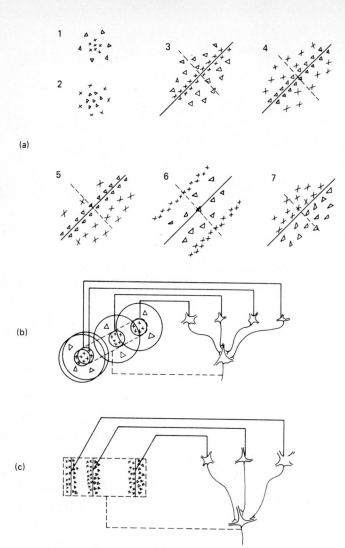

(a)

(b)

(c)

Figure 10-43

*Receptive fields of lateral geniculate and cortical cells. (a) Fields 1 and 2 are typical of lateral geniculate cells (as of ganglion cells), where × represents areas giving excitatory responses and Δ areas that give inhibitory responses ("off" responses). Fields 3 to 7 are typical of simple cortical units. All are shown oriented along the same axis, although each type of field exists in all orientations. (b) A possible synaptic arrangement whereby the convergence of excitatory input from a group of lateral geniculate cells onto a single cortical cell could explain the shape of simple receptive fields. (c) Another possible neural circuit in which several "simple" cortical cells could contribute to the receptive field of a "complex" cortical unit. Any stimulus having a vertical edge falling across the rectangle, regardless of its position, would excite some cells having simple receptive fields, and they in turn could excite the higher order cell. [From D. Hubel, and T. Wiesel. J. Physiol. (London), **160**, 106–154 (1962).]*

cal column of cortical tissue above and below it, forming *ocular dominance columns*, although there is enough lateral interaction between the cells in adjacent columns to ensure that the large majority of neurons receive input from both eyes. It is this binocular input that permits stereoscopic vision. There are two ocular dominance columns devoted to each small part of the visual field (one for each eye), and dozens of such alternating pairs of columns cover the cortex in a map of visual space.

The cortex consists not merely of such ocular dominance columns, however. Each cortical neuron has a preferred stimulus orientation. It turns out that if one samples the orientation specificities of neurons in a given ocular dominance column, penetrating exactly perpendicular to the surface, all neurons prefer the same stimulus orientation. If one samples along an electrode penetration parallel to the cortical surface and perpendicular to the ocular dominance columns, the cells show a regular shift in the preferred orientations (Figure 10-44). Every 40–45 μm along this axis there is a shift of about 12° in the preferred orientation, and there are approximately 15 such shifts, covering the whole 180° of possible orientations, in about 0.5–1 mm. The two ocular dominance columns

Figure 10-44

Plot of receptive field orientation selectivity versus electrode track distance for an oblique penetration through parts of two ocular dominance columns in the striate cortex of a 17 day old monkey that had had virtually no visual experience. The pattern is nevertheless very adultlike. Lesions were made at points L1 and L2 to identify and calibrate tracks. Open circles represent units driven most strongly by the contralateral eye, closed circles, the ipsilateral eye. The open bar, near lesion 1 (in layer IV) was an area where receptive fields were contralaterally dominated and unoriented. The inset diagrammatically shows the track of the electrode, as determined by subsequent histology. [From T. H. Wiesel and D. H. Hubel. J. Comp. Neurol., 158, 307–318 (1974).]

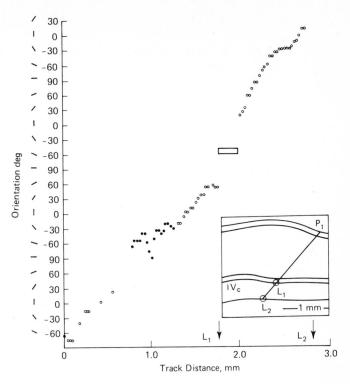

(about 1 mm wide in one axis) and the 15 orientation columns within each ocular dominance column receiving input from the same part of the visual field (about 1 mm deep in the other axis) form a "hypercolumn," all the cells of which are devoted to analyzing visual input from the same small part of the visual field. (Figure 10-45 summarizes this organization.)

Binocular convergence of input is probably essential to stereoscopic vision and depth perception. Most neurons receive input from exactly the same visual fields via the two pathways. However, sometimes there are slight visual field disparities, such that the two inputs will be exactly in register only if the visual image is closer or farther away than the plane of focus for most of the cells. These disparities, and the exact amount of parallax necessary to bring them into register, are thought to be used for depth perception. (Stretch receptors in the ciliary muscles, which govern the degree of visual accommodation, may also provide major cues to depth perception.)

Also involved in these neural integrative pro-

cesses is the discrimination of color. This has been little studied to date at higher centers, but it is apparent that retinal ganglion cells connected to cones may respond in a variety of complicated ways to light of different wavelengths. In the simplest case, there are the usual concentric excitatory and inhibitory areas, but with the center excited by one wavelength and inhibited by another, while the surround is influenced oppositely. Diffuse monochromatic light may be inhibitory in one part of the spectrum, excitatory in another, whereas white light, which triggers both effects, has little or no effect on the cell.

10.6.4 Arthropod Eye

Figure 10-29 shows the two commonest types of arthropod ommatidia, both consisting of a precise geometric array of separate sensory units, each having a separate corneal lens, crystalline focusing cone, and circular array of sensory elements, surrounded by masking pigment that can migrate outward to help isolate each unit in bright light, or

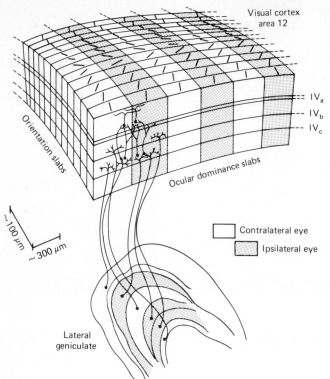

Figure 10-45
Diagram showing the apparent relationship between ocular dominance columns and orientation specificity columns in the macaque monkey. Measurements show that each ocular dominance column is about 300 nm wide, whereas in the other axis approximately 1 mm of cortex contains, in orderly sequence, cells with orientation specificities completing a full 360°. Complex cells in an upper layer are shown receiving input from two ocular dominance columns but from the same orientation column. [From T. H. Bullock, R. Orkand, and A. D. Grinnell. Introduction to Nervous Systems, W. H. Freeman, San Francisco, 1977; after D. H. Hubel and T. Wiesel. J. Comp. Neurol., **146,** 421–450 (1972).]

retract to allow light to spread to several ommatidia in dim light. There are usually eight sensory elements (retinula cells), one or two of which (eccentric cells) are often much larger than the others, with larger axons joining the dense layer of nerve fibers (neuropile) below the eye. In many insects the inner edges of these cells fuse to form a compact structure (rhabdome) consisting of a dense array of microvillar membranes. In Diptera, each retinula cell forms its own rhabdomere. The visual angle of each ommatidium is considerably larger than was originally thought, so that one or more of the sensory cells in each of several ommatidia view the same point in space. In flies, these in-register retinula cells send their axons to common second-order neurons. In this case, at least, it is not strictly correct to view the ommatidium as a single functional unit; instead, the functional units involve parts of six or more ommatidia. How input is processed in ommatidia having a fused rhabdome is not known.

Many arthropods, especially insects, have excellent image vision, although the major emphasis of this type of eye appears to be the detection of movement. In bees, with perhaps the best vision among the arthropods studied, the absolute sensitivity of vision is about 1/100 that of man, and the ability to discriminate between intensities is 20–50 times less. On the other hand, diurnal insects with fast flight speed usually have much faster temporal resolution than the vertebrate eye, showing "flicker fusion" rates up to 300 per sec, compared with a maximum of about 50 per sec in man. Many arthropods appear to be able to detect the polarization of light, although the mechanism of doing this is not understood.

As was mentioned, the visual pigments of arthropods and mollusks contain 11-*cis* retinal. In many species this is coupled with different proteins to provide color vision. This has been particularly thoroughly studied in bees, which are found to have relatively poor vision in the red but a pigment maximally sensitive far in the ultraviolet (340 nm). A survey of common flowers shows that several seemingly dull ones reflect strong ultraviolet components invisible to us but attractive to worker bees.

10.7 MIGRATION AND HOMING

The study of behaviour has frequently revealed capabilities that—at the time they are discovered—cannot be easily explained by any known sensory mechanisms. Familiar examples are orientation by bats and by electric fishes. In most instances further investigation has shown a reasonable sensory basis for these capabilities. In at least one case, however, the mystery remains. This is the ability of animals to travel long distances to the correct destination through totally unfamiliar territory during migration or homing. Migrations are best known in birds, but also occur in insects (for example, monarch butterflies), fish (for example, salmon), reptiles (for example, sea turtles), and many mammals (bats, whales, and caribou, among others). Most dramatic, probably, are the annual migrations of birds like golden plovers that spend the summer months in Arctic North America, the winters in southern South America. Adult golden plovers migrate from North to South America over about 2000 miles of the open Atlantic, returning in the spring via the land bridge of Central America. The immature, inexperienced golden plovers migrate the 8000 miles south via Central America, separate from the experienced adults. The knowledge of where to go is presumably genetically "built in."

The problem is most clearly delineated in the phenomenon of homing. Many birds, if removed from their home territory and transported long distances into totally unfamiliar territory (even inside closed boxes), are able to find their way back. In a few cases, this can be explained by random search, for example flying in an increasing spiral until familiar territory is encountered and then returning directly. In others, there is a uniform tendency, within the species, to head always in one direction, no matter what the true home direction. This behavior may be adaptive in leading them to a shore which can then be followed to familiar territory. In any case, it implies at least the ability to use natural cues to know compass directions. In many instances, however, there is true navigation: A high percentage of the animals released return in a short time, indicating that they took approximately the most direct route.

The phenomenon has been best studied in a few species of birds, notably homing pigeons, European warblers, and some North Atlantic shore birds. For example, of 42 Manx shearwaters released at 300–400 miles distance in several directions from their home colony near Wales, 90% returned at a typical speed of 200 mi/day. One of the major difficulties in the study of homing is the lack of information about what route the animal took. Some successful attempts have been made to follow homing birds with airplanes, but most observations are restricted to release time, initial heading, and recapture time. Fortunately, it has been found that certain birds, in cages, will show intention movements or preferences for certain parts of the cage that are in the home or migratory direction, making study of the phenomenon somewhat easier.

Celestial visual cues are strongly implicated as sources of information for homing and migration. In almost all cases studied, diurnal homing animals (many birds, bees, turtles) need to be able to see the sun or a patch of clear sky. Even for simple determination of compass direction, however, it is necessary to know not only the direction of the sun but the exact time. That the "internal clocks," or circadian rhythms, of the birds are used in this way has been demonstrated by resetting their clocks (for example, by 6 hr) with artificially displaced day-night cycles, and showing that the homing or migration headings are displaced by a corresponding amount (90°). "Sun-compass orientation" thus requires knowing the direction of the sun and accurately knowing the time.

This information is still not sufficient for homing, for the sun, at any given altitude, could be moving along different arcs, depending upon the bird's longitudinal and latitudinal displacement from home. It has been proposed, by Matthews in England, that homing birds, after watching the sun for a few minutes (no longer is required, for they have made the correct choice within a few minutes), can extrapolate the sun's movement to determine its full arc. This, when compared with the known arc in the home location, tells the bird how far to the north or south it has been displaced (by altitude at the highest point on the arc), and how far east or west (by the position along its arc at a given time). The difficulty is explaining how a bird, or any other animal, can accurately obtain this information when it has only a few minutes (or less) to observe the sun and the displacement may be only a few miles.

In addition, recent research indicates that birds can maintain correct migration headings even when flying at night through dense cloud cover, and experienced homing pigeons have been found to be able to "home" without being able to see the sun or blue sky. It appears that some other cue, or combination of cues, must be available in such cases. Several other sensory cues have been postulated: inertial guidance, use of the Coriolus effect (essentially a miniscule difference in a bird's weight, depending on whether it is flying west or east, and at what velocity), use of constant air currents or gradients of thermal radiation, use of low-frequency acoustic cues, and use of the earth's magnetic field. Of these, perhaps greatest interest now is in the possible use of magnetic fields. To date, using classical or operant conditioning techniques, it has not been possible to show detection by birds of the presence, absence, or change of magnetic fields. On the other hand, experienced homing birds that can orient even on sunless days have been shown to be disoriented when strong magnets are fixed to their backs. Moreover, some birds that show migration headings in cages have been shown to change their headings in a predictable way when the surrounding field is changed.

These findings are made more believable by the very convincing evidence that elasmobranch fishes (see Sections 10.3.7 and 10.4) and honey bees are able to detect and utilize the earth's magnetic field. Lindauer and Martin, using a series of very ingenious experiments, have shown that both the directional information transmitted from one bee to others in their "dances" (see Section 11.2) and the angle at which they orient cells in the combs they build can be altered by altering the magnetic field. The receptor organ is not known in this case, but both bees and pigeons (as well as certain bacteria and molluscs) have now been shown to have in their bodies small amounts of the magnetic material, magnetite (Fe_3O_4). The precise location and relationship to any nervous structures has yet to be determined, but it seems likely that there are receptors capable of monitoring the orientation of arrays of these molecules, or—by analogy with elasmobranchs and electric fish—of detecting the minuscule electric currents resulting from moving charged structures through the earth's magnetic field.

As was implied previously, many birds migrate at night. Apparently they can either use specific constellations of stars as guides to orientation or detect the axis of rotation of the night sky to determine compass direction. It is claimed that even hand-reared birds previously unexposed to the night sky can make use of star patterns in their first tests, implying highly specific, genetically built-in pattern recognition. In other cases, however, it appears that such cues are learned during initial experience.

Although bats have been much less studied than birds in this respect, certain species show both migration and homing behavior. It seems highly unlikely that their vision is adequate for celestial navigation by the stars, but it seems even less likely that echolocation can provide the necessary information.

ADDITIONAL READING

GENERAL

Brown, J. L., Jr. *Sensory Systems.* In J. R. Brobeck, (ed.). *Physiological Basis of Medical Practice*, 9th ed. Williams & Wilkens Co., Baltimore, 1973.

Burtt, E. T. *The Senses of Animals,* Wykeham Publications, London, 1974.

Case, J. *Sensory Mechanisms,* The Macmillan Company, New York, 1966.

Handbook of Sensory Physiology. Several volumes in continuing series, for example, volume I by W. R. Lowenstein (ed.)., *Principles of Receptor Physiology,* Springer Verlag, New York, 1971.

Marler, P., and W. J. Hamilton, III. *Mechanisms of Animal Behavior,* John Wiley & Sons, New York, 1966.

Mellon, D. *The Physiology of Sense Organs,* Oliver and Boyd, London, 1970. *Sensory Receptors,* vol. 30. Cold Spring Harbor Symposia on Quantitative Biology, Cold Spring Harbor, NY, 1965.

CHEMICAL

Adler, J. "Chemoreceptors in bacteria," *Science,* **166,** 1588–1597 (1969).

Amoore, J. E. *Molecular Basis of Odor,* Charles C. Thomas, Springfield, IL, 1970.

Dethier, V. G. *The Physiology of Insect Senses,* John Wiley & Sons, New York, 1963.

Moncrieff, R. W. *The Chemical Senses,* 3rd ed., Leonard Hill, London, 1967.

Pfaffman, C. (ed.). *Olfaction and Taste III,* The Rockefeller University Press, New York, 1969.

MECHANICAL

Busnel, R.-G., and J. F. Fish (eds.). *Animal Sonar Systems,* Plenum Press, New York, 1980.

Catton, W. T. "Mechanoreceptor function," *Physiol. Rev.,* **50,** 297–318 (1970).

Dijkgraaf, S. "The functioning and significance of the lateral line organs," *Biol. Rev.,* **38,** 51–106 (1963).

Griffin, D. R. *Listening in the Dark,* Yale University Press, New Haven, 1958.

Grinnell, A. D. "Comparative physiology of hearing," *Ann. Rev. Physiol.,* **31,** 545–580 (1969).

Gulick, W. L. *Hearing,* Oxford University Press, New York, 1971.

Loewenstein, W. R. "Facets of a transducer process." In *Sensory Receptors,* vol. 30, pp. 29–34, Cold Spring Harbor Symposia on Quantitative Biology, Cold Spring Harbor, NY, 1965.

Roeder, K. *Nerve Cells and Insect Behavior,* Harvard University Press, Cambridge, MA, 1963.

Stein, R. B. "Peripheral control of muscle," *Physiol. Rev.,* **54,** 215–243 (1974).

Von Bekesy, G. *Experiments in Hearing,* McGraw-Hill Book Company, New York, 1960.

———, and W. A. Rosenblith. "The mechanical properties of the ear." In S. S. Stevens (ed.), *Handbook of Experimental Psychology,* John Wiley & Sons, New York, 1951.

Yost, W. A., and D. W. Nielsen. *Fundamentals of Hearing,* Holt, Rinehart, & Winston, New York, 1977.

ELECTROMAGNETIC

Bennett, M. V. L. "Comparative physiology: electric organs," *Ann. Rev. Physiol.,* **32,** 471–528 (1970).

Dowling, J. E., and F. S. Werblin. "Organization of retina of the mudpuppy, *Necturus maculosus.* I, Synaptic structure; II, intracellular recording," *J. Neurophysiol.,* **32,** 315–355 (1969).

Galler, S. R., K. Schmidt-Koenig, G. J. Jacobs, and R. E. Belleville (eds.). *Animal Orientation and Navigation,* National Aeronautics and Space Administration Publication S-606, Washington, D.C., 1972.

Griffin, D. R. "The physiology and geophysics of bird navigation," *Quart. Rev. Biol.,* **44,** 255–276 (1969).

———. *Bird Migration,* Science Study Series, Doubleday-Anchor, New York, 1964.

Hubbell, W. C., and M. D. Bownds. "Visual transduction in vertebrate photoreceptors," *Ann. Rev. Neurosci.,* **2,** 17–34 (1979).

Hubel, D. H., and T. N. Wiesel. "Receptive fields, binocular interaction and functional architecture in the cat's visual cortex," *J. Physiol.,* **160,** 106–154 (1962).

———. "Sequence regularity and geometry of orientation columns in the monkey striate cortex," *J. Comp. Neurol.,* **158,** 267–294 (1974).

Kaneko, A. "Physiology of the retina," *Ann. Rev. Neurosci.,* **2,** 169–191 (1979).

Maturana, H. R., J. Y. Lettvin, W. S. McCulloch, and W. H. Pitts. "Anatomy and physiology of vision in the frog *(Rana pipiens),*" *J. Gen. Physiol.,* **43,** 129–175 (1960).

Toyoda, J., H. Nosaki, and T. Tomita. "Light-induced resistance changes in single photoreceptors of *Necturus* and gekko," *Vision Res.,* **9,** 453–463 (1969).

Van Essen, D. C. "Visual areas of the mammalian cerebral cortex," *Ann. Rev. Neurosci.,* **2,** 227–263 (1979).

Walls, G. L. *The Vertebrate Eye and Its Adaptive Radiation, Bulletin 19,* Cranbook Institute of Science, Bloomfield Hills, MI, 1941.

Alan D. Grinnell

11

USE OF INFORMATION: CENTRAL COORDINATION

11.1 INTRODUCTION

Previous chapters have dealt with the coding of sensory information and its transmission to the central nervous system (CNS). There it is processed and, if appropriate, action is taken. This action may be of several sorts: simple reflexes, the release of complicated but still automatic behavior patterns, or conscious perception, decision making, and learning. The sensory information in all cases is the same, but it is sent simultaneously through many different pathways. Thus a burst of spikes in a cutaneous pain receptor may do the following:

1. Through interneurons activate motoneurons controlling flexor muscles to help jerk the hand away from a hot stove.
2. Excite interneurons in the spinal cord that in turn inhibit motoneurons controlling antagonist muscles.
3. Send impulses through axon branches to other levels of the spinal cord and the cerebellum to trigger the coordinated activity of interneurons and motoneurons controlling arm and trunk muscles involved in the withdrawal movement and subsequent shaking of the hand.
4. Initiate autonomic reflexes.
5. Send impulses to cortical areas where the stimulus is perceived as painful and integrated with visual or other inputs to provide a conscious understanding of why the sensation has arisen and perhaps a memory that will prevent its happening again.

Conversely, a single motoneuron, controlling a certain group (motor unit) of muscle fibers, receives excitatory and inhibitory input from many cells within the same segment of the spinal cord, from cells elsewhere up and down the cord, from the cerebellum, from several brainstem nuclei, and from the motor cortex: more than 5000 synaptic inputs on the average.

Each successive level of sensory-to-motor processing requires more neurons with greater versatility of action. The most complex activity, conscious interpretation, is probably restricted to higher vertebrates. Indeed, it seems a safe generalization to say that differences in behavior of animals are much more the result of central processing than of differences in peripheral sensory organs. An equally valid principle, however, is that if something can be done automatically by built-in circuits, it will be. By eliminating conscious control, action is made faster and fewer cells are required, freeing the rest for more difficult problems. We would be in serious trouble if we were forced to consider every move before we made it. On the other hand, our relatively enormous nervous systems permit a degree of conscious control lacking in simpler animals.

Of the total 10^{11} or more neurons in the human brain, only about 0.02% are either sensory or motor neurons, directly involved in bringing information to the brain or instructions from the brain. The rest (99.98%) are interneurons in the spinal cord and brain—extracting information, processing it, making decisions, "learning," and "remembering."

11.2 EVOLUTION OF CENTRAL NERVOUS SYSTEM

With the evolutionary development of bilateral symmetry and a distinct "front end," animals came to possess a preponderance of exteroceptors and associated analyzing ganglia at that end. This phenomenon of *cephalization,* although apparent even in flatworms, is first seen in advanced form of annelids, where most activity is still reflexly governed within each segment by a cluster of nerve cells (ganglion), but the ganglia are connected and in many ways coordinated by the larger ganglia at the head end (Figure 11-1). All ganglia have a large proportion of unipolar cells, the axons and dendrites of which mingle in a dense central neurophile where all synaptic contacts are made. It is almost impossible to trace pathways histologically or electrophysiologically in such a matrix, so little is known about the functional organization of these ganglia, although major advances are now being made through the use of fluorescent dyes injected into single neurons.

The dominance of cephalic ganglia is increasingly strong in arthropods, cephalopod mollusks,

Figure 11-1
First stages in cephalization. (a) A distinct concentration of nervous tissue is seen in the flatworm's dorsal nerve plexus, but the development of a central nervous system is much more pronounced in annelids (b). Shown here is the dorsal view of the ventral nerve cord (Vnc) of the common earthworm, Lumbricus, *in segments I to VI. Code: Cg, cerebral ganglion; Cpc, circumpharyngeal connective; Sg, subpharyngeal ganglion. [(a) from D. Hadenfeldt. Z. Wiss. Zool.,* **133**, *586–638 (1929); (b) W. N. Hess, J. Morphol.,* **40**, *235–259 (1925).]*

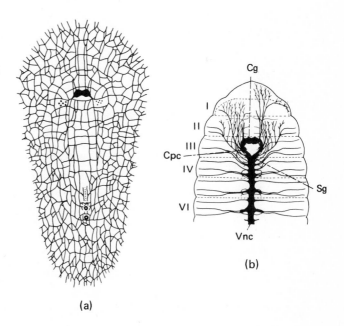

and chordates, allowing increasing flexibility and, in some cases, modification of response depending on an animal's previous experience and current physiological state. Probably the most highly developed invertebrate brains are those of hymenopteran insects and cephalopod mollusks. Honeybees are capable of assuming different behavior patterns, as needed, in the hive, and must integrate many forms of sensory information (sun-angle relative to food source, distance to source, qualitative "goodness" of source, and elapsed time) to communicate with other bees through their complicated *schwanzeldanzen,* so well studied by von Frisch. Stimulation and localized lesions of such insect brains have revealed groups of premotor neurons controlling most motor activities, largely by changing levels of tonic inhibition. Cephalopod

mollusks have extraordinarily well developed brains and flexible behavior, exhibiting a degree of learning and long-term memory not found in any other invertebrates. Implicated in this behavior is the dense neuropile of the verticle lobe, damage to which destroys much of the animals' ability to retain a memory (Figure 11-2).

Vertebrate CNSs were derived from the same beginnings and follow the same principles, although the organization is considerably changed (see Figures 11-3 to 11-5). Again there is a progressive emphasis on head ganglia, with a corresponding loss of independence of action by the spinal ganglia. The behavior of fish and amphibians, although less stereotyped than that of insects, is still quite restricted and only slightly modifiable. Nevertheless, reflexes operating within the spinal

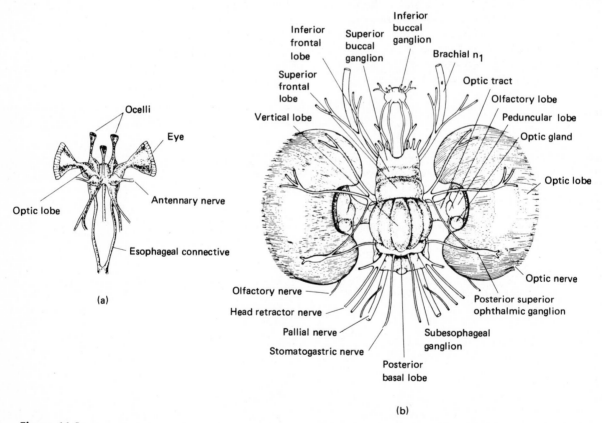

Figure 11-2
Brains of (a) an insect (locust) and (b) an octopus. The optic lobes of both are predominant, but other ganglia perform important integrative functions (see the text). [(b) from J. Z. Young. Biol. Rev., **36,** *32–96 (1961).]*

cord may be extremely complicated and adaptive. It is no wonder that early observers attributed consciousness to the spinal cord of a headless frog. Such an animal can maintain its body in a normal position and purposefully bring its leg up to wipe off its skin a piece of paper soaked in weak acetic acid, using the leg on the other side if the closer leg is held.

Sensory nerves, on entering the cord, branch many times (Figure 11-3). Some end directly on motoneurons or, more often, on chains of interneurons leading to motoneurons in the same and adjacent segments, where they help coordinate relatively simple reflexes. Other branches ascend the spinal cord to the brain via two major interneuronal pathways, one providing precise information about location and type of stimulation for higher level analysis, the other apparently converging with input from other modalities to form a good alerting system, but one that discards a great deal of information. At several levels in these interneuron chains there is controlling input from the brain, usually inhibiting responses. In this way patterns of peripheral sensitivity can be changed both in size of receptive fields and in sensitivity to different stimulus modalities.

Most (but not all) axon collaterals that ascend to the brain by way of the different spinal-cord tracts decussate at one of several levels and enter the opposite side of the brain. Such decussation is seen throughout the vertebrates, and even in invertebrates, and is paralleled by decussation of the returning motor fibers. To date there has been no acceptable, or even attractive, hypothesis to account for this decussation.

The size of the spinal cord depends largely on the complexity of movement and sensory input of trunk and limbs, and the greatest concentrations of neurons, predictably, are in the regions of the limbs, especially the hind limbs. In most tetrapods there is a prominent pelvic enlargement of the cord, which reached such a degree in dinosaurs that it has been dignified with the descriptive name *pelvic brain.*

The vertebrate CNS differs sharply from that of invertebrates in having all cell bodies within or near the central ganglia. (The only exceptions to this are the postganglionic cells of the parasympathetic nervous system, discussed in Section 11-4). The whole system is of extreme importance and delicacy, enclosed within bony chambers lined with a cushion of cerebrospinal fluid. This fluid also fills

Figure 11-3

Vertebrate spinal cord. Transverse section showing grey and white matter divisions and (left side) one of the simplest and most conspicuous reflex pathways. Sensory input from a flexor muscle stretch receptor monosynaptically excites flexor motoneurons and, via interneurons, inhibits extensor motoneurons. The same sensory information also passes to higher and lower segments and to the brain via the fiber tracts of the white matter. The right side of the diagram shows some of the principal fiber tracts and the information they carry (see also Figure 11-9). Sensory input to the cord is via the dorsal, or posterior, roots; motor outflow is via the ventral, or anterior, roots. The ratio of sensory nerve fibers to interneurons to motoneurons is approximately 2:50:1.

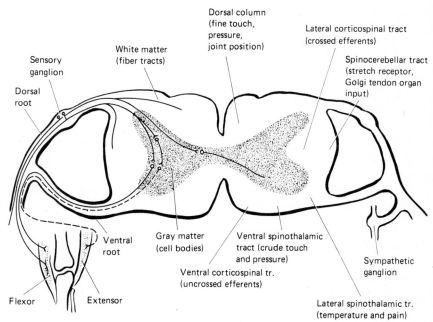

chambers within the spinal cord and brain and forms the extracellular medium for CNS cells. Mechanical isolation of the CNS is of obvious practical value because nerve cells, unlike other body cells, do not continue to develop into adult life. Each cell that is lost due to mechanical injury, accumulation of toxins (such as ethyl alcohol), or lack of proper oxygenation or nutrient substrate, is lost for good. In the adult human, even without any special source of injury, it has been estimated that thousands of central neurons die each day. This is a small number, compared with the total 10^{11} or more, but if the wrong cells are lost the damage might be irreparable, leading to loss of motor coordination and perhaps senility.

The brain is also protected by a much more carefully regulated chemical environment than the rest of the body tissues enjoy. The capillary endothelial cells are much less permeable in the brain than elsewhere, forming a *blood-brain barrier* that effectively excludes a variety of toxic substances and small biologically potent molecules, such as amino acids and small peptides, that are normally present in the blood. Any of these substances (including glucose) that are needed in the brain appear to be actively transported across the blood-brain barrier. This is an energy-expensive process, but permits rigid regulation of the fluids surrounding each neuron. Only a few specialized parts of the brain are unprotected by this barrier; these are areas involved in monitoring the composition of the blood or the levels of blood-borne hormones.

11.3 VERTEBRATE BRAIN

11.3.1 Structure

Although there has been a progressive increase in cephalic control of the vertebrate spinal cord, the cord has changed little in structure. The brain, however, has undergone dramatic changes (Figure 11-4). The primitive vertebrate brain can easily be viewed as having three divisions: hindbrain, midbrain, and forebrain. The main part of the hindbrain (the *medulla*) is simply the enlarged end of the spinal cord. Through it pass all sensory nerves, except those of vision and smell, and fibers controlling almost all motoneurons (Figure 11-5). Many

sensory nerves synapse in the hindbrain to provide information necessary, especially, for proprioceptive control of balance and movement and for simple auditory reflexes. Particularly important to these functions are inputs from the vestibular and cochlear nerves. In addition, there are nuclei controlling respiration, swallowing, vomiting, blood pressure, and heart rate. Much of the central control of equilibrium and movement takes place in the *cerebellum*, an outgrowth of the medulla. Unlike the medulla, the cerebellum increases immensely in size and importance in the course of evolution of fast-moving birds and mammals.

The midbrain has changed little in size with vertebrate evolution, but it has changed in function. In fishes and amphibians it mediates the animals' most complex behavior. In particular in submammalian vertebrates a dorsal enlargement (the *tectum*) receives most of the input from the optic nerves as well as projection from hindbrain sensory nuclei and serves as the principal integrating area of the brain. With the development of a forebrain center for visual analysis, however, much of the visual input has bypassed the midbrain to go directly to the lateral geniculate body of the thalamus, apparently leaving only such functions as reflex control of extraocular muscles, iris and eyelids in the tectal anterior colliculus. At the same time, however, another midbrain center, the posterior colliculus, assumes an increasingly important role in the analysis and relay of auditory information.

The vertebrate forebrain has two major divisions: the diencephalon and the telencephalon. The diencephalon, consisting of *thalamus, hypothalamus,* and *posterior pituitary,* is extremely important, even in the most primitive vertebrates, as the site of several regulatory centers. The function of the thalamus is little understood in fishes, where it apparently merely connects olfactory input to the midbrain. In higher vertebrates, however, it is much enlarged and becomes an integrative center for input from all the sensory systems. It also becomes progressively more important as a relay station for channeling sensory information to appropriate parts of the cerebral hemispheres, and channeling motor information in the other direction to the spinal cord. Here, particularly clearly, one can distinguish the two parallel sensory pathways mentioned in our discussion of the spinal cord: the nonspecific and the specific, or epicritic, pathways.

Figure 11-4

Evolution of the vertebrate brain. Lateral views of the brains of (a) a codfish, (b) frog, (c) alligator, (d) goose, and (e) horse. Cranial nerves are marked I to XII. Note especially the relative enlargement of the cerebellum and cerebrum. [From A. S. Romer. The Vertebrate Body, *3rd ed., W. B. Saunders Company, Philadelphia, 1962.]*

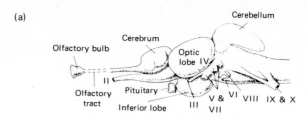

(a)

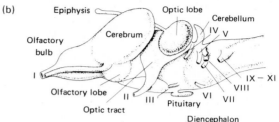

(b)

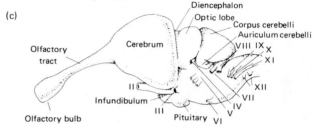

(c)

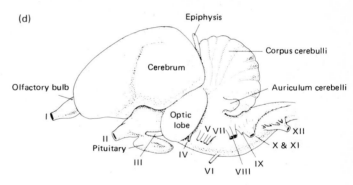

(d)

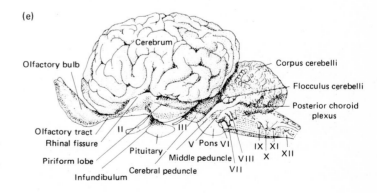

(e)

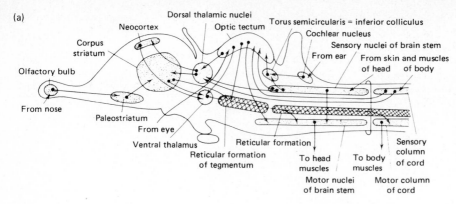

(a)

Olfactory bulb

From nose

Neocortex

Corpus striatum

Paleostriatum

From eye

Ventral thalamus

Reticular formation of tegmentum

Dorsal thalamic nuclei

Optic tectum

Torus semicircularis = inferior colliculus

Cochlear nucleus

Sensory nuclei of brain stem

From ear

From skin and muscles of head of body

Reticular formation

To head muscles

To body muscles

Motor nuclei of brain stem

Sensory column of cord

Motor column of cord

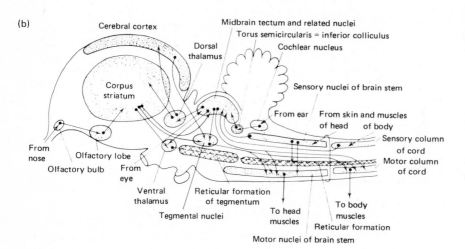

(b)

Cerebral cortex

Corpus striatum

From nose

Olfactory bulb

Olfactory lobe

From eye

Ventral thalamus

Tegmental nuclei

Dorsal thalamus

Midbrain tectum and related nuclei

Torus semicircularis = inferior colliculus

Cochlear nucleus

Sensory nuclei of brain stem

From ear

From skin and muscles of head of body

Reticular formation of tegmentum

To head muscles

To body muscles

Reticular formation

Motor nuclei of brain stem

Sensory column of cord

Motor column of cord

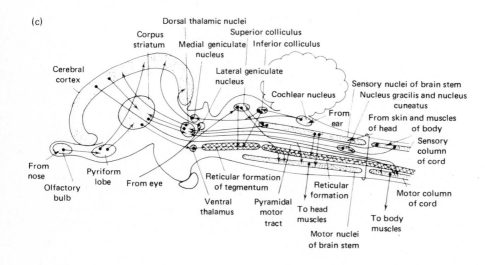

(c)

Cerebral cortex

Corpus striatum

Dorsal thalamic nuclei

Medial geniculate nucleus

Lateral geniculate nucleus

Superior colliculus

Inferior colliculus

Cochlear nucleus

From ear

Sensory nuclei of brain stem

Nucleus gracilis and nucleus cuneatus

From skin and muscles of head of body

Sensory column of cord

From nose

Olfactory bulb

Pyriform lobe

From eye

Ventral thalamus

Reticular formation of tegmentum

Pyramidal motor tract

To head muscles

Reticular formation

Motor nuclei of brain stem

To body muscles

Motor column of cord

Figure 11-5

''Wiring diagrams'' of neural connections in the evolving vertebrate brain. (a) In reptiles, the midbrain tectum is the principal integrating center, receiving input from most sensory modalities and exerting direct control over motor centers. Information is also sent to thalmic nuclei, but their role is still subordinate. (b) In birds, the corpus striatum has become dominant, although the midbrain tectum is still a critical relay center. (c) In mammals, the midbrain is important principally as a relay station for auditory information, and the neocortex of the forebrain has assumed the dominant role in integration of sensory information and motor control. Medullary and midbrain nuclei change relatively little in the three groups. [From A. S. Romer. The Vertebrate Body, 3rd ed., W. B. Saunders Company, Philadelphia, 1962.]

The nonspecific pathway is apparently designed for general alerting functions and for setting of overall levels of activation, with no accurate topographic localization, large receptive fields inhibited by even larger fields, and often more than one sensory modality represented by a single neuron. The specific pathway is characterized by accurate topographic maps, small receptive fields, localized inhibition, and restriction to one sensory modality. The information carried by this pathway mostly is processed within the cortex.

More ventral in the diencephalon is the hypothalamus. This collection of nuclei is involved in many of the regulatory functions of the body mentioned in other chapters. It contains neurosecretary cells that produce the hormones released from the posterior pituitary, and others that control the anterior pituitary. It also contains cells that sense and regulate deep body temperature and that help control osmoregulation. Stimulation of certain cells in the hypothalamus can elicit thirst, hunger, rage, sexual drive, "pleasure," or pain. Clearly this part of the brain, little changed throughout the vertebrates, is one of the highest developed primitive integrative areas, controlling many forms of complex instinctive behavior.

The remainder of the forebrain, the telencephalon, has—of all brain areas—undergone the most significant changes in the course of vertebrate evolution. In fishes and amphibians, it is little more than an olfactory center, receiving input from the olfactory bulbs and sending it, for detailed analysis, to small swellings on top of the forebrain, the cerebral hemispheres. In reptiles and birds, this area remains as an olfactory lobe and a center known as the *amygdala.* In addition, there is an elaboration of nervous tissue dorsal and lateral to this, forming the striatum (lateral) and pallium (dorsal). The striatum assumes particular importance as the locus of highest levels of integration of sensory inputs and motor responses; for example, in coordinating instinctive behavior and conditioned reflexes. The striatum of birds is particularly enlarged, approaching the relative proportion of the cerebral cortex in higher mammals. The pallium forms centers that are of still largely unknown significance to reptiles and birds (olfactory cortex, septum, hippocampus) but are known to play important roles in mammals. The hippocampus is clearly involved in short-term memory processing, and both hippocampus and septum interact with the hypothalamus, the olfactory bulb, and the amygdala in what is known as the *limbic system.* This system is important in mammals in controlling emotions, moods, and desires. Lesions or electrical stimulation in different parts of these centers, especially the amygdala, can cause intense pleasure, rage, fear (or lack thereof), excessive eating, or near-fatal lack of appetite. In a sense, this might be said to be a system critical to defining one's personality and sense of self.

Beginning in the monotreme mammals there is a progressive increase in size and importance of the bilateral *cerebral hemispheres,* growing out of the pallium. These structures expand in all directions, covering the original forebrain, the midbrain, and in some cases even the cerebellum. At the same time, they acquire even greater dominance over lower centers. This new structure, the cerebral cortex, consists of a thick surface layer of cell bodies (gray matter) below which fiber tracts (white matter) connect various parts of the cortex to each other and to other points in the CNS. A particularly prominent tract is the corpus callosum, connecting the two hemispheres. The function of different parts of the cortex, like that of other areas of the brain, has been explored in several ways: by ablation of restricted areas, followed by observations of changes in behavior; by recording activity in response to natural sensory stimulation; by electrical and chemical stimulation and observation of motor responses (or conscious reports of effects in human subjects); and, finally, by histological tracing of fiber pathways, especially in degeneration studies. It is found that in mammals, unlike lower vertebrates, there are discrete cortical projection areas for sensory input and lower areas that control specific motor functions. Within these areas the gray matter is arranged in columns of cells, each devoted to sensation or motor control of a restricted part of the body, with adjacent columns devoted to adjacent body regions—the result of maintained topographic representation at lower neural levels (Fig. 11-6). As cerebral control grows, the size of the sensory projection and motor areas grows, expanding more in area than in thickness. To provide the necessary area, the surface of the brain becomes convoluted into deep folds. There are about 10^5 neurons/mm² of surface of the cortex in man, and throughout the neocortex there is a similar segregation of cell types and interconnections into six

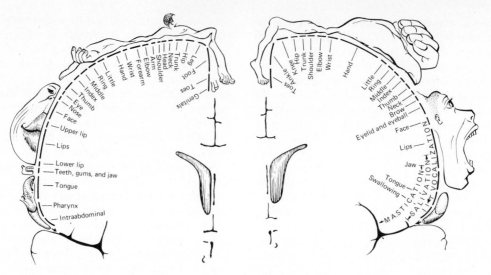

Figure 11-6

Maps of sensory (left) and motor (right) "homunculi" in the primary sensory and motor areas of the human cortex. The approximate portion of the area devoted to each part of the body is shown. Note the great importance of the lips and face in sensory function, and the disproportionate emphasis on the thumb and facial muscles in the motor projection area. [From W. Penfield, and T. Rasmussen. The Cerebral Cortex of Man, *The Macmillan Company, New York, 1957.]*

layers. The high degree of organization of the cortex has already been described in the context of the best studied part of it, the visual cortex (Section 10-6-3).

In primitive mammals, virtually the whole surface (which is relatively smooth) can be assigned some sensory or motor function. A given area either receives input from some sense organ or exercises control over a specific motor response. In higher mammals, however, and especially in primates, there come to be large unassigned areas of the brain, areas that do not appear to have an immediate sensory or motor function. Figure 11-7 shows the relative proportions of assigned and unassigned

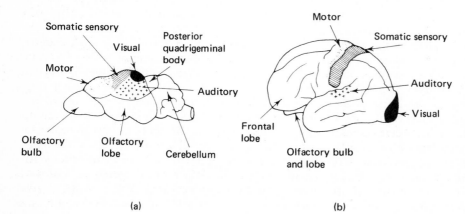

(a)

(b)

Figure 11-7

Direct sensory and motor projection areas on the brain surface of (a) a primitive mammal, a shrew; and (b) man. In the shrew, virtually all the cortical surface is assigned, that is, it has a direct sensory or motor function. In man, most of the cortical surface is unassigned, having no direct sensory or motor function. [From A. S. Romer. The Vertebrate Body, *3rd ed., W. B. Saunders Company, Philadelphia, 1962.]*

areas in representative mammals. These unassigned areas are clearly correlated with the development of versatility of response and intelligence. They serve as intersensory association areas and have many functions not yet understood. Most of the large frontal lobe, for example, has such complicated functions that the best explanation of its role is that it helps determine personality. It is so noncritical to even complicated sensory and motor functions that it was once popular to remove part of it (frontal lobotomy) to alleviate such problems, as pathologic aggressiveness, anxiety, obsessions, and schizoprenia.

An interesting aspect of neocortical function is the localization of certain complex functions to only one side of the brain. Language is perhaps the best example of this. There are two special areas, almost always restricted to the left side of the brain, that are involved specifically in verbal communication. Broca's area, located just in front of the motor cortex on the lateral side of the hemisphere, governs the ability to talk. Damage to it causes slow, ungrammatical speech and writing, although comprehension of the spoken or written word can be perfectly normal. Surprisingly, such a handicapped person can often sing familiar songs perfectly well. Between the somatic sensory cortex and visual areas of the cortex, just above the auditory cortex, is Wernicke's area; damage to this portion leads to a loss of comprehension of spoken and written words. The motor skills of speech are perfectly normal, but the spoken language is without meaningful content. Anatomical differences between the two hemispheres are correlated with these specialized left-localized functions. (Interestingly, many song birds show left hemisphere localization of brain centers for singing.) Another remarkable localization of function, in humans, which is bilateral in nature, is the recognition of faces. Damage to an area on the underside of both occipital lobes causes no other obvious deficit than the inability to recognize visually a familiar person. Voice recognition is still perfectly normal, and patients can still read and name familiar objects, just not people! A great deal of study is now being devoted to the asymmetry of structure and function in the human brain. Although this research is still in its infancy, it is felt by many that, whereas the left side of the brain is specialized for verbal and analytical tasks, the right is more important for spatial perception, mu-

sical abilities, and perhaps other skills viewed as "artistic" and "intuitive."

11.3.2 Motor Function

Just as sensory analysis is shifted and improved as higher neural centers evolve to perform the function, motor activity undergoes similar refinement. Existent reflexes and motor nuclei remain, but new centers evolve that assume control over them. It is reported that only 28 distinguishable movements can be elicited by electrical stimulation of the motor cortex of monotremes, compared with several hundred in primates. Particularly conspicuous in higher mammals are the huge corticospinal (or pyramidal) tracts. These arise in the motor cortex, partially as axons of the giant Betz cells, and pass directly to spinal motoneurons. Conscious control is so exact that with practice single motoneurons can be voluntarily controlled. Most other cortical motor outflow is interrupted in ganglia at the base of the forebrain (basal ganglia) or in deeper nuclei and reaches the spinal motoneurons through extrapyramidal tracts. Much of this control is governed by the γ motoneuron—muscle-spindle servomechanism discussed previously (Section 10-3-5). Just as most sensory input decussates to the contralateral side of the brain, motor outflow crosses back in the brainstem or spinal cord.

It is essential that movement be smooth and graded in intensity. The mechanism whereby this is achieved in vertebrates is quite different than that used by invertebrates. Vertebrate skeletal muscle fibers are controlled in all-or-none fashion by motoneurons. In most cases, a single motoneuron provides the sole innervation for several tens or hundreds of muscle fibers (a *motor unit*). When the motoneuron fires, these muscle fibers generate action potentials and contract in all-or-none fashion. Increasing the frequency of motoneuron firing increases the tension over a certain range by causing a tetanus in the fibers of that motor unit.

Vertebrate motoneurons are called the "final common pathway" because all integration happens before they fire. They receive and integrate excitatory and inhibitory inputs, and they alone send commands peripherally. Smooth movement is achieved by asynchronous activation of many or few motor units in a muscle. Recent studies have shown that there are motoneurons of different sizes and that these have different-sized motor units. The smallest motoneurons innervate the fewest muscle fibers, the largest innervate very large populations. With increasing central excitatory input to a pool of motoneurons innervating a given muscle, the smallest reach threshold first, the largest last. Consequently, very fine movements or the initiation of strong contraction involve only the small motor units, with which control is most precise. The largest, which by themselves could cause jerky movement, are brought in only when most of the rest of the muscle is active and their large contribution is superimposed on an already strong contraction.

Invertebrate muscle fibers receive multiple innervation: usually two or more excitatory nerves and one or more inhibitory nerves, which may also innervate most of the other fibers in the muscle. These muscle fibers do not produce action potentials but rather contract in proportion to their depolarization. Integration and graded activity is achieved peripherally.

A most important regulation of vertebrate central motor outflow is effected by the cerebellum, which integrates the necessary proprioceptive information about muscle and limb position with central output and, on the basis of this information, modifies outgoing commands to insure coordinated movement. The cerebellum is remarkably simple in construction, with only two types of inputs (one diffuse in its connections, the other restricted and specific, both excitatory) and five distinguishable cell types, only one of which (the Purkinje cell) sends an output from the cerebellum. This output is entirely inhibitory on subcerebellar nuclei. The Purkinje cells are gigantic cells with vast and uniformly developed dendritic trees on which more than 200,000 synapses are found per cell. The other cerebellar cell types, arranged in a highly ordered array around the Purkinje cells, appear primarily to provide restricted excitation and sharp lateral inhibition, shaping the localization within the cerebellum of an excitatory input. Because of its simplicity of organization and close association with well-known spinal circuits, the cerebellum has been the object of intense electrophysiological study. It and the visual cortex are now probably the best understood major divisions of the vertebrate brain.

Although the cerebellum is generally viewed as a motor coordination center, there is growing appreciation for its potential role in processing of sensory information. It seems almost inevitable that this is a major function in many fishes that use weak electrical signals for orientation (see Section 10-4). In some of these animals, the cerebellum is so vastly hypertrophied that it covers the entire surface of the brain and exceeds in size all the rest of the nervous system. In fact, it gives these fish the largest brain/body weight ratio of any animal studied. (This, in turn, illustrates the danger of trying to rank intelligence, or almost any other nervous system property, on the basis of a simple measurement like brain size/body size.)

11.3.3 Sensory-Motor Plasticity

The neural processes interposed between sensory output and motor outflow are almost entirely unknown. It is here that multisensory associations are made, that experience and learning play an important role. In simple animals a great deal of the interpretation of sensory input is probably instinctive. How much of human behavior is similarly determined is a subject of controversy, but it is obvious from experience that a major portion of even unconscious motor skills are largely learned; that is, they must be made smooth and automatic by practice. The judgment of distance, of size, even the simple coordination between hand and eye requires learning. Whatever neural circuits are involved, they remain plastic throughout life, for new motor patterns can be learned at any age—albeit perhaps not with the same facility as during childhood. Moreover, even the most firmly ingrained sensory-motor associations are changeable. If a man is forced to wear prismatic spectacles that displace the visual field laterally, he is at first disoriented, reaching in the wrong direction and misjudging angles by the amount of the displacement. This disorientation persists if the wearer is passively moved about in the environment. If he is free to move himself, however, and obtain feedback information from other senses (especially proprioception) constantly to acquaint him with the discrepancy between what he sees and what he expects to see, then within a short time he can adjust and automatically coordinate his movements in the displaced environment. Such adjustment probably involves the same plastic pathways used in originally acquiring coordination, and in later learning new motor skills.

11.3.4 Reticular Formation

So far we have mentioned only the classical, or primary, sensory pathways. There is, in addition, a secondary pathway, the importance of which we are only beginning to fathom. This is through the reticular formation, a large, seemingly disorganized neuropile extending from the medulla through the base of the midbrain into the thalamus. A large fraction of the cells in the reticular formation receive input from collaterals of afferent fibers of several different sensory modalities over large areas of the body and respond to moderate input on any of the channels. The reticular formation output then goes partially back down the cord, to regulate sensory and motor activity, and partially to the cortex, where it acts, in essence, as a gate to consciousness (Figure 11-8). The latter portion is called the *reticular activating system* because, when its activity is reduced by drugs, appropriate stimulation, removal, or sleep, the animal loses consciousness. When it is excited, by strong sensory input or electrical stimulation, the animal is aroused. Thus even though sensory input is reaching the cortex along its normal (primary) pathways, even during sleep, there is no conscious reaction and probably no assimilation without the tonic, nonspecific input of this diffuse nerve mass. It is noteworthy that electrical stimulation can either cause arousal or sleep, depending on the site of stimulation. Apparently there are discrete but interconnected hypnogenic regions throughout the reticular formation, which are suppressed by arousal activity originating in other parts of the system. Sleep can be induced by the implantation of ACh crystals into the same hypnogenic areas, wakefulness by implantation of atropine. The same drugs are ineffective elsewhere in the formation.

The states of sleep and attentiveness are plainly correlated with the type of slow electrical activity (*electroencephalogram*, or EEG) that can be recorded from the surface of the cortex. During wakefulness there are low-amplitude, high-frequency waves; during sleep, larger, slower waves. Indeed, several different depths of sleep are distinguisha-

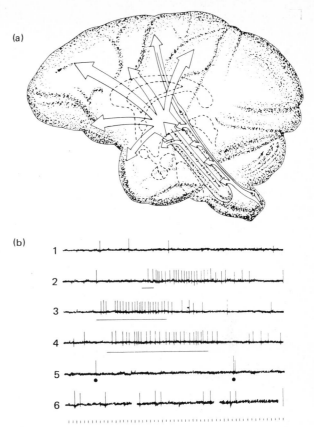

Figure 11-8

Ascending reticular formation. (a) Schematic diagram showing the ascending reticular formation receiving input from collaterals or direct afferent pathways, and projecting principally to cortical association areas. Reticular formation cells are involved in wakefulness and arousal, and are relatively nonspecific in their response to stimuli. As (b) shows, single cells in the cat reticular formation may respond to stimuli of several different modalities. (1) Spontaneous discharge; (2) tapping of ipsilateral forearm during time indicated by the bar under the trace; (3) similar rubbing of the back; (4) touching vibrissae; (5) sudden handclaps; and (6) single electric stimuli to the ipsilateral sensory cortex. [(a) From H. W. Magoun, Brain Mechanisms and Consciousness, Charles C Thomas, Publisher, Springfield, Ill., 1954; (b) M. Palestini, G. F. Rossi, and A. Zanchetti. Arch. Ital. Biol., 95, 97–109 (1957).]

ble, by EEG pattern, muscle tone, and eye movement (apparently associated with dreaming). The EEG has proved useful clinically, especially in diagnosing central disorders, such as epilepsy. What gives rise to the EEG, however, is uncertain. Although it is generally felt that these low amplitude extracellular potential changes are a reflection of synchronized activity of populations of neurons, a recent study has suggested that the most prominent and clinically useful component of the EEG, the low frequency alpha rhythm, may not be a record of nervous activity at all. The alpha rhythm was instead interpreted as being a result of eye movements at this frequency, creating a widely recordable potential by shifting of the direct-current polarized eyes in their sockets. Such movement would be comparable to the tremor of skeletal muscles recordable throughout the body. This explanation is by no means accepted by most workers in the field, but it does illustrate the fact that there is little agreement on the origin of this or other EEG components and little understanding of why any of them are correlated with clinical disorders.

11.3.5 Sleep and Rhythms

Sleep itself is entirely mysterious. It is obviously vital in most mammals, but for unknown reasons that will probably be found only after the appropriate biochemical tests are invented. Most mammals exhibit unquestionable sleep, although ungulates, like birds, manage to keep their equilibrium and proprioreceptive nervous systems constantly alert. Reptiles, amphibians, and fishes similarly display states of unresponsiveness that might be termed sleep but, in these animals, as in invertebrates, the phenomenon becomes indistinguishable from the more universal circadian rhythms (see Section 3-9). Almost all animals show daily rhythms of activity, temperature, color, and so on, and can maintain these rhythms for weeks or months with astonishing accuracy even in constant light, temperature, humidity, and acoustic conditions. The rhythms seem to be innate, but can be driven and reset by manipulating light-dark or other environmental factors. Presumably similar fundamental mechanisms are involved in these rhythms as in sleep. Recent findings suggest that a small group of neurons lying just above the optic chiasm, the suprachiasmic nucleus, may control many of these rhythms. Lesions of this nucleus, or of its output, but not the input, disrupt or eliminate many of the circadian rhythms found in mammals.

11.4 AUTONOMIC NERVOUS SYSTEM

In addition to motor control of skeletal muscle and the feedback control of receptors, there is a whole separate category of motor outflow that tonically governs the activity of smooth muscle and glands. This is the autonomic (or visceral, or involuntary) nervous system, mentioned frequently in previous chapters. Remarkably little is known about the autonomic nervous system. Most of our present knowledge is based upon work on man and a few other mammals. As more investigations are done with a wider variety of animals, the organization of the system and even many of the functional properties appear increasingly complex. For a comprehensive review of recent findings and their interpretation, the reader is referred to the excellent review by Burnstock. The description that follows is grossly oversimplified in many respects, but presents many of the more important findings.

In amniote vertebrates, the autonomic system is conspicuously different in organization and function from the somatic nervous system discussed previously, and is easily separable into two discrete functional systems: the sympathetic and the parasympathetic. Together they maintain homeostasis in the organism and effect those rapid changes in visceral function that are required in a state of emergency (see Table 11-1). The sympathetic system predominates when an animal is threatened (in the familiar "fight for flight" situations) and causes such changes as increase in blood pressure, increase in depth of respiration, accelerated heart rate, pupil dilation, unsheathing of claws, increase of blood flow to muscles, and decrease in blood flow to the digestive system. The parasympathetic system predominates in the absence of stress and in most cases has an effect antagonistic to that of the sympathetic system in the same organs. Some exceptions exist: The adrenal medulla, sweat glands, and skin and muscle capillary beds receive only sympathetic innervation, whereas in salivary glands both inputs are excitatory, causing the secretion of different types of saliva.

TABLE 11-1. Partial List of Postsynaptic Actions of Sympathetic and Parasympathetic Nervous Systems.

Structure	Sympathetic Nervous System	Parasympathetic Nervous System
Eye		
Radial muscles of iris	Dilates pupil	—
Iris sphincter muscle	—	Constricts pupil
Ciliary muscle	Relaxation	Contraction (near vision)
Lacrymal glands	—	Stimulates tearing
Salivary glands	Slight, viscous secretion	Profuse dilute secretion
Blood vessels (arterioles)	Mostly vasoconstriction* esp. in muscles, skin	Little effect, some visceral vasodilation
Heart	Increases rate and ventricular contraction force	Decreases rate
Lungs		
Bronchial lumen	Dilates	Constricts
Glands	—	Stimulates secretion
Gastrointestinal tract		
Sphincters	Contraction	Relaxation
Motility and tone	Inhibits	Stimulates
Glandular secretion (incl. gall bladder)	Inhibits	Stimulates
Urinary bladder	—	Contraction, sphincter relaxation
Liver	Increases glycogenolysis and blood sugar	—
Adrenal medulla	Secretes NE and epinephrine	—
Sex organs	Vasoconstriction, ejaculation	Vasodilation, erection
Sweat glands	Stimulated*	—

* Effect mediated by postganglionic release of ACh in some cases.

The organizational distinction between the sympathetic and parasympathetic systems is shown in Figure 11-9. The sympathetic system arises from the thoracic and lumbar segments of the spinal cord, the parasympathetic system from cranial nerves (most importantly the Xth, or vagus nerve) and the sacral spinal cord. Both differ from the somatic motor system in having a synapse interposed between the spinal cord neuron and the effector organ. In the sympathetic system, the spinal (preganglionic) neuron synapses with the postganglionic neuron in one of a chain of interconnected ganglia just outside the spinal cord. The parasympathetic preganglionic fibers normally run all the way to the effector organ before synapsing with a short postganglionic neuron. In the sympathetic system, especially, there is a wide spread of activity by means of the collaterals of stimulated fibers, so that excitation of one autonomic reflex leads to generalized activity of related fibers throughout the body.

The important functional distinction between sympathetic and parasympathetic systems is in the chemical transmitter released at the end organ. In the parasympathetic system this is ACh, in the sympathetic system, norepinephrine (NE) or epinephrine. (Curiously, sympathetic postganglionic nerves innervating sweat glands release ACh.) It has been known for some time that ATP is released from postganglionic sympathetic terminals at the same time as NE. There is now some evidence that in lower vertebrates this (or a related) nucleotide may itself serve as a transmitter. Preganglionic fibers in both systems, like somatic motoneurons, release ACh.

The mammalian autonomic nervous system is evolved from comparable systems in the earliest vertebrates and even most invertebrates. Visceral plexuses, much like those in mammals, are found in the digestive tracts of flatworms and nematodes,

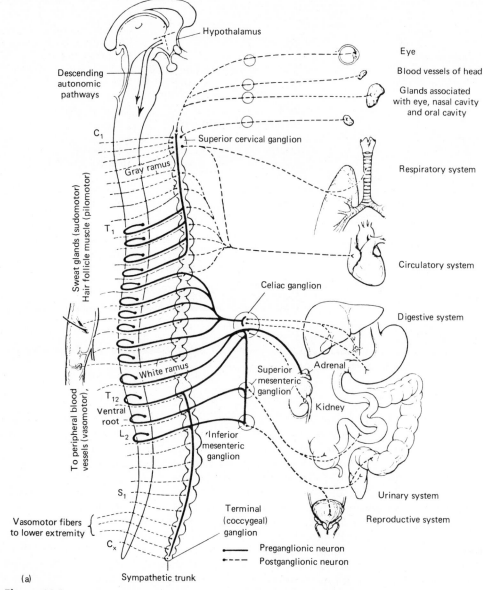

(a)

Figure 11-9

Diagrammatic representation of (a) the sympathetic and (b) the parasympathetic divisions of the autonomic nervous system. [*From C. R. Noback and R. J. Demarest.* The Nervous System, *2nd ed., McGraw Hill Book Company, New York, 1977.*]

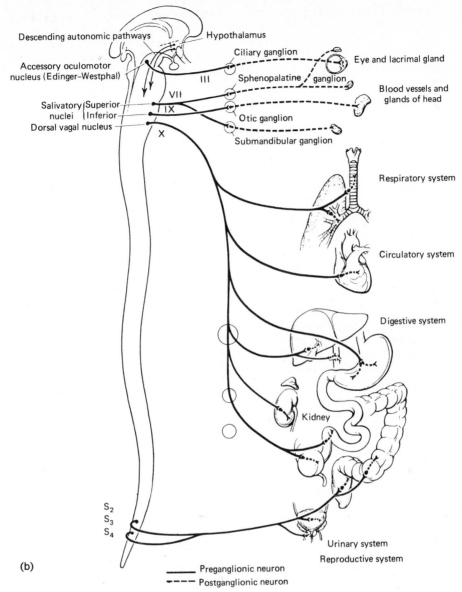

Descending autonomic pathways — Hypothalamus

Accessory oculomotor
nucleus (Edinger–Westphal)

Salivatory {Superior
nuclei {Inferior

Dorsal vagal nucleus

III — Ciliary ganglion — Eye and lacrimal gland

VII — Sphenopalatine ganglion

IX — Blood vessels and glands of head

X — Otic ganglion

Submandibular ganglion

Respiratory system

Circulatory system

Digestive system

Kidney

S_2
S_3
S_4

Urinary system

Reproductive system

(b)

——— Preganglionic neuron
------ Postganglionic neuron

Figure 11-9 (cont.)

and an increasing variety of tissues (heart, glands, chromatophores) in more advanced phyla. ACh and epinephrine are physiologically effective (not always in predictable ways) in most invertebrate groups, although another amine (5-hydroxytryptamine) is widely found and may take the place of the ACh antagonist in several forms. In all cases,

the function of these systems is the same, the maintenance of a balanced internal environment.

Among fishes, there is a wide range of evolutionary changes leading to the higher vertebrate systems. The vagus nerve is by far the most important, perhaps the only, parasympathetic nerve in the most primitive cyclostomes. Other cranial nerves sometimes do, sometimes do not, contain parasympathetic fibers in elasmobranchs and teleosts. Sympathetic fibers are first detectable in lampreys, where they emerge from both dorsal roots and ventral roots. In higher forms they emerge predominantly or entirely in the ventral roots. Throughout the fishes there seems to have been a wealth of experimentation with single, double, or no innervation of given organs, and innervation from different sources. In amphibia, however, the final system is quite well established. It is worth noting, as evidence of the primitive nature of this control system, that excitatory and inhibitory influences are integrated peripherally by the effector structure, much as in invertebrate muscle, and in marked contrast to the central inhibition characteristic of the vertebrate somatic nervous system.

11.5 NEUROPHARMACOLOGY

The development of new histochemical and microbiochemical techniques for studying the chemistry of small parts of cells has made it possible, for the first time, to identify the transmitters at many CNS synapses. (See Table 9-2.) With this has come a much greater appreciation for the role of certain brain nuclei, and the rational development of neuropharmacological tools.

11.5.1 CNS Neurotransmitters

Monoamines. One major finding is that different transmitters are often sharply localized to discrete pathways and centers. For example, many of the brain's NE-containing cells are localized to one small nucleus of the reticular formation, the *locus coeruleus.* The axons of this nuclear branch profusely and project widely, throughout the hypothalamus, cerebellum and forebrain. NE delivered by this system is thought to be instrumental in arousal, maintenance of consciousness, dreaming sleep, motivation, and mood.

Another monoamine transmitter, dopamine, is concentrated in certain cells of the midbrain that project into the forebrain, where they appear to be important in regulating emotional responses, and into the corpus striatum, where they are involved in helping coordinate complex movements.

Degeneration of the dopaminergic fibers gives rise to the motor disorder called Parkinson's disease. This disease can be quite well controlled now by treatment with one of the metabolic precursors of dopamine, L-Dopa, which can be taken up by neurons. Still another monoamine, serotonin, appears to be the transmitter in pathways regulating body temperature, the onset of sleep, and some aspects of sensory perception.

By far the commonest inhibitory transmitter in the brain is the amino acid, γ-aminobutyric acid (GABA), which is produced almost entirely in the brain and spinal cord. As many as one third of the synapses in the brain are thought to use GABA as their transmitter, in most if not all cases acting by increasing postsynaptic conductance to Cl^-. The motor coordination disease, Huntington's chorea, is characterized by a loss of GABA-containing neurons in the corpus striatum. Unfortunately neither GABA nor GABA analogs are capable of crossing the blood-brain barrier, so exogenous treatment has so far not proved possible.

In contrast to GABA, the transmitters NE, dopamine, and serotonin appear not to act directly by causing conductance changes in the postsynaptic membranes. Instead, they interact with specific receptor molecules in the postsynaptic membrane that act, by way of coupled enzymes, such as adenylate cyclase, to increase or decrease the levels of a "second messenger," normally cyclic adenosine monophosphate (cAMP) or cyclic guanidine monophosphate (cGMP), in the postsynaptic cell. This internal messenger activates specific enzymes called protein kinases, which in turn catalyze the phosphorylation of special membrane proteins to change the permeability to specific ions. This is a relatively slow process, suited especially for mediating such long-lasting actions as slow shifts in membrane potential or modulation of intracellular metabolic and synthetic pathways.

Short peptides. In addition to the monoaminergic neurotransmitters, a number of short peptides, ranging in size from 2 to 34 amino acids in length, are known to be produced by neurons and have potent effects on neurons. Some have been known for decades, such as ACTH, vasopressin, oxytocin, angiotensin, gastrin, and cholecystekinin; others have been discovered only recently, such as the hypothalamic releasing factors that govern the release of anterior pituitary hormones (see Sections 11-6-1 and 12-4-3). Perhaps the most exciting development in recent years, however, has been the discovery of short peptides that almost certainly are true neurotransmitters. Prominent among these are the enkephalins (five amino acids, found in the brain) and β-endorphin (31 amino acids, found in the general circulation), which act as endogenous opiates. In fact, they were discovered after it was determined that morphine was exerting its effect in the brain by interacting with specific receptors, and that a morphine antagonist, naloxone, blocked this specific response. If there are specific receptors, what is it that normally interacts with them? The search for this substance turned up the enkephalins and endorphins, which mimic the effects of morphine, but may be very much more potent, and which occur naturally with a widespread distribution. It seems likely that they are involved in the perception of pain, and in setting of mood, and perhaps in the hitherto unexplainable phenomena of acupuncture and hypnosis as well. Another probable short peptide neurotransmitter is substance P, 11 amino acids in length, which is concentrated in primary sensory nerves and is probably released by them. Substance P appears to excite pain pathways, and may be inhibited by presynaptic enkephalin.

11.5.2 Pharmacological Agents

With greater understanding of neurochemistry has come a greater understanding of neural disorders and the development of new pharmacological tools for treating them. We have already mentioned the use of L-Dopa in successful treatment of Parkinson's disease. A variety of other important drugs now in common use are thought to act in the following ways. The antianxiety drugs valium (diazepan) and benzodiazepan act by increasing the effectiveness of GABA at receptor sites throughout the brain. Amphetamines trigger the release of dopamine, which is implicated in arousal and activity of pleasure centers in the brain. (Overused, amphetamines induce symptoms of schizophrenia.) The antischizophrenic drugs chlorpromiazine (Thorazine) and halopuredol (Haldol) bind tightly to dopamine receptors, blocking the effects of dopamine. The antidepressant iproniazid (Marsalid) inhibits the enzyme that degrades NE, thereby re-

sulting in increased levels of NE. Other antidepressants block reuptake of NE and serotonin from synapses, so the transmitter has a prolonged effect. Cocaine probably acts by a similar mechanism. Caffeine and theophylline, from coffee and tea, act through the second messenger system; they inhibit the enzyme that degrades cAMP, and the increased levels of cAMP have different excitatory effects in a number of systems. Mescaline has effects similar to NE and dopamine, whereas LSD and psilocybin act in many ways like seratonin; but both have other actions unlike these normal neurotransmitters that complicate their effects. Neuropharmacology has become one of the most exciting and important areas of neuroscience in recent years, for obvious reasons.

11.6 NERVE FUNCTIONS NOT NECESSARILY MEDIATED BY IMPULSES

The coding, transfer, and integration of information by means of conducted impulses and synaptic potentials is not the only function of the nervous system.

11.6.1 Neurosecretion

There are many nerve fibers whose main function is the release of neurosecretory substances (neurohumors, or neurohormones) into the blood stream to help coordinate the internal chemical environment (see Section 12-5). Although there is some evidence that the release of neurohormones is accelerated by impulse activity within the fiber, this is not yet a clear correlation in most cases. These secretory substances appear to be concentrated in the nerve terminals in *secretory granules,* vesicles much larger than the synaptic vesicles of transneuronal synapses, but very possibly released in similar fashion. In invertebrates such neurohumors either directly or indirectly (via stimulation of secretory cells) control growth, maturation, molting and metamorphisis, heart rate, reproduction, and color changes. In vertebrates comparable functions are mediated by two groups of neurosecretory cells in the hypothalamus: one that releases short polypeptide "releasing factors" into the circulation of the anterior pituitary to either initiate or inhibit the production there of a wide variety of hormones (see Section 11-5-1 and Chapter 12), and another group that sends axons into the posterior pituitary (neurophypophysis) and there releases several known hormones. The chemical structures of a number of the neurosecretory productions of the posterior pituitary have been elucidated. Major parts of the hormones are octapeptides, three amino acids of which are variable (see Chapter 12). These hormones control water balance in many vertebrates and in birds and mammals can affect blood pressure and the activity of certain smooth muscles.

11.6.2 Trophic Effects of Nerve

Nerves also commonly, if not universally, have important "trophic" effects on the structures they innervate, that is, long-term effects that go beyond the immediate response to a transmitter. In some cases this is probably mediated by the neurotransmitter, or by secretion of a substance other than the transmitter that is effective on the postsynaptic cell. In other cases, it appears that the level and pattern of postsynaptic cell activity is responsible for the long-term effects.

During development, sensory innervation is necessary for the differentiation of sensory organs, such as taste buds, fish lateral line organs, specialized skin structures, and muscle spindles. Electric organs of fishes differentiate from normal muscle fibers only after they become innervated. When the nerve supply of the lateral line organs is cut, these receptors degenerate and only reappear after regeneration of the sensory axon. Taste buds, which normally degenerate within a few days of formation, even in the presence of the nerve, are not replaced in its absence. Sensory nerves not normally innervating the tongue apparently cannot induce taste-bud formation. The specificities of taste buds are thought to be determined by the tissue from which they differentiate, however, because crossing of the two nerves that innervate different parts of the tongue (glossopharyngeal and the chorda tympani) results in response patterns that remain characteristic of the part of the tongue innervated. Clearly, in receptor-cell induction by afferent endings the strong trophic influence is exerted in the absence of synaptic transmitter. Another familiar trophic influence of nerve is on

the regeneration of limbs in amphibia, which requires a substantial nerve supply in the epidermal cap, although it does not appear to be important whether the nerves are sensory or motor.

The best known trophic influence of nerve is on somatic muscle. If motor nerves are sectioned or crushed to block their function, the muscle atrophies to as little as 20% of its original weight and volume within a few days or weeks. Protein synthesis is sharply reduced and muscle filaments disappear, among other biochemical changes. (As always, there are exceptions. In certain muscles of chickens, for example, denervation causes hypertrophy instead of atrophy.) Synaptic function fails before any detectable postsynaptic changes occur. Thus, there is first a failure to evoke an EPP, after about 13 hr in a mammal, followed very shortly thereafter by a burst of minEPPs and then disappearance of both. At about 16 hr after denervation, the first postsynaptic effect is seen, a depolarization of 10–15 mV. After about 2 days (in mammals) or 2–3 weeks (in amphibia), a most interesting phenomenon occurs postsynaptically; the entire muscle membrane develops "denervation hypersensitivity" to ACh (Figure 11-10). Sensitivity to ACh applied iontophoretically from a microelectrode tip becomes as great anywhere on the membrane as it normally is at the end plate. This process reverses the effect of initial innervation, when sensitivity to ACh, originally covering the whole fetal muscle membrane, gradually recedes to become localized to the site of innervation.

Similar hypersensitivity of the muscle membrane can be accomplished by simply damaging the fibers, even in segments of nerve-free muscle. Hence the widespread ACh receptor does not literally spread out from the normal end plate. There is evidence that the ACh receptor of denervated muscle is similar, but may not be identical to that at the normal end plate (see Section 9-7). ACh causes a permeability change that displaces the membrane potential toward an equilibrium potential of about −10 mV, just as at the normal end plate, but there are some pharmacological differences, for example, atropine becomes a more effective blocking agent. Moreover, anticholinesterase drugs do not potentiate the effect of locally applied ACh on the denervated membrane, indicating that the newly exposed receptor is not associated with cholinesterase, as it is in the normal end plate. These extrajunctional

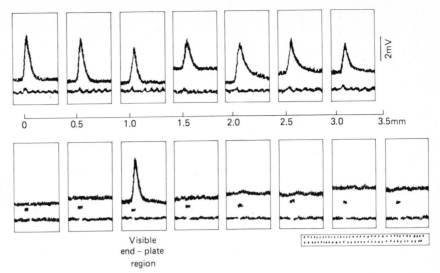

Figure 11-10
Denervation sensitivity. Effects of ACh applied electrophoretically to various points along the length of a cat tenuissimus muscle fiber. (Bottom) In a normally innervated fiber, a response is seen only at the visible end-plate region. (Top) In 14-day denervated muscle fibers, however, the response was approximately equally large at all points on the surface of the fiber. In addition to increased sensitivity to ACh, denervation results in atrophy of most muscles and increased ease of reinnervation. [From J. Axelsson and S. Thesleff. J. Physiol. (London), **149,** *178–193 (1959).]*

ACh receptors appear to be synthesized, *de novo*, after denervation, and show a much more rapid turnover rate than the subsynaptic receptors. Interestingly, in mammals (but not in birds or frogs) the development of extrajunctional sensitivity to ACh is correlated with the appearance of TTX-resistant Na^+ channels in the muscle membrane.

In addition to increasing sensitivity to transmitter, denervation causes changes that facilitate reinnervation, following which there is a gradual retraction of the sensitive area until it is again restricted to the subsynaptic membrane. However, capacity for reinnervation cannot be related directly to ACh sensitivity, for it can occur even when the receptors (both junctional and extrajunctional) are totally blocked with α-bungarotoxin.

The trophic effect of nerve on muscle includes functional properties as well as membrane sensitivity and biochemistry. During development, for example, mammalian muscle gradually increases in contraction speed. Secondarily, certain muscle fibers slow down again. If their nerve supply is cut before this happens, they remain fast. Moreover, if the nerves supplying fast and slow muscles are crossed, so that they regenerate to innervate the wrong muscles, both types partially change character, the fast becoming almost typically slow, the original slow fibers becoming fast. Correlated with this reversal of properties is a parallel change in

the ATPase activity per unit weight of myosin in the two muscle types.

In short, there are many long-term changes in skeletal muscle following denervation or cross innervation that have, until recently, been attributed to some still unidentified trophic substance or group of substances released from the nerve terminals. In the absence of these substances the observed changes occur. However, in the past few years, studies have determined that almost all of these effects can be prevented or reversed by appropriate direct stimulation of the muscle. Even a very low average rate of stimulation can entirely reverse the spread of sensitivity, restricting it again to the site of the original end plates where high sensitivity remains throughout. Moreover, the contraction speed of muscle (its slow versus fast properties) and the associated ATPase activity can be shifted back and forth by different stimulation regimes, approximating the activity pattern of the different nerves normally innervating fast and slow muscles. Thus it is clear that most or all of the trophic influences of nerve on muscle could be explained simply by the activity induced in the muscle. What aspect of activity is important is not yet clear. Action potentials in the membrane appear the most likely candidate, especially since a similar spread of sensitivity is seen in parasympathetic ganglion cells and probably other neurons where contraction is not a candidate. Nevertheless, in muscle it could be contraction that is important; or subthreshold depolarizations might be adequate in either case.

There are other observations, however, indicating that in the absence of muscle activity there still are trophic influences operative. Most compelling, perhaps, is the finding that if one blocks nerve conduction and muscle contraction completely with a TTX cuff, or injection of TTX beneath the nerve sheath, the muscle shows only partial denervation-like changes. However, if the postsynaptic sensitivity to ACh is blocked completely with α-bungarotoxin, the muscle behaves as if it had been completely denervated. This suggests that ACh itself has a significant trophic maintenance capability in the absence of muscle activity, and that the ACh that is released by leakage from the terminal, or in the form of spontaneous release of packets, even in the absence of evoked release, is adequate to exert a significant trophic influence. Additional evidence for a nerve-released trophic substance, either ACh or something else, in the absence of evoked release is the finding that colchicine or related compounds known to block axonal transport, when applied to cuffs to muscle nerves, induce denervation hypersensitivity to ACh without blocking action potentials or normal muscle function. These experiments are not as clear-cut as one would like, however, for others have found that under some conditions colchicine by itself can directly induce ACh sensitivity on muscle membranes, independent of any effect on axonal transport. On the other hand, it has been shown by many workers that the onset of denervation changes occurs at progressively greater delays, the longer the nerve stump left attached to the muscle. This is the case even though all muscle activity stops with denervation, however long the stump. Also, Miledi has performed an ingenious experiment in which he partially denervated frog sartorious muscle fibers. Each fiber is innervated twice. He simply cut the branches of the nerve that go to one end of the muscle, leaving the other sites of innervation intact. The muscles maintained apparently normal activity, but developed significant extrajunctional ACh sensitivity near the denervated end plates. These phenomena apparently reflect a second mechanism of inducing spread of ACh sensitivity, although they do not require that there be a special factor

released by the nerve. For example, an extraneous piece of nerve, or even string, laid on a muscle in the presence of its normal innervation somehow induces localized ACh sensitivity at that point. Apparently some product of nerve degeneration (which begins later, the longer the nerve stump left) or aggravation of the muscle membrane, or accumulation of phagocytes is responsible and can override the influence of activity.

There are other, quite dramatic, changes in postsynaptic properties, depending on the source of innervation, that have not been shown to be affected by activity pattern. For example, in amphibians, slow muscles, which normally do not produce action potentials and undergo prolonged contracture when exposed to high $[K^+]$ and which retain these properties after denervation, change with cross innervation by "fast" nerves to show the fast muscle characteristics of action potentials and only transient K^+ contractures. Such alterations in physiology and biochemistry may well require special trophic substances for their explanation. It appears, therefore, that muscle properties are controlled both by trophic substances (perhaps ACh itself) and by the electrical or contractile activity induced by the synaptic input.

Throughout the CNS there appear to be trophic influences of the same sort operating between nerve cells. Maintenance of nerve cells depends both on the presence of synaptic input and on the presence of normal postsynaptic cells to innervate. Only one neurotrophic substance has so far been identified, the *nerve growth factor,* which is essential to development of autonomic nerve cells and which, in excess, leads to tremendous hypertrophy of the sympathetic nervous system. The factor is a complex of three proteins, the best sources of which are certain snake venoms and the submaxillary salivary glands of male mice!

The influence of nerve on muscle (or other end organs) is not one way. During embryonic development there is extensive nerve branching and proliferation. Many of these branches fail to innervate peripheral structures and soon degenerate. Similarly, when a peripheral structure is removed, the nerve supplying it usually first attempts to innervate adjacent structures, fails if they are normally innervated, then atrophies and disappears. These changes do not take place if the nerve is simply kept from firing (when, for example, the afferent input to a group of motoneurons is removed), so the influence seems to originate somehow in the muscle.

11.6.3 Axonal Transport

These intercellular interactions occur between cells whose nuclei and synthetic machinery can be as much as several feet apart, connected only by tremendously long, fine processes. The volume of the axon, despite its small diameter, may still be 100 times that of the cell body, or more. Moreover, the axon at its end is releasing transmitter and other products mostly synthesized in the soma. Thus it is necessary, both for the maintenance of the distal regions of the cell and for intercellular chemical communication, to have a means of transporting substances in both directions. The mechanism is *axonal transport*. This is usually demonstrated by pulse-labeling a group of cells with radioactive amino acids or transmitter precursors, following which a wave of radioactivity travels down their axons. Characteristically, different substances are transported at different rates, with as many as five discrete categories of rates now recognized. These range from a slow wave traveling at 0.2–1 mm/day, associated with the gradual movement of large structural elements, such as neurofilaments or microtubules, down the axon, to a fast transport system moving substances 200–400 mm/day, carrying enzymes, other proteins, glycoproteins and lipids, usually or always packaged in membranes (Fig. 11-11). The mechanisms of transport are not known, but both Ca^{2+} and ATP are required. An actomyosin-type shuttling mechanism has been postulated, but evidence now indicates that Ca^{2+} is involved more in loading substances into membrane transport systems than in the transport process itself. It appears likely that the Ca^{2+}-activating protein, calmodulin, is an intermediary in the part of the process involving Ca^{2+} (see also Section 9.7.3). Since different waves of transport occur at different rates, it is quite possible that different mechanisms or routes of transport are involved.

Transport in the opposite direction along the axon (retrograde transport) is also fast, but apparently involves different mechanisms than rapid orthograde transport. Students of functional neuroanatomy are now taking advantage of retrograde transport to trace the distal processes of neurons. When the enzyme horseradish peroxidase is put into the region of synaptic endings, some of it is taken up pinocytotically into presynaptic terminals, from which it is transported back to the cell body.

Figure 11-11

Fast axonal transport. In experiments such as the one giving the results shown, a radioactively labelled substance (such as ^{3}H-leucine) is injected into the dorsal root ganglion (DRG). At various times thereafter, if short segments of the sciatic nerve are tested for their content of the substance, it can be seen that a wave of the substance has been carried (at much greater than diffusion rates) in both directions away from the ganglion. In the experiment shown, if curve C represents the distance the wave would normally travel in a given length of time, curve A represents the observed distribution of label when a local anoxic block has been applied to the nerve at the location shown (hatched box) for the 3 hr prior to measurement, and curve B shows the result when the block was applied for 1 hr and then removed for the final 2 hr. With sustained anoxic block, transport stops at the block, and transported substances accumulate at the proximal side. The block is reversible, however, as curve B shows. [After S. Ochs, Proc. Nat. Acad. Sci., 68, 1279–1282 (1971).]

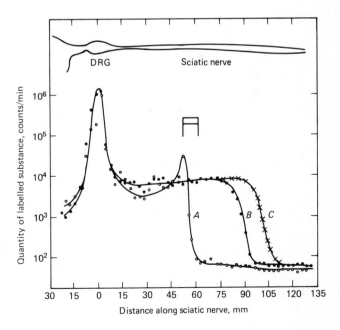

The enzyme can then be used to catalyze a reaction that clearly marks the cells whose processes picked up the marker. The degree of uptake is correlated with the presynaptic activity, so selected endings can be marked. This is one of the most popular techniques at present for tracing pathways.

11.6.4 Specificity and Plasticity of Neural Connections

Specificity of Neural Connections. The most exciting aspect of transneuronal trophic influence is the remarkable precision and reproducibility of synaptic connections. Throughout development of the nervous system, cells migrate and their processes grow via circuitous routes to make highly specific connections with the appropriate target cells. In lower vertebrates, much of this specificity remains in the adult and can be investigated in regenerating systems. For example, in fishes and urodele amphibians (but not adult anurans or higher vertebrates), if the nerve to a fin or limb is cut, completely normal function is regenerated. Even if the nerves to a given muscle are purposely misdirected, appropriate function is restored. From these results, it was once postulated that the nerves grow out randomly but that each muscle somehow confers its own specificity on the nerve fibers reaching it, causing central rearrangement of connections to form the appropriate circuits. Now the evidence instead favors the interpretation that the correct nerves are somehow finding their way to their original muscles. Almost certainly, however, some nerves do grow into the inappropriate muscles and form functional synapses. There is some suggestive evidence that these synapses can be repressed, or turned off, in some way by the appropriate nerve as it regenerates and forms connections.

Regenerating motor nerves, whether appropriate to that muscle or not, show a very strong tendency to terminate in the synaptic gutters left by the original nerve. Through the work of McMahan and his associates, it is now known that the necessary identifying cues for this selective reinnervation are located in the basement membrane, the noncellular material that forms a tightly adherent coat surrounding each muscle fiber even following the contours of the synaptic folds. Through ingenious experiments, the muscle fiber can be removed leaving only the basement membrane (and Schwann cells).

A regenerating nerve selectively forms differentiated terminals at exactly those spots where the basement membrane retained the conformation of old junctional folds. Similarly, in the absence of a nerve, as muscle fibers regenerate, they form postsynaptic specializations, including ACh receptor concentrations, at the same location adjacent to the surviving basement membrane.

One of the best studied and most specific systems is that of the connections between the amphibian optic nerve and the midbrain tectum. During development, optic-nerve axons grow in some unknown way to synapse with appropriate midbrain cells to permit accurate pattern vision. Fibers of characteristic behavior with receptive fields in a certain part of the eye always end on cells in the same part of the higher order nucleus. If the eye is rotated 180°, vision is reversed and the animal is predictably and permanently disoriented. If the optic nerve in a normal animal is cut and twisted, however, the optic-nerve fibers somehow find their way back via improbable channels to synapse in the same part of the midbrain, possibly with the same cells they originally joined, because nearly perfect vision is restored. If the eye is rotated 180° and the optic nerve cut, upside-down and backward vision is regenerated. The explanation for such phenomenally accurate regeneration must involve some form of chemical signature, of highly specific chemical guidance of regenerating nerves (if not the selective attraction proposed initially by Cajal). A proposed working hypothesis is that each regenerating axon seeks a particular point along two (or three) trophic gradients occurring in perpendicular planes in the postsynaptic nucleus, although the specificity may be even more exact than this. It is known that the specificity appears at approximately that time in development when optic ganglion cell replication stops and gap junctions between ganglion cells in the retina are lost.

Effect of Use and Disuse: Plasticity of Neural Connections. A particularly elegant demonstration of the fundamental changes occurring with use and disuse in a sensory pathway is that by Hubel and Wiesel in the mammalian visual system. They have found that in newborn kittens the receptive fields of cells in the retina, lateral geniculate, and cortex are essentially the same as in the corresponding nuclei of the adult, although responses are more

11 USE OF INFORMATION: CENTRAL COORDINATION

sluggish. Thus the appropriate connections are already made at birth. Experience is not needed to make the right connection, but is involved in the loss of incorrect connections.

When vision was prevented in one eye of a kitten for the first 3 months of life (either by an opaque occluder or suturing of the eyelids together), that eye became functionally blind, despite the presence of a pupillary reflex. Several questions are raised: (1) Is the visual failure due to a failure at some point in the pathway to the cortex, or is it in the readout of cortical information? (2) Is the defect caused by light or pattern deprivation, or both? (3) Is this defect dependent on the age of the animal? (4) What is the mechanism of failure? (5) Is it reversible?

Recordings of receptive fields in the lateral geniculate of the monocularly deprived kittens showed virtually no physiological changes. The layers receiving input from the occluded eye were about 40% thinner and cell size was reduced, but the cells were driven in the normal fashion by the deprived eye. Cortical-cell receptive fields, on the other hand, were of the normal shape, but virtually none could be driven by the occluded eye. Where normally most cells receive about equal input from each eye, all activity was now governed by the "good" eye.

Surprisingly, when both eyes were occluded for 3 months, responses at the geniculate level were essentially normal and 40% of the cortical cells showed normal response patterns, even though the animals were functionally blind. In this case failure apparently occurs "central" to the primary visual cortex, and disuse, by itself, is not sufficient for total synaptic failure.

When a translucent rather than opaque occluder was used to eliminate pattern vision without much affecting light intensity, the deprived eye was again functionally blind, but now there was less change in the size of the lateral geniculate layers driven by that eye. The mean cross-sectional area was reduced by only 10–15%. Cortical cells again responded only to input from the normal eye, having apparently lost their original input from the other eye. This result is consistent with the observation that most geniculate cells respond, if sometimes weakly, to diffuse light, whereas cortical cells normally do not.

Even in the presence of full light and pattern vision in both eyes, if both cannot be focused on the same object, one becomes functionally dominant. When this condition (squint; see Section 9.8) was produced in kittens by cutting one of the oculomotor nerves, the cortical receptive fields again came to be dominated by only one eye or the other. The same result was obtained by occluding first one eye and then the other, on alternate days, so that the eyes both had visual experience and could focus correctly on the same point but could not be used simultaneously. Thus the deficit is not so much in the disuse of synapses as in the lack of synchrony of use. Without simultaneous and "in register" activation of synapses driven by both eyes, one population takes over the cell and the other is lost completely. In recent experiments, a more accurate measurement has been made of how simultaneous the inputs need to be. When the input to the two eyes in a kitten was alternated at a frequency of 10 Hz, binocularity was normal. At 1 Hz, however, a higher than normal percentage of neurons were driven only by one eye, and a frequency of 0.1 Hz had the same effect as 24-hr alteration.

Hubel and Wiesel have found that the critical period for this plasticity in kittens is from the fourth through the seventh or eighth week after birth. In cats older than 3–4 months, none of the changes in structure or function can be induced. During the critical period, occlusion of vision for a period of only 2–3 days can cause severe loss of influence of that eye, with the first cells to be affected being those that are normally driven most nearly equally by both eyes. Curiously, during the first 3 weeks after birth, keeping the eyes closed has no damaging effect; responses become less sluggish and more stimulus orientation-specific even without visual experience.

The reversibility of changes due to visual deprivation depends on the degree to which each eye has been used during the critical period. If the deprivation is discontinued before the period of plasticity is over, a few cortical cells recover their input from the deprived eye; however, few if any are found that can be driven binocularly, and the vast majority continue to be affected only by the undeprived eye. If deprivation begins only midway through the critical period, the input from the deprived eye is still lost, but less permanently. If, before the end of the critical period, the normal eye is blindfolded and the previously deprived eye used, the ocular dominance can shift completely. Interestingly, it has been reported that, with monocular deprivation, when one eye is almost totally dominant at the level of the cortex, treatment of the kitten with bicucculine, which blocks synapses mediated by γ-aminobutyric acid (GABA), causes the inputs from the deprived eye to become effective in driving cortical cells.

Comparable effects of deprivation are known in human children, and have been demonstrated experimentally in rhesus monkeys. In monkeys, however, susceptibility begins at birth, is maximal for a few weeks, and decreases only slowly over a period of many months or years. Probably this plasticity of connections is common to mammals during development and infancy and is true of all sensory systems, if not all brain function.

The implications of such plasticity are now being explored. Of particular interest is a series of experiments showing that experience during the critical period not only can govern the optical dominance of cortical cells but can strongly influence the stimulus specificities of cells. For example, if a kitten is raised during its critical period under conditions in which it can see only vertically oriented black and white stripes, the vast majority of cortical cells, when tested later, show response only to visual stimuli with borders of approximately that orientation. For a kitten otherwise raised in the dark throughout the critical period, apparently only a few hours exposure to one orientation at the height of the critical period (day 28) is necessary to achieve such a change in the system. This probably does not reflect the formation of new connections on the basis of the experience but rather the selective survival of some synapses and loss or suppression of others. This can occur during the critical period (perhaps with the effect of adjusting the visual system for any developmental anomalies) but plasticity is then lost. Much current research is directed at trying to learn what is occurring at the synapses on cortical cells during the critical period.

11.7 MEMORY AND LEARNING

The subject of neural plasticity leads inevitably to the most mysterious and impressive of brain func-

tions, the phenomena of consciousness, learning, and memory. We are far from an understanding of any of these. Certainly all, and especially the capacities for learning and memory, are associated with the development of large areas of association and unassigned cortex. Many experiments tend to indicate that there are two processes involved in learning and memory: one that operates during the first few seconds or minutes after an experience, another that semipermanently assimilates the experience. The early process can be disrupted by electroconvulsive shock or anesthesia, whereas the latter is imprinted so indelibly that shock, anesthesia, coma, extreme cold, and various drugs cannot destroy it.

Early theories that learning is a dynamic process, involving continuously circulating patterns of impulses in specific neural chains, clearly cannot explain long-term memory. Much more promising is the suggestion that activity of certain networks of synapses, representing a given sensory input (coupled with conscious attention, whatever that is), can effect some lasting change in their functional efficacy. It is difficult even to speculate how a single representation of a stimulus, and presumed single utilization of a given pathway, could modify such synapses, but this might conceivably be the function of the attentive process, even by simply recirculating the chain of activity several times to help establish the pattern. The nature of the synaptic change is equally speculative, perhaps involving spatial movement of endings to more effective sites on the postsynaptic cell, a change in area of synaptic contact, a change in amount of transmitter released or receptor present, or even an induced specificity of receptor to a particular transmitter.

Much recent thought (and some sensational reporting) has gone into theories of how RNA (ribonucleic acid) might be used to code transmitter-receptor systems at specific junctions. Although there is some evidence for an increase in RNA synthesis in cells that are involved in conditioned responses, there is little of a convincing nature to suggest what could be happening in these cases. The chemical transmitters are not known, but one of the most universally effective depolarizing agents for brain neurons is glutamate, suggesting the (still remote) possibility that specifically coded short polypeptides might be of some significance.

Among the most dramatic studies of memory

have been those of Penfield, done mostly in the 1930s and 1940s. Penfield stimulated various points on the cortex of epilepsy patients, searching for the site of the pathological epileptic focus, stimulation of which gave rise first to an aura, then to epileptic seizure. In so doing, he found that stimulation of certain areas on the temporal cortex could elicit highly specific and detailed memories, repeatable on stimulation of the same point but totally different on stimulation a few millimeters away. Subsequent research (usually taking the form of observations on human patients with pathologic lesions in the temporal lobes or following surgical damage designed to alleviate severe clinical symptoms) has implicated especially the medial lobe of the temporal cortex and, within this, the hippocampal areas. These are joined bilaterally by the fornix, a fiber tract that continues posteriorly to terminate in the mammillary bodies at the posterior end of the hypothalamus. This hippocampus-fornix-mammillary body complex appears to be involved in motivation and emotional states and is critical to the embedding of memory, although long-established memories appear to be located elsewhere. Thus the temporal lobes of the cortex are viewed either as principal sites of memory "engrams" or as part of the pathway to their recall. The mechanism of effect of the stimulus and the cellular circuits stimulated are unknown.

As in so many other fields of biology, great advances in understanding of neurophysiological principles have depended on finding the right experimental preparation. A more complete understanding of memory and learning must probably await the discovery of a simple, perhaps invertebrate, system that exhibits conditionability. In such a system, with few cells and rigorously controlled input, the subtle changes in synaptic function that are probably responsible for these phenomena may finally be recognized.

ADDITIONAL READING

Anderson, H., J. S. Edwards, and J. Palka. "Developmental neurobiology of invertebrates," *Ann. Rev. Neurosci.,* **3,** 97–139 (1980).

The Brain, A Scientific American book, W. H. Freeman & Co., San Francisco, 1979.

Bullock, T. H., R. Orkand, and A. D. Grinnell. *Introduction to Nervous Systems,* W. H. Freeman & Co., San Francisco, 1977.

Burnstock, G. "The evolution of the automatic innervation of visceral and cardiovascular systems in vertebrates," *Pharmacol Rev.,* **21,** 247–324 (1969).

Cooper, J. R., F. E. Bloom, and R. H. Roth. *The Biochemical Basis of Neuropharmacology,* 3rd. ed., Oxford University Press, New York, 1978.

Eccles, J. C., M. Ito, and J. Szentagothai. *The Cerebellum as a Neuronal Machine,* Springer-Verlag, New York, 1967.

_____ *The Understanding of the Brain,* 2nd ed., McGraw-Hill Book Co., New York, 1977.

Fambrough, D. M. "Control of acetylcholine receptors in skeletal muscle," *Physiol. Rev.,* **59,** 165–227 (1979).

Gaze, R. M. *The Formation of Nerve Connections,* Academic Press, New York, 1970.

Grafstein, B., and D. Forman. "Intracellular transport in neurons," *Physiol. Rev.,* **60,** 1167–1283 (1980).

Green, L. A., and E. M. Shooter. "The nerve growth factor: biochemistry, synthesis and mechanism of action," *Ann. Rev. Neurosci.,* **3,** 353–402 (1980).

Hall, A., J. Hildebrand, and E. Kravitz. *The Chemistry of Synaptic Transmission,* Chiron Press, Newton, MA, 1974.

Harris, A. J. "Inductive functions of the nervous system," *Ann. Rev. Physiol.,* **36,** 251–306 (1974).

Hubel, D., and T. Wiesel. Three papers on responses of visual cells of visually inexperienced and deprived kittens, *J. Neurophysiol.,* **26,** 978–1017 (1963).

_____ and S. LeVay. "Plasticity of ocular dominance columns in monkey striate cortex," *Phil. Trans. Roy. Soc. London,* **B278,** 377–409 (1977).

Hughes, A. F. W. *Aspects of Neural Ontogeny,* Academic Press, New York, 1968.

Jacobsen, M. *Developmental Neurobiology,* 2nd ed., Plenum Press, New York, 1978.

Kandel, E. R. (ed.). *Handbook of Physiology. Neurophysiology,* vol. I, *Cellular Neurobiology,* American Physiological Society, Washington, D.C., 1976. (Contains chapters by J. Rosenthal on nerve trophic influences, and by A. D. Grinnell on the specificity of nerve connections.)

_____ *Cellular Basis of Behavior,* W. H. Freeman & Co., San Francisco, 1976.

Kuffler, S. W., and J. G. Nicholls. *From Neuron to Brain,* Sinauer Associates, Sunderland, MA, 1976.

Landmesser, L. T. "The generation of neuromuscular specificity," *Ann. Rev. Neurosci.,* **3,** 279–302 (1980).

Luce, G. G. *Current Research on Sleep and Dreams,* Publ. no. 1389, U. S. Government Printing Office, Washington, D.C., 1965.

Lund, R. D. *Development and Plasticity of the Brain,* Oxford University Press, New York, 1978.

Mark, R. F. "Synaptic repression at neuromuscular junctions," *Physiol. Rev.,* **60,** 355–395 (1980).

Noback, C. R., and R. J. Demarest. *The Nervous System,*

Introduction and Review, 2nd ed., McGraw-Hill Book Co., New York, 1977.

Penfield, W. "The role of temporal cortex in recall of past experience and intrepretation of the present," pp. 149–174. In *Neurological Basis of Behavior,* Ciba Foundation Symposium, J. B. A. Churchill Ltd., London, 1958.

Purves, D. "Long-term regulation in the vertebrate peripheral nervous system," *Int. Rev. Physiol.,* **10,** 125–177 (1976).

Romer, A. *The Vertebrate Body,* 4th ed., Saunders, Philadelphia, 1970.

Sarnat, H. S., and M. G. Netsky. *Evolution of the Nervous System,* Oxford University Press, New York, 1974.

Schmidt, F. O. (ed.). *The Neurosciences,* Second Study Program, The Rockefeller University Press, New York, 1970; and *The Neurosciences,* Third Study Program, The Rockefeller University Press, New York, 1974.

Schwartz, J. H. "Axonal transport: components, mechanisms and specificity," *Ann. Rev. Neurosci.,* **2,** 467–504 (1979).

Shepherd, G. M. *The Synaptic Organization of the Brain,* Oxford University Press, New York, 1974.

Snyder, S. H., and S. R. Childers. "Opiate receptors and opoid peptides," *Ann. Rev. Neurosci.,* **2,** 35–64 (1979).

The Synapse, vol. 40, Cold Spring Harbor Symposia on Quantitative Biology, Cold Spring Harbor, NY, 1976.

Vrbova, G., T. Gordon, and R. Jones. *Nerve-Muscle Interaction,* John Wiley & Sons, 1978.

Wells, M. J. *Brain and Behavior in Cephalopods,* William Heinemann, Ltd., London, 1962.

C. Barker Jørgensen

12

HORMONAL AND OTHER CHEMICAL COORDINATION

12.1 INTRODUCTION

The preceding three chapters have outlined the role of the nervous system in coordinating and integrating animal activities. Animals also possess other systems of coordination. These other systems are characterized by diversity and heterogeneity in both structure and function. They may serve to stimulate or inhibit specific functions of cells, tissues, or organs of the organism; to control simple or complex metabolic processes; to stimulate or inhibit growth and differentiation of specific cells, tissues, or organs, or of the whole organism; or to influence behavior.

During the long and still incomplete process of exploring the nature of nonnervous coordination between the parts of the body, emphases and concepts have been continuously changing. The existence of nonnervous coordination remained speculative long after nervous coordination had become well established. However, by the end of the last century sufficient observations had accumulated to convince many people that tissues and organs of

the body release substances to the circulating blood that stimulate or depress functions in other parts of the body.

Originally this concept of chemical coordination between parts of the body was very broad, and it included substances such as glucose, secreted by the liver, waste products, and even heat. Chemical coordination merged with the concept of internal secretion, especially after Bayliss and Starling's (1902) conclusive demonstration of the role played by secretin in the regulation of secretion of pancreatic juice (see Chapter 2).

The concept of internal secretion soon became closely connected with the ductless or endocrine glands, long known to the anatomists. It followed from the definition that the ductless glands might release products directly into the blood. In the early part of this century endocrinology became established as a subject in its own right, with the experimental demonstration that ductless glands may indeed act by means of substances given off into the blood circulation.

The standard procedure for demonstrating endocrine function of a ductless gland was to extirpate the gland to see whether the organism exhibited abnormalities in the absence of the gland. If deficient functions that developed could then be renormalized by grafting the gland to a site within the organism remote from the original site, or by injection of extracts of the gland, this was taken as evidence for an endocrine function of the gland. The factors secreted by endocrine glands became known as hormones (from Greek *hormaein,* to excite). Bayliss and Starling had coined the term for the blood-borne messengers they had demonstrated to be secreted by the gut. Thus, ironically, the first hormone to be demonstrated was not produced by an endocrine gland but by an organ. It seems a further bit of irony that the first hormones to be chemically identified were adrenaline (epinephrine) and noradrenaline (norepinephrine) from the medulla of the adrenal glands, in 1901 and 1904. Only later was adrenaline shown also to be a nerve transmitter.

During the first half of the twentieth century endocrinology flourished, primarily based on studies of the ductless glands and their hormones. The endocrine system gained recognition as the system for chemical communication within the body, acting through chemical messengers, the hormones, which reached their target organs by the systemic blood circulation.

The concept thus developed of a dual communication system involving both a nervous system that acts directly upon target cells via nerve terminals and synapses, predominantly by means of electrical transmission (see Chapter 9), and an endocrine system that acts at a distance, via blood borne chemical messengers.

This clear concept soon began to blur as new facts came to light. In the nervous system synaptic transmission turned out to be predominantly chemical, with electrical transmission the exception. Moreover, many neurons were also found to act as endocrine cells and to release hormones into the blood circulation (see Section 12.4). It gradually became clear that "classical endocrinology" only represented one end of a spectrum of types of chemical coordinations between organs, tissues, and cells of organisms. Many tissues and cells secrete specific substances that act over short distances on adjacent tissues or cells. These types of chemical communication do not involve the blood circulation and take place by diffusion of substances in intracellular or interstitial spaces, or by other local transport mechanisms.

At the biochemical level it was originally assumed that neurotransmitters were chemical entities typical of nervous function and distinct from the hormones produced by the endocrine glands. It is now clear that no such distinction exists. Representatives of classical hormones may play roles in the control of synaptic transmission within the nervous system (see Chapter 11), and classical neurotransmitters, such as ACh and serotonin, seem to act as local hormones in the control of gastrulation in the embryos of sea urchins, amphibians, and birds. The chemical natures of the molecules that have been used to act as neurotransmitters, local hormones, or systemically acting hormones are apparently determined by properties other than the types of functions they came to serve (see Section 12.7).

During the broadening of the concepts of endocrine function the inadequacy of the term "hormone" became clear, because hormones do more than excite or stimulate. Various only partly successful attempts have been made to establish a more appropriate terminology. Thus growth inhibiting factors have been termed chalones (Greek *chalinos,* curb). The little-used term autacoid (Greek *autos,* self; *akos,* remedy; *eidos,* form) was coined for internal secretions generally, to cover both hormones and chalones.

The lack of success in finding a common term for the chemical coordinations arises from the diversity and heterogeneity of functions involved. Endocrinologists seem to have acquiesced in using the term hormone, and they speak about "local hormones" when the agent does not act via the blood circulation. Another convention is that "hormone" has become reserved for chemically identified agents. Until identified chemically, hormonally active substances are termed "factors." This is a consequence of the great advances in biochemical techniques. Whereas previously the physiological functions exerted by hormones were usually well established before the chemical nature of the hormones was known, today the situation is reversed. Many substances have been isolated, chemically identified, and even synthesized, that are suspected to be hormones but with no definite evidence concerning their functions. Such substances are termed "putative hormones."

The volume of recent work on hormonal mechanisms and functions prohibits a complete coverage of the field. Instead the present chapter first provides condensed data on the main vertebrate ductless glands, their hormones and functions, on main groups of local hormones, and on some invertebrate endocrine functions. Second, an attempt is made to characterize hormonal functions by dealing with selected examples in greater detail. The examples have been chosen to illustrate hormonal functions at various levels of biological organization. Finally, the nature of the hormones and their functions is taken up again for a more detailed examination.

Other examples of hormonal functions can be found in their proper context in other chapters of the book. Thus, the mechanisms that control secretion of hydrochloric acid in the stomach are described in Chapter 2. They provide an especially illuminating example of the complexity of integrative control mechanisms that involve interactions between nerves, local hormones, and systemically acting hormones. Chapter 7 mentions the role of hormones of the neurohypophysis and the adrenal cortex in regulating the composition and volume of the extracellular fluid, including the blood. Chapter 8 includes the role thyroid hormones play in control of metabolic rate and body temperature in endothermic vertebrates. Chapter 13, on reproduction, further illustrates the basic roles hormones play in the lives of animals.

12.2 ENDOCRINE GLANDS AND HORMONES

12.2.1 Classical Hormones

Table 12-1 lists important vertebrate ductless glands, their hormones, and the main functions of these hormones. These glands and their hormones constitute the basis for much of classical endocrinology. Many of their functions are dealt with in detail in later sections of this chapter and in Chapter 13 (see references in the table).

12.2.2 Local Hormones

Our knowledge of the functions of the chemical substances that have been lumped as local hormones is still incomplete and vague, and in some groups our knowledge of the chemistry, biosynthesis, and pharmacology of the hormones has progressed far ahead of our understanding of their physiology.

Prostaglandins. As the name implies, these substances are produced by the prostate gland and were first demonstrated in semen, but they occur widely in the tissues and organs of the body. A large number of prostaglandins and associated compounds have been chemically identified during recent years (Figure 12-1). The substances are derived from the unsaturated fatty acids in the food, whose character as vitamins may thereby be explained (Table 2-5). The prostaglandin system has been implicated in physiological responses of most tissues of the body, such as blood pressure, body water homeostasis, regulation of food intake, ovulation of ovarian follicles, regression of corpora lutea, and parturition (see Chapter 13). However, the precise functions of the various prostaglandins and their derivatives mostly remain obscure. They have been characterized as agents that serve to modulate and modify physiological events.

Vasoactive Peptides. The vasoactive peptides constitute another group of substances widely distributed in the body. They may act both as local hormones and via the blood circulation. The vasoactive peptides include the kinins and the angiotensins.

The kinins dilate arterioles, increase permeability of capillaries, and cause contraction of smooth muscles, such as those in the gut and uterus. They are the most powerful vasodilator agents known. It has been suggested that they are part of the mechanisms that control blood pressure and blood perfusion of organs and glands. The vasodilator effects of kinins are probably mediated by prostaglandins.

Two angiotensins are known. The decapeptide angiotensin I is released from angiotensinogen (see Figure 12-2), a plasma globulin, by the proteolytic enzyme renin, which is secreted by the juxtaglomerular cells in the kidneys (see Section 6.7.2). Another

TABLE 12-1. Main Tetrapod Ductless Glands, their Hormones and Functions

Gland	Hormone	
	Name	Function
Hypophysis (pituitary gland)		
Adenohypophysis	Adrenocorticotropin (ACTH)	Stimulates growth and hormone secretion of adrenal cortex
	Melanotropin (MSH)	Stimulates pigment synthesis and dispersion in melanophores (amphibians, reptiles)
	Prolactin (PRL)	Various functions (see Section 12.7)
	Growth hormone (GH)	Stimulates growth and participates in general metabolism
	Follicle-stimulating hormone (FSH)	} Gonadotropins, stimulate production of gametes, secretion of sex hormones, and ovulation (see Chapter 13)
	Luteinizing hormone (LH)	
	Thyrotropin (TSH)	Stimulates growth and hormone secretion of thyroids
Neurohypophysis		
Pars nervosa	Vasopressins, vasotocin	Involved in regulation of water balance (see Section 12.4.1)
	Oxytocin	Causes milk ejection (see Section 12.4.1) and increases uterine contractions (mammals) (see Section 13.6.3)
Eminentia mediana (median eminence)	TSH-releasing hormone (TRH)	} Control release of adenohypophysial hormones (see Section 12.4.3)
	LH-releasing hormone (LRH)	
	Somatostatin, dopamine, and others not yet chemically identified	
Thyroids	Thyroxine (T_4), triiodothyronine (T_3)	Involved in growth and differentiation (especially amphibian metamorphosis, see Section 12.3.3), and energy metabolism in endotherms
Parathyroids	Calcitonin	Decreases blood calcium } (see Section 12.4.2)
	Parathyroid hormone	Increases blood calcium
Pancreatic islets	Insulin, glucagon, somatostatin	Involved in regulation of intermediary metabolism (see Section 12.4.2)
Adrenals		
Medulla	Adrenaline, noradrenaline	Raise blood pressure and blood metabolites
Cortex	Corticosterone, cortisol, aldosterone	Control carbohydrate and mineral metabolism
Gonads:		
Ovaries	Progesterone	} Involved in sexual maturation and reproduction (see Chapter 13)
	Estradiol and other estrogens	
Testes	Testosterone and other androgens	

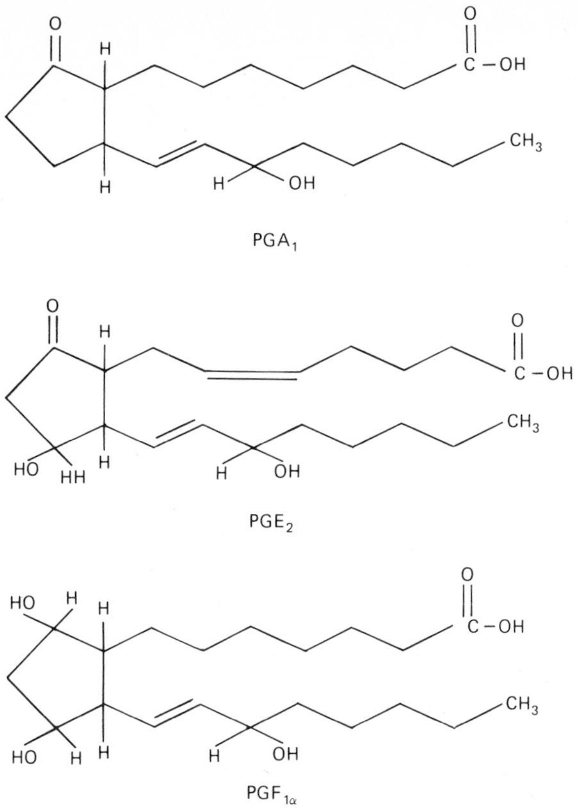

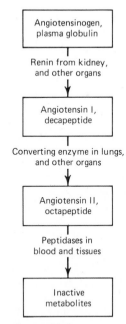

Figure 12-1
Examples of commonly occurring prostaglandins.

PGA$_1$

PGE$_2$

PGF$_{1\alpha}$

they are synthesized in organs such as the kidneys, pancreas, salivary glands, and the gut. The other type is blood plasma kallikrein, which is produced in the liver in an inactive form as prekallikrein, and released into the blood. The glandular kallikreins split the kinin kallidin from the kininogens produced by the organs. Kallidin is a decapeptide, which is further hydrolyzed by peptidase to the nonapeptide bradykinin. Kininogens that circulate in the blood are hydrolyzed directly to bradykinin in a process that involves activation of prekallikrein to plasma kallikrein. It is noteworthy that this process of activation of plasma kallikrein also constitutes part of the chemical processes associated with blood coagulation, an example of a particular bit of biochemical machinery being adopted to serve several widely differing physiological functions.

The kinins and angiotensins are short-lived, and they are further hydrolyzed by peptidases into inactive fragments. The half-life of bradykinin circulating in the blood is about 20 sec and of angiotensin II, 1–2 min.

proteolytic enzyme hydrolyzes angiotensin I to the octapeptide angiotensin II, which is biologically more potent. Angiotensin II causes arteriolar constriction, and is the most powerful vasoconstrictor known. Angiotensin II also stimulates aldosterone secretion, induces thirst, and may play a key role in the homeostatic regulation of the composition of the body fluids (see Section 12.4.2).

Both the kinins and the angiotensins are produced by similar biochemical mechanisms. The angiotensin mechanism was described in the preceding paragraph. The kinins are present as precursor proteins, kininogens, in the various tissues and organs of the body or circulating in the blood (Figure 12-3). The kinins are split from the kininogens by means of proteolytic enzymes known as kallikreins. Two types of kallikreins can be distinguished. One type is known as the glandular kallikreins because

Angiotensinogen, plasma globulin

↓ Renin from kidney, and other organs

Angiotensin I, decapeptide

↓ Converting enzyme in lungs, and other organs

Angiotensin II, octapeptide

↓ Peptidases in blood and tissues

Inactive metabolites

Figure 12-2
Formation and metabolism of angiotensins.

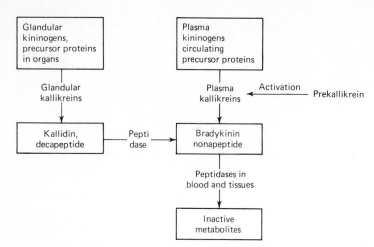

Figure 12-3
Main features in formation of a kinin, bradykinin.

Biogenic Amines. Roles as local hormones have been attributed to several biogenic amines, including serotonin and histamine. The role played by histamine in the control of secretion of hydrochloric acid was mentioned in Section 2.4.6. Serotonin and histamine (Figure 12-10) are both widely distributed in the tissues of the body.

Inducers and Growth Factors. These substances represent a heterogenous group of locally acting factors, most of which are poorly defined chemically. Growth and differentiation in early embryological development is controlled by specific substances that are secreted by certain cells or tissues and induce neighboring cells or tissues to develop according to their inherent potentialities. The chemical correlation between parts of the embryo has especially been studied in amphibians, starting with the pioneer work of H. Spemann and his school half a century ago. The inducers presumably reach their target cells by diffusion through intercellular or interstitial spaces. They act before a vascular system and blood circulation have been established in the embryo. Inducer molecules may even reach their target cell through gap junctions in the cell membrane that are especially adapted for the direct transfer from cell to cell of large molecules.

Local hormonal control of growth and differentiation is not restricted to the embryological and early development, but may persist into adult life. Specific factors have thus been found that control growth and differentiation of nerves, fibroblasts, and epidermis. These factors have been identified as polypeptides and have been named nerve growth factor, fibroblast growth factor, and epidermal growth factor. Other as yet undetected factors apparently promote and control cell replication and differentiation in their respective tissues.

It is of special interest that "classical" hormones, such as the hypophysial hormones, in addition to stimulating secretion in their target organs, thyroids, adrenals, and gonads, also play major roles as growth factors in maintaining growth and differentiation of their target structures. The broad functional spectra of some polypeptide hormones provide good examples on the lack of sharp delineations among the various types of chemical communication that have evolved between cells and organs in multicellular organisms.

12.2.3 Invertebrate Endocrinology

Invertebrate endocrinology developed independently of vertebrate endocrinology. It originated as insect endocrinology, which eventually became a discipline of its own. During the last few decades invertebrate endocrinology has expanded to include most other groups of invertebrates, but especially crustaceans, cephalopods, polychaetes, and echinoderms. During the same period, relations have become closer between the comparative endocrinology of the invertebrates and that of the vertebrates, furthered by the establishment of a journal and societies that cover all fields of animal endocrinology. Table 12-2 lists examples of endocrine mechanisms in invertebrates. Some of the examples are treated in greater detail in the text later in this chapter.

12.3 HORMONES IN SOME FACETS OF ANIMAL LIFE

In this section we will describe several biological activities in which hormones play key roles at the highest levels of structural organization. The subjects range from how hormones serve in animal communication and how they act in adapting the colors of animals to the environment, to the hormonal control of highly integrated functions such as metamorphosis and molting.

12.3.1 Pheromones

Research on how animals communicate has traditionally centered on sensory modalities such as sight and hearing, modalities that dominate communication between members of our own species. However, in most groups of animals the primary sensory modality in communication is usually olfactory or gustatory, that is, chemical. Chemical communication between animals has attracted increasing attention in recent decades, starting with chemical communication between members of groups of social insects. The field has since expanded to include fishes, mammals, and other groups of animals.

The messenger substances used in chemical communication between animals were originally termed "ectohormones," but this term has been almost completely replaced by the term *pheromones* (Greek, *pherein*, to carry, and *hormaein*, to excite).

Chemical communication may influence activities over the entire range of animal behavior and function, such as aggression, flight, social behavior, reproduction, and so on. This vast field will be illustrated by a few examples of pheromonal control of reproduction in insects and mammals.

Insects. Release of olfactory sex attractants is widely distributed among insects and has been found, for instance, in Hemiptera, numerous Lepi-

doptera, Hymenoptera, Coleoptera, and Diptera. Most sex attractants are species-specific. A number of them have been isolated and chemically identified. The first to be identified and subsequently synthesized was the sex attractant of females of the silk moth, *Bombyx mori*, by Butenandt and coworkers. They extracted 12 mg of the pure substance from the abdominal scent glands of 500,000 virgin female moths. The substance was named bombykol. It is an alcohol with the formula shown in Figure 12-4. Almost simultaneously, in 1960, the sex attractant of the female gypsy moth (*Parthetria dispar*), "gyptol," was identified and synthesized. It was the alcohol 10-acetoxyhexadec-*cis*-7-enol (Figure 12-4). The honeybee sex attractant, which is secreted by the mandibular gland of virgin queen honeybees, was identified in 1964 as a fatty acid, *trans*-9-ketodecenoic acid (Figure 12-4).

Two species of lepidopterans, the red-banded leaf roller moth (*Argyrotaenia velutinana*) and the oblique-banded leaf roller moth (*Choristoneura rosacana*) secrete identical sex attractants, *cis*-11-tetradecenyl acetate (Figure 12-4). The two species have overlapping seasonal and diurnal cycles and share the same host plants. The reproductive isolation in the field of the two species therefore requires specific means, perhaps the emission of secondary compounds by the females along with the tetradecenyl acetate.

Some groups of insects use aphrodisiac pheromones as releasers in the control of mating behavior. Usually the aphrodisiac is secreted by the male, as in several species of cockroaches. When a male of such a species has been attracted to a female, he will raise his wings and expose his aphrodisiac abdominal gland. The secretion from the gland attracts the female who mounts the male and feeds on the secretion prior to copulation (see Section 13.3).

In some gregarious and many social insects, pheromones have been adopted in the control of gonad development. Pheromones that both inhibit and accelerate gonad maturation are known. Thus, in the desert locust (*Schistocerca gregaria*) the sexually mature males present in a crowded group of animals will secrete pheromones that accelerate maturation in both male and female immature individuals. The volatile pheromone is perceived over short distances by means of antennal sense organs. The function of the pheromones is presumably to syn-

TABLE 12-2. Endocrine Mechanisms in Invertebrates

Systematic group and species	Functions of glands and hormones	Source and chemical nature of hormones
Hydrozoa *Hydra*	Growth-activating and gonad-inhibiting	NSC* in subhypostomal growth zone
Turbellaria Planarians *Dugesia* spp.	Growth-activating and gonad-inhibiting	NSC
Nermertina *Lineus* spp.	(1) Gonad-inhibiting	NSC in cerebral ganglia
	(2) Androgenic, causing sex reversal in parabiotic female partner	Testis
	(3) Osmoregulatory (diuretic) function in hypotonic media	NSC in cerebral ganglia
Polychaeta Nereidae	Gonad-inhibiting and growth (regeneration)-stimulating h.*	NSC in cerebral ganglia, "juvenile hormone," water soluble and of low molecular weight
Eulalia sp. *Eunice* sp. *Harmothoë* sp. etc.	Gonadotropic, stimulating oogenesis and/or vitellogenesis	NSC in cerebral ganglia
Oligochaeta Earthworms (lumbricids)	(1) Stimulates gametogenesis and secondary sex characters	NSC in cerebral ganglia
	(2) Hyperglycemic	
Hirudinea Leeches (*Hirudo medicinalis,* and others)	(1) Stimulates proliferation of spermatogonia and oogonia	NSC in cerebral ganglia
	(2) Chromatophorotropic	
Gastrododa Prosobranchia *Crepidula fornicata*	Stimulates vitellogenesis, spermatogenesis, and female and male accessory sex organs	NSC in cerebral ganglia
Pulmonata *Lymnaea stagnalis*	(1) Growth stimulating	NSC in cerebral ganglia
	(2) Stimulates spermatogenesis, ovulation and oviposition	
	(3) Stimulates oocyte maturation and female accessory sex organs	Dorsal bodies
Helix aspersa	(1) Stimulates spermatogenesis	NSC in cerebral ganglia
	(2) Stimulates vitellogenesis	Dorsal bodies
	(3) Stimulates male accessory sex organs	Testis
	(4) Stimulates female accessory sex organs	Ovary

TABLE 12-2. *Continued*

Systematic group and species	Functions of glands and hormones	Source and chemical nature of hormones
Cephalopoda *Octopus; Sepia*	Stimulates oogenesis, proliferation of spermatogonia, and secondary sex characters	Optic gland hormone (see Section 13.4.3)
Echinodermata All classes	(1) Stimulates ovulation and spermiation (gonad stimulating substance, GSS)	Hypodermal cell layer of radial nerves
	(2) Action on ovary indirect via second principle	Follicle cells in ovary; 1-methyladenine
Crustacea Decapoda, and other higher crustaceans	(1) Gonad-inhibiting h.	NSC in X-organ–sinus gland complex in eye stalk
	(2) Molting h.	Y-organs in some crustaceans; 20-hydroxyecdysone
	(3) Molt-inhibiting h.	Sinus gland
	(4) Red pigment concentrating h.	Sinus gland, octapeptide
	(5) Distal retinal pigment light-adapting h., and others, unidentified chromatophore and retinal pigment controlling h.s.	Sinus gland, octadecapeptide
	(6) Hyperglycemic h.	Sinus gland, protein
Insecta	(1) Growth, molting, and metamorphosis	See Section 12.3.3
	(2) Reproduction	See Section 13.4.3
	(3) Diuretic h.	Corpora cardiaca
	(4) Antidiuretic h.	Corpora cardiaca
	(5) Heart beat controlling h.s.	Corpora cardiaca
	(6) Hypo- and hyperglycemic h.s.	Corpora cardiaca, peptides
	(7) Adipokinetic h.	Corpora cardiaca, nonapeptide

* h. = hormone; NSC = neurosecretory cells.

chronize sexual maturation and reproduction within the group.

In many insect colonies, reproduction is usually restricted to one or a few females at a time. The rest of the females, the workers, remain sexually immature. The type of females reared, whether workers or queens, depends upon the activities of the workers, who rear other workers in small cells on a diet of low protein, whereas queens are reared in large cells on different diet rich in protein. The social life of the colony is thus largely determined by the mechanism that suppresses ovarian development in the workers and directs their rearing activities. The mechanism is pheromonal. It has been studied in detail in social bees, in termites, and in ants. The honeybee *Apis mellifera* is a well-studied example.

Queen rearing and oogenesis in worker honeybees are suppressed by a pheromone secreted by the queen of the colony. The pheromone is produced by the mandibular glands, and the queen distributes it over her body when grooming.

SEX ATTRACTANTS IN INSECTS

Bombykol
$$CH_3(CH_2)_2CH=CH\cdot CH=CH(CH_2)_8CH_2OH$$

Gyptol
$$CH_3(CH_2)_5\ CH\cdot CH_2\cdot CH=CH(CH_2)_5CH_2OH$$
$$\underset{\underset{O}{\underset{\|}{\overset{|}{O\cdot C\cdot CH_3}}}}{}$$

Honeybee sex attractant
$$CH_3\cdot CO(CH_2)_5CH=CH\cdot COOH$$

Leaf roller moth sex attractant
$$CH_3\cdot CH_2CH=CH(CH_2)_{10}-O-COCH_3$$

Figure 12-4
Chemical structure of some insect sex attractants.

Worker bees lick it from the body of the queen and distribute it to the other workers of the colony in regurgitated food. The active principle has been identified as 9-ketodecenoic acid, the same pheromone that serves as a sex attractant in the virgin queen during the nuptial flight. A second substance acts synergistically with the ketodecenoic acid to constitute the complete "queen substance."

The suppressing action of the queen substance on oogenesis in the workers is reversible. If the queen is removed from a colony, the ovaries start developing in many of the workers who eventually become sexually mature and reproduce.

Mammals. Communication by means of chemical messengers has long been known to play an important role in the life of macroosmatic mammals. This could be inferred, for instance, from the widespread uses of scent glands, urine, and feces to mark territories and trails. It also had long been realized that the males of many mammalian species distinguish estrous from anestrous females by smell. Similarly, females may distinguish between normal and castrated males. Physiological knowledge of the nature and mechanism of function of the sex pheromones in mammalian reproduction is, however, scanty.

Pheromones are known that may influence the estrous cycle in mammals. Both inhibitory and acceleratory pheromones have been described, for example, in the laboratory mouse. The estrous cycle of the mouse is very similar to that of the rat, nonmated females exhibiting regularly occurring cycles of 4 or 5 days duration (see Section 13.4.3). The cycles are, however, less regular than in the rat. Moreover, in contrast to the rat, the duration of the cycles tend to be longer when the females are kept isolated from the males. The introduction of males to previously isolated females has been found to accelerate sexual maturation, even when the male is kept inside a wire cage so that the female cannot make bodily contact with the male. Moreover, exposure of infantile female mice to male odor hastens vaginal opening and the first estrus. The nature of the active male scent is not known, but it is apparently present in the urine and may be a steroid.

When female mice live under crowded conditions in the absence of males, the estrous cycles tend to be depressed and the mice eventually become anestrous. Anestrus also results when mice are kept individually in small cages that prevent bodily contacts. The effect is therefore caused by pheromone that presumably acts through olfaction. When mice are released from a crowded condition the estrous cycles return. This return can be accelerated when males are present.

A third pheromone action has been observed in recently mated female mice. The fertilized eggs normally implant 5 days after mating and gestation ensues (Section 13.6.1). However, if a mated mouse encounters a strange male before implantation has occurred, the pregnancy is interrupted and the female becomes estrous again. The mechanism responsible for the blocking of pregnancy probably involves a factor that acts to identify the individual male; this must be in addition to the general male pheromone.

12.3.2 Color Change

The ability to change color by means of chromatophores is widely distributed in the animal kingdom. The color of animals possessing chromatophores in the skin is determined by the type and pattern of the chromatophores and by the state of dispersion of the pigment within the individual chromatophore. The control of pigment distribution may be either nervous or hormonal or both. In cephalopods, some groups of teleosts (*Cyprinidae* and others), and some reptiles (chamaeleons), the chromatophores are exclusively or predominantly under nervous control. Control seems to be exclusively or predominantly hormonal in crustaceans, elasmobranchs, other groups of teleosts (eel), amphibians, and other reptiles (lizards). In still other teleosts (for example, catfishes and flat fishes), a mixed nervous and hormonal control has been found. Color change by chromatophores is apparently a highly adaptable character that has evolved and disappeared, or has been modified independently, in several systematic groups. However, there is an obvious preference for the use of the slower hormonal control mechanisms.

The most widely distributed function of color change is the adaptation of the color of the animal to the background. Frogs and toads, which are among the best studied, may serve as examples. The most important chromatophores in anurans are the melanophores, which contain dark brown pigment granules. In the adult the degree of dispersion of the melanin granules within the melanophore, and thus the color of the animal, is controlled by melanophore-stimulating hormone (MSH) from pars intermedia of the hypophysis (Figure 12-21). MSH induces dispersion of the granules. Hypophysectomy causes extreme paleness and denervation of the pars intermedia extreme darkness, due to maximum dispersion of melanin granules. The denervation experiments have shown that liberation of MSH from the pars intermedia is controlled by inhibitory nerves arising in the brain. MSH seems to be the only (or at least the dominant) hormone in the control of the melanophores in adult amphibians. The secretion of the hormone is reflexly controlled. The major factor in controlling the rate of secretion seems to be the pattern of illumination of the retina, at least in species in which the adaptation of color to the background dominates in determining the state of the melanophores. The efferent neurons have not been identified, but the inhibitory neurons that innervate the pars intermedia appear to originate in the anterior hypothalamus. Other factors may also influence the color of frogs and toads, such as temperature, humidity, or internal factors determined by the state of the animal. Their mode of action has been little studied.

12.3.3 Metamorphosis

Animals are said to metamorphose when, during development, the body undergoes great and rapid changes in form and function. No sharp boundaries exist between animals with and without metamorphic development phases. Especially conspicuous metamorphoses are found among insects, cyclostomes, amphibians, and some teleosts.

Metamorphosis is usually correlated with profound changes in the life of the organisms, especially in choices of environment and food. Thus, prior to metamorphosis most amphibian larvae are aquatic; after metamorphosis they are terrestrial or amphibious. Anuran larvae generally simultaneously change from a predominantly vegetarian diet to a carnivorous one. In lampreys metamorphosis inaugurates a free-swimming, migratory life in place of the sessile, mud-burrowing habits of the ammocoete larvae.

Insects. Growth and development in insects, as in arthropods generally, take place in a number of stages that are interspaced by sloughing of the old cuticle. In insects with hemimetabolous development, each molt gradually changes the appearance of the larvae in the direction of that of the adult. In insects with holometabolous development, the larval stages are all alike and mainly differ in size; however, all larval stages are strikingly different from the adult phase. In these insects a special stage, the pupa, intervenes between the juvenile and the adult stages.

It therefore appears that hemimetabolous development proceeds through a series of minor metamorphoses, whereas in holometabolous development only the stages preceding the adult phase involve major metamorphic changes. It should further be stressed that, in arthropods, molting constitutes an integral part of growth.

Arthropod growth and metamorphosis are hormonally controlled, and the same hormonal mechanisms have been found at work both in hemimetabolous and holometabolous development.

Experiments on a number of insects have shown that molting, as well as pupation, is normally dependent upon the presence of the brain. An especially instructive series of experiments was performed on the blood-sucking bug *Rhodnius* by

Wigglesworth. This insect requires a meal of blood for each molt to occur. However, if the head is removed after feeding, the larva never molts although it may survive for many months. Molting may be induced if the headless larva is united with another larva that has had its head removed at a later stage after feeding (Figure 12-5). This experiment demonstrates that a factor circulating in the blood is responsible for normal molting.

The brain hormone, ecdysiotropin, has been found to exert its function by means of the protho-racic gland (Figure 12-6) rather than acting directly. In the absence of this gland the brain is unable to induce molting, whereas extracts of the protho-racic glands are capable of inducing molting even in the absence of the brain. The insect molting factor has been isolated and chemically identified as a steroid, named α-ecdysone (Figure 12-7).

Molting in insects is thus initiated by the activation of endocrine neurons in the brain, the neurosecretory cells. In *Rhodnius* the stimulus for secretion of the brain hormone appears to be the

(a)

(b)

(c)

Figure 12-5
Induction of molting by para-biosis in headless individuals of the bug Rhodnius prolixus. *(a) Fourth-stage larva decapitated after the critical period and joined by means of a capillary tube to a fourth-stage larva decapitated 1 day after molting. (b) Giant or sixth-stage larva of* Rhodnius *produced by implanting the corpus allatum of a fourth-stage larva into the abdomen of the fifth stage. (c) Molting fifth-stage larva with a decapitated first-stage larva joined to the tip of the head.* [*From V. B. Wigglesworth.* Control of Growth and Form, *Cornell University Press, Ithaca, N.Y., 1959.*]

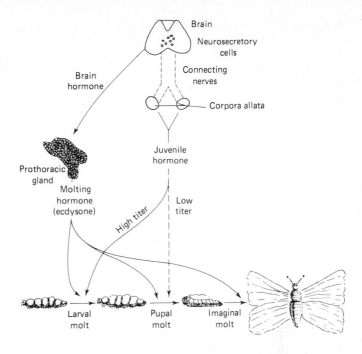

Figure 12-6

Actions of insect hormones in molting and metamorphosis. (Above) Endocrine glands and their secretions. *(Below)* Developmental stages produced by the hormones. *[Modified from P. Karlson.* Kurzes Lehrbuch der Biochemie, *Georg Thieme Verlag, KG, Stuttgart, 1961.]*

distension of the abdomen by the meal of blood. In other insects other sensory mechanisms must provide the nervous stimulation of the neurosecretory brain cells. In locusts it appears to be the process of food intake.

α-Ecdysone has no influence on the metamorphic changes that may occur in connection with the molt. Metamorphic status has been found to be determined by the juvenile hormone secreted by the corpora allata. Extirpation of the corpora allata in a young larval stage, for instance, of *Rhodnius,* results in the larva molting into a miniature adult, thus bypassing the remaining larval stages. On the other hand, implantation of corpora allata into the last larval stage will prevent the final metamorphic change into the adult stage, and molting will result in an additional larval stage. Similar results have been obtained in insects with holometabolous development (Figure 12-6). The result of molting in a caterpillar thus appears to depend upon the amount of corpus allatum hormone that is circulating in the body. Large amounts lead to the emergence of a new larval stage, smaller amounts to pupation. Below the level necessary for normal pupation, molting carries on further toward the adult stage. Extirpation of the corpora allata from a full-

grown caterpillar will result in the larva molting into a monstrous mixture of a pupa and an adult.

The corpus allatum hormone thus controls the metamorphic changes by inhibiting the differentiation of adult characters. In larger doses the hor-

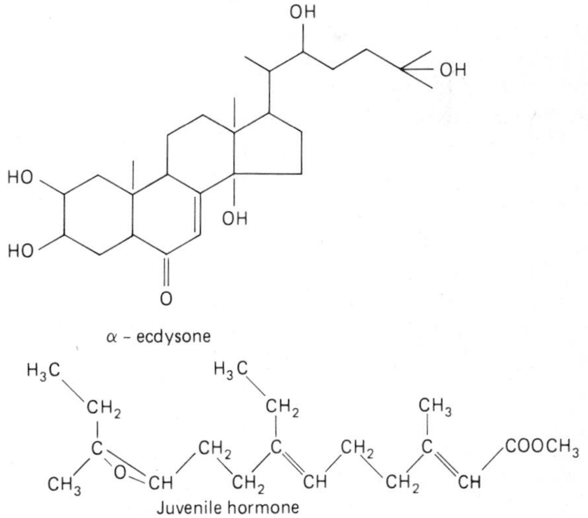

α – ecdysone

Juvenile hormone

Figure 12-7
Two important insect hormones.

mone may even reverse the metamorphic changes and thus juvenilize the molting insect—hence the name (Figure 12-7).

Ecdysone seems to act by inducing synthesis of mRNA (see Section 12.8). Presumably the genome of the insect larva consists of genes both for larval and adult properties; however, normally only the particular combination of genes that corresponds to the developmental stage to be realized through the molt becomes active. The combination of genes to be activated probably depends upon the balance of ecdysone and juvenile hormone secreted prior to metamorphosis. Absence of juvenile hormone tends to activate genes that promote the development of adult characters by the production of the proper types of RNA and enzymes presumably responsible for the character of the bodily differentiations. In the presence of juvenile hormone, ecdysone supposedly activates a somewhat different combination of genes; these consequently result in other patterns of enzymic activity, leading to the realization of juvenile traits to an extent that may vary with the relative amounts of juvenile hormone secreted.

In crustaceans, molting is controlled by the closely related β-ecdysone, which is 20-hydroxy ecdysone (Table 12-2).

Amphibians. Amphibian metamorphosis is also completely dependent upon endocrine activity. The extent to which the metamorphic changes proceed and the rate of change depend upon the amount of thyroid hormones secreted. Inadequate amounts of hormone only carry the metamorphosis to, but not beyond, a certain stage.

Biochemical changes occur in parallel with the metamorphic morphological changes. These changes have been analyzed in some detail, especially in frogs. During metamorphosis the liver starts to produce the enzymes needed for synthesis of urea, the form in which most nitrogen is excreted in the terrestrial life of amphibians (Chapter 7). Plasma proteins change in composition, with albumins being produced in increasing amounts during metamorphosis. In addition, the hemoglobin changes from a larval to an adult type. The intestinal tract increases its production of pepsin, trypsin, and so on. The effects of thyroid hormones in metamorphosis are therefore multiple and extremely diversified. This is suggestive of hormonal action at a fundamental level in the universal processes of differentiation in amphibian metamorphosis. The thyroid hormones apparently exert their action by activating the proper genes at the proper times.

Neoteny. All anurans that start their development as tadpoles normally metamorphose. In contrast, many urodeles do not metamorphose or metamorphose only partly. These forms continue to grow and reach sexual maturity despite the maintenance of typical larval characters. Examples of such permanent larvae are the American mud puppy (*Necturus*), the European cave salamander (*Proteus*), and the axolotl (*Ambystoma mexicanum*). In other species neoteny may occur occasionally in nature, for instance, in *Ambystoma tigrinum*. also known as the Western axolotl (Table 12-3). *A. tigrinum* may or may not metamorphose depending on subspecies and the conditions of the environment.

Failure to metamorphose might be due to (1) reduced sensitivity of larval tissues to thyroid hormones; (2) primarily reduced thyroid glands; or (3) secondarily deficient activity of the thyroid glands due to deficient secretion of thyroid-stimulating hormone from the hypophysis. The first condition probably applies to the families Sirenidae, Proteidae, Cryptobranchidae, and Amphiumidae (Table 12-3). Even large doses of thyroxine have been found ineffective in inducing metamorphosis in species of these families, for instance, *Necturus*, *Amphiuma, Siren,* and *Cryptobranchus*.

Several of the consistently neotenous species of the families Plethodontidae and Ambystomatidae can be induced to metamorphose by treatment with thyroxine. Some species respond only to high levels of thyroxine, such as the cavernicolous salamanders *Typhlomolge* and *Haideotriton*. Other species metamorphose in response to what seems to be physiological levels of thyroxine, such as *Eurycea tynerensis, E. neotenes,* and the axolotl *A. mexicanum*. All the species that occasionally exhibit neoteny in nature apparently react readily to small doses of thyroxine. Several of the species, including the axolotl and *Gyrinophilus palleucus,* also metamorphose in response to treatment with thyrotropin. It would, therefore, appear that the absence of metamorphosis in those species that do metamorphose in response to small doses of thyroxine may be due to a deficient secretion of thyrotropin. The ultimate reason for the deficient thyrotropin secretion may

TABLE 12-3. Degrees of Neoteny Among Urodeles

	Family	Species
1. Permanent larvae	Sirenidae	*Siren lacertina* *S. intermedia* *Pseudobranchus striatus*
	Proteidae	*Necturus maculosus* *N. punctatus* *Proteus anguinus*
	Cryptobranchidae	*Cryptobranchus alleganiensis* *Megalobatrachus japonicus* *M. davidianus*
	Amphiumidae	*Amphiuma means* *A. tridactylium*
2. Consistently neotenous in natural habitat, but metamorphosis can be induced	Plethodontidae	*Typhlomolge rathbuni* *Haideotriton wallacei* *Eurycea tynerensis* *E. neotenes* *E. nana* *E. troglodytes* *Gyrinophilus palleucus*
	Ambystomatidae	*Ambystoma mexicanum*
3. Occasionally neotenous in nature	Ambystomatidae	*Dicampton ensatus* *Ambystoma tigrinum* *A. talpoideum* *A. gracile* *Rhyacosiredon*
	Salamandridae	*Notophthalmus perstriatus* *N. viridescens* *Triturus alpestris* *T. cristatus* *T. helveticus* *T. taeniatus*
	Plethodontidae	*Eurycea multiplicata*
	Hynobidae	*Hynobius lichenatus*

Source: Dent, J. N. "Survey of Amphibian Metamorphosis." In W. Etkin and L. I. Gilbert (eds.), *Metamorphosis,* Appleton-Century-Crofts, New York, 1968,

be found in the failure of thyrotropic hypothalamic structures to become active, as they do in normal metamorphosis (Section 12.4.3).

Ammocoetes. The absolute dependency of amphibian metamorphosis on the activity of the thyroid gland has led quite naturally to the expectation that metamorphic changes in other vertebrate groups may also be induced by thyroid hormones. The role of thyroid hormones in mammalian development, where they are indispensable for normal growth, differentiation and maturation, made such an expectation reasonable. However, metamorphosis in the ammocoete larva appears not be depend upon thyroid activity. No metamorphosis-promoting activity has been observed for thyroxine in ammocoetes and extirpation of the endostyle does not prevent metamorphosis. Still, metamorphosis seems to be hormonally controlled because both pinealectomy and hypophysectomy have been found to inhibit metamorphosis.

Teleosts. The ontogenies of several diverse teleosts include more or less pronounced metamorphic changes. Salmon, eels, and mudskippers (*Periophthalmus*) undergo metamorphic changes, changes that, in the mudskipper, adapt them for a life largely out of the water. In these fishes, however, available evidence appears to indicate that the thyroid hormones do not act as specific metamorphosis-inducing hormones, at least not in ways that are comparable to their function in amphibians.

12.3.4 Molting

The keratinization of the outmost part of the epidermis plays an important role in the adaptation of tetrapods to a terrestrial life. The keratinized layer protects the body against the impact of the environment, which may be both of mechanical and chemical nature. These functions are poorly developed in amphibians, in which the cornified epidermis generally includes only one cell layer. In the squamate reptiles, the lizards and snakes, about ten cell layers may keratinize to produce one continuous layer. In birds and mammals keratinization likewise includes a number of cell layers, but additional, especially active cell proliferation and keratinization processes take place in specialized epidermal structures, the follicles, which produce feathers and hairs.

The plumage of birds and the pelage of mammals have functions in addition to those served by the integument in reptiles. The heat-insulating properties of plumage and fur are important in temperature regulation, and color and other properties may play a role as secondary sex characters.

Keratinized integumentary structures are renewed from time to time, most often according to a regular and characteristic pattern. In amphibians and squamate reptiles, the outer keratinized layer loosens from the underlying parts of the epidermis and the slough is shed. Special behavior for getting rid of the slough has generally been developed, and the slough is often eaten. Molting frequency varies with the temperature, and molting may be inhibited at low temperatures during winter. At normal summer temperatures several amphibians have been found to molt about once to several times a week, whereas reptiles molt at monthly or longer intervals. Recurrent molting

helps maintain an intact stratum corneum, which may be the main or sole function of molting in amphibians and reptiles.

In birds and mammals the renewal of the stratum corneum takes place continuously, generally as an inconspicuous desquamation. In contrast, plumage and pelage are molted, often in rather complex patterns that vary with species. Variations occur in the order in which feathers and hairs are shed. Molting often proceeds as waves spreading over the body. The duration of the molt may vary as well as its periodicity. Molting usually occurs once or twice a year, both in birds and mammals, but more frequent molting may occur. Molting may also be continuous and asynchronous. Some animals (for example, elephant seals) include stratum corneum along with the shedding of the hairs, reminiscent of reptilian molting.

Molting is under hormonal control or influence. In amphibians the hormonal mechanisms are different in urodeles and anurans. In urodeles normal molting depends upon thyroid function. Extirpation of the thyroid, or hypophysis, inhibits shedding of the slough but not its formation. Proliferation and keratinization of new cornified layers continue. These new layers consequently pile up as a thick keratinized layer on the surface of the animal. Injection of thyroid hormone will normalize the skin.

The thyroid hormones further strongly influence the frequency with which molts occur. Injection of a large dose of thyroxine in metamorphosed *Ambystoma* thus speeded the frequency of molting until the resting periods between the molts almost disappeared.

In toads, and perhaps in other anurans, similar abnormalities appear in molting in the absence of adrenocortical activity as appear in urodeles in the absence of the thyroid, whereas in the toad molting is normal in the absence of the thyroid.

Periodic formation of new cornified layers in amphibians appears to be an autonomous function of the skin; however, its normal course requires the presence of thyroid hormones in the blood of urodeles and of corticosteroids in at least some anurans. Little is known about the role of hormones in molting in the squamate reptiles, although thyroid function is presumably involved. Thus, in the shovel-nosed snake, *Chionactis occipitalis,* thyroidectomy increases the frequency of molting, whereas treatment with thyroxine permanently inhibits molting. Thyroxine apparently acts by inhibiting the proliferation of the stratum germinativum, which normally gives rise to a new generation of epidermal cells, the differentiation and keratinization of which is followed by the sloughing of the old keratinized epidermis.

In birds, growth and development of the feathers and the normal pattern of the molt depend upon hormonal control, but the mechanisms are complex and presumably varying between the species. Gonadal and thyroid hormones seem to be the most important. Molting is usually adapted to occur outside the breeding season, often as a postnuptial molt. In these instances the onset of molting is probably causally related to the decrease in internal secretion of the gonads that follows the end of the breeding season. In many, but not all, species of birds treatment with thyroxine will precipitate a molt also outside the normal molting period. The importance of gonadal activity in synchronizing the molt is indicated by the finding that gonadectomized birds molt continuously.

Mammalian hair growth and molting cycles are generally less influenced by elimination of the endocrine system than are the equivalent processes in birds and lower vertebrates. The cyclic activity of the hair follicles presumably constitutes an autonomous epidermal activity. The activity may, however, be greatly influenced by hormones, and the molting patterns typical of the various mammalian species appear to be largely determined by an interplay between a number of endocrine glands. Generally, growth of hairs and molting are stimulated by thyroxine and retarded by estrogens and corticosteroids. A molt often occurs in connection with the breeding season. There is some indication that, in the varying hare, this molt may be induced by direct action of gonadotrophic hormones and not indirectly by means of gonadal secretions.

It thus appears that, in recurrent molting that proceeds rather independently of external conditions, the role of the endocrine system is of a permissive character. Molts adapted to seasonal rhythms, however, appear to be directly induced by increased secretion of one or another hormones or by changes in the hormonal balance in the blood.

12.4 HORMONAL MECHANISMS

In the preceding sections emphasis was on the role of hormones in the lives of animals, without going into details about the hormonal mechanisms involved. In this section emphasis will be on hormonal mechanisms, but considered from a broader biological context. Hormonal mechanisms vary in complexity with respect to the degree of complexity of the physiological functions they serve. This section describes hormonal mechanisms in a sequence based upon increasing complexity of the functions.

12.4.1 "Hormonal Reflexes"

Comparisons between nervous coordination and hormonal coordination often stress that the former appears to have been adopted in functions that call for fast and precisely graded activity of the effector structure. Conversely, the endocrine system usually controls more slowly developing, and often protracted, activities. Further, nervous control is restricted to the structure or organ innervated, whereas the effects of a hormone are typically exerted everywhere in the organism where cells or structures that are capable of responding to the particular hormone are present. Hormone effects are, therefore, diffuse or localized depending upon the distribution in the organism of target structures. Hormone effects can, moreover, be relatively fast and short-lived, and may thus acquire properties that usually characterize nervously regulated mechanisms. Some examples of localized and relatively fast endocrine mechanisms will illustrate hormonal correlations of this type.

Milk Let-Down. In nursing mammals, the suckling of the young reflexly elicits milk let-down. The phenomenon is produced by the contraction of muscle fibers that surround the secretory lobules of the mammary glands. Milk stored in the glandular ducts is driven towards the nipples. It is only through this reflex that the milk becomes available to the suckling young. The reflex involves an afferent pathway from sensory cells in the nipples to the CNS. The efferent pathway is represented by hormone-secreting hypothalamic neurons that terminate in the pars nervosa of the neurohypophysis. The suckling thus reflexly stimulates the secretion of the hormone oxytocin (Table 12-1) from these neurons. By way of the systemic circulation, the hormone reaches the mammary glands and induces the muscle cells to contract. In addition, oxytocin participates in other hormonal reflexes in the organism, reflexes connected with uterine functions. The muscles of the uterus are sensitive to oxytocin and they contract in response to oxytocin that is reflexly liberated during parturition and coitus (see Section 13.6).

12.4.2 Hormones in Homeostatic Mechanisms

An important group of regulatory mechanisms in the body aims at maintaining the normal composition and volume of the body fluids. It was Claude Bernard who in the middle of the last century realized the importance of the constancy of the internal milieu to the survival and normal functioning of the cells of the body. Claude Bernard was mainly concerned with the organic composition of the blood as the nutritional medium of the cells. Later studies on the chemical composition of the body fluids and their regulations placed equal emphasis on the inorganic constituents (Chapter 7).

The mechanisms preserving the stability of the body's internal environment are often termed homeostatic mechanisms, an expression introduced by W. B. Cannon in the 1920s. Cannon was led to his concept of homeostasis in the organism mainly from his studies on the role played by the sympathetic nervous system in regulating and stabilizing the functions of the body. More recently numerous instances of hormonal homeostatic mechanisms have been found.

Bernard's concept of the fixed internal environment merged into Cannon's concept of homeostasis. In the same way, the latter merged into the wider concept of cybernetics, which, in the 1940s, was introduced by the mathematician Norbert Wiener to cover the entire field of control and communication theory, whether in machines or in animals. The term is derived from the Greek word for steersman.

Cybernetics originated from the observation that the nervous control of the voluntary movements showed striking formal similarities to technologists' servosystems. Negative feedback control in such systems turned out to be a useful analog for the neurophysiologist in describing normal and defective functioning of the muscular system. It was soon realized that the feedback concept had much wider implications for physiology and could be used to describe a great number of diverse regulatory mechanisms in living organisms. The concept was soon adopted by endocrinologists to describe the reciprocal relationships between the hypophysis and the peripheral glands (see Sections 12.4.3 and 13.5). It has now become the basic concept in the large group of hormonally mediated mechanisms that serve to maintain the steady states of organisms. The mechanisms vary from relatively simple to highly complex, as will be seen from the following examples.

Regulation of Calcium Ions in the Blood. Deviations of calcium ion concentrations in the blood from the normal range are harmful to the organism. Hypocalcemia may cause nerve fibers to fire spontaneously and may thus produce tetany. Hypercalcemia may result in calcification of nonosseous tissues. The parathyroid glands maintain blood calcium within normal limits by means of a mechanism that is diagrammatically depicted in Figure 12-8. In the control system, the parathyroid cell measures the concentration of calcium in the blood, compares the measured concentration with that wanted, and adjusts for differences. If the measured concentration of calcium is lower than the one "set" by the organism, the parathyroid cell increases its secretion of hormone to the blood and induces the effector organ, the bones, to give off calcium to the blood. If the concentration of calcium is higher than the normal set-point, the secretion of parathyroid hormone is reduced by the negative feedback action of calcium on the parathyroid cell. That the parathyroid cell acts as receptor and regulator in the servosystem has been shown in perfusion experiments. If the isolated parathyroid gland is perfused with blood of varying calcium content, the secretion of parathyroid hormone to the perfusate varies inversely with the concentration of calcium.

The parathyroid hormone was the first known endocrine regulator of blood calcium, and it was long thought to be the only one. Now several others have been added. The second one to be recognized was calcitonin produced by specific cells, the C cells, usually embedded in the tissue of the thyroid gland. The secretion of calcitonin is stimulated by

Figure 12-8
Diagrammatic representation of main features in the homeostatic regulation of calcium in the blood.

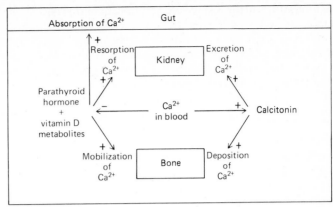

+ stimulation effect
− inhibitory effect

increased concentration of calcium in the blood. Calcitonin reduces blood calcium by enhancing its deposition in the bones. The parathyroid hormone and calcitonin thus cooperate in securing calcium homeostasis. The parathyroid hormone seems to be the prime mediator of calcium homeostasis. Extirpation of the parathyroids causes hypocalcemia, which cannot be prevented by the additional extirpation of the calcitonin-secreting tissue. On the other hand, blood calcium remains normal in the absence of calcitonin, except for a passing abnormal rise after feeding of starving animals. It is thus suggested that calcitonin represents a control mechanism that is especially adapted to handle the discontinuous introduction of calcium into the blood that results from a discontinuous food uptake.

More recently metabolites of vitamin D have been found to exert important functions in the calcium homeostasis, especially 25-hydroxycholecalciferol produced in the liver and 1,25-dihydroxycholecalciferol produced in the kidneys. These hormones are produced in amounts that vary inversely with the amounts of calcium in the diet, and they seem to be closely integrated with the parathyroid hormone in their activities in the body (see Figure 12-8).

Osmotic and Volume Regulation of Body Fluids. A diversity of mechanisms keep both the volumes and the osmotic concentrations of the body fluids within ranges that differ among species (Chapter 7). In vertebrates, hormones play crucial roles in these homeostatic mechanisms. The endocrine mechanisms are reasonably well understood only in mammals, but major features of the mechanisms probably also apply to other vertebrate classes, at least those belonging to the tetrapods. The hormones involved are common to vertebrates except cyclostomes. They include neurohypophysial antidiuretic hormone (ADH), angiotensin, and a corticosteroid hormone, aldosterone.

The function of ADH was the first to be established. It had long been known that, in mammals, extirpation of the neurohypophysis renders the kidney incapable of producing a urine hyperosmotic to the blood in response to dehydration or salt-loading of the organism (Chapter 7). The kidneys will continue to produce large volumes of dilute urine, a diuretic condition known as diabetes insipidus. The loss of ADH is the primary cause of this condition.

Dehydration or salt-loading increases the osmotic pressure of the blood. It was previously generally accepted that it was the increased osmotic pressure itself that acted as a stimulus for the secretion of ADH into the blood, and that osmoreceptors were located in the hypothalamus close to the neurons that produce ADH. More recent evidence, however, indicates that the receptors are not true osmoreceptors, but instead are sensitive to the concentration of sodium ions. They are located in the anterior wall of the third ventricle of the brain, in areas outside the blood-brain barrier.

Stimulation of the osmo- or sodium-receptors apparently activates nerves that terminate on the cell bodies of the hormone-producing cells. These neurosecretory cells also act as neurons, and action potentials propagated along the nerve fibers trigger the release of ADH, stored in the nerve terminals of the neurohypophysis, in proportion to the impulse frequency.

A reduction of blood volume, such as may be induced by bleeding, without any change in osmotic or sodium concentration of the blood also stimulates secretion of ADH. The secretion seems to be reflexly stimulated via volume receptors within the vascular system.

Fluid depletion, whether caused by bleeding or dehydration, produces physiological responses that cannot be completely explained by the ADH mechanism. The renal renin-angiotensin system is apparently also involved (see Section 12.2.2). A reduction in blood volume releases renin from the kidney into the blood circulation, and thus results in the production of angiotensins. Angiotensin II coordinates a variety of activities in the organism, all serving the common goal of maintaining or restoring the blood volume. These activities include stimulation of thirst, secretion of ADH, and renal reabsorption of sodium. Angiotensin probably stimulates thirst by acting directly upon the hypothalamic nervous structures that regulate drinking. In rats, injection of minute amounts of angiotensin into this area of the hypothalamus will cause excessive drinking even in already hydrated animals. The renal effect is indirect. Angiotensin stimulates the adrenal cortex to secrete the mineralocorticoid aldosterone, which in turn stimulates the reabsorption of sodium ions from the fluid passing the renal tubules.

The function of angiotensin in the control of water balance may have evolved quite early in the history of the vertebrates. Angiotensin has been found to be a potent agent in stimulating drinking in teleosts.

The pars nervosa plays a role in the homeostatic regulation of water balance in all terrestrial vertebrates. However, the nature of the regulatory mechanisms varies between the groups. In amphibians the antidiuretic hormone (vasotocin) diminishes urinary water loss primarily by reducing glomerular filtration rather than by increasing tubular reabsorption of water (see Chapter 7). Additionally it may act on the skin and urinary bladder to increase their permeability to water, resulting in increased water absorption. The efficiency of the water-conserving action of vasotocin in amphibians varies greatly with the ecology of the species. The hormone may exert little or no influence in purely aquatic forms (for example, *Xenopus*) but it produces maximum effects in primarily terrestrial groups, such as toads (Bufonidae). In freshwater turtles and water snakes vasotocin first decreases the glomerular filtration rates by stopping the blood flow through some of the glomeruli; with increasing dose it also increases the tubular reabsorption of water. In arid-adapted lizards, like *Sceloporus cyanogenys* and *Varanus gouldii*, vasotocin exhibits little antidiuretic effect at the level of the kidneys, but it appears to stimulate water uptake from the urine within the cloaca.

The significance of the pars nervosa in the regulation of the water balance of an animal may also be evaluated by observation of the effects of elimination of the gland. Such experiments showed that even in *Bufo* water balance is nearly normal in the absence of the pars nervosa. No diabetes insipidus ensues from elimination of the vasotocin-producing neurons, and dehydration in the absence of vasotocin effectively reduces urine flow and increases permeability of the skin to water. The pars nervosa, therefore, probably became integrated into the homeostatic water balance mechanisms of the body early in the adaptation to terrestrial life in vertebrates. However, the dominant position of the gland in the regulatory mechanisms appeared later. Birds develop polyuria after extirpation of the pars nervosa.

Regulation of Blood Sugar. Animals that eat and digest discontinuously absorb fuel independently of actual metabolic needs. In the absence of regulatory mechanisms, fuels circulating in the blood, such as glucose, amino acids, and lipids, would fluctuate greatly concurrently with the pattern of feeding and muscular activity. A pioneer in the field, Claude Bernard, realized the importance of maintaining a constant organic composition of the blood, and he was especially concerned with the regulation of the blood sugar, which he studied in both vertebrates and invertebrates.

Glucose occupies a central role as a primary fuel in most animals, and therefore deserves special attention. More recently, however, it has become clear that the amino acids and lipids are also major components in the homeostatic regulation of the circulating fuels. Glucose has been studied much more intensively for a much longer period of time, largely because techniques for the determination of glucose in the blood and tissues of animals became available to physiologists as long ago as the middle of the last century. Routine techniques for determining amino acids, free fatty acids, and glycerol in blood are of considerably more recent origin.

The mechanisms used by vertebrates to maintain blood glucose within normal concentration ranges involve two hormones secreted by the pancreatic islets: insulin, produced by the B cells, and glucagon, produced by the A cells. The relative importance of these hormones varies between vertebrate classes. Thus, glucagon seems to be the most important in birds, whereas insulin is more important in mammals and in some lower vertebrates. Recently, a third hormone, somatostatin, produced by the D cells of the pancreatic islets, has been shown to play a role in the regulation of blood glucose in mammals, in which the mechanisms of regulation are much better understood than they are in nonmammalian vertebrates and invertebrates. The diagram in Figure 12-9 depicts the main features of present concepts of the complex interplay between the hormones produced by the pancreatic islets that are active in regulation of blood glucose in mammals.

In the 1920s Banting and Best found that injection of extracts of the endocrine pancreas normalized (reduced) the high blood glucose in dogs that had been made diabetic by extirpation of the pancreas. The hormone insulin was responsible for the effect. It is the only hormone that reduces the concentration of glucose in the blood, whereas several hormones are known to increase blood glucose. Insulin is indispensable for the maintenance of normal glucose concentrations in the extracellular fluid in mammals. Insulin was first thought to reduce glucose concentrations by facilitating the uptake, and thus the catabolic breakdown, of glucose in the tissues and organs of the body, especially the muscles. Now several more mechanisms have been accepted, including stimulation of glycogen synthesis from glucose in the liver and lipogenesis in the fatty tissues.

Insulin secretion is directly controlled by the level of glucose in the blood. Increasing blood glucose levels, which normally prevail during the absorptive phase after a meal, will stimulate the B cells to increase secretion of insulin. This initiates or intensifies the array of processes that tend to

Figure 12-9
Diagram of function of pancreatic islets in regulating level of glucose in the blood. A, Glucagon producing cell; B, insulin producing cell; D, somatostatin producing cell.

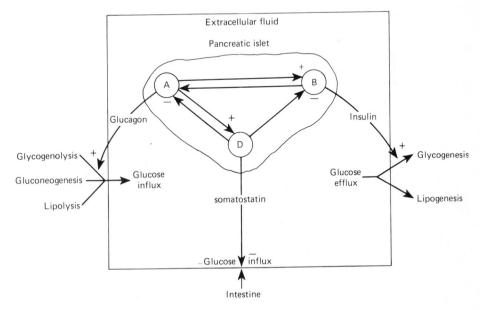

curb the increase in blood glucose and maintains the concentration within the normal range. Increased blood levels of amino acids also act to stimulate insulin secretion, as do some gastrointestinal hormones secreted in connection with the meal. Simultaneously with the increase in secretion of insulin the secretion of glucagon is reduced. Insulin seems to act on the A cells to inhibit the secretion of glucagon.

When blood glucose levels are decreasing, as they tend to in the intervals between meals, the secretory pattern is reversed, insulin being secreted at decreasing rates, glucagon at increasing rates. Glucagon acts to increase blood glucose by several mechanisms, including stimulation of glycogenolysis in the liver and, when the glycogen reserves have been depleted, stimulation of gluconeogenesis from amino acids in muscles and glycerol from adipose tissues.

The role of somatostatin in the regulation of the blood glucose is less clear than are the roles of insulin and glucagon. The secretion of somatostatin is stimulated by the same factors that stimulate the secretion of insulin, that is high levels of blood glucose, amino acids, and certain gastrointestinal hormones. The hormone seems to inhibit secretion of both glucagon and insulin. It also seems to inhibit the influx of glucose and amino acids that normally follows digestion of a meal. In the absence of D cells, the organism develops a state of diabetes mellitus that is characterized by high levels of secretion of both glucagon and insulin. The pancreatic islets apparently act as functional units in the homeostatic regulation of glucose and amino acids in the blood. The regulation is based on an interplay between the three cell types A, B, and D, probably involving local mechanisms acting among the cells.

12.4.3 Hierarchical Mechanisms

The regulation of the levels of thyroidal, adrenocortical, and gonadal hormones in the blood is not directly related to the processes that these hormones control. The regulation is indirect, by means of a superior link, the hypophysis.

The regulatory mechanism of thyroid hormones is shown diagrammatically in Figure 12-10. Thyrotropic hormone from the hypophysis stimulates the thyroid gland to secrete hormones, thyroxine and

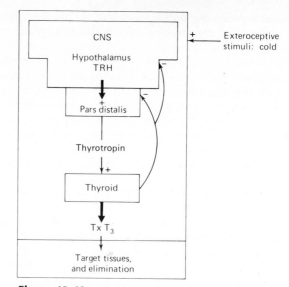

Figure 12-10
Diagram of neuroendocrine mechanisms of control of thyroid function.

triiodothyronine, to the blood. In the absence of the hypophysis the vertebrate thyroid gland secretes only at low rates. Elimination of thyroid hormones by the liver and kidneys constantly tends to reduce the level of hormones in the blood. The concentration of thyroid hormones in the blood, however, determines the rate at which thyrotropin is being secreted by the hypophysis. Decreasing levels of thyroid hormones in the blood enhance the secretion of thyrotropin, whereas increased levels of thyroid hormones depress thyrotropin secretion. The diagram suggests that the thyrotropin-producing cells of the hypophysis are also the receptors that measure the concentration of thyroid hormones in the blood. However, an additional receptor is often located outside the hypophysis in the brain. The CNS thus becomes superior in relation to the hypophysis, and a hierarchical neuroendocrine control system builds up.

The importance of the CNS in the control of thyroid function can be assessed by determining the effect of interrupting the normal connections between hypophysis and brain, for instance, by transplanting the hypophysis. The effect of the operation varies greatly. In some species, the thyroid continues to secrete hormones at normal rates, in other species the secretion is strongly reduced. In the first instance the negative feedback regulation of the concentration of thyroid hormone is apparently exerted at the level of the hypophysis; in the second the principal control is exerted by the brain, where receptors sensitive to thyroid hormones have been located in the hypothalamus. The hypothalamic receptors, in turn, regulate the rate of secretion of the thyrotropin-releasing hormone (TRH) from neurons terminating in the median eminence. The TRH finally reaches the thyrotropin-secreting cells of the pars distalis via the hypothalamic-hypophysial portal circulation (Figure 12-11).

Homeostatic regulation by means of negative feedback may secure the daily basic needs for thyroid hormones in the tetrapods and perhaps in other vertebrates. However, other mechanisms of control can become superimposed on the homeostatically working mechanisms under situations of special demands for increased thyroidal activity, as in the following examples.

Amphibian metamorphosis depends upon the presence of sufficient amounts of thyroid hormones in the circulating blood (see Section 12.3.3). Normally metamorphosis appears to be induced by increased thyroid activity in response to increased rates of secretion of TSH. However, the larvae of *Ambystoma tigrinum* have been found not to metamorphose if the hypophysis is transplanted ectopically. The primary stimulus to amphibian metamorphosis is probably an increase in the central nervous stimulation of the TSH-secreting hypophysial cells.

Mammalian temperature control includes, among other mechanisms, heat production due to increased rates of secretion of thyroid hormones (see

Figure 12-11
Hypothalamic-hypophysial portal circulation in a tetrapod (Bufo bufo). *Code: 1, optic chiasma; 2, median eminence; 3, portal vessels; 4, pars distalis; 5, endolymphatic sacs.* [*From K. G. Wingstrand, original.*]

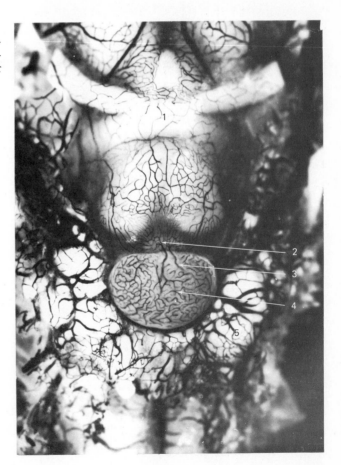

also Chapter 8). Thus, exposure of a mammal to cold environments may result in the immediate liberation of TSH from the hypophysis, presumably due to activation of central nervous TSH-stimulating structures. Cooling of skin receptors, by means of reflexes to hypothalamic structures, appear to be responsible for the increased secretion of TSH and thus for secretion of thyroid hormone, which finally results in increased heat production (Figure 12-10). Hypothalamic structures acting as centers of temperature regulation can also be directly thermally stimulated. The structure termed the *heat-loss center* seems to react upon direct cooling by stimulating TSH secretion from the hypophysis (Figure 12-12).

The features that characterize the integrated function of brain, hypophysis, and thyroid can also be recognized in the interactions between brain, hypophysis, and adrenocortex or gonads. The incorporation of central nervous mechanisms in the neuroendocrine machinery has probably evolved as an adaptation to changes in the environment that may demand fluctuations in the secretory activity of the peripheral endocrine glands, the thyroid, adrenals, and gonads. In the above mentioned example, the environmental factor was the temperature acting via receptors in the skin. Other factors are the diurnal or annual rhythms of light acting via the eyes, chemical stimuli acting via olfactory or gustatory receptors, and so on. Other important neuroendocrine mechanisms, which involve rhythmic hormone secretion from the peripheral glands, may proceed in the absence of exteroceptive stimuli; these are based on inherent rhythmic activity in the brain structures that control the secretion of the trophic hormones of the pars distalis. The most complex of these mechanisms are probably those that control reproduction. This is true of both vertebrates and invertebrates (see Chapter 13).

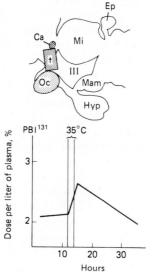

Figure 12-12
*Effect of cooling of hypothalamic heat-loss center on the rate of TSH secretion. (Above) Median sagittal section of diencephalon of the goat showing the placement of the thermode whose temperature can be adjusted by the circulation of water of different temperatures. (Below) Increase in TSH secretion upon 2 hr cooling at 35°C. The rate of TSH secretion is evaluated from the increase in protein-bound iodine, labeled with ^{131}I, in the plasma. Code: Ca, anterior commissure; Ep, epiphysis; Hyp, hypophysis; Mam, mammillary body; Mi, massa intermedia; Oc, optic chiasma; III, third ventricle. [After B. Andersson, et al. Acta Physiol. Scand., **63**, 189 (1965), modified.]*

12.5 PROPERTIES OF ENDOCRINE CELLS AND THE NATURE OF THEIR HORMONES

The majority of hormones that have been chemically identified, both in invertebrates and vertebrates, are either derived from amino acids or are steroids. Many endocrine glands produce hormones that are proteins or peptides. Some of these glands are the adenohypophysis, the thyroid and parathyroid glands, the gastric and intestinal epithelium, and the pancreatic islets (Table 12-1). All these glands differentiate from the epithelium of the gut.

Endocrine neurons also constitute an important group of cells that produce peptide hormones. These neurons are termed *neurosecretory neurons,* and are widely distributed in the animal kingdom. They dominate endocrine systems in invertebrates. In vertebrates, hypothalamic neurons produce the pars nervosa hormones and the hormones that control the secretion of hormones from the pars distalis (Table 12-1).

It is noteworthy that the endocrine neurons that terminate in the pars nervosa also seem to act as typical neurons in propagating action potentials (see Section 12.4.2). This combination of endocrine and nervous activities may be characteristic of neurosecretory neurons.

Free amino acids or their derivatives can also act as hormones. An important group is the *biogenic amines* (see Section 12.2.2). They include epinephrine, norepinephrine, histamine, serotonin, and melatonin (Figure 12-13). *Epinephrine* and *norepinephrine,* both derived from the amino acid phenylalanine, are produced by the adrenal medulla. *Histamine* is derived from the amino acid histidine and is a widely distributed hormone that acts locally in the tissues of the body. *Serotonin,* 5-hydroxytryptamine, is derived from tryptophan and is formed in many tissues of the body, including the CNS. It is of widespread occurrence in inverte-

brates. It may act both as a local hormone and as a neurotransmitter. *Melatonin* is derived from serotonin and synthesized in the pineal gland. In some species it seems to be involved in the light-regulated control of the reproductive cycle (see Section 13.5).

Norepinephrine and serotonin are well established examples of amino acid derivatives that can function as hormones or as nerve transmitters, depending upon the site of their release. More recent studies have indicated that such dual functions may be widespread among the well-known hormones of the amino acid group. Thus, a number of peptide hormones, including ACTH, gastrin, and cholecystokinin, have been localized in the brain, where they seem to function as transmitter substances or modifiers of transmitter action. Other peptides that were first isolated from the brain, such as the gonadotropin releasing hormone, somatostatin, and the morphinelike enkephalins have later been isolated from the gut.

Steroid hormones include the hormones of the adrenal cortex and the gonads in vertebrates (Figure 12-14) and the molting hormones in arthropods (Section 12.3.3). Steroids and their synthesizing and metabolizing enzymes probably occur generally in invertebrates, but conclusive demonstrations of their biological significance in nonarthropod invertebrates are lacking.

The pathways for the synthesis of the hormones

Figure 12-13
Biogenic amine hormones.

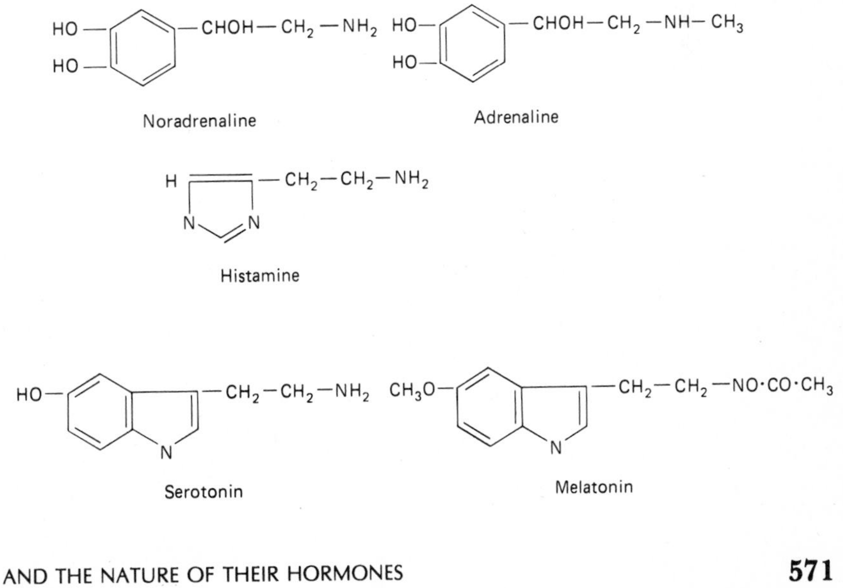

Noradrenaline

Adrenaline

Histamine

Serotonin

Melatonin

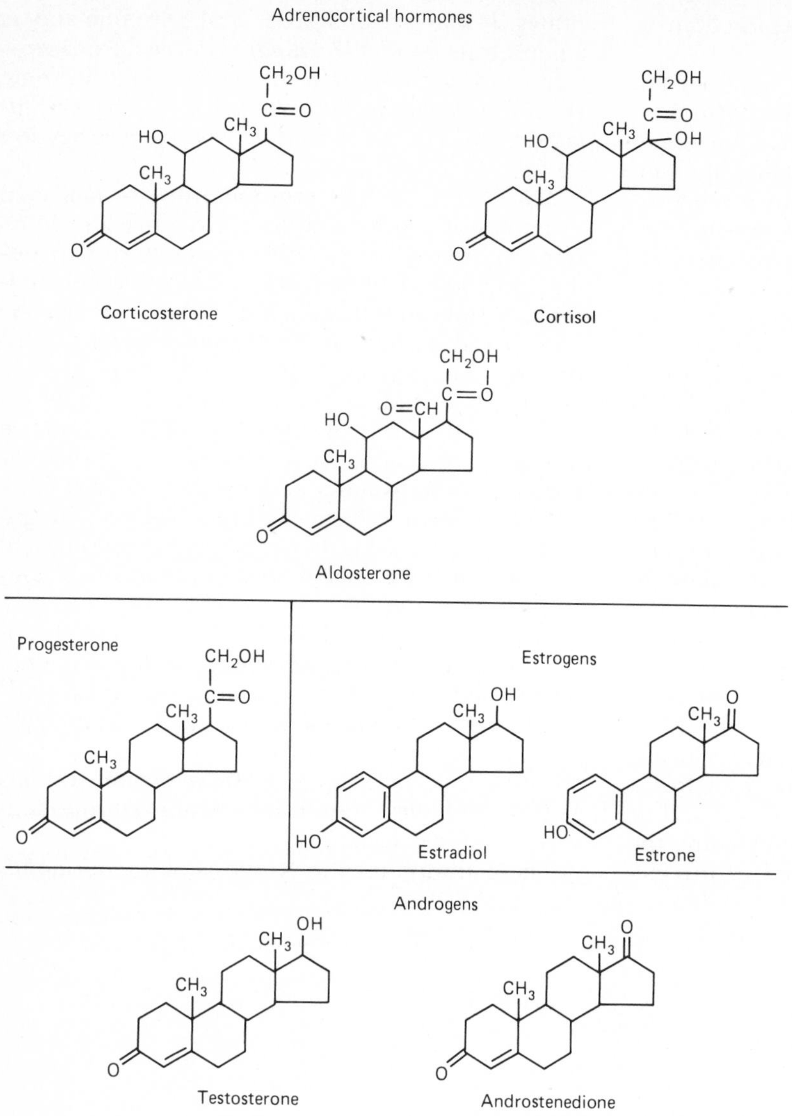

Adrenocortical hormones

Corticosterone

Cortisol

Aldosterone

Progesterone

Estrogens

Estradiol

Estrone

Androgens

Testosterone

Androstenedione

Figure 12-14
Some important steroid hormones from the adrenals and the gonads.

are fundamentally similar in all vertebrate steroidogenic cells. Production of the specific steroids appears to be established by rather slight differences in the synthetic and secretory machinery. Generally, cholesterol constitutes a starting point in the production of steroid hormones, and progesterone appears as an intermediary product that, in addition, has acquired independent function as a hormone of the corpus luteum. Testosterone, too, may function both as an intermediary product, for example, in ovarian cells, or as the final product, in the interstitial cells of the testes (Figure 12-15).

12.6 EVOLUTION OF ENDOCRINE ORGANS

The evolution of endocrine organs and structures is especially well illustrated by the glandular structures that are derived from the endoderm of the

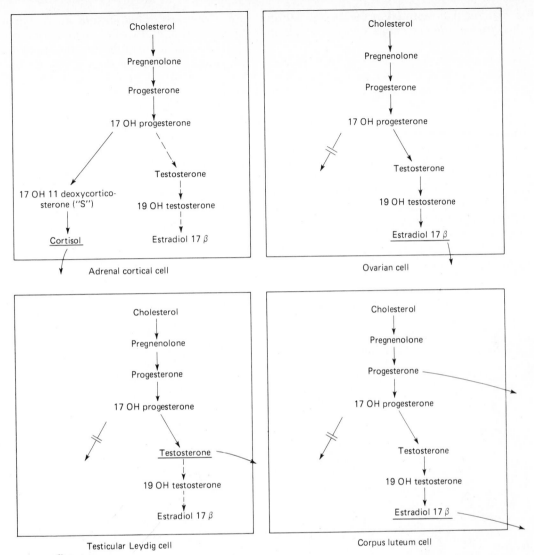

Figure 12-15
Secretory pathways in different steroid-hormone-producing cells: adrenal cortex, ovarian follicle, corpus luteum, and testicular Leydig cell. The diagram is designed to suggest that a common enzymatic pattern underlies the capability of all of these cells and that biochemical differentiation involves deletion or deemphasis of certain pathways, with the result that one or another steroid becomes the predominant secretory product. [*From C. P. R. Keele, and E. Neil.* Samson Wright's Applied Physiology, *Oxford University Press, Fair Lawn, NJ, 1965.*]

branchial region of the anterior gut: the thyroid and the parathyroids.

Thyroid Glands. The various components of the vertebrate endocrine system are characteristic of this class and not found outside the vertebrates, not even in the nonvertebrate groups of the phylum Chordata. However, in one instance, the thyroid gland, it has been possible to trace the origin back to prevertebrate stages and get an idea of the adaptational changes that resulted in the incorporation of the structure in the vertebrate endocrine system [Barrington, 1964].

Nonvertebrate chordates such as the tunicates (Urochordata) and the lancelets (Cephalochordata), as well as the ammocoete larvae of the lampreys (Cyclostomata), all possess an endostyle, whose secretion of mucus plays an important role in feeding mechanisms. During metamorphosis the endostyle of ammocoetes is known to be transformed into the thyroid gland of the adult. Those parts of the endostyle that secrete the mucus atrophy, and other parts of the endostylar epithelium give rise to the formation of the thyroid tissue. It is significant that the ability to bind iodine and to synthesize thyroxine and the other iodinated tyrosine derivatives is already present in the endostyle or the epithelium that differentiates into the adult thyroids. This epithelium can be visualized by autoradiography of endostylar tissue after addition of radioactive iodine to the medium in which the animals are living (Figure 12-16). Iodine binding and synthesis of iodinated tyrosine derivatives, even the thyroid hormones (Figure 12-17), have also been demonstrated in the endostyles of *Amphioxus* and of tunicates, thus supporting the other evidence on the homology of the endostyles in all the chordate groups. Studies with radioactive iodine in various other invertebrate groups have shown that iodine binding is not restricted to the protochordates. However, in invertebrates the synthesis of the biologically active derivatives triiodothyronine and thyroxine seems to be absent or inconspicuous.

Nothing seems to be known about the possible functions of the iodinated compounds produced

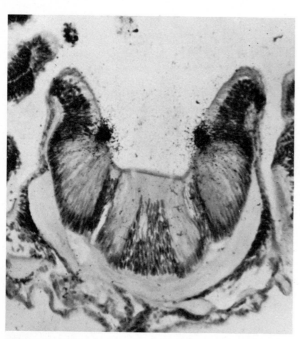 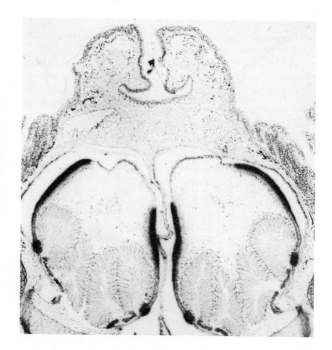

Figure 12-16
Left: *autoradiogram of a transverse section through the endostyle of an amphioxus which had been immersed in seawater containing radioactive iodide. Protein-bound* [131]*I is visible as two black areas.*
Right: *autoradiogram of a transverse section through the anterior region of the endostyle of an ammocoete larva which had received an intraperitoneal injection of radioactive iodide. Note the absence of protein-bound* [131]*I from the glandular tracts. [From E. J. W. Barrington.* An Introduction to General and Comparative Endocrinology, *Clarendon Press, Oxford, 1963, p. 183, and M. Clements-Merlini.* J. Morphol., **106**, *337 (1960).]*

Figure 12-17
Hormones of the thyroid gland.

HO — ⟨⟩ — O — ⟨⟩ — CH₂· CH(NH₂) COOH

Thyroxine (3,5,3′,5′ – Tetraiodothyronine)

HO — ⟨⟩ — O — ⟨⟩ — CH₂· CH(NH₂) COOH

3,5,3′ – Triiodothyronine

by the endostyle. Presumably they do not act as hormones; they appear to be secreted into the pharynx along with the other secretions of the endostyle to be ingested and eventually assimilated by the animal. The iodinated tyrosine derivatives were thus the secretory products of a structure with alimentary function prior to its transformation into a structure with endocrine function. We do not know when in the course of evolution this transformation took place.

In the other vertebrate groups, the thyroid gland differentiates from the epithelium of the floor of the buccal cavity. It is of similar structural appearance as in the adult lampreys. The functional unit is the thyroid follicle (Figure 12-18), the wall of which consists of one layer of squamous to cuboidal cells. The follicles vary greatly in size, depending upon the amounts of colloid they contain. The colloid consists mainly of thyroglobulin secreted by the follicle cells. In thyroglobulin, tyrosine residues are iodinated and converted into thyroxine and triiodothyronine residues. The secretion of the hormones into the blood circulation appears to be initiated by the return of thyroglobulin into the cells of the epithelium, which take up the follicular colloid by pinocytosis. In the cells the droplets containing thyroglobulin undergo proteolytic digestion by means of lysosomal cathepsin with subsequent release of free thyroxine and triiodothyronine (Figure 12-19).

Parathyroid Glands. The parathyroid glands are derived from the epithelium of the embryonic branchial pouches. They exemplify endocrine glands that appeared late in vertebrate history, being known from tetrapods only. One or two pairs are present, often embedded in the thyroid gland (Figure 12-18). The glands produce the parathyroid hormone involved in the control of calcium ions in the blood (see Section 12.4.2).

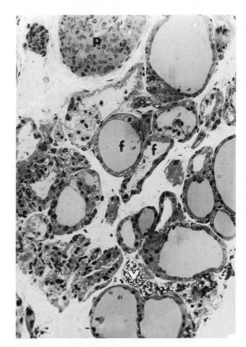

Figure 12-18
Section of thyroid and parathyroid glands in mouse (175 ×). Code: f, follicles; p, parathyroid tissue: [From J. Carstensen Egeberg, original.]

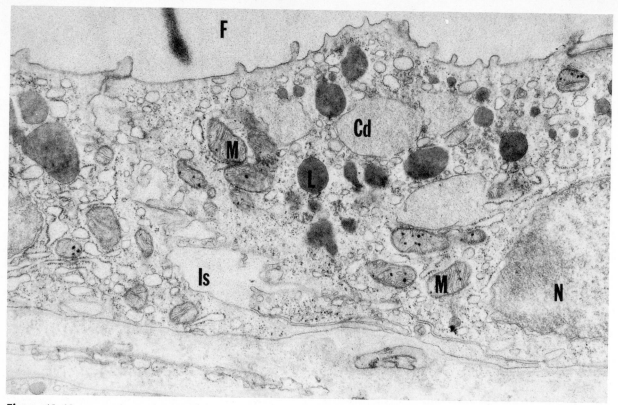

Figure 12-19
Electron micrograph of human thyroid gland (13,000×) Code: Cd, colloid droplets; F, follicle lumen; Is, intercellular space; L, lysosomes; M, mitochondriae; N, nucleus. [From J. Carstensen Egeberg, original.]

12.7 EVOLUTION OF HORMONES

Evolution of hormones embraces two concepts, chemical evolution and evolution of function of the hormones. These two concepts are not always clearly distinguished.

Evolution depends upon mechanisms inherent in the process of replication of DNA. These mechanisms occasionally incorporate incorrect base units in replicate DNA strands, or DNA sequences may become duplicated. These and other irregularities in the replication process constitute the basis for changes in either or both the amino acid compositions and sequences in the proteins, including protein and peptide hormones, synthesized according to the instructions encoded in the DNA units making up the genes. As a result of this mutational process the various types of proteins in living organisms differ to varying degrees in their amino acid compositions and sequences.

During the last 30 years or so a great number of proteins have been isolated, and their compositions and sequences of amino acids have been determined. The technique of isolating and sequencing proteins and peptides has become an important tool in the elucidation of animal phylogenies. The numbers of amino acid substitutions within homologous protein molecules can be used to estimate minimum numbers of gene mutations needed to account for the substitutions, and thus can be used to indicate the phylogenetic relatedness of the animal species that produced the homologous proteins.

The sequencing of proteins, including enzymes

and hormones, disclosed the important fact that amino acids are not substituted to the same extent, that is, at the same rate, in different parts of the molecules. The conservative amino acid sequences of enzymes and protein and peptide hormones presumably constitute parts of the molecules that are essential to the normal function of the molecule. Base substitutions during DNA replications that result in amino acid substitutions within these parts of the peptide chain are less likely to become established within the gene pool of an animal population than are base substitutions that cause exchanges of amino acids within less essential parts of the molecule. The hypophysial adrenocorticotropic hormone (ACTH) may serve as an example. This hormone has been isolated from a number of vertebrates; it is a polypeptide made up of 39 amino acid residues. Amino acid compositions differ among species, but only in the last part of the chain, which does not add to the corticotropic effect of the molecule. The biological activity resides in the initial part of the peptide including the first 24 amino acid residues (Figure 12-20). Substitutions of amino acids in homologous hormones thus constitute examples of evolution of hormone molecules, resulting from mutations of genes that did not affect the basic function of the hormone.

Protein and peptide hormones usually occur in chemically related groups, indicating that they share a common ancestry. Such hormone families have been found among the peptide hormones in the gut, brain, and hypophysis. They have been extensively studied in recent years.

Adenohypophysial Hormone Families. The hormones of the adenohypophysis fall into three families of chemically related molecules. The best known are the polypeptide hormones ACTH, melanotropin (MSH), and fat mobilizing hormone (lipotropin, LPH). Figure 12-21 shows that significant sequences of the molecules of these hormones are identical. The hormones vary in amino acid composition among animal species, and the hormones may occur in variants even in the same hypophysis. The occurrence of a family of molecules suggests a molecular evolution by extensive point mutations, gene duplications, and chromosomal rearrangements.

The hormones secreted need not represent direct translational products of the corresponding genes. Hormones may be synthesized as larger precursor proteins from which the active molecules are cleaved by proteolytic enzymes. This seems to apply to both ACTH and LPH. Moreover, the melanotropins seem to be further cleavage products of ACTH and LPH, produced by the cells of the pars intermedia of the hypophysis. α-MSH constitutes the amino acid residues 1–13 of the ACTH molecule, and β-MSH a chain of residues contained within the LPH molecule (Figure 12-21). LPHs have more recently been found also to act as precursors for the opiatelike endorphins and enkepha-

Figure 12-20
Amino acid sequences in various mammalian adrenocorticotropic hormones.

SPECIES DIFFERENCES IN AMINO ACID SEQUENCES IN CORTICOTROPIC HORMONE IN MAMMALS

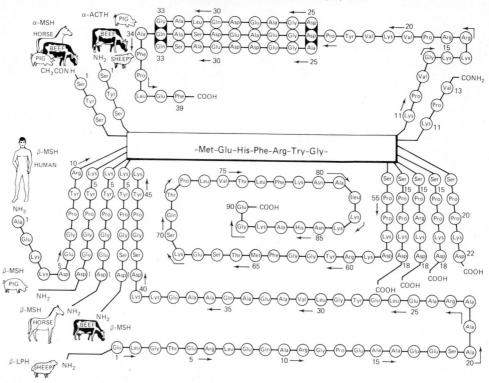

Figure 12-21
An example of a family of hormones from the adenohypophysis. α-ACTH, adrenocorticotropic hormones; α-MSH and β-MSH, melanocyte-stimulating hormones; β-LPH, lipotropic hormones. [From C. H. Li. La specificité zoologique des hormones hypophysaires et de leurs activités. M. Fontaine, (ed.) Centre National de la Recherche Scientifique, Paris, 1969.]

lins isolated from the CNS. It is thus suggested that peptide molecules constituting potential new hormones may have arisen through posttranslational events in protein synthesis, as represented by production of active molecules by enzymatic cleavage of precursor proteins.

Section 12.2.2 mentioned other examples of production of hormones by the splitting of smaller peptide chains from precursor proteins: kinins from kininogens and angiotensins from angiotensinogen.

The second family of adenohypophysial peptide hormones includes the growth hormones and prolactins. Growth hormones and prolactins from several mammalian species show great similarities in amino acid sequences, indicating the origin of the hormones from a common ancestral molecule. Moreover, it has been found that the same sequences of amino acids are repeated within the molecules, indicating that the large molecules are derived, by one or more gene reduplications, from a smaller primordial peptide. The large molecule was established early in vertebrate evolution, as indicated by the recent finding that teleost prolactin is chemically closely related to mammalian prolactin.

The third family of pars distalis hormones includes the glycoprotein hormones, the gonadotropins (FSH and LH) and thyrotropin (TSH). All three hormones consist of two subunits, the α subunit and the β subunit, of which the α subunits appear to be virtually identical in all three hormones. This can be demonstrated by separating the two subunits of, for example, LH and TSH and cross combining the subunits to establish molecules consisting of a TSH α subunit and a LH β subunit, and vice versa. Cleavage into subunits strongly reduces the potency of the hormones. Reconstitution restores function according to the origin of the β subunit. It is thus indicated that the α subunit serves to bind the hormone to the site of action, whereas the specific function the hormone exerts is determined by the β subunit. Recombinations with similar results have also been made with FSH.

Gut-Brain Peptides. The finding that identical, or chemically closely related, peptides may act both as hormones in the gut and as nerve transmitters, or modulators, in the brain strongly stimulated interest in the evolution of peptide hormones and neurotransmitters. One outcome of studies within this field has been the discovery that members of the several hormone families were present in a variety of other organs as well, and that they may have a wide phylogenetic distribution. Most or perhaps all of the secreted peptides are synthesized as part of large precursor molecules, and apparently distinct peptides may be derived from a single precursor.

One family of peptides has been termed the *gastrin-CCK-family.* Its members regulate secretion of digestive fluids in the vertebrate gut (see Section 2.4.6). A member of this family, the decapeptide *caerulein* was first isolated from the skin of the anurans *Hyla* and *Xenopus.* Caerulein is chemically closely related to CCK, and it exhibits a similar spectrum of biological actions and potencies. Caeruleinlike peptides are also present in cells in the amphibian gut. CCK-like peptides occur widely in the gut and brain of all vertebrates studied so far, from cyclostomes to mammals. Gastrinlike peptides have even been found in the neurosecretory system of an insect *(Manduca sexta)* and in the guts of certain molluscs.

The *secretin family* includes secretin, glucagon, vasoactive intestinal peptide (VIP), and gastric inhibitory peptide (GIP). The functions of the secretinlike peptides appear to range from control of blood sugar (glucagon), regulation of exocrine (secretin) and endocrine (GIP) to functions as modulators of synaptic transmission in the CNS (VIP).

The existence of families of peptide hormones provides evidence for the establishment of new hormonal functions by means of known genetic mechanisms, such as mutations, gene duplications, and so on. The lack of relationships between the chemical nature of hormones and their functions, as well as the finding that the same molecule may exert widely different functions in various tissues and organs of the body, indicate that evolution of hormonal functions may take place by adoption of existing biochemical machinery, and that the chemical nature of hormones is primarily determined by the origin and history of the endocrine cells, rather than the type of functions in which the hormones are involved.

Evolution of new hormonal mechanisms may also take place independently of chemical evolution by a change in function of an old hormone. Examples of this type of functional evolution of hormones are numerous. Prolactin probably provides the most spectacular example. Prolactin is apparently produced in the hypophysis of all vertebrates, but its function is not known in several of the lower vertebrate groups. In freshwater teleosts it seems to secure a low permeability of the gill epithelium, thus reducing the loss of ions from the body fluids. In various cichlid fishes, treatment with prolactin induces a variety of activities that may collectively be described as parental care, including digging of nests for the eggs, fanning the eggs, and causing secretions of skin glands as food for the young. In amphibians (newts such as *Diemictylus*) the hormone is responsible for the "water drive," which causes the animals to migrate to water in the breeding season. In birds prolactin has been adapted to several functions in connection with breeding in general and, in the pigeon, with the production of crop milk. The crop milk consists of a secreted crop epithelium that is regurgitated and fed to the young. Both the male and the female birds secrete the milk (see also Section 13.3). In mammals, finally, the hormone forms part of the complex control of growth and secretion of the mammary gland during pregnancy and after delivery, hence the name prolactin. In some species it apparently also supports the maintenance and secretion of the corpus luteum, hence its designation as the luteotropic hormone (see also Sections 13.6 and 13.7).

12.8 MECHANISMS OF HORMONE ACTION

A vast and rapidly increasing literature deals with the mechanisms by which hormones exert their functions. A key function has been ascribed to protein molecules of the target cells, the receptor molecules, which more or less specifically bind the hormones. The receptors can be located on the external surface of the cell membrane, in the cytosol, or they may be found attached to the chromatin of the cell nucleus. Catecholamine and peptide hormones usually bind to surface receptors, and steroid hormones to receptors in the cytosol. The thyroid hormones appear to bind directly to chromatin receptors.

Most studies of the mechanisms of hormone action have used hormones that bind to surface receptors. Figure 12-22 is a simplified model of a mechanism of this type. The location of the receptor molecules can be imagined from the diagram in Figure 9-7 of the cell membrane, which indicates the various types of protein molecules that float within the lipid bilayer constituting the membrane.

The binding of hormone to the receptor molecule initiates processes that result in activation of the enzyme adenylate cyclase, which is located on the internal surface of the cell membrane. The activated adenylate cyclase converts ATP to adenosine-3′5′-monophosphate (cAMP, Figure 12-23). cAMP initiates further processes that are specific for the hormone, and it has therefore been termed the "second messenger." The processes may involve activation of enzymes within the cell. An example

is the activation of phosphorylase by adrenalin in the liver, which results in the splitting of glycogen into free glucose, and thus to an increase in blood glucose. cAMP may also initiate a chain of only partly understood events that, via synthesis of messenger RNA (mRNA), results in the synthesis of specific enzymes and eventually leads to the biochemical and physiological "actions" of the hormone. ACTH, the gonadotropins, the releasing hormones, and glucagon are examples of hormones that appear to stimulate protein synthesis and growth with cAMP acting as a second messenger.

Many other peptide hormones do not use cAMP as an intermediary. Examples are growth hormone, prolactin, insulin, nerve growth factor, somatostatin, oxytocin, and angiotensin. In some cases other mediators may be used, such as cyclic guanosine monophosphate or calcium ions, but the steps that follow binding of the hormone to the receptors are not as well understood as the cAMP mechanism.

In most models of peptide hormone function the hormones have remained outside the target cells. They act as triggers of series of events by binding reversibly to surface receptors. However, examples have also been described where the hormone-receptor complex is taken up by the cell. The same hormone may even act both externally and internally. Thus, nerve growth factor has been ascribed this dual mechanism of action. During embryonic life, the factor stimulates growth and differentiation of the neuronal network by binding to one type of receptor on the cell surface, and the hormone-receptor complex transmits the stimulus through a process that is still undefined. After growth of

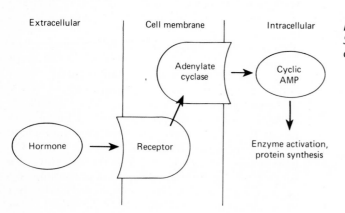

Figure 12-22
Simplified model of hormone action on cells through cAMP.

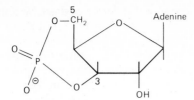

Figure 12-23
Chemical structure of cAMP.

the neurons is completed, a second type of receptor at the nerve terminals binds the hormone. This complex is taken up, internalized, and the hormone is transported from the periphery to the cell body to exert its anabolic activities.

Steroid hormones typically penetrate the cell membrane and bind to intracellular receptor molecules. Penetration may be by passive diffusion of the lipid-soluble molecules, but specific uptake mechanisms have also been suggested. By little understood mechanisms, the hormone-receptor complex reaches the nucleus, binds to the chromatin, activates genes, and thus causes the synthesis of proteins that are specific to the hormone action. For example, corticosteroid hormones, such as cortisol or cortisone, may stimulate the production of aminotransferase enzymes in the liver. The enzymes induce increased production of glucose, and thus lead to an increase in blood glucose, which is the physiological function of these so-called glucocorticoid hormones.

The thyroid hormones appear to reach receptors on the nuclear chromatin directly, but ultimately the chain of events elicited is comparable to that of the other hormones that stimulate protein synthesis.

Mechanisms that involve activation of genes and synthesis of proteins are typical of hormonal regulation of metabolic and growth processes. Other types of hormonal activities are of a more direct nature, and the hormones may act on cell structures outside the nucleus. Vasopressin and oxytocin are examples of hormones that act on the cell membrane to regulate its permeability and other properties. Hormone actions that show little delay are probably characterized by relatively simple sequences of events between the binding of the hormone to its receptor site and the final effects of the hormone.

The concept of hormone-receptor binding implies that a target cell for the hormone contains specific receptor molecules, probably located on the outer surface of the cell membrane. The number of receptor sites constitutes the capacity of the cell for binding hormone molecules (Figure 12-24). The efficiency, or affinity, with which hormone binds to the receptors is expressed by the concentration of hormone at which half the receptors bind to hormone molecules, the *half-saturation concentration* (Figure 12-24). The affinities found for different hormones roughly correspond to the normal levels of these hormones in the blood.

The sensitivity of the target cells to the hormones also depends upon the number of receptors, *the capacity*. The capacity often varies with the concen-

Figure 12-24
Diagram to illustrate relationship between hormone concentration and binding of hormone to specific receptors on target cells.

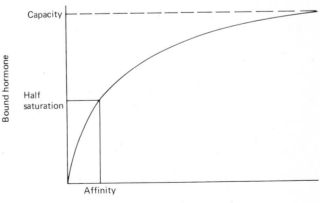

tration of hormone in the blood. Usually, a hormone will act on the target cell to decrease the number of receptor molecules, which presumably reflects a balance between rates at which new molecules are synthesized and old molecules are broken down. Such regulation of sensitivity of target cells by reduction of receptor molecules seem to be typical for hormones that act through cAMP. This desensitization process explains why treatment with large doses of hormone, for example insulin, may cause insensitivity of the body tissues toward the hormone.

Other hormones, for example prolactin and angiotensin, which do not have cAMP as an intermediary, increase the number of receptor molecules, and thus increase the sensitivity of the target cells to the hormone.

Dose-Response Relations. The effects exerted by a hormone vary in strength with the amount of hormone acting upon the target cells, but the patterns of dose-response relations differ. The magnitude of the hormone response may increase in proportion to the numbers of receptors occupied by hormone. This situation is typical of steroid hormones binding to receptors in the cytosol, and thyroxin binding to nuclear structures.

In the case of peptide and other hormones that bind to receptors in the cell membrane, maximum response may be reached with only a small fraction of the receptors occupied. The functional significance of the apparent surplus of receptors seems to be to enhance the rate at which hormone-receptor complexes are formed, and, at low concentrations of hormone, to increase the amount of complexes formed and, thus, the stimulus strength.

Hormone-Receptor Specificity. In order for the hormone-receptor mechanism to function in a specific way, the receptor should bind only the hormone and not other molecules circulating in the blood. Mostly, this hormone-receptor specificity is not absolute, but receptors will more or less efficiently bind molecules that are chemically, or structurally, related to the hormone.

The complex between the receptor and the foreign molecule may act in the same way as the hormone-receptor complex and set in train the processes expressing the normal action of the hormone. But sometimes the complex with the for-

eign molecule is inactive, and by occupying the receptor molecules the foreign molecules will prevent the hormone from acting. Foreign molecules may thus act as blockers of the normal hormonal action. A large number of substances are known that act as antihormones by blocking the hormone receptors, and they have been used extensively in studies on hormone actions and functions. They are also used clinically.

Hormones often occur in chemically related families. The receptors for the various members of such families may bind to other members, but less efficiently. This lack of absolute specificity may result in overlapping actions, which have been observed for many hormones, such as follicle stimulating hormone and luteinizing hormone, mineralocorticoids and glucocorticoids, and insulin and somatomedin. The possible physiological significance of such overlap, or redundancy, in hormone activities is not well understood.

Homologous and Heterologous Hormones. The fact that hormone receptors exhibit only relative specificity has had great practical and theoretical implications. Peptide hormones, except the smallest, are species specific. Most available peptide hormones originate from slaughtered animals, especially cows, pigs, and sheep. Most treatments with peptide and protein hormones, both clinically and experimentally, have therefore used heterologous hormones, that is, hormones that are chemically related to, but different from, the hormone produced by the organism treated (see Section 12.7). This raises the question as to whether observed activities are representative of functions exerted by the homologous hormone, the hormone produced by the animal itself. The answer depends upon many factors, the most important of which probably is: how active is the heterologous hormone?

Heterologous hormones tend to be more effective the more closely related are the donor and the recipient of the hormone. But this general rule has numerous exceptions, and the efficiency of a heterologous hormone is strikingly unpredictable. A closely related hormone may show little activity, whereas a remotely related one may be highly active.

The efficiency of a heterologous peptide hormone, or a synthetic analog of the hormone, depends both upon the affinity of the receptor for the hormone or analog, and upon the ability of the hormone-receptor complex to initiate the chain of processes that result in the final hormonal action. If a heterologous hormone is effective in physiological doses (at a blood concentration that is similar to that of the homologous hormone), the heterologous hormone is likely to simulate the mode of action of the homologous hormone. Under such circumstances experiments performed with heterologous hormones may, therefore, serve to elucidate the functions of the homologous hormone. An example is human chorionic gonadotropin acting on the gonads of the toad *Bufo bufo*. This human gonadotropin induces normal spermatogenesis and vitellogenesis in hypophysectomized male and female toads at concentrations of hormone in the blood around 10^{-10} M, which presumably are physiological doses. In hypophysectomized controls, spermatogenesis and vitellogenesis are discontinued. The heterologous gonadotropin therefore seems to act as a substitute closely mimicking the mechanism of action of the homologous gonadotropin.

Heterologous peptide hormones usually exert no effects in small physiological amounts but may act when administered in large, pharmacological, doses. In such cases it may be difficult to interpret the results in physiological terms, both with respect to mechanism of action of the unphysiologically high doses, and with respect to the effects observed. Sometimes it is, however, possible to conclude that such effects do not represent normal actions of the hormone. An example is the finding that large doses of a number of pars distalis hormones of mammalian origin, including thyrotropin, prolactin, and growth hormone, may stimulate interrenal function in the hypophysectomized teleost *Poecilia latipinna*.

ADDITIONAL READING

HISTORY

Biedl, A. *Innere Sekretion*, 3. Aufl., Urban und Schwarzenberg, Berlin und Wien, 1916.

Brooks, C. McC., and H. A. Levey. "Humorally-transported integrators of body function and the development of endocrinology." In Chandler McC. Brooks and Paul F. Cranefield (eds.), *The Historical Development of Physiological Thought*, The Hafner Publishing Company, New York, 1959.

Jørgensen, C. B. "John Hunter, A. A. Berthold, and the origins of endocrinology," *Acta Hist. Sci. Nat. Med.*, **24,** 1–54 (1971).

Rolleston, H. D. *The Endocrine Organs in Health and Disease, with an Historical Review*, Oxford University Press, London, 1936.

GENERAL AND COMPARATIVE ENDOCRINOLOGY

Assenmacher, I., and D. S. Farner (eds.). *Environmental Endocrinology*, Springer-Verlag, Berlin and New York, 1978.

Barrington, E. J. W. *An Introduction to General and Comparative Endocrinology*, 2nd ed., Oxford University Press, New York, 1975.

_____ (ed.). *Trends in Comparative Endocrinology*, Amer. Zool. Suppl. 1975.

_____ and C. B. Jørgensen (eds.). *Perspectives in Endocrinology: Hormones in the Lives of Lower Vertebrates*, Academic Press, London, 1968.

Bentley, P. J. *Comparative Vertebrate Endocrinology*, Cambridge University Press, New York, 1976.

Benzon, G. K., and J. G. Phillips (eds.). *Hormones and the Environment*, Cambridge University Press, New York, 1970.

Frieden, E., and H. Lipner. *Biochemical Endocrinology of the Vertebrates*, Prentice-Hall, Englewood Cliffs, NJ, 1971.

Gaillard, P. J., and H. H. Boer (eds.). *Comparative Endocrinology*, Elsevier, Amsterdam, 1978.

Hanke, W. *Vergleichende Wirkstoffphysiologie der Tiere*, Gustav Fischer, Jena, 1973.

Martini, L. (ed.). *Comprehensive Endocrinology*, vols. I–XII, Raven Press, New York, 1979–1981.

Williams, R. H. (ed.). *Textbook of Endocrinology*, 5th ed., Saunders, Philadelphia, 1974.

VERTEBRATE ENDOCRINOLOGY

Beers, R. F., Jr., and E. G. Bassett (eds.). *Polypeptide Hormones*, Raven Press, New York, 1980.

Chester Jones, I. (ed.), *General, Clinical and Comparative Endocrinology of the Adrenal Cortex*, vols, 1–3, Academic Press, New York, 1976–1980.

Fitzsimons, J. T. "The role of renin-angiotensin system in the regulation of extracellular fluid volume," pp. 100–115. In C. Barker Jørgensen and Erik Skadhauge (eds.). *Osmotic and Volume Regulation*, Alfred Benzon Symposium XI, Munksgaard, Copenhagen, 1978.

Glass, G. B. J., Jr., (ed.). *Gastrointestinal Hormones*, Raven Press, New York, 1980.

Harris, G. W., and B. T. Donovan (eds.). *The Pituitary Gland*, vols. 1–3, Butterworth & Co., London, 1966.

Holmes, R. L., and J. N. Ball. *The Pituitary Gland, a Comparative Account*, Cambridge University Press, New York, 1974.

Idler, D. R. (ed.). *Steroids in Nonmammalian Vertebrates*, Academic Press, New York, 1972.

Krieger, D. T. (ed.). *Endocrine Rhythms*, Raven Press, New York, 1980.

LOCAL HORMONES

Ajmone Marsan, C., and W. Z. Traczyk (eds.). *Neuropeptides and Neural Transmission*, Raven Press, New York, 1980.

Bloom, F. E. (ed.). *Peptides: Integrators of Cell and Tissue Function*, Raven Press, New York, 1980.

Bullough, W. S. "Mitotic control in adult mammalian tissues," *Biol. Rev.*, **50**, 99–127 (1975).

Gross, F., and H. G. Vogel (eds.). *Enzymatic Release of Vasoactive Peptides*, Raven Press, New York, 1980.

Hay, E. D., T. J. King, and J. Papaconstantinou (eds.). *Macromolecules Regulating Growth and Development*, Academic Press, New York, 1974.

Horton, E. W. *Prostaglandins*, Springer-Verlag, Berlin-Heidelberg-New York, 1972.

Jimenez de Asua, L., R. Levi-Montalcini, R. Shields, and S. Iacobelli (eds.). *Control Mechanisms in Animal Cells*, Raven Press, New York, 1980.

Kadowitz, P. J., P. D. Joiner, and A. L. Hyman. "Physiological and pharmacological roles of prostaglandins," *Ann. Rev. Pharmacol.*, **15**, 285–306 (1975).

LoBue, J., and A. S. Gordon (eds.). *Humoral Control of Growth and Differentiation*, vols. 1–2, Academic Press, New York, 1973.

INVERTEBRATE ENDOCRINOLOGY

Fingerman, M. "Endocrine mechanisms in marine invertebrates," *Life Sci.*, **14**, 1007–1018 (1974).

Golding, D. W. "A survey of neuroendocrine phenomena in nonarthropod invertebrates," *Biol. Rev.*, **49**, 161–224 (1974).

Highnam, K. C., and L. Hill. *The Comparative Endocrinology of the Invertebrates*, 2nd ed., Edward Arnold, London, 1977.

Sláma, K., M. Romaňuk, and F. Šorm. *Insect Hormones and Bioanalogues*, Springer-Verlag, Wien and New York, 1974.

Tombes, A. S. *An Introduction to Invertebrate Endocrinology*, Academic Press, New York, 1970.

Wigglesworth, V. B. *Insect Hormones*, Oliver & Boyd, Edinburgh, 1970.

PHEROMONES

Barrington, E. J. W. "Chemical communication," *Proc. Roy. Soc. London*, **B199**, 361–375 (1977).

Birch, M. C. (ed.). *Pheromones*, North Holland, Amsterdam, 1974.

Dunham, P. J. "Sex pheromones in crustacea," *Biol. Rev.*, **53**, 555–583 (1978).

Hinde, R. A. (ed.). *Non-Verbal Communication*, Cambridge University Press, New York, 1972.

Johnston, J. W., D. G. Moulton, and A. Turk. *Communication by Chemical Signals*, Appleton-Century-Crofts, New York, 1970.

Shorey, H. H. *Animal Communication by Pheromones*, Academic Press, New York, 1976.

COLOR CHANGE

Bagnara, J. T., and M. E. Hadley. *Chromatophores and Color Change*, Prentice-Hall, Inc., Englewood Cliffs, NJ, 1973.

Burtt, E. H., Jr., (ed.). *The Behavioral Significance of Color*, Garland Press, New York, 1979.

Needham, A. E. *The Significance of Zoochromes*, Springer-Verlag, Berlin-Heidelberg-New York, 1974.

Waring, H. *Color Change Mechanisms of Cold-Blooded Vertebrates*, Academic Press, New York, 1963.

METAMORPHOSIS AND MOLTING

Etkin, W., and L. I. Gilbert (eds.). *Metamorphosis*, Appleton-Century-Crofts, New York, 1968.

Larsen, L. O. "Physiology of molting," pp. 53–100. In B. Lofts (ed.). *Physiology of Amphibia*, vol. 3, Academic Press, New York, 1976.

Ling, J. K. "Adaptive functions of vertebrate molting cycles," *Am. Zool.*, **12**, 77–93 (1972).

NEUROENDOCRINOLOGY

Ball, J. N. "Hypothalamic control of the pars distalis in fishes, amphibians and reptiles," *Gen. Comp. Endocrinol.*, **44**, 135–170 (1981).

Greep, R. O., and E. B. Astwood (eds.). "The pituitary gland and its neuroendocrine control," *Handbook of Physiology*, vol. IV, sec. 7, Amer. Physiol. Soc., Bethesda, MD, 1974.

Harris, G. W. *Neural Control of the Pituitary Gland*, Edward Arnold, London, 1955.

Jeffcoate, S. L., and J. S. M. Hutchinson (eds.). *The Endocrine Hypothalamus*, Academic Press, New York, 1978.

Lederis, K., and W. L. Veale (eds.). *Current Studies of Hypothalamic Function 1978*, vols. 1–2, Karger, Basel, 1978.

Martini, L., and W. F. Ganong (eds.). *Neuroendocrinology*, vols. 1–2, Academic Press, New York, 1966–1967.

Meites, J., B. T. Donovan, and S. M. McCann (eds.). *Pioneers in Neuroendocrinology*, vols. 1–2, Plenum Press, New York, 1975, 1978.

EVOLUTION OF HORMONES

Barrington, E. J. W. *Hormones and Evolution*, English Universities Press Limited, London, 1964.

——— (ed.). *Hormones and Evolution*, vols. 1–2, Academic Press, New York, 1979–1980.

Li, C. H. (ed.). *Hormonal Proteins and Peptides*, vols. I–II, Academic Press, New York, 1973.

van Noorden, S., and S. Falkmer. "Gut-islet endocrinology—some evolutionary aspects," *Invest. Cell Pathol.*, **3,** 21–35 (1980).

Pang, P. K. T., and A. Epple (eds.). *Evolution of Vertebrate Endocrine Systems,* Texas Tech Press, Lubbock, Texas, 1980.

Wallis, M. "The molecular evolution of pituitary hormones," *Biol. Rev.,* **50,** 35–98 (1975).

MECHANISMS OF HORMONE ACTION

Baxter, J. D., and J. W. Funder. "Hormone Receptors," *New Engl. J. Med.,* **301,** 1149–1161 (1979).

Birnbaumer, L., and B. W. O'Malley (eds.). *Receptors and Hormone Action,* vols. I–III, Academic Press, New York, 1978.

Litwack, G. (ed.). *Biochemical Actions of Hormones,* vols. 1–6, Academic Press, New York, 1970–1979.

C. Barker Jørgensen

13

REPRODUCTION

13.1 INTRODUCTION

Organisms produce offspring and thereby maintain their kinds. The means by which organisms reproduce vary remarkably between species and constitute central fields in several branches of biology, including genetics, embryology, phylogeny, ecology, population dynamics, husbandry, and ethology, apart from physiology. The comparative animal physiologist will study types and mechanisms of animal reproduction, whereas the adaptational physiologist may try to survey and classify breeding patterns, with special reference to the environmental factors that control breeding in the various species and individuals. Classification of breeding patterns is made difficult by the extreme diversity, and also because the patterns are often highly variable within the species, depending on external factors.

Most animals reproduce sexually, but asexual reproduction also occurs as an alternative reproductive mechanism. In sexual reproduction a zygote results from the fertilization of an ovum by a spermatozoan. Fertilization may be internal, connected with copulation, or external. The eggs may be laid soon after ovulation (*oviparity*), or laying may be delayed and the eggs may develop within the maternal organism. In the latter situation, the young emerge from the eggs at about the time of egg deposition and we speak of *ovoviviparity*. If development does not involve hatching, and the young are born, we speak of *viviparity*. In viviparous species, the embryo may be nourished by yolk present in the egg or by maternal secretions, or the embryo may establish placental contact with the mother inside a uterus (see Section 13.6).

13.2 BREEDING PATTERNS AND NATURE OF CONTROL

Animals may breed more or less continuously at all times of the year, or they may breed discontinuously, at irregular or regular intervals.

When considering continuous breeding, we should distinguish between the individuals and the population. Breeding may go on continuously within a population without all individuals breeding continuously. But continuous breeding of the individuals constituting a population is also widespread throughout the animal kingdom.

As might be expected, continuous breeding is usually correlated with constant environments, such as the oceans, tropical rain forests, man's dwellings and stores, intestinal tracts, and other special niches in animals that serve as hosts for parasites. However, inhabitants of constant environments need not be continuous breeders. Thus, of 60 species of birds from a Colombian rain forest, 22 were found to breed throughout the year, whereas the remaining 38 species had restricted seasons.

Discontinuous breeding *may* thus be found among animals inhabiting constant environments. However, it is *typical* of animals living under changing environmental conditions. If the environmental conditions change unpredictably, the inhabitants may become opportunistic breeders that adapt to utilize favorable conditions for breeding when they become available. Thus, many animals, for example, birds or amphibians that inhabit desert or semi-arid areas, are such opportunistic breeders that they take advantage of unpredictable and sporadic rainfalls to produce young. Breeding is often postponed for long periods, which may last for more than 1 year if there is no rain; the populations may breed several times in succession after a heavy rainfall.

By far the best known type of discontinuous breeding is that which can be correlated with the seasons of the year. This is probably due to the fact that the large majority of studies on breeding patterns have been made on animals inhabiting the temperate regions at higher latitudes where light and temperature vary substantially and regularly through the year.

Breeding in animal populations is closely adapted to the environments in which the population live. A breeding pattern encompasses both adaptive traits that have become genetically fixed and traits that are physiologically flexible. The genetically fixed adaptations evolved through natural selection. Alleles within the genomes of individuals that breed during less favorable periods tend to be eliminated because of heavier mortality among the offspring and perhaps also the parents. Such selection eventually leads to the evolution of species-specific periodicities in breeding. The physiological adaptations enable the individuals within a population to adjust their breeding according to actual environmental factors. A distinction can be made between the environmental factors that caused the evolutionary adaptations in breeding and the factors that cause the physiological adjustments to the environment. The factors that caused the genetically fixed traits in breeding patterns have been termed *ultimate factors,* those acting as physiological cues have been termed *proximate factors.* The terms ultimate and proximate factors have also been used to distinguish between evolutionary and physiological adaptations of functions other than breeding.

An example may help to elucidate the concepts of ultimate and proximate factors in animal breeding. Temperate zone ungulates, like deer, bear their young in spring or early summer when vegetation is fresh and abundant and nutritional conditions are good for both the mother and for the production of milk. Being born in spring therefore implies that the fawn can become well nourished and is exposed to a mild climate, so the young deer can grow big and sturdy before winter. Parturition in spring must be prepared for many months in advance, corresponding to the long pregnancy in deer. Mating thus occurs in the autumn, and autumn is the only time of the year when deer mate. Environmental factors that cause parturition in deer to occur in spring are ultimate factors. They include food supply and temperature. Proximate factors ensure that the deer mate at the proper time of year. They include environmental cues that are characteristic of autumn, especially the declining lengths of the days.

For breeding cycles in general, food probably ranks as the predominant ultimate factor, but other factors may operate, including temperature, availability of suitable breeding sites, and predation pressure. The most important proximate factors are daylength (photoperiod), temperature, water, food, and social stimuli. Several examples follow.

13.3 COURTSHIP AND MATING

In many animal species a specific behavior pattern precedes spawning of eggs and sperm or mating. Such courtship behavior is the rule in mammals and birds, and it occurs widely among lower vertebrates and invertebrates. Ethologists are analyzing courtship and mating behavior in a rapidly increasing number of species. They attempt to explain the functional significance, the mechanisms, and the origins of the often very peculiar and complex behavioral patterns. Courtship behavior acts as a species-specific signal for individuals of the opposite sex, thus serving to further and to secure the meeting of the right mates.

Spawning and mating imply contacts between individuals that would not be tolerated outside the breeding seasons; at other times such contacts would result in hostilities between the individuals. Consequently, much courtship behavior serves to counteract aggressive responses from the approached partner. In some instances it has been possible to resolve courtship behavior into components, some of which can be recognized as parts of other behavioral patterns, such as aggressive behavior. The zig-zag dance of the male stickleback is a well-known example. Here the "zig" has been interpreted as a component of the aggressive behavioral pattern, whereas the "zag" serves to lead the female towards the nest for egg-laying.

Male courtship behavior may also serve to stimulate the final maturation and ovulation of the oocytes, to initiate the estrous condition in the female, and thus to synchronize the breeding activities of the two sexes. Courtship therefore belongs among the proximate factors in controlling reproduction.

The various aspects of courtship and mating behavior may be illustrated by some well-analyzed examples.

Cockroaches. Courtship and mating behavior in cockroaches and many other insects encompass a complex sequence of interactions between the sexes. These interactions are controlled by a number of chemical and tactile stimuli that act as releasers of the appropriate behavior (Figure 13-1). Copulation is preceded by the female mounting the male. The stimuli that induce the receptive female to mount therefore play a central position in the mating behavior. Mounting is closely associated with feeding motions of the female over the dorsal surface of the male's abdomen, elicited by chemical stimuli produced by the courting male cockroach.

Stickleback. The complex pattern of breeding in the three-spined stickleback, *Gasterosteus aculeatus,* and hormonal control has been analyzed in great detail. Complete breeding comprises several

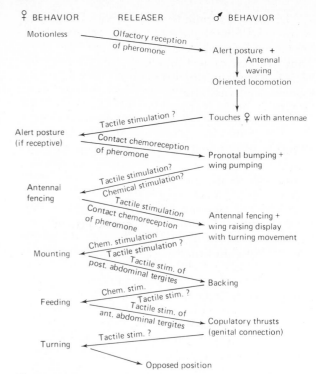

Figure 13-1
A summary of the mating behavior of the cockroach, Bursotria fumigata, *indicating the probable releasers for each step in the sequence.* [*From R. H. Barth, Jr.* Advan. Reprod. Physiol., **3,** *167 (1968).*]

phases. Initially, the male establishes a territory and builds a nest. The establishment of the territory rests upon the development of aggressive behavior towards other sticklebacks that enter the territory. Nest building is succeeded by the development of courtship behavior, which differentiates the attitude of the male towards intruders of the territory. Other males are still attacked; however, gravid females elicit courting by the male, which eventually causes the female to enter the nest to deposit her eggs and the male to fertilize them. When spawning is completed the behavioral pattern changes again. The female is attacked and driven away, and the male assumes parental behavior, which consists of fanning the nest to aerate the eggs. These various activities, aggressive (agonistic), sexual, next building, and fanning activities, presumably represent innate activity patterns in the CNS that can be elicited by various stimuli from the environment. The expression of the activities, however, depends upon the CNS being exposed to the proper hormones. Thus aggressive, nest-building, and courtship behavior depend upon the presence of testis hormones. After castration these activities decline or disappear, but they can be restored by treatment of the castrate fish with testosterone.

Ring-dove. In female birds social stimuli from the courting males are normally needed for the final maturation of the oocytes. The interactions between breeding male and female birds may be illustrated by the ring-dove, *Streptopelia risoria,* which has been especially thoroughly studied (Lehrman). The female dove will build a nest only after some days of courtship with the male. After nest building the female mates for some days followed by egg laying and brooding. However, if treated with gonadotropin the female dove will build a nest, lay eggs, and brood in the absence of a male. Courtship may, therefore, induce the secretion of gonadotropin, which completes oocyte development and also enhances secretion of estrogen from the ovary. These hormones presumably facilitate nest-building behavior and may induce the ovulatory surge in the secretion of luteinizing hormone (see Section 13.4.3). A resulting increased secretion of progesterone may be of importance in facilitating brooding. Brooding eventually elicits the secretion of prolactin, which is responsible for proliferation of the crop sac epithelium and thus the secretion of "pigeon milk" on which the young are fed. The male partakes in the feeding of the young. Prolactin secretion in the male may be initiated by stimuli he receives during nest building or by the sight of the brooding female.

Thus, it is normally the courting male that stimulates ovarian development in the female dove and synchronizes the reproductive performances of the two sexes. However, castrated, noncourting males, other female doves, or even the female's own mirror image, can also stimulate ovarian development and ovulation in females at advanced ovarian stages. In earlier ovarian stages stimulation requires courting males.

Mammals. It appears from the previous examples that courtship and mating behavior in the animal kingdom depends upon the nervous system being exposed to the appropriate hormone milieu. In mammals hormone-dependent neural structures have been localized to the hypothalamus. It has been found that ovariectomized anestrous cats can be brought into heat by implantation of small pellets of estrogen into the mammillary region of the hypothalamus. The estrogen is effective in amounts that do not influence the genital tract. Implantation of the pellets in other regions of the brain does not induce estrous behavior. Similar results have been obtained in the rabbit. Presumably, therefore, certain hypothalamic structures must be exposed to estrogenic hormone in order that estrous behavior can be elicited. Interestingly, testosterone-containing pellets likewise elicited estrous behavior and not malelike mounting activities. Male behavior can be induced in the female rabbit by the action of testosterone in other areas of the hypothalamus. Such observations from both invertebrates and vertebrates strongly support the suggestion that, in the animal kingdom generally, hormones do not organize the behavioral patterns of male or female sexual activities but rather activate innate behavioral patterns.

13.4 MECHANISMS OF REPRODUCTION

We have up till now been concerned with sexual reproduction as the interplay between the entire organism and its environment, resulting in breeding. In the following sections we shall analyze how the various parts inside the organism interact in order to prepare the organism for breeding. These parts consist primarily of the gonads, the CNS, the endocrine organs that control the functions of the gonads, and the sensory organs that mediate the control exerted by factors in the environment or by internal factors.

13.4.1 Gonads

In most animal species the gonad primordium develops to function either as an ovary or a testis, the sex of the individual being genetically determined. Gonadal function is, however, characterized by its plasticity and adaptability, as exemplified in the numerous exceptions to this general rule. Hermaphroditism occurs widely throughout the animal kingdom. It may be genetically fixed or may be purely phenotypic. In some species the individuals contain functioning ovaries and testes simultaneously, whereas in other species the functional status of the gonads may change. Often the gonad begins as a testis to produce spermatozoa, and later in the life of the individual switches to producing eggs, a pattern known as *protandric hermaphroditism.* Examples of *protogynic* (first-female) *hermaphroditism* are also known. In some species sex reversal occurs only once in a lifetime, in other species repeatedly. Sex reversals may be genetically fixed, or the sex of an individual may be environmentally determined.

13.4.2 Ovary

Oogenesis is basically similar in all animals. The primordial germ cells differentiate into oogonia, which divide and multiply mitotically. Eventually oogonia differentiate into oocytes that grow, finally to undergo reduction division, or meiosis, and to mature. In the morphologically most primitive condition, egg formation is diffuse, as in various coelenterates, turbellarians, and other lower invertebrate groups. Usually, however, egg formation is localized in an ovary or an ovotestis. Eggs may develop solitarily, as in various lower worms, molluscs and echinoderms, but often the oocytes become surrounded by follicle cells, as in a great number of invertebrate groups and in vertebrates. The ovarian follicle then becomes the functional unit of the ovary. Oogenesis usually continues throughout the life of the individual, but in some forms it is restricted to part of the life cycle. Thus, in most mammals studied formation of new oocytes ceases in late embryonic life or shortly after birth.

Ovaries produce eggs in a number and of a size that is largely genetically determined. The sizes of eggs vary between species from less than one-twentieth of a millimeter to several centimeters, in birds. The volume of eggs thus varies by a factor of about 10^9 from the smallest to the largest known. Egg size is primarily determined by the amounts of yolk accumulated. Increasing the stores of energy for the embryo has apparently been of adaptive value during evolution within several taxonomic groups. Yolk may be synthesized within the oocyte itself. In animals that produce small eggs,

all yolk may originate in this way. It seems that the synthetic capacity of the oocytes puts narrow limits on final oocyte size. At any rate, animals that produce large eggs do so by means of yolk precursors that are synthesized outside the oocyte itself. Thus, in insects the fat bodies synthesize the yolk precursors, and in vertebrates, the liver performs this function.

The mechanism of vitellogenesis is strikingly similar in insects and nonmammalian vertebrates and represents a good example of parallel evolution. In both insects and vertebrates the extraovarian synthesis of yolk precursor depends on hormonal stimulation. The precursor molecules circulate in the body fluid and reach the oocyte surface by passing between the follicle cells into the space formed between these cells and the oocyte plasma membrane, which takes up the yolk precursors by pinocytosis.

13.4.3 Ovarian Cycles and Mechanisms of Control

Ovarian functional patterns throughout the animal kingdom are characterized by extreme diversity. At one extreme, oogonia continue to divide throughout the period of ovarian development simultaneously with the asynchronous growth and maturation of oocytes. The breeding period of the population is usually protracted and the individual female may spawn repeatedly during the breeding period. This type is common among lower marine invertebrates that produce numerous small eggs. At the other extreme, egg formation proceeds synchronously from the oogonial stage throughout the oocyte growth stages until final size and maturation. Ovulation and spawning may occur simultaneously within a whole population, triggered by a specific environmental factor. The Japanese crinoid *Comanthus japonica* is a striking example of synchronized oogenesis and spawning. Spawning takes place between 3 and 4 A.M. during the night of the first quarter moon in the month of October. If the moon happens to be hidden behind clouds during that particular night, the local populations of crinoids fail to ovulate and spawn that year, and the eggs degenerate and are resorbed.

Synchronous development of eggs may be more common in species that exhibit only one spawning during their lifetime, such as octopus and lampreys.

However, synchronous oocyte development is also widespread among animals that breed repeatedly.

An important variant of synchronous development is one in which the synchrony is postponed until late stages of oocyte growth. In its basic pattern, this type of cyclic ovarian function is characterized by specific numbers of small, slowly growing oocytes simultaneously entering the final growth phase and undergoing rapid growth, maturation, and ovulation all before the next population of small oocytes leaves the pool of reserve oocytes to repeat the ovarian cycle. This pattern is typical of ovarian function in vertebrates (Figure 13-2), but it has also been found within different groups of invertebrates. It seems to be correlated with the evolution of yolky eggs. The pattern has reached its highest development and complexity with respect to regulatory mechanisms in cephalopods, insects, and vertebrates.

Cephalopods. The female octopus (*Octopus vulgaris*) represents the widespread, but uncommon type of reproduction in which the ovaries mature only once. When the individual has spawned, it dies. The octopus ovary is characterized by the synchronous development and growth of its population of oocytes. Ovarian function in octopus is also remarkable in that all stages of oogenesis are hormonally dependent. Gonadotropin produced in the optic glands, two small bodies situated on the stalks of the optic lobes (Table 12-2 and Figure 13-3), is necessary for all stages in oogenesis, including oogonial multiplication, differentiation and early vitellogenic growth of oocytes, proliferation of follicle cells, as well as final vitellogenic growth and maturation of oocytes. The secretion of gonadotropin from the optic glands is controlled by inhibitory nerves arising from the subpedunculate lobe of the brain (Figure 13-3). In the cuttlefish, *Sepia*, reduced daylength accelerates ovarian development. Daylength seems to control the activity of the subpedunculate lobe through the eyes and optic nerves.

The optic-gland hormone also stimulates growth of the oviduct. The octopus ovary does not seem to possess endocrine function (Figure 13-4).

Insects. A rapidly growing, extensive literature deals with ovarian function and its mechanisms of control in insects. Some insects have a very short adult life and vitellogenesis is often almost complete when the adult emerges from the last molt. In these insects the gonads behave like other somatic tissues of the body, growing and developing during the larval stages. However, in the majority of insects vitellogenic growth and maturation of oocytes are postponed until after the last molt and proceed only under specific hormonal and neuroendocrine control.

As mentioned in Section 13.4.2, the vitellogenic growth and maturation of oocytes in the insect

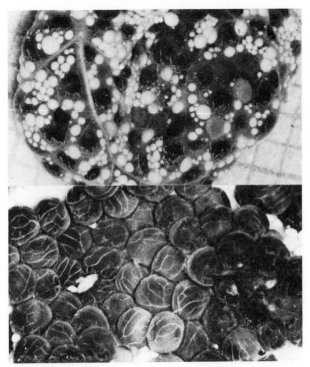

Figure 13-2
Photographs of Bouin-fixed, maturing ovaries of the European toad Bufo bufo *seen from the outside* (above) *and the inside of the ovarian wall* (below). *Numerous small unpigmented oocytes of various sizes are clearly seen from the outside through the membraneous tunica surrounding the ovaries. The large pigmented oocytes bulge into the interior of the ovarial sac, as can be seen in the lower half of the figure. The oocytes in the phase of rapid vitellogenesis grow synchronously, all oocytes being of about the same size. The fast-growing follicles can be seen to be surrounded by dense capillary nets. The squares of the background are 1 mm.*

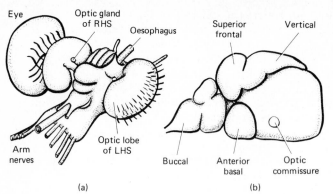

Figure 13-3
Diagrams of (a) the brain of Octopus *from above and in front. The optic glands lie on the stalks of the optic lobes; (b) longitudinal vertical section through the central supraesophageal part of the brain. The subpedunculate lobe, origin of the nerve supply controlling secretion by the optic glands, is shaded. [From M. J. Wells and J. Wells.* Nature, **222**, *293 (1969).]*

ovary is based on the pinocytotic uptake of yolk precursors, which must have direct access to the plasma membrane of the oocyte. The enveloping follicle cells prevent this access in the small oocytes of the immature ovary. A change in configuration of the follicle cells therefore constitutes an early stage in oocyte growth. The follicle cells separate under the influence of gonadotropin and thus expose the oocyte surface to the hemolymph.

The secretion of gonadotropin from the corpora allata is under inhibitory control from the brain.

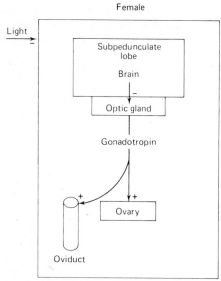

Figure 13-4
Diagram of the neuroendocrine mechanisms of control of the ovary in Octopus.

This inhibitory control is mediated through the neurosecretory system originating in the procerebrum (see Figure 12-6). Ovarian development thus depends upon release from the central nervous inhibition. Feeding is one of the most important of a variety of external or internal stimuli that have been adopted to act as the factor that abolishes this inhibition of gonadotropin secretion and thus triggers off the series of processes leading to ovarian maturation.

Lower vertebrates. The ovaries of lower vertebrates typically contain oogonia that multiply throughout life and a large pool of small oocytes, some of which enter the phase of rapid vitellogenic growth from time to time. Eventually, the follicles mature and ovulate. We may use the common European toad *Bufo bufo* as an example to illustrate a type of ovarian cycle that occurs widely in the lower vertebrates (Figures 13-2 and 13-5).

This toad spawns in spring, as do many other lower vertebrates inhabiting higher latitudes. During spawning the ovaries completely discharge their population of mature eggs. The ovaries return to a juvenilelike condition, dominated by the small oocytes. After a resting period that may last from one to a few months some oocytes from the pool of small oocytes begin to grow and within some weeks the complete clutch of oocytes that are destined to be ovulated during the following breeding period is recruited. This clutch numbers several thousand oocytes, about one tenth of the oocytes present in the reserve pool. The vitellogenic oocytes grow synchronously to reach full size within 2–3 months. The growth period is followed by an-

Figure 13-5
Diagram of annual cycle in behavior and reproductive state in a temperate zone anuran, the toad Bufo bufo. *Full drawn line, ovary mass; dashed line, number of oocytes.*

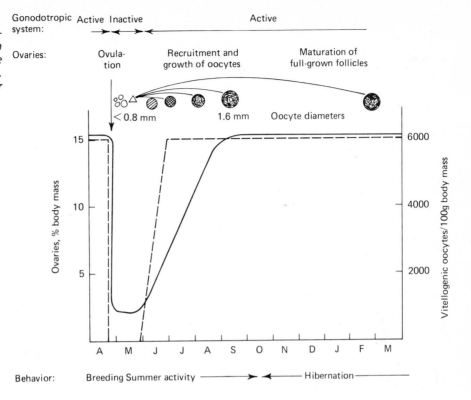

other resting period, which lasts from the autumn, throughout the period of hibernation, and until spawning the following spring. No further oocytes enter the final vitellogenic growth stage until the previous clutch of eggs has been shed, or has degenerated. Some intraovarian mechanism appears to assure that only the normal number of eggs develops in the ovaries at any one time. Growth and ovulation of the large follicles depend upon gonadotropin secreted from the adenohypophysis. Secretion of gonadotropin in amounts needed to maintain a normal ovarian cycle in the toad, as well as in other vertebrates, again depends upon stimulation from the CNS. The relations between the CNS and gonadotropin secretion are dealt with in greater detail in Section 13.5.

Reptiles and Birds. In teleosts and amphibians the number of eggs produced during the breeding season usually amounts to thousands per 100 g of body mass, with sizes around 1 mm in diameter. But there are many species that produce fewer and larger eggs. The reptiles are characterized by diversity in ovarian functional patterns. There is a trend towards production of larger eggs, which in its extreme leads to the production of one egg at a time.

Ovarian function in birds can be considered as further development of the reptilian patterns. In most species of birds only the left ovary becomes functional, the right ovary remaining undeveloped. Most studies on follicle growth and ovulation cycles in the bird ovary have been done in the domestic hen, which is selected for long sequences of egg-laying and is therefore not representative of wild birds. However, basic principles are probably common to both wild and domesticated birds.

The ovary of the hen and other birds during the egg laying period contains several large vitellogenic follicles of different sizes. As the largest follicle grows, matures, and ovulates, it is replaced by the second-largest, which is replaced by the third-largest, and so on. This growth pattern appears to be controlled by an intraovarian mechanism. The hierarchical organization of follicles seems to extend down into the numerous population of small oocytes.

A continuous secretion of gonadotropin maintains vitellogenic growth of the oocytes; ovulation of the mature follicles results from increased secretion of progesterone and luteinizing hormone. Both hormones increase to peak values some hours before ovulation. Studies indicate that the peak values arise by a positive feedback type of mechanism, where increasing levels of the one hormone cause increased secretion of the other until the peak values needed to produce ovulation have been reached (see also Section 13.5).

Mammals. The mammalian ovary exhibits distinct follicular growth patterns that can be related to the lack of yolk deposition in the follicles and the discontinuation of production of new follicles around the time of birth (Section 13.4.2). At birth the ovaries of female mammals contain large populations of small follicles; these number, for example, about 2 million in human ovaries. Already at birth, or even before, these populations of follicles exhibit growth patterns that continue throughout life until the populations are depleted. The pattern is characterized by the continuous recruitment of follicles to the growth phase from the pool of resting primary follicles (Figure 13-6). The number of follicles recruited to the growth phase is roughly proportional to the number of remaining primary follicles. This mechanism rapidly reduces the number of small follicles. Thus, in the human at puberty only 10% of the population at birth remains, and at the age of 40 the population is reduced to less than 1%.

The primary follicles that begin further follicular growth and development may reach stages around the beginning of antrum formation; that is, the stage when vacuoles start to develop within the zona granulosa (Figure 13-6), before they start to degenerate and become atretic. This pattern of follicular growth and atresia appears to be determined by mechanisms that are inherent in the ovaries and independent of the functional state of the organism, such as age, sexual maturity, estrous or menstrual cycles, and pregnancy. The pattern continues until the pool of small follicles is empty.

At puberty the pattern of growth of the follicles becomes modified under the influence of a cyclic pattern of gonadotropin secretion. Under the influence of this secretion some growing follicles escape atresia and continue to grow and develop until they reach the mature state as Graafian follicles (Figure 13-6). The numbers of follicles that are selected for final growth and maturation during an ovarian cycle varies among species. It ranges from one in

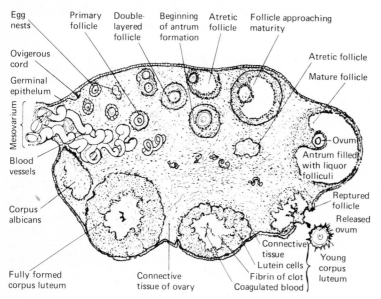

Figure 13-6
Mammalian ovary, showing the sequence of events in the origin, growth, and rupture of the Graafian follicle and the formation and retrogression of the corpus luteum. Follow clockwise around the ovary, starting at the mesovarium. [*From C. E. Corliss. Patten's Human Embryology. McGraw-Hill Book Company, New York, 1976.*]

humans and other primates to 10 or more in other groups of mammals. This inherent basic pattern of continuous recruitment of follicles to the growth phase ensures that large follicles are always present that can respond to gonadotropin with final growth and maturation.

The process that leads from gonadotropin-independent to gonadotropin-dependent growth of follicles has been carefully studied in rat ovaries. The process depends upon the follicles becoming sensitive to luteinizing hormone (LH) as a result of the formation of receptors that bind LH (see Section 12.8). The LH-binding cells are the theca cells, which differentiate peripherally in the follicles. LH stimulates the theca cells to produce testosterone, which by local transport mechanisms reaches the nearby granulosa cells. The granulosa cells convert the testosterone enzymatically into estradiol (Figure 13-10). Estradiol promotes its own rate of secretion by increasing the sensitivity of the follicles toward LH and FSH (follicle stimulation hormone) and by increasing the enzymic activity (within the follicle) that produces estradiol from testosterone. This self-enforcing production of estradiol is a central factor in the final growth and maturation of the follicles. Estradiol also controls the ovulatory surge of LH secretion (see Rat Estrous Cycle). Estradiol thus acts as the primary factor that couples follicle maturation and ovulation in the rat ovary, and probably also in the ovaries of other mammals. Ovarian cycles vary widely among different groups of mammals. In some groups the ovarian cycles are short, for example, 4 or 5 days in small rodents. Short cycles have very short functional periods of the corpora lutea. Long ovarian cycles are usually of 2–3 weeks' duration, mainly due to more prolonged luteal phases. They are found widely among larger mammals. The primate menstrual cycle may be considered as a further development of the long ovarian cycles with long luteal phases.

Ovarian cycles are also correlated with functional changes in the female genital tract and with behavioral changes, both controlled by estrogen and progesterone from the ovaries. The functional changes in the genital tract can be considered as preparatory for copulation, and, in the uterus, for implantation of the fertilized eggs. The behavioral changes imply cyclicly recurring sexual receptiveness of sexual urge, heat, hence the term *estrous cycles.*

The best studied ovarian cycles are the estrous cycle in the white laboratory rat and the menstrual cycle in the primates. In addition, they represent extremes, and may serve to characterize the range of ovarian cycles within the eutherian (placental) mammals.

Rat Estrous Cycle. The estrous cycle of the white rat is usually studied under constant and standardized conditions in the laboratory, including a day and night schedule consisting of 12 hr of light alternating with 12 hr of darkness (Figure 13-7). The estrous cycle under these conditions lasts 4 or 5 days, depending on the strain of rats used. The stages in the cycle can be estimated from an examination of the vaginal smears, which reflect the secretion of hormones from the ovaries from day to day in the cycle (Figure 13-7).

Luteinizing hormone (LH) is present in the blood at a low level during most of the cycle. However, even at this low level LH plays an important role in the timing of the estrous cycle. It is during metestrus that the large follicles become responsive to LH in the blood, and the chain of events starts that results in the increase in the secretion of estradiol. During metestrus there is also an increase in the secretion of progesterone, resulting in a low peak of this hormone. This peak probably represents autonomous activity of the corpora lutea that developed from the ovulated follicles. Progesterone in the blood declines to a low level during diestrus, indicating the regression of the corpora lutea. The high levels of estradiol reached in diestrus cause the surges of LH, FSH, and prolactin seen in proestrus.

The timing of these events seems to be as follows: High levels of circulating estradiol activate hypothalamic structures that control the secretion of gonadotropin (see Section 13.5). This activation only takes place if the high level of estradiol coincides with the sensitive phase of the gonadotropin-controlling hypothalamic structures. These structures exhibit a daily rhythm in sensitivity that is endogenous, but is timed by the daily photoperiod. The structures thus act as a biological clock (see Section 13.5.1). At the schedule of day and night shown in Figure 13-7, the sensitive period falls in the afternoon of proestrus. It lasts about 2 hr.

The activity of the hypothalamic structures triggers the surge of LH secretion in the early evening that is mainly responsible for the ovulation of the

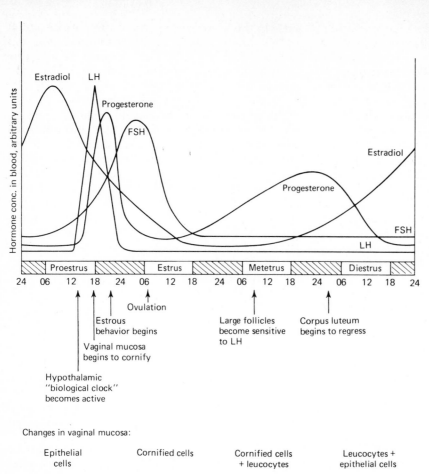

Figure 13-7
Diagram of 4-day estrous cycle in the white rat. [Based on N. B. Schwarz, Recent Prog. Horm. Res., **25,** 1 (1969); and L. G. Nequin, J. Alvares, and N. B. Schwarz, Biol. Reprod., **20,** 259–270 (1979).]

Changes in vaginal mucosa:

| Epithelial cells | Cornified cells | Cornified cells + leucocytes | Leucocytes + epithelial cells |

follicles the following night. An immediate effect of LH on the follicles is to induce the peak secretion of progesterone. The effects of LH on follicle maturation and ovulation seem to be mediated through progesterone, which leads to the enzymatic breakdown of connective tissue in the follicle and thus to its rupture and the release of the oocyte. Simultaneously with the increase in progesterone levels in the blood, the levels of estradiol decline. This change in ratio of the two hormones causes cornification of the vaginal epithelium and induces mating behavior (Figure 13-7).

The rat estrous cycle may last 4 or 5 days. There has been much debate concerning which hormonal mechanisms determine this variation in the length of the cycle. It seems now that the length is determined by the survival of the corpora lutea, and thus by the continued secretion of progesterone. In the 5-day cycle, the longer duration of progesterone secretion, which depends upon prolactin secretion, may delay by one day the effect of estradiol in triggering the surge of LH, thus introducing two days of diestrus in the cycle.

Menstrual cycle. Long estrous cycles and menstrual cycles are characterized by the long persistence of the corpora lutea formed from ovulated follicles. The designation "long estrous cycles" indicates that the typical behavior of the female in heat is still the predominant feature of the ovarian cycle and connected processes. The designation "menstrual cycle" indicates that the conspicuous feature is menstrual bleeding. It is customary to count the beginning of the cycle from the first day of menstrual bleeding.

The main features of the human menstrual cycle at the hormonal, ovarian, and uterine levels are depicted in Figure 13-8. The figure represents a standard cycle lasting 4 weeks. Early in the cycle, the gonadotropins and ovarian hormones are present at basal levels in the blood. At the end of the first week estradiol levels start to increase, simultaneously with follicular growth. These coinciding events apparently correspond to the phase in the rat estrous cycle in which the large follicles become sensitive to the basal levels of LH. During the second week of the cycle a follicle becomes selected for final growth and differentiation. Estradiol in the blood increases steeply to reach levels that initiate the processes that result in the surge of LH secretion and ovulation around midcycle. FSH also rises to a peak value, but the peak is less pronounced than the LH peak, and both hormones soon return to basal levels. Estradiol levels also decrease rapidly but increase again when the ovulated follicle has developed into a corpus luteum. The corpus luteum secretes some estradiol, but more progesterone. The ovarian cycle is terminated by regression of the corpus luteum, with a concurrent strong decline in the ovarian secretion of estradiol and progesterone.

The changing patterns of circulating estradiol and progesterone strongly affect the structure and function of the uterus. Early in menstruation the upper layer of the mucosa is shed; the mucosa is maximally reduced at the end of the menstrual pe-

Figure 13-8
Diagram of menstrual cycle. [Hormone data from B.-M. Landgren, A.-L. Unden, and E. Diczfalusy. Acta Endocrinol., **94**, 89–98 (1980); and A. Miyake, Y. Kawamura, T. Aono, and K. Kurachi. Acta Endocrinol., **93**, 257–263 (1980).]

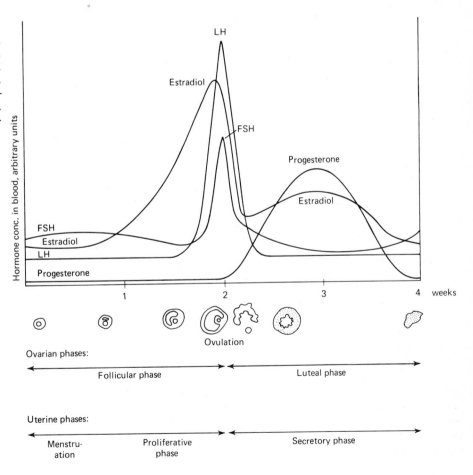

riod. The increasing levels of estradiol stimulate growth of the mucosa, including the differentiation of numerous serous and mucous glands. This proliferative phase extends to the time of ovulation and formation of a corpus luteum, when the ratio of estradiol to progesterone in the blood changes from high to low. This change causes the uterine mucosa to enter the secretory phase, when the glands become active and the mucosa fit for implantation of a blastocyst resulting from a fertilized egg (see Section 13.6.1). If the ovulated egg is not fertilized and the corpus luteum regresses, the fall in estrogen and especially progesterone induces the processes that lead to the shedding of the mucosa and the menstrual bleeding.

Termination of the Luteal Phase. In many mammals a specific factor, *luteolysin,* has been found to be responsible for the regression of the corpus luteum, and thus for the termination of the luteal phase. In several instances the factor has been identified as the prostaglandin $F_{2\alpha}$ (see Section 12.2). Prostaglandin $F_{2\alpha}$ seems to act by desensitizing the luteal cells toward their trophic hormone LH, perhaps by inhibiting the LH activation of adenylate cyclase (see Section 12.8).

Spontaneous and Reflex Ovulation. One crucial element in the estrous cycle is the synchronization of ovulation and mating in order to secure fertilization. In the rat and other "spontaneously" ovulating mammalian species, it is presumably the rise in estrogen secretion that triggers off two chains of events that lead to the simultaneous ovulation and mating behavior. In other mammals, ovulation is directly coupled to mating, which serves as a stimulus for the processes that eventually result in ovulation. This latter type has been termed reflex, or induced, ovulation.

The main difference between spontaneous and reflex ovulation seems to be that, in the reflex ovulators, the neurogenic stimulation elicited during mating takes over the function of the biological clock. Afferent nerve tracts from the genital region impinge on the hypothalamic, gonadotropic structures to enhance secretion of gonadotropin-releasing hormone and thus to induce ovulation in ovaries containing mature follicles.

Reflex ovulation has long been known in the rabbit, cat, and ferret, which were believed to represent exceptions amongst mammals. More recently, however, reflex ovulation has been demonstrated in a number of additional groups. Thus most of all Insectivora, Lagomorpha, and Carnivora may be reflex ovulators, as well as several Rodentia. Reflex ovulation appears to be absent amongst Marsupialia, Primates, Cetacea, Proboscidea, Perissodactyla, and Artiodactyla.

It is noteworthy that a spontaneous ovulator like the rat can be turned into a reflex ovulator. Treatment with constant light has been found to block spontaneous ovulation (see Section 13.5.1). In such rats mating, or artificial stimulation of the vagina or uterine cervis, will produce ovulation.

Coitus-induced nonspontaneous ovulation may represent the original eutherian pattern from which the various types of spontaneous cycles have developed. It may even be that under special circumstances ovulation can still be reflexly elicited in spontaneous ovulators. There are indications that, in the rat, mating early in the estrous cycle may accelerate ovulation, that is, cause reflex ovulation.

General Conclusions on Ovarian Function and Control. Despite all of the conspicuous diversity, it is possible to discern basic patterns that may represent ancient features in ovarian structure and function and also trends common to taxonomically unrelated groups that may represent convergent evolution.

It is a basic pattern that animal ovaries that undergo more than one cycle maintain a pool of small oocytes from which specific numbers more or less synchronously start to grow, mature, and ovulate. Two of the basic, but unsettled, questions still remaining are what controls the numbers of oocytes to be recruited and what prevents further recruitment to the rapid, final growth phase from taking place until the clutch of large oocytes has been eliminated, either by ovulation or by degeneration. In several species, representing both invertebrates and vertebrates, it seems that the mechanisms of control are largely to be found within the ovary itself.

It is also typical of ovarian cycles that the earlier phases, including production and early growth of oocytes, are largely controlled locally, from within the ovarian tissue, whereas the later growth stages are controlled by factors from the outside, especially by hormones acting as gonadotropins. It thus seems to be a general feature of vertebrates that the last stages in follicular function are initiated by a surge in gonadotropin secretion, since such a surge has been demonstrated in several species of mammals, in birds, and in lower vertebrates. This preovulatory surge of gonadotropin causes maturation of the eggs and ovulation.

The animal ovary may become a part of systems that serve to adjust both the later follicular growth stages and ovulation to specific environmental conditions, by mediation through sensory organs. A proper ovarian cyclicity may also become secured within internal, integrated systems that have become more or less independent of direct control from the environment, for example, in some insects and in some vertebrates. One important feature in the establishment of integrated reproductive mechanisms has been the potential of the CNS for establishing partly autonomous patterns of activity, patterns that constitute a central element in the concept of biological clocks (see Section 13.5.1).

The daily rhythm in sensitivity toward estradiol of the gonadotropic structures within the CNS of rats and other small rodents constitutes such a biological clock.

It was previously believed that endogenous cyclic activity of gonadotropic structures also controlled the other types of ovarian cycles. No such biological clock has, however, been demonstrated to control the menstrual cycle. The mechanisms that control corpus luteum survival also determine the duration of a cycle, and thus the initiation of the next cycle.

13.4.4 Testicular Function and Control

The process of the production of spermatozoans is remarkably uniform throughout the animal kingdom. Spermatogonia divide mitotically to maintain a stock of male germ cells, or they may differentiate into spermatocytes to initiate the process of spermatogenesis. During spermatogenesis the spermatocytes undergo meiotic divisions, each spermatocyte giving rise to four haploid spermatids that

Figure 13-9
(a) Transverse section of rat testis showing a seminiferous tubule in full spermatogenetic activity. (b) Transverse section of civet cat testis. This is a gelatine-embedded section stained with Sudan black to show the secretory interstitial Leydig cells. [Courtesy of Dr. Brian Lofts.]

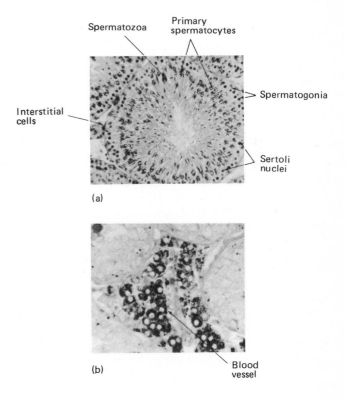

differentiate into spermatozoans. This last differentiation is often termed spermiogenesis. Spermatogenesis typically proceeds within a specific microenvironment of the body, such as cysts in many invertebrates or seminiferous tubules in most vertebrates. Typically, spermatogenesis requires the participation of specific somatic cells, such as the Sertoli cells of the vertebrate seminiferous tubules (Figure 13-9). The fundamental units, either cysts or seminiferous tubules, usually develop within embryonic structures that differentiate into a distinct organ, the testis.

In vertebrates, the testes have two main functions. Besides producing spermatozoans they constitute the primary source of the main male sex hormone, testosterone. Testosterone is needed for the development and maintenance of the secondary sexual characters secreted by interstitial cells, or Leydig cells, which are located between the seminiferous tubules (Figure 13-9).

Spermatogenesis in adult vertebrate males depends upon the secretion of gonadotropins from the hypophysis, but the hormonal control of spermatogenesis is indirect, acting through the Sertoli cells, which enclose the developing germ cells. Research during recent years has greatly advanced our knowledge of the biochemical and molecular events elicited by the gonadotropins in the testes, especially in the Leydig and the Sertoli cells.

Both Leydig cells and Sertoli cells play key roles in present models of testis function. The two cell types act as target cells of the gonadotropins (Figure 13-10). LH stimulates the Leydig cells to secrete testosterone, whereas FSH primarily acts upon the Sertoli cells. These cells are unique in the range of their responses toward FSH. These responses include synthesis and secretion of general and specific cell proteins, such as androgen-binding protein and inhibin, conversion of testosterone to estradiol, and secretion of fluid into the lumen of the seminiferous tubules. The testosterone that is converted to estradiol originates from the nearby Leydig cells (Figure 13-10).

Testosterone secreted by the Leydig cells, and the several hormones, androgen-binding protein, and other proteins secreted by the Sertoli cells accumulate within the seminiferous tubules to build up the microenvironment needed for spermatogenesis. However, little definite is known about how the various biochemical factors stimulate spermato-

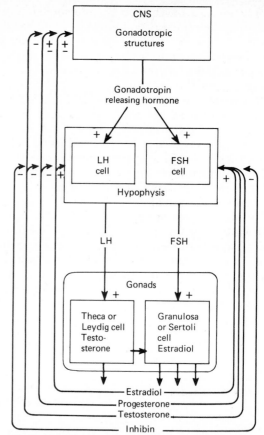

Figure 13-10
Diagram of CN-hypophysial gonadal relationships. +, indicates stimulatory action; −, indicates inhibitory action. See text for further explanation.

genesis or about which developmental stages depend upon these factors.

13.4.5 Female and Male Central Nervous (CN) Activity Patterns

In sexually mature males, the CN gonadotropic structures apparently maintain a tonic stimulation of the secretion of gonadotropins from the hypophysis. Tonically stimulated gonadotropin secretion seems also to characterize the female in species of mammals that exhibit reflex ovulation. As mentioned in Section 13.4.3, in some spontaneously ovulating species an additional, cyclical activity appears to stimulate the tonic center in order to induce the ovulationary surge of gonadotropin secretion. In such spontaneous ovulators the gonadotropin secretion thus differs between males and females due to differences in the pattern of CN control of the secretion.

The male or female pattern of CN activity may be fixed in prenatal or early postnatal life, depending upon the species. In the rat it is possible to interfere with normal sexual differentiation of the brain by administering small amounts of testosterone to newborn (1–5 days old) female rats. At sexual maturity such animals become sterile. Apparently, in these animals the secretion of FSH and LH suffice to support the growth and development of Graafian follicles, but the surge of gonadotropin secretion necessary for ovulation fails to appear. The secretory pattern of gonadotropins is therefore incompatible with normal ovarian function.

A testis implanted in adult female rats that received testosterone in their early postnatal life may function normally both with respect to spermatogenesis and secretory activity of the Leydig cells. It thus appears that early testosterone treatment causes the adult brain to maintain acyclic gonadotropin secretion.

If the testes are extirpated from the newborn rat, the brain of the adult rat becomes fixed according to the female type: ovaries implanted in the castrate will show normal cyclic activity with follicular maturation followed by ovulation. Extirpation of the ovaries in the newborn female rat, however, has no effect on CN stimulatory pattern. Ovaries implanted later in the life of ovariectomized animals will assume normal cyclic function.

The female pattern of CN activity thus appears normally to be suppressed in the male rat in early postnatal life by testosterone secreted by the infantile testes. This assumption is supported by the finding of increased concentrations of testosterone in the plasma of male rats in early postnatal life. On days 1 and 5, the plasma has been found to contain 0.27 and 0.21 μg per l as compared with 0.09–0.1 on days 10 and 15. In sexually mature rats, concentrations increase to 1–2 μg per l plasma.

In order for the treatment to exert any effect on the pattern of activity, the exposure of the brain to androgen must occur within a critical period of time. Testosterone treatment of the female rat fetus or young before and after this period has no effect on ovarian function at the onset of puberty.

13.5 NEUROENDOCRINOLOGY OF REPRODUCTION

The previous sections have provided examples of reproductive patterns from various parts of the animal kingdom. It seems that some basic elements or concepts can be recognized as parts of the mechanisms that regulate gonadal function independent of the taxonomic or ecological status of particular types of animals. The most prominent of these basic elements is the functional complex formed by a gland that produces gonadotropin and a CNS that controls the rate of gonadotropin secretion. This functional unit can act as the mediator between the environment and the gonads, and the CNS can act as the integrator in the interplay with the gonad. But the CNS may also act autonomously, according to inherent patterns of activity. These various activities of the CNS have attracted much interest, and the neuroendocrinology of reproduction has become a vast and diversified branch of physiology, concerned mainly with the white rat, but also some other mammals and birds. We will now consider aspects of this field that may help to further elucidate the relationships that exist between the CNS, the hypophysis, and the gonads, as well as the ways in which environmental factors may act upon these systems.

The CN-hypophysial-gonadal system is a member of the group of neuroendocrine systems that were described in Section 12.4.3, with the CN-hypophysial-thyroid system as an example. The systems serve to maintain the secretory patterns of hypophysial hormones that are required for the normal function of the target organs. CN structures control the secretion of hormones from the various types of hypophysial cells. This control is mediated by the neurosecretory neurons that terminate in the median eminence to release their hormones into the primary capillary plexus of the hypothalamic-hypophysial portal system (Figure 12-11). The portal vessels convey the hypothalamic-releasing hormones to the pars distalis of the hypophysis where they act upon their specific cells to stimulate or inhibit hormone secretion.

It has long been known that gonadotropin secretion is controlled by CN stimulation. If the hypophysis is disconnected from its normal hypothalamic relations, for example by ectopic grafting, gonadotropin secretion decreases. It has also long been known that hormones produced by the gonads control gonadotropin secretion. Removal of ovaries or testes increases gonadotropin secretion. These two sets of findings led to the concept that the gonads control gonadotropin secretion by a negative feedback mechanism. In the female, estradiol and progesterone presumably controlled gonadotropin secretion by inhibiting gonadotropic structures within the CNS, especially the hypothalamus, or by acting directly upon the hypophysial cells that produce LH and FSH. In the male testosterone served in the negative feedback control of gonadotropin secretion.

These original elements are still contained in modern concepts of the mechanisms that control gonadotropin secretion, but during recent years there has been a recognition that neuroendocrine mechanisms are vastly more complex. As indicated in the model diagrammed in Figure 13-10 a variety of gonadal hormones may act both on CN gonadotropic structures and on the LH and FSH cells of the hypophysis. Moreover, the gonadal hormones may stimulate or inhibit depending upon the level of hormones and the physiological state of the organism. The relative importance of the various elements in the regulatory processes mostly remains unsettled. A few important features of the model should be pointed out.

There is a striking similarity in the general patterns of female and male regulatory mechanisms. In both ovaries and testes neighboring cells cooperate in the production of testosterone and its enzymatic conversion into estradiol. Moreover, the peptide hormone inhibin, which was first found to be secreted by the Sertoli cells, also seems to be secreted by the granulosa cells of the ovarian follicles. Inhibin acts on the hypophysis to check, as the name implies, the secretion of gonadotropin, primarily FSH.

The ability of central structures to be inhibited or stimulated by gonadal hormones has become of special importance in the control of ovulation. Low levels of circulating estradiol inhibit the gonadotropic structures, and thus gonadotropin secretion. Higher estradiol levels alter the response

of the gonadotropic structures, and estradiol induces the ovulatory surge in LH secretion (see Section 13.4.3).

LH and FSH may be secreted in diverging ratios under different physiological conditions, both in males and females. This has been difficult to explain, because only one gonadotropin-releasing hormone is known. However, divergent secretory patterns of LH and FSH could result from differential effects of the circulating sex steroids on the sensitivity of the LH and FSH cells toward the gonadotropin-releasing hormone. Usually, the LH cells respond more strongly to the releasing hormone than do the FSH cells (Figure 13-10).

The CN structures that control gonadotropin secretion and, therefore, reproductive behavior are ill defined anatomically. Gonadotropic structures are probably concentrated within the hypothalamus. Numerous experiments have attempted to localize and identify neuronal elements involved in the control of either gonadotropin secretion or sexual behavior. In the rat, and perhaps other mammals with autonomous ovarian cycles, the early evidence indicated that the neuronal circuitry that maintains the cycles includes two well-defined hypothalamic centers: a tonic center responsible for the maintenance of a basal level of gonadotropin needed to support follicular growth, and a cyclic center whose activity pattern was responsible for the pattern of ovulatory LH surges from the hypophysis.

The rigid concepts of anatomically and functionally well-defined gonadotropic hypothalamic centers and specific neurosecretory pathways to the median eminence were, however, too simplistic, and they are being replaced by more dynamic concepts of increasing complexity. Neural regulation of reproduction may be of a more diffuse and complex nature anatomically. This can be exemplified by the neurons that produce gonadotropin-releasing hormone (GnRH).

Originally, GnRH neurosecretory neurons were believed to function only as mediators of the control the hypothalamic gonadotropic centers exerted on the gonadotropic cells in the hypophysis. GnRH-producing neurons can be visualized in histological sections by treating the sections with fluorescent antibodies against GnRH. By means of this technique GnRH-producing neurons have been demonstrated in the brains of vertebrates down to the fishes. It turned out that the neurons have projections that not only contact the cells which release gonadotropin from the hypophysis, but also make connections in many other areas of the brain in addition to the median eminence. The functions of these other neurons remain uncertain. They may act in the control of reproductive behavior, which GnRH has been found to stimulate in female rats. GnRH thus seems to constitute another example of a peptide molecule that may act both as a hormone and as a neurotransmitter or modulator of neuronal activity.

The activity of the gonadotropic neural structures may also be controlled by external, environmental factors. These factors are those listed previously as proximate factors (Section 13.2), for example daylength (photoperiod), sight or scent of another individual, and tactile stimuli such as those during mating.

The photoperiod is the environmental factor that has been most intensively studied as a stimulus in the neuroendocrine control of reproduction, especially in mammals, birds, teleosts, and insects. In most animals where the photoperiod is the factor that times and controls reproduction, it is increasing daylength which acts as the stimulus that activates the neuroendocrine processes leading to gonadal development and reproduction, but it may also be decreasing daylength, for example, in ruminating ungulates that mate in the fall.

In mammals the photoreceptors are located in the eyes since blind animals do not respond to changing photoperiods by changes in gonadal development. The afferent pathways connecting the peripheral receptors with the hypothalamic gonadotropic structures are only incompletely known. During recent years much interest has been focused on the possible role of the pineal gland in mammalian reproduction.

In the Syrian hamster, *Cricetus auratus*, the gonads and the other parts of the reproductive systems in both sexes atrophy during winter, when the animals hibernate. Gonadal atrophy also takes place in the laboratory in hamsters that are exposed to artificial illumination schedules, which mimic those during winter in nature. If the pineal gland is extirpated, the gonads become active again and the animals reproduce. This reactivation of the CN-hypophysial-gonadotropic system by pinealectomy can be prevented by treating the pinealectomized hamsters with melatonin, produced by the pineal gland. The synthesis of melatonin is controlled by the photoperiod. The light environment thus acts via the pineal gland in controlling the annual reproductive cycle in the Syrian hamster and perhaps in other seasonally breeding mammals living with the temperature zones.

In birds and insects light may affect the CN gonadotropic structures not only by acting through the eyes but also by acting directly on specific areas of the brain. In birds this has been demonstrated by placing small discs emitting radioluminescence in various parts of the brain in animals kept on artificial short days. In such birds (for example, Japanese quail, *Coturnix japonica*) testes developed when the disc was placed in the orbital cavity, in the base of the hemispheres of the brain, in the olfactory lobes, and in thalamic and hypothalamic areas. In other areas of the brain the discs were ineffective in eliciting testis development. The retinal and the deep photoreceptors may differ in their sensitivity towards spectral colors. The retina responds only to orange and red rays, whereas the deep photoreceptors also respond to the short wavelengths of the visible spectrum. At least in some birds the deep photoreceptors seem to be the most important. Thus, in house sparrows, *Passer domesticus,* exposed to long days the testes developed normally even in eyeless birds.

Birds, and other photoperiodically stimulated animals, differ greatly with regard to the duration of the daily light period that can stimulate development of the gonads. The thresholds for photostimulation are closely related to the environment and the habits of the birds, for example, whether they are stationary or migratory, and if so where they migrate.

Most birds that have been studied stop breeding before the daylength falls below the critical values necessary for stimulation of gonad development. The ovaries and testes regress, due to discontinued secretion of gonadotropin. The CN-hypophysial gonadotropic system centers in the brain seem to become insensitive to the stimulus elicited by long days. In some birds this ***refractory period*** starts early in the breeding season, permitting the rearing of only one generation of young in the year. In other birds the refractory period appears late in the season, thus permitting prolonged breeding. A refractory period may even be absent, for instance, in some pigeons. The refractory period may be considered as an adaptation to prevent breeding during periods of the year when the lengths of the days are still above the critical values but food supplies are decreasing or failing.

The mechanisms that cause photorefractoriness in birds probably reside at the CN-hypophysial level, because the gonads remain normally sensitive toward gonadotropin. It has recently been shown that in the white-crowned sparrow, *Zonotrichia leucophrys*, the hypophysis of photorefractory male birds is normally sensitive toward gonadotropin-releasing hormone. It thus seems that in the male white-crowned sparrow, and perhaps other birds, the photorefractory state depends upon processes within the CNS.

The length of the refractory period varies, and its length can be controlled by the light and darkness schedule to which the refractory birds are exposed. Thus it has been found that refractory female canaries, *Serinus canarius*, required exposure to short days (8 hr light daily) for 4–6 weeks in order to regain normal sensitivity to long daily photoperiods.

13.5.1 Biological Clocks in Reproduction

Both in photostimulated birds and teleost fishes (stickleback, *Gasterosteus aculeatus*), the absolute length of the light period does not determine its stimulatory or inhibitory effect. The effective stimulus is an exposure to light during a limited period of time each day when the brain mechanisms are sensitive to light.

The nature of such a *circadian rhythm* in light sensitivity has been studied by exposing animals (for example, birds) to daily light schedules mimicking a short day (for example, 6 hr of light) combined with a short period of light (for example, 15 min) given at various times during the night (the 18-hr-period of darkness following the 6-hr exposure to light). Experiments of this type have shown that the short day followed by brief exposures to light during the night can stimulate gonadotropin secretion and normal growth of the regressed testis, but only if these brief exposures fall within a definite period of the night.

The name circadian implies that the rhythm is not precisely a 24-hr rhythm. When the rhythm in experiments of the type just described presents itself as a 24-hr rhythm it is because of the light schedule to which the birds were exposed. The environmental light schedule thus acts to entrain the biological clock to a precise 24-hr rhythm. The environment acts as the so-called *Zeitgeber* (time giver) or synchronizer. In birds, and other species exhibiting circadian rhythmicity, but kept isolated from the environmental synchronizer, the biological clock becomes "free-running" and the sensitive period may occur progressively earlier or later during the daily 24-hr period.

This approximate nature of the periodicity of biological clocks seems to be basic to their adaptability to shifts in the environmental synchronizer, for example, the schedule of light and darkness. Adaptations to such shifts can be observed in the laboratory by exposing animals to schedules of light and darkness that are different from those of the normal or previous environment. This can be shown for example, in the rat, whose estrous cycle is timed by the daily rhythm of estradiol sensitivity (Section 13.4.3). Complete reversal of the light-dark schedule changed the timing of the estrous cycles correspondingly within a few cycles.

Entrained rhythms may persist for some time even in the absence of the entraining agent, the synchronizer. Thus, in the white rat normal estrous cycles may continue in animals kept in constant darkness. However, when exposed to continuous light the rats stop ovulating, presumably because the ovulatory surge of gonadotropin fails to appear.

13.6 VIVIPARITY

Courtship and mating behavior normally result in release of the gametes and in fertilization. In aquatic animals, release of gametes to the water is the rule. However, many purely aquatic groups have evolved mechanisms for internal fertilization. Internal fertilization is an almost universal concomitant of terrestrial life. Fertilization must necessarily take place immediately in aquatic animals that discharge eggs and sperm into the water, before water currents have scattered the sperm. Discharge of sperm into the female genital tract improves the chances of contact between eggs and spermatozoa and thus permits a greater economy in the production of sperm. It also permits fertilization to be delayed. The survival of the spermatozoa is usually short, in mammals normally only a few days. However, in forms exhibiting delayed fertilization the life span of the spermatozoa may be greatly prolonged. Thus, in social insects a single insemination may suffice for the fertilization of eggs ovulated during the entire fertile life of the animal, in the viviparous teleost *Lebistes reticulatus,* for the fertilization of several broods. In mammals, too, examples are known of prolonged survival of sperm. In hibernating bats that mate in autumn, ovulation and fertilization do not occur until the following spring.

In most animals internal fertilization is combined with the immediate release of the eggs to the exterior. However, among many invertebrates and vertebrates further developments have led to the fertilized eggs being retained for shorter or longer periods in the body of the mother, resulting in viviparity or ovoviviparity, that is, in delivery of live young or of eggs that are ready to hatch. These adaptations have resulted in reduction of the numbers of offspring, apparently a consequence of the protection against the environment, including predation, that the mother affords the embryos.

Both viviparity and the production of large yolky eggs may be considered as adaptations that improve the chances that the individual embryo will survive until reproductive age. The first step in the evolution of viviparity in the various taxonomic groups of animals may have been the retention of yolky eggs in the genital tract of the female. In this form, viviparity primarily serves to protect the developing embryo, which still feeds upon yolk. In later evolutionary stages the mother supplies additional food for the embryo. This trend has reached its highest development in placental mammals, which have reverted to producing eggs practically without yolk.

In many vertebrate and invertebrate groups ovoviviparity or viviparity are of scattered occurrence. Among the vertebrates examples are known from the elasmobranches, teleosts, amphibians, and reptiles. In mammals viviparity dominates and is absent only in the egg-laying monotremes. Viviparity is not known in birds, in which brooding of the eggs constitutes an alternative to gestation.

Lower Vertebrates. The scattered instances of viviparity that have evolved independently in the dif-

ferent groups of lower vertebrates have attracted interest because of the opportunities they provide for comparisons with the types of mammalian gestation and their mechanisms of control. In mammals the evolution of viviparity is closely correlated with the development and function of corpora lutea, which usually persist during the period of pregnancy. Studies of viviparity in lower vertebrates have therefore paid special attention to the development of corpora lutea from the ovulated follicles and their possible role in the maintenance of gestation.

The formation of corpora lutea has often been described both in ovoviviparous and viviparous lower vertebrates. Experimental evidence concerning the functional significance of the corpora lutea in lower vertebrates has been conflicting. It seems, though, that the corpora lutea in some species are able to synthesize progesterone, as do the mammalian corpora lutea. Such synthesis has been observed in *in vitro* experiments with ovaries of the viviparous toad, *Nectophrynoides occidentalis.* In other species, such as the viviparous urodele *Salamandra salamandra,* histochemical studies indicate that the corpora lutea produce the enzymes necessary for the conversion of precursors to progesterone. The relations between gestation and corpora lutea have been most thoroughly studied in *Nectophrynoides,* which may serve as an example of viviparity in lower vertebrates.

N. occidentalis inhabits the top of Mount Nimba in West Africa. It spends the dry season, from November to March, estivating in the soil, being active only during the rainy season, from April through October. The breeding season is in autumn when the sexually mature female ovulates 5–14 eggs. Egg number depends upon body size, which varies from 20 to 30 mm in length. The number of eggs is thus strongly reduced in comparison with oviparous anurans of the same size. The ovulated eggs are unusually small, about 0.5 mm in diameter or about the size of previtellogenic oocytes in other amphibians. Ovulation is not dependent upon mating. Isolated females ovulate at the same time as females in amplexus. The ovulated and fertilized eggs remain in the uterus for about nine months. Shortly after breeding, the animals start estivating; during this period embryonic development is suspended, beginning again only when the females resume an active life with the onset of the next rainy season. The young are delivered fully metamorphosed in early summer.

Following ovulation the ovarian follicles are converted into well-developed corpora lutea, which maintain their size during the period of estivation and start regressing when the females emerge in spring, concurrently with the rapid development of the embryos. Some interaction presumably exists between embryos and corpora lutea. Thus, the corpora lutea regress at a faster rate in females carrying unfertilized eggs than in gravid females. Moreover, ovariectomy during early pregnancy in October has been found to cause abortion, but only during the first pregnancy. Ovariectomy during the second to fourth pregnancy is compatible with delivery of normal young, but embryonic development is accelerated and parturition is significantly advanced. It has been suggested that the function of the progesterone secreted by the corpora lutea is to stimulate growth and differentiation of the oviducts, to make them suitable for accommodating the developing embryos.

In the other viviparous amphibian, *S. salamandra,* possessing corpora lutea that may produce progesterone, extirpation of the ovaries is without effect on the intrauterine development of the embryos, even when the operation is performed early in the period of gestation.

13.6.1 Implantation

In mammals, the initiation of gestation is regulated by complex interactions between the fertilized eggs and the maternal organism. Upon fertilization the egg becomes an independent individual whose continued existence depends upon the proper timing of processes in the egg and its surroundings leading up to the implantation of the embryo in the uterus wall.

The first critical event in the life of the young embryo is its recognition by the maternal organism. The moment of recognition varies. Recognition is early in the mare, where it apparently occurs during the passage of the fertilized egg along the Fallopian tube because it has been observed that only fertilized eggs pass the Fallopian tube. Unfertilized eggs remain in the tubes for several months before they finally degenerate. Maternal recognition is more often postponed till the egg arrives in the lumen of the uterus; here recognition may be elicited by

interactions between the fertilized egg, now in the blastocyst stage, and the uterus epithelium. Recognition may, however, come even later. In marsupials it occurs after parturition, when the young settle in the pouch and start suckling. Gestation itself proceeds in marsupials without any change in the ovarian cycle. This and other features distinguish reproduction in marsupials from that in eutherian mammals, suggesting that viviparity has evolved separately in marsupials and eutherians after their derivation from a common oviparous ancestor.

Marsupial reproductive pattern is determined by the length of the gestation period and the absence or presence of suckling young. In nonmacropodid marsupials with short gestational periods, birth occurs before the onset of the next estrous period. If the young have settled in the pouch and started suckling, ovulation and estrus will fail to appear. In macropodids gestation usually lasts about as long as the estrous cycle, in most species a little shorter in some a little longer. The former group mate postpartum, the latter group prepartum. The fertilized egg is accommodated in the nonpregnant uterus. (Marsupials possess two completely separate uteri.) However, no implantation takes place and development of the young embryo is arrested at the blastocyst stage during the period of parturition and/or lactation of the previous young (Figure 13-11). When the young is ready to leave the pouch, or if it is removed, embryonic development is resumed. The quiescent state of the embryo probably results from reflex inhibition (caused by the suckling) of corpus luteum function, that is secretion of progesterone is inhibited. The reflex inhibition acts through the hypophysis, which seems to secrete an inhibitory factor that may be prolactin or oxytocin (see Table 12-1).

In eutherian mammals postpartum estrus and mating, followed by delayed implantation of the fertilized eggs, is known from some insectivores and rodents. In rodents corpora lutea develop only if the animals are mated. In these forms it has been possible to resolve the events of early gestation into major components, which has greatly facilitated the analysis of the integrating mechanisms. The studies have mainly been performed in the mouse and the rat.

In estrous rats and mice the uterus is extended by secretion ("ballooned") but the walls are thin and not yet prepared for implantation of the ferti-

Figure 13-11

Reproductive patterns in two groups of diprotodont marsupials. Sides 1 and 2 refer to the completely separated right and left uteri. [Modified from G. B. Sharman, et al. In I. W. Rowlands (ed.), Comparative Biology of Reproduction in Mammals, Academic Press, Inc., New York, 1966, p. 221.]

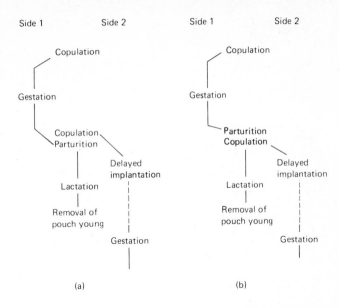

lized eggs. The necessary further development is controlled by ovarian secretion of progesterone and estrogen. If newly mated rats are ovariectomized no implantation takes place. However, implantation occurs if the operated animals are treated with progesterone and estrogen. Progesterone alone is ineffective.

Prior to implantation the blastocysts are distributed in epithelial crypts at regular intervals along the uterine horns. They adhere to the epithelium by their embryonic pole and remain in this state of preattachment if implantation is delayed (Figure 13-12a). However, if estrogen has been secreted along with progesterone, the epithelium differentiates, decidua formation and implantation start. The first step is an increase in the permeability of the uterine capillaries around the blastocyst, which results in the development of local edema. Subsequently, the uterine connective tissue proliferates and completely encloses the blastocyst (Figure 13-12b). The basic effects of the steroids in implantation thus seem to be decidua formation and a stimulation of DNA replication, and of cell division (see Section 12.8).

Sensitization of the uterine epithelium need not result in decidua formation. Thus, if female rats or mice are mated with sterile, vasectomized males, the epithelium starts proliferating but decidua are not formed and pseudopregnancy ensues. The de-

cidual processes presumably result from the interaction between the blastocysts and uterine epithelium that has previously been sensitized under the influence of estrogen.

The mouse becomes estrous soon after parturition. In nature, mice therefore normally become pregnant after delivery. Implantation, however, is delayed. The length of the delay increases with the number of young the mother nurses. The suckling of the young apparently reflexly inhibits the secretion of LH from the hypophysis and, thus, of estrogen. The uterus epithelium is not sensitized and does not respond to contact with the blastocyst by decidua formation. If the litter is removed, the central inhibition of gonadotropin secretion is released and implantation ensues.

13.6.2 Gestation

The implanted blastocyst differentiates into the fetus and the trophoblast. The trophoblast differentiates further into the placenta by the continued process of decidua formation, as a response of the uterine epithelium to the trophoblast invading the endometrium. The more invasive the trophoblast the larger the area of the decidual epithelium through which both the fetus and the placenta obtain oxygen, nutrients, water, ions, hormones, and so on, from the mother, and discharge waste prod-

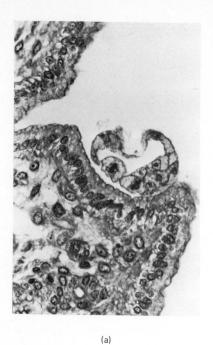

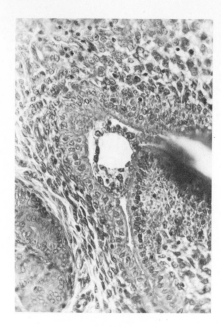

Figure 13-12
Longitudinal sections of the uterus of pregnant rats. (a) The blastocyst is attached to the epithelium by its embryonic pole at the fifth day of pregnancy. (b) Implanted blastocyst. [From A. Psychoyos. Advan. Reprod. Physiol., **2**, 257–277 (1967).]

(a)

(b)

ucts. The human trophoblast is the most invasive known, and it elicits the most pronounced uterine decidual response. This development depends basically upon the continued functioning of the corpora lutea, which again depends upon secretion of gonadotropins or prolactin (rat).

In the rabbit and some other species, ovariectomy or hypophysectomy performed at any time during the period of gestation results in abortion or in resorption of the fetuses. In other mammals, gestation may continue after one or both of these operations. Thus, in the rat and mouse, hypophysectomy is compatible with continued gestation if the operation is performed after a certain date during gestation. In such species it has been found that the secretion of the necessary gonadotropins is taken over by the placenta. Rats and mice, however, consistently abort if the ovaries are extirpated, indicating that gestation depends upon the continued function of the corpora lutea. In other mammals, however, gestation becomes independent of the ovaries. Thus, in the ewe, gestation may proceed after removal of the ovaries if the operation is performed after about 50 days of gestation. Presumably, at this time the placenta replaces the cor-

pora lutea in the secretion of progesterone. In keeping with this assumption the progesterone level in the blood starts to increase after 50 days, and placental secretion of progesterone in late pregnancy in the ewe may be about 5 times greater than the ovarian secretion. In women anlyses of progesterone in plasma and studies on the effect of ovariectomy on gestation have indicated that the main secretion of progesterone changes from the corpus luteum to the placenta around the ninth week of gestation.

Placental secretion of hormones is widespread among eutherian mammals. It has reached its highest development in the primates. The human placenta synthesizes both "human chorionic gonadotropin" and "human placental lactogen" besides progesterone and several estrogens, estradiol, estrone, and estriol. Chorionic gonadotropin is the first hormone to be synthesized, and it has been found to coat the differentiating trophoblast. It is believed that the hormone at this early stage acts as an immunologically protective layer that prevents the rejection of the blastocyst. The hormone also secures the survival of the corpora lutea throughout the period of gestation.

13.6.3 Parturition

During the period of gestation the myometrium of the uterus grows and builds up the contractile system. The synthesis of the contractile proteins is stimulated by estradiol or other estrogens secreted by the ovaries and, in some species, the placenta. At parturition the contractile system becomes active and expels the fetuses. Understanding of the mechanisms that control the termination of gestation and initiate parturition, therefore, ultimately rests upon knowledge of the electrophysiological and contractile properties of the myometrial muscle fibers and their reactions to circulating hormones.

The uterus in immature animals is quiescent. Estradiol matures the muscles and eventually starts spontaneous discharge of action potentials, which may propagate and result in regular rhythmic contractions. The administration of progesterone to an estradiol-primed uterus counteracts the effect of estradiol. Muscular activity is greatly reduced. There is no propagation of action potentials and contractions become local and uncoordinated. It has, therefore, been assumed that a major function of progesterone during the later phases of pregnancy is to prevent waves of muscular contractions from arising in the uterus. In support of this theory it has been observed, for example, in rabbits, rats, and sheep, that progesterone levels in the blood decline sharply immediately after parturition starts.

The mechanisms that terminate gestation and initiate parturition probably vary among species, but mostly they remain obscure. Major advances have, however, been made in understanding the mechanisms that control parturition, at least in the sheep. The series of events that leads to term in the sheep can be traced back to the maturation of the fetal adrenocorticotropic-adrenal system. During the last part of gestation the fetal hypophysis starts secreting adrenocorticotropic hormone (ACTH), but the high rates of ACTH secretion do not stimulate the fetal adrenal to secrete adrenocortical hormones, such as cortisol, until the last days of gestation. Presumably, this increase in cortisol secretion from about 1 mg/24 hr 5 days before term to 2–6 mg/24 hr at the onset of labor results from the maturation and consequently enhanced sensitivity of the fetal cortex toward circulating ACTH. The high rate of cortisol secretion apparently initiates parturition. This is supported by the finding that fetal hypophysectomy or adrenalectomy resulted in prolonged gestation, whereas injections of ACTH or corticosteroids caused premature parturition.

The prepartum fetal cortisol secretion acts on the placenta to reduce progesterone secretion and to increase estradiol secretion. This change in progesterone-estradiol ratio initiates the uterine contractions. The increased estradiol seems not to act directly, but through the liberation of a prostaglandin, $PGE_{2\alpha}$ (see Section 12.2), which thus acts as a messenger (Figure 13-13). Prostaglandins may also act as messengers in parturition in other mammals and in women. Certain placental cells are major sites for the synthesis of these prostaglandins. Prostaglandins can induce abortion in the later part of gestation, whereas agents that inhibit synthesis of prostaglandins can extend gestation.

Parturition also involves the oxytocin mechanism. Distension of the cervix by a fetus during parturition reflexly stimulates the secretion of oxytocin from the pars nervosa of the hypophysis (Section 12.4.1). Oxytocin specifically stimulates the myometrial muscle fibers, and the sensitivity of the uterus toward oxytocin becomes maximal during parturition as a result of the decline in progesterone levels.

ENDOCRINE REGULATION OF PARTURITION

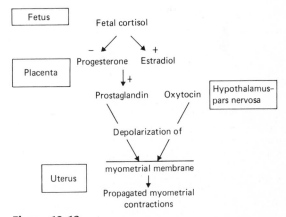

Figure 13-13
Diagram of major steps in the endocrine mechanisms leading to propagated myometrial contractions and thus to parturition in sheep.

The importance of the oxytocin mechanism in normal parturition varies among species. In species such as the rabbit, in which parturition normally proceeds quickly and is completed within 15–30 min, lack of oxytocin causes delayed or protracted parturition and a high fetal mortality. In other species, such as man and the domestic goat, parturition is a slow process. In its initial stages oxytocin has not been observed to be secreted. During the final stages of labor, however, large amounts of oxytocin appear in the blood and contribute to the myometrial contractions. In some species, parturition may proceed more or less normally even in the absence of oxytocin secretion.

13.7 LACTATION

In most animals, care of the young is restricted to supplying food for the developing embryo in the form of yolk deposited in the egg. Posthatching or postnatal parental care has evolved sporadically within several animal groups, including invertebrates, such as several insects. It has reached its highest levels in mammals, which nourish their young on a special secretion, milk. Pigeons also nourish their young on a secretion, produced by the epithelium of the crop sac, but they represent exceptions among the birds. Milk secretion and its mechanisms of regulation exhibit certain basic features but—as was the case with the other phases of mammalian reproduction such as breeding pattern, mating behavior, ovarian cycles, gestation, and parturition—the mechanisms that control the development of the mammary glands and their secretion have repeatedly been modified during evolution.

Monotremes. Lactation preceded viviparity in mammalian evolution. In the egg-laying monotremes, mammary glands are fully differentiated and ready for secretion when the animals reach sexual maturity. The glands develop both in males and females, but they secrete only in the female nursing the young.

Marsupials. In marsupials the mammary glands also reach the lactational stage at sexual maturity, but only in females. Development depends upon secretions from the ovaries, ovariectomy resulting in involution of the glands.

Gestation does not influence the mammary glands. Thus, in the phalanger *Trichosures,* the mulgara *Dasyurus,* and the opossum *Didelphis,* the weight of the mammary glands and the differentiation of secretory alveoli was found to be the same in females that went through a pregnancy and delivered young as it was in nonmated females in the equivalent postestrous stage. If young were transferred to the pouch of the nonpregnant postestrous animals, they would suckle normally. Thus, in marsupials lactation as well as gestation is characterized by the relatively slight development of neuroendocrine interactions among uterus, ovaries, and mammary glands.

The mammary glands do not start to secrete milk until the young attach themselves to the teats. It is apparently the suckling of the teat rather than the withdrawal of milk from the alveoli that acts as the main stimulus for the initiation and maintenance of secretion of milk *(lactogenesis).*

Eutheria. In eutherian mammals the prepuberal development of the mammary glands varies greatly among species. The ferret represents one extreme, with the mammary glands starting to develop only after the female has become pregnant. The rabbit exhibits slight ductal growth at sexual maturity, with little difference between males and females. In most species, though, the ductal system becomes well developed with the onset of puberty in the female and remains rudimentary in the males. In the sexually mature female, the ductal system often fluctuates synchronously with the estrous or menstrual cycle. Differentiation of the secretory part of the mammary gland, the alveoli which bud from the branching ductal system, is usually slight in nonpregnant females. Differentiation takes place during pregnancy. At the end of gestation mammogenesis is complete, and the gland is ready for lactogenesis, which is initiated at parturition.

We may thus distinguish four phases in mammary gland development: the prepuberal, puberal, gestational, and lactational phases. Each of these phases is determined by the endocrine status of the organism.

Endocrine control of the development of the mammary gland has been studied by means of extirpation of the endocrine glands, especially the hypophysis and ovaries but also the thyroid gland, the adrenal cortex, and the pancreatic islets. Operated animals were treated with hormones in various combinations in order to see the effect on growth and differentiation of ducts and alveoli and on secretion of milk. Much valuable information has also been obtained from *in vitro* studies in which isolated mammary gland tissue, taken from immature, puberal, pregnant, or lactating animals, was kept in controlled media to which hormones were added. The experiments have shown that a great number of hormones may stimulate mammary gland development. However, the effects of hormone combinations and substitutions may vary among species, as well as intraspecifically, depending upon the developmental stage of the gland at the time of the experiment. It is often difficult to draw conclusions from the experimental results as to the roles played by the hormones in the intact organism. Some trends in endocrine control have, however, emerged (Figure 13-14).

The growth of ducts in puberal animals apparently results mainly from the secretion of estrogen by the ovaries. In ovariectomized rats and rabbits, normal growth of ducts can be induced by treatment of the animals with estrogen. Additional treatment with progesterone causes alveolar growth as well. Alveolar development during pregnancy seems to result mainly from the secretion of pro-

gesterone from the corpora lutea, in some species supplemented or superceded by progesterone secretion from the placenta (see Section 13.6.2). Progesterone given alone to ovariectomized animals has no stimulating effect on the mammary glands.

Estrogen and progesterone may also cause duct growth and alveolar differentiation in ovariectomized-hypophysectomized animals (rabbit), but the effect is smaller than in animals only ovariectomized. Replacement therapy by hypophysial hormones or hormones of the target organs showed that thyroid hormone (thyroxine) enhanced the effects of the ovarian steroid hormones, but only if given along with a corticosteroid hormone such as cortisone. Corticosteroid hormones must thus be present in order that the other hormones can exert their effects on the mammary gland tissue. (Hence the designation "permissive" about the effect of corticosteroid hormones in this and many other situations.) In the rat and several other species, but not the rabbit, insulin could replace thyroxine.

Prolactin and growth hormone also seem to act synergistically with estrogen and progesterone in establishing the prelactational state of the mammary gland during gestation.

After parturition prolactin stimulates milk secretion. During pregnancy progesterone inhibits the effect of prolactin on lactogenesis, presumably

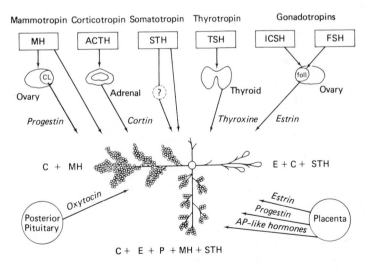

Figure 13-14
*Diagram showing some of the hormones that influence mammary growth and lactation. Upper part of the mammary diagram, rudimentary gland; right, prepuberal or puberal gland; lower, gland of pregnancy (prolactational); left, lactating gland. Mammotropin (MH) = prolactin; corticotropin (ACTH) = adreno corticotropic hormone; somatotropin (STH) = growth hormone; Thyrotropin (TSH) = thyroxins (T$_3$ and T$_4$); ICSH = interstitial-cell-stimulating hormone = luteinizing hormone; FSH = follicle-stimulating hormone; estrin (E) = estrogen; cortin (C) = corticosteroid hormones; progestin (P) = progesterone and other progestational hormones; AP-like hormones = anterior-pituitary (pars distalis)-like hormones (for example, gonadotropins and luteotropin). [From W. R. Lyons, C. H. Li and R. E. Johnson, Recent Progr. Horm. Res., **14**, 219 (1958).]*

by blocking the formation of receptors for prolactin on the alveolar cells. The prolactin binding activity of the mammary glands increases greatly after parturition, and prolactin can exert its wide spectrum of activities in stimulating lactogenesis, including synthesis of milk proteins, metabolism of lactate and lipid, ion transport, and production of the antibodies in the milk.

The role of prolactin in maintaining lactogenesis varies among species. In the cow prolactin is crucial for lactation at parturition but not later. In women prolactin maintains milk secretion during the period of lactation, but it is especially important during the first weeks after parturition. Suckling reflexly elicits secretion of prolactin and thus serves to stimulate milk secretion (see Section 12.4.1).

ADDITIONAL READING

REPRODUCTIVE BEHAVIOR
Bermont, G., and J. M. Davidson. *Biological Bases of Sexual Behavior,* Harper & Row, New York, 1974.

Friedman, R. C., R. M. Richart, and R. L. Vande Wiele (eds.). *Sex Differences in Behavior,* John Wiley & Sons, New York, 1974.

Montagna, W., and W. A. Sadler (eds.). *Reproductive Behavior,* Plenum Press, New York, 1974.

Truman, J. W., and L. M. Riddiford. "Hormonal mechanisms underlying insect behavior," *Adv. Insect Physiol.,* **10,** 297–352 (1974).

GENERAL
Browning, H. C. "The evolutionary history of the corpus luteum," *Biol. Reprod.,* **8,** 128–157 (1973).

Greep, R. O. (ed.). *Reproductive Physiology,* University Park Press, Baltimore, 1974.

Jones, R. E. (ed.). *The Vertebrate Ovary,* Plenum Press, New York, 1979.

Nalbandov, A. V. *Reproductive Physiology of Mammals and Birds,* 3rd ed. W. H. Freeman & Company, San Francisco, 1976.

Parkes, A. S. (ed.). *Marshall's Physiology of Reproduction,* vols. 1–3, Longmans, Green and Co., London, 1952–1966.

Roosen-Runge, E. C. *The Process of Spermatogenesis in Animals,* Cambridge University Press, New York, 1977.

Zuckerman, Sir Solly, and B. Weir (eds.). *The Ovary,* vols. 1–3, Academic Press, New York, 1977.

MAMMALS
Austin, C. R., and R. V. Short (eds.). *Reproduction in Mammals,* vols. 1–7, Cambridge University Press, New York, 1972–1979.

Gilmore, D., and B.Cook (eds.). *Environmental Factors in Mammal Reproduction,* Macmillan, London, 1981.

Grumbach, M. M., G. D. Grave, and F. E. Meyer (eds.). *The Control of the Onset of Puberty,* John Wiley & Sons, New York, 1974.

Hafez, E. S. E. (ed.). *Reproduction in Farm Animals,* 4th ed., Lea and Febiger, Philadelphia, 1980.

Henderson, K. M. "Gonadotrophic regulation of ovarian activity," *Brit. Med. Bull.,* **35,** 161–166 (1979).

Karim, S. M. M. *Prostaglandins and Reproduction,* University Park Press, Baltimore, 1975.

Lincoln, G. A. "Pituitary control of testicular activity," *Brit. Med. Bull.,* **35,** 167–172 (1979).

Means, A. R., J. R. Dedman, J. S. Tash, D. J. Tindall, M. van Sickle, and M. J. Welsh. "Regulation of the testis Sertoli cell by follicle stimulating hormone," *Ann. Rev. Physiol.,* **42,** 59–70 (1980).

Richards, J. S. "Hormonal control of ovarian follicular development: a 1978 perspective," *Recent Progr. Hormone Res.,* **35,** 343–378 (1979).

Steinberger, A., and E. Steinberger (eds.). *Testicular Development, Structure, and Function,* Raven Press, New York, 1980.

Tyndale-Biscoe, C. H., J. P. Hearn, and M. B. Renfree, "Control of reproduction in macropod marsupials," *J. Endocrinol.,* **63,** 589–614 (1974).

NONMAMMALIAN VERTEBRATES
Breder, Jr. C. M., and D. E. Rosen. *Modes of Reproduction in Fishes,* The Natural History Press, New York, 1966.

Farner, D. S. (ed.). *Breeding Biology of Birds,* National Academy of Science, Washington, DC, 1973.

Farner, D. S., J. R. King, and K. C. Parkes. *Avian Biology,* vols. I–III, Academic Press, New York, 1972.

Jørgensen, C. B., L. O. Larsen, and B. Lofts. "Annual cycles of fat bodies and gonads in the toad *Bufo bufo bufo* (L)," *Biol. Skr. Dan. Vid. Selsk.,* **22**(5), 1–37 (1979).

Murton, R. K., and N. J. Westwood. *Avian Breeding Cycles,* Clarendon Press, Oxford, 1977.

Taylor, D. H., and S. H. Guttman (eds.). *The Reproductive Biology of Amphibians,* Plenum Press, New York, 1977.

INVERTEBRATES
Adiyodi, K. G., and R. G. Adiyodi. "Comparative physiology of reproduction in arthropods," *Adv. Comp. Physiol. Biochem.,* **5,** 37–107 (1974).

Giese, A. C., and J. S. Pearse (eds.). *Reproduction in Marine Invertebrates,* vols.1–2, Academic Press, New York, 1974.

Wigglesworth, V. B. *Insect Hormones,* Oliver and Boyd, Edinburgh, 1970.

Wilde, J. de, and A. de Loof. "Reproduction," pp. 11–157. In M. Rockstein (ed.). *The Physiology of Insecta,* 2nd ed., vol. 1, Academic Press, New York, 1973.

NEUROENDOCRINOLOGY OF REPRODUCTION

Gorski, R. A. "The neuroendocrinology of reproduction: an overview," *Biol. Reprod.*, **20**, 111–127 (1979).

Harlan, R. E., J. H. Gordon, and R. A. Gorski. "Sexual differentiation of the brain: implications for neuroscience," *Rev. Neurosci.*, **4**, 31–71 (1979).

Jeffcoate, S. L., and J. S. M. Hutchinson (eds.). *The Endocrine Hypothalamus*, Academic Press, New York, 1978.

Labrie, F., P. Borgeat, J. Drouin, M. Beaulieu, L. Lagacé, L. Ferland, and V. Raymond. "Mechanism of action of hypothalamic hormones in the adenohypophysis," *Ann. Rev. Physiol.*, **41**, 555–569 (1979).

Meites, J., B. T. Donovan, and S. M. McCann (eds.). *Pioneers in Neuroendocrinology*, vols. 1–2, Plenum Press, New York, 1975, 1978.

Reiter, R. J. "The pineal and its hormones in the control of reproduction in mammals," *Endocrine Rev.*, **1**, 109–131 (1980).

BIOLOGICAL CLOCKS AND RHYTHMS

Beck, S. D. *Insect Photoperiodism.* 2nd ed., Academic Press, New York, 1980.

Bünning, E. *The Physiological Clock: Circadian Rhythms and Biological Chronometry*, Springer-Verlag, New York, 1974.

Ferin, M., F. Halberg, R. M. Richart, and R. L. Vande Wiele (eds.). *Biorhythms and Human Reproduction*, John Wiley & Sons, New York, 1974.

Hedlund, L. W., J. M. Franz, and A. D. Kenny (eds.). *Biological Rhythms and Endocrine Function*, Plenum Press, New York, 1975.

Murton, R. K., and J. Kear. "Photoperiodism in waterfowl: phasing of breeding cycles and zoogeography," *J. Zool.*, **186**, 243–283 (1978).

Pengelley, E. T. (ed.). *Circannual Clocks: Annual Biological Rhythms*, Academic Press, New York, 1974.

GESTATION AND LACTATION

Davies, I. J., and K. J. Ryan. "Comparative endocrinology of gestation," *Vitamins and Hormones*, **30**, 223–279 (1972).

Larson, B. L., and V. R. Smith (eds.). *Lactation. A Comprehensive Treatise*, vols. 1–III, Academic Press, New York, 1974.

Shiu, R. P. C., and H. G. Friesen. "Mechanism of action of prolactin in control of mammary gland function," *Ann. Rev. Physiol.*, **42**, 83–96 (1980).

Steven, D. H. (ed.). *Comparative Placentation*, Academic Press, New York, 1976.

Wolstenholme, G. E. W., and Maeve O'Connor (eds.). *Foetal Autonomy*, J. and A. Churchill, London, 1969.

Index

Blood calcium
   regulation of, 564
Blood flow
   forms of energy involved in, 209
   kidney, 243
   liver, 244
   renal portal system, 242
   resistance to, 207–208
     calculation of, 208
   to various organs, 206
   tissues of external respiration, 239
   turbulence, 210
Blood pressure, 213
   pulmonary and systemic contrasted, 240–241
Blood vessels
   geometry of, 200
   physical properties of, 211
   series and parallel arrangements of, 206
   types of, 206
Blood volume, 201, 217–218
   determination of, 217
   diving animals in, 257
   diving and nondiving species, 257
   hypovolemia, 216
   in the pulmonary vessels, 242
   of various vertebrates, 218
   of venous system, 226
BMR, see metabolic rate, basal
Body hydration, 282
Body spaces, 282
Body temperature, 66
   daily rhythms, 66
Body water, 281
Bohr effect, 268
Boiling-point elevation, 275
Bombykol, 461, 553
Bombyx mori, 553
Boyle, R., 252
Brachiopods, 17
Bradykinin, 551
   effect on skin blood flow, 246
   and pain, 456
Brain
   evolution of, 521
   hormone, 558
   light sensitivity in, 497
   organization of language centers, 529
   oxygen consumption of, 428
   structure of, 524–529
Brain temperature, 386
Branchial heart, 230
Brandt, 20
Brocas area, 529

Brody, S., 59
Brooding, 589, 605
Budgerigar, 79
Buffers
   blood, 177–178
   water, 137
Bufo, 76, 583, 592
   water balance in, 566
Bufonidae, 566
αbungarotoxin, 540
   and ACh receptors, 433
Butenandt, 553

C

CA²⁺
   action potentials, role of, 428–429
   and presynaptic inhibition, 445
   role in axonal transport, 540
   role in photoreceptors, 507
   role in PTP, 444
   role in synapses facilitation, 443
   role in transmitter release, 439–440
Calanida, 18
Calcium, 26, 38
   homeostasis, 565
Calcium hypothesis, of transmitter release, 443
Calcitonin, 564
Callorhinus, 373
Calmodulin
   role in transmitter release, 443
Calorimetry, direct, 52
Calorimetry, indirect, 52
   definition, 52
   methods of, 52
Calorie, definition of, 48
Calypte, 56, 67, 394
Camels, 22
Cameloids, 262
Capacitance, 419, 421
   of biological membranes, 419
   of membranes, 421
Capacitance coefficient, gases, 137, 139
Capacity for growth, 40
Capillary, 200, 214
   blood flow, 216
     ascending vasodilation and, 217
     local control of, 216
   function, 214–217, 241
     fluid movements across, 215–216
     pulmonary, 241
   surface area, 200
   transfer of fluids across, 215

Capra, 381
Caprimulgiformes, 369
Carassius, 343
Carbohydrases, 28 (*see also* Glycosidases)
Carbohydrate absorption, 37
Carbohydrates, 28–29
Carbon dioxide
   acid-base regulation and, 179–184
   buffering of, 177–180
   dissociation curve for, 178
   in external media, 134
   hemoglobin transport of, 178
   respiratory control by, 188–189
   transport of, 177–180
   in water, 134–135
Carbonic anhydrase, 171, 178
Carboxypeptidase, 20
Cardiac cycle, 225–228
   filling and emptying sequence, 226–227
   mechanics of, 225
Cardiac muscle (*see also* Muscle, cardiac)
   intercolated disks of, 445
Cardiac output, 222
   blood convection requirement and, 224
   in exercise, 231
   measurement of, 224
   in various vertebrates, 222
Cardiovascular functions, 198, 247–251
   baroreceptors, and, 247
   control, 247–248
     in hypotensive states, 248
   general nature of, 247
   presoreceptors and, 247
   sympathetic effects, 248
Cardiovascular performance, 198
Cationic pumps, 428
Carnivora, 598
Carnivores, 40
Carotenoids, and vision, 496
Carotid bodies, respiratory control by, 188
Carrier-mediated transport, 37
Cat, 598
Catch muscles, 128
Catecholamines, 580
Cathartes, 389
Cation pump, 428
Caudal hearts, 230
Cecum, 22
Cellular energy metabolism, 46, 47
Cellulose, 29
Central nervous system (CNS), 409
Cephalization in evolution of brain, 521

Filtration, 278
Filtration fraction (FF), 279
Fishes, 230–233, 602
  blood pressure drop across gills, 241
  circulation in exercise, 251
  electroreception in, 491
  hearts of, 230
  heart weights of, 231–232
Flat fishes, 557
Flight, 92, 128, 401
  cruising range and size, 92
  and heat production, 401
  muscle, 128
    in insects, 401
Fluorine, 26, 27
Flying foxes, 79
Focusing, mechanisms in different animals, 497
Follicle cells, 592
Follicle-stimulation hormone, 595
Food chains, 421
Food energy, partitioning of, 39
Food webs, 14, 24
Foramen panizzae, 237
Fovea, 499, 506
Freezing-point depression, 275
Frequency resolution
  in invertebrates, 477
  mechanism of, 483
Freshwater sponges, 20
Freshwater turtles, 566
Frogs, 557
Fructose, 28
FSH, 578, 597, 600, 601, 602
Fundulus, 177
Fungi, 21, 23

G

GABA, and synaptic inhibition, 445
Gadus, 143
Galacturonic acid, 28
Galapagos marine iguana, 181
Galen, 185
Gall bladder, 33
Galvani, 95
Gammarus, 39
Ganglia, autonomic, 533
Ganglion cells
  of retina, 510
    response types, 510
Gap junctions, and electrical transmission, 445
Gas law, 137

Gases, in external environment, 134 (*see also* Respiration)
Gasterosteus oculeatus, 588, 604
Gastric fistula, 31
Gastric inhibitory peptide, 33, 579
Gastric juice, 31
Gastrin, 32, 536, 571
Gastrin-CCK-family, 579
Gastrointestinal
  blood flow, 245
  hormones, 568
Gastropods, 17, 42
Gazella, 381
Generator potential, 416, 422–423
  in stretch receptors, 467
Genetic makeup of experimental animals, 9
Geococcyx, 400
Geological time scale, 3
Gerrhonotus, 352
Gessamen, J. A., 54
Gestation, 605, 606, 607–609 (*see also* Viviparity)
Giant clams, 20, 21
Gibbs-Donnan equilibrium, 284, 293, 417
Gills
  blood, 239
  blood flow through, 149–153, 281
  blood and water flow, 240
    timing of, 240
  as heat exchangers, 340
  water flow through, 142, 148–149, 163–165
Giraffe, 263–265
  heart and blood vessels of, 263–265
  high blood pressure in, 263
Glands, autonomic control of, 532
Glia, 411, 507 (*see also* Neuroglia)
  in retina, 507
Glomerular filtration rate (GRF), 279
Glomeruli, 278
Glucagon, 33, 566, 579
Gluconlogenisis, 567, 568
Glucocorticoid hormones, 581
Glucose, 28, 29, 47
  energy yield of, 47
β-Glucoside, 29
Glucuronic acid, 28
Glycine, 278
Glycogen, 28, 42
Glycogendysis, 29, 568
Glycolysis, 69 (*see also* Anaerobiosis)
  aerobic and anaerobic, 69
Glycolytic enzymes in lizards, 71

Glycoprotein hormones, 578
Goats, 22
Goldfish, 37
Goldman-Hodgkin-Katz equation, 419
Golgi tendon organs, 467
Gonadotropin, 43, 578, 589, 591, 592, 593, 599, 600, 601, 608
  secretion, control of, 601
Gonadotropin-releasing hormone, 602
Gonads, 589–590
Graafian follicles, 594, 601
Granulosa cells, ovarian follicles, 602
Green turtle, 22
Growth, 40
Growth factors, 43, 522
Growth hormone, 43, 578, 611
Guanaco, see *Lama*
Gular flutter, 387
Gut-brain peptides, 281, 282, 579
Gymnemic acid, 460
Gymnodinium microadriaticum, 20, 21
Gypsy moth, 553
Gyptol, 553
Gyrinophilus palleucus, 560

H

Hagfish, accessory hearts of, 199
Haideotriton, 560
Hair cells, 470–473, 485
  electrical responses of, 473
  inner and outer, in cochlea, 480–481
  mechanisms of function in cochlea, 481
  mechanoreceptors, 470
  organization in cochlea, 485
  synaptic mechanisms of, 473
Hair plate, sensory organ in invertebrate, 470, 476
Hall, M., 185
Hamburger shift, 178
Haplochromis, 143
Harmothoe, 554
Harvey's, E. N. equation, 197
Harvey, W., 198
Hazel, J. R., 343
Hearing, 476–491
  absolute sensitivity of, 485
  in birds, 487, 488
  dynamic range in, 485
  evolution of, 478
  frequency resolution, 481–485
  in insects and evasion by bats, 490–491

Weis-Fogh, T., 79
Wernicke's area, 529
Western axolot, 560
Whales, 259
White-crowned sparrow, 604
White laboratory rat, 595
White muscle vs. red in fishes, 375 (*see also* Muscle)
Weiner, N., 564
Wigglesworth, V. B., 558
Wing loading and body temperature, 403
Work, 77
    definition and units, 77

Work of heart, 228
    sympathetic nervous influence on, 228
Work of ventricle, 227

X

Xantusia, 352
X-cells of retina, 510
Xenon clearance, 354
Xenopus, 566, 579
Xiphas, 375
Xylan, 28, 29
Xylose, 28

Y

Y-cells, of retina, 510
Yolk synthesis, 590

Z

Zeitgeber, 66, 604
Zinc, 26
Zona granulosa, 594
Zonotrichia, 369, 604
Zooanthellae, 20, 21
Zoochlorella, 20
Zooplankton, crustaceans, 17